Amazon.co.uk, 5 Dec 2013 £51-95

Birthday gift from Stephen.

Selected Titles in This Series

48 **Yoav Benyamini and Joram Lindenstrauss,** Geometric nonlinear functional analysis, Volume 1, 2000
47 **Yuri I. Manin,** Frobenius manifolds, quantum cohomology, and moduli spaces, 1999
46 **J. Bourgain,** Global solutions of nonlinear Schrödinger equations, 1999
45 **Nicholas M. Katz and Peter Sarnak,** Random matrices, Frobenius eigenvalues, and monodromy, 1999
44 **Max-Albert Knus, Alexander Merkurjev, and Markus Rost,** The book of involutions, 1998
43 **Luis A. Caffarelli and Xavier Cabré,** Fully nonlinear elliptic equations, 1995
42 **Victor Guillemin and Shlomo Sternberg,** Variations on a theme by Kepler, 1990
41 **Alfred Tarski and Steven Givant,** A formalization of set theory without variables, 1987
40 **R. H. Bing,** The geometric topology of 3-manifolds, 1983
39 **N. Jacobson,** Structure and representations of Jordan algebras, 1968
38 **O. Ore,** Theory of graphs, 1962
37 **N. Jacobson,** Structure of rings, 1956
36 **W. H. Gottschalk and G. A. Hedlund,** Topological dynamics, 1955
35 **A. C. Schaeffer and D. C. Spencer,** Coefficient regions for Schlicht functions, 1950
34 **J. L. Walsh,** The location of critical points of analytic and harmonic functions, 1950
33 **J. F. Ritt,** Differential algebra, 1950
32 **R. L. Wilder,** Topology of manifolds, 1949
31 **E. Hille and R. S. Phillips,** Functional analysis and semigroups, 1957
30 **T. Radó,** Length and area, 1948
29 **A. Weil,** Foundations of algebraic geometry, 1946
28 **G. T. Whyburn,** Analytic topology, 1942
27 **S. Lefschetz,** Algebraic topology, 1942
26 **N. Levinson,** Gap and density theorems, 1940
25 **Garrett Birkhoff,** Lattice theory, 1940
24 **A. A. Albert,** Structure of algebras, 1939
23 **G. Szegö,** Orthogonal polynomials, 1939
22 **C. N. Moore,** Summable series and convergence factors, 1938
21 **J. M. Thomas,** Differential systems, 1937
20 **J. L. Walsh,** Interpolation and approximation by rational functions in the complex domain, 1935
19 **R. E. A. C. Paley and N. Wiener,** Fourier transforms in the complex domain, 1934
18 **M. Morse,** The calculus of variations in the large, 1934
17 **J. M. Wedderburn,** Lectures on matrices, 1934
16 **G. A. Bliss,** Algebraic functions, 1933
15 **M. H. Stone,** Linear transformations in Hilbert space and their applications to analysis, 1932
14 **J. F. Ritt,** Differential equations from the algebraic standpoint, 1932
13 **R. L. Moore,** Foundations of point set theory, 1932
12 **S. Lefschetz,** Topology, 1930
11 **D. Jackson,** The theory of approximation, 1930
10 **A. B. Coble,** Algebraic geometry and theta functions, 1929
9 **G. D. Birkhoff,** Dynamical systems, 1927
8 **L. P. Eisenhart,** Non-Riemannian geometry, 1927
7 **E. T. Bell,** Algebraic arithmetic, 1927

(Continued in the back of this publication)

Geometric Nonlinear Functional Analysis

Volume 1

American Mathematical Society
Colloquium Publications
Volume 48

Geometric Nonlinear Functional Analysis

Volume 1

Yoav Benyamini
Joram Lindenstrauss

American Mathematical Society
Providence, Rhode Island

1991 *Mathematics Subject Classification.* Primary 46–XX;
Secondary 22–XX, 28–XX, 47–XX, 52–XX, 54–XX.

ABSTRACT. This volume is devoted to the study of uniformly continuous and, in particular, Lipschitz functions between Banach spaces. Also studied is the related question of classifying Banach spaces and their subsets (mainly spheres) with respect to uniform or Lipschitz classification.

Library of Congress Cataloging-in-Publication Data

Benyamini, Yoav, 1943-
Geometric nonlinear functional analysis / Yoav Benyamini, Joram Lindenstrauss.
p. cm. — (American Mathematical Society colloquium publications, ISSN 0065-9258 ; v. 48)
Includes bibliographical references.
ISBN 0-8218-0835-4 (hardcover : alk. paper)
1. Nonlinear functional analysis. I. Lindenstrauss, Joram, 1936– . II. Title. III. Series: Colloquium publications (American Mathematical Society) ; v. 48.
QA321.5.B46 2000
515'.7–dc21

99-17734
CIP

∞ The paper used in this book is acid-free and falls within the guidelines
established to ensure permanence and durability.
Visit the AMS home page at URL: http://www.ams.org/

10 9 8 7 6 5 4 3 2 1   05 04 03 02 01 00

Dedicated to A. Dvoretzky, our teacher, colleague and friend
Founder of the functional analysis school in Israel

Contents

Introduction

1. The Subject Matter

The main theme of this book is the study of uniformly continuous and, in particular, Lipschitz functions between Banach spaces. Such functions appear in many contexts, and their study involves a rich interplay between topology, geometry of Banach spaces, geometric measure theory, classical analysis, probability and even combinatorics. The study of uniformly continuous and Lipschitz maps is naturally connected to the classification of Banach spaces and their subsets (especially balls) with respect to uniform homeomorphisms and bi-Lipschitz maps. It also leads to the development of notions and tools which are of interest in the applications of nonlinear functional analysis to other branches of mathematics.

To put the theory of uniform classification in context, we quote two well-known results. The first, due to Mazur and Ulam [**407**], is that a surjective isometry between two Banach spaces which takes 0 to 0 is necessarily linear. The second (and much deeper) result, due to Kadec [**303**], is that all separable infinite-dimensional Banach spaces are mutually homeomorphic. (And, as shown by Anderson [**16**], they are homeomorphic to the product space $\mathbb{R}^{\aleph_0}$.) The first result means that the linear structure of a Banach space is completely determined by its structure as a metric space, while the second says that its structure as a topological space, which is very simple, contains no information whatsoever on its linear structure.

The uniform classification lies between these two extremes. Even when the isometry assumption is slightly weakened, and instead of isometries we consider mappings which "almost" preserve distances, we already encounter difficult and very interesting questions (see Chapters 14 and 15, and Theorem 6.14). The situation becomes even more involved (and interesting) when we move further away from isometries and consider bi-Lipschitz maps and uniform homeomorphisms.

In addition to uniformly continuous maps, we also discuss some other topics in the nonlinear theory which are closely related to questions in the geometry of Banach spaces.

We treat several topics in the context of finite-dimensional Banach spaces. However most of the discussion concerns infinite-dimensional spaces. In many classical topics, once one leaves the locally compact setting, one immediately encounters the type of problems which are the main subject of this book.

We shall now outline the highlights of the various chapters.

In Chapter 1 we consider uniformly continuous (and Lipschitz) retractions, extensions and selections. We prove that some important metric spaces are Lipschitz retracts of any metric space containing them. These include the spaces of continuous functions on compact metric spaces (with the usual sup norm) and the space of all closed, bounded and convex subsets of a given convex set in a Banach space (equipped with the Hausdorff metric). We then study the question of extending uniformly continuous functions, defined on a subset of a Banach space E and taking values in a Banach space F, to the whole space E. This question is of special interest when E or F are Hilbert spaces. In the last section we study selections of set-valued maps. The problems of retraction and of extension can both be formulated as selection problems, and our main interest here is in obtaining uniformly continuous selections whose moduli of continuity are as good as possible.

In Chapter 2 we consider several independent topics related to the material of Chapter 1. We mention just three of these topics. The first is the question of approximating a general uniformly continuous function by functions with a prescribed modulus of continuity. In particular we obtain the exact exponent α so that every uniformly continuous function from the unit ball of L_p into L_q can be approximated arbitrarily well by functions of Hölder class α. Another subject is the nearest point map onto convex sets in a uniformly convex and uniformly smooth Banach space. We obtain precise estimates on the modulus of continuity of this map. These estimates yield, in particular, that a space whose moduli of convexity and smoothness are of the best possible order must be isomorphic to a Hilbert space. The third topic is the Steiner point map, a classical method to select a distinguished point in a bounded convex subset in $\mathbb{R}^n$. We show, among other results, that its modulus of continuity on the set of closed bounded convex subsets in $\mathbb{R}^n$ (equipped with the Euclidean norm) is the smallest possible among all such selection maps.

Chapter 3 is devoted to fixed point theory. Besides the basic standard results of Banach and Schauder, we also consider Lipschitz and nonexpansive maps of closed convex sets in Banach spaces. The result in this chapter which is most closely related to the rest of the book is that if C is a closed convex noncompact set in a Banach space, then there is a Lipschitz map f from C into itself with no approximate fixed point, i.e., $\inf\{\|f(x) - x\| : x \in C\} > 0$. In particular it follows that if E is an infinite-dimensional Banach space, then there is a Lipschitz retraction from the unit ball of E onto its unit sphere. We also discuss in some detail the rather delicate problems of the existence of fixed points and of the nonlinear ergodic theorem for nonexpansive maps.

One of the main tools of nonlinear functional analysis is the "linearization" of maps. A natural way to do so is to use derivatives. In the first section of Chapter 4 we present the two most natural concepts of differentiability, the Gâteaux and the Fréchet derivatives. We also present the basic calculus results that carry over with no difficulty from the finite-dimensional setting to infinite dimensions. The study of the existence of derivatives in infinite-dimensional spaces is much more involved than in finite-dimensional ones, and the main part of the chapter is devoted to the question of differentiation of convex real-valued functions. This subject is quite well understood today. There is even a complete characterization of sets in separable Banach spaces that can be the sets of points of nondifferentiability (in the Gâteaux sense) of some convex continuous function.

In Chapter 5 we study Banach spaces which have the Radon-Nikodým property (RNP). This class of spaces appears naturally in a variety of contexts and has many equivalent characterizations involving, for example, vector-valued measures, vector-valued martingales, or the extremal structure of closed bounded convex sets. The characterization most relevant to the subject of this book is that spaces with RNP are exactly those spaces E for which every Lipschitz function from $\mathbb{R}$ to E is differentiable almost everywhere.

In Chapter 6 we turn to Gâteaux differentiability of Lipschitz functions. To study this topic we need, in addition to the concept of RNP spaces, also the notion of "almost everywhere" in infinite-dimensional Banach spaces. In spite of the fact that infinite-dimensional Banach spaces do not carry a natural measure, one can still define in a meaningful way the notion of negligible sets. Actually, this can be done in several nonequivalent ways. We give some examples which illustrate these notions and also exhibit their delicate nature. With these tools (RNP and negligible sets) at hand, we proceed to prove the existence almost everywhere of Gâteaux derivatives of Lipschitz functions between certain Banach spaces. This generalizes in a very useful way the classical theorem of Rademacher on the differentiability of Lipschitz functions between finite-dimensional spaces. In a separate section we show that the "almost everywhere" results do not hold for Fréchet differentiation even for real-valued convex and continuous functions on a separable Hilbert space. The chapter ends with a section which summarizes and gives an overview of the various classes of negligible sets which appear in this volume.

In Chapter 7 we apply the Gâteaux differentiability results of Chapter 6 in order to linearize Lipschitz retractions and bi-Lipschitz maps. This procedure reduces in many important cases the study of Lipschitz embeddings and retractions to the linear theory. It follows from the discussion in this chapter that some concrete separable spaces which are not linearly isomorphic to each other also fail to be Lipschitz equivalent. On the other hand, we show that every separable Banach space is Lipschitz equivalent to a subset of c_0. This demonstrates the decisive role of the existence of derivatives in some of the results. We also present an example of two nonisomorphic spaces which are Lipschitz equivalent. (This example is nonseparable. No separable example of this type is known.) We end this chapter with examples of Lipschitz equivalences of l_2 onto itself with some surprising properties.

In Chapter 8 we turn to uniform embeddings; i.e., uniformly continuous maps whose inverses are uniformly continuous on their domain of definition. In this chapter we study uniform embeddings into Hilbert space and give a complete linear characterization of the normed (even quasi-normed) spaces that embed uniformly into Hilbert space. They are exactly the spaces which are linearly isomorphic to a subspace of $L_0(\mu)$, the space of all the measurable functions on some measure space with the topology of convergence in measure.

In Chapter 9 we characterize the Banach lattices whose unit spheres are uniformly homeomorphic to the unit sphere of Hilbert space. Examples (related to the Krivine-Maurey stability theory) show that for spaces which are not lattices the situation is more complicated. By using various methods (starting with the classical explicit map of Mazur, and proceeding via lattice factorization, complex interpolation and the Pełczyński decomposition method), we prove that the unit spheres of a very large class of spaces are uniformly homeomorphic to the unit sphere of a Hilbert space. We also show that the uniform classification of unit balls of Banach spaces is essentially the same as the classification of their spheres.

In Chapter 10 we study uniform homeomorphism between Banach spaces. It turns out that for several classes of Banach spaces, the structure of the space as a uniform space determines its linear structure. Moreover, for all Banach spaces the uniform structure already determines the finite-dimensional linear subspaces of the space (up to an isomorphism constant independent of the dimension of the subspace). Comparing these results with those of Chapter 9, one sees that there is a big difference between the uniform classification of spheres and that of spaces. We also give some examples that put the results of this chapter into perspective. For example, there are separable uniformly convex spaces which are uniformly homeomorphic but are not linearly isomorphic. There is even a separable uniformly convex space E with the following property: Any Banach space which is uniformly homeomorphic to E is either isomorphic to E or to $E \oplus H$ (where H is an infinite-dimensional Hilbert space), and E and $E \oplus H$ are uniformly homeomorphic but not linearly isomorphic. Another topic discussed in this chapter is the structure of discrete nets in Banach spaces. We show that knowing a discrete net in a Banach space up to Lipschitz equivalence gives much information on the space (and sometimes even determines it uniquely up to linear isomorphism). Infinite-dimensional Banach spaces determine their nets up to Lipschitz equivalence, but this is no longer true for finite-dimensional spaces (except, of course, when the dimension is 1).

In Chapter 11 we study quotient maps, a notion dual to that of embedding in the context of the linear theory. The direct analog, namely, surjective Lipschitz or uniformly continuous maps, turns out to be insufficient from our point of view. Indeed, for each pair of separable infinite-dimensional Banach spaces E and F there is a Lipschitz (and C^1) surjective map from E onto F. We introduce another notion of uniform (or Lipschitz) quotient map and study its properties. With this new notion we get, for example, that a uniform quotient space of a Hilbert space must be isomorphic to a Hilbert space. In finite dimensions the notion of Lipschitz quotient maps is related to the theory of quasi-regular maps.

In Chapter 12 we show that a real-valued uniformly continuous function on the unit sphere of a finite-dimensional Banach space is essentially constant on large sections of the sphere. More precisely, for every n, modulus of continuity φ, and $\varepsilon > 0$, there is a $k = k(n, \varepsilon, \varphi)$ (with $\lim_{n\to\infty} k(n, \varepsilon, \varphi) = \infty$ for fixed ε and φ), such that if f is a uniformly continuous real-valued function on the unit sphere of an n-dimensional Banach space E, and if the modulus of continuity of f is bounded by φ, then there is a k-dimensional subspace F of E such that the oscillation of f on the unit sphere of F is smaller than ε. We actually prove two results of this type. The first gives concrete estimates of k (which are sometimes asymptotically sharp), but does not show explicitly how F is located in E. The second gives much smaller k (and we do not even attempt to get any explicit estimate), but produces a subspace F which is nicely located in E: If E has a basis, then F is spanned by a "block basis". The heart of the first approach is the theorem of Dvoretzky on the existence of almost spherical sections in high-dimensional convex bodies. The second approach also reduces essentially to a result of a similar nature: For every k and ε there is an n such that every n-dimensional Banach space with a basis contains a block basis of length k which is ε-close to the unit vectors of l_p^k for some $1 \le p \le \infty$.

In Chapter 13 we consider the analogous question in an infinite-dimensional setup. We consider a uniformly continuous real-valued function f on the unit sphere of an infinite-dimensional Banach space E and ask whether there is an infinite-dimensional subspace F of E such that the restriction of f to the unit sphere of F has a small oscillation (prescribed in advance). The results of the previous chapter ensure the existence of finite-dimensional subspaces F of arbitrarily high dimension, but it turns out that in general there does not have to be an infinite-dimensional subspace F with the required property. The main example here is that of a distorted norm on a Hilbert space; i.e., an equivalent norm on a separable Hilbert space H (which we consider as a Lipschitz function on the unit sphere on

H) for which there is a $\delta > 0$ such that the oscillation of the new norm on the unit sphere of any infinite-dimensional subspace F of H is at least δ. The construction and the proof are quite involved, and the norm is not constructed directly on a Hilbert space. The first step is the construction of a rather complicated Banach space. Results on the structure of this Banach space are then transferred to Hilbert space via an appropriate nonlinear map. We also show in this chapter that every uniformly continuous function from the unit sphere of c_0 to $\mathbb{R}$ has arbitrarily small oscillation when restricted to a suitable infinite-dimensional subspace of c_0. It turns out that c_0 is essentially the only Banach space with this property.

Chapter 14 considers maps which are close to isometries. After reviewing the main results concerning isometries, we present the theory developed by F. John in the framework of the theory of elasticity. Consider a map f defined on a bounded open set Ω which is locally almost an isometry; then one asks whether f is injective on all of Ω, or even close to an actual isometry there. Some of these questions can be answered for general Banach spaces, while others have a nice answer only in Hilbert space (actually only in finite-dimensional Euclidean spaces). In finite-dimensional spaces, the maps we consider are differentiable almost everywhere, and it is natural to study their derivative. It is in the framework of this study that BMO functions were introduced for the first time.

In Chapter 15 we also consider the approximation of maps which are close to isometries by isometries, but an essential feature here is that the maps are globally defined and surjective. This issue is now well understood, and we present two quite precise results concerning it. This study is closely related to that of functions which are almost linear in the sense that $\|f(x+y) - f(x) - f(y)\| = o(\|x\| + \|y\|)$ as $\|x\|, \|y\| \to \infty$.

In Chapter 16 we study functions which are close to being linear in a weaker sense, i.e., $\|f(x+y) - f(x) - f(y)\| = O(\|x\| + \|y\|)$. It turns out that these functions (the so-called quasi-linear functions) come up naturally in the study of "twisted sums" of Banach spaces. (A space G is a twisted sum of E and F if $G \supset E$ and G/E is F.) The natural setting for this study is not of Banach spaces, but that of quasi-Banach spaces. Quasi-linear functions are used in particular in order to produce nontrivial twisted sums of Hilbert space with itself. (The existence of such a twisted sum is far from obvious.) The properties of twisted sums of Hilbert space with itself are studied in some detail. They give examples of Banach spaces which are interesting from many different points of view such as the study of unconditional structure or in the framework of complex Banach spaces, where they are used to produce two complex Banach spaces which are linearly isometric as real spaces but fail to be isomorphic as complex spaces.

In the final chapter, 17, we study questions related to Hilbert's fifth problem (concerning Lie groups) in the setting of groups modeled on an infinite-dimensional Banach space. We first examine whether there exist nontrivial group structures on a Banach space. It is shown that under some metric and other natural conditions, the only way to put a commutative group structure on a Banach space is to transfer to it the addition operations from another Banach space via a uniform homeomorphism. We also consider the analogue of Hilbert's problem concerning local groups. We study how far one can go in the construction of a Lie structure without changing the underlying model space. It is clear that without change of the model space one cannot solve the problem completely, and it seems to be unknown whether one can solve the infinite-dimensional analogue of Hilbert's problem if a change in the model space is allowed. In all these considerations it is essential that one assumes that the group operation is uniformly continuous, and thus the subject is intimately connected to the uniform classification of Banach spaces.

A more detailed outline of the material in chapters 7-12 is given in the survey [**362**] by Lindenstrauss. Note however that some of the open problems mentioned in [**362**] have already been solved; see, e.g., Examples 7.17 and 11.24.

Volume 2 of this book will treat topics which complement those considered here. The three main topics in Volume 2 will be

(i) Fréchet differentiability of Lipschitz functions between Banach spaces.
(ii) Lipschitz maps from the discrete point of view.
(iii) Connections with analytic functions.

Let us briefly describe these topics:

(i) In this volume we present existence theorems for Gâteaux differentiability of Lipschitz maps between Banach spaces. Fréchet differentiability is considered only for convex (real-valued) functions. The question of existence of Fréchet derivatives of Lipschitz functions between Banach spaces is much harder. Presently only one general result (due to Preiss [**471**]) is known in this direction: Every real-valued Lipschitz map defined on a space E with a separable dual is Fréchet differentiable on a dense set. Easy examples show that a Lipschitz map from ℓ_2 into itself may be nowhere Fréchet differentiable. This also holds for Lipschitz maps on other infinite-dimensional spaces. In Volume 2 we shall present the above-mentioned result on real-valued functions and also discuss several weakened notions of Fréchet differentiability. In terms of these weakened versions several positive results are available for Lipschitz maps between pairs of Banach spaces E and F (when both may be infinite-dimensional). The discussion involves also a new type of exceptional sets and a notion related to density points in infinite dimensions.

(ii) It is proved in Chapter 10 of this volume that uniformly homeomorphic Banach spaces have the same finite-dimensional subspaces. This means that all Banach space notions which involve only finitely many vectors (i.e., the entire so-called "local theory of Banach spaces"; see e.g. Appendix G) are invariant under uniform homeomorphism. It follows that at least in principle, these notions can be defined using only distances and without any use of vector space operations. We can even go one step further and use these notions in their metric version and apply them to general metric spaces. Naturally, there are major problems in such a program. Finding explicit metric formulations for linear notions is not always easy, and the metric formulation can sometimes be done in several different ways. Nevertheless, there is an emerging theory in this spirit for general (and in particular for finite) metric spaces.

We shall present this theory in Volume 2. It involves, e.g., such questions as the possibility of embedding metric spaces with small distortion into Banach spaces (and in particular Hilbert spaces); metric notions corresponding to type, cotype and superreflexivity; extension of Lipschitz maps defined on finite metric spaces and taking values in a Banach space, etc. There is even a metric version of Dvoretzky's theorem. The main tools in this study are of probabilistic and combinatorial nature. Some connections have already been found between this direction and theoretical computer science. Discrete problems also enter, to a very limited extent, into some discussions of the present volume (e.g., in the study of Banach groups in Chapter 17).

(iii) As one example of the connection of analytic functions to the subject matter of this book we mention the notion of Analytic Radon-Nikodým Property (ARNP). A Banach space E has the ARNP if every bounded E-valued analytic function on the open unit disc has radial limits a.e. on the unit circle. There is a theory of this class of spaces which is analogous to the theory of RNP spaces presented in Chapter 5 of this volume. The notions of convexity and convex functions that appear in the RNP theory are replaced in this theory by notions of complex convexity such as harmonic and plurisubharmonic functions. Also, general martingales are replaced by "analytic" martingales.

Another topic that falls under this heading and will be treated in Volume 2 is the biholomorphic classification of some subsets in complex Banach spaces. For example, it turns out that when the unit balls of two complex Banach spaces are biholomorphically equivalent, then the spaces are necessarily linearly isometric.

Besides these three major directions, Volume 2 will also contain some other topics. We mention just one such additional topic. A nice and useful result (e.g. in infinite-dimensional manifold theory) is that the general linear group of l_2 is contractible. This will be proved in Volume 2, and we shall also consider which other Banach spaces can replace l_2 in the statement above.

After reviewing the contents of the two volumes, it should be clear that large parts of what is usually called nonlinear functional analysis are not treated at all in this book. In particular we do not cover the one aspect of nonlinear functional analysis which is treated most often, namely, the theory of nonlinear differential equations. We also do not treat here the related subjects of infinite-dimensional manifold theory, degree theory, index theories, bifurcations results, etc. There are, of course, many books which treat this material. The books by Krasnoselskii [**336**], Vainberg [**567**], Schwartz [**526**], Nirenberg [**438**], Browder [**96**], Berger [**55**], Zeidler [**593**], Deimling [**139**], and Ambrosetti and Prodi [**12**] comprise just a sample of the books in which the general theory is presented. See also the collections of expository articles [**393, 394**] edited by Matzeu and Vignoli.

We also do not treat here questions related to purely topological homeomorphism, i.e., the theory which is usually called infinite-dimensional topology. Although several arguments in this theory use the geometry of Banach spaces, the final results are, as we have already mentioned above, not connected to the linear structure of the spaces. We refer the interested reader to the books by Bessaga and Pełczyński [**62**] and by van Mill [**576**], where this subject is covered in detail.

2. The Book. Conventions, Prerequisites, Etc.

We tried to write the book so that it will be possible to read each chapter independently of the other chapters. Thus each chapter gives a rather complete picture of one aspect of the subject, and cross-references between the chapters are few and are usually restricted to quoting a theorem or a definition.

Each chapter ends with a Notes and Remarks section in which we refer to the literature, present additional comments and results in the same direction, and often also mention some open problems.

The end of a proof is marked with a □ or a ◊. The □ is used when the proof comes directly after the statement of the result. If an auxiliary result or a discussion separates the proof of a result from its statement, we use the ◊ symbol.

The book ends with a detailed subject index. We point out, however, that the list of theorems in the subject index is *not* the list of the main results proved in the book. Rather, it is a list of theorems which have commonly used names attached to them, and most of them are only quoted here without a proof. The bibliography serves also as an enhanced author index; every item contains a reference to the page numbers in the text where the item is quoted.

The prerequisites for reading this book are a knowledge of material usually covered in the first year of graduate studies (in real analysis, functional analysis and also some complex analysis). Some further more specialized topics, which are used in a few places in the book, are presented in the various appendices at the end. In these appendices we present the necessary definitions and basic results. Some of these results are proved, but in most cases we refer to books and to journal articles where proofs can be found. We emphasize that the appendices are not a prerequisite for the main body of the book. They should be consulted only at the places in the text where a specific reference is made.

Two general remarks concerning the Banach spaces which appear in this book: For the large majority of the results, it does not matter whether the spaces are taken over the real or the complex field. In these cases we always work with real scalars without further notice. Complex scalars are used, again with no further notice, when the context requires it, e.g., when analytic functions or spectral theory are used, or when we apply the complex interpolation method. Our main interest is in separable Banach spaces. When separability makes no difference, we treat general Banach spaces; but when nonseparability poses additional difficulties, our general policy is to treat the separable case only. We do, however, state specifically that a space is separable whenever this is assumed.

We use standard notation and terminology. In particular we use the following notation for the classical Banach spaces. If K is a compact Hausdorff space, we denote by $C(K)$ the space of scalar-valued continuous functions on K, equipped with the supremum norm. The scalar field is usually assumed to be the real field and we do not indicate this in the notation. If K is a convergent sequence, we denote $C(K)$ by c. The hyperplane in c of sequences converging to 0 is denoted by c_0. For an infinite set Γ we denote by $c_0(\Gamma)$ the space of all scalar -valued functions f on Γ such that $\{\gamma : |f(\gamma)| > \varepsilon\}$ is finite for every $\varepsilon > 0$, equipped with the supremum norm. If Γ is countable, then $c_0(\Gamma)$ is just the space c_0.

Let $(\Omega, \mathcal{B}, \mu)$ be a measure space, i.e., $\mathcal{B}$ is a σ-algebra of subsets of Ω, and μ is a σ-additive nonnegative measure on $\mathcal{B}$. We denote by $L_p(\Omega, \mathcal{B}, \mu)$ for $1 \le p < \infty$ (respectively $L_\infty(\Omega, \mathcal{B}, \mu)$ for $p = \infty$) the space of equivalence classes of scalar-valued measurable functions f such that $\|f\|_p = \big(\int_\Omega |f|^p d\mu\big)^{1/p} < \infty$ (respectively $\|f\|_\infty = \sup\big\{t : \mu\{|f| > t\} > 0\big\} < \infty$). Once again the scalar field is usually assumed to be the real field. We usually suppress Ω and $\mathcal{B}$ from the notation and write $L_p(\mu)$. If μ is the Lebesgue measure on $[0,1]$, we denote $L_p(\mu)$ by $L_p[0,1]$ or simply by L_p. If μ is the counting measure on an abstract set Γ (i.e., $\mu(\gamma) = 1$ for every γ, and $\mathcal{B}$ is the σ-algebra of all subsets of Γ), we denote $L_p(\mu)$ by $l_p(\Gamma)$. If Γ is countable, or finite with cardinality n, we write l_p or l_p^n respectively.

Many other Banach spaces are defined throughout the book.

The vast majority of the material presented in this book appears for the first time in book form. A small part of the material in Chapters 1–5 is standard and is covered in many books. A larger part of the material in these chapters, well-known to experts but perhaps not to functional analysts in general, is covered in some specialized books (e.g. Deville, Godefroy and Zizler [**144**], Diestel and Uhl [**146**], Goebel and Kirk [**217**], Phelps [**461**]). However, much of the material even in these chapters also appears here for the first time in book form. Most of the material in Chapter 12 appears in Milman and Schechtman [**420**], and most of the material in the first section of Chapter 16 is contained in Castillo and Gonzalez [**114**].

A considerable amount of information is now known on the structure of Lipschitz and uniformly continuous functions, and on the relation of this subject to the geometry of Banach spaces. Many natural and fundamental problems in the area are, however, still open and some natural directions have not been studied at all. We hope that this book will help in making the results of geometric nonlinear functional analysis better known to the general community of functional analysts and that it will stimulate further research in the area.

3. Acknowledgements

In writing this book we were helped by many friends and colleagues who made suggestions, supplied references, gave us unpublished manuscripts or pointed out and corrected some mistakes in the various preliminary versions. These include, in particular, J. Castillo, P. Enflo, T. Gelander, G. T. Gowers, P. Hájek, A. Ioffe, W. B. Johnson, N. J. Kalton, A. Koldobsky, E. Matoušková, A. Leibmann, B. Maurey, S. Mendelson, R. Nahum, A. Naor, E. Odell, R. R. Phelps, A. Pełczyński, S. Reich, Th. Schlumprecht and I. Vestfrid. We thank them all.

We owe very special thanks to G. Schechtman, who encouraged and helped us throughout the process of writing this book. In particular he arranged for us to stay together at the Weizmann Institute for three days a week during two semesters at the decisive stages in the planning and writing of this book.

We also owe special thanks to D. Preiss, who made very many suggestions for improvements in the text. He also conducted a seminar in University College, London, on a preliminary version of a part of the book and transmitted to us the comments of the participants of the seminar.

The technical part of writing the book was facilitated considerably by Simcha Kojma of the Hebrew University. She helped us in setting up the system we used for writing the book and typed a large part of it. Her work was supported by the Edmund Landau Center for Research in Mathematical Analysis at the Hebrew University (supported by the Minerva Foundation, Germany).

We are also very grateful to all the people of the AMS who helped in publishing this book: the Editors of the Colloquium Publications series; S. Gelfand, director of the publishing department, who handled all our requests promptly and efficiently; the copy-editor, Deborah Smith, who made careful corrections and comments; and Barbara Beeton of the technical support staff, who helped with adjusting and solving software problems.

Y. Benyamini was partially supported by the Fund for the Promotion of Research at the Technion - Israel Institute of Technology.

Yoav Benyamini and Joram Lindenstrauss
Haifa and Jerusalem, October 1999

CHAPTER 1

Retractions, Extensions and Selections

1. Absolute Lipschitz Retracts

Let X and Y be metric spaces and let $f : X \to Y$. The modulus of continuity of f is the function

$$\omega_f(t) = \sup\{d(f(x), f(y)) : x, y \in X \quad \text{and} \quad d(x,y) \leq t\}.$$

The map f is said to be uniformly continuous if there is a $t_0 > 0$ such that $\omega_f(t) < \infty$ for $t < t_0$, and if $\lim_{t\to 0^+} \omega_f(t) = 0$. When X is a convex subset of a Banach space, or more generally when it is metrically convex (see Definition 1.3 below), the triangle inequality shows that $\omega_f(t+s) \leq \omega_f(t) + \omega_f(s)$, i.e., ω_f is subadditive. For general metric spaces this is not necessarily true, and we shall mainly be interested in maps f whose modulus of continuity is dominated by a nondecreasing subadditive function $\omega \geq \omega_f$ with $\lim_{t\to 0^+} \omega(t) = 0$. If $\omega_f(t) \leq ct$ for some constant c, we say that f is a Lipschitz map, and the smallest possible constant c, i.e., $\sup\{d(f(x), f(y))/d(x,y) : x \neq y\}$ is called the Lipschitz constant of the map f. If $\omega_f(t) \leq ct^\alpha$ for some $\alpha > 0$, we say that f is a Hölder map, or more precisely a Hölder(α) map with Hölder constant c.

In this chapter we introduce and study uniformly continuous retractions, uniform retracts, uniformly continuous extensions of maps, and uniformly continuous selections of set-valued maps. The next chapter is a continuation of this chapter, and we present there several topics of more specialized nature.

Let Y be a metric space and let X be a subset of Y. A Lipschitz map $r : Y \to X$ is called a Lipschitz retraction if it is the identity on X. When such a Lipschitz retraction exists, we say that X is a Lipschitz retract of Y. A metric space X is called an absolute Lipschitz retract if it is a Lipschitz retract of every metric space containing it. Uniform retractions, uniform retracts, and absolute uniform retracts are defined in a similar way.

In the first section we study absolute Lipschitz retracts and give some examples of such spaces. In the second section we consider the extension of uniformly continuous maps on a Hilbert space.

Selection of set-valued maps appear in many branches of mathematics. In the third section we present Michael's theorem, a very useful result in the topological category, along with some of its consequences. In the last section we study the selection of set-valued maps in the uniform category, and we give some examples of absolute uniform retracts (which are not absolute Lipschitz retracts).

The spaces $l_\infty(\Gamma)$ play a special role in the study of absolute Lipschitz and uniform retracts.

Lemma 1.1. (i) *Every metric space X is isometric to a subset of $l_\infty(\Gamma)$ for some set Γ.*

(ii) *Let Z be a subset of the metric space Y, and let ω be a nondecreasing subadditive function satisfying* $\lim_{t\to 0}\omega(t)=0$. *Assume that $f : Z \to l_\infty(\Gamma)$ is uniformly continuous and that $\omega_f \le \omega$. Then f can be extended to a uniformly continuous function $F : Y \to l_\infty(\Gamma)$ whose modulus of continuity ω_F is also dominated by ω. In particular, if f is Lipschitz so is F, and with the same Lipschitz constant.*

Proof. (i) Fix any $x_0 \in X$, and take Γ to be the set X itself. The embedding $\phi : X \to l_\infty(\Gamma)$ is defined by $\phi(x)(y) = d(x,y) - d(x_0,y)$.

(ii) By considering each coordinate separately, it suffices to prove (ii) for $\mathbb{R}$ as the target space instead of $l_\infty(\Gamma)$, i.e., for real-valued functions f. This is done by the explicit formula

$$F(y) = \inf\{f(z) + \omega(d(z,y)) : z \in Z\}.$$

To see that $F(y)$ is finite for every y, fix any $z_0 \in Z$. Then

$$f(z) + \omega(d(z,y)) \ge f(z_0) - \omega(d(z,z_0)) + \omega(d(z,y)) \ge f(z_0) - \omega(d(y,z_0))$$

for every $z \in Z$. It is also clear that F is an extension of f. To estimate the modulus of continuity of F, fix $x, y \in Y$ and choose z_0 such that $F(x)$ is essentially equal to $f(z_0) + \omega(d(z_0,x))$. Since $F(y) \le f(z_0) + \omega(d(z_0,y))$, it follows from the triangle inequality in Y and from the subadditivity of ω that $F(y) - F(x) \le \omega(d(x,y))$. A similar argument shows that $F(x) - F(y) \le \omega(d(x,y))$. □

Remark. The subadditivity of ω is essential for the existence of a uniformly continuous extension. Indeed, if ω_f is not bounded by such a function, then f cannot be extended to a uniformly continuous function on any metrically convex space Y containing Z.

The next proposition characterizes absolute Lipschitz retracts by the possibility to extend Lipschitz maps which are either defined on or take values in them.

Proposition 1.2. *Let X be a metric space. The following are equivalent:*

(i) *X is an absolute Lipschitz retract.*

(ii) *For every metric space Y and for every subset $Z \subset Y$, every Lipschitz function $f : Z \to X$ can be extended to a Lipschitz function $F : Y \to X$.*

(iii) *For every metric space Y containing X and for every metric space Z, every Lipschitz function $f : X \to Z$ can be extended to a Lipschitz function $F : Y \to Z$.*

Proof. By taking $Z = X$ and f to be the identity map on X, it is clear that either (ii) or (iii) implies (i). Also (i) implies (iii), because if $r : Y \to X$ is a Lipschitz retraction, then $F = f \circ r$ is the extension.

It remains to see that (i) implies (ii). By Lemma 1.1(i) X can be regarded as a subset of some $l_\infty(\Gamma)$, and since it is an absolute Lipschitz retract there is a Lipschitz retraction r from $l_\infty(\Gamma)$ onto X. Considering f as a function from Z to $l_\infty(\Gamma)$, Lemma 1.1(ii) yields a Lipschitz extension $G : Y \to l_\infty(\Gamma)$ and $F = r \circ G$ is the desired function. □

Remarks. (i) There is an analogous characterization of uniform absolute retracts in term of extensions of uniformly continuous functions. One should note, however, that (i) or (iii) only implies a restricted version of (ii): The modulus of continuity of f should be assumed to be dominated by a nondecreasing subadditive function satisfying $\lim_{t\to 0}\omega(t) = 0$.

(ii) It follows from Proposition 1.2 that if $Z \subset X$ is a Lipschitz retract of the absolute Lipschitz retract X, then Z itself is also an absolute Lipschitz retract. A similar remark holds for uniform retracts.

(iii) The definition of absolute retracts can be quantified: If X is an absolute Lipschitz retract, then there is, in fact, a constant λ such that it is an absolute λ-Lipschitz retract; i.e., there is a λ-Lipschitz retraction onto X from any containing metric space. Indeed, it follows from the previous results that $l_\infty(\Gamma)$ is an absolute 1-Lipschitz retract, and we can thus take λ to be the Lipschitz constant of any retraction from $l_\infty(\Gamma)$ onto any isometric copy of X. The extensions in (ii) and (iii) of Proposition 1.2 will then satisfy the condition that the Lipschitz constant of F will be bounded by λ times the Lipschitz constant of f.

Similarly, if X is an absolute uniform retract, then there is a subadditive function $\omega(t)$ satisfying $\lim_{t\to 0^+}\omega(t) = 0$ such that there is a retraction $\varphi : Y \to X$ with $\omega_\varphi \le \omega$ for every metric space Y containing X. The extensions in (ii) and (iii) of Proposition 1.2 will then satisfy $\omega_F \le \omega \circ \omega_f$.

We now proceed to give a geometric characterization of absolute 1-Lipschitz retracts.

Definition 1.3. (i) *A metric space X is said to be metrically convex if for every $x_0, x_1 \in X$ and for every $0 < t < 1$ there is a point $x_t \in X$ such that $d(x_0, x_t) = td(x_0, x_1)$ and $d(x_1, x_t) = (1-t)d(x_0, x_1)$. Equivalently, two closed balls $B(x, s)$ and $B(y, t)$ in X intersect if and only if $d(x, y) \le s + t$.*

(ii) *A metric space X is said to have the binary intersection property if every collection of mutually intersecting closed balls in X has a common point.*

The real line has the binary intersection property because every collection of mutually intersecting closed intervals in $\mathbb{R}$ has a common point. By considering each coordinate separately, it follows that the same is true for $l_\infty(\Gamma)$.

Proposition 1.4. *A metric space X is an absolute 1-Lipschitz retract if and only if it is metrically convex and has the binary intersection property.*

Proof. Assume that X is an absolute 1-Lipschitz retract. To see that it is metrically convex, fix two points $x_0, x_1 \in X$, define $f(0) = x_0$ and $f(1) = x_1$, and extend it to a function $F : [0,1] \to X$ with the same Lipschitz constant $d(x_0, x_1)$ as f. Then take $x_t = F(t)$.

To see that X has the binary intersection property, embed X isometrically in some $l_\infty(\Gamma)$, and let φ be a 1-Lipschitz retraction from $l_\infty(\Gamma)$ onto X. Let $\{B(x_i, r_i)\}_{i\in I}$ be a collection of mutually intersecting closed balls in X. Then the balls in $l_\infty(\Gamma)$ with the same centers and with the same radii intersect there (because $l_\infty(\Gamma)$ has the binary intersection property). If $x \in l_\infty(\Gamma)$ is in this intersection, then $\varphi(x)$ is a point in X which is in the intersection of the original collection.

Conversely, we show that condition (ii) of Proposition 1.2 holds with F having the same Lipschitz constant λ as f. A standard application of Zorn's lemma shows that it is enough to consider the case where Y is obtained from Z by adding just one new point, say y. Then the closed balls $\{B(f(z), \lambda d(z, y))\}_{z\in Z}$ in X are mutually intersecting. Indeed, for each $z_1, z_2 \in Z$

$$d(f(z_1), f(z_2)) \le \lambda d(z_1, z_2) \le \lambda d(z_1, y) + \lambda d(z_2, y)$$

and the two balls $B(f(z_j), \lambda d(z_j, y))$ intersect because X is metrically convex.

It thus follows that there is a point x common to all these balls, and we put $F(y) = x$. Then $d\big(F(y), F(z)\big) = d\big(x, f(z)\big) \le \lambda d(y, z)$ for each $z \in Z$ as required. □

Example 1.5. *There is a Lipschitz retraction from l_∞ onto c_0. Consequently, the space c_0 is an absolute Lipschitz retract.*

For each $x = (x_n) \in l_\infty$ put $d(x) = d(x, c_0) = \limsup |x_n|$, and define

$$\big(r(x)\big)_n = \begin{cases} 0 & \text{if } |x_n| < d(x) \\ \big(|x_n| - d(x)\big)\operatorname{sign}(x_n) & \text{if } |x_n| \ge d(x). \end{cases}$$

It is easy to check that r is a retraction with Lipschitz constant 2. Notice that there is no retraction with a smaller Lipschitz constant. Indeed, let $Y = c_0 \cup \{e\}$, where e is the function on $\mathbb{N}$ which is identically 1. Y is considered as a subset of l_∞ with its usual norm. Let e_n be the standard basis in c_0. If ρ is a retraction from Y to c_0 with Lipschitz constant λ, then

$$\|\rho(e) - \rho(2e_n)\| \le \lambda \|e - 2e_n\| = \lambda.$$

But the coordinates of $\rho(e)$ converge to zero and $\rho(2e_n) = 2e_n$, thus the left-hand side converges to 2.

Before passing to a more general example, we introduce some notation and make some preparations.

Let X be a metric space, and let $C_u(X)$ denote the space of real-valued, bounded, uniformly continuous functions on X, equipped with the sup norm. This class contains, of course, the $C(K)$ spaces for compact metric spaces K. We denote by $D(f)$ the distance of a function $f \in l_\infty(X)$ to $C_u(X)$.

Given $f \in l_\infty(X)$ and a positive constant λ, put

$$\psi(f,\lambda)(x) = \inf\big\{g(x) : \ g \ge f \text{ and } \ g \ \text{ is } \lambda\text{-Lipschitz}\big\}.$$

We leave it to the reader to check that the function $\psi(f, \lambda)$ is itself λ-Lipschitz and that it is given by the explicit formula

$$\psi(f,\lambda)(x) = \sup\big\{f(y) - \lambda d(x,y) : \ y \in X\big\}. \tag{1.1}$$

Note that $\psi(f, \lambda)$ is 1-Lipschitz as a function of f for each fixed λ. Indeed, assume that $\|f - g\| = \tau$. Then $f \le g + \tau$, and the minimality of ψ implies that $\psi(f, \lambda) \le \psi(g + \tau, \lambda) = \psi(g, \lambda) + \tau$. Similarly $\psi(g, \lambda) \le \psi(f, \lambda) + \tau$.

In (1.1) we clearly need to consider only points y for which $d(x, y) \le 2\|f\|/\lambda$. If $f \in C_u(X)$ has modulus of continuity ω, then for each such y

$$f(y) - \lambda d(x,y) - f(x) \le f(y) - f(x) \le \omega\big(d(x,y)\big) \le \omega(2\|f\|/\lambda).$$

Thus $0 \le \psi(f, \lambda) - f \to 0$ as $\lambda \to \infty$; i.e., the functions $\psi(f, \lambda)$ converge uniformly to f as $\lambda \to \infty$. (See also Proposition 2.1 in the next chapter for another simple argument that shows that uniformly continuous functions can be uniformly approximated by Lipschitz functions.)

It follows that for every $f \in l_\infty(X)$ and for every $\delta > 0$ there is a Lipschitz function h such that $\|f - h\| < D(f) + \delta$. The inequality $f < h + D(f) + \delta$ and the definition of ψ imply that when λ is large enough, then $\psi(f, \lambda) \le h + D(f) + \delta$, and thus $\psi(f, \lambda) - f \le 2D(f) + 2\delta$.

We are now ready to prove

Theorem 1.6. *Let X be a metric space. Then $C_u(X)$ is an absolute Lipschitz retract.*

Proof. For each $f \notin C_u(X)$ and for each $t > 0$, put

$$K(f,t) = \inf\big\{\lambda :\ \psi(f,\lambda) - f \le (2+t)D(f)\big\}.$$

(Such λ's exist by the remarks above.) The retraction is then defined for $f \notin C_u(X)$ by

$$\varphi(f)(x) = \int_0^1 \psi\big(f, K(f,t)\big)(x)dt$$

and $\varphi(f) = f$ when $f \in C_u(X)$.

Fix $0 < t_0 < 1$. Since $\|\psi(f,K)\| \le \|f\|$ for every K, and since $\psi\big(f, K(f,t)\big)$ is Lipschitz with constant $K(f,t_0)$ for every $t \ge t_0$, it follows that

$$|\varphi(f)(x) - \varphi(f)(y)| \le K(f,t_0)d(x,y) + 2t_0\|f\|$$

for all $x, y \in X$. Thus $\varphi(f)$ is, indeed, a uniformly continuous function.

By its definition φ is continuous at points of $C_u(X)$. Hence, to show that φ is Lipschitz it is enough to show that it is 20-Lipschitz in the neighborhood of every point not in $C_u(X)$. We shall show that $\|\varphi(f) - \varphi(g)\| \le 20\|f-g\|$ whenever $f, g \notin C_u(X)$ satisfy $10\|f-g\| \le \min\big(D(f), D(g)\big)$.

Put $\|f-g\| = \varepsilon$. Since $\psi(g,\lambda) \le \psi(f,\lambda) + \varepsilon$ for each fixed λ, it follows that for every $t \le 1$

$$\begin{aligned}
\psi\big(g, K(f,t)\big) - g &\le \psi\big(f, K(f,t)\big) - f + 2\varepsilon \le (2+t)D(f) + 2\varepsilon \\
&\le (2+t)D(g) + 5\varepsilon = (2 + t + \frac{5\varepsilon}{D(g)})D(g).
\end{aligned}$$

By the minimality of $K(g, t + \frac{5\varepsilon}{D(g)})$, it follows that $K(f,t) \ge K(g, t+\frac{5\varepsilon}{D(g)})$, and hence that $\psi\big(g, K(f,t)\big) \le \psi\big(g, K(g, t + \frac{5\varepsilon}{D(g)})\big)$. Thus

$$\begin{aligned}
\varphi(f) &= \int_0^1 \psi\big(f, K(f,t)\big)dt \le \varepsilon + \int_0^1 \psi\big(g, K(f,t)\big)dt \\
&\le \varepsilon + g + \int_0^1 \Big(\psi\big(g, K\big(g, t + \frac{5\varepsilon}{D(g)}\big)\big) - g\Big)dt \\
&\le \varepsilon + g + \int_0^{1+\frac{5\varepsilon}{D(g)}} \Big(\psi\big(g, K(g,t)\big) - g\Big)dt \\
&\le \varepsilon + \varphi(g) + \int_1^{1+\frac{5\varepsilon}{D(g)}} (2+s)D(g)ds \\
&\le \varepsilon + \varphi(g) + \frac{5\varepsilon}{D(g)}(3 + \frac{5\varepsilon}{D(g)})D(g) \le \varphi(g) + 20\varepsilon
\end{aligned}$$

because the restrictions on f and g imply that $\frac{5\varepsilon}{D(g)} \le 1/2$. A similar argument shows that $\varphi(g) \le \varphi(f) + 20\varepsilon$, i.e., $\|\varphi(f) - \varphi(g)\| \le 20\|f-g\|$. □

Let E be a Banach space, and let A be a closed convex subset of E. Another important example of an absolute Lipschitz retract is given by the space $H(A)$ of all nonempty, closed, convex, and bounded subsets of A. The Hausdorff metric

$$d_H(C,D) = \inf\big\{\varepsilon :\ C \subseteq (D)_\varepsilon \text{ and } D \subseteq (C)_\varepsilon\big\}$$

(where $(C)_\varepsilon = \{x \in A : d(x, C) < \varepsilon\}$, and similarly for $(D)_\varepsilon$) is a complete metric on $H(A)$.

Theorem 1.7. *Let A be a closed convex subset of a Banach space E. Then $H(A)$ is an absolute 8-Lipschitz retract.*

For the proof we need the following simple geometric lemma, which will also be used in Section 4 below.

Lemma 1.8. *Let A be a convex subset of a normed space E, and let $\beta > \alpha > 0$. Assume that $A \cap B(z, \alpha) \neq \emptyset$, and let $x \in A$. Then there is a point y in $A \cap B(z, \beta)$ satisfying*

$$\|x - y\| \leq \big(\|x - z\| - \beta\big)^+ (\beta + \alpha)/(\beta - \alpha).$$

Proof. By translation and scaling assume that $z = 0$ and that $\beta = 1$. In this case the lemma is trivial unless $\|x\| > 1$.

Fix any point u in $A \cap B(0, \alpha)$, and take y to be the unique point on the segment joining x and u satisfying $\|y\| = 1$. The convexity of A and the choice of y imply that $y \in A \cap B(0, 1)$.

Let $0 < \lambda < 1$ be such that $y = \lambda x + (1 - \lambda)u$. It follows from the estimate $\|x\| - 1 = \|x\| - \|y\| \geq (1 - \lambda)\big(\|x\| - \|u\|\big)$ that

$$\begin{aligned} \|x - y\| &= (1 - \lambda)\|x - u\| \leq \big(\|x\| - 1\big)\big(\|x\| - \|u\|\big)^{-1}\|x - u\| \\ &\leq \big(\|x\| - 1\big)\big(\|x\| - \|u\|\big)^{-1}\big(\|x\| + \|u\|\big) \\ &\leq \big(\|x\| - 1\big)(1 + \alpha)/(1 - \alpha) \end{aligned}$$

because $\|x\| > 1$ and $\|u\| \leq \alpha$. □

Corollary 1.9. *Let M be a closed convex subset of a Banach space E, and let $\gamma > 1$. Then the map $F : E \to H(M)$, given by $F(x) = M \cap B\big(x, \gamma d(x, M)\big)$, is Lipschitz with Lipschitz constant at most $(1 + \gamma)^2/(\gamma - 1)$. (In particular, taking $\gamma = 3$, we obtain a map with Lipschitz constant 8.)*

Proof. Fix $z, w \in E$, and let $x \in F(w)$. Using the lemma with $A = M$, any $\alpha > d(z, M)$ and $\beta = \gamma\alpha$, we find a point $y \in M \cap B(z, \beta)$ for which

$$\|x - y\| \leq \big(\|x - z\| - \beta\big)^+ (1 + \gamma)/(\gamma - 1).$$

Since

$$\|x - z\| \leq \|x - w\| + \|w - z\| \leq \gamma d(w, M) + \|w - z\|$$

and since $\beta > \gamma d(z, M)$, it follows that

$$\begin{aligned} \|x - y\| &\leq \Big(\gamma\big(d(w, M) - d(z, M)\big)^+ + \|w - z\|\Big)(1 + \gamma)/(\gamma - 1) \\ &\leq \|w - z\|(1 + \gamma)^2/(\gamma - 1). \end{aligned}$$

As β can be made arbitrarily close to $\gamma d(z, M)$, and since a similar estimate holds when we start with an arbitrary point $x \in F(z)$, the corollary follows. □

Proof of Theorem 1.7. Let Γ be the unit sphere of E^*. Recall that the support functional of a closed convex bounded set B in E, denoted by h_B, is the function on Γ given by

$$h_B(x^*) = \sup\big\{\langle x, x^* \rangle : x \in B\big\}.$$

Then h_B is continuous, and it follows immediately from the definitions and from the separation theorem that $d_H(B,C) = \|h_B - h_C\|_\infty$, that $B \subset C$ if and only if $h_B \le h_C$, and that $h_{B+C} = h_B + h_C$. We check the first identity for example. The separation theorem implies that for every $x \in B$ there is a functional $x^* \in \Gamma$ such that

$$\begin{aligned}\operatorname{dist}(x,C) &= \langle x^*, x\rangle - \sup\left\{\langle x^*, y\rangle : y \in C\right\}\\ &= \langle x^*, x\rangle - h_C(x^*) \le h_B(x^*) - h_C(x^*) \le \|h_C - h_C\|_\infty.\end{aligned}$$

The same estimate holds for $\operatorname{dist}(x,B)$ when $x \in C$, hence $d_H(B,C) \le \|h_B - h_C\|_\infty$, and the other direction is trivial.

Thus the map $B \to h_B$ identifies $H(A)$ (with its metric, order and affine structures) with the set $M = \{h_B : B \in H(A)\}$ in $l_\infty(\Gamma)$, which is closed and convex.

By the corollary there is an 8-Lipschitz map G from $l_\infty(\Gamma)$ to $H(M)$ such that $G(h) = h$ for all $h \in M$. It thus suffices to show that there is a 1-Lipschitz retraction u from $H(M)$ onto M. Indeed, the composition $u \circ G$ will then be a Lipschitz retraction from $l_\infty(\Gamma)$ onto M. As $l_\infty(\Gamma)$ is an absolute 1-Lipschitz retract, it will follow that M, hence also $H(A)$, is an absolute 8-Lipschitz retract.

The retraction u is given by an explicit formula: For a set $\mathcal{B}$ in $H(M)$, define $u(\mathcal{B})(x^*) = \sup\{h(x^*) : h \in \mathcal{B}\}$. Geometrically, the function $u(\mathcal{B})$ is the support function of the closure of $\cup\{B \in H(A) : h_B \in \mathcal{B}\}$. This latter set is a convex subset of A because $\mathcal{B}$ is convex. Thus $u(\mathcal{B})$ belongs to M. Clearly $u(\{h\}) = h$ when $h \in M$, and it follows directly from the definitions of the Hausdorff metric and the support function that u is 1-Lipschitz. $\diamondsuit$

Hölder maps into absolute Lipschitz retracts have a nice characterization in terms of approximation by Lipschitz maps.

Proposition 1.10. *Let X and Y be metric spaces, and let $0 < \alpha < 1$. Let f be a map from X into Y.*

(i) *Assume that there is a constant $C > 0$ such that for every $0 < t < \infty$ there is a Lipschitz map $f_t : X \to Y$ whose Lipschitz constant is bounded by $Ct^{1-\alpha^{-1}}$ and $\sup_{x\in X} d\big(f_t(x), f(x)\big) \le t$. Then f is a Hölder(α) map.*

(ii) *If Y is an absolute Lipschitz retract, then the converse also holds; i.e., if f is Hölder(α) then there are C and f_t as in* (i).

Proof. (i) Fix $x \ne y$ in X and put $t = d(x,y)^\alpha$. Then

$$\begin{aligned}d\big(f(x), f(y)\big) &\le d\big(f(x), f_t(x)\big) + d\big(f_t(x), f_t(y)\big) + d\big(f_t(y), f(y)\big)\\ &\le t + d(x,y)Ct^{1-\alpha^{-1}} + t \le (C+2)d(x,y)^\alpha.\end{aligned}$$

(ii) We may assume that f is Hölder(α) with constant 1.

Assume first that Y is the real line with its usual metric, and define

$$f_t(x) = \inf\left\{f(y) + d(x,y)t^{1-\alpha^{-1}} \,:\, y \in X\right\}.$$

Since $f(x) \le f(y) + d(x,y)^\alpha \le f(y) + d(x,y)t^{1-\alpha^{-1}}$ whenever $d(x,y) \ge t^{\alpha^{-1}}$, we can also write

$$f_t(x) = \inf\left\{f(y) + d(x,y)t^{1-\alpha^{-1}} \,:\, d(x,y) \le t^{\alpha^{-1}}\right\}.$$

By its definition f_t is Lipschitz with constant bounded by $t^{1-\alpha^{-1}}$, and it satisfies $f(x) - t \le f_t(x) \le f(x)$ as required.

If $Y = l_\infty(\Gamma)$ the result follows by applying the argument above to each component of f separately. If Y is a general λ-absolute Lipschitz retract, it can be isometrically embedded in some $l_\infty(\Gamma)$ and there is a retraction $r : l_\infty(\Gamma) \to Y$ which is Lipschitz with constant λ. Considering f as map into $l_\infty(\Gamma)$ there are maps $g_t : X \to l_\infty(\Gamma)$ satisfying the requirements of the theorem. The maps $f_t = r \circ g_t$ are Lipschitz with constant bounded by $\lambda t^{1-\alpha^{-1}}$ and $d\big(f_t(x), f(x)\big) \le \lambda t$. The result follows now by renormalization. □

Part (ii) is false for general metric spaces Y. For example, take $X = \mathbb{R}$ with the usual metric and $Y = \mathbb{R}$ with the metric $|x - y|^\alpha$. Then the identity from X to Y is Hölder(α), but there are no nonconstant Lipschitz maps from X to Y. Part (ii) may also fail when Y is a Banach space. Theorem 2.3 shows that this is the case for $Y = L_q$ when $q < \infty$. On the other hand (ii) holds when Y is a Hilbert space and X is a subset of a Hilbert space, see Theorem 2.1(ii).

We finish this section with a simple proposition that will be used repeatedly in the book.

Proposition 1.11. *Let X be a metrically convex space. Then every uniformly continuous map on X is Lipschitz for large distances; i.e., for every $d > 0$ there is a K_d such that $d\big(f(x), f(y)\big) \le K_d d(x, y)$ whenever $d(x, y) \ge d$.*

The same is true if instead of metric convexity we only assume that there is a $\lambda > 0$ such that every two points $x, y \in X$ can be connected by a curve whose length is smaller than $\lambda d(x, y)$.

Proof. Let $C = \sup\{d\big(f(x), f(y)\big) : d(x, y) \le d\}$. If $d(x, y) \ge d$, then we can find $n \le 2d(x, y)/d$ points $x = x_0, x_1, \dots, x_n = y$ such that $d(x_i, x_{i-1}) \le d$ for every $1 \le i \le n$. Thus

$$d\big(f(x), f(y)\big) \le \sum_{i=1}^{n} d\big(f(x_i), f(x_{i-1})\big) \le nC \le K_d d(x, y)$$

where $K_d = 2C/d$. The proof of the second statement is similar. □

2. Extension of Maps on Hilbert Space

In this section we consider the problem of extending uniformly continuous maps between Hilbert spaces.

Theorem 1.12. *Let H_1 and H_2 be two Hilbert spaces. Assume that A is a subset of H_1 and that ω is a nonnegative concave function on $\mathbb{R}^+$ satisfying $\lim_{t\to 0^+} \omega(t) = 0$. Then every uniformly continuous function $f : A \to H_2$ whose modulus of continuity is bounded by ω extends to a uniformly continuous function $F : H_1 \to H_2$ defined on all of H_1 and whose modulus of continuity is also bounded by ω.*

Remark. To be able to extend f to a uniformly continuous function defined on all of H_1, it is necessary, of course, that ω_f be dominated by a subadditive function η satisfying $\lim_{t\to 0^+} \eta(t) = 0$. The theorem shows that this condition is also sufficient. Indeed, it is easy to check that the function ω whose subgraph is the convex hull of the subgraph of η is concave and satisfies $\omega \le 2\eta$. Thus the theorem can be applied using this ω.

As we have already seen in the previous section, extensions of uniformly continuous functions are related to intersection properties of balls. The proof of Theorem 1.12 exhibits this principle again. We start with

Proposition 1.13. *Assume that* $\{B(x_i, r_i)\}_{i\in I}$ *and* $\{B(y_i, r_i)\}_{i\in I}$ *are two systems of closed balls in the Hilbert spaces* H_1 *and* H_2 *respectively and that*

$$\|y_i - y_j\| \le \|x_i - x_j\| \quad for\ all \quad i, j \in I.$$

If $\bigcap_I B(x_i, r_i) \neq \emptyset$, *then also* $\bigcap_I B(y_i, r_i) \neq \emptyset$.

Proof. Since closed balls in a Hilbert space are weakly compact, it is enough to prove the theorem for finite collections of, say, n balls in the same Hilbert space H. We can then also assume that H is finite-dimensional.

Fix $x \in \bigcap_{i=1}^n B(x_i, r_i)$ and assume, as we may, that $x \neq x_i$ for all i. Since H is finite-dimensional, the function

$$f(z) = \max_{i\le n} \frac{\|z - y_i\|}{\|x - x_i\|} \tag{1.2}$$

attains its minimum value λ at a point y, say. We shall show that $\lambda \le 1$, and hence that $y \in \bigcap_{i=1}^n B(y_i, r_i)$.

Let J be the set of indices $i \le n$ such that

$$\|y - y_i\| = \lambda\|x - x_i\| \quad \text{for} \quad i \in J$$

and

$$\|y - y_i\| < \lambda\|x - x_i\| \quad \text{for} \quad i \notin J.$$

It follows that y belongs to the convex hull of $\{y_i : i \in J\}$. Indeed, otherwise there would be a hyperplane L separating y from $\{y_i : i \in J\}$. Moving y a little in the direction of L would then turn the equalities on J to strict inequalities, while preserving the strict inequalities off J, and this would contradict the assumption that $f(y)$ is minimal. Thus fix $\{\alpha_i\}_{i\in J}$ such that $\alpha_i \ge 0$, $\sum \alpha_i = 1$ and $y = \sum \alpha_i y_i$.

For each $i, j \in J$ write

$$\begin{aligned} \|x_i - x_j\|^2 &\ge \|y_i - y_j\|^2 = \|y_i - y\|^2 + \|y_j - y\|^2 - 2\langle y_i - y, y_j - y\rangle \\ &= \lambda^2\|x_i - x\|^2 + \lambda^2\|x_j - x\|^2 - 2\langle y_i - y, y_j - y\rangle. \end{aligned}$$

Multiplying this inequality by $\alpha_i\alpha_j$, summing over $i, j \in J$, and using the fact that $\sum_{i,j\in J} \alpha_i\alpha_j\langle y_i - y, y_j - y\rangle = 0$, we obtain that

$$\begin{aligned} 0 &\ge \sum_{i,j\in J} \alpha_i\alpha_j\big(\lambda^2\|x_i - x\|^2 + \lambda^2\|x_j - x\|^2 - \|x_i - x_j\|^2\big) \\ &= \sum_{i,j\in J} \alpha_i\alpha_j\big(\|x_i - x\|^2 + \|x_j - x\|^2 - \|x_i - x_j\|^2\big) \\ &\quad +(\lambda^2 - 1)\sum_{i,j\in J} \alpha_i\alpha_j\big(\|x_i - x\|^2 + \|x_j - x\|^2\big). \end{aligned} \tag{1.3}$$

Since the first sum in the last term is $2\|\sum_{i\in J} \alpha_i(x_i - x)\|^2 \ge 0$, we obtain that $\lambda^2 - 1 \le 0$, i.e., $\lambda \le 1$. □

Theorem 1.12 for $\omega(t) = Kt$, i.e., for extending Lipschitz functions with preservation of the Lipschitz constant, follows immediately from Proposition 1.13. Indeed, by normalization, we can assume that $\omega(t) = t$, i.e., that f is 1-Lipschitz. By Zorn's lemma it suffices to show that f can be extended to one more point, say z, not in

A. Consider the balls $\{B(x, \|x-z\|)\}_{x\in A}$ and $\{B(f(x), \|x-z\|)\}_{x\in A}$. They satisfy the assumptions of Proposition 1.13. Indeed $\|f(x_1)-f(x_2)\| \le \|x_1-x_2\|$ for all $x_1, x_2 \in A$, and z clearly belongs to all the $B(x, \|x-z\|)$'s. If y is any point in $\bigcap B(f(x), \|x-z\|)$, then $f(z)=y$ gives the desired extension.

Note also that Theorem 1.12 for $\omega(t)=t$ and Proposition 1.13 are actually equivalent. Indeed, assuming that Lipschitz functions extend with the same Lipschitz constant, consider two systems of balls as in the proposition. The 1-Lipschitz map $f(x_i)=y_i$ extends from $\{x_i\}_{i\in I}$ to all of H_1. Fix $x\in\bigcap B(x_i, r_i)$ and put $y=f(x)$. Then $\|y-y_i\| = \|f(x)-f(x_i)\| \le \|x-x_i\| \le r_i$ for all i. Thus $y\in\bigcap B(y_i, r_i)$, and the intersection is nonempty.

To prove Theorem 1.12 for a general concave function ω, we introduce the following notation. Denote by $\mathcal{F}$ the class of all positive functions $\delta(t)$ on $\mathbb{R}^+$ such that

$$\sum \alpha_i\alpha_j \left(\delta(\|x_i\|)^2 + \delta(\|x_j\|)^2 - \delta(\|x_i-x_j\|)^2\right) \ge 0 \tag{1.4}$$

for all finite systems $\{x_i\}_{i\in I}$ in some Hilbert space and for all nonnegative numbers $\{\alpha_i\}_{i\in I}$.

The function $\delta(t)=t$ belongs to $\mathcal{F}$ because the left-hand side is then just $2\|\sum\alpha_i x_i\|^2$. This was the point that was used in estimating the first sum in the last term of inequality (1.3). So the definition of the class $\mathcal{F}$ is tailor-made in order that the same proof as that of Proposition 1.13 will also prove the following variant (when we replace the function f in (1.2) by $f(z)=\max \frac{\|z-y_i\|}{\delta(\|x-x_i\|)}$).

Proposition 1.13′. *Assume that the function δ belongs to the class $\mathcal{F}$, and let $\{B(x_i, r_i)\}_{i\in I}$ be a system of closed balls in a Hilbert spaces H_1. Let $\{y_i\}_{i\in I}$ be points in another Hilbert space H_2 such that*

$$\|y_i-y_j\| \le \delta(\|x_i-x_j\|) \quad \text{for all} \quad i,j\in I.$$

If $\bigcap_I B(x_i, r_i)\ne\emptyset$, then also $\bigcap_I B\big(y_i, \delta(r_i)\big)\ne\emptyset$.

Once we prove the next lemma, Theorem 1.12 will follow from Proposition 1.13′ by applying Zorn's lemma in the same way as its proof for Lipschitz maps followed from Proposition 1.13.

Lemma 1.14. *Let ω be a concave and nondecreasing function on $\mathbb{R}^+$ satisfying* $\lim_{t\to0^+}\omega(t)=0$. *Then $\omega\in\mathcal{F}$.*

Proof. We start with the simple observations that $\mathcal{F}$ is closed with respect to pointwise convergence and that the set $\mathcal{F}'=\{\delta^2 : \delta\in\mathcal{F}\}$ is a convex cone.

For each $\lambda>0$ the function $\varphi_\lambda(t)=\min(\lambda, t)$ belongs to $\mathcal{F}$. Indeed, the sum in (1.4) over the terms i,j for which both $\|x_i\|\le\lambda$ and $\|x_j\|\le\lambda$ is nonnegative because $\delta(t)=t\in\mathcal{F}$. Each of the other terms is itself nonnegative. Note also that $\varphi_\lambda(t)^2=\varphi_{\lambda^2}(t^2)$. It follows that if η is any nondecreasing concave function which satisfies $\lim_{t\to0^+}\eta(t)=0$, then $\delta(t)=\eta^{1/2}(t^2)\in\mathcal{F}$. Indeed, as $\mathcal{F}$ is pointwise closed, it is enough to check this when η is a bounded nondecreasing piecewise linear concave function, i.e., a finite linear combination $\eta(t)=\sum\alpha_i\varphi_{\lambda_i}(t)$ with positive coefficients. But then $\delta^2(t)=\eta(t^2)=\sum\alpha_i\varphi_{\lambda_i}(t^2)=\sum\alpha_i\varphi_{\sqrt{\lambda_i}}(t)^2\in\mathcal{F}'$ because each φ_λ^2 belongs to $\mathcal{F}'$ and $\mathcal{F}'$ is a convex cone. Thus $\delta\in\mathcal{F}$.

Finally, given ω as in the lemma, we see that the function $\eta(t)=\omega^2(t^{1/2})$ is concave and nondecreasing too. Indeed, if $\omega\in C^1$, then $\eta'(t)=\big(\omega(t^{1/2})/t^{1/2}\big)\omega'(t^{1/2})$, and both factors are nonnegative and nonincreasing. A general ω can be approximated by differentiable ones. Thus $\omega(t) = \eta^{1/2}(t^2) \in \mathcal{F}$ by the previous discussion.

This finishes the proof of the lemma, hence also of Theorem 1.12. □

Corollary 1.15. *Let H be a Hilbert space, and let X be a subset of the metric space Y. Then every Hölder(1/2) map $f : X \to H$ extends to a Hölder(1/2) map defined on all of Y and with the same constant.*

Proof. Again, by Zorn's lemma, it suffices to consider the case where Y is obtained from X by adding just one more point y, and we may also assume that f is Hölder(1/2) with constant 1. Consider the Hilbert space $l_2(X)$ of all square summable real functions on X, and denote by e_x the function with value 1 at x and zero elsewhere. The two systems of balls $\{B(d(x,y)^{1/2}e_x, d(x,y)^{1/2})\}_{x\in X}$ in $l_2(X)$ and $\{B(f(x), d(x,y)^{1/2})\}_{x\in X}$ in H satisfy the conditions of Proposition 1.13. Indeed, by the triangle inequality in $l_2(X)$

$$\|f(x_1) - f(x_2)\| \le d(x_1,x_2)^{1/2} \le \|d(x_1,y)^{1/2}e_{x_1} - d(x_2,y)^{1/2}e_{x_2}\|_{l_2(X)}.$$

Since 0 belongs to all the balls of the first system, it follows that there is a common point z in all the balls of the second. An extension is then given by $f(y) = z$. □

Remark. The corollary also holds for Hölder(α) functions when $\alpha < 1/2$. This can be shown by either modifying the proof or by replacing the metric d on Y by $d^{2\alpha}$ and applying the corollary with respect to this new metric.

The analogous result fails for $\alpha > 1/2$. Indeed, as we shall see in Corollary 1.29, if $\alpha > 1/2$, then there is a metric space Y containing the closed unit ball $B(H)$ of a Hilbert space such that there is no Hölder(α) retraction from Y onto $B(H)$.

3. Michael's Selection Theorem

Let ϕ be a set-valued function from a set X to the subsets of a set Y. A selection φ of ϕ is a mapping $\varphi : X \to Y$ such that $\varphi(x) \in \phi(x)$ for all x in X.

Selections appear in a variety of mathematical problems, and, depending on the context, one usually looks for selections with some specified additional properties: measurable, continuous, etc.

Many types of problems can be rephrased as selection problems. For example, a retraction from a space X onto its subset Y is just a selection for the set-valued map

$$\phi(x) = \begin{cases} x & \text{if } x \in Y \\ Y & \text{if } x \notin Y. \end{cases}$$

A similar formula describes an extension as a selection of an appropriate set-valued function.

Before embarking on our main topic of uniformly continuous selections and of their applications, we present in this section one of the classical selections theorems in the topological category.

Let X and Y be topological spaces, and let ϕ be a set-valued function on X whose values are subsets of Y. We say that ϕ is lower semi-continuous (l.s.c.) if the set $\{x : \phi(x) \cap G \neq \emptyset\}$ is open in X for every open subset G of Y. When the spaces are metrizable, ϕ is l.s.c. if and only if $x_i \to x$ in X and $y \in \phi(x)$ imply that there are $y_i \in \phi(x_i)$ such that $y_i \to y$. For the definition and properties of covers, paracompact spaces, and partitions of unity, see Appendix B. We just mention here that metric spaces and compact Hausdorff spaces are paracompact.

Theorem 1.16. *Assume that X is a paracompact space, that E is a Banach space, and that ϕ is a set-valued function from X to subsets of E. Assume that ϕ is l.s.c. and that $\phi(x)$ is a closed and convex nonempty subset of E for every $x \in X$. Then ϕ admits a continuous selection; i.e., there is a continuous function $\varphi : X \to E$ such that $\varphi(x) \in \phi(x)$ for every $x \in X$.*

The theorem will follow from an "approximate selection" lemma.

Lemma 1.17. *Let X, E and ϕ be as above, except that the sets $\phi(x)$ are not assumed to be closed any more. Then for every $\varepsilon > 0$ there is a continuous function $\varphi : X \to E$ such that $d\big(\varphi(x), \phi(x)\big) < \varepsilon$ for every $x \in X$.*

Proof. For each $v \in E$ let $B(v, \varepsilon)$ be the *open* ball in E with center v and radius ε, and set $G_v = \{x \in X : \phi(x) \cap B(v, \varepsilon) \neq \emptyset\}$. Then each G_v is open because ϕ is l.s.c., and the collection $\{G_v : v \in E\}$ is an open cover of X. Let $\{\psi_\alpha\}$ be a partition of unity subordinated to this open cover. For each α choose v_α such that ψ_α is supported in G_{v_α}, and define

$$\varphi(x) = \sum \psi_\alpha(x) v_\alpha.$$

The sum is locally finite; hence φ is a well-defined continuous function. Also, if $\psi_\alpha(x) \neq 0$, then $d\big(v_\alpha, \phi(x)\big) < \varepsilon$. Thus $\varphi(x)$ is a convex combination of points in the ε-neighborhood of $\phi(x)$. Since $\phi(x)$ is convex, $\varphi(x)$ also belongs to this neighborhood. □

Proof of Theorem 1.16. The theorem follows from the lemma by an iteration procedure.

Put $\phi_0 = \phi$, and let φ_0 be the map obtained from the lemma for $\varepsilon = 1$. Set $\phi_1(x) = \phi_0(x) \cap B(\varphi_0(x), 1)$. The values of ϕ_1 are convex, and by the choice of φ_0 they are nonempty. To see that ϕ_1 is l.s.c., fix an open set $G \subset E$ and a point $z \in \{x : \phi_1(x) \cap G \neq \emptyset\}$. There is an $r < 1$ such that the open set $\{x : \phi_0(x) \cap (B(\varphi_0(z), r) \cap G) \neq \emptyset\}$ contains z. The continuity of φ_0 implies that the set $\{x : \phi_0(x) \cap B(\varphi_0(x), 1) \cap G \neq \emptyset\}$ also contains an open neighborhood of z. Thus ϕ_1 satisfies the assumptions of the lemma, and we let φ_1 be the map obtained from the lemma for ϕ_1 and for $\varepsilon = 1/2$.

Continuing in the same way, we obtain a sequence of continuous mappings $\varphi_n : X \to E$ for which $\|\varphi_n(x) - \varphi_{n-1}(x)\| < 2^{-n+1} + 2^{-n}$ and $d\big(\varphi_n(x), \phi(x)\big) < 2^{-n}$ for all $x \in X$. Let φ be the limit of the φ_n's. Then φ is continuous and $\varphi(x) \in \phi(x)$ for all x because each $\phi(x)$ is a closed set. Thus φ is a continuous selection. ◇

We now pass to some applications of Theorem 1.16. Recall that a real-valued function f is called lower semi-continuous (respectively upper semi-continuous) if the set $\{x : f(x) > \alpha\}$ (resp. $\{x : f(x) < \alpha\}$) is open for every $\alpha \in \mathbb{R}$. The connection with l.s.c. set-valued functions is that f is lower semi-continuous if and only if the set-valued map $x \to (-\infty, f(x)]$ is l.s.c.

The oscillation of f at the point $x \in X$ is defined by

$$\operatorname{osc}(f)(x) = \inf_U \sup_{y,z \in U} |f(y) - f(z)|$$

where the infimum is taken over the neighborhoods U of x.

The next two propositions are typical applications of Theorem 1.16. (Note, however, that Proposition 1.18 actually holds under the weaker assumption that X is normal.)

Proposition 1.18. *Let X be a paracompact space.*

(i) *Let $f_1 \le f_2$ be two real-valued functions on X such that f_1 is upper semi-continuous and f_2 is lower semi-continuous. Then there is a continuous function f on X such that $f_1(x) \le f(x) \le f_2(x)$ for every $x \in X$.*

(ii) *Let $g \in l_\infty(X)$. Then the distance in $l_\infty(X)$ of g from the subspace of bounded and continuous functions on X is given by $\frac{1}{2}\sup\{\mathrm{osc}(g)(x) : x \in X\}$.*

Proof. (i) One checks easily that the set-valued function $\phi(x) = [f_1(x), f_2(x)]$ is l.s.c.; hence Theorem 1.16 applies.

(ii) Put $\delta = \frac{1}{2}\sup\{\mathrm{osc}(g)(x) : x \in \mathrm{X}\}$. It is clear that the distance is at least δ. To prove the other direction, put

$$f_1(x) = \inf_U \sup\{g(y) : y \in U\} - \delta$$
$$f_2(x) = \sup_U \inf\{g(y) : y \in U\} + \delta$$

where the infimum and supremum are taken over the neighborhoods U of x.

It is easy to check that f_1 is upper semi-continuous, that f_2 is lower semi-continuous, and that $f_1 \le f_2$. By part (i) there is a continuous function f between them. It follows that $f - \delta \le f_2 - \delta \le g \le f_1 + \delta \le f + \delta$, and hence that $\|f - g\| \le \delta$. □

Proposition 1.19. (i) *Let A be a closed and convex subset of a Banach space. Then A is an absolute retract; i.e., there is a continuous retraction on A from any metric space containing it as a closed subset.*

(ii) *Let F be a closed linear subspace of the Banach space E. Then the quotient map $\pi : E \to E/F$ admits a continuous positive homogeneous lifting; i.e., there is a continuous map $\varphi : E/F \to E$ such that $\pi \circ \varphi$ is the identity on E/F and such that $\varphi(\lambda x) = \lambda\varphi(x)$ for every $x \in E/F$ and $\lambda > 0$.*

Proof. (i) Assume that X is a metric space containing A as a closed subset, and define

$$\phi(x) = \begin{cases} x & \text{if } x \in A \\ A & \text{if } x \notin A. \end{cases}$$

It is easy to check that ϕ is lower semi-continuous (because A is closed in X). The values of ϕ are closed and convex subsets of a Banach space; hence Theorem 1.16 applies and ϕ admits a continuous selection which is the desired retraction.

(ii) Fix any $\varepsilon > 0$, and let B be the *open* ball $\{y \in E : \|y\| < 1 + \varepsilon\}$. We first show that the set-valued map $\Phi(x) = B \cap \pi^{-1}x$ is a lower semi-continuous map from the unit sphere $S(E/F)$ of E/F into the convex subsets of B. Indeed, assume that $x_i \to x$ in $S(E/F)$, and that $y \in \Phi(x)$. By the definition of the quotient norm there are $z_i \in E$ so that $\pi z_i = x_i - x$ and $\|z_i\| \le (1 + \varepsilon)\|x_i - x\|$. In particular $z_i \to 0$, and we assume that $\|z_i\| < 1 + \varepsilon - \|y\|$ for every i. Then $y_i = y + z_i \to y$ and $y_i \in \Phi(x_i)$, because $\pi y_i = x_i$ and $\|y_i\| < 1 + \varepsilon$.

Put now $\Psi(x) = \overline{\Phi(x)}$. Then it is easy to see that Ψ is also lower semi-continuous, and since its values are closed Theorem 1.16 applies and Ψ admits a continuous selection $\psi : S(E/F) \to (1 + \varepsilon)B(E)$. Finally, define φ to be the positive homogeneous extension of ψ to all of E/F. The continuity of φ at zero follows from the boundedness of ψ on the unit sphere, and the continuity elsewhere is obvious. □

There are no analogues of the last proposition in the uniform category. We shall see later (see Corollary 7.3) that there are situations (e.g., when E is a reflexive Banach space) in which the existence of a uniformly continuous retraction from E onto a closed subspace F of E even implies that F is actually linearly complemented. Thus (i) certainly does not hold universally as in the topological category. Similarly, there are generally no uniformly continuous liftings for quotient maps. If $\varphi : E/F \to E$ is such a lifting, then $x \to x - \varphi(\pi x) + \varphi(0)$ is a uniformly continuous retraction from E onto F, and as remarked above, such retractions do not necessarily exist.

If E is uniformly convex, then Corollary 1.25 below ensures the existence of liftings which are uniformly continuous on bounded subsets of E/F. But this is no longer true for general Banach spaces; see Remark (ii) after Example 1.30.

We now present an explicit example where a uniformly continuous lifting does not exist.

Example 1.20. *Let D be the space of bounded right-continuous functions f on $[0,1]$ with limit from the left at every point, and such that the only jumps of f may occur at rational points in the open interval $(0,1)$. D is equipped with the supremum norm.*

Then the quotient map $\pi : D \to D/C[0,1]$ admits no uniformly continuous lifting.

We first note that for every $\varepsilon > 0$ a function in D can have only finitely many jumps of size larger than ε, for otherwise at least one of the one-sided limits will fail to exist at any accumulation point of the points where these jumps occur. It follows that the quotient $D/C[0,1]$ is isometric to c_0. More precisely, recalling that the jumps can occur only at rational points q, index the unit vectors of c_0 as $\{e_q\}$. By applying Proposition 1.18(ii) locally we see that the qth coordinate of πf as an element of c_0 is half the jump of f at q.

Assume that $\psi : c_0 \to D$ is a uniformly continuous lifting. We shall obtain a contradiction by a "midpoint" argument, a method that will be presented and used systematically in Section 2 in Chapter 10.

Since ψ is uniformly continuous, it satisfies a Lipschitz condition for large distances. The Lipschitz constants K_ρ for distances larger than ρ are a nonincreasing function of ρ; hence their limit exists and satisfies $K = \lim_{\rho\to\infty} K_\rho \geq 1$. Fix $0 < \varepsilon < 1/100$, and let ρ be so large that $\|\psi(u) - \psi(v)\| \leq (K+\varepsilon)\|u-v\|$ whenever $\|u - v\| > \rho$.

Choose now $x, y \in c_0$ such that $\|x - y\| > 2\rho$ and such that

$$\|\psi(x) - \psi(y)\| > (K - \varepsilon)\|x - y\|.$$

Since elements in D are right continuous, there is an open set G in $[0,1]$ such that

$$|\psi(x)(t) - \psi(y)(t)| > (K - \varepsilon)\|x - y\| \tag{1.5}$$

for every $t \in G$. Moreover, we can also assume that the oscillations of both $\psi(x)$ and $\psi(y)$ are smaller than $\varepsilon\|x - y\|$ on G.

On the other hand the choice of ρ implies that if $z \in c_0$ is a metric midpoint between x and y, i.e., if z satisfies $\|x-z\| = \|y-z\| = \|x-y\|/2 > \rho$, then

$$\|\psi(x) - \psi(z)\|, \|\psi(y) - \psi(z)\| < (K+\varepsilon)\|x-y\|/2. \tag{1.6}$$

Comparison of (1.6) and (1.5) shows that

$$\left|\psi(z)(t) - \frac{\psi(x)(t) + \psi(y)(t)}{2}\right| \leq 3\varepsilon\|x-y\|$$

for all $t \in G$. In particular, the oscillation of $\psi(z)$ on G is smaller than $7\varepsilon\|x-y\|$.

Finally, choose a rational point $q \in G$ such that the modulus of the qth coordinates of x and of y are smaller than $\|x-y\|/8$. Then $z = (x+y)/2 + \frac{1}{4}\|x-y\|e_q$ is a metric midpoint between x and y. But then the qth coordinate of z is at least $\frac{1}{8}\|x-y\|$, and being half the jump of $\psi(z)$ at q, the latter is at least $\frac{1}{4}\|x-y\|$, a contradiction.

Remark. Let F be a closed linear subspace of the Banach space E. Then F is linearly complemented in E if and only if there is a linear lifting for the quotient map $\pi : E \to E/F$. This is no longer true in the nonlinear case, and although Theorem 1.6 implies that there is a Lipschitz retraction from D onto $C[0,1] = \operatorname{Ker} \pi$, there is no uniformly continuous lifting for the quotient map. Note, however, that if π admits a uniformly continuous lifting φ, then a uniform retraction from E to F is given by $r(x) = x - \varphi(\pi(x)) - \varphi(0)$.

One can sometimes use Theorem 1.16 when the set-valued map ϕ is not apriori defined into the subsets of a Banach space with its norm topology. We describe one such application.

Let K be a compact Hausdorff space, and let H be a closed subset of K. A simultaneous extension operator from $C(H)$ to $C(K)$ is a bounded linear operator $T : C(H) \to C(K)$ such that for each $f \in C(H)$ the function Tf is an extension of f to all of K.

Theorem 1.21. *Let K and H be as above, and assume that H is metrizable. Then there is a simultaneous extension operator $T : C(H) \to C(K)$ such that $\|T\| = 1$ and such that $T1 = 1$.*

Proof. Since H is metrizable, $C(H)$ is separable. Let $\mathcal{P}$ be the set of probability measures on H, considered as a subset of the unit ball of $C(H)^*$. Equipped with the the w^*-topology, $\mathcal{P}$ is compact and is affinely homeomorphic to a (clearly convex) subset of a Banach space: The map $\mu \to \left(2^{-i}\langle \mu, f_i\rangle\right) \in l_2$, where $\{f_i\}$ is a countable dense subset in the unit ball of $C(H)$, is such a homeomorphism into the Hilbert space l_2. It follows that we can use Theorem 1.16 for set-valued maps into $\mathcal{P}$. Let $\delta_h \in \mathcal{P}$ denote the point mass at h; i.e., it is the functional in $C(H)^*$ defined by $\delta_h(f) = f(h)$. The set-valued map on K given by

$$\phi(k) = \begin{cases} \delta_k & \text{if } k \in H \\ \mathcal{P} & \text{if } k \notin H \end{cases}$$

satisfies the assumptions of Theorem 1.16, so let φ be a continuous (with respect to the w^*-topology) selection of ϕ. The linear extension operator is defined by the formula

$$Tf(k) = \langle \varphi(k), f\rangle.$$

The continuity of φ with respect to the w^*-topology ensures that Tf is continuous on K, and the other required properties are immediate. □

4. Lipschitz and Uniform Selections

The intersection $F(s) \cap G(s)$ of two Lipschitz functions from a metric space S into $H(E)$ is usually not a Lipschitz function. The next proposition describes an important situation where it is.

Proposition 1.22. *Let E be a Banach space, and fix a number $\delta < 1$. Let F and G be two Lipschitz maps from a metric space (S, d) into $H(E)$, such that G is of the form $G(s) = B\big(f(s), g(s)\big)$ for two Lipschitz functions $f : S \to E$ and $g : S \to \mathbb{R}^+$.*

If the intersection $F(s) \cap B\big(f(s), \delta g(s)\big)$ is nonempty for every $s \in S$, then the map $F \cap G$ is also Lipschitz.

Proof. Fix s and t in S. We shall show that for every u in $F(s) \cap G(s)$ there is a point $y \in F(t) \cap G(t)$ such that $\|u - y\| \leq K d(s,t)$, where the constant K depends on δ and the Lipschitz constants L_F, L_f, and L_g of F, f and g respectively.

If $g(t) = 0$, then $y = f(t)$ gives $\|u - y\| \leq \|u - f(s)\| + \|f(s) - f(t)\|$. But the first term is bounded by $g(s) = g(s) - g(t) \leq L_g d(s,t)$, and the second by $L_f d(s,t)$.

Assume that $g(t) > 0$. We shall simplify and assume that the infimum in the Hausdorff distance and the distance between $f(t)$ and $F(t)$ are both attained (thus saving some necessary yet cumbersome ε's).

Choose a point $x \in F(t)$ such that $\|u - x\| \leq d_H\big(F(s), F(t)\big)$. If x is in $G(t)$, take it as y and we are done. If not, then $\|x - f(t)\| > g(t)$, and we use Lemma 1.8 with this x, $A = F(t)$, $z = f(t)$, $\alpha = d\big(f(t), F(t)\big)$, and $\beta = g(t)$. Since by the assumptions of the proposition $\alpha/\beta \leq \delta$, and thus $(\beta + \alpha)/(\beta - \alpha) \leq (1+\delta)/(1-\delta)$, it follows that there is a point y with $\|x - y\| \leq \big(\|x - f(t)\| - g(t)\big)(1+\delta)/(1-\delta)$ in $A \cap B(z, \beta) = F(t) \cap G(t)$.

Note also that

$$\|x - f(t)\| - g(t) \leq \|x - u\| + \|u - f(s)\| - g(t) + \|f(s) - f(t)\|$$

and since $\|x - u\| \leq d_H\big(F(s), F(t)\big)$ and $\|u - f(s)\| \leq g(s)$ we obtain that

$$\|x - f(t)\| - g(t) \leq (L_F + L_g + L_f) d(s,t).$$

Finally,

$$\begin{aligned}
\|u - y\| &\leq \|u - x\| + \|x - y\| \\
&\leq d_H\big(F(s), F(t)\big) + \big(\|x - f(t)\| - g(t)\big)(1+\delta)/(1-\delta) \\
&\leq \Big(L_F + (L_F + L_g + L_f)(1+\delta)/(1-\delta)\Big) d(s,t).
\end{aligned}$$

□

Definition 1.23. *Let A and B be two subsets of the Banach space E, and assume that A is bounded.*

(i) *The radius of A around the point $x \in E$, denoted by $r_x(A)$, is defined by*

$$r_x(A) = \sup\big\{\|x - y\| : \, y \in A\big\}.$$

(ii) *The relative radius of A with respect to B, denoted by $\mathrm{rad}_B(A)$, is defined by*

$$\mathrm{rad}_B(A) = \inf\big\{r_x(A) : \, x \in B\big\}.$$

When $B = E$ is the whole space, we suppress the E in the notation and write simply $\operatorname{rad}(A)$.

(iii) *The relative Chebyshev center of A with respect to B is the set of all $x \in B$ such that $r_x(A) = \operatorname{rad}_B(A)$. When the relative Chebyshev center of A contains exactly one point, we call this point simply the relative center of A and denote it by* $c_B(A)$.

Remarks. (i) When E is reflexive, the relative Chebyshev center of every bounded set A with respect to any weakly closed set B is nonempty. Indeed, it is the intersection of the decreasing sequence of nonempty weakly compact sets $K_n = \{x \in B : r_x(A) \leq \operatorname{rad}_B(A) + 1/n\}$.

(ii) When E is uniformly convex and A and B are as in (i), then the Chebyshev center of A with respect to B is actually a single point.

A subset $\mathcal{A}$ of $H(E)$ is said to have a uniformly normal structure if there is a $\gamma < 1$ such that $\operatorname{rad}(A) \leq \gamma \operatorname{diam}(A)$ for every $A \in \mathcal{A}$. If this holds for $\mathcal{A} = H(E)$, we say that E has uniformly normal structure. By Proposition A.9 uniformly convex spaces have a uniformly normal structure.

Let $\mathcal{A}$ be a subset of $H(E)$. A mapping $F : \mathcal{A} \to E$ is called a selector if $F(A) \in A$ for every $A \in \mathcal{A}$. The set $\mathcal{A}$ is said to be hereditary if $H(A) \subseteq \mathcal{A}$ for every $A \in \mathcal{A}$.

Theorem 1.24. *Let $\mathcal{A}$ be a hereditary subset of $H(E)$ which has a uniformly normal structure. Then there is a selector $\phi : \mathcal{A} \to E$ which is uniformly continuous on $\{A \in \mathcal{A} : \operatorname{diam}(A) \leq r\}$ for every $r > 0$.*

Proof. Let γ be the constant ensured by the uniform normality, and choose $\alpha > 1$ such that $k = \alpha\gamma < 1$. As in the proof of Theorem 1.7, we denote the unit sphere of E^* by Γ and identify E with a subspace of $l_\infty(\Gamma)$. Define a map T from $\mathcal{A}$ to $\mathcal{A}$ by

$$T(A) = A \cap B\big(h_A, \alpha \operatorname{rad}(A)\big)$$

(where h_A is the support function of A, and the ball is, of course, in $l_\infty(\Gamma)$). $T(A) \in \mathcal{A}$ because $\mathcal{A}$ is hereditary.

Proposition 1.22 now applies: The maps $A \to h_A$ and $A \to \operatorname{rad}(A)$ are Lipschitz (with Lipschitz constant 1), and for any fixed $1 < \eta < \alpha$ the definitions imply that $A \cap B\big(h_A, \eta \operatorname{rad}(A)\big) \neq \emptyset$. Thus T is a Lipschitz map.

We claim that $\operatorname{diam}\big(T(A)\big) \leq \alpha \operatorname{rad}(A)$. Indeed, if $x, y \in T(A)$, then $y \in A$ implies that $h_A(x^*) \geq \langle x^*, y\rangle$ for every $x^* \in \Gamma$, and thus

$$\begin{aligned} \alpha \operatorname{rad}(A) &\geq \|h_A - x\|_\infty = \sup\big\{h_A(x^*) - \langle x, x^*\rangle : x^* \in \Gamma\big\} \\ &\geq \sup\big\{\langle y, x^*\rangle - \langle x, x^*\rangle : x^* \in \Gamma\big\} = \|x - y\| \end{aligned}$$

as required.

Since $\mathcal{A}$ has uniform normal structure with constant γ, it follows from the choice of α that $\operatorname{diam}\big(T(A)\big) \leq k \operatorname{diam}(A)$ for all $A \in \mathcal{A}$. Thus

$$\operatorname{diam}\big(T^n(A)\big) \leq k^n \operatorname{diam}(A) \to 0$$

and $\phi(A) = \bigcap T^n A$ is a singleton. The function ϕ is a selector on $\mathcal{A}$, and it is uniformly continuous on $\{A \in \mathcal{A} : \operatorname{diam}(A) \leq r\}$ for every $r > 0$. Indeed, given $\varepsilon > 0$, choose $n = n(\varepsilon)$ such that $k^n r < \varepsilon$, and let $C = C(\varepsilon)$ be the Lipschitz constant of T^n. If $d_H(A, B) < \varepsilon/C$, then $d_H(T^n A, T^n B) < \varepsilon$. Since the diameters of A and B are less than r, it follows from the choice of n that the diameters of $T^n A$ and $T^n B$ are smaller than ε. Thus $\|\phi(A) - \phi(B)\| < 2\varepsilon$. □

As an immediate consequence of the theorem we obtain that linear quotient maps from uniformly convex spaces, and more generally from spaces with uniformly normal structure, admit liftings which are uniformly continuous on bounded sets.

Corollary 1.25. *Let E be a Banach space with a uniformly normal structure, and let $\pi : E \to F$ be a surjective linear map. Then there is a homogeneous map $\varphi : F \to E$ which is uniformly continuous on bounded sets and such that $\pi \circ \varphi$ is the identity on F.*

Proof. By the open mapping theorem there is a constant K such that the set $A_z = \{x \in E : \|x\| \leq K \text{ and } \pi x = z\}$ is nonempty for every z in the unit sphere of F. Let ϕ be the uniformly continuous selector on the closed and convex subsets of $KB(E)$ given by Theorem 1.24, and take φ to be the homogeneous extension of the map $z \to \phi(A_z)$. □

A metric space X is said to be an absolute neighborhood Lipschitz retract if for every metric space Y containing X there is an $r > 0$ such that X is a Lipschitz retract of the r-neighborhood $\{y \in Y : d(y, X) < r\}$ of X in Y. Absolute neighborhood uniform retracts are defined similarly.

Corollary 1.26. *Let E be a Banach space with a uniformly normal structure, and let A be a closed and convex subset of E. Then A is an absolute neighborhood uniform retract. If A is bounded, it is an absolute uniform retract.*

Proof. Assume that Y is a metric space containing A. Since $H(A)$ is an 8-absolute Lipschitz retract, there is a Lipschitz function $f : Y \to H(A)$ (with Lipschitz constant at most 8) extending the natural isometric embedding of A into $H(A)$ as the set of singletons. Fix any $r > 0$, and consider the neighborhood $Y_r = \{y \in Y : d(y, A) < r\}$. Since $d_H(B, x) \geq \text{rad}(B)$ for every $x \in A$ and $B \in H(A)$, it follows that $\text{rad}\big(f(y)\big) \leq 8r$ for every $y \in Y_r$. Thus $\phi \circ f$ (with ϕ as in Theorem 1.24) is a uniformly continuous retraction from Y_r onto A.

When A is bounded, its radius bounds the radii of all the sets in $H(A)$; thus $\phi \circ f$ is uniformly continuous on all of Y. □

For uniformly convex spaces it is possible to construct retractions as in the last corollary explicitly. The estimates for the resulting modulus of continuity are better than those obtained from a quantitative analysis of the proof of the corollary. We shall show in Lemma 1.28 and Corollary 1.29 that these estimates are sharp for $E = L_p$. Recall that $c_A(D)$ is the center of the set D relative to A, and that it is a singleton when E is uniformly convex and A is convex.

Theorem 1.27. *Let A be a closed and convex subset of a uniformly convex Banach space E. Then the map $D \to c_A(D)$ is a retraction from $H(A)$ onto A which is uniformly continuous on $\{D \in H(A) : \text{rad}_A(D) \leq r\}$ for every $r > 0$. Its modulus of continuity on this set satisfies $\omega(r; \varepsilon) \leq 2r\delta_E^{-1}(\varepsilon/2r)$ whenever $\varepsilon \leq r$.*

Proof. Fix two sets $B, D \in H(A)$. To simplify the notation, denote their radii and centers relative to A by r_B, r_D, c_B and c_D respectively. Assume that $r_B \leq r_D \leq r$ and that $d_H(B, D) \leq \varepsilon$.

Since r_D is the minimal radius of a ball with center in A that contains D, there is a point $z \in D$ such that $\|z - (c_B + c_D)/2\| \geq r_D$. On the other hand, $\|z - c_B\| \leq r_B + d_H(B, D) \leq r_D + \varepsilon$ and also $\|z - c_D\| \leq r_D \leq r_D + \varepsilon$.

By the definition of the modulus of convexity δ_E (see Appendix A for its definition and basic properties)

$$t\delta_E^{-1}(\varepsilon/t) = \sup\big\{\|x - y\| : \|x\|, \|y\| \leq t \text{ and } \|x + y\|/2 \geq t - \varepsilon\big\}$$

and $t\delta_E^{-1}(\varepsilon/t)$ is monotone increasing in t. Using this for $\varepsilon \leq r$ with $x = z - c_B$, $y = z - c_D$ and $t = r_D + \varepsilon$ gives that

$$\|c_B - c_D\| \leq (r_D + \varepsilon)\delta_E^{-1}\big(\varepsilon/(r_D + \varepsilon)\big) \leq 2r\delta_E^{-1}(\varepsilon/2r).$$

□

Lemma 1.28. *Let $1 < p < \infty$ and consider any isometric embedding of l_p^n into l_∞. Let $r > 0$, and let φ be a retraction from $\{y \in l_\infty : d(y, l_p^n) \leq r\}$ onto l_p^n. Then the modulus of continuity ω of φ satisfies*

$$\omega(r^{1-p}\varepsilon^p) + 4\omega(r/2)n^{-1/p} \geq \varepsilon.$$

Proof. Fix ε, r and n. For every subset A of $\{1, \dots, n\}$, let $x_A \in l_p^n$ be the element whose coordinates are equal to ε on A and are equal to 0 off A. Thus $\|x_A - x_B\| = \varepsilon\Delta(A, B)^{1/p}$, where $\Delta(A, B)$ denotes the cardinality of the symmetric difference between A and B.

Since $t \leq r + r^{1-p}t^p$ for $t > 0$, it follows that $\|x_A - x_B\| \leq r + \varepsilon^p r^{1-p}\Delta(A, B)$ for every $A, B \subset \{1, \dots, n\}$. We can thus choose, for every such set A, a point $y_A \in l_\infty$ such that

$$\|y_A - x_A\| \leq r/2 \text{ and } \|y_A - y_B\| \leq \varepsilon^p r^{1-p}\Delta(A, B). \tag{1.7}$$

Indeed, the ith coordinate of y_A is defined by the formula

$$y_A(i) = \min_C \big\{x_C(i) + r/2 + \varepsilon^p r^{1-p}\Delta(A, C)\big\}.$$

We check the second inequality for example. Assume that $y_A(i) \geq y_B(i)$, and choose a set C such that $y_A(i) = x_C(i) + r/2 + \varepsilon^p r^{1-p}\Delta(A, C)$. Then

$$\begin{aligned} 0 &\leq y_A(i) - y_B(i) \\ &\leq \big(x_C(i) + r/2 + \varepsilon^p r^{1-p}\Delta(A, C)\big) - \big(x_C(i) + r/2 + \varepsilon^p r^{1-p}\Delta(B, C)\big) \\ &\leq \varepsilon^p r^{1-p}\Delta(A, B). \end{aligned}$$

Denote the group of permutations on $\{1, \dots, n\}$ by Σ. The action of a permutation σ on a set A is denoted by σA, and the isometry that σ induces on l_p^n is denoted by U_σ. Thus $U_\sigma x(i) = x(\sigma^{-1}i)$, $U_{\tau\sigma} = U_\tau U_\sigma$ and $U_\sigma x_A = x_{\sigma A}$.

To show that the modulus of continuity of φ satisfies the estimate of the lemma, we shall use only the fact that φ maps $Y = l_p^n \cup \big\{y_A : A \subset \{1, \dots, n\}\big\}$ into l_p^n and satisfies $\varphi(x_A) = x_A$ for every set A. Fix A and put

$$z_A = \frac{1}{n!}\sum_{\sigma \in \Sigma} U_\sigma \varphi(y_{\sigma^{-1}A}).$$

Direct computation shows that $U_\tau z_A = z_{\tau A}$ for every permutation τ. In particular it follows that $U_\tau z_A = z_A$ whenever A is invariant under τ. Also, since the U_σ's act as isometries on l_p^n

$$\begin{aligned} (1.8) \qquad \|z_A - x_A\| &\leq \frac{1}{n!}\sum_{\sigma\in\Sigma}\|U_\sigma\varphi(y_{\sigma^{-1}A}) - x_A\| \\ &= \frac{1}{n!}\sum_{\sigma\in\Sigma}\|\varphi(y_{\sigma^{-1}A}) - x_{\sigma^{-1}A}\| \leq \omega(r/2). \end{aligned}$$

Similarly $\Delta(\sigma A, \sigma B) = \Delta(A, B)$ implies that

$$(1.9) \qquad \|z_A - z_B\| \leq \omega\big(\varepsilon^p r^{1-p}\Delta(A, B)\big)$$

for all $A, B \subset \{1, \dots, n\}$.

Let $m = [\frac{n}{2}]$, and put $C = \{1, \dots, m+1\}$. Since $U_\sigma z_C = z_C$ for all permutations that leave C (and the complement of C) invariant, it follows that $z_C(i)$ is a constant, say α, when $i \in C$, and $z_C(i)$ is identically β if $i \notin C$. By (1.8) it then follows that

$$(m+1)|\alpha - \varepsilon|^p + (n-m-1)|\beta|^p = \|z_C - x_C\|^p \leq \omega(r/2)^p.$$

Hence $\varepsilon - \alpha \leq \omega(r/2)(m+1)^{-1/p}$. Similarly, if $D = \{1, \dots, m\}$, then z_D is constant on D and on its complement; and if we denote the constant value of z_D on the complement of D by γ, then $\gamma \leq \omega(r/2)(n-m)^{-1/p}$.

Equation (1.9) and the choice of C and D finally give that

$$\begin{aligned} \omega(\varepsilon^p r^{1-p}) &\geq \|z_C - z_D\| \geq \big(z_C - z_D\big)(m+1) \geq \alpha - \gamma \\ &\geq \varepsilon - \omega(r/2)\big((m+1)^{-1/p} + (n-m)^{-1/p}\big) \geq \varepsilon - 4\omega(r/2)n^{-1/p} \end{aligned}$$

by the choice of m. □

Remark. Analysis of the proof shows that if instead of a retraction we only assume that $\|\varphi x_A - x_A\| < \varepsilon/10$, then $\omega(r^{1-p}\varepsilon^p) + 4\omega(r/2)n^{-1/p} \geq \varepsilon/2$.

Corollary 1.29. *Let $1 < p < \infty$, and put $\gamma = \min(1/2, 1/p)$ and $E = L_p(\mu)$.*

(i) *Assume that Y is a metric space containing E such that $d(y, E) \leq r$ for all $y \in Y$. Then there is a retraction from Y onto E whose modulus of continuity satisfies $\omega(\varepsilon) \leq C_p r^{1-\gamma}\varepsilon^\gamma$.*

(ii) *Assume that Y is a metric space containing the closed unit ball $B(E)$ of E. Then there is a retraction from Y onto $B(E)$ whose modulus of continuity satisfies $\omega(\varepsilon) \leq C_p \varepsilon^\gamma$.*

(iii) *The estimates of* (i) *and* (ii) *are sharp. In particular $L_p(\mu)$ is not an absolute uniform retract.*

Proof. Parts (i) and (ii) follow from Theorem 1.27 and the fact (see Appendix A) that the modulus of convexity of $E = L_p(\mu)$ is equivalent to ε^2 for $1 < p \leq 2$ and to ε^p for $2 \leq p < \infty$.

Part (iii) follows from Lemma 1.28. Letting $n \to \infty$ shows that the estimate in (i) is sharp for $p \geq 2$. For $p \leq 2$ the result follows from the case $p = 2$ and the fact that every $L_p(\mu)$ space contains uniformly complemented copies of l_2^n. (See the remark after Proposition D.7.)

The sharpness of (ii) for $p < 2$ follows again from the estimate for $p = 2$, so assume $p \geq 2$. If $\varepsilon = n^{-1/p}$, then all the points $\{x_A\}$ of Lemma 1.28 belong to $B(E)$. Taking r in the lemma such that $4\omega(r/2) < 1/2$ and $\varepsilon = n^{-1/p}$, we obtain that $\omega(r^{1-p}\varepsilon^p) \geq \varepsilon/2$, yielding the lower estimate on ω.

The final remark also follows from Lemma 1.28. Fixing $\varepsilon > 0$ and letting $n \to \infty$ and then $r \to \infty$ give $0 \geq \varepsilon$, which is impossible. □

Remark. The arguments above do not work for $p = 1$. However, by Theorem 9.1 the unit ball of $L_1(\mu)$ is uniformly homeomorphic to the unit ball of $L_p(\mu)$ with a homeomorphism which is Lipschitz in one direction and Hölder$(1/p)$ in the other. Thus (ii) holds for $E = L_1(\mu)$ with $\omega(\varepsilon) < C_\alpha \varepsilon^\alpha$ for every $\alpha < 1/2$.

Being an absolute uniform retract is, of course, preserved under uniform homeomorphisms. It thus follows from the results of Chapter 9 that the unit balls of many nonuniformly convex spaces are also absolute uniform retracts. We shall see in the second volume that the unit balls of l_∞ and of c_0, on the other hand, are not uniformly homeomorphic to the unit ball of a uniformly convex space, and yet, being Lipschitz retracts of their respective spaces, they are certainly absolute Lipschitz retracts. This discussion may suggest that $B(E)$ is always an absolute uniform retract, but this is not so. The next example shows that $B(E)$ need not even be an absolute neighborhood uniform retract.

Example 1.30. *Let $E = (\sum \oplus l_{p_k})_2$. If $p_k \to \infty$, then $B(E)$ is not an absolute neighborhood uniform retract.*

Indeed, assume that there is a retraction from a ρ-neighborhood of $B(E)$ in l_∞ onto $B(E)$ with modulus of continuity ω. Since each l_{p_k} is norm one complemented in E, it follows that there is a uniformly continuous retraction with modulus ω onto the unit ball of each l_{p_k} from its ρ-neighborhood in l_∞. As in the proof of the corollary, choose $r < \rho$ such that $4\omega(r/2) < 1/2$. Fix now ε such that $\varepsilon < r$, and choose n_k such that $\varepsilon \sim n_k^{-1/p_k}$. We then obtain that $\omega(r^{1-p_k}\varepsilon^{p_k}) \geq \varepsilon/2$, and letting $k \to \infty$, one finds that the left-hand side converges to 0, a contradiction.

Remarks. (i) Since $B(E)$ is a Lipschitz retract of E, it follows that the whole space $E = (\sum \oplus l_{p_k})_2$ is also not an absolute neighborhood uniform retract.

(ii) Embed $E = (\sum \oplus l_{p_k})_2$ isometrically as a linear subspace of l_∞. Then there is no lifting for the quotient map $\pi : l_\infty \to l_\infty/E$ which is uniformly continuous on the unit ball of l_∞/E. (Recall that such liftings do exist for quotient maps on spaces with a uniformly normal structure; see Corollary 1.25.) Indeed, assume such a lifting φ exists, and then $\psi(x) = x - \varphi(\pi x) + \varphi(0)$ is a uniformly continuous map from $B(l_\infty)$ into E, which is the identity on $B(E)$. Thus, if $\tau : E \to B(E)$ is a Lipschitz retraction, then $\tau \circ \psi$ is a uniformly continuous retraction of $B(l_\infty)$ onto $B(E)$, contradicting the fact that the latter is not an absolute neighborhood uniform retract.

5. Notes and Remarks

A Banach space E is called injective if there is a linear projection on E from any Banach space containing it. In this case there is a λ such that the norms of these projections are bounded by λ, and then E is called a $\mathcal{P}_\lambda$-space. For $\lambda = 1$ a real Banach space E is $\mathcal{P}_1$ if and only if it has the binary intersection property for balls, hence if and only if it is an absolute 1-Lipschitz retract. The work of Nachbin, Goodner, Kelley and Hasumi characterizes real and complex $\mathcal{P}_1$ spaces as the $C(K)$ spaces for extremally disconnected compact Hausdorff spaces K. For details see Day [**138**], pp. 123–125.

We shall use in the next chapter, and therefore prove here, that a real $L_\infty(\mu)$ space with μ finite has the binary intersection property, and hence is a $\mathcal{P}_1$-space. To this end we first show that under its usual lattice structure $L_\infty(\mu)$ is a complete lattice; i.e., every order-bounded set in $L_\infty(\mu)$ has a supremum. (A set A is order-bounded if there is a $g \in L_\infty(\mu)$ such that $g \geq f$ for every $f \in A$. A function g_0 is a supremum for A if $g_0 \geq f$ for every $f \in A$ and $g_0 \leq g$ for every g which satisfies $g \geq f$ for every $f \in A$.) Note that this requires some care because the sets are not necessarily countable, and measurability is possibly lost if the supremum is taken pointwise. Thus let A be an order-bounded set in $L_\infty(\mu)$, and let $\tilde{A}$ be the set of all suprema of countable subsets of A. Then $\tilde{A}$ is closed under countable suprema, and therefore there is a $g_0 \in \tilde{A}$ such that $\int g_0 d\mu = \sup\{\int f d\mu : f \in \tilde{A}\}$. One checks easily that g_0 is the required supremum. Assume now that $\{B(f_\alpha, r_\alpha) : \alpha \in \Gamma\}$ is a family of mutually intersecting balls in $L_\infty(\mu)$; hence $f_\alpha - r_\alpha \leq f_\beta + r_\beta$ for all $\alpha, \beta \in \Gamma$. Then $\sup\{f_\alpha - r_\alpha : \alpha \in \Gamma\}$ belongs to all these balls.

Isbell [**267**] introduced a general construction of absolute 1-Lipschitz retracts, which, in general, are not Banach spaces. Let (X, d) be a metric space. A function $f : X \to \mathbb{R}$ is called extremal if $f(x) + f(y) \geq d(x, y)$ for all $x, y \in X$ and if it is minimal in the sense that the only function $g : X \to \mathbb{R}$ which satisfies the same inequality and $g(x) \leq f(x)$ for every $x \in X$ is $g = f$. The set of all extremal functions on X, equipped with the metric $\rho(f, g) = \sup\{|f(x) - g(x)| : x \in X\}$, is denoted by ϵX. It was called by Isbell the injective envelope of X. It is easy to check that $(\epsilon X, \rho)$ is metrically convex and has the binary intersection property, hence an absolute 1-Lipschitz retract. The map $x \to f_x$, where $f_x(y) = d(x, y)$, is an isometry of X into ϵX, and ϵX is minimal in the sense that there is no proper subset of ϵX which is an absolute 1-Lipschitz retract and contains the natural copy of X. This property characterizes ϵX. If $Y \supset X$ is an absolute 1-Lipschitz retract which is minimal as above, then there is an isometry of Y onto ϵX whose restriction to X is the natural isometry from X to ϵX.

If X is compact, then so is ϵX. If E is a Banach space, then ϵE can be given (a unique) linear structure which makes it an injective Banach space; see Cohen [**123**] and Isbell [**266**]. The article [**267**] contains other interesting facts concerning the properties of ϵX for general metric spaces.

There are no separable infinite-dimensional injective Banach spaces. This was proved by Grothendieck [**232**], who showed that every operator T from $l_\infty(\Gamma)$ to a separable space is weakly compact. If T is a projection, it then follows from the Dunford-Pettis theorem (see Dunford and Pettis [**158**] or Dunford and Schwartz [**159**], p. 494) that $T = T^2$ is even compact, and hence that T has a finite-dimensional range.

Every real $\mathcal{P}_\lambda$ space is an absolute λ-Lipschitz retract. Indeed, it is linearly isometric to a subspace of some $l_\infty(\Gamma)$ space and is the image of a λ-Lipschitz retraction (even linear projection) there. Example 1.5 shows that the converse is false, and Banach spaces which are absolute Lipschitz retracts do not have to be $\mathcal{P}_\lambda$ spaces.

The complex line is a complex $\mathcal{P}_1$-space by the complex version of the Hahn-Banach theorem. As a metric space it is isometric to the Euclidean plane and its projection constant is $4/\pi$; see Grünbaum [**237**]. It follows that a complex $l_\infty(\Gamma)$ is a $4/\pi$-absolute Lipschitz retract, and hence every complex $\mathcal{P}_\lambda$-space is a $4\lambda/\pi$-absolute Lipschitz retract.

Example 1.5 and Theorem 1.6 are due to Lindenstrauss [**357**]. (A weaker version of Theorem 1.6 was proved earlier by Isbell [**265**].) It is unknown if $C(K)$ is an absolute Lipschitz retract for every compact Hausdorff space K, but if Y is a metric space containing $C(K)$, and if $Y \setminus C(K)$ is finite, then there is a retraction from Y onto $C(K)$ whose Lipschitz constant is at most 2 (see [**357**]). Since the second dual of a $C(K)$ space is injective (see Theorem F.2(v)), this problem is a special case of the question whether every Banach space E is a Lipschitz retract of E^{**}.

Isbell [**265**] proved that $H(E)$ is an absolute uniform retract, and Lindenstrauss observed in [**357**] that the proof actually shows that it is an absolute Lipschitz retract. The proof of Theorem 1.7 that we gave here is due to Przesławski and Yost [**480**]. Lemma 1.8 was used earlier by Le Donne and Marchi [**153**] and by Artstein [**25**] in a similar context. Let $H_\alpha\big(B(E)\big)$ denote the set of all convex subsets A of $B(E)$ for which there are two concentric balls B_1 and B_2 with $B_1 \subset A \subset B_2$ such that the ratio of their radii is bounded by α. R. Nahum [**432**] has modified the proof of Theorem 1.7 and showed that $H_\alpha\big(B(E)\big)$ is an absolute Lipschitz retract for every $\alpha > 1$.

Proposition 1.10 is closely related to the real interpolation method (see Bergh and Löfström [**56**]) and to the representation of the space of Hölder(α) functions from X to $\mathbb{R}$ as an interpolation space between the spaces of Lipschitz and of bounded continuous functions from X to $\mathbb{R}$. A detailed discussion of the proposition and its applications for $Y = \mathbb{R}$ appears in Appendix B of the book [**230**] which was written by Semmes.

Proposition 1.11 was observed by several authors; see e.g. [**125**]. A sphere S in a Banach space is a typical example of a set which is not metrically convex but for which there is a λ such that every $x, y \in S$ are connected by an arc in S whose length is smaller than $\lambda d(x, y)$. A detailed study of length of arcs in spheres and of their relation to the structure of Banach spaces is contained in Schäffer [**514**].

Theorem 1.12 for $\omega(t) = \lambda t$, i.e., for extensions of Lipschitz functions, is due to Kirszbraun [**327**]. The general case and the introduction of the class $\mathcal{F}$ are due to F. Grünbaum and Zarantonello [**238**].

Corollary 1.15 is due to Minty [**424**]. Instead of using Proposition 1.13 he introduces a general scheme from which Corollary 1.15 and Proposition 1.13 both follow.

Theorem 1.16 is due to Michael [**412**]. Part (ii) of Proposition 1.19 is due to Bartle and Graves [**38**], and its proof here is also from [**412**]. (Some of the ingredients of the proof of Theorem 1.16 already appear in [**38**].)

The article of Preiss and Vilímovský [**473**] contains several results in the spirit of Proposition 1.18(i) for uniformly continuous functions.

Example 1.20 is taken from Aharoni and Lindenstrauss [**3**]. It can be shown that D is isometric to the space of continuous functions on the Cantor set Δ, and we thus obtain an example of a noncomplemented embedding of $C[0,1]$ into $C(\Delta)$. The first example of this type was given by Amir [**13**].

Theorem 1.21 is one variant of the extension theorems of Borsuk [**79**], Kakutani [**307**], and Dugundji [**157**]. An immediate consequence of it is that $C(H)$ is isometric to a norm-one complemented subspace of $C(K)$. Theorem 1.21 does not hold without the assumption that H is metrizable; see for example Arens [**19**],

Corson and Lindenstrauss [**126**] and Benyamini [**47**]. For a detailed study of extension operators see Pełczyński [**451**].

Theorem 1.27 is due to Lindenstrauss [**357**]. Theorem 1.24, which extends the qualitative part of Theorem 1.27 to families of sets with uniformly normal structure, is due to Skaletskii [**536**]. The proof we give here, as well as Proposition 1.22 is due to Przesławski and Yost [**480**].

Amir observed in [**14**] that Theorem 1.27 actually characterizes uniformly convex spaces. If E is a Banach space such that the Chebyshev center $c(A)$ of every bounded convex set $A \in H(E)$ is a singleton, and if the map $A \to c(A)$ is uniformly continuous on bounded sets, then E is uniformly convex.

Lemma 1.28, Corollary 1.29 and Example 1.30 are due to Lindenstrauss [**357**].

CHAPTER 2

Retractions, Extensions and Selections (Special Topics)

1. Approximation of Uniformly Continuous Functions

In this chapter we present several topics related to the material of Chapter 1. The various sections are independent of each other.

We start by studying the approximation of uniformly continuous functions by functions with a given modulus of continuity. The next proposition lists some specific cases of a general principle: An extension theorem for a class of functions implies an approximation theorem by functions in this class.

Proposition 2.1. *Let X be a metric space, and let E be a Banach space. Fix $\varepsilon > 0$, and let $f : X \to E$ be a uniformly continuous map whose modulus of continuity ω_f is dominated by a subadditive function ω satisfying $\lim_{t\to 0} \omega(t) = 0$.*

(i) *Let $E = l_\infty(\Gamma)$ for some set Γ (hence, in particular, E could be $\mathbb{R}$). Then there is a Lipschitz function $g : X \to E$ with Lipschitz constant $2\omega(\varepsilon)/\varepsilon$ such that $\|g(x) - f(x)\| \le 3\omega(\varepsilon)$ for every $x \in X$.*

(ii) *Let X be a subset of a Hilbert space, and let E be a Hilbert space. Then there is a Lipschitz function $g : X \to E$ with Lipschitz constant $2\omega(\varepsilon)/\varepsilon$ such that $\|g(x) - f(x)\| \le 3\omega(\varepsilon)$ for every $x \in X$.*

Proof. (i) Let $\mathcal{A}$ be a maximal ε-separated set in X; i.e., $d(x,y) \ge \varepsilon$ for every $x \ne y$ in $\mathcal{A}$ and $\mathcal{A}$ is maximal with this property. The subadditivity of ω implies that $\|f(x) - f(y)\| \le \omega\big(d(x,y)\big) \le 2\omega(\varepsilon)d(x,y)/\varepsilon$ for all $x, y \in \mathcal{A}$. By Lemma 1.1 the restriction of f to $\mathcal{A}$ extends to a Lipschitz function g on all of X with the same Lipschitz constant $2\omega(\varepsilon)/\varepsilon$. Given $x \in X$, choose $y \in \mathcal{A}$ such that $d(x,y) < \varepsilon$. Then $g(y) = f(y)$, and thus

$$\|g(x) - f(x)\| \le \|g(x) - g(y)\| + \|f(y) - f(x)\| \le 3\omega(\varepsilon).$$

Part (ii) follows similarly from Theorem 1.12. □

The next proposition gives a result of a similar nature for approximating bounded uniformly continuous functions into a Banach space whose unit ball is an absolute uniform retract. If a metric space X is an absolute uniform retract, then there is a function η such that in every metric space containing X there is a retraction on X whose modulus of continuity is bounded by η. We shall say that X is an η-absolute uniform retract.

Proposition 2.2. *Let ω be a subadditive function with $\lim_{t\to 0} \omega(t) = 0$, and let F be a Banach space such that $B(F)$ is an η-absolute uniform retract. Then for every $\delta > 0$ and $r > 0$ there are constants a and b (depending on δ, r and ω) with the following property: Let X be a metric space, and let $f : X \to rB(F)$ be a uniformly* *continuous function whose modulus of continuity is bounded by* ω. *Then there is a uniformly continuous function* $h : X \to rB(F)$ *whose modulus of continuity is bounded by* $a\eta(b\varepsilon)$ *and such that* $\|f(x) - h(x)\| \le \delta$.

If F *is uniformly convex, then* f *can be approximated by Hölder*(α) *functions for some* $\alpha > 0$. *If* $F = L_q(\mu)$ *with* $1 \le q < \infty$, *then* α *can be taken to be* $\min(1/2, 1/q)$.

Proof. Assume that $r = 1$. Embed F in l_∞, and use the previous proposition to approximate f by a Lipschitz function $g : X \to l_\infty$. Let $\rho : l_\infty \to B(F)$ be a retraction whose modulus of continuity is bounded by η. The desired approximating function is then $\rho \circ g$. We leave the details to the reader.

If F is uniformly convex we renorm it so that $\delta_F(\varepsilon) \ge C\varepsilon^p$ for some $p < \infty$ by Theorem A.6, and then use Theorem 1.27 (and Theorem 1.7).

If $1 < q < \infty$ we use the known estimate of the modulus of convexity of $L_q(\mu)$.

For $q = 1$ the Mazur map $\phi : B(L_1(\mu)) \to B(L_2(\mu))$ is Hölder(1/2) and its inverse is Lipschitz (see Proposition 9.2). By the case $q = 2$ the map $\phi \circ f$ can be approximated by Hölder(1/2) maps $\{h_n\}$, and then $\phi^{-1} \circ h_n$ are also Hölder(1/2), and they approximate f. □

Denote the unit ball of $L_p(\mu)$ by B_p. We now compute the exact exponent α for which every uniformly continuous function from B_p into $L_q(\nu)$ can be approximated by Hölder(α) functions. To avoid technicalities we take $\mu = \nu$ (although, as the proof shows, the results also hold for different measures), and to simplify the notation we suppress the measure altogether.

Theorem 2.3. *Let* $1 \le p, q \le \infty$. *Every uniformly continuous function from* B_p *to* L_q *can be uniformly approximated by Hölder*(α) *functions where*

$$\alpha = \begin{cases} 1 & \text{if } p \ge 2 \ge q \text{ or } q = \infty \\ \min\{1/2, 1/q\} & \text{if } p = \infty \text{ and } q < \infty \\ 2/q & \text{if } p, q \ge 2 \\ p/2 & \text{if } p, q \le 2 \\ p/q & \text{if } p \le 2 \le q \end{cases}$$

and this is the best possible value of α *(when the spaces are infinite-dimensional).*

Proof. Let $1 \le r, s < \infty$. The Mazur map

$$\phi_{s,r}(x)(t) = |x(t)|^{s/r}\operatorname{sign}(x(t))$$

from L_s to L_r satisfies $\phi_{s,r}^{-1} = \phi_{r,s}$, and we shall see in Proposition 9.2 that the restriction of $\phi_{s,r}$ to balls in L_s is Lipschitz for $s \ge r$ and Hölder(s/r) for $s \le r$.

To prove the first part of the theorem for $p, q < \infty$, consider the uniformly continuous map $\phi_{q,2} \circ f \circ \phi_{2,p} : B_2 \to L_2$. By Proposition 2.1 it is the uniform limit of a sequence $g_n : B_2 \to L_2$ of Lipschitz functions. The functions $\phi_{2,q} \circ g_n \circ \phi_{p,2}$ converge uniformly to f, and direct inspection shows that they are in the appropriate Hölder(α) classes for the different values of p and q.

For $q = \infty$ the estimate follows from Proposition 2.1(i) because L_∞ is a $\mathcal{P}_1$-space, and for $p = \infty$ it follows from Proposition 2.2.

The following claim is the main tool in showing that these values of α are best possible.

Assume that $1 \le p < q < \infty$, *and let* $\alpha > p/q$. *If* $\psi : B(l_p^n) \to l_q^n$ *is a Hölder*(α) *map, then* $\|\psi - \phi_{p,q}\|_\infty \ge 1/2 - \delta_n$, *where* $\delta_n \to 0$ *as* $n \to \infty$.

Proof of the Claim. Denote the ith coordinate of x by $x(i)$. For every permutation σ of $\{1,\dots,n\}$ and for every choice of signs $\theta=(\theta_1,\dots,\theta_n)$ the operator

$$U_{\sigma,\theta}x(i)=\theta_i x(\sigma^{-1}i)$$

is an isometry on both l_p^n and l_q^n. Thus the operator

$$V_{\sigma,\theta}f(x)=U_{\sigma,\theta}\left(f(U_{\sigma,\theta}^{-1}x)\right)$$

which is a norm-one operator on bounded functions from l_p^n to l_q^n preserves the modulus of continuity of f. The average

$$V=\frac{1}{n!2^n}\sum_{\sigma,\theta}V_{\sigma,\theta}$$

commutes with the $V_{\sigma,\theta}$'s. Thus $\varphi=V\psi$ (which is a Hölder(α) function with the same constant C as ψ) satisfies $V_{\sigma,\theta}\varphi=\varphi$, i.e., $U_{\sigma,\theta}\varphi(x)=\varphi(U_{\sigma,\theta}x)$ for all x,σ and θ. In particular, if χ_A is the indicator function of a subset A of $\{1,\dots,n\}$ and if $\delta>0$, then $\varphi(\delta\chi_A)=\lambda\chi_A$ where the constant λ depends only on the cardinality of A and on δ. Also $V_{\sigma,\theta}\phi_{p,q}=\phi_{p,q}$ implies that $\|\varphi-\phi_{p,q}\|_\infty\le\|\psi-\phi_{p,q}\|_\infty$.

Assume that $n=2m$ is even, and put $x_k=n^{-1/p}\chi_{\{k,\dots,k+m-1\}}$ for $k\le m+1$. As noted above there is a λ independent of k such that $\varphi(x_k)=\lambda\chi_{\{k,\dots,k+m-1\}}$. Thus the vectors $\varphi(x_{k+1})-\varphi(x_k)$ for $1\le k\le m$ are disjointly supported and

$$\begin{aligned}\|\varphi(x_{m+1})-\varphi(x_1)\|_q &= \Big\|\sum\big(\varphi(x_{k+1})-\varphi(x_k)\big)\Big\|_q\\ &= \Big(\sum\|\varphi(x_{k+1})-\varphi(x_k)\|_q^q\Big)^{1/q}\\ &\le m^{1/q}C(2^{1/p}n^{-1/p})^\alpha\to 0\end{aligned}$$

as $n\to\infty$ because $\alpha>p/q$. But the definitions of $\phi_{p,q}$ and of the x_k's give

$$\|\phi_{p,q}(x_{m+1})-\phi_{p,q}(x_1)\|_q=1.$$

Thus

$$\begin{aligned}&2\sup\{\|\varphi(x)-\phi_{p,q}(x)\|_q:\|x\|_p\le 1\}\\ &\quad\ge\Big\|\big(\phi_{p,q}(x_{m+1})-\phi_{p,q}(x_1)\big)-\big(\varphi(x_{m+1})-\varphi(x_1)\big)\Big\|_q\\ &\quad\ge\|\phi_{p,q}(x_{m+1})-\phi_{p,q}(x_1)\|_q-\|\varphi(x_{m+1})-\varphi(x_1)\|_q\end{aligned}$$

which converges to 1.

We now complete the proof of the theorem.

When $p\ge 2\ge q$ or $q=\infty$; i.e., when $\alpha=1$, there is nothing to prove. The result for $p=\infty$ follows from $p=1$ because B_∞ contains a subset isometric to B_1 and B_1 is an absolute uniform retract.

The case $p\le 2\le q$ follows directly from the claim. Indeed, for each n represent l_p^n as the subspace of L_p consisting of functions that are constant on some fixed n disjoint sets (and similarly for l_q^n). Then $\phi_{p,q}:L_p\to L_q$ restricts to the natural Mazur map from l_p^n into l_q^n. Since l_q^n is complemented in L_q by a norm-one projection, $\phi_{p,q}$ cannot be approximated better as a map from L_p to L_q than the corresponding map between l_p^n and l_q^n.

For $p,q\ge 2$ we use the facts that L_p contains a subspace F isomorphic to L_2 and that the unit ball of L_q is an absolute uniform retract (Corollary 1.29(ii)), and hence that every uniformly continuous map from $B(F)$ to L_q extends to a uniformly continuous map from B_p into L_q. Thus the estimate cannot be better than the estimate $\alpha \le 2/q$ for approximation from B_2 to L_q.

If $p, q \le 2$ and $q > 1$, we use the fact that L_q contains a complemented copy of L_2 (see the remark after Proposition D.7). It follows that we cannot obtain a better estimate (up to a constant depending only on q) than for the approximation of maps from B_p to L_2.

Finally, fix $p \le 2$ and let α be the best possible exponent for $q = 1$. Let $s > 1$. Factoring functions from B_p to L_s through L_1 (using the Hölder(s) map $\phi_{1,s}$) shows that they can be approximated by Hölder(α/s) functions. On the other hand, by the previous case, the best possible exponent for approximation from B_p to L_s is $p/2$. Thus $\alpha/s \le p/2$, and since $s > 1$ was arbitrary $\alpha \le p/2$. □

Remarks. (i) Except for the case of $p \le 2$ and $q = 1$ the proof actually shows that the result is best possible in a stronger sense. It is impossible to approximate with functions whose moduli of continuity satisfy $\liminf_{t\to 0} \omega(t)/t^\alpha = 0$ for the critical indices α. We do not know if this is also true for $p \le 2$ and $q = 1$.

(ii) For $q \ge 2$ the estimate $\alpha \le 2/q$ actually holds for every infinite-dimensional Banach space E. Indeed, B_q is an absolute uniform retract, and by Dvoretzky's theorem E contains l_2^n's uniformly. Thus α cannot be bigger than for functions from B_2 to L_q.

We now discuss the approximation of continuous maps by differentiable maps. Section 1 of Chapter 4 presents the basic facts on differentiability.

Let E and F be Banach spaces with E^* separable, and let U be an open subset of E. We shall see in Corollary 4.14 that every continuous map $f : U \to F$ can be uniformly approximated by C^1 maps. The derivatives of the approximating maps are, however, only continuous, and approximation by maps with uniformly continuous derivatives is not necessarily possible even when U is the unit ball of E and when f is uniformly continuous on U. Indeed, a map with a uniformly continuous derivative on the unit ball would be Lipschitz, but by Theorem 2.3 approximation by Lipschitz maps is not always possible. We shall show that there is even a Lipschitz map from the unit ball of a Hilbert space into itself which cannot be approximated by Fréchet differentiable maps with uniformly continuous derivatives. The difference between C^1 maps and maps with a uniformly continuous derivative will also play a role in Section 1 of Chapter 11. Denote the unit ball of $L_2(\mathbb{R})$ by B_2.

Proposition 2.4. *Consider the map $f\big(x(t)\big) = |x(t)|$ from B_2 to itself. Then $\|g - f\|_\infty \ge 1/16$ for every Fréchet differentiable map g with a uniformly continuous derivative.*

Proof. Assume that g satisfies $\|g - f\|_\infty < 1/16$ and is Fréchet differentiable with a uniformly continuous derivative, and we shall obtain a contradiction.

The map f is positive homogeneous, commutes with measure preserving transformation of $\mathbb{R}$, depends only on the absolute value of $x \in B_2$, and commutes with scaling of $\mathbb{R}$, i.e., with the operators $U_a x(t) = a^{1/2} x(at)$ for $a > 0$. In the first part of the proof we show that we may assume that to a large extent g shares these properties with f. We shall do this in four steps, and we leave it to the reader to check that in each of these steps we do not destroy the invariance properties achieved in the previous steps.

(i) Changing 1/16 to 1/8, we may assume that $g(x) = \|x\| g(x/\|x\|)$ whenever $\|x\| \geq 1/2$. Indeed, just replace g by $h(x) = \tau(\|x\|) g(x/\|x\|)$ for $x \neq 0$ (and put $h(0) = 0$), where $\tau : [0,1] \to [0,1]$ is C^2 with $\tau(0) = \tau'(0) = \tau''(0) = 0$, $|\tau(t) - t| < 1/16$ for every $0 \leq t \leq 1$, and $\tau(t) = t$ for $t \geq 1/2$. Verifying that h satisfies the required properties is routine. We only check that it approximates f. Since f is positive homogeneous, it follows that for every x with $\|x\| \leq 1$

$$\begin{aligned}&\|f(x) - h(x)\| \\ &\leq \Big| \|x\| - \tau(\|x\|) \Big| \big\| f(x/\|x\|) \big\| + \tau(\|x\|) \Big\| f(x/\|x\|) - g(x/\|x\|) \Big\| < 1/8.\end{aligned}$$

In the next steps we use averaging procedures. These steps do not increase the modulus of continuity of the derivative of the map g from step (i). We denote this modulus of continuity by ω, and recall that by (4.3)

$$\|g(x+y) - g(x) - D_g(x)y\| \leq \omega(\|y\|)\|y\| \tag{2.1}$$

whenever $x, x+y \in B_2$.

(ii) We may replace g by a map which commutes with the measure preserving transformations of $\mathbb{R}$ which interchange pairs of disjoint intervals of the same length.

Fix N, and let G_N be the group of transformations of $\mathbb{R}$ which permute the $2N2^N$ intervals $\{[i2^{-N}, (i+1)2^{-N})\}_{-N2^N \leq i < N2^N}$. For each $\sigma \in G_N$ and $y \in B_2$ put $y_\sigma(t) = y(\sigma(t))$. Define $g_\sigma : B_2 \to B_2$ by $g_\sigma(x) = \big(g(x_\sigma)\big)_{\sigma^{-1}}$, and consider their average $g_N = |G_N|^{-1} \sum_{\sigma \in G_N} g_\sigma$. The derivative D_{g_N} is uniformly continuous with modulus of continuity bounded by ω; hence (2.1) holds with g replaced by g_N. The map g_N also commutes with G_N and satisfies $\|g_N - f\|_\infty \leq \|g - f\|_\infty$. If D_g is bounded on B_2 by K, then so is D_{g_N}. We denote the K-ball in the space of bounded linear operators on $L_2(\mathbb{R})$ by A. Note that g, and hence also g_N, maps B_2 into $2B_2$.

Equipped with the weak operator topology A is compact, and so is $2B_2$ under the weak topology. Thus the sequence $\{(g_N, D_{g_N})\}_{N=1}^\infty$, considered as maps from B_2 into $2B_2 \times A$ has a limit point (h, D) in the topology of pointwise convergence. The lower semi-continuity of the norm implies that $\|h - f\|_\infty \leq \|g - f\|_\infty$ and that (2.1) continues to hold when g is replaced by h and D_g by the function D. It follows that h is differentiable and that its derivative at x is $D(x)$. Similarly, the inequality $\|D_{g_N}(x) - D_{g_N}(y)\| \leq \omega(\|x - y\|)$ for $x, y \in B_2$ is also preserved in the limit. Now replace g by h.

(iii) We may assume that $g(x) = g(|x|)$ for every $x \in B_2$. Indeed, consider each measurable function σ with $|\sigma(t)| = 1$ a.e. as a unitary operator on $L_2(\mathbb{R})$ defined by $\sigma x = \sigma(t)x(t)$, and let S be the (abelian) group of such operators. Let M be an invariant mean on $l_\infty\big(S, L_2(\mathbb{R})\big)$ (see Corollary C.2), and replace g by $h(x) = M_\sigma\big(g(\sigma x)\big)$. The S-invariance of f and $\|M\| = 1$ imply that the S-invariant map h also satisfies $\|h - f\|_\infty < 1/8$ and (2.1). It is thus differentiable, and its derivative $D_h(x) = M_\sigma\big(D_g(\sigma x)\big)$ satisfies $\|D_h(x) - D_h(y)\| \leq \omega(\|x - y\|)$.

(iv) We may assume that $g\big(U_a(x)\big) = U_a\big(g(x)\big)$ for every $a > 0$ and $x \in B_2$, where $U_a x(t) = a^{1/2} x(at)$. Just replace g by $h(x) = M_a\big(U_a^{-1}(U_a x)\big)$, where M is an invariant mean on $l_\infty\big(\mathbb{R}^+, L_2(\mathbb{R})\big)$, and the abelian group $(\mathbb{R}^+, \cdot)$ acts on $L_2(\mathbb{R})$ as the unitary operators $\{U_a\}$.

We now pass to the second part of the proof. Let Δ be an interval in $\mathbb{R}$. By (ii) $g(\chi_\Delta)$ is constant on Δ and on its complement. Since it belongs to $L_2(\mathbb{R})$, it must vanish off Δ. Putting $x_a = \chi_{[0,a]}$ for $1/2 \le a \le 1$, we see that $g(x_a) = \lambda_a x_a$ for some constant λ_a. If $1/2 \le a < b \le 1$, then $U_{a/b}(x_a) = (a/b)^{1/2} x_b$. By (iv)

$$g\big(U_{a/b}(x_a)\big) = U_{a/b}\big(g(x_a)\big) = (a/b)^{1/2}\lambda_a x_b$$

and since $\|x_b\| \ge \|(a/b)^{1/2}x_b\| = a^{1/2} > 1/2$, we see by (i) that

$$g\big((a/b)^{1/2}x_b\big) = (a/b)^{1/2}g(x_b) = (a/b)^{1/2}\lambda_b x_b.$$

Thus $\lambda_a = \lambda_b$. Put $y = x_b - x_a$, and denote the derivative of g at x_a by D. Then $\|g(x_a \pm y) - g(x_a) - D(\pm y)\| \le \|y\|\omega(\|y\|)$ by (2.1), and $g(x_b) = g(x_a \pm y)$ by (iii); hence

$$\begin{aligned}\|g(x_b) - g(x_a)\| &= \frac{1}{2}\Big\|\big(g(x_a + y) - g(x_a) - Dy\big) + \big(g(x_a - y) - g(x_a) + Dy\big)\Big\| \\ &\le \|y\|\omega\big(\|y\|\big).\end{aligned}$$

Fix N, and put for $x_i = x_{1/2+i/2N}$ for $0 \le i \le N$ and $y_i = x_i - x_{i-1}$. Then $\|y_i\| = (2N)^{-1/2}$ and the functions $g(x_i) - g(x_{i-1})$ are disjointly supported, hence orthogonal. Thus

$$\begin{aligned}\|g(x_N) - g(x_0)\| &= \Big(\sum_{i=1}^{N} \|g(x_i) - g(x_{i-1})\|^2\Big)^{1/2} \\ &\le N^{1/2}(2N)^{-1/2}\omega\big((2N)^{-1/2}\big) \to 0\end{aligned}$$

as $N \to \infty$. But $\|g(x_N) - g(x_0)\|$ is bounded from below by $\|f(x_N) - f(x_0)\| - 1/4$ and $\|f(x_N) - f(x_0)\| = \|\chi_{[1/2,1]}\| > 1/4$, a contradiction. □

2. The Nearest Point Map

Let A be a closed convex subset of the Banach space E. If for every point $x \in E$ there is a point in A nearest to x, and if this point is unique, we denote this point by $P_A x$ and call the mapping P_A the metric projection from E to A.

In a Hilbert space P_A exists for every closed and convex set A, and it is non-expansive, i.e., $\|P_A x - P_A y\| \le \|x - y\|$ for all $x, y \in E$. When E is reflexive and strictly convex, P_A exists for every closed convex subset of E. It turns out that under suitable geometric conditions on E the metric projection on such sets is even uniformly continuous on bounded sets, and we obtain sharp estimates on the modulus of continuity of P_A in spaces that are both uniformly convex and uniformly smooth. The result will be applied to prove that spaces that are simultaneously "very" convex and "very" smooth are isomorphic to a Hilbert space.

Let A be a closed convex set, and let x be a point in E for which there is a $z \in A$ nearest to x. It is convenient to characterize z in terms of support functionals. The point $z \in A$ is nearest to x if and only if there is a norm-one functional $x^* \in E^*$ such that $\langle x^*, x - z\rangle = \|x - z\|$ and such that $\langle x^*, u\rangle \le \langle x^*, z\rangle$ for all $u \in A$. Indeed, if z is nearest to x, take x^* as the (normalized) functional separating A from the ball $B(x, \|x - z\|)$. The other direction is trivial. Note that z is also a nearest point in A to any point y on the ray through x emanating from z. Indeed, since $y - z$ is a positive multiple of $x - z$ we obtain $\langle x^*, y - z\rangle = \|y - z\|$.

Lemma 2.5. *Let A be a closed and convex subset of the uniformly convex Banach space E. Then P_A is uniformly continuous on bounded neighborhoods of A.*

Proof. Fix x. By translation we may assume that $P_A x = 0$, and since we deal with bounded neighborhoods we may also assume that $\|x\| = 1$. Let x^* be a norm-one functional such that $\langle x^*, x\rangle = 1$ and $\langle x^*, v\rangle \le 0$ for all $v \in A$.

Fix $\varepsilon > 0$ and a point u such that $\|u - x\| < \varepsilon$. Then $\|u - P_A u\| \le \|u\| \le 1+\varepsilon$. Since also $\langle x^*, P_A u\rangle \le 0$ and $\langle x^*, u\rangle = \langle x^*, x\rangle - \langle x^*, x - u\rangle \ge 1 - \varepsilon$ we obtain $\|(u - P_A u) + u\| \ge \langle x^*, (u - P_A u) + u\rangle \ge 2\langle x^*, u\rangle \ge 2 - 2\varepsilon$. By the definition of δ_E

$$\|P_A x - P_A u\| = \|P_A u\| = \|(u - P_A u) - u\| \le (1+\varepsilon)\delta_E^{-1}\big(2\varepsilon/(1+\varepsilon)\big).$$

□

Assume now that E is both uniformly convex and uniformly smooth. We introduce some notation. Fix a point x such that $\|x\| = 1$, and let $x^* \in E^*$ be the unique functional supporting $B(E)$ at x. The main tool we shall use is the geometry of the "cross-sections"

$$B_\alpha = \big\{u \in B(E) : \langle x^*, u\rangle = 1 - \alpha\big\}$$

obtained for $0 < \alpha < 1$ by intersecting the unit ball of E with parallel translates of the kernel of x^*. We shall consider two geometric parameters associated with the cross-sections, their diameter and inner radius given by

$$\begin{aligned} \operatorname{diam}(B_\alpha) &= \sup\big\{\|u - v\| : u, v \in B_\alpha\big\} \\ ir(B_\alpha) &= \inf\big\{\|u - x_\alpha\| : u \in B_\alpha,\ \|u\| = 1\big\} \end{aligned}$$

where $x_\alpha = (1-\alpha)x$. Since $\|u\| \le (1-\alpha) + \|u - x_\alpha\|$ for every u it follows that $\alpha \le ir(B_\alpha)$. Also, if $u \in B_\alpha$ and if $\|u - x_\alpha\| \le ir(B_\alpha)$, then $2x_\alpha - u$ is also in B_α; hence $ir(B_\alpha) \le \operatorname{diam}(B_\alpha)/2$.

The diameter and the inner radius give information on the "shape" of the cross-section B_α. If $ir(B_\alpha)$ is close to $\operatorname{diam}(B_\alpha)/2$, then B_α is "round" in the sense that the lengths of the radius vectors are similar in all directions. On the other hand, if $ir(B_\alpha)$ is much smaller than $\operatorname{diam}(B_\alpha)/2$, then there are directions in which it is much narrower than in others. It turns out that the metric projection has better continuity properties when the cross-sections B_α are more "round".

Let A be a closed convex set, and assume that $d(x, A) = 1$. By translation we may assume that $P_A x = 0$; hence $\|x\| = 1$. The cross-sections and the values of the associated parameters depend on the point x, but the estimates that follow are independent of x and depend only on the fact that $\|x\| = d(x, A) = 1$.

The continuity of P_A at x is measured by $\sup\{\|P_A u\| : \|u - x\| \le \varepsilon\}$ as a function of ε (because $P_A x = 0$). This is estimated in two steps. In the first step (Lemma 2.6) we show that for a suitable value of α, which is given in terms of the inner radius of B_α, this supremum can be estimated by $\sup\{\|P_A u\| : u \in B_\alpha\}$. In uniformly smooth spaces the inner radius cannot increase too fast and α is controlled by the modulus of smoothness of E. In the second step (Lemma 2.7) we show that the supremum over B_α is bounded by $\operatorname{diam}(B_\alpha)$ and that the latter is controlled by the values of the modulus of convexity of E.

Lemma 2.6. (i) *There is a $0 < \theta < 1/20$ such that if $0 < \varepsilon < \theta$ and if α is such that $ir(B_\alpha) = 3\varepsilon$, then*

$$\sup\big\{\|P_A u\| : \|u - x\| \le \varepsilon\big\} \le \sup\big\{\|P_A u\| : u \in B_\alpha\big\}.$$

(ii) *For every $0 < \beta < 1$*

$$ir(B_\beta) \ge (1-\beta)\rho_E^{-1}(\beta/2). \tag{2.2}$$

In particular, if α *is chosen as in* (i), *then* $\alpha \le 2\rho_E(6\varepsilon)$.

Proof. (i) P_A is uniformly continuous on bounded neighborhoods of A, so choose $\theta < 1/20$ such that $\|P_A u\| \le 1/20$ whenever $\|u - x\| < \theta$. (Note that the choice of θ is independent of x.)

Since $P_A u = P_A v$ whenever v lies on the ray emanating from $P_A u$ in the direction of u, we need to show only that this ray intersects B_α for each u satisfying $\|u - x\| \le \varepsilon$.

Fix such u, let v be the point where the ray intersects the unit sphere and put $\langle x^*, v\rangle = 1 - \tau$. We shall show that $\langle x^*, v\rangle \ge 1 - \alpha$, i.e., that $\tau \le \alpha$.

Let w be the (unique) point in the segment $[u, P_A u]$ that belongs to the kernel of x^*, and write $v = w + (1-\gamma)(u - w)$ where $1 - \gamma > 0$. Then

$$\begin{aligned} \tau &\le ir(B_\tau) \le \|v - x_\tau\| = \big\|w + (1-\gamma)(u-w) - (1-\tau)x\big\| \\ &\le \big\|w + (1-\tau)(u-w) - (1-\tau)x\big\| + |\gamma - \tau|\,\|u - w\| \\ &\le \tau\|w\| + (1-\tau)\|u - x\| + |\gamma - \tau|\,\|u - w\|. \end{aligned}$$

We shall estimate the terms on the right-hand side separately. Choose $0 \le \beta \le 1$ such that $w = \beta u + (1-\beta)P_A u$. The choice of θ gives

$$0 = \langle x^*, w\rangle = \beta\langle x^*, u\rangle + (1-\beta)\langle x^*, P_A u\rangle \ge \beta(1-\varepsilon) - (1-\beta)/20$$

hence $\beta < 1/20$. Using this estimate and the choice of θ again, we obtain

$$\|w\| \le \beta\|u\| + (1-\beta)\|P_A u\| \le \beta(1+\varepsilon) + (1-\beta)/20 < 1/4.$$

The definitions of τ and of γ give that $1 - \tau = \langle x^*, v\rangle = (1-\gamma)\langle x^*, u\rangle$. Since also $\big|\langle x^*, u\rangle - 1\big| \le \varepsilon$ we obtain that $\big|\gamma - \tau\big| \le \varepsilon(1-\gamma) \le \varepsilon/\langle x^*, u\rangle \le \varepsilon/(1-\varepsilon)$. Finally

$$\|u - w\| \le \|u - P_A u\| \le \|u\| \le 1 + \varepsilon.$$

Inserting these estimates into the inequality above, we obtain

$$\tau \le ir(B_\tau) \le \tau/4 + (1-\tau)\varepsilon + \varepsilon(1+\varepsilon)(1-\varepsilon)^{-1} \tag{2.3}$$

and in particular $3\tau/4 \le \varepsilon + \varepsilon(1+\varepsilon)(1-\varepsilon)^{-1} = 2\varepsilon/(1-\varepsilon)$.

When we insert this estimate of τ in the right-hand side of (2.3), we obtain $ir(B_\tau) \le 3\varepsilon$. Since $ir(B_\alpha) = 3\varepsilon$ it follows that $\tau \le \alpha$ as required.

(ii) Fix any point $u \in B_\beta$ with $\|u\| = 1$, and let $v = u - x_\beta$. To simplify the notation put $(1-\beta)^{-1} = \eta$. Then $\|x + \eta v\| = \eta\|x_\beta + v\| = \eta\|u\| = \eta$, while $\langle x^*, v\rangle = 0$ implies that $\|x - \eta v\| \ge \langle x^*, x - \eta v\rangle = \langle x^*, x\rangle = 1$. Hence

$$\begin{aligned} \rho_E\big(\eta\|v\|\big) &\ge \big(\|x + \eta v\| + \|x - \eta v\| - 2\big)/2 \\ &\ge (\eta + 1 - 2)/2 = \beta(1-\beta)^{-1}/2 > \beta/2 \end{aligned}$$

and $\|v\| \ge (1-\beta)\rho_E^{-1}(\beta/2)$. Since u was arbitrary this proves (2.2).

We now use $\alpha \le ir(B_\alpha) = 3\varepsilon$ in (2.2). Recalling that $\varepsilon < 1/20$, we see that $1 - \alpha \ge 1 - 3\varepsilon \ge 1/2$ and thus $3\varepsilon \ge \rho_E^{-1}(\alpha/2)/2$. Hence $\alpha \le 2\rho_E(6\varepsilon)$. □

Lemma 2.7. *For each* $0 < \alpha < 1$

$$\sup\big\{\|P_A u\| : u \in B_\alpha\big\} \le \operatorname{diam}(B_\alpha) \le \delta_E^{-1}(\alpha).$$

Proof. To prove the first inequality let $u \in B_\alpha$. Let w be the point of intersection of the kernel of x^* and the segment $[u, P_A u]$, and fix $0 \le \beta \le 1$ such that $w = (1-\beta)P_A u + \beta u$. Then $\|u - w\| = (1-\beta)\|u - P_A u\| \le (1-\beta)\|u\| \le (1-\beta) \le 1$, and since $\langle x^*, u - w\rangle = \langle x^*, u\rangle = 1 - \alpha$ it follows that $u - w \in B_\alpha$. Since actually

$\|u - w\| \le (1-\beta)$, all the points in the hyperplane $(x^*)^{-1}(1-\alpha)$ whose distance to $u - w$ is less than β also belong to B_α. Since also $\|P_A u - w\| \le \beta$ we obtain

$$\begin{aligned}\operatorname{diam}(B_\alpha) &\ge \|u - (u-w)\| + \beta = \|w\| + \beta \\ &\ge \|P_A u\| - \|P_A u - w\| + \beta \ge \|P_A u\|.\end{aligned}$$

To prove the second inequality fix any two points $u, v \in B_\alpha$. It follows that $\|(u+v)/2\| \ge \langle x^*, (u+v)/2\rangle = 1 - \alpha$ and $\|u - v\| \le \delta_E^{-1}(\alpha)$ by the definition of δ_E. Since u and v were arbitrary the result follows. □

Theorem 2.8. *Let E be a uniformly convex and uniformly smooth Banach space, and denote its moduli of uniform convexity and uniform smoothness by δ_E and ρ_E respectively. Then for every closed convex subset A of E the metric projection P_A is uniformly continuous on bounded neighborhoods of A and its modulus of continuity ω on the set $\{x : d(x, A) \le 1\}$ satisfies $\omega(\varepsilon) \le \delta_E^{-1}\big(2\rho_E(6\varepsilon)\big)$.*

Proof. Fix a point x such that $d(x, A) = 1$. By translation we may assume that $P_A x = 0$ and thus $\|x\| = 1$. Let u be such that $\|u - x\| \le \varepsilon$. By Lemma 2.6 $\|P_A u\| \le \sup\{\|P_A v\| : v \in B_\alpha\}$ where $\alpha \le 2\rho_E(6\varepsilon)$. By Lemma 2.7 this supremum is bounded by $\delta_E^{-1}(\alpha) \le \delta_E^{-1}\big(2\rho_E(6\varepsilon)\big)$. □

Remarks. (i) Since $\rho_E(\tau) = o(\tau)$ as $\tau \to 0$ the estimate here improves the estimate obtained in the proof of Lemma 2.5.

(ii) The estimate in Theorem 2.8 is sharp in L_p (up to a constant). We check this for $2 < p < \infty$. In this case $\delta_{L_p}(\varepsilon) \simeq \varepsilon^p$ and $\rho_{L_p}(\tau) \simeq \tau^2$, so the theorem gives that $\omega(\varepsilon) \le c_p \varepsilon^{2/p}$. We now show that this cannot be improved.

We work for convenience in $L_p[0,2]$. Denote the Rademacher functions on $[0,1]$ and on $[1,2]$ by $\{r_n\}_{n=1}^\infty$ and $\{s_n\}_{n=1}^\infty$ respectively. Fix a sequence $\{\varepsilon_n\}$ of positive numbers converging to zero, and put $\varphi_n = \varepsilon_n r_n + \varepsilon_n^{2/p} s_n$. Let F be the closed linear subspace spanned by the φ_n's in $L_p[0,2]$. By Khintchine's inequality every function in F has an expansion as a sum $\sum a_n \varphi_n$ where $\sum |a_n \varepsilon_n^{2/p}|^2 < \infty$. Let f_0 be the indicator function of $[0,1]$, and put $f_n = f_0 + \varepsilon_n r_n$. It is easy to check that $P_F f_0 = 0$. Also $\|f_n - t\varphi_n + \sum_{j\ne n} a_j \varphi_j\| = \|f_n - t\varphi_n - \sum_{j \ne n} a_j\varphi_j\|$ and the strict convexity of L_p imply that $P_F f_n = t_n \varphi_n$ where t_n minimizes $\|f_n - t\varphi_n\|$. Direct computation using the formula

$$\|f_n - t\varphi_n\|^p = \frac{1}{2}\Big(|1 + (1-t)\varepsilon_n|^p + |1 - (1-t)\varepsilon_n|^p\Big) + |t|^p \varepsilon_n^2$$

shows that $-(p-1)(1-t_n) + t_n^{p-1} = O(\varepsilon_n^2)$. It follows that there is a constant $a_p > 0$ such that $t_n \ge a_p$ when ε_n is small enough. Thus $\|f_n - f_0\| = \varepsilon_n$ and $\|P_F f_n - P_F f_0\| = \|t_n \varphi_n\| \ge a_p \varepsilon_n^{2/p}$. Finally, note that if $\varepsilon_n \ge \varepsilon_{n-1}/2$, say, it follows from the monotonicity of ω that $\omega(\varepsilon) \ge c_p \varepsilon^{2/p}$ for all small ε.

Theorem 2.9. *Let E be a uniformly smooth and uniformly convex Banach space, and assume that its moduli satisfy $\delta_E(\varepsilon) \ge \varepsilon^2/C$ and $\rho_E(\tau) \le C\tau^2$ for some constant C. Then E is isomorphic to a Hilbert space and the isomorphism constant depends only on C.*

Sketch of Proof. By Theorem 2.8 the metric projection from E onto any of its closed subspaces F is a retraction whose modulus of continuity on the unit ball is bounded by $\delta_E^{-1}\big(2\rho_E(6\varepsilon)\big) \le K\varepsilon$ where the constant K depends only on C. Since

$P_F(\lambda x) = \lambda P_F x$ for every λ it follows that P_F is a Lipschitz retraction from E onto F with a constant K independent of F.

If E is separable, then by Corollary 7.3 (to be proved in Chapter 7) every closed subspace of E is complemented by a linear projection with norm bounded by K. By a theorem of Lindenstrauss and Tzafriri [**370**] E is isomorphic to a Hilbert space and the isomorphism constant depends only on K (hence only on C). Since the argument above holds whenever E is replaced by any of its separable subspaces, they are all isomorphic to a Hilbert space with a constant depending only on C. It follows that the same holds for E itself. ♢

3. The Contraction Extension Property

Definition 2.10. *A Banach space E is said to have the Contraction Extension Property (CEP) if every Lipschitz map from a subset of E into E extends to a Lipschitz map, with the same Lipschitz constant, defined on all of E.*

More generally, if E and F are Banach spaces such that every Lipschitz map from a subset of E into F extends to a Lipschitz map with the same Lipschitz constant defined on all of E, we say that the pair (E, F) has the CEP.

It is enough, of course, to consider only 1-Lipschitz maps in the definition.

Hilbert space has the CEP (Theorem 1.12); and if E is a $\mathcal{P}_1$-space, then (F, E) has the CEP for any Banach space F. It turns out that the CEP provides a joint characterization of $\mathcal{P}_1$ and Hilbert spaces (at least for separable spaces).

Banach spaces whose duals are isometric to an $L_1(\mu)$ space for some measure space μ will be called L_1-preduals.

Theorem 2.11. *Let E be a real Banach space with the CEP. If E is strictly convex, then it is a Hilbert space. Otherwise it is an L_1-predual.*

As mentioned above Hilbert spaces actually have the CEP. Not all L_1-preduals, however, have the CEP. In the Notes and Remarks section we point out that the separable L_1-preduals with the CEP are finite-dimensional, i.e., l_∞^n for some n. Also, a dual space E which is an L_1-predual is actually a $\mathcal{P}_1$-space (see below). It is quite conceivable that every L_1-predual with the CEP is a $\mathcal{P}_1$-space without any additional assumptions on E.

Proof of Theorem 2.11 for E strictly convex. This case is easy, and we actually show more, namely, that if there is a Banach space F with $\dim F \geq 2$ such that (F, E) has the CEP, then E is a Hilbert space.

Assume that E is not Hilbertian. By Proposition A.2 there are $x, y \in E$ with $\|x\| = \|y\| = 1$ and $\|x+y\|^2 + \|x-y\|^2 > 4$. Take $\varepsilon = \|x-y\|$ in Proposition A.1, and find $x', y' \in F$ with $\|x'\| = \|y'\| = 1$ and $\|x'-y'\| = \varepsilon$ so that $\|x'+y'\|^2 \leq 4-\varepsilon^2$. Then the isometry that takes $0, x', y'$ to $0, x, y$ respectively cannot be extended as a 1-Lipschitz map to the point $z' = (x'+y')/2$. Indeed, $\|x'-z'\| = \|y'-z'\| = \varepsilon/2$, but $z = (x+y)/2$ is the only point in E with $\|x-z\| = \|y-z\| \leq \varepsilon/2$ (because E is strictly convex) and $\|z-0\| = \|z\| > \|z'\| = \|z'-0\|$. ♢

Remark. A similar proof shows that if (E, F) has the CEP and if F is strictly convex with $\dim F \geq 2$, then E is a Hilbert space. Just choose $x, y \in E$ with $\|x\| = \|y\| = 1$ such that $\|x+y\|^2 + \|x-y\|^2 < 4$ and then $x', y' \in F$ with $\|x'\| = \|y'\| = 1$ such that $\|x'-y'\| = \|x-y\|$ and $\|x'+y'\|^2 + \|x'-y'\|^2 \geq 4$.

The case of nonstrictly convex spaces is deeper. We consider only the case $\dim E > 2$. For references on the two-dimensional case as well as on the properties of L_1-preduals stated in Lemma 2.12 below, see the Notes and Remarks section.

As we have already seen in Proposition 1.4 and in Section 2 of Chapter 1, there is a close relationship between extension properties of maps and intersection properties of balls. The proof for spaces which are not strictly convex exhibits this principle again. All the balls in this section will be closed balls.

Lemma 2.12. (i) *A Banach space E is an L_1-predual if and only if E^{**} is such a space.*

(ii) *A Banach space E is an L_1-predual if and only if every collection of four mutually intersecting balls in E has a common point.*

The next lemma will not be used explicitly, but together with the remark following it, it clarifies the role of the number 4 in part (ii) above.

Lemma 2.13. *Let E be a Banach space such that every collection of four mutually intersecting balls in E has a common point. Then every finite collection of mutually intersecting balls in E has a common point. If E is a dual space, then E has the binary intersection property (i.e., E is a $\mathcal{P}_1$-space).*

Proof. We prove by induction for $n \geq 4$ that if collections of n mutually intersecting balls have a common point, then the same is true for collections of $n+1$ balls.

Let $\{B_i\}_{i=1}^{n+1}$ be a collection of $n+1$ mutually intersecting balls. Fix any point $x_0 \in E$, and put $d_0 = \max d(x_0, B_i)$. We shall show that there is a point x_1 such that $d_1 = \max d(x_1, B_i) \leq \frac{4}{5} d_0$ and such that $\|x_1 - x_0\| \leq d_0$. Iterating this procedure we obtain a sequence $\{x_k\}$ such that $\|x_{k+1} - x_k\| \leq (\frac{4}{5})^k d_0$ and such that $d_k = \max d(x_k, B_i) \leq (\frac{4}{5})^k d_0$. It follows that $\{x_k\}$ is a Cauchy sequence, and since $d_k \to 0$ the limit of the x_k's belongs to all the balls as required.

Let $A \subset \{1, \dots, n+1\}$ be a subset of cardinality $n-1$. By the induction hypothesis the collection of n balls consisting of $B(x_0, d_0)$ and $\{B_i\}_{i \in A}$ has a common point y_A. We take $x_1 = \frac{2}{n(n+1)} \sum y_A$, the average of the y_A's.

Since the ball $B(x_0, d_0)$ is convex and contains all the y_A's, it also contains x_1 and hence $\|x_1 - x_0\| \leq d_0$. To estimate the distance of x_1 from B_i, note first that $d(y_A, B_i) \leq \|y_A - x_0\| + d(x_0, B_i) \leq 2d_0$ for every i. Since there are exactly $n(n-1)/2$ sets A for which $i \in A$, and since $n \geq 4$, it follows that

$$
\begin{aligned}
d(x_1, B_i) &\leq \frac{2}{n(n+1)} \sum_{\{A : i \notin A\}} d(y_A, B_i) \\
&\leq \frac{2}{n(n+1)} \left(\frac{n(n+1)}{2} - \frac{n(n-1)}{2} \right) 2d_0 \\
&= \frac{4d_0}{n+1} \leq \frac{4}{5} d_0.
\end{aligned}
$$

The finite intersection property for a mutually intersecting family of balls in a dual space implies that they have a common point because they are w^*-compact. □

Remark. The number four in the lemma cannot be reduced to three. It is not hard to check that in $L_1(\mu)$ spaces any three mutually intersecting balls have a point in common.

Lemma 2.14. *Fix $n > 2$. Let E be a Banach space such that every collection of n mutually intersecting balls in E, all having the same radius, has a common point. Then every collection of n mutually intersecting balls in E with arbitrary radii has a common point.*

Proof. We shall use only the lemma for a dual space E, so we give only the simpler proof of this case.

Assume that $\{B(x_i, r_i)\}$ is a collection of n mutually intersecting balls such that $\bigcap B(x_i, r_i) = \emptyset$, and put $r = \max r_i$. We shall obtain a contradiction by finding n balls $B(y_i, r)$ with $B(y_i, r) \supset B(x_i, r_i)$ (so they are certainly mutually intersecting) and such that $\bigcap B(y_i, r) = \emptyset$.

The choice will be inductive. Assume that the $j-1$ points $y_1, \dots, y_{j-1}$ satisfy $B(y_i, r) \supset B(x_i, r_i)$ and $\left(\bigcap_{i<j} B(y_i, r)\right) \cap \left(\bigcap_{i \geq j} B(x_i, r_i)\right) = \emptyset$, and put

$$K = \Big(\bigcap_{i<j} B(y_i, r)\Big) \cap \Big(\bigcap_{i>j} B(x_i, r_i)\Big)$$

(with the usual conventions for $j = 1$ or n). We shall simplify the notation and assume that $x_j = 0$ and $r_j = 1$.

Since $B(0, 1)$ and K are disjoint and w^*-compact they can be strictly separated by a w^*-continuous functional x^*. Normalize so that $\|x^*\| = 1$ and $\langle x^*, x\rangle > 1$ for every $x \in K$, and choose z with $\|z\| = 1$ and $\langle x^*, z\rangle = -1$. We show that the point $y_j = (r-1)z$ has the required properties.

If $\|x\| \leq 1$, then $\|x - y_j\| \leq \|x\| + \|y_j\| \leq 1 + (r-1) = r$; hence $B(0, 1) \subset B(y_j, r)$. If $x \in B(y_j, r)$, then

$$\langle x^*, x\rangle = \langle x^*, x - y_j\rangle + \langle x^*, y_j\rangle \leq r + (r-1)\langle x^*, z\rangle = 1$$

by the choice of z. By the choice of x^* this means that $x \notin K$. □

The next lemma shows that the CEP is transferred from E to E^{**}. Since by part (i) of Lemma 2.12 the space E is an L_1-predual whenever E^{**} is, we shall be able to assume whenever convenient that E is a dual space.

Lemma 2.15. *If E satisfies the CEP, then so does E^{**}.*

Proof. Let A be a subset of E^{**}, and assume that $f : A \to E^{**}$ is a Lipschitz map with Lipschitz constant 1. For every choice of finite subsets $B \subset A$, $C \subset E^{**}$ and $D \subset E^*$, put $\varepsilon = (|B| + |C| + |D|)^{-1}$. By the principle of local reflexivity (Proposition F.1) there is a map $F = F_{B,C,D} : B \cup f(B) \cup C \to E$ such that

(i) $(1+\varepsilon)^{-1}\|x^{**} - y^{**}\| \leq \|F(x^{**}) - F(y^{**})\| \leq (1+\varepsilon)\|x^{**} - y^{**}\|$ for all $x^{**}, y^{**} \in B \cup f(B) \cup C$.

(ii) $\langle x^{**}, x^*\rangle = \langle F(x^{**}), x^*\rangle$ for all $x^{**} \in B \cup f(B) \cup C$ and $x^* \in D$.

By (i) the map $F \circ f \circ F^{-1}$ is a $(1+\varepsilon)^2$-Lipschitz map from $F(B)$ to E and thus extends to a $(1+\varepsilon)^2$-Lipschitz map $g = g_{B,C,D} : E \to E$. In particular,

$$\|g(F(x^{**})) - g(F(y^{**}))\| \leq (1+\varepsilon)^3\|x^{**} - y^{**}\| \tag{2.4}$$

for all $x^{**}, y^{**} \in B \cup C$.

Consider E as a subset of E^{**}. The net $g_{B,C,D} \circ F_{B,C,D}$ (which is ordered by $(B, C, D) \preceq (B', C', D')$ if and only if $B \subset B'$, $C \subset C'$ and $D \subset D'$) has a subnet which converges pointwise in the w^*-topology on E^{**}. Denote its limit by h. By (ii) h extends f, and by (2.4) h is a Lipschitz map with Lipschitz constant 1. □

Lemma 2.16. *Assume that E has the CEP. Then*

(i) *Let F be a two-dimensional subspace of E, and let $e \in F$ be an extreme point of the unit ball of E. Then there is a point $y \in F$ with $\|y\| = 1$ such that $\|e + sy\| = \|e - sy\|$ for every $s \in \mathbb{R}$.*

(ii) *Assume that E is not strictly convex, and let α be the supremum of the lengths of segments in the unit sphere of E. Let F be a two-dimensional subspace of E containing an extreme point e of the unit ball of E. Then e is the endpoint of two segments on the unit sphere of F each of length α.*

Proof. (i) Fix $\varepsilon > 0$. A simple continuity argument shows that there is a point $z = z_\varepsilon \in F$ with $\|z\| = 1$ such that $\|e + \varepsilon z\| = \|e - \varepsilon z\|$. We shall show that $\|te + \varepsilon z\| = \|te - \varepsilon z\|$ for every $0 \le t \le 1$. We then take y as any limit point of the z_ε's as $\varepsilon \to 0$.

Consider the map f defined by $f(e + \varepsilon z) = e - \varepsilon z$, $f(-e + \varepsilon z) = -e - \varepsilon z$ and $f(0) = 0$. Then f is an isometry by the choice of z. Fix $0 < t < 1$ and extend f to the point $te + \varepsilon z$ as a 1-Lipschitz map. When we put $x = f(te + \varepsilon z)$ we obtain the two inequalities

$$\|x - (e - \varepsilon z)\| \le \|te + \varepsilon z - (e + \varepsilon z)\| = 1 - t,$$
$$\|x - (-e - \varepsilon z)\| \le \|te + \varepsilon z - (-e + \varepsilon z)\| = 1 + t.$$

Since $2 = \|(x - e + \varepsilon z) - (x + e + \varepsilon z)\| \le 1 - t + 1 + t = 2$, the two inequalities are actually equalities.

Since e is extreme, the equalities $\|e + (x + \varepsilon z)\| = 1 + t$ and $\|e - (x + \varepsilon z)\| = 1 - t$ imply that $x + \varepsilon z = te$, i.e., $x = te - \varepsilon z$. We now use $f(0) = 0$ and obtain the inequality $\|te - \varepsilon z\| \le \|te + \varepsilon z\|$. The reverse inequality follows by symmetry.

(ii) Fix e and F and let y be as in (i). The map that fixes e and maps y to $-y$ induces an isometry of F, so it suffices to show that e is an endpoint of one segment of length α.

Let $[u, v]$ be a segment on the unit sphere of E (hence $\|u + v\| = 2$), and put $\|u - v\| = \beta$. By a continuity argument there is a point w in F such that $\|e - w\| = \beta$ and such that $\|e + w\| = 2$. Thus the map $f(u) = e$, $f(-u) = -e$ and $f(v) = w$ is an isometry, and it extends to a 1-Lipschitz map defined at 0. Since e is extreme the condition $\|f(0) \pm e\| \le 1$ implies that $f(0) = 0$. But then $\|w\| = \|f(v) - f(0)\| \le 1$ and $\|w\| = 1$ because $\|e + w\| = 2$. The condition $\|e + w\| = 2$ also implies that $[e, w]$ is a segment (of length β) on the unit sphere of F. Since β was arbitrary the maximal segment emanating from e is actually of length α. □

Proof of Theorem 2.11 for E not strictly convex. We shall find four points $\{x_i\}_{i=1}^4$, a fifth point x, and a number $\beta > 0$ such that $\|x_i - x_j\| = 2\beta$ and $\|x_i - x\| = \beta$. This will prove that every four mutually intersecting balls with the same radius r have a common point. Indeed, if $\{B(y_i, r)\}$ is such a collection, then $f(x_i) = y_i$ is a Lipschitz map with constant r/β. It extends to x with the same constant; hence $f(x) \in \bigcap B(y_i, r)$. By Lemmas 2.14 and 2.12(ii) E is an L_1-predual.

We start by showing that we may assume that there are three extreme points e_1, e_2, e_3 in the unit ball of E so that the plane passing through them intersects the unit ball in a subset of a nontrivial face.

Indeed, replacing if necessary E by E^{**} (which is permissible by Lemma 2.15), we may assume that the unit ball of E has an extreme point e. We claim that there is a maximal segment $[x, e]$ on the unit sphere of E so that x is not an extreme point. Indeed, by Lemma 2.16 there is a neighborhood of e in the unit sphere of E which is composed entirely of segments of length α emanating from e. To see that not all the other endpoints of these segments can be extreme, let e' be such an endpoint which is extreme. If x is another such endpoint and if $\|x - e'\| < \alpha$, then by applying Lemma 2.16 again x is not even extreme in the two-dimensional space spanned by e' and x.

Now let $x^* \in E^*$ satisfy $\|x^*\| = \langle x^*, (x+e)/2\rangle = 1$. Since x is not extreme the dimension of the face $\{y \in B(E) : \langle x^*, y\rangle = 1\}$ is at least two. By the Krein-Milman Theorem the w^*-compact face $\{y^{**} \in B(E^{**}) : \langle x^*, y^{**}\rangle = 1\}$ of the unit ball of E^{**} must contain at least three extreme points. Passing again from E to E^{**} if necessary, we may assume that these three points belong to E as required.

Denote the intersection of the plane passing through e_1, e_2, e_3 with the unit ball of E by P, and distinguish two cases.

(i) P is the triangle generated by the three extreme points. Then for every x which lies on the side of the triangle opposite to e_i the segment $[e_i, x]$ is maximal on the unit sphere of E; hence $\|e_i - x\| = \alpha$ by Lemma 2.16. Direct computation then shows that if we put $x_1 = e_1$, $x_2 = e_2 + e_3 - e_1$ and $x_3 = 2e_2 - e_3$, then $\|x_i - x_j\| = 2\alpha$ and also $\|x_i - e_2\| = \alpha$. By the same argument as in the beginning of the proof with the points $\{x_i\}_{i=1}^3$ and e_2, the fact that E has the CEP implies that every collection of three mutually intersecting balls in E all having the same radius must intersect.

Choose an extreme point e and a point y as in part (i) of Lemma 2.16, and let $\lambda > 0$ be such that $\|e+\lambda y\| = 2$. Then the three balls $B(e, 1)$, $B(-e, 1)$ and $B(\lambda y, 1)$ are mutually intersecting and hence have a common point. But since e is extreme the only point common to the first two is 0. Thus $\|\lambda y\| \le 1$ and $\|e + \lambda y\| = 2$ imply that $\lambda = 1$. The required four points are then $\pm e$ and $\pm y$ (with $\beta = 1$ and with 0 as the fifth point).

(ii) Assume that P contains properly the triangle generated by e_1, e_2, e_3, and let x be an extreme point of P different from the e_i's. (Note that x is not necessarily extreme in the unit ball of E.) The four points e_1, e_2, e_3 and x are the four required points (with $\beta = \alpha/2$). Indeed, the segments $[e_i, x]$ and $[e_i, e_j]$ are maximal in the unit sphere of the two-dimensional spaces spanned by e_i and x or e_j respectively. By Lemma 2.16 they all have length α. Since all the sides and the two diagonals of the quadrilateral with vertices e_1, e_2, e_3 and x have the same length α, it follows from the triangle inequality that all the vertices are at distance $\alpha/2$ from the point w where the diagonals meet. This w is the fifth required point. ♢

Remark. Both parts of the proof of Theorem 2.11 used only a restricted form of the extension property. We extended maps defined only on sets with three or four points to maps with the same Lipschitz constant on one additional point.

4. The Steiner Point

In this section we study selections for closed bounded convex subsets of finite-dimensional Euclidean spaces. By $\|\cdot\|$ we denote the usual Euclidean norm on $\mathbb{R}^n$. Let $A \subset \mathbb{R}^n$ be closed convex and bounded. The Steiner point of A is defined by

$$s(A) = n \int_{S^{n-1}} h_A(u) u \, d\sigma(u)$$

where $h_A(u) = \sup\{\langle u, x\rangle : x \in A\}$ is the support functional of the set A and where σ denotes the normalized Lebesgue measure on the unit sphere S^{n-1} in $\mathbb{R}^n$. We shall see that from several natural points of view the Steiner point is the "best" selection map on $H(\mathbb{R}^n)$.

We shall use the same notation h_A for the support functional as a function on $\mathbb{R}^n$ and for its restriction to the sphere S^{n-1}. It is a continuous, positive-homogeneous and subadditive function. Denote the space of continuous functions on S^{n-1} with the supremum norm by $C(S^{n-1})$. The map that identifies a set $A \in \mathbb{R}^n$ with $h_A \in C(S^{n-1})$ is affine (i.e., $h_{\alpha A+\beta B} = \alpha h_A + \beta h_B$ for $\alpha, \beta \geq 0$), isometric (i.e., $d_H(A, B) = \|h_A(u) - h_B(u)\|_\infty$) and order preserving (i.e., $A \subset B$ if and only if $h_A \leq h_B$).

Lemma 2.17. *The set of differences of support functions is dense in* $C(S^{n-1})$.

Proof. Observe first that every positive-homogeneous and convex function h on $\mathbb{R}^n$ is the support function of some closed convex and bounded set, namely the set $A = \bigcap_{u\in S^{n-1}}\{x : \langle x, u\rangle \leq h(u)\}$. To see this we shall show that $A \neq \emptyset$ and that $h_A \geq h$. Since clearly $h_A \leq h$, they are equal.

Fix $0 \neq v \in \mathbb{R}^n$. Since the epigraph of h is a convex cone in $\mathbb{R}^{n+1}$, the linear function whose graph supports the epigraph at $\big(v, h(v)\big)$ takes the value 0 at zero. It thus has the form $\langle y, \cdot\rangle$ for some $y \in \mathbb{R}^n$. Then $y \in A$, and the claim follows from

$$h(v) = \langle y, v\rangle \leq \sup\big\{\langle z, v\rangle :\ z \in A\big\} = h_A(v).$$

Let f be a positive-homogeneous function on $\mathbb{R}^n$ which is C^2 on $\mathbb{R}^n \setminus \{0\}$, and write $f(x) = \big(f(x) + k\|x\|\big) - k\|x\|$. The function $k\|x\|$ is the support function of the ball with radius k centered at zero; and we shall show that if k is large enough, then the first term is also a support function. Since the restrictions of such functions f to S^{n-1} are dense in $C(S^{n-1})$, this will prove the lemma.

By homogeneity it is enough to check that the Hessian of $f(x)+k\|x\|$ is positive definite on S^{n-1}. But the Hessian of the norm is positive definite, f is C^2, and S^{n-1} is compact. Thus the Hessian of $f(x) + k\|x\|$ is indeed positive definite provided k is large enough. $\square$

Lemma 2.18. *Let h be a positive-homogeneous and subadditive function on $\mathbb{R}^n$. Then*

$$n\int_{S^{n-1}} \langle u, v\rangle h(u)\, d\sigma(u) \leq h(v)$$

for every $v \in \mathbb{R}^n$.

Proof. For every $\lambda \in \mathbb{R}$

$$\lambda h(w + \lambda v) \leq \lambda h(w) + \lambda^2 h(v).$$

Indeed, for $\lambda \geq 0$ this follows directly from the homogeneity and subadditivity of h and for $\lambda < 0$ from $h(w) = h\big((\lambda v + w) + (-\lambda)v\big) \leq h(\lambda v + w) - \lambda h(v)$.

Assuming $\|v\| = 1$ and taking $\lambda = \langle u, v\rangle$ and $w = u - \langle u, v\rangle v$, we obtain

$$\begin{aligned} &n\int_{S^{n-1}} \langle u, v\rangle h(u)\, d\sigma(u) \\ &= n\int_{S^{n-1}} \langle u, v\rangle h\Big(\big(u - \langle u, v\rangle v\big) + \langle u, v\rangle v\Big)\, d\sigma(u) \\ &\leq n\int_{S^{n-1}} \langle u, v\rangle h\big(u - \langle u, v\rangle v\big)\, d\sigma(u) + nh(v)\int_{S^{n-1}} \langle u, v\rangle^2\, d\sigma(u). \end{aligned}$$

The first integral is zero because σ is invariant under the reflection with respect to the hyperplane $H = \{w : \langle w, v\rangle = 0\}$, the point $u - \langle u, v\rangle v$ is in H, and $\langle u, v\rangle$ changes its sign under this reflection.

The invariance of σ implies that $\int \langle u, v\rangle^2 \, d\sigma(u)$ is independent of v; thus

$$n \int \langle u, v\rangle^2 \, d\sigma(u) = \sum_{i=1}^{n} \int \langle u, e_i\rangle^2 \, d\sigma(u) = \int \sum_{i=1}^{n} \langle u, e_i\rangle^2 \, d\sigma(u) = \int d\sigma(u) = 1.$$

□

Remark. Let $v \in \mathbb{R}^n$. If h in the lemma is a linear function, then the inequalities in the proof above are actually equalities. Thus

$$v = n \int_{S^{n-1}} \langle u, v\rangle u \, d\sigma(u).$$

We are now ready to prove the basic properties of the Steiner point map s.

Proposition 2.19. (i) $s(\alpha A + \beta B) = \alpha s(A) + \beta s(B)$ *for all* $A, B \in H(\mathbb{R}^n)$ *and nonnegative* α, β.

(ii) s *commutes with affine isometries of* $\mathbb{R}^n$*; i.e.,* $s(\rho A) = \rho s(A)$ *for each such isometry* ρ.

(iii) s *is Lipschitz with respect to the Hausdorff distance on* $H(\mathbb{R}^n)$*, and its Lipschitz constant as* $n \to \infty$ *is*

$$2\pi^{-1/2}\Gamma(n/2 + 1)/\Gamma\big((n+1)/2\big) = 2^{1/2}\pi^{-1/2}n^{1/2} + O(1).$$

(iv) s *is a selection; i.e.,* $s(A) \in A$ *for all* $A \in H(\mathbb{R}^n)$.

Proof. (i) This follows trivially from the fact that $h_{\alpha A+\beta B} = \alpha h_A + \beta h_B$ for all $A, B \in H(\mathbb{R}^n)$ and $\alpha, \beta \geq 0$.

(ii) By the remark above we obtain $s(\{v\}) = n \int \langle u, v\rangle u \, d\sigma(u) = v$ and thus $s(A + v) = s(A) + s(\{v\}) = s(A) + v$; i.e., s commutes with translations. If ρ is an orthogonal transformation, then $h_{\rho A}(u) = h_A(\rho^{-1}u)$, and the invariance of σ shows that $s(\rho A) = \rho s(A)$.

(iii) Define a linear operator $P : C(S^{n-1}) \to \mathbb{R}^n$ by

$$\langle Pf, v\rangle = n \int_{S^{n-1}} f(u)\langle v, u\rangle \, d\sigma(u)$$

for every v. If $\|v\| = 1$, then

$$\begin{aligned} \|P\| &= n \int_{S^{n-1}} |\langle v, u\rangle| \, d\sigma(u) \\ &= 2\pi^{-1/2}\Gamma(n/2 + 1)/\Gamma\big((n+1)/2\big). \end{aligned}$$

The first equality holds because we can take functions f with $\|f\| = 1$ which converge pointwise to the sign of $\langle v, u\rangle$. The resulting integral is independent of v, and the second equality follows by direct computation.

Identifying A with h_A, the formula $s(A) = n \int_{S^{n-1}} h_A(u)u \, d\sigma(u) = Ph_A$ shows that the Steiner map is the restriction of P to support functions. It follows that $\|s\|_{Lip} \leq \|P\|$. For the reverse inequality note that if $f = h_A - h_B$, then

$$\begin{aligned} \|Pf\| &= \|Ph_A - Ph_B\| = \|s(A) - s(B)\| \\ &\leq \|s\|_{Lip}\|h_A - h_B\| = \|s\|_{Lip}\|f\| \end{aligned}$$

and $\|P\| \le \|s\|_{Lip}$ because differences of support functions are dense in $C(S^{n-1})$.

(iv) By Lemma 2.18 $\langle v, s(A)\rangle = n\int_{S^{n-1}} h_A(u)\langle v,u\rangle\, d\sigma(u) \le h_A(v)$ for every v. Thus $s(A) \in A$ by the separation theorem. □

The properties exhibited in Proposition 2.19 above, and even somewhat weaker ones, are sufficient to characterize the Steiner point. Note that we do not assume in the next theorem that ϕ is a selection. Also, homogeneity follows from additivity for continuous maps. Indeed, $\phi(A) = n\phi(\frac{1}{n}A)$ for every positive integer n; hence also $\phi(\frac{m}{n}A) = m\phi(\frac{1}{n}A) = \frac{m}{n}\phi(A)$, and ϕ is positive-homogeneous by continuity.

Theorem 2.20. *Let $\phi : H(\mathbb{R}^n) \to \mathbb{R}^n$ be a continuous additive function which commutes with the affine isometries of $\mathbb{R}^n$. Then ϕ is the Steiner point map.*

Proof. Let $O(n)$ be the compact group of orthogonal transformations on $\mathbb{R}^n$. The restriction of ϕ to $\mathbb{R}^n$ (i.e. to singletons) is an additive transformation on $\mathbb{R}^n$ that commutes with $O(n)$, hence a multiple of the identity. Since it commutes with translations it is actually the identity, and ϕ agrees with s on the singletons. Assume now that A is a convex bounded set with $\dim A > 0$. Since ϕ commutes with translations we can assume that the origin is in the relative interior of A.

Fix $x \in \mathbb{R}^n$, $v \in S^{n-1}$ and a number $c > \|x\|$, and put

$$B = \int_{O(n)} \big(c + \langle x, \rho v\rangle\big)\rho^{-1}A\, d\mu(\rho)$$

where μ is the Haar measure on $O(n)$. The integrand is a continuous function from $O(n)$ to $H(\mathbb{R}^n)$ for every fixed A, hence integrable. (The integral of a continuous set-valued function is defined in a natural way as the limit of finite Riemann sums. Note that we can take only linear combinations with positive coefficients in $H(\mathbb{R}^n)$, and the extra constant c was chosen to ensure that $c + \langle x, \rho v\rangle > 0$.)

We shall prove below that B is a ball. Once this is proved the proof of the theorem is completed as follows. Let z be the center of the ball B. Since ϕ commuted with affine isometries it follows that $\phi(B) = z$. On the other hand, the continuity and additivity of ϕ, together with the fact that $\phi(\rho^{-1}A) = \rho^{-1}\phi(A)$, show that

$$\begin{aligned} z &= \phi(B) = \int_{O(n)} \big(c + \langle x, \rho v\rangle\big)\rho^{-1}\phi(A)\, d\mu(\rho) \\ &= \int_{O(n)} \langle x, \rho v\rangle \rho^{-1}\phi(A)\, d\mu(\rho) \end{aligned}$$

because the $O(n)$ invariance of μ implies that the vector $\int_{O(n)} \rho^{-1}\phi(A)\, d\mu(\rho)$ is also invariant under $O(n)$, hence 0.

Since this holds for every ϕ satisfying the assumptions of the theorem, it also holds for the Steiner map s. Hence

$$\int_{O(n)} \langle x, \rho v\rangle \rho^{-1}\big(\phi(A) - s(A)\big)\, d\mu(\rho) = 0$$

for all $x \in \mathbb{R}^n$ and $v \in S^{n-1}$. Taking $x = \phi(A) - s(A)$ and using the orthogonality of ρ, we obtain

$$\begin{aligned} \int \langle x, \rho v\rangle^2 d\mu(\rho) &= \int \langle x, \rho v\rangle \langle \rho^{-1}x, v\rangle\, d\mu(\rho) \\ &= \left\langle \int \langle x, \rho v\rangle \rho^{-1}x\, d\mu(\rho), v \right\rangle = 0. \end{aligned}$$

Thus $\langle x, \rho v\rangle = 0$ for all v and ρ; hence $x = 0$ and $\phi(A) = s(A)$.

To show that B is indeed a ball we compute its support function. Since $h_{\alpha K+\beta L} = \alpha h_K + \beta h_L$ and $h_{\rho^{-1}K}(u) = h_K(\rho u)$ for all $\alpha, \beta \geq 0$ and $\rho \in O(n)$, it follows that

$$h_B(u) = \int_{O(n)} (c + \langle x, \rho v\rangle) h_A(\rho u)\, d\mu(\rho).$$

By the invariance of μ the integral $c\int_{O(n)} h_A(\rho u)\, d\mu(\rho)$ is a constant independent of u, and we denote it by r. The assumptions that $\dim A > 0$ and that 0 is in the relative interior of A imply that h_A is strictly positive almost everywhere; hence $r > 0$. We claim that there is a vector z such that

$$\int_{O(n)} \langle x, \rho v\rangle h_A(\rho u)\, d\mu(\rho) = \langle z, u\rangle$$

for all $u \in S^{n-1}$. Indeed, given u, choose $\tau \in O(n)$ which interchanges u and v, i.e., $\tau u = v$ and $\tau v = u$. By the invariance of μ

$$\begin{aligned}\int_{O(n)} \langle x, \rho v\rangle h_A(\rho u)\, d\mu(\rho) &= \int_{O(n)} \langle x, \rho\tau u\rangle h_A(\rho\tau v)\, d\mu(\rho)\\ &= \int_{O(n)} \langle \rho^{-1}x, u\rangle h_A(\rho v)\, d\mu(\rho)\end{aligned}$$

so take $z = \int_{O(n)} \rho^{-1}x\, h_A(\rho v)\, d\mu(\rho)$.

Thus $h_B(u) = \langle z, u\rangle + r$; i.e., it is the support function of the ball $B(z, r)$. □

By Proposition 2.19 the Steiner map is a Lipschitz map with Lipschitz constant $c_n = 2\pi^{-1/2}\Gamma(n/2+1)/\Gamma\big((n+1)/2\big)$. It turns out that this is the smallest possible modulus of continuity for retractions from $H(\mathbb{R}^n)$ to $\mathbb{R}^n$. (We consider only the Euclidean norm on $\mathbb{R}^n$ here. Recall, for example, that with the l_∞^n norm there is even a nonexpansive retraction.)

Proposition 2.21. *Let ϕ be a retraction from $H(\mathbb{R}^n)$ onto $\mathbb{R}^n$. Then the modulus of continuity ω_ϕ of ϕ satisfies $\omega_\phi(t) \geq \omega_s(t)$ for every $t > 0$.*

Proof. Let $M_A(\cdot)$ be an invariant mean on the bounded $\mathbb{R}^n$-valued functions defined on the abelian semigroup $H(\mathbb{R}^n)$, and let $M_x(\cdot)$ be an invariant mean on the bounded $\mathbb{R}^n$-valued functions defined on $\mathbb{R}^n$ (see Appendix C). Given ϕ define

$$\psi(B) = \int_{O(n)} \rho^{-1} M_A\Big(M_x\Big(\phi(A + \rho B + x) - \phi(A + x)\Big)\Big)\, d\mu(\rho).$$

Since d_H is an invariant metric on $H(\mathbb{R}^n)$, i.e., $d_H(A+C, A+B) = d_H(B, C)$ for all $A, B, C \in H(\mathbb{R}^n)$, we obtain $\|\phi(A+\rho B+x) - \phi(A+x)\| \leq \omega_\phi\big(d_H(B, \{0\})\big)$, and the "integrand" is a bounded function. Similar arguments show that it is continuous as a function of ρ (so ψ is well defined) and also that ψ is uniformly continuous and satisfies $\omega_\psi(t) \leq \omega_\phi(t)$ for every $t > 0$. We shall show that ψ is additive and commutes with affine isometries of $\mathbb{R}^n$. By Theorem 2.20 it is the Steiner map.

Fix $y \in \mathbb{R}^n$ and a set A. Using $M_x(y) = y$, $\phi(x) = x$ and $\phi(x+y) = x+y$ gives

$$\begin{aligned}&M_x\Big(\phi(A + x + y) - \phi(A + x)\Big)\\ &= y + M_x\Big(\phi(A + x + y) - \phi(x + y)\Big) - M_x\Big(\phi(A + x) - \phi(x)\Big) = y\end{aligned}$$

because the second "integrand" is obtained from the first by translation by y and M_x is invariant under such translations. It follows that $\psi(y) = y$.

Fix $B, C \in H(\mathbb{R}^n)$. Then

$$\begin{aligned} &M_A\Big(M_x\big(\phi(A+B+C+x) - \phi(A+x)\big)\Big) \\ &= M_A\Big(M_x\big(\phi(A+B+C+x) - \phi(A+B+x)\big)\Big) \\ &\quad + M_A\Big(M_x\big(\phi(A+B+x) - \phi(A+x)\big)\Big). \end{aligned}$$

Thus $\psi(B+C) = \psi(B) + \psi(C)$; i.e., ψ is additive. Since $\psi(y) = y$ we see that ψ commutes with translations. Commuting with the orthogonal group is obvious. □

Corollary 2.22. *Let E be an infinite-dimensional Banach space. Then there is no uniformly continuous selection on $H(E)$.*

Proof. By Dvoretzky's theorem E contains nearly Euclidean subspaces of all dimensions. Since the Lipschitz constant of the Steiner map increases with dimension, the corollary follows from the proposition. □

Note that there is a difference between the existence of selections and of general Lipschitz (or uniformly continuous) retractions from $H(E)$ onto E. Indeed, retractions exist if (and only if) E is an absolute Lipschitz (or uniform) retract, but selections never exist for infinite-dimensional spaces (although selections which are uniformly continuous on bounded sets do exist for certain spaces, e.g., for uniformly convex spaces; see Theorem 1.27).

5. Simultaneously Continuous Maps

Let E be an infinite-dimensional Banach space. We consider it as simultaneously equipped with two topologies: the norm topology and the weak topology. (When E is a dual space, we may also consider the w^*-topology.)

There is a natural norm-Lipschitz retraction from E onto its unit ball, namely, $r(x) = x/\|x\|$ for $\|x\| > 1$. This map is, however, never continuous with respect to the weak topology. Indeed, it follows from the definition of the weak topology that the unit sphere is weakly dense in the unit ball. If $\{x_\alpha\}$ is a net in the unit sphere which converges weakly to a point x with $\|x\| = 1/2$, then $r(2x_\alpha) = x_\alpha$ converges weakly to $x \neq 2x = r(2x)$.

In nonseparable spaces weakly continuous retractions on $B(E)$ may fail to exist altogether (e.g., for Hilbert space, see Corson and Lindenstrauss [**126**]). When E^* is separable, or when we consider E^* with the w^*-topology for separable E, such retractions do exist by Michael's theorem (see Theorem 1.16).

In the first part of this section we consider retractions from E onto $B(E)$ that are simultaneously continuous with respect to both topologies. In the second part we characterize reflexivity by the possibility of extending simultaneously continuous real-valued functions from weakly closed subsets to all of E.

The existence of simultaneously continuous retractions does not seem to follow from general selection theorems, and their construction requires ad hoc methods. When such retractions do exist one may also look for weakly continuous retractions that are uniformly continuous with respect to the norm, and the following theorem describes one instance where this is possible.

Theorem 2.23. *Let E be a separable uniformly convex Banach space. Then there is a retraction from E onto its unit ball which is weakly continuous and norm uniformly continuous. Its norm modulus of continuity is bounded by δ_E^{-1}.*

Proof. Fix a decreasing sequence $\{E_n\}_{n=1}^{\infty}$ of codimension n subspaces of E such that $\bigcap_{n=1}^{\infty} E_n = \{0\}$. Put $F_n = E/E_n$, and let $q_n : E \to F_n$ be the quotient maps. By a slight abuse of notation we shall also denote the natural quotient map from F_m onto F_n (for $m > n$) by q_n. If a net $\{x_\alpha\}$ is weakly convergent to x, then $\|q_n x_\alpha - q_n x\| \to 0$ for all n. Since $\bigcap_{n=1}^{\infty} E_n = \{0\}$ the converse is also true provided the net is bounded. For each $x \in F_n$ the uniform convexity of E implies that there is a unique point in E, which we denote by $\psi_n(x)$, such that $q_n\psi_n(x) = x$ and $\|\psi_n(x)\| = \|x\|$. For each $x \in E$ put

$$n(x) = \inf\{n : \|q_n x\| \geq 1\}$$

(if no such n exists, which can only happen if $\|x\| \leq 1$, put $n(x) = \infty$).

Fix x with $\|x\| > 1$ and such that $n = n(x) > 1$. The point $x_n = q_n\psi_{n-1}(q_{n-1}x)$ is the unique point in F_n with $q_{n-1}x_n = q_{n-1}x$ such that $\|x_n\| = \|q_{n-1}x\|$. Since $n(x) = n$ it follows that $\|x_n\| < 1$, and thus there is a unique $0 < \lambda \leq 1$ such that the point $z = \lambda q_n x + (1-\lambda)x_n$ satisfies $\|z\| = 1$. Of course, $q_{n-1}z = q_{n-1}x$. With these notations we can now define the retraction by

$$\phi(x) = \begin{cases} x & \|x\| \leq 1 \\ \psi_1(q_1 x/\|q_1 x\|) & \|x\| > 1 \text{ and } n(x) = 1 \\ \psi_n(z) & \|x\| > 1 \text{ and } n(x) = n > 1 \end{cases}$$

where z was defined above.

We first check the continuity of ϕ with respect to the norm. Let $x, y \in E$, and put $\varepsilon = \|x - y\|$. Assume that $n(x) = n \leq m = n(y)$ and consider two possibilities.

(i) $n < m$. Write $\phi(x) = \psi_n(z)$ where $z = \lambda q_n x + (1-\lambda)x_n$ and $\|z\| = 1$. (When $n = 1$ take $z = q_1 x/\|q_1 x\|$.) Choose $z^* \in F_n^*$ with $\|z^*\| = \langle z^*, z\rangle = 1$, and notice that $\langle z^*, q_n x\rangle \geq 1$ because $\langle z^*, x_n\rangle < 1$.

Identifying F_n^* as a subspace of E^*, consider z^* as a functional on E, and note that by the construction $q_n\phi(y) = q_n y$. Thus

$$\langle z^*, \phi(y)\rangle = \langle z^*, q_n y\rangle \geq \langle z^*, q_n x\rangle - \|q_n x - q_n y\| \geq 1 - \varepsilon$$

and clearly $\langle z^*, \phi(x)\rangle = \langle z^*, z\rangle = 1$. Hence $\|\phi(x) + \phi(y)\| \geq \langle z^*, \phi(x) + \phi(y)\rangle \geq 2 - \varepsilon$, and by the definition of the modulus of convexity $\|\phi(x) - \phi(y)\| \leq \delta_E^{-1}(\varepsilon)$.

Before passing to the next case, notice that the condition $m > n$ was used only to ensure that $q_n\phi(y) = q_n y$.

(ii) $n = m$. We treat only the case $1 < n < \infty$. The case $n = m = 1$ requires only minor modifications, and $n = m = \infty$ is of course trivial.

Fix a point $u \in E$ such that $q_n u \neq 0$ and $q_{n-1}u = 0$. We shall replace x and y by $x_t = x + tu$ and $y_t = y + tu$ for an appropriate t. Clearly $\|x_t - y_t\| = \|x - y\|$, $q_{n-1}y_t = q_{n-1}y$ and $q_{n-1}x_t = q_{n-1}x$ for all t.

Recall that $x_n = q_n\psi_{n-1}(q_{n-1}x)$. Since the kernel of q_{n-1} in F_n is the one-dimensional space generated by $q_n u$ and since $q_{n-1}x = q_{n-1}x_n$, it follows that there is a real scalar τ such that $q_n x = x_n + \tau q_n u$. Thus $\phi(x_t) = \phi(x)$ as long as $t + \tau$ has the same sign as τ and $\|q_n x_t\| \geq 1$. The same argument holds for y_t. Interchanging x and y if necessary, we can choose t such that

$$\phi(x_t) = \phi(x), \; \phi(y_t) = \phi(y), \; \|q_n y_t\| = 1, \; \|q_n x_t\| \geq 1.$$

But now the argument of the previous case holds. Indeed, as remarked at the end of case (i) we needed only that $q_n\phi(y_t) = q_n y_t$, which holds here because $\|q_n y_t\| = 1$. Thus $\|\phi(x) - \phi(y)\| = \|\phi(x_t) - \phi(y_t)\| \leq \delta_E^{-1}(\varepsilon)$ because $\|x_t - y_t\| = \|x - y\| = \varepsilon$.

It remains to check that ϕ is weakly continuous. Fix a net $\{x_\alpha\}$ that converges weakly to a point x; hence $q_j x_\alpha \to q_j x$ for all j. Since $\{\phi(x_\alpha)\}$ is a bounded net it is enough to check that $q_j\phi(x_\alpha) \to q_j\phi(x)$ for all j.

If $n(x) = \infty$, then $\|q_j x\| < 1$ for all j. Fix j and find $\alpha(j)$ such that $\|q_j x_\alpha\| < 1$ for $\alpha > \alpha(j)$. The definition of ϕ implies that $q_j\phi(x) = q_j x$ and $q_j\phi(x_\alpha) = q_j x_\alpha$ for $\alpha > \alpha(j)$. Since $q_j x_\alpha \to q_j x$ it follows that $q_j\phi(x_\alpha) \to q_j\phi(x)$.

Assume now that $n = n(x) < \infty$. If $\|q_n x\| > 1$, then $n(x_\alpha) = n(x)$ for large enough α, so assume this holds for all α. The definitions of $\phi(x_\alpha)$ and of $\phi(x)$ then depend only on $q_n x_\alpha$ and $q_n x$; hence $\phi(x_\alpha)$ actually converge in norm to $\phi(x)$.

If $\|q_n x\| = 1$, choose $\lambda > 1$ which is close enough to 1 so that $n(\lambda x) = n$. Put $y = \lambda x$ and $y_\alpha = \lambda x_\alpha$, and then $\|\phi(y_\alpha) - \phi(x_\alpha)\|, \|\phi(y) - \phi(x)\| \leq \delta_E^{-1}(\lambda - 1)$. Since $\|q_n y\| > 1$, it follows that $\phi(y_\alpha) \to \phi(y)$ weakly; i.e., all the weak limit points of the $\phi(x_\alpha)$'s belong to the ball of radius $2\delta_E^{-1}(\lambda - 1)$ around $\phi(x)$. Since λ was arbitrary it follows that $\phi(x_\alpha) \to \phi(x)$ weakly. □

The norm modulus of continuity of a simultaneously continuous retraction ϕ on the unit ball in l_p $(1 < p < \infty)$ is bounded from below by $c_p t^{1/p}$. Indeed, let $\{e_n\}$ be the standard basis in l_p, fix $0 < t < 2/p$, and let $\delta = (pt/2)^{1/p}$. Put $x_n = e_1 + \delta e_n$ and $y_n = x_n/\|x_n\|$. Then

$$\|x_n - y_n\| = \|x_n\| - 1 = (1 + \delta^p)^{1/p} - 1 \leq 2\delta^p/p = t.$$

On the other hand, $x_n \to e_1$ weakly; hence $\phi(x_n) \to \phi(e_1) = e_1$ weakly. Since $\|\phi(x_n)\| \leq 1$ and l_p is uniformly convex, the convergence is actually in norm. Since $\phi(y_n) = y_n$, estimating the norm by the value of the nth coordinate gives

$$\lim \|\phi(x_n) - \phi(y_n)\| \geq \delta/(1 + \delta^p)^{1/p} \geq \delta/2 \geq c_p t^{1/p}.$$

Let $p \geq 2$. Since L_p contains a norm-one complemented subspace isometric to l_p and since the modulus of convexity of both l_p and L_p is equivalent to ε^p, the estimate in the theorem is best possible (up to constants) for both $L_p[0,1]$ and l_p.

For $p < 2$ the modulus of convexity of L_p is equivalent to ε^2, and we need a different computation to show that the result of the theorem is still best possible. In l_p, on the other hand, the correct estimate for $p < 2$ is $Ct^{1/p}$. See Benyamini [**50**] for proofs of both claims.

We now turn to the extension of real-valued functions.

A dual Banach space E^* is normal in its w^*-topology because it is σ-compact, hence Lindelöf (see Kelley [**325**], page 113). It follows that if A is a w^*-closed subset of E^* and if f is a bounded real-valued function on A which is w^*-continuous, then f extends with the same bounds to a w^*-continuous function on all of E^*. Also, by Lemma 1.1(ii) every norm-Lipschitz function f on A extends with the same Lipschitz constant to a norm-Lipschitz function on all of E^*. It turns out that it is possible to extend functions in such a way that both the w^*-continuity and the norm-Lipschitz condition are preserved simultaneously.

Theorem 2.24. *Let A be a w^*-closed subset of E^*, and let f be a bounded real-valued function on A which is both w^*-continuous and norm-Lipschitz. Then*

f extends to a w^*-continuous function on all of E^* with the same bounds and with the same norm-Lipschitz constant as f.

Proof. The proof is a modification of the usual proof of Urysohn's lemma, where we also need to take care of the Lipschitz condition. By normalizing and scaling we can assume that $0 \le f(x^*) \le 1$ and that $|f(x^*) - f(y^*)| \le \|x^* - y^*\|$ for all $x^*, y^* \in A$.

The sets $U_t = \{x^* \in A : f(x^*) < t\}$ are relatively w^*-open F_σ subsets of A with $U_t = \emptyset$ for $t \le 0$ and $U_t = A$ for $t > 1$. They satisfy $U_t = \bigcup_{r<t} U_r$ and $\overline{U_t} \subset U_s$ for every $t < s$. Also, since f is norm-Lipschitz with constant 1, it follows that $d(U_t, A \setminus U_s) \ge s - t$ for all $t < s$. (Here and below the bar denotes closure with respect to the w^*-topology.)

We shall construct w^*-open F_σ subsets V_t of E^* satisfying $V_t \bigcap A = U_t$ for all t and such that

(i) $V_t = \emptyset$ for $t \le 0$ and $V_t = E^*$ for $t > 1$

(ii) $\overline{V_t} \subset V_s$ for every $t < s$

(iii) $d(V_t, E^* \setminus V_s) \ge s - t$ for all $t < s$.

Conditions (i) and (ii) and the fact that $V_t \bigcap A = U_t$ imply that the function $g(x^*) = \inf\{t : x^* \in V_t\}$ is a w^*-continuous extension of f to all of E^* with $0 \le g(x^*) \le 1$. Also, if $x^*, y^* \in E^*$ and if $g(x^*) < g(y^*)$, say, choose s, t such that $g(x^*) < t < s < g(y^*)$. The definition of g implies that $x^* \in V_t$ and $y^* \notin V_s$. By (iii) $s - t \le d(V_t, E^* \setminus V_s) \le \|x^* - y^*\|$. Since s, t were arbitrary, g satisfies a Lipschitz condition with constant 1.

Put $V_t = E^*$ for $t > 1$ and $V_t = \emptyset$ for $t \le 0$. It then suffices to construct the sets V_t for any dense subset of $(0, 1)$. Fix such a dense sequence $T = \{t_n\}$, and construct V_{t_n} by induction on n. We shall describe only the induction step, the construction of V_{t_1} being similar and even somewhat simpler.

To simplify the notation, we denote U_{t_j} by U_j and V_{t_j} (for $j < n$) by V_j. We also strengthen the inductive hypothesis. In addition to the condition

(iii′) $d(V_i, E^* \setminus V_j) \ge t_j - t_i$ for $i, j < n$ such that $t_j > t_i$

we shall also assume that

(iv) $d(U_t, E^* \setminus V_j) \ge t_j - t$ for all $j < n$ such that $t < t_j$

and that

(v) $d(V_j, A \setminus U_t) \ge t - t_j$ for all $j < n$ such that $t > t_j$.

For each set $C \subset E^*$ and $\varepsilon > 0$ we denote the set $\{x^* \in E^* : d(x^*, C) \le \varepsilon\}$ by $C(\varepsilon)$. It follows from the w^*-compactness of closed balls in E^* that $C(\varepsilon)$ is w^*-closed whenever C is. Put

$$W = \Big(\bigcup_{t \in T,\ t \le t_n} U_t(t_n - t) \Big) \cup \Big(\bigcup_{j<n,\ t_j < t_n} V_j(t_n - t_j) \Big)$$

and

$$B = \Big(\bigcup_{t \in T,\ t \ge t_n} (A \setminus U_t)(t - t_n) \Big) \cup \Big(\bigcup_{j<n,\ t_j > t_n} (E^* \setminus V_j)(t_j - t_n) \Big).$$

We first show that B is w^*-closed. We need to check only the first term, so assume that $x^*_\alpha + y^*_\alpha \to z^*$ where $x^*_\alpha \in A \setminus U_{t_\alpha}$ and $\|y^*_\alpha\| \le t_\alpha - t_n$ (and $t_\alpha \ge t_n$). Passing to a subnet, we may assume that $t_\alpha \to s \ge t_n$ and that $y^*_\alpha \to y^*$ with $\|y^*\| \le s - t_n$.

Then $x^*_\alpha \to x^* = z^* - y^*$. Hence $f(x^*) = \lim_\alpha f(x^*_\alpha) \geq s$; i.e., $x^* \in A \setminus U_s$ and $z^* \in \big(A \setminus U_s\big)(s - t_n) \subset B$.

The sets B and W satisfy $B \bigcap W = \emptyset$, $W \bigcap A = U_n$ and $B \bigcap A = A \setminus U_n$. We check, for example, that $B \bigcap W = \emptyset$ (the other relations are proved similarly).

Fix $t \leq t_n \leq s$ in T, and assume that $t_j < t_n < t_i$ for some $i, j < n$. Hence $U_t(t_n - t) \cap \big(A \setminus U_s\big)(s - t_n) = \emptyset$ because $U_t = \bigcup_{r<t} U_r$, and since $r < t$ implies $U_r(t_n - t) \cap \big(A \setminus U_s\big)(s - t_n) = \emptyset$ by the assumption that $d(U_r, A \setminus U_s) \geq s - r$. Similarly $U_t(t_n - t) \cap \big(E^* \setminus V_i\big)(t_i - t_n) = \emptyset$ by (iv). That $\big(A \setminus U_s\big)(s - t_n) \cap V_j(t_n - t_j) = \emptyset$ and $V_j(t_n - t_j) \cap \big(E^* \setminus V_i\big)(t_i - t_n) = \emptyset$ follows from (v) (with t replaced by s) and from (iii′) respectively, when we note that V_j is norm open, and hence that for every $x^* \in V_j(t_n - t_j)$ there is a $y^* \in V_j$ with $\|x^* - y^*\| < t_n - t_j$.

We now use the normality of E^* in its w^*-topology. Since U_t is w^*-F_σ for every t, there is a w^*-F_σ set $\tilde{W} \subset W$ which contains U_n and which is norm-dense in W. (Note that if $D = \bigcup_j F_j$ is a w^*-F_σ with w^*-closed sets F_j, then for every $\varepsilon > 0$ the set $\bigcup_j F_j(\varepsilon)$ is a w^*-F_σ set which is norm-dense in $D(\varepsilon)$.) Write $\tilde{W} = \bigcup W_k$ with closed W_k's. Since $W_k \bigcap B = \emptyset$ there are w^*-continuous functions $0 \leq h_k \leq 1$ such that $h_k(x^*) = 0$ for $x^* \in B$, and $h_k(x^*) = 1$ for $x^* \in W_k$. Then $h = \sum 2^{-k} h_k$ is also w^*-continuous, and $V_n = \{x^* : h(x^*) > 0\}$ is the desired w^*-open F_σ set. Indeed, $V_n \cap A = U_n$ because $U_n \subset \tilde{W}$ and $A \setminus U_n \subset B$. That (iii′) holds (with $i = n$ or $j = n$) is obvious. To check (iv) fix $t < t_n$. Then $E^* \setminus V_n \subseteq E^* \setminus \tilde{W}$, the norm-denseness of $\tilde{W}$ in W and $U_t(t_n - t) \subset W$ imply that $d(U_t, E^* \setminus V_n) \geq d(U_t, E^* \setminus \tilde{W}) = d(U_t, E^* \setminus W) \geq t_n - t$. Condition (v) is proved similarly. □

Let E be a general (not necessarily dual) Banach space, and consider the same extension problem with the w^*-topology replaced by the weak topology. It turns out that the possibility of extending functions which are simultaneously weakly continuous and norm-Lipschitz characterizes reflexivity.

Theorem 2.25. *A Banach space E is reflexive if and only if every weakly continuous and norm-Lipschitz bounded function $f : A \to \mathbb{R}$, defined on a weakly closed subset $A \subset E$, extends to a weakly continuous and norm-Lipschitz function on all of E.*

Proof. If E is reflexive, then the previous theorem applies and extensions exist. Thus assume E is nonreflexive, and fix $0 < \delta < 1/3$. By Proposition E.7 there are $u_j \in E$ and $u^*_j \in E^*$ with $\|u_j\| = \|u^*_j\| = 1$ such that $\langle u^*_n, u_j\rangle = 0$ for $j \leq n$ and $\langle u^*_n, u_j\rangle > 1 - \delta$ for $j > n$.

It follows that the set $A_\delta = \big\{u_j - (1 - 3\delta)u_i : i < j\big\}$ is weakly closed. Indeed, assume that z is a weak limit point of A_δ. Then z belongs to the closed linear span of the u_j's, and there is a finite linear combination $v = \sum_{k=1}^n \lambda_k u_k$ such that $\|z - v\| < \delta$. If $j > n$, then u^*_n separates $u_j - (1 - 3\delta)u_i$ from z. Indeed,

$$\begin{aligned}\langle u^*_n, u_j - (1 - 3\delta)u_i - z\rangle &= \langle u^*_n, u_j\rangle - (1 - 3\delta)\langle u^*_n, u_i\rangle - \langle u^*_n, z - v\rangle - \langle u^*_n, v\rangle \\ &> 1 - \delta - (1 - 3\delta) - \delta = \delta.\end{aligned}$$

Hence z is necessarily in the closure of the finite set $\big\{u_j - (1 - 3\delta)u_i : i < j \leq n\big\}$ and must then belong to this set.

Note that $\|u_j - (1 - 3\delta)u_i\| \geq \langle u^*_i, u_j - (1 - 3\delta)u_i\rangle > 1 - \delta > 2/3$, i.e., $\operatorname{dist}(0, A_\delta) > 2/3$, and that $A_\delta \subset B(0, 2)$. On the other hand, 0 belongs to the weak closure of $A_\delta(3\delta)$. Indeed, $u_j - u_i \in A_\delta(3\delta)$ for all $i < j$, and 0 is an accumulation point of $\{u_j - u_i : i < j\}$. (If $V = \{x : |\langle x_l^*, x\rangle| < \varepsilon,\ l \le n\}$ is a basic open neighborhood of 0, find $i < j$ such that $|\langle x_l^*, u_i\rangle - \langle x_l^*, u_j\rangle| < \varepsilon$ for all $l \le n$ and then $u_j - u_i \in V$.)

Define a function f to be $2/3$ on A_δ and $f(0) = 0$. Then f is weakly continuous on the weakly closed set $A_\delta \bigcup\{0\}$ and norm-Lipschitz with Lipschitz constant at most 1. If g is a weakly continuous extension of f to E, then there are $x \in A_\delta$ and $\|y\| \le 3\delta$ such that $g(x) = 2/3$ while $g(x+y)$ is arbitrarily close to 0. Thus g cannot satisfy a Lipschitz condition with constant smaller than $2/9\delta$.

We now "glue" together a sequence of such examples with $\delta \to 0$.

By Proposition E.7 there are $z_n \in E$ and $z_n^* \in E^*$ with $\|z_n\| = 6$ and $\|z_n^*\| = 1$ and such that $\langle z_n^*, z_j\rangle = 0$ for $j \le n$ and $\langle z_n^*, z_j\rangle > 5$ for $j > n$. Let A_n be the set A_δ for $\delta = 1/3n$. Then $A_n \subset B(0,2)$ implies that $d(A_n + z_n, A_m + z_m) > 1$ for every $n \ne m$, and an argument similar to the one used above to show that A_δ is weakly closed shows that both sets $\{z_n\}$ and $\bigcup(A_n + z_n)$ are weakly closed. Hence

$$A = \{z_n\} \cup \left(\bigcup (A_n + z_n)\right)$$

is also closed. The function f defined to be $2/3$ on $\bigcup(A_n + z_n)$ and $f(z_n) = 0$ for all n is weakly continuous and norm-Lipschitz, and by the argument above it does not have a weakly continuous extension which is also norm-Lipschitz. □

Remark. The impossibility of extending the function is really a consequence of the interplay between the two topologies. Indeed, we may assume in the proof above that E is separable. But separable Banach spaces are certainly Lindelöf in their weak topology, hence normal. Thus when only the weak topology is considered, weakly continuous functions do extend.

6. Notes and Remarks

Proposition 2.2 and Theorem 2.3 are due to Tsaŕkov [**562**] (see also the announcement [**560**]). The final argument in the proof of Theorem 2.3 showing that $\alpha = p/2$ for $1 \le p \le 2$ and $q = 1$ was overlooked in [**562**] and was observed by Assaf Naor.

Proposition 2.4 is due to Tsaŕkov [**561**]. (The formulation in [**561**] refers to approximation by maps with Hölder derivatives, but the proof, which we reproduced here, does not use this stronger condition.)

The results of Section 2 are due to Björnestål [**68**]. He proved the results for closed linear subspaces of E, but only minor modifications of the original proof were necessary to treat general closed convex subsets.

The metric projection does not have to be continuous even when E is reflexive and strictly convex. See, e.g., Brown [**97**] and Veselý [**577**].

Theorem 2.8 is of a local nature: Given a point $x \notin A$, it gives good estimates for $\|P_A y - P_A x\|$ when y is close to x. This gives, of course, estimates for $\|P_A x - P_A y\|$ also when y is far from x, but it is sometimes possible to obtain better estimates that do not follow from the local situation. See for example Alber [**6**].

Theorem 2.9 is originally due to Figiel and Pisier [**191**] (cf. also Lindenstrauss and Tzafriri [**372**], 1.e.16). Their approach is entirely different, showing that a space satisfying the assumption of the theorem is of type 2 and of cotype 2 (see Theorem A.7 for a simple proof of this fact). Hence it is isomorphic to a Hilbert space by a theorem of Kwapien [**343**]. In this approach it suffices that the space has one norm which is uniformly convex and another which is uniformly smooth (with appropriate moduli), and it is not necessary to have one norm with both properties.

Several other interesting isomorphic characterizations of Hilbert space, involving C^2 norms or other special smooth functions, follow from Theorem 2.9. For these results and for references to the original literature, see Section V3 of the book [**144**] by Deville, Godefroy and Zizler (as well as the notes and remarks to that section).

Theorem 2.11 was proved for two-dimensional spaces by Grünbaum [**236**] and for general finite-dimensional spaces by Schönbeck [**524**]. Our proof of Theorem 2.11 mostly follows the original proof of Schönbeck, with some arguments by Wells and Williams [**581**].

To give a proper infinite-dimensional formulation, we had to introduce, however, Lemma 2.15. The formulation in [**581**] is vacuous. There are no separable infinite-dimensional spaces satisfying the CEP which are not strictly convex. Indeed, by a theorem of Zippin [**595**] every separable infinite-dimensional L_1-predual contains a subspace isometric to c_0. If E were a separable L_1-predual with the CEP, choose a sequence $\{e_n\}_{n=1}^{\infty} \subset E$ which is isometrically equivalent to the unit vector basis of c_0. For each sequence of signs $\theta = \{\theta_n\}$ with $\theta_n = \pm 1$ consider the map f_θ defined by $f_\theta(x_n) = 2\theta_n e_n$ where $x_1 = e_1$ and $x_n = e_n - \sum_{i<n} e_i$ for $n > 1$. Since each f_θ is an isometry it extends as a 1-Lipschitz map to 0, so put $f_\theta(0) = y_\theta$ and then $\|y_\theta - 2\theta_n e_n\| \leq 1$ for all n. Since $\theta \neq \eta$ implies that there is an n such that $\|2\theta_n e_n - 2\eta_n e_n\| = 4$, it follows that $\|y_\theta - y_\eta\| \geq 2$. But this contradicts the separability of E.

de Figueiredo and Karlovitz [**192**] also give an infinite-dimensional version of Theorem 2.11 (which assumes that the unit ball of E has an extreme point).

The nontrivial part of Lemma 2.12(i) is due to Grothendieck [**233**]. Part (ii) is due to Lindenstrauss [**356**]. For a systematic study of intersection properties of balls and of their relations to various extension properties, see Aronszajn and Panitchpakdi [**24**], Lindenstrauss [**356**] and Lima [**352**]. Lemmas 2.13 and 2.14 are due to Lindenstrauss [**356**]. In the finite-dimensional case they were proved by Hanner [**248**] by a different method.

For historical background on the Steiner point, see the Notes to sections 3.4 and 5.4 in Schneider [**521**].

For convex bodies A with C^2 boundary the Steiner point is the center of gravity of the boundary of A with respect to the probability measure on the boundary whose density is proportional to the Gaussian curvature; see [**521**], page 305.

The computation of the Lipschitz constant of the Steiner map appears, e.g., in Positsel'skiĭ [**469**]. We mention in passing that c_n is the same as the projection constant of l_2^n; see Grünbaum [**237**].

There are several proofs (e.g. Meyer [**411**], Schneider [**519**]) of Theorem 2.20 with various sets of assumptions on ϕ. The proof here is by Positsel'skiĭ [**470**]. By a little more care in the handling of the integrals he actually proves the theorem under the assumption that ϕ is continuous only at balls rather than at all sets. Some continuity assumptions are, however, needed; see Sallee [**507**] and Schneider [**520**].

Another weakening of the assumptions by a very small change in the proof was observed by Schneider; see [**521**] pages 367–369. It is enough to assume that ϕ commutes with proper isometries (i.e. translations and orientation-preserving orthogonal transformations). For $n \geq 3$ the proof is the same, being careful to choose τ in the final argument to be in $SO(n)$. For $n = 2$ the only possible τ is a reflection, so a different argument is needed at this point.

Vitale [**578**] showed that the Steiner map does not even have a merely continuous extension to infinite-dimensional Hilbert spaces. He showed that if A is a compact convex and infinite-dimensional set in a Hilbert space, then not only do the Steiner points of nearby finite-dimensional sets not converge to a single point in A, they actually "cover" it: $A = \cap_{\varepsilon>0} S_\varepsilon$ where S_ε is the closure of the set of all Steiner points of finite-dimensional sets B satisfying $d_H(A, B) < \varepsilon$.

Proposition 2.21 is due to Positsel'skiĭ [**469**]. It was rediscovered by Przesławski and Yost [**479**].

Theorem 2.23 is due to Benyamini [**50**]. The study of simultaneously continuous retractions on the unit ball of E^* was motivated by Benyamini [**49**] where such maps were used to prove that linear into almost-isometries between separable $C(K)$ spaces can be approximated by linear into isometries.

We mention two more results from [**50**]. The first is that for separable $C(K)$ spaces there is a w^*-continuous retraction from $C(K)^*$ onto its unit ball which is 2-Lipschitz with respect to the norm. The second is that the existence of simultaneously continuous retraction from E onto its unit ball depends on the norm of E. For example, Hilbert space can be renormed so that there is no weakly continuous retraction onto the new unit ball which is uniformly norm-continuous.

The constructions in Schachermayer [**510**] and in Johnson and Wolfe [**287**] show that there is no retraction from $L_\infty[0,1]$ onto its unit ball which is simultaneously w^*-continuous (as the dual of $L_1[0,1]$) and norm-uniformly continuous.

Theorems 2.24 and 2.25 are due to Matoušková [**387**]. Note that the same proofs yield the analogous results with the Lipschitz condition replaced by any other given modulus of continuity. It is also shown in [**387**] that the assumption that f is bounded is essential in these results. There is an unbounded weakly continuous and norm-Lipschitz function on a weakly closed subset of a Hilbert space which cannot be appropriately extended to the whole space.

It should be remarked that there are Banach spaces which are not normal in their weak topology; see Corson [**124**].

CHAPTER 3

Fixed Points

1. Continuous Maps

Brouwer's Theorem in $\mathbb{R}^n$ has several equivalent forms. For example:

(i) There is no continuous retraction from the unit ball onto the unit sphere.

(ii) A continuous map of a closed bounded convex set into itself has a fixed point.

(iii) The sphere is not contractible.

(Recall that a metric space X is contractible if there is a point $x_0 \in X$ and a homotopy $h : X \times [0,1] \to X$ such that $h(x,0) = x$ and $h(x,1) = x_0$ for all $x \in X$.)

In infinite dimensions the analogous results are false (see Theorem 3.3 below), and we need to impose additional conditions either on the map or on the set where the map is defined. The following theorem of Schauder and Tychonoff is classical.

Theorem 3.1. *Let K be a compact convex set in a locally convex Hausdorff linear topological space E. Then every continuous map of K into itself has a fixed point.*

Since compactness implies that the map, initially assumed to be just continuous, is actually uniformly continuous, Theorem 3.1 raises the following natural question: Let f be a *uniformly continuous* map on a closed convex and bounded set in a Banach space. Does it have to have a fixed point? (Analogous questions can be asked for the other equivalent forms of Brouwer's Theorem.)

The main result of Section 2 will be that the answer is negative: On every noncompact closed and convex set in a Banach space there is a Lipschitz map without a fixed point.

Another classical fixed point theorem is Banach's contraction principle.

Theorem 3.2. *Let X be a complete metric space, and assume $f : X \to X$ is a Lipschitz map with Lipschitz constant strictly smaller than 1. Then f has a (unique) fixed point.*

In Sections 3 and 4 we consider nonexpansive maps rather than strict contractions, i.e., Lipschitz maps whose Lipschitz constant is equal to 1. In contrast to the universal nature (positive or negative) of the results in the first two sections, the results here depend on the geometry of the space or of the set.

Proof of Theorem 3.2. Let $\lambda < 1$ be the Lipschitz constant of f, and fix any $x \in X$. Then $d\big(f^{n+1}(x), f^n(x)\big) \le \lambda d\big(f^n(x), f^{n-1}(x)\big) \le \ldots \le \lambda^n d\big(f(x), x\big)$. Thus $\{f^n(x)\}$ is Cauchy, and its limit is a fixed point of f. The uniqueness is obvious. $\diamondsuit$

Proof of Theorem 3.1. Since K is compact, we can find for each convex symmetric neighborhood U of zero in E another convex symmetric neighborhood $V \subset U$

of zero such that $f(x) - f(y) \in U$ whenever $x - y \in V$. By compactness again there is a finite set $A \subset K$ such that $\{V + z : z \in A\}$ is an open cover of K. Let $\{\varphi_z : z \in A\}$ be a partition of unity subordinated to this cover (see Appendix B).

The map $f_U(x) = \sum_{z \in A} \varphi_z(x) f(z)$ is a continuous function from K into the convex hull of a finite number of points in K. By Brouwer's theorem it has a fixed point which we denote by x_U. Note that $f_U(x)$ is a convex combination of points $f(z)$ with $x - z \in V$. Hence $f(z) - f(x) \in U$ for each of them, and since U is convex it follows that $f_U(x) - f(x) \in U$ for every $x \in K$. In particular, $f(x_U) - x_U = f(x_U) - f_U(x_U) \in U$. Pass now to a convergent subnet of $\{x_U\}$, and let x be its limit. Given a neighborhood U_0 of 0, consider only sets U with $U \subset U_0$. Then $f(x_U) - x_U \in U \subset U_0$, and passage to limit gives $f(x) - x \in \overline{U_0}$. Since U_0 was an arbitrary neighborhood of zero it follows that $f(x) = x$. ♢

Theorem 3.3. *Let K be a noncompact closed convex subset of a Banach space. Then there is a continuous map of K into itself with no fixed points.*

Proof. We shall consider only bounded K. (Refer to the proof of Theorem 3.4 to see how to handle the unbounded case.) The proof has four steps.

(i) K contains a closed subset R which is homeomorphic to the real line. Indeed, since K is not compact there are a $\delta > 0$ and an infinite sequence $\{x_n\}_{n=-\infty}^{\infty}$ in K such that $\operatorname{dist}\big(x_n, \operatorname{span}\{x_i : |i| \le n, i \ne n\}\big) > \delta > 0$ for all n. Set R to be the polygonal line $\bigcup [x_n, x_{n+1}]$ with the obvious homeomorphism onto $\mathbb{R}$.

(ii) The real line is an absolute retract (i.e., there is a continuous retraction on $\mathbb{R}$ from every metric space containing it as a closed set). This is just a reformulation of Tietze's theorem. Thus there is a continuous retraction $r : K \to R$.

(iii) There is a map $\psi : R \to R$ without a fixed point. Take, for example, the image in R of the map $x \to x + 1$ in the real line.

(iv) The map $\varphi = \psi \circ r : K \to K$ does not have a fixed point. □

Remark. The natural homeomorphism between the real line and R is a local Lipschitz homeomorphism with a uniform constant on each interval of length not exceeding 1, say. But, at least when K is bounded, there is no Lipschitz retraction from K onto R. Indeed, let $l(x, y)$ be the intrinsic (path) distance on R, given by

$$\sup_{\varepsilon > 0} \inf \left\{ \sum_{i=1}^{n} \|u_i - u_{i-1}\| : u_0 = x, \dots, u_n = y;\ u_i \in R \text{ and } \|u_i - u_{i-1}\| \le \varepsilon \right\}$$

and assume that there is a Lipschitz retraction $r : K \to R$. For any $x, y \in R$ choose a chain $\{u_i\}$ that almost realizes $l(x, y)$. Since the segment $[x, y]$ is mapped by r to a connected set in R containing x and y, there are points $\{z_i\}$ in $[x, y]$ such that $r(z_i) = u_i$. Then

$$\begin{aligned} l(x, y) &\approx \sum \|u_i - u_{i-1}\| = \sum \|r(z_i) - r(z_{i-1})\| \\ &\le \|r\|_{Lip} \sum \|z_i - z_{i-1}\| = \|r\|_{Lip} \|x - y\| \\ &\le \|r\|_{Lip} \operatorname{diam}(K). \end{aligned}$$

But $l(x, y)$ is unbounded on R; thus $\|r\|_{Lip}$ cannot be finite.

2. Lipschitz Maps

It is very easy to give examples of Lipschitz maps on closed bounded and convex sets without fixed points. For example, the map $x \to (1 - \|x\|, a_1, a_2, \dots)$ is such a map on the unit ball of l_p for every $1 \le p < \infty$ and for c_0 (where a_i are the coordinates of the vector x). However, this map has an approximate fixed point, i.e., a sequence $\{x_n\}$ such that $\|f(x_n) - x_n\| \to 0$. For example, one can take $x_n = (\sum_{i \le n} e_i)/n^{1/p}$ (for l_p), and $x_n = \frac{1}{2}\sum_{i \le n}(1 - \frac{i}{n})e_i$ (for c_0).

The main result of this section strengthens Theorem 3.3 to Lipschitz maps.

Theorem 3.4. *Let K be a noncompact closed and convex subset of a Banach space. Then there is a Lipschitz map $f : K \to K$ without approximate fixed points; i.e., there is a $\delta > 0$ such that $\|f(x) - x\| > \delta$ for all $x \in K$.*

Proof. The proof follows the scheme of the proof of Theorem 3.3, and we first consider a bounded K. The role of $\mathbb{R}$ will be played by another model-set, which we denote by $\mathbb{F}$. Denote the unit vectors in l_∞ by e_n, and set

$$\mathbb{F} = \bigcup_{n \ge 0} \operatorname{co}\{0, e_n, e_{n+1}\}.$$

We shall show that

(i) K contains a closed subset F which is Lipschitz equivalent to $\mathbb{F}$.

(ii) $\mathbb{F}$ is an absolute Lipschitz retract.

(iii) There is a Lipschitz map $\psi : \mathbb{F} \to \mathbb{F}$ without approximate fixed points.

Once these are proved, fix a Lipschitz retraction r from K onto F, and let $\varphi : F \to F$ be the image on F of the map ψ on $\mathbb{F}$. Then $\varphi \circ r$ is a Lipschitz map on K, and it is easy to check that it has no approximate fixed points.

(i) Assume that $0 \in K$. Using the one-sided subsequence $\{x_n\}_{n=0}^{\infty}$ of the sequence from the proof of Theorem 3.3, we set

$$F = \bigcup_{n=0}^{\infty} \operatorname{co}\{0, x_n, x_{n+1}\}.$$

We leave it to the reader to check that F is, indeed, Lipschitz equivalent to $\mathbb{F}$.

(ii) We construct explicitly a Lipschitz retraction from l_∞ onto $\mathbb{F}$. Since l_∞ is an absolute Lipschitz retract (see Lemma 1.1), so is $\mathbb{F}$.

Let e be the vector whose coordinates are all equal to 1, and define for $x \in l_\infty$

$$\varepsilon(x) = \inf\{\varepsilon : (x - \varepsilon e) \vee 0 \in \mathbb{F}\}.$$

Then $\varepsilon(x) - \varepsilon(y) \le \|x - y\|$, i.e., $\|\varepsilon\|_{Lip} = 1$; hence $r(x) = (x - \varepsilon(x)e) \vee 0$ is a retraction onto $\mathbb{F}$ with $\|r\|_{Lip} \le 2$.

(iii) The subset $\bigcup[e_n, e_{n+1}]$ of $\mathbb{F}$ is naturally homeomorphic to $[0, \infty)$, and we denote by $g(t)$ the point corresponding to t, i.e., $g(t) = (n + 1 - t)e_n + (t - n)e_{n+1}$ when $n \le t \le n + 1$. This gives natural "coordinates" in $\mathbb{F}$, and we can represent each x in $\mathbb{F}$ as

$$x = \lambda g(t)$$

where $0 \le t < \infty$, $0 \le \lambda \le 1$, and the representation is unique except for $x = 0$.

The definition of ψ will be given separately for $t \le 1$ and for $t \ge 1$.

For $0 \le t \le 1$, put

$$\psi\big(\lambda g(t)\big) = \begin{cases} (1-3\lambda t)e_0 & \lambda \le 1/3 \\ (1-t)e_0 + (3\lambda - 1)t(e_0+e_1)/2 & 1/3 \le \lambda \le 2/3 \\ g(t/2\ + 3\lambda - 2) & 2/3 \le \lambda. \end{cases}$$

When $n \ge 1$ and $n \le t \le n+1$, put

$$\psi\big(\lambda g(t)\big) = \begin{cases} (1-3\lambda)e_0 & \lambda \le 1/3 \\ (3\lambda - 1)g(t-\ 1/2) & 1/3 \le \lambda \le 2/3 \\ g\big(t-\ 1/2\ + (3\lambda - 2)\big) & 2/3 \le \lambda. \end{cases}$$

A simple direct computation shows that ψ is Lipschitz. To see that it has no approximate fixed points, observe that if $t \ge 1$ and $\lambda \ge 1/3$, and if we put $\psi\big(\lambda g(t)\big) = \mu g(s)$, then $\psi\big(\lambda g(t+1)\big) = \mu g(s+1)$. It follows that it is enough to check that ψ does not have a fixed point in the compact set where $t \le 2$. We leave the easy checking to the reader.

Assume now that K is unbounded. If K contains a noncompact closed and bounded subset, then this subset contains a copy of $\mathbb{F}$, and we are done. If not, assume that $0 \in K$, and choose for every n a point $x_n \in K$, $\|x_n\| = n$. By the compactness of $\{x \in K : \|x\| \le 1\}$, we may pass to a convergent subsequence of x_n/n and obtain a point y such that the infinite ray $\mathbb{R}^+ y$ is contained in K. This ray is an absolute Lipschitz retract, so fix a Lipschitz retraction $r : K \to \mathbb{R}^+ y$. Then the map $x \to (\|r(x)\| + 1)y$ has no approximate fixed point. □

Corollary 3.5. *Let E be an infinite-dimensional Banach space. Then*

(i) *There is a Lipschitz retraction from the unit ball $B(E)$ of E onto the unit sphere $S(E)$.*

(ii) *The unit sphere $S(E)$ is Lipschitz contractible.*

Proof. (i) We shall work with the ball of radius 2. Starting with a map $f_1 : B(E) \to B(E)$ with no approximate fixed points, extend it to a map on $2B(E)$ by setting $f(x) = (2 - \|x\|)f_1(x/\|x\|) + (\|x\| - 1)x/\|x\|$ for $1 \le \|x\| \le 2$. The map f is a Lipschitz map of $2B(E)$ into $B(E)$ with no approximate fixed points, and $f(x) = x/2$ when $\|x\| = 2$. Put $\delta = \inf \|f(x) - x\| > 0$. Then $\|x + 3\big(x - f(x)\big)/\delta\| \ge 3 - \|x\| \ge 1$, and thus

$$r(x) = 2\frac{x + 3\big(x - f(x)\big)/\delta}{\|x + 3\big(x - f(x)\big)/\delta\|}$$

is a Lipschitz map from $2B(E)$ into its boundary. If $\|x\| = 2$, then $f(x) = x/2$ and hence $r(x) = x$.

(ii) If $r : B(E) \to S(E)$ is a Lipschitz retraction, then a Lipschitz homotopy of the identity on $S(E)$ to a point $x_0 \in S(E)$ is given by

$$h(x,t) = r\big(tx_0 + (1-t)x\big).$$

□

Remark. There is a standard argument to deduce that the unit sphere in $\mathbb{R}^n$ is a retract of the unit ball once we assume that there is a fixed point free mapping on the unit ball. If f is such a fixed point free map, put $r(x) = x + \lambda\big(x - f(x)\big)$, where $\lambda \ge 0$ is chosen such that $\|r(x)\| = 1$. The analogous mapping is not necessarily continuous in a general Banach space. We demonstrate the nature of the difficulty by an example of a partially defined function on the unit square in the plane. Define a function f on the segment $[0,1]$ on the x-axis by $f(t,0) = (1,t)$ for $0 \le t \le 1$. Then $\|f(x) - x\| \ge 1/2$. Yet, if we use the formula above (with respect to the unit square) to define r, then $\lim_{t\to 1} r(t,0) = (1,-1) \ne (1,0) = r(1,0)$. In infinite-dimensional spaces such phenomena can occur for maps that are defined on the whole ball. In the proof of (i) we made sure that $f\big(2B(E)\big) \subset B(E)$, and this eliminated the problem because f does not take values near the boundary.

It is perhaps instructive to give at least one example of a Banach space where it is possible to give an explicit formula for a Lipschitz retraction from the unit ball onto the unit sphere. Let $\varphi : C[0,1] \to C[0,1]$ be the map

$$\varphi(x)(t) = \big|x(t) + G(x,t)\big| - G(x,t)$$

where $G(x,t) = 1 - 2\big(1 - \|x\|\big)t$. Then φ is 5-Lipschitz, $\varphi(x) = x$ when $\|x\| = 1$, and $\inf\{\|\varphi(x)\| : x \in C[0,1]\} > 0$. Hence $f(x) = \varphi(x)/\|\varphi(x)\|$ is a Lipschitz retraction from the unit ball of $C[0,1]$ onto the unit sphere.

3. Nonexpansive Maps

In this section we impose on the map f the stronger condition that it is nonexpansive, i.e., that it is a Lipschitz map whose Lipschitz constant is one. We then study geometric conditions on the space E or on the set K which ensure that f has a fixed point. In the next section we shall describe some constructive ways to approximate fixed points. In both sections we consider only bounded sets.

Nonexpansive maps on bounded sets do not have to have fixed points. However, as the next proposition shows, they always have approximate fixed points, i.e., a sequence $\{x_n\}$ such that $\lim \|f(x_n) - x_n\| \to 0$. Moreover, the set of approximate fixed points is "metrically convex" in a natural sense.

Proposition 3.6. *Let K be a closed bounded and convex subset of a Banach space, and let $f : K \to K$ be nonexpansive. Then*

(i) *f has an approximate fixed point.*

(ii) *Let $\{x_n\}$ and $\{y_n\}$ be approximate fixed points such that $\lim \|x_n - y_n\| = d$ exists. Then for every $0 < \alpha < 1$ there is an approximate fixed point $\{z_n\}$ such that $\lim \|x_n - z_n\| = \alpha d$ and $\lim \|z_n - y_n\| = (1-\alpha)d$.*

Proof. (i) Assume that $0 \in K$. For each $t > 0$ the Lipschitz constant of the map $f_t(x) = f(tx)$ is t. If $t < 1$, then f_t has a fixed point x_t by Theorem 3.2 and

$$\|f(x_t) - x_t\| = \|f(x_t) - f_t(x_t)\| = \|f(x_t) - f(tx_t)\| \le (1-t)\|x_t\| \to 0$$

as $t \to 1$ because K is bounded.

(ii) Choose $\varepsilon_n \to 0$ such that $\max\big(\|f(x_n) - x_n\|, \|f(y_n) - y_n\|\big) < \varepsilon_n^2$, and put $w_n = (1-\alpha)x_n + \alpha y_n$. Then $f_n(z) = (1-\varepsilon_n)f(z) + \varepsilon_n w_n$ maps K into itself and has Lipschitz constant $1 - \varepsilon_n < 1$. Hence f_n has a fixed point z_n; i.e.,

$$z_n = (1-\varepsilon_n)f(z_n) + \varepsilon_n w_n.$$

It follows that $f(z_n) - z_n = \varepsilon_n(f(z_n) - w_n) \to 0$, so $\{z_n\}$ is an approximate fixed point. Also

$$\begin{aligned} \|z_n - x_n\| &= \|(1-\varepsilon_n)f(z_n) + \varepsilon_n w_n - x_n\| \\ &\le (1-\varepsilon_n)\Big(\|f(z_n) - f(x_n)\| + \|f(x_n) - x_n\|\Big) + \varepsilon_n\|w_n - x_n\| \\ &\le (1-\varepsilon_n)\Big(\|z_n - x_n\| + \varepsilon_n^2\Big) + \varepsilon_n\alpha\|x_n - y_n\|. \end{aligned}$$

It follows that $\|z_n - x_n\| \le \varepsilon_n + \alpha\|x_n - y_n\|$; hence $\limsup\|x_n - z_n\| \le \alpha d$. Similarly, $\limsup\|y_n - z_n\| \le (1-\alpha)d$. The reverse inequalities then follow from the triangle inequality. □

Not much can be said on the existence of fixed points without further assumptions on the set K, and from now on our standing assumption will be that K is weakly compact. This enables us to introduce some topological considerations while still allowing the sets to be "large" with respect to the norm. One immediate consequence of this assumption (and Zorn's lemma) is that K contains a minimal closed and convex subset, invariant under f, and we shall look for conditions that ensure that such a minimal set is, in fact, a single point, i.e., a fixed point.

A point $x \in K$ is called diametral if $\sup\{\|x - y\| : y \in K\} = \operatorname{diam}(K)$.

Proposition 3.7. *Let K be a weakly compact convex subset of a Banach space, and assume that $f : K \to K$ is nonexpansive. Assume that K does not have a proper closed convex subset which is f-invariant. Then*

(i) *Every point in K is diametral.*

(ii) *For every $x \in K$ and for every approximate fixed point $\{x_n\}$ of f, the limit $\lim\|x_n - x\|$ exists and is equal to* $\operatorname{diam}(K)$.

Proof. (i) The function $\psi(x) = \sup\{\|x - y\| : y \in K\}$ is a convex and continuous function, and thus the set M where it attains its minimum is closed and convex. The lower semi-continuity of the norm with respect to the weak topology implies that ψ is also weakly lower semi-continuous, and since K is weakly compact M is nonempty. Using the minimality of K and the triangle inequality, we obtain $\psi(x) = \sup\{\|x - f(y)\| : y \in K\}$. Thus

$$\psi\big(f(x)\big) = \sup\big\{\|f(x) - f(y)\| :\ y \in K\big\} \le \sup\big\{\|x - y\| :\ y \in K\big\} = \psi(x)$$

and M is f-invariant. Using the minimality of K again gives $M = K$, and hence ψ is constant. This constant value is clearly equal to $\operatorname{diam}(K)$.

(ii) Since K is weakly compact, minimality implies that it is weakly separable, hence also norm separable. By diagonalization we can pass to a subsequence (still denoted by $\{x_n\}$) so that $\varphi(x) = \lim\|x - x_n\|$ exists for every $x \in K$. We shall show that $\varphi(x) = \operatorname{diam}(K)$ independently of the initial sequence. It then follows that the original limit exists and that there was no need to pass to a subsequence.

Since $\{x_n\}$ is an approximate fixed point, we obtain

$$\varphi\big(f(x)\big) = \lim\|f(x) - x_n\| = \lim\|f(x) - f(x_n)\| \le \lim\|x - x_n\| = \varphi(x)$$

and the same argument as in the proof of part (i) shows that the convex and continuous function φ is actually constant, say equal to d. Clearly $d \le \operatorname{diam}(K)$.

By passing to a further subsequence, we may assume that $\{x_n\}$ converges weakly to some point $y \in K$, and thus $d = \lim\|x - x_n\| \ge \|x - y\|$ for all $x \in K$. But by (i) y is diametral; hence $d \ge \operatorname{diam}(K)$. □

A bounded convex set K is said to have normal structure if every convex subset S of K which does not reduce to a single point contains a nondiametral point, i.e., a point x such that $\sup\{\|x-y\| : y \in S\} < \operatorname{diam}(S)$. Bounded convex subsets of uniformly convex and of uniformly smooth Banach spaces have normal structure (even uniformly so); see Proposition A.9.

Corollary 3.8. *Let K be a weakly compact convex set with normal structure, and let f be a nonexpansive map on K. Then f has a fixed point.*

Proof. We may assume that K is minimal. The corollary follows directly from Proposition 3.7(i) and the definition of normal structure. □

A typical example of a weakly compact set in which every point is diametral is the closed convex hull of 0 and of the unit vectors in c_0. Nevertheless, c_0 has the fixed point property.

Theorem 3.9. *Let K be a weakly compact convex subset of c_0, and let f be a nonexpansive map on K. Then f has a fixed point.*

Proof. Assume that K is minimal and that $\operatorname{diam}(K) = 1$, and fix an approximate fixed point $\{x_n\}$. By passing to a subsequence and by translation we may assume that the x_n's converge weakly to 0 and, by passing to a further subsequence, that they have essentially disjoint supports. Since $\|x_n\| \to 1$ (this holds for every approximate fixed point by Proposition 3.7(ii)), the essential disjointness of the supports implies that $\|x_n - x_{n+1}\| \to 1$. By Proposition 3.6(ii) (with $y_n = x_{n+1}$) there is an approximate fixed point $\{z_n\}$ such that $\lim \|x_n - z_n\| = \lim \|x_{n+1} - z_n\| = 1/2$ (and of course $\|z_n\| \to 1$). But this is impossible. Indeed, if n and j are chosen such that $\|x_n - z_n\|, \|x_{n+1} - z_n\| < 5/8$, and $|z_n(j)| > 6/8$, then $|x_n(j)|, |x_{n+1}(j)| > 1/8$, contradicting the fact that x_n and x_{n+1} have essentially disjoint supports. □

In the next two results we consider weakly compact subsets of $L_1(\mu)$ for some probability measure μ. We give an example of a nonexpansive map on a weakly compact and convex set without a fixed point, and we show that convex weakly compact subsets of *reflexive* subspaces of $L_1(\mu)$ do have the fixed point property for nonexpansive maps.

Recall that a closed convex subset of $L_1(\mu)$ is weakly compact if and only if it is uniformly integrable; i.e., for every $\varepsilon > 0$ there is a $\delta > 0$ such that $\int_A |f| < \varepsilon$ whenever $\mu(A) < \delta$. (See Dunford and Schwartz [**159**], page 294.)

Example 3.10. *There is a convex weakly compact subset K of $L_1(\mu)$ and a nonexpansive map on K without a fixed point.*

Consider the interval $I = [0, 1]$ with the normalized Lebesgue measure, and let (Ω, μ) be a probability space for which there is a one-one measure preserving transformation $\tau : I \times \Omega \to \Omega$. The set

$$K = \left\{ f \in L_1(\Omega, \mu) : \ 0 \le f \le 1 \ \text{ and } \ \int f \, d\mu = 1/2 \right\}$$

is convex and weakly compact.

Denote the subgraph $\{(t, \omega) : 0 \le t \le f(\omega)\}$ of $f \in K$ by $G(f)$, and define $\varphi : K \to K$ by $\varphi(f) = \chi_{\tau(G(f))}$, the indicator function of the set $\tau(G(f)) \subset \Omega$. It is immediate that φ is an isometry on K and that $f \in K$ is a fixed point of φ if and only if $f = \chi_A$ for a set A satisfying $\tau(I \times A) = A$ and $\mu(A) = 1/2$.

Now take Ω to be the product $I^{\aleph_0}$ with the product Lebesgue measure. If $\omega = (s_1, s_2 \dots) \in \Omega$ and $t \in I$, denote the point $(t, s_1, s_2, \dots) \in \Omega$ by $t \triangleleft \omega$ and define the measure preserving map $\tau : I \times \Omega \to \Omega$ by $\tau(t, \omega) = t \triangleleft \omega$. If $\tau(I \times A) = A$ and $\omega \in A$, then $t \triangleleft \omega \in A$ for every $t \in I$. Inductively $t_1 \triangleleft \dots t_n \triangleleft \omega \in A$ for every $t_1, \dots, t_n \in I$. It follows that A is a "tail" event; hence its measure is either zero or one by the zero-one law. Thus $\chi_A \notin K$ and φ has no fixed point.

Theorem 3.11. *Let E be a reflexive subspace of $L_1(\mu)$, and let K be a convex weakly compact subset of E. Then every nonexpansive map $f : K \to K$ has a fixed point.*

Proof. We may assume that K is minimal and that it is contained in the unit ball of $L_1(\mu)$ with $\operatorname{diam}(K) = 1$. Fix a countable dense set D in K. To avoid ambiguities that may arise from the fact that elements of $L_1(\mu)$ are defined only up to a set of measure zero, we fix, once and for all, everywhere-defined representatives of the elements of D. Put

$$v(t) = \sup\big\{y(t) : \ y \in D\big\} \ ; \quad u(t) = \inf\big\{y(t) : \ y \in D\big\}$$

and then $u \le x \le v$ almost everywhere for every $x \in K$.

Let $\{x_n\}$ be an approximate fixed point of f. We first show that for almost all t the only accumulation points of the sequence of numbers $\{x_n(t)\}$ are $u(t)$ and $v(t)$. Fix $x, y \in D$. By Proposition 3.7(ii) applied to the sequence $\{x_n\}$ and to each of the three points $x, y, (x + y)/2 \in K$ we obtain

$$\int \big(|x_n - x| + |x_n - y| - |2x_n - x - y|\big)$$
$$= \|x_n - x\| + \|x_n - y\| - 2\|x_n - (x + y)/2\| \to 0.$$

Since the integrands are nonnegative functions, we may assume, by passing to a subsequence, that they converge to zero almost everywhere. Since D is countable, there is a set A with $\mu(A) = 1$ such that

$$|x_n(t) - x(t)| + |x_n(t) - y(t)| - |2x_n(t) - x(t) - y(t)| \to 0$$

for every $t \in A$ and for every $x, y \in D$. Fix $t \in A$, and let α_t be an accumulation point of $\{x_n(t)\}$. Then $|\alpha_t - x(t)| + |\alpha_t - y(t)| = |2\alpha_t - x(t) - y(t)|$ for all $x, y \in D$; hence $\alpha_t - x(t)$ and $\alpha_t - y(t)$ must have the same sign. Since x and y were arbitrary, α_t must coincide with either $v(t)$ or $u(t)$.

We now show that u and v are finite almost everywhere. Set for example $I = \{t \in \Omega : \ v(t) = +\infty\}$, and fix an approximate fixed point $\{x_n\}$ of f. Since K is contained in the unit ball of $L_1(\mu)$

$$\mu\big\{t \in \Omega : \ |x(t)| > 2^n\big\} < 2^{-n}$$

for all $x \in K$ and all n. Since $v(t)$ is the only nonnegative accumulation point of $\{x_n(t)\}$, it follows that $\lim_{n\to\infty} \mu\{t \in I : 1 \le x_n(t) \le 2^m\} = 0$ for every m. By passing to a subsequence we may assume that

$$\mu\{t \in I : 1 \le x_n(t) \le 2^n\} < 2^{-n}$$

for every n. Combining this with the previous estimate (for $x = x_n$), we obtain $\mu\{t \in I : 1 \le x_n(t)\} < 2^{-n+1}$. By the Borel-Cantelli lemma $x_n(t) \ge 1$ only finitely many times for almost all $t \in I$. Hence $\mu(I) = 0$ by the definition of I.

Now fix N. We shall show that there are N functions $\{h^i\}$ in E and N disjoint sets $\{C_i\}$ such that $\int_{C_i} |h^i| \ge 3\|h^i\|/4$ for all $1 \le i \le N$. Since the measure of at least one of the C_i's must be smaller than $1/N$, this will show that the unit ball of E is not uniformly integrable, hence not weakly compact, contradicting the assumption that E is reflexive.

Let $\{x_n\}$ be an approximate fixed point of f. Applying Proposition 3.7(ii) and a diagonal argument, we can pass to a subsequence of $\{x_n\}$, denoted by $\{y_n\}$, such that $\|x_n - y_n\| \to 1$. The sequence $\{y_n\}$ is, of course, also an approximate fixed point. Writing $x_n^0 = x_n$ and $x_n^N = y_n$, apply the "metric convexity" of the set of approximate fixed points (Proposition 3.6(ii)) to find approximate fixed points $\{x_n^j\}$ for $j = 1, \ldots, N-1$ such that $\lim_{n\to\infty} \|x_n^i - x_n^j\| = |i-j|/N$ for all $0 \le i, j \le N$ with $i \ne j$.

Fix $\varepsilon > 0$. By the uniform integrability of the unit ball of E there is a $\delta > 0$ such that $\int_W |h| < \varepsilon\|h\|$ for every $h \in E$ and for every set W satisfying $\mu(W) < \delta$. Since for almost every t the only accumulation points of the $x_n(t)$'s are the finite values $u(t)$ and $v(t)$, it follows that the only accumulation points of $|x_n^i(t) - x_n^j(t)|$ are $v(t) - u(t)$ and 0. We can thus choose n large enough so that

$$\|x_n^i - x_n^{i-1}\| \le (1+\varepsilon)/N \text{ for } i = 1, \ldots, N \text{ and } \|x_n^N - x_n^0\| \ge 1 - \varepsilon$$

and so that there is a set B with $\mu(B) > 1 - \delta$ and subsets $A_1, \ldots, A_N$ and A of B with

$$\big|(v-u) - |x_n^i - x_n^{i-1}|\big| < \varepsilon \text{ on } A_i, \text{ and } |x_n^i - x_n^{i-1}| < \varepsilon \text{ on } B \setminus A_i.$$

Similarly $\big|(v-u) - |x_n^N - x_n^0|\big| < \varepsilon$ on A, and $|x_n^N - x_n^0| < \varepsilon$ on $B \setminus A$.

To simplify the notation, we denote $x_n^i - x_n^{i-1}$ by h^i, $x_n^N - x_n^0$ by h, and $v - u$ by g. Hence

$$\|h^i\| \le (1+\varepsilon)/N, \ \|h\| \ge 1 - \varepsilon, \text{ and } h = \sum h^i \tag{3.1}$$

$$\big|g - |h^i|\big| < \varepsilon \text{ on } A_i, \ |h^i| < \varepsilon \text{ on } B \setminus A_i, \text{ and } \int_{\Omega\setminus B} |h^i| < \varepsilon \tag{3.2}$$

$$\big|g - |h|\big| < \varepsilon \text{ on } A, \ |h| < \varepsilon \text{ on } B \setminus A, \text{ and } \int_{\Omega\setminus B} |h| < \varepsilon. \tag{3.3}$$

To motivate the final computations, consider the simplified situation where $\varepsilon = 0$ and $\mu(B) = 1$. Since $h = \sum h^i$, and since for each fixed t the only possible values of all these functions are $\pm g(t)$ and 0, the equality $\|h\| = \sum \|h^i\|$ implies that the h^i's necessarily have disjoint supports. The computations below quantify this idea when we have only approximate equalities.

By (3.3), (3.1) and (3.2)

$$\int_A g \ge \int_A |h| - \varepsilon = \|h\| - \int_{\Omega\setminus B} |h| - \int_{B\setminus A} |h| - \varepsilon \ge 1 - 4\varepsilon \tag{3.4}$$

and

$$\sum \int_{A_i} g \le N\varepsilon + \sum \int_{A_i} |h^i| \le N\varepsilon + \sum \|h^i\| \le 1 + (N+1)\varepsilon. \tag{3.5}$$

Note that if $t \in A$ and $|h(t)| \ge N\varepsilon$, then the identity $h(t) = \sum h^i(t)$ implies that there is an i such that $|h^i(t)| \ge \varepsilon$, i.e., $t \in A_i$. It follows that $|h| < N\varepsilon$ on $A \setminus \cup A_i$, and thus

$$\int_{A\setminus\cup A_i} g \le \varepsilon + \int_{A\setminus\cup A_i} |h| \le (N+1)\varepsilon.$$

Combining this estimate with (3.4) and (3.5), we find that if we put

$$U = \{t : t \text{ belongs to at least two } A_i'\text{s}\}$$

then

$$\int_U g \leq \sum \int_{A_i} g - \int_{\cup A_i} g \leq \sum \int_{A_i} g - \int_A g + \int_{A \setminus \cup A_i} g \leq (2N+6)\varepsilon.$$

The required disjoint sets are $C_i = A_i \setminus U$. Indeed, $\big|g - |h^i|\big| < \varepsilon$ on A_i, hence on $U \cap A_i$. Thus

$$\begin{aligned} \int_{C_i} |h^i| &\geq \|h^i\| - \int_{\Omega \setminus B} |h^i| - \int_{B \setminus A_i} |h^i| - \int_{U \cap A_i} g - \int_{U \cap A_i} \big|g - |h^i|\big| \\ &\geq \|h^i\| - (2N+9)\varepsilon > 3\|h^i\|/4 \end{aligned}$$

provided ε is small enough. □

4. Approximation of Fixed Points

Up to this point we have studied only the existence of fixed points. In this section we consider the problem of constructively approximating fixed points of nonexpansive maps. We shall consider two approaches. The first was actually used already in the construction of approximate fixed points: For a fixed $z \in K$ and $0 < t < 1$, the map $x \to tf(x) + (1-t)z$ has Lipschitz constant strictly smaller than 1, and thus has a unique fixed point x_t. One can now look for conditions (on E, K or f) that will ensure that $\lim_{t \to 1} x_t$ exists.

The second approach is motivated by the observation that if the iterates $f^n(x)$ of f converge for some $x \in K$, then the limit is a fixed point of f. This was already used in the proof of Theorem 3.2. Even when the iterates do not converge, one can look for conditions (on E, K or f) so that some useful weaker asymptotic behavior of the iterates $f^n(x)$ holds, for example, convergence of the orbit of $f_\lambda = \lambda f + (1-\lambda)I$ (where I is the identity on E and $0 < \lambda < 1$), weak convergence, or some mode of convergence of the averages $\frac{1}{n}\sum_{i=0}^{n-1} f^i(x)$.

To streamline the presentation, we shall occasionally impose conditions on the space E which are stronger than what we actually need. Some results may hold under a variety of weaker conditions, but we shall not pursue this direction.

We first introduce some notation. Assume that the norm on E is Fréchet differentiable (see Section 1 of Chapter 4). Then for each $y \neq 0$ in E there is a unique functional, which we denote by $j(y)$, such that $\|j(y)\| = 1$ and $\langle j(y), y\rangle = \|y\|$. The functional $j(y)$ is the derivative of the norm at the point y. More precisely, there is a function θ_y satisfying $\theta_y(t)/t \to 0$ as $t \to 0^+$, such that

$$\big|\|y+z\| - \|y\| - \langle j(y), z\rangle\big| \leq \theta_y(\|z\|)$$

for every $z \in E$. When E is uniformly smooth, θ_y can be replaced by one function θ independent of y on the unit sphere of E, hence on any bounded subset of E which is bounded away from zero. If E is uniformly smooth, then the map $j : E \setminus \{0\} \to E^*$ is continuous (see Proposition A.5).

Theorem 3.12. *Let E be a uniformly smooth Banach space, and let f be a nonexpansive map on a closed bounded and convex subset K of E. Fix a point $z \in K$ and $0 < t < 1$, and let x_t be the unique point in K satisfying $x_t = tf(x_t) + (1-t)z$. Then $\lim_{t \to 1} x_t$ exists and is a fixed point of f.*

Proof. Assume that $z = 0$ (by translation). Replacing K by the minimal closed convex set containing 0 and invariant under f, and then E by the closed linear space spanned by K, we may also assume that E is separable. It is also clear that if the limit in the theorem exists, then it is a fixed point of f.

Let t_j be a sequence tending to 1. By a standard diagonalization procedure (using the separability of E) we can pass to a subsequence $\{x_n\}$ of $\{x_{t_j}\}$ for which

$$\varphi(y) = \lim_{n\to\infty} \|x_n - y\|$$

exists for all $y \in E$. Using the separability of E again, we can assume, by passing to a further subsequence, that $j(x_n - y)$ is w^*-convergent for every $y \in E$. We denote this limit by $\psi(y)$.

The main step in the proof is to show that $\{x_n\}$ has a convergent subsequence. Assuming this is not the case, we first show that φ is Fréchet differentiable and that its derivative is given by $D_\varphi(y) = -\psi(y)$. Indeed, fix y. Since $\{x_n\}$ does not have a convergent subsequence, the bounded sequence $\|x_n - y\|$ is also bounded away from zero for $n > n(y)$. Thus there is a function θ, independent of n, satisfying $\theta(t)/t \to 0$ as $t \to 0^+$ such that

$$\Big| \|x_n - y - z\| - \|x_n - y\| - \langle j(x_n - y), -z\rangle \Big| \le \theta(\|z\|)$$

for sufficiently large n. Letting $n \to \infty$ we see that

$$\big|\varphi(y+z) - \varphi(y) - \langle -\psi(y), z\rangle\big| \le \theta(\|z\|)$$

as claimed.

The function φ is continuous and convex. Since K is weakly compact, the closed and convex set M where φ attains its minimum is nonempty. It follows from $\|f(x_n) - x_n\| \to 0$ and $\|f(x_n) - f(y)\| \le \|x_n - y\|$ that

$$\varphi(f(y)) \le \limsup \big(\|x_n - f(x_n)\| + \|f(x_n) - f(y)\|\big) \le \lim \|x_n - y\| = \varphi(y).$$

Thus M is also f-invariant. By Corollary 3.8 and by Lemma A.9 it contains a fixed point y_0 of f.

We observe next that for every fixed point y_0 of f

$$\langle j(x_t - y_0), x_t\rangle \le 0. \tag{3.6}$$

Indeed, this follows from $t < 1$ and the inequality

$$\begin{aligned} \langle j(x_t - y_0), x_t - ty_0\rangle &\le \|x_t - ty_0\| = t\|f(x_t) - f(y_0)\| \\ &\le t\|x_t - y_0\| = \langle j(x_t - y_0), tx_t - ty_0\rangle. \end{aligned}$$

On the other hand, since φ attains its minimum on K at y_0 it follows that

$$-\lim\langle j(x_n - y_0), y - y_0\rangle = \langle -\psi(y_0), y - y_0\rangle = \langle D_\varphi(y_0), y - y_0\rangle \ge 0 \tag{3.7}$$

for every $y \in K$. Taking $y = 0$ and comparing with (3.6) (with x_t replaced by x_n) we see that $\limsup \|x_n - y_0\| = \limsup\langle j(x_n - y_0), x_n - y_0\rangle \le 0$, i.e., $x_n \to y_0$ contrary to the assumption they have no convergent subsequence.

To conclude the proof we need to show that $\{x_t\}$ have at most one limit point as $t \to 1$, so assume that $x'_n \to y_1$ and $x''_n \to y_2$. By (3.6) $\langle j(x'_n - y_2), x'_n\rangle \le 0$. Since j is norm continuous $\langle j(y_1 - y_2), y_1\rangle \le 0$ and similarly $\langle j(y_2 - y_1), y_2\rangle \le 0$. Adding the two inequalities gives $\|y_1 - y_2\| = \langle j(y_1 - y_2), y_1 - y_2\rangle \le 0$, i.e., $y_1 = y_2$. □

We now turn to the study of the orbits $\{f^n(x)\}$ of f. The orbits do not have to converge even in finite-dimensional spaces: A nontrivial rotation of the unit disk in the Euclidean plane even satisfies $\|f^{n+1}(x) - f^n(x)\| \not\to 0$ for $x \neq 0$. However, the fixed points of f and of $f_\lambda = \lambda f + (1-\lambda)I$ coincide, and it seems conceivable that f_λ behaves "better" than the original f. (This is obvious for the rotation above.) The next proposition shows that the behavior of the iterates is, indeed, somewhat improved, but the example following it shows that the iterates of f_λ still do not need to converge even in Hilbert space. To simplify the notation we shall consider only $\lambda = 1/2$.

Proposition 3.13. *Let f be a nonexpansive map on a closed bounded and convex subset K of a Banach space. Then $\|f_{\frac{1}{2}}^{n+1}(x) - f_{\frac{1}{2}}^n(x)\| \to 0$ for every $x \in K$.*

Proof. Fix x, and let $x_n = f_{\frac{1}{2}}^n(x)$. Hence $x_{n+1} = \big(x_n + f(x_n)\big)/2$, and we set $u_n = x_{n+1} - x_n$ and $v_n = f(x_{n+1}) - f(x_n)$. Then $\|v_n\| \le \|u_n\|$ and

$$u_{n+1} = (v_n + u_n)/2. \tag{3.8}$$

Thus $\|u_n\|$ is nonincreasing and the limit $L = \lim \|u_n\|$ exists, and $\lim \|v_n\| = L$ too. We need to show that $L = 0$, so assume $L \neq 0$ and fix an integer p. We shall show that $\|f(x_{n+p+1}) - f(x_{n+1})\| > pL/2$ provided n is large enough (depending on p). Since $\|f(x_{n+p+1}) - f(x_{n+1})\| \le \operatorname{diam}(K)$ and since p was arbitrary, this contradicts $L \neq 0$.

Put $\alpha_0 = 2^{-p}$ and $\alpha_j = 2^{j-p-1}$. Then $\sum \alpha_j = 1$, and iteration of (3.8) gives

$$u_{n+p+1} = \alpha_0 u_{n+1} + \sum_{j=1}^{p} \alpha_j v_{n+j}. \tag{3.9}$$

Fix $\delta > 0$ and a large enough n so that $\|u_{n+1}\| \le L + \delta$. Choose a functional $\|x^*\| = 1$ such that $\langle x^*, u_{n+p+1}\rangle = \|u_{n+p+1}\| \ge L$. Then $\langle x^*, u_{n+1}\rangle \le L + \delta$ and also $\langle x^*, v_{n+j}\rangle \le \|v_{n+j}\| \le \|u_{n+j}\| \le \|u_{n+1}\| \le L + \delta$. Applying x^* to (3.9) gives

$$\begin{aligned}
\alpha_i \langle x^*, v_{n+i}\rangle &= \langle x^*, u_{n+p+1} - \alpha_0 u_{n+1} - \sum_{j\neq i} \alpha_j v_{n+j}\rangle \\
&\ge L - (1-\alpha_i)(L+\delta) > \alpha_i L/2
\end{aligned}$$

for every $i \le p$, provided $\delta = \delta(p, L)$ is small enough. Thus

$$\begin{aligned}
\|f(x_{n+p+1}) - f(x_{n+1})\| &= \Big\|\sum_{i=1}^{p} \big(f(x_{n+i+1}) - f(x_{n+i})\big)\Big\| \\
&= \Big\|\sum_{i=1}^{p} v_{n+i}\Big\| \ge \sum_{i=1}^{p} \langle x^*, v_{n+i}\rangle > pL/2.
\end{aligned}$$

□

Example 3.14. *There is a closed bounded and convex subset K in a Hilbert space, a nonexpansive map $f : K \to K$, and a point $x_0 \in K$, such that the sequence $f_{\frac{1}{2}}^n(x_0)$ does not converge.*

We shall construct a nonconvergent sequence $\{x_j\}$ in the unit ball of a Hilbert space H and define the values $f(x_j)$ of f at these points. The definition will be such that $x_{j+1} = f_{\frac{1}{2}}(x_j) = f_{\frac{1}{2}}^{j+1}(x_0)$ and such that f is nonexpansive on $\{x_j\}$. Once this is done we take K to be the closed convex hull of the sequence and extend f to a nonexpansive map from K to K by Theorem 1.12. (Formally the theorem gives only an extension from K to H, but we then compose it with the nearest point retraction from H onto K.)

The sequence $\{x_j\}$ will be defined in "blocks", and we describe the inductive choice of the nth block.

Let $\{e_i\}_{i=0}^{\infty}$ be an orthonormal sequence in H. Starting with $x_0 = e_0$, the choice of the first $(n-1)$ blocks yields vectors $\{x_i\}_{i=0}^{k+1}$ with nonnegative coordinates such that:

(i) The first k vectors $\{x_i\}_{i=0}^{k}$ and the images $\{f(x_i)\}_{i=0}^{k-1}$ of the first $k-1$ of them belong to $\operatorname{span}\{e_i : i < n\}$.

(ii) $f(x_k) \in \operatorname{span}\{e_{n-2}, e_{n-1}, e_n\}$ and $x_{k+1} \in \operatorname{span}\{e_{n-1}, e_n\}$.

(iii) $\|x_i\| \geq \|x_{i+1}\| > 1/2$ and $\|f(x_i)\| < \|x_i\|$ for all i.

In the nth block we define $f(x_{k+1})$ and m more vectors $\{y_j\}_{j=1}^{m}$ with nonnegative coordinates such that:

(i′) $f(x_{k+1})$ and y_j belong to $\operatorname{span}\{e_{n-1}, e_n\}$ for every $j < m$.

(ii′) $f(y_{m-1}) \in \operatorname{span}\{e_{n-1}, e_n, e_{n+1}\}$ and $y_m \in \operatorname{span}\{e_n, e_{n+1}\}$.

(iii′) $\|y_j\| \geq \|y_{j+1}\| > 1/2$ and $\|f(y_j)\| < \|y_j\|$ for all j.

To simplify the notation we shall denote x_{k+1} by y_0.

Let P be the plane spanned by e_{n-1} and e_n. Let α be the angle between y_0 and e_n, and fix an integer m (to be specified later). Let V be the rotation of the plane P around the origin by the angle $2\alpha/(2m-1)$. The vectors $\{y_j\}_{j=1}^{m-1}$ are defined inductively in the plane P by

$$f(y_j) = V^2 y_j; \quad y_{j+1} = \big(y_j + f(y_j)\big)/2.$$

Then $\|f(y_j)\| = \|y_j\|$, the angle between y_j and $f(y_j)$ is $4\alpha/(2m-1)$, and the angle between y_j and y_{j+1} is $2\alpha/(2m-1)$. Also $\|f(y_i) - f(y_j)\| = \|y_i - y_j\|$ for all $i, j < m$, and $\|y_j\| = \|y_{j-1}\| \cos\big(2\alpha/(2m-1)\big) < \|y_{j-1}\|$ for $j < m$.

For each $j < m-1$ the two pairs of vectors $Vy_{m-1}, f(y_j)$ and y_{m-1}, y_j have the same lengths respectively, and the angle between the first two vectors is smaller than the angle of the second pair; hence $\|Vy_{m-1} - f(y_j)\| < \|y_{m-1} - y_j\|$.

Put $y_{m-1} = \beta e_{n-1} + \gamma e_n$. Then $Vy_{m-1} = -\beta e_{n-1} + \gamma e_n$. It follows that if we choose small enough ε and define

$$f(y_{m-1}) = -\beta e_{n-1} + (\gamma - \varepsilon) e_n + \varepsilon e_{n+1}$$

then $\|f(y_{m-1})\| < \|y_{m-1}\|$ and $\|f(y_{m-1}) - f(y_j)\| < \|y_{m-1} - y_j\|$ for all $j < m-1$. We also define

$$y_m = \big(y_{m-1} + f(y_{m-1})\big)/2$$

and then $y_m \in \operatorname{span}\{e_n, e_{n+1}\}$ and $\|y_m\| > \|y_{m-1}\| \cos\big(2\alpha/(2m-1)\big)$. Hence

$$\|y_m\| > \|y_0\| \big(\cos\big(2\alpha/(2m-1)\big)\big)^m > 1/2$$

provided m is large enough.

This completes the choice of the vectors in the nth block, and it remains to show that if f is nonexpansive on $\{x_i\}_{i=0}^{k}$ and if m is large enough, then it will also be nonexpansive on $\{x_i\}_{i=0}^{k} \cup \{y_j\}_{j=0}^{m-1}$. The only inequalities that still need to be checked are $\|f(y_j) - f(x_i)\| \leq \|y_j - x_i\|$ for $0 \leq i \leq k$ and $0 \leq j < m$.

For all such i and j the value $\|y_j - f(x_i)\|$ is strictly smaller than $\|y_j - x_i\|$. Indeed, if x_i belongs to the $(n-1)$st block, this follows from the construction and from the fact that $\|(u+v)/2 - w\| > \|u-w\|$ for every u, v, w in a Hilbert space satisfying $\|u\| = \|v\| \geq \|w\|$ and $\langle v, w\rangle > \langle v, w\rangle$. (For x_k this is true because x_k and $f(x_k)$ are at the same distance to $\text{span}\{e_{n-1}, e_n\}$, and the orthogonal projection of $f(x_k)$ on this plane is closer to the y_j's than the projection of x_k.) If x_i belongs to one of the earlier blocks, then x_i and $f(x_i)$ are both orthogonal to $\text{span}\{e_{n-1}, e_n\}$ which contains all the y_j's, and the inequality follows because $\|f(x_i)\| < \|x_i\|$.

Choose $\delta > 0$ so that $\|y_j - f(x_i)\| + \delta < \|y_j - x_i\|$ for all $0 \leq j < m$ and $0 \leq i \leq k$. By choosing m large enough, we can ensure that $\|y_j - f(y_j)\| < \delta$ for all $j < m$, and hence that $\|f(y_j) - f(x_i)\| < \|y_j - x_i\|$ for all j and i.

A standard way to regularize a nonconvergent sequence is to pass to its averages. For linear operators this leads to the mean ergodic theorem (on reflexive spaces, say). Unfortunately $\frac{1}{n}\sum_{i=0}^{n-1} f^i(x)$ does not need to converge in norm even on a Hilbert space. To see this we construct another (simpler) version of the example above. In the nth block define $y_j = V^j e_{n-1}$ for $1 \leq j \leq m_n$, where V is the rotation by the angle $\pi/2m_n$ in the plane spanned by e_{n-1} and e_n.

The sequence $\{x_i\}$ is the union of all the blocks, and f is defined on the x_i's by $f(x_i) = x_{i+1}$. The map f is nonexpansive on the sequence, and it can be extended to be nonexpansive on its closed convex hull. Also, if we put $N_k = \sum_{n \leq k} m_n$ and assume that the m_n's tend fast to infinity so that $N_{k-1}/m_k \to 0$ as $k \to \infty$, then $\frac{1}{N_k}\sum_{i=0}^{N_k-1} f^i(x)$ is essentially the average over the kth block, which is close to a multiple bigger than $1/4$ of $(e_{k-1} + e_k)$. Hence the averages do not converge.

Note, however, that the sequence above does converge weakly, and it turns out that a nonlinear version of the mean ergodic theorem (on "nice" spaces) does hold when we replace the norm convergence by weak convergence.

Theorem 3.15. *Let E be a uniformly convex space with a Fréchet differentiable norm, and let K be a closed bounded and convex subset of E. Let $f : K \to K$ be nonexpansive. Then the sequence $\frac{1}{n}\sum_{i=0}^{n-1} f^i(x)$ converges weakly to a fixed point of f.*

We start with some preparations for the proof. Denote the set of fixed points of f by $F(f)$. When E is strictly convex, $F(f)$ is convex and f is affine on it. Indeed, assume that $x, y \in F(f)$. Then $\|f\big((x+y)/2\big) - f(x)\| \leq \|(x+y)/2 - x\| = \|x-y\|/2$ and similarly $\|f\big((x+y)/2\big) - f(y)\| \leq \|x-y\|/2$. Since $f(x) = x$ and $f(y) = y$, the strict convexity implies that $f\big((x+y)/2\big) = (x+y)/2$. (Note, however, that without strict convexity the fixed points set need not be convex. For example, the fixed point set of the map $f(s,t) = (s, |s|)$ from the unit ball of the two-dimensional l_∞ space to itself is not convex.) It follows that in strictly convex spaces which are also reflexive $F(f)$, being closed and convex, is weakly compact.

The first lemma generalizes these observations (in uniformly convex spaces) to the set $F^\varepsilon(f)$ of ε-fixed points, i.e., points where $\|f(x) - x\| < \varepsilon$ for some $\varepsilon > 0$. Heuristically, part (i) implies that the set of nearly fixed points is nearly convex and that f is nearly affine on this set. (See also Theorem 3.20 below for a stronger variant.) Parts (ii) and (iii) show that the sets $F^\varepsilon(f)$ "approach" $F(f)$ as $\varepsilon \to 0$.

Lemma 3.16. *Let E be a uniformly convex Banach space with modulus of convexity δ_E, and let K be a closed bounded and convex subset of E with diameter d. Let f be a nonexpansive mapping from K into E. Then*

(i) *For every $x, y \in K$ and for every $\lambda, \mu \geq 0$ with $\lambda + \mu = 1$*

$$\frac{1}{2} d \cdot \delta_E\Big(\frac{4}{d}\|\lambda f(x) + \mu f(y) - f(\lambda x + \mu y)\|\Big) \leq \|x - y\| - \|f(x) - f(y)\|.$$

(ii) *If f has an approximate fixed point in K, then it actually has a fixed point there.*

(iii) *For every weak neighborhood W of the set $F(f)$ there is an $\varepsilon > 0$ such that $\|x - f(x)\| < \varepsilon$ implies that $x \in W$.*

Proof. (i) By the definition of the modulus of convexity

$$1 - \|\lambda u + \mu v\| \geq 2\min(\lambda, \mu)\, \delta_E(\|u - v\|) \geq 2\lambda\mu\, \delta_E(\|u - v\|)$$

for every u and v with $\|u\|, \|v\| \leq 1$. Taking $u = \big(f(y) - f(\lambda x + \mu y)\big)/\lambda\|x - y\|$ and $v = \big(f(\lambda x + \mu y) - f(x)\big)/\mu\|x - y\|$, we obtain

$$\begin{aligned} &2\lambda\mu\|x - y\|\ \delta_E\Big(\|\lambda f(x) + \mu f(y) - f(\lambda x + \mu y)\| \,/\, \lambda\mu\|x - y\|\Big) \\ &\quad \leq \|x - y\| - \|f(x) - f(y)\| \end{aligned}$$

and the result follows from $\lambda\mu\|x - y\| \leq d/4$ and from the fact that $\delta_E(t)/t$ is nondecreasing. (See Proposition A.4.)

(ii) For each $r > 0$ put $\eta(r) = \inf\{\|f(x) - x\| : x \in K, \|x\| \leq r\}$. By our assumption $s = \inf\{r : \eta(r) = 0\}$ is well-defined. If $s = 0$, then zero is fixed by f, so assume $s > 0$ and find a sequence $x_n \in K$ such that $\|x_n\| \to s$ and such that $\varepsilon_n = \|f(x_n) - x_n\| \to 0$. If x_n has a convergent subsequence, then its limit is a fixed point as required. So assume, by passing to a subsequence, that there is $\tau > 0$ such that $\|x_n - x_m\| > \tau$ for all n, m. We now estimate

$$\begin{aligned} \|f\big((x_n + x_m)/2\big) - x_n\| &\leq \|f\big((x_n + x_m)/2\big) - f(x_n)\| + \|f(x_n) - x_n\| \\ &\leq \|x_n - x_m\|/2 + \varepsilon_n \end{aligned}$$

and similarly $\|f\big((x_n + x_m)/2\big) - x_m\| \leq \|x_n - x_m\|/2 + \varepsilon_m$. Since $\varepsilon_n \to 0$, it follows from the uniform convexity that $\|f\big((x_n + x_m)/2\big) - (x_n + x_m)/2\| \to 0$. But using the uniform convexity again, there is an $s_0 < s$ (depending on τ and s) such that $\|(x_n + x_m)/2\| \leq s_0$ for large n and m, contradicting the minimality of s.

(iii) It is enough to show that if a sequence $\{x_n\}$ satisfies $\|f(x_n) - x_n\| \to 0$, then every weak limit point of $\{x_n\}$ belongs to $F(f)$. To see this, assume, by passing to a subsequence, that the sequence x_n converges weakly to a point x. By (ii) applied to the set $\overline{\text{co}}\{x_j : j \geq n\}$ the function f has a fixed point y_n in that set. But the sequence $\{y_n\}$ must also converge weakly to x. Thus x, being a weak limit of points in the weakly closed set $F(f)$, also belongs to $F(f)$. □

In addition to points that are almost fixed we shall also need to consider sequences which are approximately the orbits of a single point under f. More precisely, we call a sequence $\{x_n\} \subset K$ an almost-orbit if it satisfies

$$\lim_{n \to \infty} \sup_{m \geq 0} \|f^m(x_n) - x_{n+m}\| = 0.$$

Lemma 3.17. *Let E be a uniformly convex Banach space, and let K be a closed bounded and convex subset of E. Let f be a nonexpansive mapping from K into E. Assume that $\{x_n\}$ and $\{y_n\}$ are almost-orbits of f. Then* $\lim \|x_n - y_n\|$ *exists, and for every $0 < \lambda, \mu < 1$ with $\lambda + \mu = 1$ the sequence $z_n = \lambda x_n + \mu y_n$ is also an almost-orbit.*

Proof. Choose $\varepsilon_n \to 0$ such that $\|f^m(x_n) - x_{n+m}\|, \|f^m(y_n) - y_{n+m}\| \le \varepsilon_n$ for every $m \ge 0$. For each n and m

$$\begin{aligned}
&\|x_{n+m} - y_{n+m}\| \\
&\le \|x_{n+m} - f^m(x_n)\| + \|y_{n+m} - f^m(y_n)\| + \|f^m(y_n) - f^m(x_n)\| \\
&\le 2\varepsilon_n + \|x_n - y_n\|
\end{aligned}$$

and thus $\lim \|x_n - y_n\|$ exists. Applying Lemma 3.16(i) we obtain

$$\begin{aligned}
\|z_{n+m} - f^m(z_n)\| &\le \lambda\|x_{n+m} - f^m(x_n)\| + \mu\|y_{n+m} - f^m(y_n)\| \\
&\quad + \|\lambda f^m(x_n) + \mu f^m(y_n) - f^m(z_n)\| \\
&\le \varepsilon_n + \frac{d}{4}\delta_E^{-1}\Big(\frac{2}{d}(\|x_n - y_n\| - \|f^m(x_n) - f^m(y_n)\|)\Big) \\
&\le \varepsilon_n + \frac{d}{4}\delta_E^{-1}\Big(\frac{2}{d}(\|x_n - y_n\| - \|x_{n+m} - y_{n+m}\| + 2\varepsilon_n)\Big)
\end{aligned}$$

and the last term converges to zero because $\lim \|x_n - y_n\|$ exists. □

Proposition 3.18. *Let E be a uniformly convex Banach space, and let f be a nonexpansive map from a closed convex and bounded subset K of E into itself. Fix $x \in K$. Then for every weak neighborhood V of $F(f)$ there is an N such that $\frac{1}{n}\sum_{i=0}^{n-1} f^i(x) \in V$ for all $n \ge N$.*

Proof. Assume, without loss of generality, that K is contained in the unit ball $B(E)$ of E. Since $F(f)$ is convex and weakly compact there is a $\delta > 0$ and a *convex* weak neighborhood W of $F(f)$ such that $W + \delta B(E) \subset V$. Fix $\varepsilon > 0$ as in (iii) of Lemma 3.16 and an integer p such that $2/p < \varepsilon/2$, and set

$$w_n = \frac{1}{p}\sum_{i=0}^{p-1} f^{i+n}(x).$$

Since $\{f^{i+n}(x)\}$ is an orbit for each fixed i, it follows from Lemma 3.17 that $\{w_n\}$ is an almost-orbit. In particular there is an M such that $\|f(w_n) - w_{n+1}\| < \varepsilon/2$ for all $n \ge M$. Since clearly $\|w_n - w_{n+1}\| \le 2/p$, it follows from the choice of p that $\|f(w_n) - w_n\| < \varepsilon$, and hence by the choice of ε that $w_n \in W$. Since W is convex, it follows that if $m \ge M$, then

$$\frac{1}{kp}\sum_{i=m}^{m+kp-1} f^i(x) = \frac{1}{k}\sum_{j=0}^{k-1} w_{pj+m} \in W.$$

Fix N such that $2(M+p)/N < \delta$, and represent each $n > N$ as $n = M + kp + r$ where $0 \le r < p$. We then write $\frac{1}{n}\sum_{i=0}^{n-1} f^i(x)$ as a sum of four terms

$$\frac{1}{n}\sum_{i=0}^{M-1} f^i(x) + \frac{1}{kp}\sum_{i=M}^{M+kp-1} f^i(x) + \Big(\frac{1}{n} - \frac{1}{kp}\Big)\sum_{i=M}^{M+kp-1} f^i(x) + \frac{1}{n}\sum_{i=M+kp}^{n} f^i(x).$$

Then the second term is in W, and the sum of the norms of the other three is smaller than $M/N + (M+p)/N + p/N < \delta$. Thus $\frac{1}{n}\sum_{i=0}^{n} f^i(x) \in W + \delta B(E) \subset V$ by the choice of W and δ. □

Let $f : K \to K$. We denote the set of all weak accumulation points of the orbit $\{f^n(x)\}$ by $\omega_\omega(x)$.

Proposition 3.19. *Let E be a uniformly convex space with a Fréchet differentiable norm, and let f be a nonexpansive map on a closed convex and bounded subset of E. Then $\langle j(y_1 - y_2), z_1 - z_2\rangle = 0$ for every $y_1, y_2 \in F(f)$ and for every $z_1, z_2 \in \overline{\mathrm{co}}(\omega_\omega(x))$. In particular, $F(f) \cap \overline{\mathrm{co}}(\omega_\omega(x))$ contains at most one point.*

Proof. Fix $0 < \tau < 1$. The constant sequences y_1 and y_2 are orbits. Hence by Lemma 3.17 the sequence $\tau f^n(x) + (1-\tau)y_1$ is an almost-orbit and the limit of $\|\tau f^n(x) + (1-\tau)y_1 - y_2\| = \|y_1 - y_2 + \tau(f^n(x) - y_1)\|$ exists.

Since the norm is Fréchet differentiable at $y_1 - y_2$, there is a function θ with $\theta(t)/t \to 0$ such that

$$\Big| \|y_1 - y_2 + \tau(f^n(x) - y_1)\| - \|y_1 - y_2\| - \tau\langle j(y_1 - y_2), f^n(x) - y_1\rangle \Big| \le \theta(\tau).$$

The first term converges and $\langle j(y_1 - y_2), y_1\rangle$ and $\|y_1 - y_2\|$ are constants; hence

$$\limsup \langle j(y_1 - y_2), f^n(x)\rangle \le \liminf \langle j(y_1 - y_2), f^n(x)\rangle + 2|\theta(\tau)|/\tau.$$

Letting $\tau \to 0$ we see that $L = \lim\langle j(y_1 - y_2), f^n(x)\rangle$ exists. Thus $\langle j(y_1 - y_2), z\rangle = L$ for every $z \in \omega_\omega(x)$, hence also in $\overline{\mathrm{co}}(\omega_\omega(x))$. This proves the proposition. □

Proof of Theorem 3.15. By Proposition 3.18 all the weak accumulation points of the averages $\frac{1}{n}\sum_{i=0}^{n-1} f^i(x)$ are in $F(f)$. Since these points are certainly in $\bigcap \overline{\mathrm{co}}\{f^k(x) : k \ge n\}$ the theorem follows from Proposition 3.19 once we check that $\overline{\mathrm{co}}(\omega_\omega(x)) = \bigcap \overline{\mathrm{co}}\{f^k(x) : k \ge n\}$.

That $\overline{\mathrm{co}}(\omega_\omega(x))$ is contained in the intersection is clear. To prove the converse, fix $y \in \bigcap \overline{\mathrm{co}}\{f^k(x) : k \ge n\}$ and a functional x^*. For each n we obtain

$$\begin{aligned} \langle x^*, y\rangle &\le \sup\big\{\langle x^*, z\rangle : z \in \overline{\mathrm{co}}\{f^k(x) : k \ge n\}\big\} \\ &= \sup\big\{\langle x^*, z\rangle : z \in \{f^k(x) : k \ge n\}\big\}. \end{aligned}$$

Hence $\langle x^*, y\rangle \le \sup\{\langle x^*, z\rangle : z \in \omega_\omega(x)\}$, and $y \in \overline{\mathrm{co}}(\omega_\omega(x))$ by the separation theorem. ◇

We finish this section with a stronger version of Lemma 3.16(i), which is close in spirit to the material in Chapters 14 and 15. It shows that a nonexpansive mapping in a uniformly convex space which is "almost isometric" on a set is necessarily "almost affine" on its convex hull.

Theorem 3.20. *Let K be a closed bounded and convex subset of a uniformly convex Banach space E, and let $f : K \to E$ be a nonexpansive map. Then for every $\varepsilon > 0$ there is a $\delta > 0$ (which depends on ε and on the diameter of K) such that*

$$\Big\| \sum \lambda_i f(x_i) - f\Big(\sum \lambda_i x_i\Big) \Big\| \le \varepsilon$$

for every choice of points $x_i \in K$ with $\max_{i,j}\big(\|x_i - x_j\| - \|f(x_i) - f(x_j)\|\big) \le \delta$ and for every $\lambda_i \ge 0$ with $\sum \lambda_i = 1$.

We need some preparations. The lemma below considers sets of m points and gives estimates that depend on m. The proposition that follows it measures the approximation of a general convex combination by convex combinations of m points.

Lemma 3.21. *Let K be a convex and bounded subset of a uniformly convex Banach space E, and let $f : K \to E$ be a nonexpansive map. Then for every integer*

$m \geq 2$ there is a positive, continuous, strictly increasing function γ_m satisfying $\gamma_m(0) = 0$ such that

$$\gamma_m\Big(\Big\|\sum_{i=1}^{m}\lambda_i f(x_i) - f\Big(\sum_{i=1}^{m}\lambda_i x_i\Big)\Big\|\Big) \leq \max_{i,j\leq m}\big(\|x_i - x_j\| - \|f(x_i) - f(x_j)\|\big)$$

for every m points $x_i \in K$ and for every $\lambda_i \geq 0$, $\sum \lambda_i = 1$.

Proof. For $m = 2$ this is exactly Lemma 3.16(i). We proceed by induction assuming that the lemma holds for m.

Fix $x_1, \dots, x_{m+1} \in K$ and $\{\lambda_i\}_{i=1}^{m+1}$ such that $\lambda_i \geq 0$ and $\sum \lambda_i = 1$, and assume that $0 < \lambda_{m+1} < 1$. Put $t = \max_{i,j\leq m+1}\big(\|x_i - x_j\| - \|f(x_i) - f(x_j)\|\big)$.

In the computations below the indices i, j will range between 1 and $m+1$, and k, l between 1 and m. Put

$$\begin{aligned}\mu_k &= \lambda_k/(1-\lambda_{m+1}),\\ u_k &= (1-\lambda_{m+1})x_k + \lambda_{m+1}x_{m+1},\\ v_k &= (1-\lambda_{m+1})f(x_k) + \lambda_{m+1}f(x_{m+1}).\end{aligned}$$

It follows that $\|f(u_k) - v_k\| \leq \gamma_2^{-1}(t)$ and also that

$$\|u_k - u_l\| - \|v_k - v_l\| = (1-\lambda_{m+1})\big(\|x_k - x_l\| - \|f(x_k) - f(x_l)\|\big) \leq t$$

and hence that

$$\begin{aligned}&\|u_k - u_l\| - \|f(u_k) - f(u_l)\|\\ &\leq \|u_k - u_l\| - \|v_k - v_l\| + \|f(u_k) - v_k\| + \|v_l - f(u_l)\| \leq t + 2\gamma_2^{-1}(t).\end{aligned}$$

Thus $\|f(\sum \mu_k u_k) - \sum \mu_k f(u_k)\| \leq \gamma_m^{-1}\big(t + 2\gamma_2^{-1}(t)\big)$ and

$$\begin{aligned}&\Big\|f\Big(\sum \lambda_i x_i\Big) - \sum \lambda_i f(x_i)\Big\|\\ &= \Big\|f\Big(\sum \mu_k u_k\Big) - \sum \mu_k v_k\Big\|\\ &\leq \Big\|f\Big(\sum \mu_k u_k\Big) - \sum \mu_k f(u_k)\Big\| + \sum \mu_k\|f(u_k) - v_k\|\\ &\leq \gamma_m^{-1}\Big(t + 2\gamma_2^{-1}(t)\Big) + \gamma_2^{-1}(t).\end{aligned}$$

The required function is then defined so that the final term above is $\gamma_{m+1}^{-1}(t)$. □

Let A be a subset of a Banach space E, and set

$$\mathrm{co}_m(A) = \Big\{\sum_{i=1}^{m}\lambda_i a_i : a_i \in A, \lambda_i \geq 0, \sum \lambda_i = 1\Big\}.$$

It turns out that in spaces with nontrivial type the sets $\mathrm{co}_m(A)$ uniformly approximate the full convex hull of A.

Proposition 3.22. *Let A be a subset of a Banach space E with type $p > 1$. Then for every $\varepsilon > 0$ there is an m (depending only on ε, the type-p constant of E and the diameter of A) such that* $\mathrm{co}(A) \subset \mathrm{co}_m(A) + \varepsilon B(E)$.

Proof. Let c be the type-p constant of E, and put $\mathrm{diam}(A) = d$. Fix a convex combination $x = \sum \mu_i a_i$ with $a_i \in A$. The random variable X which takes the value $a_i - x$ with probability μ_i has zero expectation and is bounded by d. Let

$X_1, \ldots, X_m$ be m independent copies of X. Then (as noted after the definition of type in Appendix G)

$$\mathbb{E}\,\Big\|\sum_{j=1}^{m} X_j\Big\|^p \leq (2c)^p \sum_{j=1}^{m} \mathbb{E}\,\|X_j\|^p \leq (2cd)^p m.$$

Thus there is a ω for which $\|\sum_{j=1}^m X_j(\omega)\| \leq 2cdm^{1/p}$. Setting $X_j(\omega) = b_j - x$ (where each b_j is one of the a_i's), we see that $(\sum_{j=1}^m b_j)/m \in \mathrm{co}_m(A)$ and that $\|x - (\sum_{j=1}^m b_j)/m\| = \|\sum_{j=1}^m X_j(\omega)\|/m \leq 2cdm^{1/p-1} < \varepsilon$ for large enough m. □

Proof of Theorem 3.20. Fix $\varepsilon > 0$. By Proposition 3.22 applied to $(E \oplus E)_\infty$ (which has type p for some $p > 1$ because E is uniformly convex), we can find m with $\mathrm{co}(A) \subset \mathrm{co}_m(A) + \varepsilon/3\, B\big((E \oplus E)_\infty\big)$ for every $A \subset K \times f(K) \subset (E \oplus E)_\infty$. Choose γ_m by Lemma 3.21. The required $\delta > 0$ is then chosen so that $\gamma_m^{-1}(\delta) < \varepsilon/3$.

Fix $\sum_{j=1}^n \lambda_j x_j \in \mathrm{co}(A)$. By the choice of m there is a set $I \subset \{1, \ldots, n\}$ of cardinality m, and for each $i \in I$ nonnegative μ_i with $\sum_{i\in I} \mu_i = 1$, such that $\|f(\sum_{j=1}^n \lambda_j x_j) - f(\sum_{i\in I} \mu_i x_i)\| \leq \|\sum_{j=1}^n \lambda_j x_j - \sum_{i\in I} \mu_i x_i\| < \varepsilon/3$ and at the same time $\|\sum_{j=1}^n \lambda_j f(x_j) - \sum_{i\in I} \mu_i f(x_i)\| < \varepsilon/3$. Thus

$$\begin{aligned}
&\Big\|\sum_{j=1}^{n} \lambda_j f(x_i) - f\Big(\sum_{j=1}^{n} \lambda_j x_j\Big)\Big\| \\
&\leq 2\varepsilon/3 + \Big\|\sum_{i\in I} \mu_i f(x_i) - f\Big(\sum_{i\in I} \mu_i x_i\Big)\Big\| \\
&\leq 2\varepsilon/3 + \gamma_m^{-1}\left(\max_{i,j\in I}\big(\|x_i - x_j\| - \|f(x_i) - f(x_j)\|\big)\right) < \varepsilon
\end{aligned}$$

provided the term $\max_{i,j\in I}\big(\|x_i - x_j\| - \|f(x_i) - f(x_j)\|\big)$, which is smaller than $\max_{i,j\leq n}\big(\|x_i - x_j\| - \|f(x_i) - f(x_j)\|\big)$, is less than δ. ◇

5. Notes and Remarks

Fixed point theory is, of course, an important and well-developed field with many ramifications and many applications. We have presented only a few results related to the geometry of Banach spaces.

Theorem 3.1 is due to Schauder [**516**] and Tychonoff [**566**]. Theorem 3.2 is due to Banach [**35**]. Both theorems are basic tools in the study of differential and integral equations (as well as in many other areas), and it was this study that motivated their original formulation.

Theorem 3.3 is due to Klee [**329**]. The first example of a continuous map of the unit ball of a Hilbert space into itself without fixed points was given by Kakutani [**308**].

Theorem 3.4 is due to Lin and Sternfeld [**353**]. Our presentation of the construction of ψ follows Goebel and Kirk [**217**]. Previously, Nowak [**440**] proved Corollary 3.5 for certain Banach spaces having some additional structure (e.g. for L_p spaces and for $C[0,1]$). Benyamini and Sternfeld [**53**] pushed Nowak's ideas further and proved the corollary for all infinite-dimensional Banach spaces. The explicit example at the end of Section 2 is taken from Goebel and Kaczor [**215**].

Corollary 3.5 is related to the problem, which we discuss in Chapter 9, of whether balls and spheres in infinite-dimensional Banach space can be Lipschitz (or uniformly) homeomorphic. The corollary shows that there is no obvious obstruction for the existence of a uniform homeomorphism (and in the topological category they are actually homeomorphic; see Bessaga and Pełczyński [**62**] or van Mill [**576**]). The answer to this problem is unknown even in Hilbert space.

There is a vast literature on nonexpansive maps, continuous semigroups of such maps, their fixed points, concrete schemes for approximating fixed points, asymptotic behavior of the semigroups, etc. We refer the reader to the books by Browder [**96**], Goebel and Reich [**218**], Zeidler [**593**, Vol. I] and Goebel and Kirk [**217**], and to the articles by Reich [**487**], Reich and Shafrir [**488**] and Shafrir [**531**] for more information and for further references. We have restricted ourselves to maps on bounded sets only. The references above contain information also on the unbounded case.

Part (ii) of Proposition 3.6 is due to Maurey [**398**]. Our proof follows Besbes [**57**].

Proposition 3.7(i), as well as the notions of a diametral point and of normal structure and the fact that uniformly convex spaces have normal structure, are all due to Brodskii and Milman [**93**]. Part (ii) is due to Karlovitz [**322**] and to Goebel [**213**]. That closed bounded and convex subsets of uniformly convex spaces have the fixed point property was proved independently by Browder [**94**], Göhde [**220**] and Kirk [**326**]. The uniformly smooth case was first proved by Baillon [**32**].

Theorem 3.9 is from Maurey [**398**]. Maurey also used the same technique in other spaces which have bases with some properties similar to those of the c_0 basis. See also Elton, Lin, Odell and Szarek [**167**] for a presentation of Maurey's work and for more applications.

Example 3.10, due to Alspach [**10**], solved a long-standing problem. Schechtman [**517**] used Alspach's idea and constructed, for each n, a convex weakly compact set and a family of n (respectively, countably many) commuting nonexpansive maps on the set, such that the whole family has no common fixed point, but every proper (respectively, finite) subfamily does have one. The presentation here follows [**517**].

Theorem 3.11 is due to Maurey [**398**]. His proof uses ultra products and random measures. A proof that follows Maurey's ideas without using these techniques is presented in the article [**167**] by Elton, Lin, Odell and Szarek. Our proof uses some ideas from Besbes [**57**]. Using these ideas, Besbes also extends the theorem to subsets of $L_1(E)$ and $c_0(E)$ for uniformly convex spaces E.

In connection with Theorem 3.11 we mention that Dowling and Lennard [**155**] noted that if $E \subset L_1(0,1)$ is nonreflexive, then there is a closed bounded and convex subset K of E and a nonexpansive map f on K without a fixed point.

It is unknown if a nonexpansive map on a convex weakly compact subset of a reflexive (or even superreflexive) space has to have a fixed point.

The article [**167**] also contains the proof of another theorem of Maurey: If K is a convex weakly compact subset of a superreflexive space, and if f is an *isometry* of K into itself, then f has a fixed point.

There are also fixed point theorems for maps on closed convex and bounded sets whose iterates satisfy a uniform Lipschitz condition, i.e., maps f for which there is a constant $k > 1$ such that $\|f^n(x) - f^n(y)\| \leq k\|x - y\|$ for all x, y and for all positive integers n. Goebel and Kirk [**216**] proved that if E is uniformly convex and if k is smaller than the unique solution of the equation $t\big(1-\delta_E(t^{-1})\big)=1$, then the map has a fixed point. In Hilbert space the result actually holds for any $k<\sqrt{2}$. The book [**217**] by Goebel and Kirk contains a discussion of this direction. It also contains several results that use various measures of noncompactness. These notions combine in a certain sense Schauder's theorem with the Banach contraction principle. A more detailed treatment of both topics can be found in the book [**29**] by Ayerbe Toledano, Dominguez Benavides and López Acedo.

Theorem 3.12 is due to Reich [**486**]. We followed the proof in Bruck [**101**].

It is of some interest to note (see Baillon, Bruck and Reich [**33**]) that if K is a closed convex set symmetric around zero in a uniformly convex space, and if f is odd ($f(-x)=-f(x)$), then $\|f^{n+1}(x)-f^n(x)\|\to 0$ actually implies the convergence of $f^n(x)$. Indeed, the oddness implies that $f(0)=0$; hence $\|f^n(x)\|$ decreases to a limit d. Also, $f^m(-x)=-f^m(x)$ for every m implies that $\|f^{n+i}(x)+f^n(x)\|$ is monotone nonincreasing in n for each fixed i, hence convergent. Since $\|f^{n+i}(x)-f^n(x)\|\to 0$ for each fixed i, this limit is necessarily equal to $2d$. Thus $\|f^{n+i}(x)+f^n(x)\|\geq 2d$, and by the uniform convexity $\|f^{n+i}(x)-f^n(x)\|\to 0$ uniformly in i.

Krasnoselskii [**335**] was the first to consider the iterates of $f_\lambda=\lambda f+(1-\lambda)I$. He proved their convergence when E is uniformly convex and K is compact. Proposition 3.13 is due to Ishikawa [**268**], who also considered other iteration schemes, and to Edelstein and O'Brien [**163**]. The proof here is due to Bruck [**101**].

A standard "soft" argument shows that Proposition 3.13 gives that for every λ and d there is a sequence $\varepsilon_n\to 0$ (independent of the Banach space and the set K) such that if $\operatorname{diam}(K)\leq d$, then $\|f_\lambda^{n+1}(x)-f_\lambda^n(x)\|\leq\varepsilon_n$. Baillon and Bruck showed that one can actually take $\varepsilon_n=C(\lambda,d)n^{-1/2}$. The case $\lambda=1/2$, which is significantly simpler than that of general λ, is presented in Bruck [**102**], where there are also historical remarks and references for the general case.

Example 3.14 is from Genel and Lindenstrauss [**200**].

Baillon proved Theorem 3.15 for Hilbert space in [**30**] and for L_p in [**31**]. Bruck and Reich modified his arguments to cover uniformly convex spaces with a Fréchet differentiable norm (unpublished). We followed Bruck [**99**]. Another proof for odd maps appears in Wittmann [**585**]. For another example that the averages do not converge in norm, see Krengel and Lin [**337**]. Lemmas 3.16(i) and 3.17 are from [**99**]. Parts (ii) and (iii) of Lemma 3.16 are due to Browder [**95**].

Lemma 3.21 and Theorem 3.20 are due to Bruck [**100**]. Theorem 3.20 was proved earlier for Hilbert space by Zarantonello [**592**]. The proof in this case is elementary: Put $z=\sum\lambda_i x_i$, expand $\|\sum\lambda_i f(x_i)-f(z)\|^2=\sum\lambda_i\lambda_j\langle f(x_i)-f(z),f(x_j)-f(z)\rangle$, and substitute $\langle f(x_i),f(x_j)\rangle=\big(\|f(x_i)\|^2+\|f(x_j)\|^2-\|f(x_i)-f(x_j)\|^2\big)/2$ and $\langle f(x_i),f(z)\rangle=\big(\|f(x_i)\|^2+\|f(z)\|^2-\|f(x_i)-f(z)\|^2\big)/2$. Using the estimate $\|f(x_i)-f(z)\|\leq\|x_i-z\|$ and the identity $\sum\lambda_i\|x_i-z\|^2=\big(\sum\lambda_i\lambda_j\|x_i-x_j\|^2\big)/2$, we find that the expression is bounded by

$$\begin{aligned}&\sum \lambda_i\lambda_j\big(\|x_i - z\|^2 + \|x_j - z\|^2 - \|f(x_i) - f(x_j)\|^2\big)/2\\ &= \sum \lambda_i\lambda_j\big(\|x_i - x_j\|^2 - \|f(x_i) - f(x_j)\|^2\big)/2\\ &\le \max\big(\|x_i - x_j\|^2 - \|f(x_i) - f(x_j)\|^2\big)/2\\ &\le d \max\big(\|x_i - x_j\| - \|f(x_i) - f(x_j)\|\big)\end{aligned}$$

where d is the diameter of the set K.

Proposition 3.22 was proved independently by Maurey (see Pisier [**465**]) and by Bruck [**100**]. It was used by Bruck to show that the rate of convergence in Theorem 3.15 is actually uniform and does not depend on x. Indeed, in the proof of Proposition 3.18 we can approximate, given $\varepsilon > 0$, the average ω_n of $p = p(\varepsilon, x)$ elements by an average of only $m = m(\varepsilon)$ of them. See Theorem 1.3 in [**100**] for details. One consequence of this uniformity is that under the above conditions $f^n_{\frac{1}{2}}(x)$, which does not have to converge in norm even in Hilbert space, is weakly convergent. This follows from Proposition 3.13 and from the simple fact that if $\{a_n\}$ is a sequence of real numbers such that $a_{n+1} - a_n \to 0$ and $\frac{1}{N}\sum_{n=k+1}^{k+N} a_n \to 0$ uniformly in k, then $a_n \to 0$. This line of reasoning is due to Reich [**485**], who applied it to more general situations.

Nonexpansive maps that satisfy the inequality of Lemma 3.21 for $m = 2$ were introduced by Bruck in [**99, 100**] and are called maps of type (γ). The proof of Theorem 3.20 holds (and was so formulated in [**100**]) for nonexpansive maps of type (γ) in Banach spaces with nontrivial type.

Let E be uniformly convex so that the weak limit of $\frac{1}{n}\sum_{i=0}^{n-1} f^i(x)$ exists for every $x \in K$. Then in view of Proposition 3.18 the limit defines a nonexpansive retraction from K onto $F(f)$. For more information on nonexpansive retractions on convex sets (not necessarily related to fixed point sets of mappings), see de Figueiredo and Karlovitz [**193**], Bruck [**98**], Beauzamy and Maurey [**46**], Goebel and Reich [**218**], Davis and Enflo [**133**], Enflo [**176, 177**].

CHAPTER 4

Differentiation of Convex Functions

1. Basic Definitions and Results

In this chapter we start to study the differentiability of functions on infinite-dimensional Banach spaces. The first section presents simple and straightforward generalizations of results from elementary calculus. Genuine problems resulting from the infinite-dimensional setting appear only when we consider the existence of derivatives. In the second section of this chapter we consider real-valued convex functions on Banach spaces. This subject is relatively well-understood, at least for separable Banach spaces (and we are mainly interested in this case).

The natural context, in some sense, in which differentiation should be studied is that of Lipschitz functions. We shall study Gâteaux differentiability of Lipschitz functions in Chapter 6, and the much more difficult question of Fréchet differentiability of Lipschitz functions in the second volume.

Definition 4.1. *Let f be a function defined on an open set in a Banach space E with values in a Banach space F. The function f is said to be Gâteaux differentiable at a point x_0 if there is a bounded linear operator $T : E \to F$ such that*

$$Tu = \lim_{t\to 0} \big(f(x_0 + tu) - f(x_0)\big)/t \tag{4.1}$$

for every $u \in E$. The (uniquely determined) operator T is called the Gâteaux derivative of f at x_0 and is denoted by $D_f(x_0)$. If the limit in (4.1) exists uniformly for u in the unit sphere of E, we say that f is Fréchet differentiable at x_0, and $D_f(x_0)$ is then called the Fréchet derivative of f at x_0.

We make some simple observations concerning these notions. Another way to write that $D_f(x_0)$ is the Fréchet derivative of f at x_0 is

$$f(x_0 + u) = f(x_0) + D_f(x_0)u + o(\|u\|). \tag{4.2}$$

A function which is Fréchet differentiable at a point is continuous there. This is not the case for Gâteaux differentiable functions even in finite dimensions. For example, the function $f : \mathbb{R}^2 \to \mathbb{R}$ defined by $f(0,0) = 0$ and

$$f(x,y) = \frac{x^4 y}{x^6 + y^3}, \qquad x^2 + y^2 > 0$$

has 0 as its Gâteaux derivative at the origin, but fails to be continuous there.

There are some obvious differentiation rules, and we mention just a few. If f is continuous and linear, then f is its own Fréchet derivative. If the composition $f \circ g$ is defined and if g and f are Fréchet differentiable at x_0 and at $g(x_0)$ respectively, then $f \circ g$ is Fréchet differentiable at x_0 and the usual chain rule formula holds

$$D_{f\circ g}(x_0) = D_f\big(g(x_0)\big) \circ D_g(x_0).$$

If g is Gâteaux differentiable at x_0 and if f is Fréchet differentiable at $g(x_0)$, then $f \circ g$ is Gâteaux differentiable at x_0 and the same formula holds. In general, Gâteaux differentiability of f at $g(x_0)$ is not sufficient for the differentiability of $f \circ g$, but it is sufficient if f is Lipschitz. If f and g map an open subset of E into F and into F^* respectively, and if both are Fréchet (respectively Gâteaux) differentiable at x_0, then so is $\langle g(x), f(x)\rangle$, and the derivative is given by

$$D_{\langle g,f\rangle}(x_0)u = \langle D_g(x_0)u, f(x_0)\rangle + \langle g(x_0), D_f(x_0)u\rangle.$$

The following version of the mean value formula will be very useful. If f is Gâteaux differentiable on the interval I connecting x_0 and y_0, then

$$\|f(y_0) - f(x_0)\| \le \|x_0 - y_0\| \sup\big\{\|D_f(x)\| : x \in I\big\}.$$

Indeed, choose $u^* \in F^*$ with $\|u^*\| = 1$ and $\langle u^*, f(y_0) - f(x_0)\rangle = \|f(y_0) - f(x_0)\|$. The result follows by applying the one-dimensional mean value formula to the function $h(t) = \big\langle u^*, f\big(x_0 + t(y_0 - x_0)\big)\big\rangle$.

Proposition 4.2. *Assume that f is Gâteaux differentiable at a neighborhood of a point x_0 and that $D_f(x)$ is continuous at x_0, i.e., $\lim_{x\to x_0} \|D_f(x) - D_f(x_0)\| = 0$. Then f is, in fact, Fréchet differentiable at x_0.*

Proof. Fix a vector u, and put $g(t) = f(x_0 + tu) - f(x_0) - tD_f(x_0)u$. Then $g(0) = 0$, and by the mean value formula

$$\begin{aligned}
&\|f(x_0 + u) - f(x_0) - D_f(x_0)u\| \\
&= \|g(1)\| = \sup\big\{\big|\langle x^*, g(1)\rangle\big| : \|x^*\| = 1\big\} \\
&\le \sup\big\{\big|\frac{d}{dt}\langle x^*, g(t)\rangle\big| : \|x^*\| = 1,\ 0 \le t \le 1\big\} \\
&\le \|u\| \sup\big\{\|D_f(x_0 + tu) - D_f(x_0)\| : 0 \le t \le 1\big\} = o(\|u\|).
\end{aligned}$$

□

A function f is said to be of class C^1 in an open set if it is continuously differentiable there. In view of Proposition 4.2 it does not matter whether we take in this context Fréchet or Gâteaux derivatives.

A Fréchet differentiable function f from an open set U in E to a Banach space F is said to be uniformly Fréchet differentiable on a subset $V \subset U$ if the $o(\|u\|)$ in (4.2) is uniform in $x_0 \in V$. If U is open and convex and $f : U \to F$ is Fréchet differentiable such that the map $x \to D_f(x)$ is uniformly continuous with modulus of continuity ω, then by the mean value formula

$$\|f(x + u) - f(x) - D_f(x)u\| \le \omega(\|u\|)\|u\| \tag{4.3}$$

for x and u with $x, x + u \in U$; hence f is uniformly Fréchet differentiable on U.

Another easy result connecting Gâteaux and Fréchet differentiability is

Proposition 4.3. *Let E be a finite-dimensional Banach space, and let f be a Lipschitz function from an open set in E to a (possibly infinite-dimensional) Banach space F. If f is Gâteaux differentiable at some x_0, then it is, in fact, Fréchet differentiable at that point.*

Proof. Fix $\varepsilon > 0$, and choose $\{u_i\}_{i=1}^m$ in the unit sphere of E such that for any u in this sphere there is an i with $\|u - u_i\| < \varepsilon$. Let $t_0 > 0$ be such that

$$\|f(x_0 + tu_i) - f(x_0) - tD_f(x_0)u_i\| \le \varepsilon|t|$$

for all $|t| \leq t_0$ and for all $i \leq m$. It follows that for every u with $\|u\| = 1$

$$\|f(x_0 + tu) - f(x_0) - tD_f(x_0)u\| \leq (C + \|D_f(x_0)\| + 1)\varepsilon|t|$$

for all $|t| \leq t_0$, where C is the Lipschitz constant of f. □

We conclude this section by quoting the basic theorems on inverse functions and on implicit functions. The proofs are identical to the usual proofs in finite dimensions and are therefore omitted.

Theorem 4.4 (The inverse function theorem). *Let f be a C^1 function from an open set in a Banach space E into a Banach space F. Assume that $D_f(x_0)$ is invertible. Then f is a homeomorphism from a neighborhood of x_0 onto a neighborhood of $f(x_0)$. The inverse function f^{-1} is a C^1 function near $f(x_0)$.*

Theorem 4.5 (The implicit function theorem). *Let E, F and G be Banach spaces, and let f be a C^1 function from an open neighborhood of a point (x_0, y_0) in $E \oplus F$ into G. Assume that the derivative of $f(x, y_0)$ at x_0 is an invertible operator from E onto G and that $f(x_0, y_0) = 0$. Then there are open neighborhoods $U \subset E$ of x_0 and $V \subset F$ of y_0 and a C^1 function $g : U \to V$ so that if $(u, v) \in U \times V$, then $f(u, v) = 0$ if and only if $v = g(u)$.*

2. Convex Functions

In this section we study the differentiability of real-valued convex functions defined on open convex sets in a Banach space. Note that in finite-dimensional spaces such functions are always locally bounded.

Proposition 4.6. *A convex function which is defined and bounded on an open convex set is locally Lipschitz.*

Proof. Assume that f is bounded by M on the ball $B(x_0, 2r)$, and fix points $x, y \in B(x_0, r)$ such that $f(x) \leq f(y)$. Put $z = y + r(y - x)/\|y - x\| \in B(x_0, 2r)$, and write $y = (1 - \lambda)x + \lambda z$ where $\lambda = \|x - y\|/(r + \|x - y\|) \leq \|x - y\|/r$. By convexity $f(y) \leq (1 - \lambda)f(x) + \lambda f(z)$; hence

$$|f(y) - f(x)| = f(y) - f(x) \leq \lambda|f(z) - f(x)| \leq 2M\|y - x\|/r.$$

□

From now on we shall consider only continuous (hence locally Lipschitz) convex functions defined on open convex sets. Recall that for a convex function f on $\mathbb{R}$ the right derivative f'_+ and left derivative f'_- of f exist at every point, the derivative itself exists except at a countable set of points, and it is monotone increasing. Moreover, $f'_+(s) \leq \big(f(t) - f(s)\big)/(t - s) \leq f'_-(t)$ for every $s < t$.

Let f be a convex function on an open convex subset G of E containing a point x_0. The subdifferential of f at x_0, denoted by $\partial_f x_0$, is the set of all functionals $y^* \in E^*$ satisfying $\langle y^*, x - x_0 \rangle \leq f(x) - f(x_0)$ for all $x \in G$. Geometrically, $\partial_f x_0$ is the set of all linear functionals on E whose graph supports the epigraph $\{(x, \lambda) \in E \oplus \mathbb{R} : x \in G, \lambda > f(x)\}$ of f at $\big(x_0, f(x_0)\big)$. It is nonempty by the separation theorem and it is w^*-closed. Since G is open, $\partial_f x_0$ is bounded by the local Lipschitz constant of f at x_0. The function f is Gâteaux differentiable at x_0 if and only if $\partial_f x_0$ consists of a single point. This single point yields a unique supporting hyperplane to the graph of f at $\big(x_0, f(x_0)\big)$; i.e., it is exactly $D_f(x_0)$.

Indeed, if there were a direction u such that the function $h(t) = f(x_0 + tu)$ fails to be differentiable at $t_0 = 0$, then the epigraph of $h(t)$ would have more than one supporting line at $\big(0, h(0)\big) = \big(0, f(x_0)\big)$. Separating each of these lines from the epigraph of f by a linear functional gives more than one point in $\partial_f x_0$.

The subdifferential $\partial_f x$ satisfies

$$\langle u^* - v^*, x - y\rangle \geq 0 \tag{4.4}$$

for all $u^* \in \partial_f x$ and $v^* \in \partial_f y$. Indeed, (4.4) follows by taking the sum of the two inequalities $\langle u^*, y - x\rangle \leq f(y) - f(x)$ and $\langle v^*, x - y\rangle \leq f(x) - f(y)$. (A set-valued map τ from a subset of E into E^* which satisfies inequality (4.4) for all $u^* \in \tau(x)$ and $v^* \in \tau(y)$ is called monotone. Thus the subdifferential is a monotone map.)

Gâteaux differentiability of a convex function is determined by the behavior of its second-order differences; f is Gâteaux differentiable at x_0 if and only if

$$f(x_0 + tu) + f(x_0 - tu) - 2f(x_0) = o(t) \tag{4.5}$$

for all $u \in E$. Indeed, (4.5) implies that f is Gâteaux differentiable at x_0 by the one-dimensional case. The converse is obvious.

The left-hand side of (4.5) is nonnegative. Hence if f and g are convex and $f + g$ is Gâteaux differentiable at x_0, then so are f and g separately.

The function f is Fréchet differentiable at x_0 if and only if the $o(t)$ in (4.5) is uniform with respect to u when $\|u\| = 1$. One direction is trivial, so assume that (4.5) holds uniformly. Then f is Gâteaux differentiable at x_0, and the convexity of f implies that $t^{-1}\big(f(x_0 + tu) - f(x_0)\big) - D_f(x_0)u \geq 0$ for every $t > 0$ and also that $t^{-1}\big(f(x_0) - f(x_0 - tu)\big) - D_f(x_0)u \leq 0$ for $t > 0$. Thus

$$\begin{aligned}&f(x_0 + tu) + f(x_0 - tu) - 2f(x_0)\\ &\quad = \big(f(x_0 + tu) - f(x_0) - tD_f(x_0)u\big) - \big(f(x_0) - f(x_0 - tu) - tD_f(x_0)u\big)\end{aligned}$$

is a sum of two nonnegative expressions, both of which have to be $o(t)$ uniformly in u for $\|u\| = 1$ if this holds for (4.5).

It follows that if f and g are two convex functions such that $f + g$ is Fréchet differentiable at a point x_0, then so is each of the functions f and g.

Proposition 4.7. *Assume that f is convex and Fréchet differentiable at x_0. Then $\|u_n^* - D_f x_0\| \to 0$ whenever $u_n^* \in \partial_f(v_n)$ and $v_n \to x_0$. In particular, a convex function which is Fréchet differentiable on an open set is C^1 there.*

Proof. Put $w^* = D_f(x_0)$. For every $\varepsilon > 0$ there is a $\delta > 0$ such that

$$f(x_0 + y) - f(x_0) - \langle w^*, y\rangle \leq \varepsilon\|y\| \tag{4.6}$$

whenever $\|y\| \leq \delta$. Since $u_n^* \in \partial_f(v_n)$

$$\langle u_n^*, x_0 + y - v_n\rangle \leq f(x_0 + y) - f(v_n) \tag{4.7}$$

for every y and for all n. Choose $\|y_n\| = \delta$ so that $\langle u_n^* - w^*, y_n\rangle \geq \delta\|u_n^* - w^*\|/2$. Then (4.6) and (4.7) with $y = y_n$ give the two inequalities

$$\begin{aligned}-\varepsilon\delta &\leq -f(x_0 + y_n) + f(x_0) + \langle w^*, y_n\rangle,\\ 0 &\leq f(x_0 + y_n) - f(v_n) - \langle u_n^*, x_0 + y_n - v_n\rangle.\end{aligned}$$

Adding these inequalities gives by the choice of y_n

$$-\varepsilon\delta \leq -\delta\|u_n^* - w^*\|/2 + \langle u_n^*, v_n - x_0\rangle + f(x_0) - f(v_n)$$

hence $\limsup \|u_n^* - w^*\| \leq 2\varepsilon$. □

We specialize now to the case where $f(x) = \|x\|$. The subdifferential $\partial_f x_0$ at a point x_0 with $\|x_0\| = 1$ is exactly the set of functionals $u^* \in E^*$ such that $\|u^*\| = 1$ and $\langle u^*, x_0\rangle = 1$; i.e., $x_0 \to \partial_f x_0$ is just the standard duality map j. The norm is Gâteaux differentiable at x_0 if and only if there is a unique supporting functional x_0^* to the unit ball at x_0. By Proposition 4.7 the norm is Fréchet differentiable at x_0 if and only if whenever $v_n \to x_0$ and $\|v_n^*\| = 1$ satisfy $\langle v_n^*, v_n\rangle = \|v_n\|$, then v_n^* tends in norm to the (unique) supporting functional at x_0. In particular, the norm in E is Fréchet differentiable at every $x \neq 0$ if and only if the (single-valued) duality map j is norm to norm continuous from E into E^*.

Proposition 4.8. *The norm in a Banach space E is uniformly Fréchet differentiable on the unit sphere $S(E)$ if and only if E is uniformly smooth.*

Proof. If E is uniformly smooth, then the norm is Fréchet differentiable on $E\backslash\{0\}$, and for every $r > 0$ its derivative, the duality map j, is uniformly continuous on $\{x : \|x\| > r\}$ by Proposition A.5. By (4.3) the norm is uniformly Fréchet differentiable on this set.

Conversely, when the norm is uniformly Fréchet differentiable on $S(E)$, then $\rho_E(\tau) = o(\tau)$ because $|x\| = 1$ and $\|y\| = \tau$ give that $\|x + y\| + \|x - y\| - 2\|x\|$ is

$$\big(\|x + y\| - \|x\| - \langle j(x), y\rangle\big) + \big(\|x - y\| - \|x\| - \langle j(x), -y\rangle\big) = o(\tau).$$

□

Proposition 4.9. *Let E be a separable space such that its norm is Fréchet differentiable at every $x \neq 0$. Then E^* is separable.*

The proposition follows from the preceding discussion and the following theorem.

Theorem 4.10. *Let C be a nonempty, closed, convex and bounded set in a Banach space E. Then the set of functionals which attain their maximum on C is dense in E^*. In particular, the set of functionals that attain their norm (i.e., their maximum on the unit ball) is dense in E^*.*

The proof is based on the following lemma.

Lemma 4.11. *Let f be a continuous real-valued function which is bounded from above on a complete metric space X. Then for every $\varepsilon > 0$ there is an $x_0 \in X$ such that $f(x) \leq f(x_0) + \varepsilon d(x, x_0)$ for all $x \in X$.*

Proof. Starting with any $y_1 \in X$, define inductively a sequence $\{y_n\}$ in X by choosing y_{n+1} such that

$$f(y_{n+1}) - \varepsilon d(y_n, y_{n+1}) \geq \sup\big\{f(x) - \varepsilon d(y_n, x) : x \in X\big\} - 2^{-n}.$$

(The right-hand side is finite because f is bounded from above.)

The sequence $\{f(y_n)\}$ is "almost monotone nondecreasing" in the sense that $f(y_{n+1}) \geq f(y_{n+1}) - \varepsilon d(y_n, y_{n+1}) \geq f(y_n) - 2^{-n}$, hence convergent. Summing the inequalities $\varepsilon d(y_n, y_{n+1}) \leq f(y_{n+1}) - f(y_n) + 2^{-n}$ over n, we find from the convergence of $\{f(y_n)\}$ that $\sum d(y_n, y_{n+1})$ converges, and hence that $\{y_n\}$ is Cauchy. The required point x_0 is its limit. Indeed, for any $x \in X$ just let $n \to \infty$ in $f(y_{n+1}) \geq f(x) - \varepsilon d(y_n, x) - 2^{-n}$. □

Proof of Theorem 4.10. We shall prove more generally that for every continuous concave function f on C which is bounded from above and for every $\varepsilon > 0$ there is a linear functional $u^* \in E^*$ such that $\|u^*\| \leq \varepsilon$ and such that $f + u^*$ attains its maximum in C.

Use Lemma 4.11 to find a point $x_0 \in C$ such that $f(x) \leq f(x_0) + \varepsilon\|x - x_0\|$ for all $x \in C$, and consider the following two convex sets in $E \oplus \mathbb{R}$:

$$K_1 = \{(x, t) : \ x \in C, \ t \leq f(x)\},$$
$$K_2 = \{(x, t) : \ x \in E \text{ and } \ t \geq f(x_0) + \varepsilon\|x - x_0\|\}.$$

The interior of K_2 is nonempty and disjoint from K_1. Hence there is a nonzero functional $(u^*, \alpha) \in E^* \oplus \mathbb{R}$ that separates the two sets; i.e., there is a scalar β such that $\langle u^*, x\rangle + \alpha t \leq \beta$ if $(x, t) \in K_1$ and $\langle u^*, x\rangle + \alpha t \geq \beta$ if $(x, t) \in K_2$.

It is evident that α cannot be negative, and $\alpha = 0$ would give $\langle u^*, x\rangle \geq \beta$ for every $x \in E$. But this would mean that $u^* = 0$, which is impossible. So $\alpha > 0$, and we normalize so that $\alpha = 1$. Since $(x_0, f(x_0)) \in K_1 \cap K_2$ it follows that $\langle u^*, x_0\rangle + f(x_0) = \beta$. Consequently

$$\langle u^*, x\rangle + f(x) \leq \langle u^*, x_0\rangle + f(x_0)$$

for all $x \in C$; i.e., $u^* + f$ attains its maximum on C at x_0. On the other hand,

$$\langle u^*, x\rangle + f(x_0) + \varepsilon\|x - x_0\| \geq \langle u^*, x_0\rangle + f(x_0)$$

for all $x \in E$, and thus $\|u^*\| \leq \varepsilon$. ♢

The norm in l_1 is Gâteaux differentiable on the dense set of vectors, all of whose coordinates are different from 0, and is nowhere Fréchet differentiable. It turns out that this is typical of separable spaces with a nonseparable dual (although we may need to renorm properly).

Proposition 4.12. *Let E be a separable Banach space with E^* nonseparable. Then there is an equivalent norm on E which is nowhere Fréchet differentiable.*

Proof. Let B^* be the closed unit ball in E^* (with respect to the given norm). Since E^* is nonseparable there is an $\eta > 0$ and an uncountable subset A of B^* such that $\|u^* - v^*\| > \eta$ for every $u^* \neq v^*$ in A. Since B^* is separable and metrizable in its w^*-topology, the Cantor-Bendixson theorem gives an uncountable subset A_0 of A such that A_0 has no isolated point in the w^*-topology. Let $C = \overline{\text{co}}(A_0 \cup -A_0)$ (the closure is taken with respect to the w^*-topology), and set $U^* = B^* + C$.

Since U^* is a w^*-compact symmetric body in E^*, it is the unit ball of the dual norm of some equivalent norm $|||\cdot|||$ on E, and $U^* \subset 2B^*$ implies that $|||x^*||| \geq \|x^*\|/2$ for every $x^* \in E^*$.

Let $x \in E$ with $|||x||| = 1$, and assume that the norm $|||\cdot|||$ is Fréchet differentiable at x. Hence there is an $r = r(x) > 0$ such that $|||x + u||| + |||x - u||| \leq 2 + \eta|||u|||/4$ whenever $|||u||| \leq r$. Fix $0 < \varepsilon < \eta r/8$, and consider the w^*-open "slice"

$$S = \{x^* \in U^* : \ \langle x^*, x\rangle > 1 - \varepsilon\}$$

of U^*. By the definition of U^*, the w^*-open set S contains a finite convex combination of points of the form $x^* + a$ with $x^* \in B^*$ and $a \in A_0 \cup -A_0$, and at least one of these points must belong to S because it is a slice. But A_0 does not have w^*-isolated points and S is w^*-open; thus S must actually contain two different points $x_1^* = x^* + a_1$ and $x_2^* = x^* + a_2$ of this form. If $\|u\| \le r$, then

$$\begin{aligned} 2(1-\varepsilon) + \langle x_1^* - x_2^*, u\rangle &\le \langle x_1^* + x_2^*, x\rangle + \langle x_1^* - x_2^*, u\rangle \\ &= \langle x_1^*, x+u\rangle + \langle x_2^*, x-u\rangle \\ &\le \|x+u\| + \|x-u\| \le 2 + \eta\|u\|/4 \le 2 + \eta r/4 \end{aligned}$$

i.e., $2(1-\varepsilon) + r\|x_1^* - x_2^*\| \le 2 + \eta r/4$, and thus $\|x_1^* - x_2^*\| < \eta/2$ because $\varepsilon < \eta r/8$. But this contradicts $\|x_1^* - x_2^*\| \ge \|x_1^* - x_2^*\|/2 = \|a_1 - a_2\|/2 > \eta/2$. □

Remark. The proof shows that there is an $\eta > 0$ such that $\|x\| = 1$ implies $\liminf_{\|u\|\to 0} \big(\|x+u\|+\|x-u\|-2\big)/\|u\| > \eta/4$. A norm which satisfies this condition is called a rough norm.

Theorem 4.13. (i) *Every separable Banach space E has an equivalent norm which is Gâteaux differentiable at every $x \neq 0$.*

(ii) *If E^* is separable, then E has an equivalent norm which is Fréchet differentiable at every $x \neq 0$.*

Proof. (i) Let $\{x_i\}_{i=1}^\infty$ be a dense sequence in the unit ball of E. The operator $Tu^* = \big(\langle u^*, x_1\rangle, \dots, \langle u^*, x_n\rangle/n, \dots\big)$ is bounded and injective from E^* into l_2. Define a norm $\|\cdot\|$ on E^* by $\|u^*\| = \|u^*\| + \|Tu^*\|_2$. Since its unit ball is w^*-closed it is the dual of an equivalent norm (also denoted by $\|\cdot\|$) on E. The new norm on E^* is strictly convex; hence the $\|\cdot\|$ unit ball of E admits a unique supporting functional at every $x \in E$ with $\|x\| = 1$. Thus $\|\cdot\|$ is Gâteaux differentiable at every $x \neq 0$ in E.

(ii) Let E^* be separable, and fix an increasing sequence $\{F_n\}_{n=1}^\infty$ of finite-dimensional subspaces of E^* whose union is norm dense in E^*. Let $T : E^* \to l_2$ be as above, and define a nonlinear homogeneous map $S : E^* \to l_2$ by

$$S(u^*) = \Big(d(u^*, F_1), \dots, d(u^*, F_n)/n, \dots\Big)$$

where the distances are with respect to the original norm in E^*. Then the components of S are w^*-lower semi-continuous (because the F_n's are finite-dimensional) and $\|S(u^* + v^*)\|_2 \le \|Su^*\|_2 + \|Sv^*\|_2$. Thus the norm $\|\cdot\|$ on E^* defined by

$$\|u^*\|^2 = \|u^*\|^2 + \|Tu^*\|_2^2 + \|S(u^*)\|_2^2 \tag{4.8}$$

is an equivalent norm with a w^*-closed unit ball; i.e., it is dual to an equivalent norm on E (also denoted by $\|\cdot\|$).

We first show that $(E^*, \|\cdot\|)$ is locally uniformly convex; i.e., if $u^*, u_k^* \in E^*$ satisfy $\|u^*\| = \|u_k^*\| = 1$ and $\lim_{k\to\infty} \|u^* + u_k^*\| = 2$, then $\|u_k^* - u^*\| \to 0$.

Indeed, each of the three expressions whose squares appear in (4.8) satisfy the triangle inequality and all three are w^*-lower semi-continuous. It follows that $\|u_k^*\| \to \|u^*\|$ and similarly for the other two terms. A similar argument shows that $\lim_{k\to\infty} \|T(u^* + u_k^*)\|_2 = 2\|Tu^*\|_2$ and thus $\|T(u_k^* - u^*)\|_2 \to 0$. Since $\{x_i\}$ is dense in the unit ball of E, it follows that u_k^* tends to u^* in the w^*-topology. By considering S we obtain similarly $\lim_{k\to\infty} d(u_k^*, F_n) = d(u^*, F_n)$ for each fixed n. Let $\varepsilon > 0$, and pick n such that $d(u^*, F_n) < \varepsilon$. Then $d(u_k^*, F_n) < \varepsilon$ for large enough k, so find $z_k^* \in F_n$ such that $\|u_k^* - z_k^*\| = d(u_k^*, F_n) < \varepsilon$. Assume (by passing to a subsequence) that the z_k^*'s converge to a point z^*. Then $\|u_k^* - z^*\| \le 2\varepsilon$ for large k, and $\|u^* - z^*\| \le 2\varepsilon$ because u_k^* is w^*-convergent to u^*. Thus $\limsup \|u_k^* - u^*\| \le 4\varepsilon$, and since ε was arbitrary $\|u_k^* - u^*\| \to 0$.

We now apply the criterion of equation (4.5) to show that whenever a dual norm $\|\cdot\|$ on a dual space E^* is locally uniformly convex, then the norm on E is Fréchet differentiable at every $x \neq 0$. Indeed, assume that $\|x\| = 1$, and choose $u^* \in E^*$ with $\|u^*\| = \langle u^*, x\rangle = 1$. For every $y \in E$ choose norm-one functionals $v_y^*, w_y^* \in E^*$ such that $\langle v_y^*, x+y\rangle = \|x+y\|$ and $\langle w_y^*, x-y\rangle = \|x-y\|$. Then

$$\|v_y^* + u^*\| \geq \langle v_y^* + u^*, x\rangle = 1 + \|x+y\| - \langle v_y^*, y\rangle \to 2$$

as $y \to 0$. By local uniform convexity $\|v_y^* - u^*\| \to 0$ as $y \to 0$, and $\|w_y^* - u^*\| \to 0$ in a similar way. Thus $\|v_y^* - w_y^*\| \to 0$, and hence

$$\begin{aligned} \|x+y\| + \|x-y\| - 2 &= \langle v_y^*, x\rangle + \langle w_y^*, x\rangle + \langle v_y^* - w_y^*, y\rangle - 2 \\ &\leq \|v_y^* - w_y^*\| \, \|y\| = o(\|y\|). \end{aligned}$$

□

Remark. The same proof as in the first part of (ii) above shows that every separable Banach space can be given an equivalent locally uniformly convex norm.

Once the norm in E is Fréchet differentiable (away from 0) it is C^1 there by Proposition 4.7. It follows that E admits a very rich collection of C^1 functions, namely, functions of the form $\varphi(\|x\|^2)$ where $\varphi : \mathbb{R} \to \mathbb{R}$ is C^1. In particular, E admits C^1 bump functions (i.e., functions with bounded support). If E^* is separable, then standard use of C^1 partitions of unity (see Appendix B) gives

Corollary 4.14. *Let E be a Banach space with a separable dual. Then every continuous function from an open set U in E into a Banach space F can be uniformly approximated by C^1 functions.*

Proof. Fix $\varepsilon > 0$, and choose, for each $y \in U$, a neighborhood $U_y \subset U$ such that the oscillation of f on U_y is smaller than ε. Let $\{\varphi_\alpha\}$ be a C^1 partition of unity subordinated to this cover (see Proposition B.3), and fix points x_α with $\varphi_\alpha(x_\alpha) \neq 0$. Put $g(x) = \sum \varphi_\alpha(x) f(x_\alpha)$. Then $\|f(x) - g(x)\| \leq \sum \varphi_\alpha(x)\|f(x) - f(x_\alpha)\| < \varepsilon$ because $\sum \varphi_\alpha(x) = 1$, and a term in the sum is nonzero only when x and x_α belong the some U_y, and hence $\|f(x) - f(x_\alpha)\| < \varepsilon$. □

The following theorem, which also uses the existence of C^1 bump functions, produces maximum attaining nonlinear perturbations in the spirit of Theorem 4.10.

Theorem 4.15. *Let E be a separable Banach space, and let f be an upper semi-continuous function which is bounded from above on E. (We allow f to take the value $-\infty$, but not on all of E.) Then for every $\varepsilon > 0$ there is a continuous Gâteaux differentiable g on E such that $\|g\|_\infty < \varepsilon$, $\|D_g\|_\infty < \varepsilon$ and $f - g$ attains its maximum on E. If E^* is separable, then g can be taken to be a C^1 function.*

Proof. The proof of the two cases is similar. We shall give the proof only when E^* is separable, and we shall find g in C^1.

Let $C^1(E)$ be the Banach space of all C^1 functions h on E for which the norm $\|h\|_1 = \|h\|_\infty + \|D_h\|_\infty$ is finite. Let $b \in C^1(E)$ be a bump function such that $b(0) = 1$ and $b(x) = 0$ for $\|x\| > 1$. Fix n and set

$$U_n = \Big\{g \in C^1(E) : \exists z \in E \text{ with } (f-g)(z) > \sup\{(f-g)(x) : \|x - z\| > 1/n\}\Big\}.$$

Then U_n is open and is dense in $C^1(E)$. Indeed, fix $h \in C^1(E)$, and let $\varepsilon > 0$. Since $f - h$ is bounded from above there is a point z such that

$$(f-h)(z) > \sup\{(f-h)(x) : x \in E\} - \varepsilon.$$

Put $g(x) = h(x) - \varepsilon b(n(x-z))$. Then $\|g-h\|_1 \le n\varepsilon\|b\|_1$, and since $b(0) = 1$ and $b(n(x-z)) = 0$ for $\|x-z\| > 1/n$ it follows that

$$(f-g)(z) = (f-h)(z) + \varepsilon > \sup\{(f-g)(x) : \|x-z\| > 1/n\}$$

i.e., $g \in U_n$.

By the Baire category theorem there is a $g \in \bigcap U_n$, and we shall show that $f-g$ attains its maximum. Indeed, $g \in U_n$ implies that there are $z_n \in E$ such that

$$(f-g)(z_n) > \sup\{(f-g)(x) : \|x-z_n\| > 1/n\}.$$

Thus $\|z_p - z_n\| \le 1/n$ for every $p > n$, and hence $\{z_n\}$ converges to a point z, say. Finally, $f-g$ attains its maximum at z because f is upper semi-continuous. □

Remark. The proof shows that $\operatorname{diam}\{x : f(x) - g(x) > f(z) - g(z) - \alpha\} \to 0$ as $\alpha \to 0$. In the terminology used in the next chapter (see Theorem 5.16) this means that $f-g$ attains a strong maximum at z.

We now study the structure of the set of points of differentiability of a general convex function on a separable Banach space. We start with an easy result.

Proposition 4.16. *Let f be a continuous convex function on an open set in a Banach space E. The set where f is Fréchet differentiable is a (possibly empty) G_δ set. If E is separable, this also holds for points of Gâteaux differentiability.*

Proof. Since f is locally Lipschitz, the sets

$$G_n = \Big\{x : \exists\delta > 0 \text{ so that } \sup_{\|y\|\le 1} \big|f(x+\delta y) + f(x-\delta y) - 2f(x)\big| < \delta/n\Big\}$$

are open, and it follows from the monotonicity of $\big(f(x+ty) + f(x-ty) - 2f(x)\big)/t$ as a function of t for fixed x and y and from the uniform form of the criterion (4.5) that the set of points of Fréchet differentiability of f is exactly $\bigcap_{n=1}^\infty G_n$.

To prove the second part, let $\{y_m\}$ be a dense sequence in the unit sphere of E. Then the points of Gâteaux differentiability of f is $\bigcap_{n,m} G_{n,m}$ where

$$G_{n,m} = \Big\{x : \exists\delta > 0 \text{ so that } \big|f(x+\delta y_m) + f(x-\delta y_m) - 2f(x)\big| < \delta/n\Big\}.$$

□

Theorem 4.17. (i) *A convex continuous function on an open subset of a separable Banach space is Gâteaux differentiable on a dense G_δ subset of its domain of definition.*

(ii) *A convex continuous function on an open subset of a space E with separable dual is Fréchet differentiable on a dense G_δ subset of its domain of definition.*

The separability of E^* in part (ii) is essential in view of Proposition 4.12. We shall prove below stronger versions of both parts of the theorem and show that the set of nondifferentiability points is not only residual but is actually much smaller. We just note here that part (i) as stated follows easily from the previous proposition because for fixed x and y the function $f(x+ty)$ is differentiable for a dense set of t's; hence each of the sets $G_{n,m}$ above is dense.

We now introduce a new concept of "small sets".

Definition 4.18. *A set A in a Banach space E is called porous if there is a number $0<\lambda<1$ with the following property. For every $x\in A$ and for every $\delta>0$ there is a $y\in E$ such that $0<\|y-x\|<\delta$ and $A\cap B(y,\lambda\|y-x\|)=\emptyset$. A countable union of porous sets is called σ-porous.*

Porous sets are nowhere dense, and hence σ-porous sets are of first category. Lebesgue's density theorem implies that σ-porous sets in a finite-dimensional space have measure 0. We mention (although this will not be used) that the class of σ-porous sets in a finite-dimensional space is much smaller than the class of sets which have measure 0 and are of first category. Also, the class of σ-porous sets is the same as the class of countable union of λ-porous sets with some fixed λ (say $\lambda=1/2$). The next result is a stronger form of Theorem 4.17(ii).

Theorem 4.19. *Let E be a Banach space with a separable dual, and let f be a continuous convex function defined on an open subset of E. Then the set of points of non-Fréchet differentiability of f is σ-porous.*

Proof. Since f is locally Lipschitz, we may assume that it is actually Lipschitz with Lipschitz constant C. For each m cover E^* by a family of balls $\{B_{k,m}\}_{k=1}^{\infty}$ of radius $1/8m$, and consider the sets $A_{k,m}$ defined by

$$\Big\{x:\ \exists u^*\in\partial_f(x)\cap B_{k,m}\ \text{and}\ \limsup_{y\to 0}\big(f(x+y)-f(x)-\langle u^*,y\rangle\big)>\|y\|/m\Big\}.$$

It follows directly from the definitions of the subdifferential and of Fréchet differentiability that the set where f fails to be Fréchet differentiable is $\bigcup_{k,m}A_{k,m}$, and we shall show that each $A_{k,m}$ is $1/6Cm$-porous. Fix $x\in A_{k,m}$ and $\delta>0$. By the definition of $A_{k,m}$ we can find y with $\|y\|<\delta$ and $u^*\in B_{k,m}\cap\partial_f(x)$ such that $f(x+y)-f(x)-\langle u^*,y\rangle>\|y\|/m$. We claim that $A_{k,m}\cap B(x+y,\|y\|/6Cm)=\emptyset$. Assume that z belongs to this intersection, and let v^* be the element of $\partial_f(z)$ guaranteed by the fact that $z\in A_{k,m}$. Then $\|v^*-u^*\|\le 1/4m$ (because the radius of $B_{k,m}$ is $1/8m$).

Since $v^*\in\partial_f(z)$ implies that $f(x)-f(z)\ge\langle v^*,x-z\rangle$, and since $\|u^*\|\le C$ and $\|x-z\|\le 2\|y\|$, we obtain

$$\begin{aligned}
f(x+y)-f(z) &= f(x+y)-f(x)+f(x)-f(z)\\
&\ge \|y\|/m+\langle u^*,y\rangle+\langle v^*,x-z\rangle\\
&= \|y\|/m+\langle u^*,x+y-z\rangle+\langle v^*-u^*,x-z\rangle\\
&\ge \|y\|/m-\|u^*\|\,\|x+y-z\|-\|v^*-u^*\|\,\|x-z\|\\
&\ge \|y\|/m-C\|y\|/6Cm-2\|y\|/4m=\|y\|/3m.
\end{aligned}$$

And this contradicts $|f(x+y)-f(z)|\le C\|x+y-z\|\le\|y\|/6m$. □

We now characterize subsets of the set of non-Gâteaux differentiability of continuous convex functions on separable Banach spaces.

A subset M of E will be called a hypersurface if it is the graph of a function, defined on a hyperplane in E, which is the difference of two convex Lipschitz functions. More precisely, E decomposes as a direct sum $E=F\oplus\mathbb{R}v$, and there are convex Lipschitz real-valued functions h and g on F such that

$$M=\big\{u+\big(h(u)-g(u)\big)v:\ u\in F\big\}.$$

The next theorem generalizes the fact that a convex function on $\mathbb{R}$ has at most countably many points of nondifferentiability. The generalization is to infinite-dimensional spaces, but it is not any easier on $\mathbb{R}^n$ when $n \geq 2$.

Theorem 4.20. *Let A be a subset of a separable Banach space E. Then there is a continuous convex function f on E which is nowhere Gâteaux differentiable on A if and only if A is contained in a countable union of hypersurfaces as above.*

Proof. Observe first that if the hypersurface M is determined by F, v, h and g, then the convex Lipschitz function f on E defined by

$$f(u + \lambda v) = \max\Big(h(u), g(u) + \lambda\Big), \qquad u \in F,\ \lambda \in \mathbb{R}$$

is not Gâteaux differentiable at the points of M. Indeed, a point $u + \lambda v$ belongs to M if and only if $\lambda = h(u) - g(u)$, and hence $f\big(u + (\lambda + t)v\big) = h(u) + t$ for $t > 0$, while $f\big(u + (\lambda + t)v\big) = h(u)$ for $t < 0$. Thus f is not differentiable at $u + \lambda v$ in the direction of v. Rescaling and adding a constant, we find a 1-Lipschitz function which vanishes at 0 and which is not Gâteaux differentiable at the points of M.

Assume now that $\{M_i\}$ is a sequence of hypersurfaces, and let $\{f_i\}_{i=1}^{\infty}$ be a sequence of convex 1-Lipschitz functions satisfying $f_i(0) = 0$ for all i so that f_i is not Gâteaux differentiable at the points of M_i. The function $\phi = \sum_{i=1}^{\infty} f_i/2^i$ is convex with Lipschitz constant 1, and since $\phi = f_{i_0}/2^{i_0} + \sum_{i \neq i_0} f_i/2^i$ for every i_0 it follows from the observation following (4.5) that ϕ is not Gâteaux differentiable at any point where at least one f_i is not differentiable, i.e., on $\bigcup M_i$.

To prove the converse direction, let f be a continuous convex function defined on an open set in E. We shall show that for every v with $\|v\| = 1$ the set of all points $x \in E$ such that f is not differentiable at x in the direction of v can be represented as a subset of a union of countably many hypersurfaces. To see that this suffices, let $\{v_i\}$ be a dense sequence in the unit sphere of E. Then, since f is locally Lipschitz, it follows that if it is not Gâteaux differentiable at some point, then it must also fail to be differentiable at that point in at least one of the countable number of directions v_i.

Let F be a closed hyperplane in E such that $E = F \oplus \mathbb{R}v$. Parameterize the points of E by the values at v attained by the supporting functionals to the graph of f; i.e., for each real scalar α put

$$B(\alpha) = \big\{x \in E :\ \exists u^* \in \partial_f(x) \text{ so that } \langle u^*, v\rangle = \alpha\big\}.$$

Then f is nondifferentiable at a point x in the direction v if and only if there are $\alpha > \beta$ such that $x \in B(\alpha) \cap B(\beta)$. In this case $x \in B(\gamma)$ for all $\alpha > \gamma > \beta$, so we can assume that α and β are rational.

Consider a point $z = w + \mu v \in B(\alpha)$, and let $z^* \in \partial_f(z)$ satisfy $\langle z^*, v\rangle = \alpha$. By the definition of the subdifferential $f(x) - f(z) \geq \langle z^*, x - z\rangle$ for all x in the domain of definition of f. Writing $x = u + \lambda v$, we see that

$$f(u + \lambda v) - \alpha\lambda \geq \langle z^*, u - w\rangle + f(w + \mu v) - \alpha\mu. \tag{4.9}$$

Denote the affine function on the right-hand side of (4.9) by $a_{z,z^*}(u)$, and define

$$g_n^{\alpha}(u) = \sup\Big\{a_{z,z^*}(u) :\ z = w + \mu v \in B(\alpha),\ \langle z^*, v\rangle = \alpha,\ z^* \in \partial_f(z),\ \|z^*\| \leq n\Big\}.$$

Then g_n^{α} is a convex function on F, and its Lipschitz constant is at most n. Also, for any $x = u + \lambda v \in B(\alpha)$ there is an n such that $g_n^{\alpha}(u) = f(x) - \alpha\lambda$. Indeed,

$g_n^\alpha(u) \le f(x) - \alpha\lambda$ for all n by (4.9); and if $x^* \in \partial_f(x)$ satisfies $\langle x^*, v\rangle = \alpha$, then a_{x,x^*} appears in the definition of g_n^α whenever $n \ge \|x^*\|$.

If f is not differentiable at x in the direction v, then, as we saw earlier, there are rational numbers $\alpha > \beta$ such that $x \in B(\alpha) \cap B(\beta)$. Write $x = u + \lambda v$, and choose integers $n, m \in \mathbb{N}$ such that $g_n^\alpha(u) = f(x) - \alpha\lambda$ and $g_m^\beta(u) = f(x) - \beta\lambda$. Then $\lambda = \big(g_m^\beta(u) - g_n^\alpha(u)\big)/(\alpha - \beta)$, and the set of nondifferentiability in the direction v is contained in the countable union of the graphs of $(g_m^\beta - g_n^\alpha)/(\alpha - \beta)$ where $n, m \in \mathbb{N}$ and α, β are rational. □

The following theorem is proved by using the same argument. Note that the theorem is interesting even in finite-dimensional spaces when $k < \dim E$.

Theorem 4.20′ *Let k be a positive integer, and let A be a subset of a separable Banach space E. Then there is a continuous convex function on E which is not differentiable at each point of A in at least k linearly independent directions if and only if A is contained in the union of countably many sets of the form*

$$\Big\{u + \sum_{i=1}^{k} \big(h_i(u) - g_i(u)\big)v_i \,:\, u \in F\Big\}$$

where $\{v_i\}_{i=1}^k$ are linearly independent, $E = F \oplus \operatorname{span}\{v_i\}_{i=1}^k$, and h_i and g_i are convex Lipschitz functions on F.

The last result of this section is not connected to differentiation. It relates the approximation of general Lipschitz functions by differences of convex functions to the geometry of the space.

Theorem 4.21. *A Banach space E is superreflexive if and only if every real-valued Lipschitz function on E can be approximated, uniformly on bounded sets, by differences of convex functions which are Lipschitz on bounded sets.*

Proof. Assume first that E is superreflexive. By Theorem A.6 we can assume that it is uniformly convex. Let $f : E \to \mathbb{R}$ be Lipschitz with constant 1, say. Put

$$f_n(x) = \inf\big\{f(y) + n\big(2\|x\|^2 + 2\|y\|^2 - \|x+y\|^2\big) \,:\, y \in E\big\}$$

and represent $f_n(x) = h_n(x) - g_n(x)$ where $h_n(x) = 2n\|x\|^2$ and

$$g_n(x) = \sup\big\{n\|x+y\|^2 - 2n\|y\|^2 - f(y) \,:\, y \in E\big\}.$$

Then g_n and h_n are convex, f_n, g_n and h_n are Lipschitz on bounded sets, and $f_n(x) \le f(x)$ for all x. We shall show that the f_n's converge to f uniformly on $\{x : \|x\| \le R\}$ for every $R \ge 1$. Consider the set

$$A_n = \Big\{y :\, f(y) + n\big(2\|x\|^2 + 2\|y\|^2 - \|x+y\|^2\big) \le f(x)\Big\}.$$

In the definition of $f_n(x)$ we need to consider only points $y \in A_n$, so fix such a y. By the triangle inequality and since f is Lipschitz with constant one,

$$\begin{aligned} n\big(\|x\| - \|y\|\big)^2 &\le n\big(2\|x\|^2 + 2\|y\|^2 - \|x+y\|^2\big) \\ &\le f(x) - f(y) \le \|x - y\|. \end{aligned} \tag{4.10}$$

We thus need to show that $\sup\{\|x-y\| : y \in A_n\} \to 0$ uniformly in $\{\|x\| \le R\}$.

It follows from (4.10) that if $R \ge 1$ and $n \ge 3$, then $\|y\| \le 2R$ for every $y \in A_n$. (If $\|y\| > 2R$, then the left-hand side is bigger than $3\big(\|y\| - R\big)^2$, while the right-hand side is at most $\|y\| + R$.) Thus $\big|\|x\| - \|y\|\big| \le (3R/n)^{1/2}$. Put $z = y\|x\|/\|y\|$;

then $\|z\| = \|x\|$ and $\|z-y\| = \big|\|x\| - \|y\|\big|$. A simple computation, using the second and the last terms of (4.10), gives $\|x+z\|^2 \geq 4\|x\|^2 - C(R)/n^{1/2}$. By the definition of the modulus of convexity $\|x-z\| \leq \|x\|\delta_E^{-1}(C_1(R)/n^{1/2}\|x\|^2)$. Distinguishing the two cases $\|x\| \geq n^{-1/8}$ and $\|x\| \leq n^{-1/8}$, we obtain, respectively, the two estimates

$$\|x-y\| \leq \|x-z\| + \|z-y\| \leq R\delta_E^{-1}(C_1(R)n^{-1/4}) + (3R/n)^{1/2},$$
$$\|x-y\| \leq 2\|x\| + \big|\|x\| - \|y\|\big| \leq 2n^{-1/8} + (3R/n)^{1/2}.$$

Hence $\sup\{\|x-y\| : y \in A_n\} \to 0$ uniformly in $\{\|x\| \leq R\}$ as required.

Conversely, if E is not superreflexive, use Theorem A.6 and the remark following it to find for every n a 1/4-tree of height n in the unit ball of E; i.e., a set

$$T_n = \big\{x(\theta_1, \theta_2, \ldots, \theta_k) : \ 0 \leq k \leq n, \theta_i = \pm 1\big\}$$

such that $2x(\theta_1, \ldots, \theta_k) = x(\theta_1, \ldots, \theta_k, 1) + x(\theta_1, \ldots, \theta_k, -1)$ for every $k < n$ (for $k = 0$ we mean the point 0) and the distance between any two different vectors in T_n is at least 1/4.

Recall also that for every finite set A in E there is a finite-codimensional subspace $E_0 \subset E$ such that $\|u+v\| \geq \|u\|$ whenever $u \in A$ and $v \in E_0$. (For each $u \in A$ choose $u^* \in E^*$ with $\|u^*\| = 1$ and $\langle u^*, u\rangle = \|u\|$, and take E_0 to be the intersection of the kernels of the u^*'s.) We can thus construct inductively a sequence of trees $T^{(n)}$ of height n and vectors u_n with $\|u_n\| = 2$ so that any two different points in the set $\bigcup\{u_n + T^{(n)}\} \subset B(0,3)$ are of distance at least 1/4 from each other. A point in $u_n + T^{(n)}$ is denoted by $u_n + x^{(n)}(\theta_1, \ldots, \theta_k)$ with $0 \leq k \leq n$ (and $k = 0$ corresponds to the point u_n).

For each $x \in B(0,3)$ let $f(x)$ be the distance of x from the "even" levels of the trees, i.e., from

$$\bigcup_n \big\{u_n + x^{(n)}(\theta_1, \ldots, \theta_k) : 0 \leq k \leq n \text{ is even}\big\}.$$

Then f is a Lipschitz function with constant 1, and we show that if h and g are convex function, such that $|f(x) - \big(h(x) - g(x)\big)| \leq \delta < 1/8$ for all $x \in B(0,3)$, then h is unbounded on $B(0,3)$.

Fix an integer n. Then $g(u_n) - h(u_n) \geq -\delta$ because $f(u_n) = 0$. Since g is convex there is a sign θ_1 such that $g\big(u_n + x^{(n)}(\theta_1)\big) \geq g(u_n)$. Also $f\big(u_n + x^{(n)}(\theta_1)\big) \geq 1/4$ by the definition of f; hence

$$h\big(u_n + x^{(n)}(\theta_1)\big) \geq \frac{1}{4} + g\big(u_n + x^{(n)}(\theta_1)\big) - \delta \geq \frac{1}{4} + g(u_n) - \delta \geq h(u_n) + \frac{1}{4} - 2\delta.$$

By the convexity of h there is a θ_2 such that

$$h\big(u_n + x^{(n)}(\theta_1, \theta_2)\big) \geq h\big(u_n + x^{(n)}(\theta_1)\big) \geq h(u_n) + \frac{1}{4} - 2\delta.$$

Since $f\big(u_n + x^{(n)}(\theta_1, \theta_2)\big) = 0$ by the definition of f, we can repeat the same argument. Hence, if $2r \leq n$, then there are signs θ_k for $1 \leq k \leq 2r$ such that

$$h\big(u_n + x^{(n)}(\theta_1, \theta_2, \ldots, \theta_{2r})\big) \geq h(u_n) + r\left(\frac{1}{4} - 2\delta\right).$$

Thus h is unbounded because $\delta < 1/8$ and because n and $2r \leq n$ are as large as we wish. □

3. Notes and Remarks

The material in Section 1 is standard. Higher-order derivatives are defined in a natural way. If f is a map from an open subset U of E into F, then D_f maps U into $L(E,F)$, the space of bounded linear operators from E into F. Hence the values of the second derivative are elements in $L\big(E, L(E,F)\big)$ which can be naturally identified with the space of bounded bilinear functions from $E\times E$ into F. There is, of course, a choice at every stage whether we take Gâteaux or Fréchet derivatives. The situation becomes simpler when we consider only Fréchet derivatives, and then the classes of C^k, C^∞ and real-analytic functions are defined in the obvious way. The basic theorems of calculus (e.g. Taylor's theorem with a remainder, or the fact that differentiations with respect to x and y of $f(x,y)$ commute) remain valid (with the same proof) in infinite dimensions. Also if the given data in Theorems 4.4 and 4.5 are C^k functions, then they also yield C^k functions.

We do not consider here the many connections of higher-order derivatives to the geometry of Banach spaces. A nice presentation of this topic is available in the book [**144**] by Deville, Godefroy and Zizler (mainly Chapter V there). We just mention two results in this direction (they are discussed in detail in [**144**]). The first gives precise information on the smoothness of bump functions on infinite-dimensional $L_p(\mu)$ spaces. Direct computations show that the function $f(x)=\|x\|^p$ on $L_p(\mu)$ is C^∞ when p is an even integer, it is C^{p-1} (with a Lipschitz $(p-1)$th derivative) when p is an odd integer, and for nonintegral p it is $C^{[p]}$ (where $[p]$ is the integer part of p) and its $[p]^{\text{th}}$ derivative satisfies a Hölder condition of order $p-[p]$. If φ is a real-valued C^∞ bump function on $\mathbb{R}$, then $\varphi(\|\cdot\|^p)$ is a bump function on $L_p(\mu)$ with the same order of smoothness as $\|x\|^p$. It turns out— and this is a deeper fact— that bump functions on infinite-dimensional $L_p(\mu)$ cannot have any higher order of smoothness than those given above. Fabian and Zizler [**183**] give a proof of this fact which is different from the one in [**144**] and which is based on the variational principle of Theorem 5.16.

The second result is a striking theorem of Deville [**140**]. Assume that an infinite-dimensional Banach space E has finite cotype and that it admits a C^∞ bump function. Then $\inf\{q : E \text{ has cotype } q\}$ is necessarily an even integer.

The differentiability criteria (4.5) for convex functions and Proposition 4.7 (for $f(x)=\|x\|$) are essentially due to Šmulian [**538**].

It is unknown if Proposition 4.9 can be strengthened to say that if E is separable and if the set where the norm fails to be Fréchet differentiable is "small" in some sense, then E^* is separable. In particular, it is unknown whether l_1 has an equivalent norm which is Fréchet differentiable outside a countable union of hyperplanes.

Theorem 4.10 is due to Bishop and Phelps [**66, 67**]. They also proved that if C is a closed bounded and convex set, then the set of support points in C (i.e., points where some linear functional attains a nontrivial maximum) is dense in ∂C.

Let C be a closed bounded and convex subset of a complex Banach space E. If C is circled (i.e., $\alpha x\in C$ for every $x\in C$ and $|\alpha|=1$), then it follows easily from the real case that the set of linear functionals x^* for which $|x^*|$ attains its supremum on C is dense in E^*. This is no longer true for a general closed bounded and convex set without the assumption that it is circled; see Lomonosov [**373**].

Lemma 4.11 is due to Ekeland [**165, 166**], who extracted the general principle from an argument in the proof of Bishop and Phelps.

Proposition 4.12 is due to Leach and Whitfield [**348**]. Theorem 4.13(i) is due to Day [**137**]. Part (ii) was proved independently by Klee [**330**] and by Kadec [**302**].

Corollary 4.14 is due to Bonic and Frampton [**77**]. It characterizes spaces E with a separable dual among all separable Banach spaces. This follows, for example, from Corollary 10.44. As noted already in Chapter 2, it is interesting to compare the corollary with Theorem 2.3 and Proposition 2.4.

Theorem 4.15 is due to Deville, Godefroy and Zizler [**143**] and extends related results of Borwein and Preiss [**81**]. Results of this nature, usually called "variational principles", were initiated by Ekeland [**165, 166**]. They are very useful in various problems in nonlinear analysis, because they make it possible to apply variational methods to the perturbed function.

Theorem 4.17(i) is a classical result of Mazur [**406**]. Part (ii) is due to Lindenstrauss [**354**] and Asplund [**27**]. The proof of the stronger Theorem 4.19 is perhaps the simplest known proof of Theorem 4.17(ii).

The main object of [**354**] was to study when Theorem 4.10 extends to operators, more precisely, to study for which Banach spaces E and F the norm attaining operators from E to F are dense in the space of all linear operators from E to F. This is not always so, and we refer the reader to Johnson and Wolfe [**287**], Partington [**448**], Schachermayer [**510**], and the appendix in Gowers [**223**] for more information. We mention only that Bourgain [**82**] showed that if E has the Radon-Nikodým property, then the answer is positive for every Banach space F.

There is a connection between simultaneously continuous retractions (see Section 5 in Chapter 2) and the density of norm attaining operators into $C(K)$ spaces. This was observed independently by Schachermayer, Stegall and Wolfe; see Johnson and Wolfe [**287**]. Let K be a compact Hausdorff space. The bounded linear operators from E to $C(K)$ are in one-one correspondence with the w^*-continuous maps from K to E^*. The map $\varphi(k) = T^*\delta_k$ corresponds to the operator T (where $\delta_k \in C(K)^*$ is defined by $\delta_k(f) = f(k)$), and when the map φ is given, then the operator T is defined by $Tx(k) = \langle \varphi(k), x\rangle$. In this correspondence $\|T\| = \sup \|\varphi(k)\|$.

Proposition 4.22. *Let K be a compact Hausdorff space, and let E be a Banach space for which there is a w^*-continuous and uniformly norm-continuous retraction from E^* onto its unit ball. Then the norm-attaining operators are dense in the space of all bounded linear operators from E to $C(K)$.*

Proof. Fix $T : E \to C(K)$, and let $\varphi : K \to E^*$ be the corresponding map. Assume, as we may, that $\|T\| = 1$, and fix $\varepsilon > 0$. Choose a point k_0 such that $\|\varphi(k_0)\| > 1-\varepsilon$, and use Theorem 4.10 to find a norm attaining functional x_0^* with $\|x_0^*\| = 1$ such that $\|x_0^* - \varphi(k_0)\| < \varepsilon$. Assume that x_0^* attains its norm at x_0; i.e., $\|x_0\| = \|x_0^*\| = \langle x_0^*, x_0\rangle = 1$.

Let $r : E^* \to B(E^*)$ be the retraction, and define a map $\psi : K \to E^*$ by $\psi(k) = r\big(\varphi(k) + x_0^* - \varphi(k_0)\big)$. Then $\psi(k_0) = r(x_0^*) = x_0^*$ because $\|x_0^*\| = 1$, and ψ is w^*-continuous because r is. Let S be the operator corresponding to ψ. Then $\|S\| = 1$ and S attains its norm because $\|Sx_0\| \geq \langle \psi(k_0), x_0\rangle = \langle x_0^*, x_0\rangle = 1$. Since $1 \leq \|y^*\| \leq 1+\varepsilon$ implies that

$$\|r(y^*) - y^*\| \leq \big\|r(y^*) - r\big(y^*/\|y^*\|\big)\big\| + \big\|\big(y^*/\|y^*\| - y^*\|\big)\big\| \leq \omega_r(\varepsilon) + \varepsilon,$$

it follows that

$$\|S - T\| = \sup \|\varphi(k) - \psi(k)\| \leq \omega_r(\varepsilon) + \varepsilon + \|x_0^* - \varphi(k_0)\| \leq 2\varepsilon + \omega_r(\varepsilon).$$

□

The article [**590**] of Zajíček is a survey related to the notions of porous and σ-porous sets. Theorem 4.19 and its proof are due to Preiss and Zajíček [**477, 476**]. Theorems 4.20 and 4.20′ are due to Zajíček [**588**]. There is no known result of this nature giving precise limitations on the size of sets of non-Fréchet differentiability of convex functions. (See also Example 6.46 in this connection.)

For convex continuous functions which are everywhere Gâteaux differentiable, there are some stronger results on Fréchet differentiability. Deville, Godefroy, Hare and Zizler [**141**] characterize the separable Banach spaces E so that every convex continuous $f : E \to \mathbb{R}$ which is everywhere Gâteaux differentiable must be Fréchet differentiable on a dense set. It turns out that E^* may be nonseparable (e.g., E may be the James tree space), but E cannot contain a subspace isomorphic to l_1. We refer to [**141**] for further details and references.

The following result of Zajíček [**589**] is related to Theorem 4.20. Let E be a strictly convex separable Banach space, and let A be a closed subset of E. Let B be the set of points $x \in E$ such that x has more than one nearest point in A. Then B can be covered by countably many Lipschitz hypersurfaces, i.e., by sets of the form $M = \{u + f(u)v : u \in F\}$, where $E = F \oplus \mathbb{R}v$ and f is a real-valued Lipschitz function on F. If E is a Hilbert space, then the functions defining the hyperspaces can actually be chosen to be of the form $f = h - g$ where h and g are convex Lipschitz functions as in Theorem 4.20.

We have limited the discussion in this chapter to separable spaces (unless the results hold with the same proof in the nonseparable case). In nonseparable spaces the situation is often more complicated. For example, the norm in such a space may be nowhere Gâteaux differentiable, and the set of points of Gâteaux differentiability of a convex continuous function need not be a G_δ set. (There are still several open problems in this area.) The article [**591**] by Zajíček describes a rather simple general separable reduction method, due to Preiss and to Zajíček, which allows one to transfer results from the separable to the nonseparable setting.

A Banach space E is called an Asplund space if every convex continuous function on it is Fréchet differentiable on a dense G_δ set. It is known that E is Asplund if and only if every separable subspace of E has a separable dual.

A Banach space E is called a weak Asplund space if every convex continuous function on it is Gâteaux differentiable on a dense G_δ set. It is called a Gâteaux differentiability space if every convex continuous function on E is Gâteaux differentiable on a dense set (which need not be G_δ). It is not known if these two notions are actually different. Differentiation on nonseparable spaces is discussed in detail in Deville, Godefroy and Zizler [**144**], Phelps [**461**], and Fabian [**180**]. We refer to these books for further information.

Theorem 4.21 and its proof are due to Cepedello-Boiso [**115**]. The "if" part was obtained independently by Strömberg [**546**]. For Hilbert space it was considered earlier by Lasry and P.L. Lions [**347**]. In this case the formula for f_n takes the simpler form $f_n = \inf\{f(y) + n\|x - y\|^2 : y \in E\}$. Inspection of the proof of the "if" part presented here shows that when E is locally uniformly convex, then the sequence f_n converges to f pointwise.

CHAPTER 5

The Radon-Nikodým Property

1. Vector Measures and Integration of Vector-Valued Functions

This chapter will be devoted to the following question: Under what conditions on E is the Radon-Nikodým theorem valid for E-valued vector measures? It turns out that the validity of this theorem is equivalent to various geometric conditions on the structure of convex sets in E. Most important from our point of view is the connection to the existence of derivatives of E-valued Lipschitz functions.

In this first section we present basic facts on vector measures and on integration of vector-valued functions with respect to a scalar measure. We present these topics in the simplest possible setting since this will suffice for our purposes.

Let $(\Omega, \mathcal{B})$ be a measurable space, i.e., a set Ω and a σ-algebra $\mathcal{B}$ of subsets of Ω, and let E be a Banach space. An E-valued measure on Ω is a map $\tau : \mathcal{B} \to E$ such that $\tau(\emptyset) = 0$ and

$$\tau\left(\bigcup_{i=1}^{\infty} A_i\right) = \sum_{i=1}^{\infty} \tau(A_i)$$

whenever $\{A_i\}_{i=1}^{\infty}$ are mutually disjoint sets in $\mathcal{B}$. Note that for this definition to make sense, the series $\sum \tau(A_i)$ has to converge unconditionally.

The variation $|\tau|$ of τ is the σ-additive nonnegative measure on $(\Omega, \mathcal{B})$

$$|\tau|(A) = \sup\left\{\sum \|\tau(A_j)\|\right\}$$

where the supremum is taken over all finite partitions of A with $A_j \in \mathcal{B}$. We shall be mainly interested in measures of finite variations, i.e., measures τ for which $|\tau|(\Omega) < \infty$. To understand the scope of this rather restrictive assumption, consider the following two simple examples:

(i) Take Ω to be the set of positive integers $\mathbb{N}$, and let $\mathcal{B}$ be the algebra of all the subsets of $\mathbb{N}$. A measure τ is determined by a choice of vectors $\tau\{i\} = x_i \in E$ so that the series $\sum x_i$ converges unconditionally. This measure is of finite variation if and only if the series converges absolutely, i.e., $\sum \|x_i\| < \infty$.

(ii) Let $(\Omega, \mathcal{B}, \mu)$ be the standard Lebesgue space on $[0, 1]$, and let $E = L_p(0, 1)$ with $1 \le p \le \infty$. Define $\tau : \mathcal{B} \to E$ by $\tau(A) = \chi_A$, the indicator function of A. Then τ is σ-additive if and only if $p < \infty$. It has finite variation if and only if $p = 1$. (If $1 < p < \infty$, then actually $|\tau|(A) = \infty$ for every A with $\mu(A) > 0$.)

Let τ be a measure with finite variation. The integral of a simple function $f = \sum_{i=1}^{m} a_i \chi_{A_i}$ in $L_1(|\tau|)$ is defined by $\int f d\tau = \sum a_i \tau(A_i)$, and the definition is then extended to a general f by continuity. Clearly $\| \int_\Omega f d\tau\| \le \int_\Omega |f| d|\tau|$.

The set of characteristic functions in $L_1(|\tau|)$ is relatively weakly compact (as a bounded set in $L_2(|\tau|)$), and integration with respect to τ is a bounded linear operator from $L_1(|\tau|)$ to E; hence the range $\{\tau(A) : A \in \mathcal{B}\}$ of τ has a weakly compact closure. In example (i) above, if τ has bounded variation, then the closure of its range is actually norm-compact, while in example (ii) (with $p = 1$) the closure of the range is weakly but not norm-compact.

A measure τ of finite variation is said to be absolutely continuous with respect to a scalar-valued σ-finite positive measure μ (in notation $\tau \prec \mu$) when one of the following equivalent conditions holds,

(i) $|\tau| \prec \mu$.

(ii) $\mu(A) = 0 \Rightarrow \tau(A) = 0$.

(iii) For every $\varepsilon > 0$ there is a $\delta > 0$ so that $\mu(A) < \delta$ implies that $|\tau|(A) < \varepsilon$. In particular $\tau \prec |\tau|$.

We now discuss vector-valued measurable functions. Let $(\Omega, \mathcal{B}, \mu)$ be a complete and σ-finite measure space (with a scalar measure μ), and let E be a Banach space. A simple function $f = \sum_{i=1}^{m} x_i \chi_{A_i}$ with $x_i \in E$ is called measurable if $A_i \in \mathcal{B}$ for every i. In general $f : \Omega \to E$ is called measurable if there is a sequence $\{f_n\}$ of measurable simple functions which converges a.e. to f as $n \to \infty$. Measurable functions have the following useful characterization.

Proposition 5.1. *A function $f : \Omega \to E$ is measurable if and only if the following two assertions hold.*

(i) *$\langle x^*, f(w)\rangle$ is a scalar-valued measurable function for every $x^* \in E^*$.*

(ii) *The function f is almost separably valued; i.e., there is a subset $A_0 \in \mathcal{B}$ with $\mu(A_0) = 0$ such that $\overline{\operatorname{span}}\{f(w) : w \notin A_0\}$ is separable.*

Proof. It is clear that (i) and (ii) hold for every measurable f. To prove the converse assume that (i) and (ii) hold and also, as we may, that there is a separable subspace $F \subset E$ such that $f(w) \in F$ for all w. Let $\{x_i^*\}_{i=1}^{\infty} \in F^*$ be a sequence of norm-one functionals such that

$$\|x\| = \sup \langle x_i^*, x\rangle$$

for every $x \in F$. It follows that $\|f(w)\|$ is measurable. Similarly $\|f(w) - x\|$ is measurable for every fixed $x \in F$. Fix a dense sequence $\{x_j\}_{j=1}^{\infty}$ in F, and let $\varepsilon > 0$. Then $\Omega = \bigcup A_j$ where $A_j = \{w : \|f(w) - x_j\| \le \varepsilon\}$ is a measurable set for each j by the discussion above. Put $B_j = A_j \setminus \bigcup\{A_k : k < j\}$. Then $g = \sum x_i \chi_{B_i}$ is a countably valued function with $\|g(w) - f(w)\| \le \varepsilon$ for every w. Thus f is a uniform limit of such countably valued "simple" functions. It follows that f is a pointwise limit of genuine simple functions. □

We pass to integration of vector-valued functions. If $f = \sum x_i \chi_{A_i}$ is a simple function (with $\mu(A_i) < \infty$ for all i), put $\int_\Omega f d\mu = \sum x_i \mu(A_i)$ and note that $\|\int_\Omega f d\mu\| \le \int_\Omega \|f\| d\mu$. A measurable function f is called Bochner integrable if there is a sequence of measurable simple functions $\{f_n\}_{n=1}^{\infty}$ converging a.e. to f so that $\int_\Omega \|f_n - f_m\| d\mu \to 0$. The integral $\int_\Omega f d\mu$ (called the Bochner integral of f) is then defined as $\lim \int_\Omega f_n d\mu$. Exactly as in the scalar case one verifies that the limit exists, that it does not depend on the choice of the sequence $\{f_n\}$, that $\int \|f\| d\mu$ is finite, and that $\|\int f d\mu\| \le \int \|f\| d\mu$.

Proposition 5.2. *A function $f : \Omega \to E$ is Bochner integrable if and only if it is measurable and $\int_\Omega \|f\| d\mu < \infty$.*

Proof. Assume that f is measurable with $\int_\Omega \|f\| d\mu < \infty$. We may also assume that μ is a finite measure. Fix $\varepsilon > 0$. By the proof of Proposition 5.1 there is a function $g = \sum_{j=1}^{\infty} x_j \chi_{B_j}$ such that $\|g(w) - f(w)\| \leq \varepsilon$ for all w. Thus

$$\int_\Omega \|f(w) - \sum_{j=1}^{m} x_j \chi_{B_j}\| d\mu \leq \varepsilon\mu(\Omega) + \int_{\bigcup\limits_{j>m} B_j} \|f\| \leq 2\varepsilon\mu(\Omega)$$

if m is large enough. The other direction is obvious. □

The space of equivalence classes of Bochner integrable functions $f : \Omega \to E$ with the norm $\|f\|_1 = \int_\Omega \|f\| d\mu$ is denoted by $L_1(\mu, E)$ (or by $L_1(\Omega, \mathcal{B}, \mu, E)$ when we want to note the measure space). The spaces $L_p(\mu, E)$ for $1 < p \leq \infty$ are defined in the natural way. The properties of the Bochner integral are similar to those of the scalar integral. For example, Lebesgue's dominated convergence theorem holds for this integral. If $T : E \to F$ is a bounded linear operator, then

$$T \int_\Omega f d\mu = \int_\Omega T f d\mu$$

for every Bochner integrable $f : \Omega \to E$. If Ω is the unit interval with the Lebesgue measure m, and if f is continuous, then $\int f dm$ coincides with the Riemann integral, and the same holds in more general situations where the Riemann integral (defined as the limit of Riemann sums) makes sense. The usual results for the differentiation of the integral are also valid for Bochner integrals.

Proposition 5.3. *Let m be the Lebesgue measure on $\mathbb{R}^n$, and denote the ball of radius r in $\mathbb{R}^n$ centered at a point u by $B_u(r)$. If $f \in L_1(m, E)$, then*

$$m\big(B_u(r)\big)^{-1} \int_{B_u(r)} \|f(v) - f(u)\| dm(v) \to 0 \tag{5.1}$$

for almost all $u \in \mathbb{R}^n$ as $r \to 0$. In particular for these u's

$$f(u) = \lim_{r \to 0} m\big(B_u(r)\big)^{-1} \int_{B_u(r)} f(v) dm(v). \tag{5.2}$$

Proof. We may assume that E is separable. Let $\{x_j\}_{j=1}^{\infty}$ be a dense sequence in E. By applying the usual differentiation theorem to the scalar functions $\|f(v) - x_j\|$ we get that for almost all u

$$\|f(u) - x_j\| = \lim_{r \to 0} m\big(B_u(r)\big)^{-1} \int_{B_u(r)} \|f(v) - x_j\| dm(v)$$

for all j. It follows that for all such u and for all j

$$\begin{aligned}
&\limsup_{r \to 0} m\big(B_u(r)\big)^{-1} \int_{B_u(r)} \|f(v) - f(u)\| dm(v) \\
&\leq \lim_{r \to 0} m\big(B_u(r)\big)^{-1} \int_{B_u(r)} \big(\|f(v) - x_j\| + \|x_j - f(u)\|\big) dm(v) = 2\|x_j - f(u)\|.
\end{aligned}$$

Since $\{x_j\}_{j=1}^{\infty}$ is dense in E this implies (5.1), which in turn implies (5.2). □

The same result holds, as usual, when balls are replaced by other appropriate sets. For $n = 1$ it follows from the proposition that $g(t) = \int_0^t f(x) dx$ is differentiable almost everywhere and that $g'(t) = f(t)$ a.e. for every $f \in L_1(m, E)$.

Let $(\Omega, \mathcal{B}, \mu)$ be a measure space. For every $f \in L_1(\mu, E)$ the formula

$$\tau(A) = \int_A f d\mu = \int_\Omega f \cdot \chi_A d\mu$$

defines an E-valued measure on $(\Omega, \mathcal{B})$ which is absolutely continuous with respect to μ. The measure τ has finite variation given by $|\tau|(A) = \int_A \|f\| d\mu$. In the next section we shall study the converse question, namely, when is a vector measure τ of finite variation, which is absolutely continuous with respect to μ, necessarily obtained in this way.

Let $\mathcal{B}_0$ and $\mathcal{B}_1$ be two σ-algebras of subsets of Ω with $\mathcal{B}_0 \subset \mathcal{B}_1$, and let μ be a probability measure on $(\Omega, \mathcal{B}_1)$. The conditional expectation $g = \mathbb{E}_{\mathcal{B}_0} f$ of a function $f \in L_1(\Omega, \mathcal{B}_1, \mu, E)$ is defined as in the scalar case. To establish its existence, note that if $f = \sum x_i \chi_{B_i}$ is a simple function, then $\mathbb{E}_{\mathcal{B}_0} f = \sum x_i \mathbb{E}_{\mathcal{B}_0} \chi_{B_i}$, where $\mathbb{E}_{\mathcal{B}_0} \chi_{B_i}$ are the conditional expectations of the scalar functions χ_{B_i}, and then take limits. As in the scalar case, $\mathbb{E}_{\mathcal{B}_0}$ is a contractive projection from $L_p(\Omega, \mathcal{B}_1, \mu, E)$ onto $L_p(\Omega, \mathcal{B}_0, \mu, E)$ for every $1 \leq p < \infty$.

Vector-valued martingales are defined as in the scalar case. The same proof as that of Proposition D.9 shows that a vector-valued martingale which converges in the L_1 norm converges almost everywhere (and to the same limit).

The question whether every bounded vector-valued martingale necessarily converges (in L_1 or pointwise) will be treated in the next section.

2. The Radon-Nikodým Property

Definition 5.4. *A closed convex and bounded set C in a Banach space E has the Radon-Nikodým property (RNP) if it satisfies the following property:*

Let $(\Omega, \mathcal{B})$ be a measurable space, and let τ be an E-valued measure and μ a scalar probability measure on $(\Omega, \mathcal{B})$. Assume that $\tau(A)/\mu(A) \in C$ for all $A \in \mathcal{B}$ with $\mu(A) \neq 0$. Then there is an $f \in L_1(\mu, E)$ such that

$$\tau(A) = \int_A f(w) d\mu(w), \quad A \in \mathcal{B}. \tag{5.3}$$

The space E is said to have the RNP if its unit ball has the RNP.

Note that a closed convex subset of a set with the RNP also has this property; hence every closed bounded and convex set in a space with the RNP has the RNP.

Since C is bounded, the assumption $\tau(A)/\mu(A) \in C$ means, in particular, that τ is of finite variation, that $\tau \prec \mu$, and that the (uniquely determined) function f appearing in (5.3) actually belongs to $L_\infty(\mu, E)$. Also the range $\{\tau(A) : A \in \mathcal{B}\}$ of a vector measure τ satisfying (5.3) is relatively norm-compact. Indeed, all these properties are obvious when f is a simple function, and the general case follows by approximating f by simple functions.

The function f in (5.3) is called the Radon-Nikodým derivative of τ with respect to μ and is denoted by $d\tau/d\mu$. If E has the RNP and if τ is an E-valued vector measure with finite variation which is absolutely continuous with respect to a σ-finite nonnegative measure μ, then τ has a Radon-Nikodým derivative with respect to μ given by $d\tau/d\mu = (d|\tau|/d\mu)(d\tau/d|\tau|)$. In this general case $d\tau/d\mu$ need not be essentially bounded, and we have only that $d\tau/d\mu \in L_1(\mu, E)$.

We now give three simple (and typical) examples of vector measures which do not have Radon-Nikodým derivatives.

(i) Let $E = L_1(0,1)$, and let m be the Lebesgue measure on $[0,1]$. The E-valued measure $\tau(A) = \chi_A$ is of finite variation ($|\tau| = m$) and no f satisfies (5.3). Indeed, consider such a function as an L_1 function $f(s,t)$ on the unit square. Then (5.3) implies that $\int_A \int_B f(s,t)dsdt = m(A \cap B)$ for all measurable subsets $A, B \subset [0,1]$. Consequently f vanishes outside the diagonal, which is clearly impossible.

(ii) Take $E = c_0$. Let m be the normalized Lebesgue measure on $[0, 2\pi]$, and put

$$\tau(A) = \Big(\int_A 1dt, \int_A e^{it}dt, \dots, \int_A e^{int}dt, \dots\Big).$$

The Riemann-Lebesgue lemma shows that $\tau(A) \in c_0$ for every A. Then $|\tau| \prec m$, but τ has no Radon-Nikodým derivate with respect to m. The only possible candidate, $f(t) = (1, e^{it}, \dots, e^{int}, \dots)$, does not map into c_0.

In both examples above the fact that τ has no representation of the form (5.3) follows also from the fact that its range $\{\tau(A) : A \in \mathcal{B}\}$ is not relatively norm-compact. As the next example shows, the relative norm-compactness of the range does not imply that τ can be represented in the form (5.3).

(iii) Let $E = C(0,1)$, and let m be the Lebesgue measure on $[0,1]$. Define τ by $\tau(A)(t) = m\big(A \cap [0,t]\big)$. Then $|\tau| = m$ and the functions $\tau(A)$ are uniformly bounded by 1 and equicontinuous. Thus the range of τ has a compact closure. If $m\big(A \cap [0,t]\big) = \int_A f(s,t)ds$, then necessarily $f(s,t) = 1$ for almost every $s \le t$ and $f(s,t) = 0$ for almost every $s > t$. But then $f(s, \cdot) \notin C(0,1)$.

The RNP can be reformulated in terms of representation of operators from $L_1(\mu)$ into E. Denote the set of all probability densities in $L_1(\mu)$ by $\mathcal{P}(\mu)$, i.e.,

$$\mathcal{P}(\mu) = \big\{\varphi \in L_1(\mu) : \ \varphi \ge 0 \text{ and } \int \varphi d\mu = 1\big\}.$$

More generally, if A is a measurable set with $\mu(A) > 0$, we denote the set of all the functions $\varphi \in \mathcal{P}(\mu)$ supported in A by $\mathcal{P}(\mu, A)$.

An operator $T : L_1(\mu) \to E$ is said to be representable if there is a function $f \in L_\infty(\mu, E)$ such that

$$Tg = \int g f d\mu \tag{5.4}$$

for every $g \in L_1(\mu)$.

Proposition 5.5. *A bounded closed and convex set C in E has the RNP if and only if every bounded linear operator $T : L_1(\mu) \to E$ such that $T\varphi \in C$ for every $\varphi \in \mathcal{P}(\mu)$ is representable.*

Proof. Note first that the formula (5.4) always represents a bounded linear operator from $L_1(\mu)$ to E with $\|T\| = \|f\|_\infty$.

Assume that τ and μ satisfy $\tau(A)/\mu(A) \in C$ for all A, and define an operator $T : L_1(\mu) \to E$ by $T\chi_A = \tau(A)$. By the convexity of C

$$T\Big(\sum \lambda_i \chi_{A_i}\Big) = \sum \lambda_i \mu(A_i) \cdot \big(\tau(A_i)/\mu(A_i)\big) \in C$$

whenever $\lambda_i > 0$ and $\sum \lambda_i \mu(A_i) = 1$. Hence T has a unique extension to a bounded linear operator on all of $L_1(\mu)$ and $T\varphi \in C$ whenever $\varphi \in \mathcal{P}(\mu)$. Using (5.4) for this T and for $g = \chi_A$, we obtain (5.3) (with the same f).

Conversely, given an operator T as in the statement of the proposition, define τ by $\tau(A) = T\chi_A$ and use (5.3) to produce a bounded function f. By the linearity of T the relation (5.4) will hold (with this f) for every g. $\square$

Given $T : L_1(\mu) \to E$ and a measurable set B with $\mu(B) > 0$, put

$$\Gamma_B = T\big(\mathcal{P}(\mu, B)\big). \tag{5.5}$$

By the proposition a set C has the RNP if and only if every operator $T : L_1(\mu) \to E$ satisfying $\Gamma_\Omega = T\big(\mathcal{P}(\mu)\big) \subset C$ is representable.

Lemma 5.6. *Let $(\Omega, \mathcal{B}, \mu)$ be a probability space, and let $T : L_1(\Omega, \mathcal{B}, \mu) \to E$ be a bounded linear operator. Then T is representable if and only if for every $A \in \mathcal{B}$ with $\mu(A) > 0$ and for every $\varepsilon > 0$ there is a set $B \subset A$ with $\mu(B) > 0$ such that* $\operatorname{diam}\big(\Gamma_B\big) < \varepsilon$.

Proof. Assume that for every A and ε there is a subset $B \subset A$ with the desired property. A simple exhaustion argument shows that given $\varepsilon > 0$ we can decompose Ω into a countable disjoint union $\bigcup B_i$ such that $\operatorname{diam}\big(\Gamma_{B_i}\big) < \varepsilon$ for all i. Put $x_i = T\big(\chi_{B_i}/\mu(B_i)\big)$, and let $h = \sum x_i \chi_{B_i}$. Let $g \in L_1$ be a nonnegative function with $\|g\|_1 = 1$, and put $\alpha_i = \int_{B_i} g d\mu$. Then $g\chi_{B_i}/\alpha_i, \chi_{B_i}/\mu(B_i) \in \mathcal{P}(\mu, B_i)$; hence

$$\|Tg - \int ghd\mu\| \le \sum \alpha_i \big\|T\big(g\chi_{B_i}/\alpha_i\big) - T\big(\chi_{B_i}/\mu(B_i)\big)\big\| \le \varepsilon$$

by the choice of the B_i's and since $\sum \alpha_i = 1$. Hence $\|Tg - \int ghd\mu\| \le \varepsilon\|g\|_1$ for all $g \in L_1$. Now take $\varepsilon_n = 1/n$, say, and construct partitions $\big\{B_i^n\big\}$ as above with each partition refining its predecessor. The corresponding functions $\{h_n\}$ converge in $L_\infty(\mu, E)$ to a function f satisfying (5.4).

The other direction is trivial. $\square$

Definition 5.7. *Let C be a bounded closed and convex set in a Banach space E, and let $x^* \in E^*$ and $\alpha > 0$. The slice $S(C, x^*, \alpha)$ is defined as*

$$\Big\{y \in C : \langle x^*, y\rangle \ge \sup\big\{\langle x^*, x\rangle : x \in C\big\} - \alpha\Big\}.$$

C is called dentable if it has slices of arbitrary small diameter. A point $y \in C$ is called a denting point of C if it is contained in slices of arbitrary small diameter.

A convenient way to check that C is not dentable is to find an $\varepsilon > 0$ so that $x \in \overline{\mathrm{co}}\big(C \setminus B(x, \varepsilon)\big)$ for every $x \in C$, where $B(x, \varepsilon)$ is the ball with center x and radius ε. (If x is not in this set, then separating it from $\overline{\mathrm{co}}\big(C \setminus B(x, \varepsilon)\big)$ would produce a slice of C of diameter at most 2ε.)

Observe that if y is a denting point of C, then the identity is continuous at y as a map from C in its weak topology to C in its norm topology. We now present one of the main results on the structure of sets with the RNP.

Theorem 5.8. *Let C be a closed bounded and convex set in a Banach space E. Then the following three assertions are equivalent:*

(i) *Every closed convex subset of C is dentable.*

(ii) *C has the RNP.*

(iii) *Every martingale $\{f_n\}$ (on some probability space) such that $f_n(w) \in C$ for all n and w converges a.e. as $n \to \infty$.*

For the proof we need two lemmas.

Lemma 5.9. *Let $T : L_1(\mu) \to E$, and define the sets Γ_B as in (5.5). Fix A with $\mu(A) > 0$, an $x^* \in E^*$ and $\alpha > 0$. Then there is a $B \subset A$ with $\mu(B) > 0$ and*

$$\Gamma_B \subset S(\overline{\Gamma_A}, x^*, \alpha).$$

Proof. Let $\gamma = \sup\{\langle x^*, x\rangle : x \in \Gamma_A\} - \alpha$. We denote the set of functions in L_1 which are supported on A by $L_1(A)$ (and define $L_\infty(A)$ similarly). Since $\alpha > 0$ there is a function $\varphi \in \mathcal{P}(\mu, A)$ such that $\langle x^*, T\varphi\rangle > \gamma$. The set

$$K = \{\psi : 0 \le \psi \in L_1(A) \text{ and } \langle x^*, T\psi\rangle \le \gamma\|\psi\|_1\}$$

is a closed convex cone in $L_1(A)$ which does not contain φ. By the separation theorem there is a $g \in L_\infty(A)$ such that $\int \varphi g > \sup\{\int \psi g : \psi \in K\}$. Since K is a cone it follows that $\sup\{\int \psi g : \psi \in K\} = 0$. The set $B = \{w : g(w) > 0\}$ has positive measure because $\varphi \ge 0$ and $\int \varphi g > 0$. It follows that if $0 \le \psi \in L_1(B)$ satisfies $\int \psi = 1$, then $\int \psi g > 0$, i.e., $\psi \notin K$ and hence $\Gamma_B \subset S(\overline{\Gamma_A}, x^*, \alpha)$. □

The second lemma is a perturbation lemma. It says that a sequence which is almost a martingale is necessarily close to one.

Lemma 5.10. *Let $(\Omega, \mathcal{B}, \mu)$ be a probability space, and let $\{\mathcal{B}_n\}_{n=1}^\infty$ be an increasing sequence of sub σ-algebras of $\mathcal{B}$. Let $\{g_n\}$ be a sequence of E-valued $\mathcal{B}_n$-measurable and Bochner integrable functions such that for some $\varepsilon > 0$*

$$\sum_{n=1}^\infty \|g_n - \mathbb{E}_{\mathcal{B}_n} g_{n+1}\|_1 < \varepsilon.$$

Then there is a $\mathcal{B}_n$-adapted martingale $\{f_n\}$ such that $\|g_n - f_n\|_1 \le \varepsilon$ for all n.

Proof. Since the conditional expectations are contractive projections, it follows from the assumption that for every k

$$\|g_k - \mathbb{E}_{\mathcal{B}_k} g_{k+1}\|_1 + \sum_{n=k+1}^\infty \|\mathbb{E}_{\mathcal{B}_k} g_n - \mathbb{E}_{\mathcal{B}_k} g_{n+1}\|_1 < \varepsilon.$$

It follows that the sequence $\{\mathbb{E}_{\mathcal{B}_k} g_n\}_{n=1}^\infty$ converges in $L_1(\mu, E)$ to a limit f_k for which $\|g_k - f_k\|_1 \le \varepsilon$. It is also evident that $\{f_k\}_{k=1}^\infty$ is a martingale. □

Remark. If $f \in L_1(\mu, E)$ and $\mathbb{E}$ is a conditional expectation operator on $L_1(\mu, E)$, then the range of $\mathbb{E}f$ is contained in the closed convex hull of the range of f. (This is obvious for a simple function f, and the general case is also easy.) Hence if the ranges of all the g_n's in Lemma 5.10 are contained in some closed and convex set C, this will also be true for the f_n's.

Proof of Theorem 5.8. (i) $\Rightarrow$ (ii). Let μ be a probability measure, and let $T : L_1(\mu) \to E$ be a bounded linear operator with $T\varphi \in C$ for every $\varphi \in \mathcal{P}(\mu)$. Let A be a set with $\mu(A) > 0$. By (i) the subset $\overline{\Gamma_A} \subset C$ is dentable. It follows from Lemma 5.9 that for every $\varepsilon > 0$ there is a $B \subset A$ with $\operatorname{diam}(\Gamma_B) < \varepsilon$. By Lemma 5.6 the operator T is representable, and since T was arbitrary the set C has the RNP by Proposition 5.5.

(ii) $\Rightarrow$ (iii). Let $(\Omega, \mathcal{B}, \mu)$ be a probability space, and let $\mathcal{B}_n$ be an increasing sequence of sub σ-algebras of $\mathcal{B}$. We may assume that $\mathcal{B}$ is generated by $\bigcup \mathcal{B}_n$. Let $\{f_n\}$ be a $\mathcal{B}_n$-adapted martingale with $f_n : \Omega \to C$. If $A \in \mathcal{B}_m$ for some m, then $\int_A f_n d\mu = \int_A f_m d\mu$ for every $n \ge m$. Hence $\lim \int_A f_n d\mu = \tau(A)$ exists for all $A \in \mathcal{B}$. The vector measure τ has finite variation and $\tau(A)/\mu(A) \in C$ whenever

$\mu(A) > 0$. Since C has the RNP, it follows that there is an $f \in L_\infty(\mu, E)$ such that $\tau(A) = \int_A f d\mu$ for all A. If $A \in \mathcal{B}_m$, then $\int_A f d\mu = \tau(A) = \int_A f_m d\mu$, i.e., $f_m = \mathbb{E}_{\mathcal{B}_m} f$. By the discussion preceding Proposition D.9 and by the proposition itself, $\|f_n - f\|_1 \to 0$ and also almost everywhere.

(iii) $\Rightarrow$ (i). Assume that (i) does not hold. We may assume that C itself is not dentable, and hence there is an $\varepsilon > 0$ such that $x \in \overline{\text{co}}\big(C \setminus B(x, \varepsilon)\big)$ for every $x \in C$. We shall produce inductively an increasing sequence of finite subalgebras $\mathcal{B}_n$ of $[0, 1]$ and $\mathcal{B}_n$-measurable functions $g_n : [0, 1] \to C$ such that

(a) $|g_n(t) - g_{n+1}(t)| \geq \varepsilon$ all n and for all $t \in [0, 1]$.

(b) $\sum_{n=1}^\infty \|g_n - \mathbb{E}_{\mathcal{B}_n} g_{n+1}\| \leq \varepsilon/4$.

This will suffice. Indeed, by Lemma 5.10 we would then produce from g_n a $\mathcal{B}_n$-adapted martingale $\{f_n\}$ with $\|f_n - g_n\|_1 \leq \varepsilon/4$ for all n. Then $\|f_n - f_{n+1}\|_1 \geq \varepsilon/2$ for all n by (a), and thus $\{f_n\}$ cannot converge in norm (or a.e.).

Let $\mathcal{B}_1$ be the trivial algebra consisting of $\emptyset$ and $[0, 1]$, and put $g_1 \equiv x$ for some $x \in C$. Assume that $\mathcal{B}_n$ and g_n have already been constructed. Let $[0, 1] = \bigcup_{i=1}^m A_i$ be the decomposition of $[0, 1]$ into the atoms of $\mathcal{B}_n$, and write $g_n = \sum_{i=1}^m x_i \chi_{A_i}$ for suitable $x_i \in C$. By the choice of ε we can find for every i vectors $\{x_{i,j}\}_{j=1}^{k_i}$ in C and $\lambda_{i,j} \geq 0$, $\sum \lambda_{i,j} = 1$ such that $\|x_i - x_{i,j}\| \geq \varepsilon$ and $\|\sum_{j=1}^{k_i} \lambda_{i,j} x_{i,j} - x_i\| \leq \varepsilon/2^{n+3}$. Partition each A_i into disjoint sets $\{A_{i,j} : j \leq k_i\}$ with $m(A_{i,j}) = \lambda_{i,j} m(A_i)$ for all i and j, take $\mathcal{B}_{n+1}$ to be the algebra whose atoms are the $A_{i,j}$'s, and define $g_{n+1} = \sum_{i=1}^m \sum_{j=1}^{k_i} x_{i,j} \chi_{A_{i,j}}$. Then (a) and (b) hold with this construction. ♢

Remarks. (i) The A_i's in the last step of the construction can be chosen to be intervals and such that $\bigcup \mathcal{B}_n$ generates the σ-field of all Borel subsets. It follows that restricting Definition 5.4 and part (iii) of Theorem 5.8 to the Lebesgue measure on $[0, 1]$ does not change the notion of RNP or the assertion of the theorem.

(ii) Remark (i) shows also that a set C has the RNP if (and only if) every closed convex separable subset of C has the RNP.

(iii) Inspection of the last step of the proof shows that the $\mathcal{B}_n$-adapted martingale $\{f_n\}$ actually satisfies $\|f_{n+1}(t) - f_n(t)\| \geq \varepsilon/2$ for every $t \in [0, 1]$.

Theorem 5.11. (i) *A weakly compact convex set has the RNP.*

(ii) *A separable w^*-compact convex set in a dual space has the RNP.*

Proof. (i) Denote the set by C. By Remark (ii) above we may assume that C is separable, and by Theorem 5.8 we need to show that every closed convex subset of C is dentable. Since closed and convex subsets of C are also weakly closed, hence weakly compact, it is enough to show that the set C itself is dentable.

Let D be the weak closure of the extreme points of C. By the Krein-Milman theorem D is nonempty. Fix $\varepsilon > 0$, and let $\{x_n\}_{n=1}^\infty$ be a dense sequence in C. The closed balls $\{B(x_n, \varepsilon)\}_{n=1}^\infty$ cover C, and since D is weakly compact it follows from Baire's category theorem that there is a weakly open set V and an integer n_0 such that $\emptyset \neq V \cap D \subset B(x_{n_0}, \varepsilon)$. Let $C_1 = \overline{\text{co}}(V \cap D)$ and $C_2 = \overline{\text{co}}(D \setminus V)$. Then $\text{diam}(C_1) \leq 2\varepsilon$ and $C = \text{co}(C_1 \cup C_2)$ by the Krein-Milman theorem. Since D is the weak closure of the extreme points of C, there is a $z_0 \in D \cap V$ which is an extreme point of C. Also since $D \setminus V$ is weakly compact, all the extreme points of C_2 belong to $D \setminus V$. In particular $z_0 \notin C_2$. Now let $0 < s < 1$, and consider the set

$$K_s = \big\{\lambda z + \mu y : \; z \in C_1, \; y \in C_2, \; \lambda + \mu = 1, \; 0 \leq \lambda \leq s\big\}.$$

The set K_s is a closed convex subset of C which does not contain z_0. If $u, v \in C \backslash K_s$, then $u = \lambda_1 z_1 + \mu_1 y_1$, $v = \lambda_2 z_2 + \mu_2 y_2$ with $z_i \in C_1$, $y_i \in C_2$, $\lambda_i + \mu_i = 1$, $\lambda_i > s$ (hence $0 \le \mu_i < 1 - s$). Thus

$$\|u - v\| \le |\lambda_1 - \lambda_2| \cdot \|z_1\| + \operatorname{diam}(C_1) + (1-s)\big(\|y_1\| + \|y_2\|\big).$$

Since C is bounded, $|\lambda_1 - \lambda_2| \le 1 - s$ and $\operatorname{diam}(C_1) \le 2\varepsilon$, it follows that if s is close enough to 1, then $\operatorname{diam}(C \setminus K_s) < 3\varepsilon$. Thus separating z_0 from K_s yields a slice of C of diameter less than 3ε.

(ii) Assume that C is a w^*-compact and norm separable convex set in $E = F^*$. There is a separable subspace F_0 of F such that the canonical map $F^* \to F_0^*$ is an isometry on C; hence there is no loss of generality to assume that F is separable. Let $\{f_n\}$ be a $\mathcal{B}_n$-adapted martingale from $(\Omega, \mathcal{B}, \mu)$ into C. We show that $\{f_n\}$ converges. Let $\{y_m\}_{m=1}^{\infty}$ be a dense set in F, and assume, as we may, that $\mathcal{B}$ is generated by $\bigcup \mathcal{B}_n$. For every m the scalar martingale $\{\langle f_n, y_m\rangle\}$ converges a.e., and since C is bounded we get that $f(t) = w^* - \lim f_n(t)$ exists almost everywhere. Since C is norm separable it follows that f is separably valued. Moreover, since every norm open ball in C is a countable union of w^*-compact sets (balls of smaller radius), the separability of C implies that the w^*-Borel structure of C and its norm Borel structure coincide. Hence they also coincide with the weak Borel structure of C. Since $\langle f, y\rangle$ is measurable for every $y \in F$, it follows that $\langle y^{**}, f\rangle$ is measurable for every $y^{**} \in E^* = F^{**}$. By Proposition 5.1 f is measurable.

The equality $\int_A f d\mu = \int_A f_n d\mu$ holds for every $A \in \mathcal{B}_n$, as we see by applying both sides to elements of F. Thus $f_n = \mathbb{E}_{\mathcal{B}_n} f$, and hence $\|f_n - f\|_1 \to 0$. □

Remarks. (i) The unit ball of $l_\infty = l_1^*$ does not have the RNP because l_∞ contains c_0. Thus the norm-separability assumption in (ii) cannot be removed.

(ii) Part (i) of the theorem actually follows from (ii). Indeed, a separable weakly compact convex set in E can be considered as a separable w^*-compact subset of E^{**}. We included the direct proof of (i) because it is of interest in itself.

The next corollary is an immediate consequence of Theorem 5.11.

Corollary 5.12. (i) *Every separable conjugate Banach space has the RNP.*

(ii) *Every reflexive space has the RNP.*

In the proof of Theorem 5.11(i) we used extreme points. There is indeed a close connection between the RNP and extremal structure of convex sets. This will become even clearer from Proposition 5.13 and Theorem 5.17 below.

Proposition 5.13. *Every closed bounded and convex set C with the RNP is the norm closure of the convex hull of its extreme points.*

Note that in general C need not be compact in any reasonable topology. Thus the Krein-Milman theorem cannot even be applied to deduce the existence of an extreme point. But Proposition 5.13 gives more than the Krein-Milman theorem even if C is the w^*-compact unit ball in a separable conjugate space; C is the norm (and not just the w^*) closed convex hull of its extreme points.

The following (obvious) lemma will be used now for linear functions only, but will later be used for "slices"

$$S(C, f, \alpha) = \Big\{ y \in C : \ f(y) \ge \sup\{f(x) : \ x \in C\} - \alpha \Big\}$$

of arbitrary real-valued functions f on C which are bounded from above.

Lemma 5.14. *Let C be a set, and let f and g be real-valued functions on C which are bounded from above. If $\alpha > 0$ and if $\delta = \sup\{|f(y) - g(y)| : y \in C\} < \alpha/2$, then $S(C, g, \alpha - 2\delta) \subset S(C, f, \alpha)$.*

Proof of Proposition 5.13. Recall that a face of a convex set C is a convex subset F of C such that $x, y \in C$ and $(x + y)/2 \in F$ imply that $x, y \in F$. Nonempty intersections of faces is a face, and a face of a face is a face of the original set. The extreme points are the faces with a single point. If $x^* \in E^*$ attains its maximum on C, then the set where the maximum is attained is a closed face of C.

We show first that C has extreme points. To this end we construct faces $\{F_n\}_{n=1}^{\infty}$ so that F_1 is a closed face of C, each F_{n+1} is a closed face of F_n, and $\operatorname{diam}(F_n) \to 0$. By the observations above $\bigcap_{n=1}^{\infty} F_n$ will be an extreme point of C. The inductive construction is as follows. Since C has the RNP, the closed face F_n is dentable. Let S_n be slice with diameter smaller than $1/n$ of F_n, determined by the functional x^*. By the Bishop-Phelps theorem (Theorem 4.10) x^* can be approximated by a functional z^* which attains its maximum on F_n. Thus z^* determines a face F_{n+1} of F_n and $F_{n+1} \subset S_n$ by Lemma 5.14 provided $\|x^* - z^*\|$ is small enough.

Let K be the closed convex hull of the extreme points of C. We shall show that $C = K$. Indeed, if $x \in C \setminus K$, then it can be separated from K by a continuous linear functional. Using the Bishop-Phelps theorem and Lemma 5.14 again, we see that there is a closed face F of C disjoint from K. By the first part of the proof F has an extreme point which is necessarily also an extreme point of C. This contradicts the definition of K. ♢

It is unknown whether Proposition 5.13 characterizes sets with the RNP in the sense that if every closed convex subset of C has an extreme point, then C has to have the RNP. (See the Notes and Remarks section for some strong partial results.) Theorem 5.17 below is a stronger version of Proposition 5.13, which trivially characterizes RNP sets. We introduce first some more geometric notions.

Definition 5.15. *Let C be a closed convex set in a Banach space E.*

(i) *A point $x \in C$ is called an exposed point of C if there is an $x^* \in E$ such that $\langle x^*, x\rangle > \langle x^*, y\rangle$ for $x \neq y \in C$.*

(ii) *A point $x \in C$ is called a strongly exposed point of C if there is an x^* as above and such that $\operatorname{diam}(S(C, x^*, \alpha)) \to 0$ as $\alpha \to 0$. A functional x^* for which this holds is said to be a strongly exposing functional (at the point x).*

Every exposed point is an extreme point, and easy 2-dimensional examples show that the converse is false. In finite-dimensional spaces the notions of exposed and strongly exposed points coincide, but easy examples show that in the infinite-dimensional case exposed points need not be strongly exposed. In connection with Definition 5.7, note that while a strongly exposed point is a denting point, the converse is again false, and a denting point need not even be exposed.

Let C be a set, and assume that $f : C \to \mathbb{R}$ attains its maximum at a point $x \in C$. If $\operatorname{diam}(S(C, f, \alpha)) \to 0$ as $\alpha \to 0$, we say, in analogy to (ii) above, that f has a strong maximum at x.

Theorem 5.16. *Let C be a bounded closed and convex set with the RNP in a Banach space E, and let f be a real-valued upper semi-continuous function on C which is bounded from above. Then for every $\varepsilon > 0$ there is an $x^* \in E^*$ with $\|x^*\| \leq \varepsilon$ such that $f + x^*$ attains a strong maximum on C.*

This theorem is of the same nature as Theorem 4.15. There are, however, important differences. The assumption here is stronger; the set where f is assumed to be finite is a subset of a set with the RNP. (As stated, f is assumed to be finite on all of C, but it could just as well take the value $-\infty$ at some points of C provided it is finite somewhere. Also, as in Theorem 4.15 we can consider f as a function on all of E taking the value $-\infty$ off C.) The conclusion here is also stronger; the perturbation is by a linear functional rather than by a general differentiable function.

Proof. The proof is in two steps. The first and main step is to show that for every $\varepsilon > 0$ there are an $x^* \in E^*$ with $\|x^*\| < \varepsilon$ and an $\alpha > 0$ such that $\operatorname{diam}\big(S(C, f + x^*, \alpha)\big) \leq 2\varepsilon$. The second step is a simple iteration argument.

Step 1. Assume that there is an $\varepsilon > 0$ such that $\operatorname{diam}\big(S(C, f+x^*, \alpha)\big) > 2\varepsilon$ for every x^* with $\|x^*\| < \varepsilon$ and for every $\alpha > 0$. We shall show that this contradicts the assumption that C has the RNP.

Let n_0 be such that $2^{-n_0} < \varepsilon$, and define subsets A_n of C for $n \geq n_0$ by

$$A_n = \bigcup \big\{ S(C, f + x^*, 4^{-n}) : \ \|x^*\| \leq \varepsilon - 2^{-n} \big\}.$$

The (not necessarily convex) sets A_n are unions of slices of functions which are close to the given function f. We show that for every $n \geq n_0$ and for every $y \in E$

$$\text{(5.6)} \qquad \operatorname{co} A_n \subset \operatorname{co}\big(A_{n+1} \setminus B(y, \varepsilon)\big) + 4 \cdot 2^{-n} B$$

where, as usual, $B(y, \varepsilon)$ is the ball with center y and radius ε and $B = B(0, 1)$.

Assume that (5.6) is false for some n and for some $y \in E$. Since the set on the right-hand side of (5.6) is convex, it follows that there are an x^* with $\|x^*\| \leq \varepsilon - 2^{-n}$ and a point $x \in S(C, f + x^*, 4^{-n})$ which can be separated from the set on the right-hand side of (5.6) by a $y^* \in E^*$ with $\|y^*\| = 1$. Put $z^* = x^* + 2^{-(n+1)} y^*$. Then $\|z^*\| \leq \varepsilon - 2^{-(n+1)}$, and the choice of y^* gives $\langle y^*, x \rangle \geq \langle y^*, z \rangle + 4 \cdot 2^{-n}$ for every $z \in S(C, f + z^*, 4^{-(n+1)}) \setminus B(y, \varepsilon)$. (Such points z exist because the diameter of the slice is larger than 2ε by the assumption in the beginning of step 1.) It follows from $x \in S(C, f + x^*, 4^{-n})$ that for z as above

$$\begin{aligned} f(x) + \langle z^*, x \rangle &= f(x) + \langle x^*, x \rangle + 2^{-(n+1)} \langle y^*, x \rangle \\ &\geq f(z) + \langle x^*, z \rangle - 4^{-n} + 2^{-(n+1)} \langle y^*, x \rangle \\ &\geq f(z) + \langle x^*, z \rangle - 4^{-n} + 2^{-(n+1)} \langle y^*, z \rangle + 4 \cdot 2^{-n} \cdot 2^{-(n+1)} \\ &= f(z) + \langle z^*, z \rangle + 4^{-n} \end{aligned}$$

and this contradicts the assumption that $z \in S(C, f + z^*, 4^{-(n+1)})$.

Having established (5.6) we now verify that the set

$$A = \bigcap_{n=n_0}^{\infty} \overline{\operatorname{co}} \left(\bigcup_{m=n}^{\infty} A_m \right)$$

is a nondentable subset of C, contradicting the assumption that C has the RNP.

Since by (5.6) $\operatorname{co} A_n \subset \operatorname{co} A_{n+1} + 4 \cdot 2^{-n} B$, a simple iteration argument shows that A is nonempty. In fact

$$\text{(5.7)} \qquad \operatorname{co} A_n \subset A + 8 \cdot 2^{-n} B$$

for every $n \geq n_0$. The same argument shows also that

$$\text{co} \bigcup_{k=n}^{m} A_k \subset \text{co}\big(A_{m+1} \setminus B(y,\varepsilon)\big) + 16 \cdot 2^{-n} B \tag{5.8}$$

for every $m > n \geq n_0$ and every $y \in E$. Let $x \in A$, and fix n such that $16 \cdot 2^{-n} < \varepsilon$. By the definition of A there is an $m > n$ and a $u \in \text{co} \bigcup_{k=n}^{m} A_k$ with $\|x - u\| \leq 2^{-n}$. By (5.8) (with $y = x$) there is a $v \in \text{co}\big(A_{m+1} \setminus B(x,\varepsilon)\big)$ such that $\|v - u\| \leq 16 \cdot 2^{-n}$; hence $\|x - v\| \leq \|x - u\| + \|u - v\| \leq 17 \cdot 2^{-n}$. Represent v as a convex combination $v = \sum \lambda_j v_j$, where $v_j \in A_{m+1}$ and $\|v_j - x\| \geq \varepsilon$. By (5.7) (with n replaced by $m+1$) there are $w_j \in A$ with $\|w_j - v_j\| \leq 8 \cdot 2^{-m-1} \leq 8 \cdot 2^{-n} < \varepsilon/2$ for all j. Thus $\|w_j - x\| > \varepsilon/2$ for all j, while $\|x - \sum \lambda_j w_j\| \leq 25 \cdot 2^{-n}$. Since n is as large as we please, $x \in \overline{\text{co}}\big(A \setminus B(x,\varepsilon/2)\big)$. Thus A is not dentable.

Step 2. Note first that since f is upper semi-continuous the sets $S(C, f + x^*, \alpha)$ are closed for every x^*. Let $\varepsilon > 0$ be given. By Step 1 there are an $x_1^* \in E^*$ and $\alpha_1 > 0$ with $\|x_1^*\| < \varepsilon/2$ and such that $\text{diam}\big(S(C, f + x_1^*, \alpha_1)\big) \leq 1/2$.

Put $\varepsilon_2 = \min(\varepsilon/4, \alpha_1/2)$, and use Step 1 and Lemma 5.14 to find $x_2^* \in E^*$ and $\alpha_2 > 0$ with $\|x_2^*\| < \varepsilon_2$ so that $\text{diam}\big(S(C, f + x_1^* + x_2^*, \alpha_2)\big) \leq 1/4$ and

$$S(C, f + x_1^* + x_2^*, \alpha_2) \subset S(C, f + x_1^*, \alpha_1).$$

Continuing inductively, find $x_k^* \in E^*$ and a decreasing sequence of sets of the form $S(C, f + \sum_{k=1}^{n} x_k^*, \alpha_n) = S_k$ whose diameters tend to 0. Moreover, the $\{x_k^*\}_{k=1}^{\infty}$ can be chosen so that $\sum_{k=n}^{\infty} \|x_k^*\| < \alpha_{n-1}/2$ for every $n > 1$ and $\sum_{k=1}^{\infty} \|x_k^*\| \leq \varepsilon$. It follows (use Lemma 5.14 again) that $f + \sum_{k=1}^{\infty} x_k^*$ attains a strong maximum at the unique point of intersection of the sequence $\{S_k\}_{k=1}^{\infty}$. □

The following strengthening of Proposition 5.13 is obtained by applying Theorem 5.16 to linear functionals.

Theorem 5.17. *Let C be a closed bounded and convex set with the RNP in a Banach space E. Then C is the closed convex hull of its strongly exposed points. The functionals which strongly expose points of C form a dense G_δ set in E^*.*

Proof. We first prove the second statement. Put

$$G_n = \big\{x^* : \text{diam}\big(S(C, x^*, \alpha)\big) < 1/n \text{ for some } \alpha > 0\big\}.$$

By Lemma 5.14 the G_n's are open subsets of E^*, and they are dense in E^* by Theorem 5.16. Hence $G = \bigcap_{n=1}^{\infty} G_n$, the set of functionals which strongly expose a point of C, is a dense G_δ in E^*. In particular, C has strongly exposed points. Let K be the closed convex hull of the strongly exposed points of C. If K were a proper subset of C, we could separate a point in C from K by an element x^* of the dense set G. The point exposed by x^* is then a strongly exposed point of G which does not belong to K, a contradiction. □

Corollary 5.18. *A weakly compact convex set in a Banach space is the closed convex hull of its strongly exposed points.*

3. Differentiability, Trees and the RNP

The main theme in this section is the connection between properties related to the RNP (especially the existence of certain types of "trees" in a Banach space E) and differentiability properties of E-valued functions on an interval.

We start with two definitions. The first introduces notions related to Fréchet differentiability. They will be studied in detail in the second volume in the context of the general study of Fréchet differentiability of Lipschitz functions between Banach spaces, and we only consider them now in the special context needed here.

Definition 5.19. (i) *Let f be a function from $\mathbb{R}$ (or from an interval in $\mathbb{R}$) to a Banach space E, and let $\varepsilon > 0$. The function f is said to be ε-differentiable at t_0 if there is a $\delta > 0$ and an $x \in E$ so that*

$$\|f(t_0 + h) - f(t_0) - hx\| \leq \varepsilon |h|$$

for every h with $|h| < \delta$.

(ii) *The pair $(\mathbb{R}, E)$ is said to have the approximation by affine functions property (AAP) if for every Lipschitz function f from $\mathbb{R}$ (or an interval in $\mathbb{R}$) to E and for every $\eta > 0$ there are an interval I and an affine function $k(t) = x_0 + tx_1$ on I such that $\|f(t) - k(t)\| \leq \eta |I|$ for all $t \in I$ (where $|I|$ is the length of I).*

Note that f is differentiable at t_0 if and only if it is ε-differentiable there for every $\varepsilon > 0$. Also if E has the property that every Lipschitz function $f : \mathbb{R} \to E$ has points of ε-differentiability for every $\varepsilon > 0$, then $(\mathbb{R}, E)$ has the AAP. One of the delicate results, proved in this and in the next section, will be that the converse is false. It follows that while the standard way to construct affine approximations of a function is by taking a derivative, there are circumstances where one can find good affine approximations which are not necessarily related to derivatives.

The second definition formalizes a concept which has already appeared implicitly in the proof of Theorem 5.8.

Definition 5.20. *Let E be a Banach space, and let $\varepsilon > 0$. A set of vectors $\{x_{n,j}\}_{n \geq 0, 1 \leq j \leq m_n}$ in E is called an ε-bush if for every $n \geq 1$ there is a partition $\{A_k^n\}_{k=1}^{m_{n-1}}$ of $\{1, 2, \ldots, m_n\}$ such that*

$$\|x_{n,j} - x_{n-1,k}\| \geq \varepsilon \tag{5.9}$$

for every $j \in A_k^n$, and

$$x_{n-1,k} = \sum_{j \in A_k^n} \lambda_{n,j} x_{n,j} \tag{5.10}$$

for some $\lambda_{n,j} \geq 0$, $\sum_{j \in A_k^n} \lambda_{n,j} = 1$.

If the cardinality of all the sets A_k^n is 2 (i.e., if $m_n = 2^n$ for all n), the ε-bush is called a generalized ε-dyadic tree. If in addition $\lambda_{n,j} = 1/2$ for all j and for all $n \geq 1$, the ε-bush is called an ε-dyadic tree.

An ε-bush in E is naturally identified with an E-valued martingale on $[0, 1]$. Put $f_0(t) = x_{0,1}$ for every $t \in [0, 1]$. To define f_1, partition $[0, 1]$ into subintervals $\{B_{1,j}\}_{j \leq m_1}$ of measure $\lambda_{1,j}$ respectively, and put $f_1(t) = x_{1,j}$ for $t \in B_{1,j}$. Continue inductively. By (5.10) we obtain a martingale, and $\|f_n(t) - f_{n-1}(t)\| \geq \varepsilon$ for every t by (5.9). Inspection of the proof of Theorem 5.8 (see also Remark (iii) that follows its proof) shows that a bounded closed and convex subset of a Banach space fails to have the RNP if and only if it contains an ε-bush for some $\varepsilon > 0$.

The notion of a vector-valued absolutely continuous function is defined in the same way as for scalar functions.

Theorem 5.21. *Let E be a Banach space. Then the following three assertions are equivalent:*

(i) *E has the RNP.*

(ii) *Every absolutely continuous function $f : [0,1] \to E$ is differentiable almost everywhere.*

(iii) *Every Lipschitz function $f : [0,1] \to E$ has a point of ε-differentiability for every $\varepsilon > 0$.*

Proof. (i) $\Rightarrow$ (ii). Let $f : [0,1] \to E$ be absolutely continuous. The variation $V(t) = \sup\{\sum_{j=1}^{m} \|f(s_j) - f(s_{j-1})\| : 0 = s_0 < s_1, \ldots, < s_m = t\}$ of f is an absolutely continuous scalar function. Thus $V(t) = \mu\{[0,t]\}$ for some finite positive measure μ on $[0,1]$ which is absolutely continuous with respect to Lebesgue measure.

Define a vector measure τ on the Borel subsets of $[0,1]$ as follows: If A is open, write $A = \bigcup(a_i, b_i)$, a disjoint union, and put $\tau(A) = \sum \big(f(b_i) - f(a_i)\big)$. If A is a general Borel set, choose a decreasing sequence $\{G_n\}$ of open sets containing A so that $\mu(G_n) \to \mu(A)$. Since $\mu(G_n \setminus G_m) \to 0$ as $n, m \to \infty$ it follows that $\tau(A) = \lim \tau(G_n)$ exists. One verifies easily that τ is a well-defined σ-additive measure of finite variation, that $\|\tau(A)\| \le \mu(A)$ for all A, and that $\tau \prec \mu$. Since E has the RNP there is a Bochner integrable $g : [0,1] \to E$ such that

$$f(t) = \tau[0,t] + f(0) = \int_0^t g(x)dx + f(0).$$

Then f is differentiable by Proposition 5.3 and $f'(t) = g(t)$ almost everywhere.

(ii) $\Rightarrow$ (iii) is clear.

(iii) $\Rightarrow$ (i). Assume that E does not have the RNP, and find $\varepsilon > 0$, finite subalgebras $\mathcal{B}_n$ of $[0,1]$ consisting of intervals, and a $\mathcal{B}_n$-adapted martingale $\{f_n\}$ with values in the unit ball of E, such that $\|f_n(t) - f_{n-1}(t)\| \ge \varepsilon$ for all $n \ge 1$ and for all $t \in [0,1]$. Note that since the martingale is bounded, this implies in particular that the end points of the intervals in $\bigcup \mathcal{B}_n$ are dense in $[0,1]$.

The functions $g_n(t) = \int_0^t f_n(t)dt$ are 1-Lipschitz on $[0,1]$. By the martingale property $g_n(t) = g_k(t)$ for $n \ge k$ whenever t is an end point of an interval in $\mathcal{B}_k$. Hence $g(t) = \lim g_n(t)$ exists on a dense set of $[0,1]$, and since the g_n's are 1-Lipschitz the limit actually exists everywhere and is 1-Lipschitz.

We claim that g does not have a point of η-differentiability for any $\eta < \varepsilon/2$. Indeed, if t_0 were such a point, find $\delta > 0$ such that $\|g(t_0 + h) - g(t_0) - hx\| \le \eta|h|$ for $|h| \le \delta$. Let n be such that the atom $[a,b]$ of $\mathcal{B}_n$ which contains t_0 is contained in $[t_0 - \delta, t_0 + \delta]$, and let $[c,d]$ be the atom of $\mathcal{B}_{n+1}$ containing t_0. Then

$$\left\|\frac{g(b) - g(a)}{b - a} - x\right\| \le \eta, \quad \left\|\frac{g(d) - g(c)}{d - c} - x\right\| \le \eta.$$

On the other hand,

$$\left\|\frac{g(b) - g(a)}{b - a} - \frac{g(d) - g(c)}{d - c}\right\| = \|f_n(t_0) - f_{n+1}(t_0)\| \ge \varepsilon$$

and this is impossible if $2\eta < \varepsilon$. (Actually we should be more careful when t_0 happens to be one of the points a,b,c or d where f_n is not defined. For example, if $t_0 = a$, we should take the limit of $f_n(t)$ as $t \searrow t_0$ instead of $f_n(t_0)$.) $\square$

Proposition 5.22. *The pair $(\mathbb{R}, E)$ has the AAP if and only if the unit ball of E does not contain a generalized ε-dyadic tree for any $\varepsilon > 0$.*

Proof. Assume that the unit ball of E contains a generalized ε-dyadic tree for some $\varepsilon > 0$, and use this tree to produce a $\mathcal{B}_n$-adapted martingale $\{f_n\}$ on $[0,1]$ with values in the unit ball of E so that $\|f_n(t) - f_{n-1}(t)\| \geq \varepsilon$ for all n and for all t. Since we started with a generalized dyadic tree, each atom $[a,b]$ of $\mathcal{B}_n$ splits in the next stage to two atoms $[a,c] \cup [c,b]$. We also have an estimate on the relative size of $[a,c]$. Assume that $f_n = u$ on $[a,b]$, and $f_{n+1} = v$ (respectively w) on $[a,c]$ (respectively $[c,b]$). Then $u = \theta v + (1-\theta)w$ with $\theta = (c-a)/(b-a)$, and $\|u-v\| \geq \varepsilon$, $\|u - w\| \geq \varepsilon$, $\|v - w\| \leq 2$. It follows that $\varepsilon/2 \leq (c-a)/(b-a) \leq 1 - \varepsilon/2$.

Let g be the Lipschitz function from $[0,1]$ to E constructed from $\{f_n\}$ as in the proof of Theorem 5.21. Let I be an interval, and let $k(t) = x + ty$ be an affine function such that $\|g(t) - k(t)\| \leq \eta|I|$ on I. We shall show that η is bounded from below by some function of ε, and this will prove that E fails to have the AAP.

We claim that I contains an atom $[a,b]$ of some $\mathcal{B}_n$ such that $4(b-a)/\varepsilon \geq |I|$. Indeed, let a be the first partition point of some $\mathcal{B}_m$ that falls into I. Choose $n \geq m$ and an atom $J \in \mathcal{B}_{n-1}$ so that a is an end point of J, $|I \cap J| \geq |I|/2$, J is not contained in I, but so that when we split $J = J_1 \cup J_2$ with $J_i \in \mathcal{B}_n$, the interval $J_1 = [a,b]$ (say) is contained in I. By the estimate on the relative size of atoms $b - a = |J_1| \geq |J|\varepsilon/2 \geq \varepsilon|I|/4$.

Now let c, u, v, w be as above. It follows that $g(b) - g(a) = (b-a)u$ and that $\|g(b) - g(a) - (b-a)y\| \leq 2\eta|I|$, i.e.

$$\|u - y\| \leq 2\eta|I|(b-a)^{-1} \leq 8\eta/\varepsilon.$$

Similarly $g(c) - g(a) = (c-a)v$ and $\|g(c) - g(a) - (c-a)y\| \leq 2\eta|I|$. Hence

$$\|v - y\| \leq 2\eta|I|(c-a)^{-1} \leq 2\eta|I|(b-a)^{-1}2/\varepsilon \leq 16\eta/\varepsilon^2$$

and consequently $\varepsilon \leq \|u - v\| \leq 24\eta/\varepsilon^2$.

Assume conversely that E fails to have the AAP, and let $g : [0,1] \to E$ be a 1-Lipschitz function for which there is an $\eta > 0$ so that for every choice of an interval I and an affine function k on I there is a point $c \in I$ such that $|g(c) - k(c)| \geq \eta|I|$. In particular for every choice of $0 \leq a < b \leq 1$ there is a $c \in (a,b)$ such that

$$\left\| g(c) - \frac{c-a}{b-a}g(b) - \frac{b-c}{b-a}g(a) \right\| \geq \eta(b-a).$$

We define a generalized η-dyadic tree in the unit ball of E as follows. Take $x_{0,1} = g(1) - g(0)$. Choose c in $(0,1)$ so that $\|g(c) - cg(1) - (1-c)g(0)\| \geq \eta$, and put $x_{1,1} = \big(g(c) - g(0)\big)/c$ and $x_{1,2} = \big(g(1) - g(c)\big)/(1-c)$. We continue inductively to define algebras $\mathcal{B}_n$ and vectors $\{x_{n,k}\}$. Let $[a,b]$ be the atom corresponding to $x_{n,k}$, i.e., $x_{n,k} = \big(g(b) - g(a)\big)/(b-a)$. We then find c in (a,b) as above and define $x_{n+1,2k-1} = \big(g(c) - g(a)\big)/(c-a)$ and $x_{n+1,2k} = \big(g(b) - g(c)\big)/(b-c)$. One checks easily that this is a tree and that $\|x_{n,k} - x_{n+1,2k-1}\| \geq \eta(b-a)/(c-a) \geq \eta$ and similarly for $\|x_{n,k} - x_{n+1,2k}\|$. □

In view of these results it is natural to inquire about the existence of various types of bushes in spaces failing to have the RNP. The next theorem and corollary give a precise answer in the duals of separable spaces.

Theorem 5.23. *Let E be a separable space so that E^* is nonseparable. Then the unit ball of E^* contains an ε-dyadic tree for some $\varepsilon > 0$.*

Proof. Choose $\varepsilon > 0$ and an uncountable set $A \subset B(E^*)$ with $\|x^* - y^*\| > \varepsilon$ for all $x^* \neq y^*$ in A. Since $B(E^*)$ is a separable metric space in the w^*-topology, we can assume, by the Cantor-Bendixson theorem, that A has no w^*-isolated point. We construct inductively a sequence $\{G_n\}_{n=0}^{\infty}$ of nonempty relatively w^*-open subsets of A, vectors $x_n \in E$ with $\|x_n\| = 1$, and scalars λ_n so that

$$G_n \supset G_{2n+1} \cup G_{2n+2},$$
$$\langle x^*, x_n \rangle > \lambda_n \quad \text{for every } x^* \in G_{2n+1},$$
$$\langle x^*, x_n \rangle < \lambda_n - \varepsilon \quad \text{for every } x^* \in G_{2n+2}.$$

Indeed, pick any $u^* \neq v^*$ in G_n. Since $\|u^* - v^*\| > \varepsilon$ there are a $\rho > 0$ and an $x_n \in E$ with $\|x_n\| = 1$ such that $\langle u^* - v^*, x_n \rangle = \varepsilon + \rho$. Take $\lambda_n = \langle u^*, x_n \rangle - \rho/2$, and put $G_{2n+1} = \{x^* \in G_n : \langle x^*, x_n \rangle > \lambda_n\}$ and $G_{2n+2} = \{x^* \in G_n : \langle x^*, x_n \rangle < \lambda_n - \varepsilon\}$. (Note that $u^* \in G_{2n+1}$ and $v^* \in G_{2n+2}$; hence these sets are nonempty.)

Let C_n be the w^*-closure of $\text{co}(G_n)$. Then $\|x^* - y^*\| \geq \langle x^* - y^*, x_n \rangle \geq \varepsilon$ whenever $x^* \in C_{2n+1}$ and $y^* \in C_{2n+2}$, and $C_n \supset (C_{2n+1} + C_{2n+2})/2$.

We will now show that it is possible to choose points $x_n^* \in C_n$ so that

$$x_n^* = (x_{2n+1}^* + x_{2n+2}^*)/2 \tag{5.11}$$

for all n. With the obvious indexing these vectors will form an $\varepsilon/2$-dyadic tree.

Fix m and choose $x_n^*(m) \in C_n$ for $2^{m-1} \leq n \leq 2^m - 1$ arbitrarily. Using (5.11) backwards, this choice determines $x_n^*(m)$ for $n < 2^{m-1}$ uniquely. By a diagonal argument there is a sequence $\{m_j\}_{j=1}^{\infty}$ such that $x_n^* = \lim_{j\to\infty} x_n^*(m_j)$ exists (in the w^*-topology) for every n. These are the desired $\{x_n^*\}_{n=0}^{\infty}$. □

We shall see in the next section that a nonconjugate space may fail to have the RNP without even containing a generalized ε-dyadic tree for any $\varepsilon > 0$. The theorem and Corollary 5.12 give immediately

Corollary 5.24. *Let E be a separable Banach space. Then E^* has the RNP if and only if E^* is separable.*

4. Examples Related to the RNP

The results so far give a fairly good idea when does a Banach space have the RNP. Separable conjugate spaces have the RNP, and c_0 and $L_1(0,1)$ fail to have it. Thus spaces isomorphic to subspaces of separable conjugate spaces have the RNP, and spaces having a subspace isomorphic to c_0 or to $L_1(0,1)$ do not. In many cases these simple criteria suffice to decide whether a given separable space has the RNP. In this section we present examples which show that these criteria are not always sufficient. The first example is of a separable space with the RNP which is not isomorphic to a subspace of a separable conjugate space. The second, much more delicate example is of a Banach space which fails to have the RNP but does not contain subspaces isomorphic to either c_0 or to $L_1(0,1)$. The second example actually shows more, and this is of particular interest in view of Proposition 5.22 above: There is a separable Banach space E which fails to have the RNP and whose unit ball does not contain a generalized ε-dyadic tree for any $\varepsilon > 0$. (Comparing this example with Theorem 5.8 shows a surprising difference between the existence of trees and of general bushes.) Since both c_0 and $L_1(0,1)$ have 1-dyadic trees in their unit balls, E cannot contain isomorphic copies of either c_0 or $L_1(0,1)$.

Example 5.25. *There is a separable Banach space with the RNP which is not isomorphic to a subspace of a separable conjugate space.*

Let F be a Banach space, and assume that its unit ball contains a 1-dyadic tree $\{x_{n,j} : n \geq 0,\ 1 \leq j \leq 2^n\}$ (e.g., $F = c_0$ or $L_1(0,1)$). For each n let

$$F_n = \operatorname{span}\{x_{k,j} :\ k \leq n,\ 1 \leq j \leq 2^k\}.$$

Fix $0 < \delta < 1$, and let E_δ be the subspace of $(F \oplus \sum_{n=0}^{\infty} \oplus F_n)_1$ (the direct sum in the l_1 sense) consisting of vectors of the form

$$y = (x, \delta x_0, \delta x_1, \dots, \delta x_n, \dots), \quad x_n \in F_n,\ x = \sum_{n=0}^{\infty} x_n,\ \sum_{n=0}^{\infty} \|x_n\| < \infty.$$

The map $y \to (x_0, x_1, \dots, x_n, \dots)$ is an isomorphism of E_δ onto $(\sum_{n=0}^{\infty} \oplus F_n)_1$, and this latter space is a separable conjugate space (because each F_n is finite-dimensional). Thus E_δ has the RNP. We shall show that if T is an isomorphism from E_δ into a separable conjugate space, then necessarily $\|T\|\|T^{-1}\| \geq (4\delta)^{-1}$. Once this is proved it follows that $E = (\sum_{k=1}^{\infty} \oplus E_{1/k})_1$ does not embed isomorphically into a separable conjugate space. But every l_1-direct sum of spaces all having the RNP has the RNP. (This can be verified in various ways, e.g., by using Theorem 5.21.)

For every $m > n \geq 0$ and $1 \leq j \leq 2^n$ put

$$y_{n,j,m} = (x_{n,j}, 0, \dots, 0, \delta x_{n,j}, 0, \dots) \in E_\delta$$

where $\delta x_{n,j}$ appears in the mth coordinate, and $x_{n,j} \in F_m$ because $m > n$. Then $\|y_{n,j,m_1} - y_{n,j,m_2}\| \leq 2\delta$ and also $y_{n,j,m} = (y_{n+1,2j-1,m} + y_{n+1,2j,m})/2$ (provided $m > n+1$ so that the expression on the right is defined).

Assume first that E_δ is isometric to a subspace of a separable conjugate space W. By a standard diagonal argument we could find a sequence $\{m_k\}$ so that $w_{n,j} = w^* - \lim_{k\to\infty} y_{n,j,m_k}$ exists for every n and j. The set $\{w_{n,j}\}$ forms a dyadic tree in the ball of radius 2 in W, and $\|w_{n,j} - y_{n,j,m}\| \leq 2\delta$ for all $m > n$ and all j. It follows that the tree $\{w_{n,j}\}$ is $1 - 4\delta$ separated, which is impossible since W has the RNP. If we assume that E_δ is only isomorphic to a subspace of W by an isomorphism T, then the same argument would produce a dyadic tree $\{w_{n,j}\}$ in the ball of radius $2\|T\|$ in W which is $\|T^{-1}\|^{-1} - 4\delta\|T\|$ separated. Hence, necessarily, $\|T^{-1}\|^{-1} - 4\delta\|T\| \leq 0$, i.e., $4\delta\|T\|\|T^{-1}\| \geq 1$.

We pass to the second example. We shall use the L_1 norm (which we denote by $\|\cdot\|$) of measurable functions on a probability space as well as the distance

$$d(f,g) = \inf\left\{\varepsilon > 0 :\ m\{t : |f(t) - g(t)| \geq \varepsilon\} \leq \varepsilon\right\}$$

which induces the topology of convergence in measure.

The first lemma uses probabilistic tools.

Lemma 5.26. *Let $\varepsilon > 0$. Then there is a function $f \in L_1(0,1)$ such that*

(i) $f \geq 0$, $\|f\| = 1$, $\|f - 1\| \geq 2 - \varepsilon$.

(ii) *Let $\{f_j\}_{j=1}^{\infty}$ be a sequence of (stochastically) independent functions on $[0,1]$ with the same distribution as f. Then for every $g \in \overline{\operatorname{span}}\{f_j\}_{j=1}^{\infty}$ with $\|g\| \leq 1$ there is a constant function c such that $d(g,c) \leq \varepsilon$.*

(iii) $\|n^{-1}\sum_{j=1}^{n} f_j - 1\| \to 0$ *as $n \to \infty$.*

Proof. We refer to Appendix D and especially to Theorem D.8 for information on stable distributions.

Fix $1 < p \le 2$, and let h be a function on $[0,1]$ with a symmetric p-stable distribution and $\|h\| = 1$. Thus $\mathbb{E}h = \int_0^1 h(x)dx = 0$, and its characteristic function is $\mathbb{E}e^{ith} = \mathbb{E}\cos(th) = e^{-c_p|t|^p}$ with $c_p \to 0$ as $p \to 1$. If $\{h_j\}$ are independent copies of h, then $\|\sum \lambda_j h_j\| = (\sum |\lambda_j|^p)^{1/p}$.

By Theorem D.5(iv) if the characteristic functions $\varphi_{X_n}(t)$ of the random variables X_n converge pointwise to e^{ict}, the characteristic function of the constant c, then X_n converge in probability to c. Combined with the fact that the only random variables X whose characteristic function satisfies $|\varphi_X(t)| = 1$ for $|t| \le 1$ are the constants, we obtain that for every $\eta > 0$ there is a $\delta > 0$ such that if $\|g\| \le 1$ and if $|1 - \mathbb{E}e^{itg}| < \delta$ for all $|t| \le 1$, then $d(g, 0) < \eta$. More generally, if $\|g\| \le 1$ and if $\big|1 - |\mathbb{E}e^{itg}|\big| < \delta$ for all $|t| \le 1$, then there is a constant c such that $d(g, c) < \eta$.

We are now ready to prove the lemma.

Let $p > 1$ be sufficiently close to 1. (Depending on the given $\varepsilon > 0$. The precise requirements will be clarified below and at the end of the proof.) Let h be a p-stable function as above, and take $f(x) = |h(x)|$. Then $f \ge 0$ and $\|f\| = 1$. The third condition in (i) holds if p is close enough to 1. Indeed, we use $c_p \to 0$ to obtain that $|1 - \mathbb{E}e^{itg}| = 1 - e^{-c_p|t|^p}$ can be made as small as we wish for all $|t| \le 1$. By the remarks above we can choose p so that $m\{f > \varepsilon/4\} < \varepsilon/4$. Then direct computation, using also that $f \ge 0$ and $\|f\| = 1$, shows that $\|f - 1\| > 2 - \varepsilon$.

Assertion (iii) of the lemma is an immediate consequence of the strong law of large numbers. Note in particular that (iii) implies that $1 \in \overline{\operatorname{span}}\{f_j\}$.

We turn to (ii). Let $\{h_j\}$ be independent copies of h, and put $f_j = |h_j|$. Since $\{h_j\}$ and $\{f_j - 1\}$ are two sequences of independent random variables with mean 0, they are 2-unconditional in $L_1(0,1)$. By Khintchine's inequality in L_1 there is an absolute constant K (independent of p) so that the two inequalities

$$K^{-1}\Big\|\sum \lambda_j h_j\Big\| \le \int_0^1 \Big(\sum |\lambda_j h_j(x)|^2\Big)^{1/2} dx \le K\Big\|\sum \lambda_j h_j\Big\|,$$

$$K^{-1}\Big\|\sum \lambda_j (f_j - 1)\Big\| \le \int_0^1 \Big(\sum |\lambda_j (f_j(x) - 1)|^2\Big)^{1/2} dx \le K\Big\|\sum \lambda_j (f_j - 1)\Big\|$$

hold for all choices of scalars $\{\lambda_j\}$.

We shall show next that there is a constant M, independent of p, such that

$$\Big(\sum |\lambda_j|^p\Big)^{1/p} \le M\Big\|\sum \lambda_j (f_j - 1)\Big\|.$$

(Actually the two expressions are equivalent, but we shall need only this inequality.)

Since L_1 has cotype 2, there is a constant C such that for every choice of scalars $\|\sum \lambda_j (f_j - 1)\| \ge C^{-1}(\sum |\lambda_j|^2)^{1/2}$. Let χ_j be the indicator function of the set $\{x : |h_j(x)| \ge 2\}$. It follows that $|h_j|\chi_j \le 2|f_j - 1|$, and we obtain that

$$\begin{aligned}
K^{-1}\Big(\sum |\lambda_j|^p\Big)^{1/p} &= K^{-1}\Big\|\sum \lambda_j h_j\Big\| \le \int \Big(\sum |\lambda_j h_j|^2\Big)^{1/2} dx \\
&\le \int \Big(\sum |\lambda_j h_j|^2 \chi_j\Big)^{1/2} dx + \int \Big(\sum |\lambda_j h_j|^2 (1 - \chi_j)\Big)^{1/2} dx \\
&\le 2\int \Big(\sum |\lambda_j (f_j - 1)|^2\Big)^{1/2} dx + 2\Big(\sum |\lambda_j|^2\Big)^{1/2} \\
&\le (2K + 2C)\Big\|\sum \lambda_j (f_j - 1)\Big\|.
\end{aligned}$$

We are now ready to finish the proof of (ii). Assume that $g = \sum \lambda_j f_j$ satisfies $\|g\| \le 1$. Then $(\sum |\lambda_j|^p)^{1/p} \le 2M$ because $|\sum \lambda_j| = |\int g dx| \le 1$ and hence $\|\sum \lambda_j(f_j - 1)\| \le 2$. By the independence of the $\{f_j\}$ we obtain that

$$\begin{aligned} \left|\mathbb{E}e^{itg}\right| &= \prod \left|\mathbb{E}e^{it\lambda_j f_j}\right| = \prod \left|\mathbb{E}e^{it\lambda_j |h|}\right| \\ &\ge \prod \Re\left(\mathbb{E}e^{it\lambda_j|h|}\right) = \prod \left(\mathbb{E}\cos(t\lambda_j h)\right) \\ &= e^{-c_p|t|^p \sum |\lambda_j|^p} \ge e^{-c_p(2M)^p|t|^p}. \end{aligned}$$

Choosing p close to 1 (and hence c_p close to 0) ensures that $\left|1 - |\mathbb{E}e^{-itg}|\right|$ is as small as we wish for all $|t| \le 1$. By the remarks in the beginning of the proof it follows that there is a constant c such that $d(g, c) \le \varepsilon$ as claimed. □

Lemma 5.26 will be used in the proof of the next lemma to describe the typical step in the construction of an approximate ε-bush in L^1. We shall work with functions defined on the product space $\Omega = I^{\aleph_0}$ with the usual product measure μ. A subspace of $L_1(\Omega)$ is said to depend on finitely many coordinates if all the functions in the subspace depend only on a (common) finite set of coordinates.

Lemma 5.27. *Let G be a finite-dimensional subspace of $L_1(\Omega)$ which depends on finitely many coordinates. Let $\{u_k\}_{k=1}^m$ be a finite set of norm-one elements in G, and let $\varepsilon > 0$. Then there is a finite-dimensional subspace F of $L_1(\Omega)$ containing G which also depends on finitely many coordinates, and there are an integer n and functions $\{v_{k,j}\}_{k \le m, j \le n}$ in F such that*

(i) *$\|v_{k,j}\| = 1$ and $\|u_k - v_{k,j}\| \ge 2 - \varepsilon$ for all k and j.*

(ii) *$\|u_k - n^{-1}\sum_{j=1}^n v_{k,j}\| \le \varepsilon$ for all k.*

(iii) *For every $\varphi \in F$ with $\|\varphi\| = 1$ there is a $\psi \in G$ with $d(\varphi, \psi) \le \varepsilon$.*

Proof. Let f be the function given by Lemma 5.26 for the given ε, and let n be such that $\|1 - n^{-1}\sum_{j=1}^n f_j\| \le \varepsilon$ (by (iii) of the lemma).

Assume that G depends on the first N coordinates of Ω. For each $k \le m$ choose $\{f_{k,j}\}_{j \le n}$ which depend on the $(N+k)$th coordinate of Ω and having the same joint distribution as the functions $\{f_j\}_{j \le n}$. Put $v_{k,j} = f_{k,j}u_k$, and let F be the span of G and of $\{v_{k,j} : k \le m,\ j \le n\}$. With this choice (i) and (ii) clearly hold, and we verify (iii) (with ε replaced by $3m\sqrt{\varepsilon}$).

Let $\varphi = w + \sum_{k,j} a_{k,j}v_{k,j}$ with $w \in G$ be any element in F with $\|\varphi\| = 1$. Then

$$\varphi = \psi_0 + \sum_{k,j} a_{k,j}u_k(f_{k,j} - 1) = \psi_0 + \sum_k u_k g_k$$

where $\psi_0 = w + \sum_{k,j} a_{k,j}u_k \in G$ and $g_k = \sum_j a_{k,j}(f_{k,j} - 1)$. Let $\mathcal{B}_k$ be the σ-algebra in Ω generated by the Lebesgue σ-algebras in all the coordinates except for the $(N+k)$th. Then $\mathbb{E}_{\mathcal{B}_k} g = g$ for every $g \in G$ and $\mathbb{E}_{\mathcal{B}_k} g_l = g_l$ for $l \ne k$, while $\int g_k d\mu = 0$ implies that $\mathbb{E}_{\mathcal{B}_k} g_k = 0$. Hence $\mathbb{E}_{\mathcal{B}_k}\varphi = \psi_0 + \sum_{l \ne k} u_l g_l = \varphi - u_k g_k$ and

$$\|g_k\| = \|u_k g_k\| \le \|\varphi\| + \|\mathbb{E}_{\mathcal{B}_k}\varphi\| \le 2.$$

By Lemma 5.26(ii) (using the fact that $1 \in \overline{\text{span}}\{f_j\}$) there are constants c_k with $\mu\{t : |g_k(t) - c_k| \ge 2\varepsilon\} \le \varepsilon$. Then $d(g_k u_k, c_k u_k) \le 3\sqrt{\varepsilon}$ for every k because

$$\begin{aligned} \mu\{t : |g_k u_k - c_k u_k| \ge \sqrt{\varepsilon}\} &\le \mu\{t : |u_k| \ge 1/\sqrt{4\varepsilon}\} + \mu\{t : |g_k - c| \ge 2\varepsilon\} \\ &\le 2\sqrt{\varepsilon} + \varepsilon \le 3\sqrt{\varepsilon} \end{aligned}$$

and hence $d(\varphi, \psi_0 + \sum_{k=1}^m c_k u_k) \le 3m\sqrt{\varepsilon}$. □

The next proposition gives a procedure for constructing a class of subspaces of $L_1(0,1)$ which fail to have the RNP.

Proposition 5.28. *Assign a number $\delta(F,\varepsilon) > 0$ to every finite-dimensional subspace F of $L_1(0,1)$ and to every $\varepsilon > 0$. Then for every sequence $\{\varepsilon_n\}_{n=1}^{\infty}$ of positive numbers there is an increasing sequence $F_1 \subset F_2 \subset \dots$ of finite-dimensional subspaces of $L_1(0,1)$ such that $E = \overline{\bigcup F_n}$ satisfies*

(i) *E does not have the RNP.*

(ii) *For every $f \in E$ with $\|f\| \le 1$ and for every integer n there is a $g \in F_n$ with $\|g\| \le 1$ and $d(f,g) \le \delta(F_n, \varepsilon_n)$.*

Proof. We note first that we may add the conclusion that $\|\psi\| \le 1$ in part (iii) of Lemma 5.27. Indeed, the functions in the unit ball of G (as in any finite-dimensional space) are equi-integrable. Thus, given $\varepsilon > 0$, there is a $\delta > 0$ (which we can assume to be smaller than ε) so that whenever $\psi \in G$ and $\mu(A) \le \delta$, then $\int_A |\psi| d\mu \le \varepsilon\|\psi\|$. Hence if $\varphi \in F$ with $\|\varphi\| = 1$ and if $d(\varphi,\psi) \le \delta$ with $\psi \in G$, then, putting $A = \{w : |\varphi - \psi| \ge \delta\}$,

$$\|\psi\| = \int_A |\psi| d\mu + \int_{\Omega\setminus A} |\psi| d\mu \le \varepsilon\|\psi\| + 1 + \delta$$

or $\|\psi\| \le (1+\varepsilon)/(1-\varepsilon)$. The function $\psi_1 = (1-\varepsilon)\psi/(1+\varepsilon) \in G$ then satisfies $\|\psi_1\| \le 1$, and it is also close in the d metric to the given φ.

For the proof of the proposition we shall replace $I = [0,1]$ by the isomorphic measure space $\Omega = I^{\aleph_0}$.

Let $F_1 = \operatorname{span}\{1\}$, and pick the element $u \equiv 1$ in F_1. Let $\eta_1 = \frac{1}{2}\delta(F_1, \varepsilon_1)$, and use Lemma 5.27 (with $\varepsilon = \eta_1$) to find $F_2 \supset F_1$ and functions $\{v_{k,j}\}$ (actually $k = 1$ here). Pick $\eta_2 = 2^{-2}\min\big(\delta(F_2,\varepsilon_2), \delta(F_1,\varepsilon_1)\big)$, and use Lemma 5.27 with the $v_{k,j}$'s obtained in the first step as the finite given family of norm-one functions in F_2. This gives an $F_3 \supset F_2$. Continue inductively and put $E = \overline{UF_n}$. By the construction there is an E-valued approximate martingale (in the sense of Lemma 5.10) which is not L_1-convergent. Hence, by the same lemma, there is also a nonconvergent martingale. Thus E fails to have the RNP. Also, by Lemma 5.26(iii) and the remark in the beginning of this proof, if $\varphi \in F_l$ with $l > n$ and if $\|\varphi\| = 1$, then there is a $\psi \in F_n$ with $\|\psi\| \le 1$ and $d(\varphi,\psi) \le \eta_n + \eta_{n+1} + \dots + \eta_{l-1} \le \delta(F_n, \varepsilon_n)$. The result follows because l was arbitrary. □

Remark. Note that the space E constructed in Proposition 5.28 has the property that its unit ball is compact in the topology of convergence in measure (provided, e.g., that $\delta(F,\varepsilon) \le \varepsilon$ and that $\varepsilon_n \downarrow 0$).

The next lemma is a consequence of the fact that a bounded martingale with values in a finite-dimensional space F converges. It says (with a quantitative estimate) that if a martingale into a Banach space which contains F does not converge, then its values must eventually have relatively large distance from F. This estimate will be used later in Example 5.30 for a choice of $\delta(F,\varepsilon)$ in the application of Proposition 5.28. Put $\rho(x,F) = \min\{\|x - y\| : y \in F\}$. The norm in $L_p(\mu, E)$ is denoted by $\|\cdot\|_p$.

Lemma 5.29. *Given $\varepsilon > 0$ and an integer m, put $N(m,\varepsilon) = 36m\varepsilon^{-2} + 2$. Let F be an m-dimensional subspace of a Banach space E, and let $\{f_n\}$ be a $\mathcal{B}_n$-adapted* *martingale defined on some probability space* $(\Omega, \mathcal{B}, \mu)$ *with values in the unit ball of* E *so that* $\|f_{n+1} - f_n\|_1 \geq \varepsilon$ *for all* n. *Then*

$$\int \rho\big(f_n(\omega), F\big) d\mu(\omega) > \varepsilon/3$$

for all $n \geq N(m, \varepsilon)$.

Proof. Note first that for every martingale $\{g_n\}$ with values in a Hilbert space H, the functions $g_{n+1} - g_n$ and g_n are orthogonal in $L_2(\mu, H)$; hence

$$\|g_{n+1} - g_n\|_2^2 = \|g_{n+1}\|_2^2 - \|g_n\|_2^2.$$

It follows that if $\|g_n\|_\infty \leq 1$ for all n, then

$$\sum \|g_{n+1} - g_n\|_1^2 \leq \sum \|g_{n+1} - g_n\|_2^2 \leq 1.$$

Since the Banach-Mazur distance from an m-dimensional Banach space F to l_2^m is at most $m^{1/2}$, any martingale $\{g_n\}$ with values in the unit ball of such F satisfies

$$\sum \|g_{n+1} - g_n\|_1^2 \leq m.$$

Assume now that $\{f_n\}$ are as in the statement of the lemma and that there is an n_0 such that $\int \rho\big(f_{n_0}(\omega), F\big) d\mu(\omega) \leq \varepsilon/3$. For each $\omega \in \Omega$, choose $g_{n_0}(\omega) \in F$ such that $\|f_{n_0}(\omega) - g_{n_0}(\omega)\| = \rho\big(f_{n_0}(\omega), F\big)$. Then $\|g_{n_0}(\omega)\| \leq 2$ for every ω. We omit the easy proof that g_{n_0} can be chosen to be measurable. (We shall need only the case where f_{n_0} takes a finite number of values, and in this case this is obvious.) Define a finite martingale g_n for $n < n_0$ by $g_n = \mathbb{E}_{\mathcal{B}_n} g_{n_0}$. Then $\|f_n - g_n\|_1 \leq \varepsilon/3$ for all $n < n_0$, and thus

$$\|g_{n+1} - g_n\|_1 \geq \|f_{n+1} - f_n\|_1 - \|f_{n+1} - g_{n+1}\|_1 - \|f_n - g_n\|_1 \geq \varepsilon/3.$$

Moreover, $\|g_n\|_\infty \leq \|g_{n_0}\|_\infty \leq 2$. Applying the observation above to the (finite) martingale $\{g_n\}_{n=1}^{n_0}$, we get

$$(n_0 - 1) \cdot (\varepsilon/3)^2 \leq \sum_{n=1}^{n_0 - 1} \|g_{n+1} - g_n\|_1^2 \leq 4m$$

i.e., $n_0 < 36m\varepsilon^{-2} + 2 = N(m, \varepsilon)$. □

We are now ready to present our second example.

Example 5.30. *There is a Banach space* E *which does not have the RNP but whose unit ball does not contain a generalized* ε*-dyadic tree for any* $\varepsilon > 0$.

For every finite-dimensional subspace F of $L_1(0,1)$ and for every $\varepsilon > 0$ put

$$\delta(F, \varepsilon) = 2^{-N(m,\varepsilon)} \gamma(F, \varepsilon)$$

where $m = \dim F$, the value $N(m, \varepsilon)$ is given by Lemma 5.29, and $\gamma(F, \varepsilon) \leq \varepsilon$ is such that $\int_A |f| dx \leq \varepsilon \|f\|$ whenever $\mu(A) \leq \gamma(F, \varepsilon)$ and $f \in F$.

The space E constructed in Proposition 5.28 with this definition of $\delta(F, \varepsilon)$ and with $\varepsilon_n \to 0$ has the desired properties.

Let F_n be the finite-dimensional spaces produced by Proposition 5.28, and assume that the unit ball of $E = \overline{\cup F_n}$ contains a generalized ε-dyadic tree. Then there is a sequence $\mathcal{B}_n$ of σ-algebras of $[0, 1]$, with each $\mathcal{B}_n$ having 2^n atoms, and a

$\mathcal{B}_n$-adapted martingale $\{f_n\}$ such that $\|f_n\|_\infty \le 1$ and $\|f_{n+1} - f_n\|_1 \ge \varepsilon$ for every n. Let $\sigma = \sup_n \|f_n\|_1$. Choose an integer m and then an integer $n > m$ so that

$$(5.12) \qquad \|f_m\|_1 \ge \sigma - \varepsilon/30, \quad \varepsilon_n \le \varepsilon/30, \text{ and } \rho\big(f_m(t), F_n\big) \le \varepsilon/30 \text{ for all } t,$$

and let $r = \dim F_n$.

By Lemma 5.29, $\int \rho(f_s(t), F_n)dt \ge \varepsilon/3$ for $s = N(r, \varepsilon)$. Fix an atom of $\mathcal{B}_s$, and apply Proposition 5.28(ii) with $f = f_s(t)$ (which is constant on the atom). We thus obtain a $\mathcal{B}_s$ measurable function $g_s(t)$ with values in F_n such that $\|g_s(t)\| \le 1$ and $d\big(g_s(t), f_s(t)\big) \le \delta(F_n, \varepsilon_n)$ for all t. Put $g_m = \mathbb{E}_{\mathcal{B}_m} g_s$. The value $g_m(t) - f_m(t)$ is the average of the 2^{s-m} values of $g_s - f_s$ on the $\mathcal{B}_m$-atom containing t. Using the estimate $\mu\{|h| \ge \delta\} \le k \max \mu\{|h_i| \ge \delta\}$ whenever h is the average of the k functions $\{h_i\}$, we see that for every $0 \le t \le 1$

$$d\big(g_m(t), f_m(t)\big) \le 2^{s-m}\delta(F_n, \varepsilon_n) \le 2^{N(r,\varepsilon)}\delta(F_n, \varepsilon_n) \le \gamma(F_n, \varepsilon_n).$$

By (5.12) there is a $\mathcal{B}_m$ measurable function φ with values in F_n which satisfies $\|f_m(t) - \varphi(t)\| \le \varepsilon/30$ for every t. Thus $\|\varphi(t)\| \le \|f_m(t)\| + \varepsilon/30 \le 1 + \varepsilon/30$.

Fix $0 \le t \le 1$, and put $A(t) = \big\{x : |g_m(t)(x) - f_m(t)(x)| \ge \gamma(F_n, \varepsilon_n)\big\}$. It follows that $\mu\big(A(t)\big) \le \gamma(F_n, \varepsilon_n)$; hence $\int_{A(t)} |\varphi(t)(x)|dx \le \varepsilon_n \|\varphi(t)\|$ by the definition of $\gamma(F_n, \varepsilon_n)$. Recalling that $\gamma(F, \varepsilon) \le \varepsilon$ and combining the estimates, we obtain

$$\begin{aligned}
\|g_m(t)\| &\ge \int_{[0,1]\setminus A(t)} \big|f_m(t)(x)\big|dx - \gamma(F_n, \varepsilon_n) \\
&\ge \|f_m(t)\| - \int_{A(t)} \big|f_m(t)(x)\big|dx - \varepsilon_n \\
&\ge \|f_m(t)\| - \int_{A(t)} \big|\varphi(t)(x)\big|dx - \varepsilon/30 - \varepsilon_n \\
&\ge \|f_m(t)\| - \varepsilon_n\|\varphi(t)\| - \varepsilon/30 - \varepsilon_n \\
&\ge \|f_m(t)\| - \varepsilon_n(1 + \varepsilon/30) - \varepsilon/30 - \varepsilon_n.
\end{aligned}$$

Integrating over $0 \le t \le 1$ we get

$$\|g_m\|_1 \ge \|f_m\|_1 - 3\varepsilon_n - \varepsilon/30 \ge \sigma - \varepsilon/30 - 3\varepsilon_n - \varepsilon/30 \ge \sigma - \varepsilon/6.$$

For each $t \in [0,1]$ put $B(t) = \big\{x : |f_s(t)(x) - g_s(t)(x)| \ge \delta(F_n, \varepsilon_n)\big\}$. Since $\mu\big(B(t)\big) \le \delta(F_n, \varepsilon_n) \le \gamma(F_n, \varepsilon_n) \le \varepsilon_n$ and $\|g_s(t)\| \le 1$ for all t, we obtain that

$$\begin{aligned}
&\int_{B(t)} \big|f_s(t)(x)\big|dx \\
&\qquad \ge \|f_s(t) - g_s(t)\| - \int_{[0,1]\setminus B(t)} \big|f_s(t)(x) - g_s(t)(x)\big|dx - \int_{B(t)} \big|g_s(t)(x)\big|dx \\
&\qquad \ge \rho\big(f_s(t), F_n\big) - \delta(F_n, \varepsilon_n) - \varepsilon_n \ge \rho\big(f_s(t), F_n\big) - 2\varepsilon_n
\end{aligned}$$

and also

$$\begin{aligned}
\int_{[0,1]\setminus B(t)} \big|f_s(t)(x)\big|dx &\ge \int_0^1 \big|g_s(t)(x)\big|dx - \int_{B(t)} \big|g_s(t)(x)\big|dx - \delta(F_n, \varepsilon_n) \\
&\ge \|g_s(t)\| - \varepsilon_n - \delta(F_n, \varepsilon_n) \ge \|g_s(t)\| - 2\varepsilon_n.
\end{aligned}$$

Combining these two estimates we get

$$\int_0^1 |f_s(t)(x)|dx \geq \|g_s(t)\| + \rho(f_s(t), F_n) - 4\varepsilon/30.$$

Integrating over t we finally obtain

$$\begin{aligned} \|f_s\|_1 &\geq \|g_s\|_1 + \varepsilon/3 - 4\varepsilon/30 \\ &\geq \|g_m\|_1 + \varepsilon/3 - 4\varepsilon/30 \\ &\geq \sigma - \varepsilon/6 + \varepsilon/3 - 4\varepsilon/30 > \sigma \end{aligned}$$

and this contradicts the definition of σ.

Remark. The same proof also shows that for every given sequence of integers $\{k_n\}_{n=1}^{\infty}$ it is possible to construct a subspace E of L_1 failing the RNP and such that for any $\varepsilon > 0$ the unit ball of E does not contain an ε-bush with $m_n \leq k_n$ for all n (where m_n are the numbers appearing in Definition 5.20).

5. Notes and Remarks

The material in Section 1 is standard. Measures of finite variation suffice for our purposes. In other contexts (e.g., in spectral theory) it is essential to consider general vector measures. If τ is an E-valued vector measure on $(\Omega, \mathcal{B})$ and $x^* \in E^*$, then $\langle x^*, \tau\rangle$ is a finite scalar measure on $(\Omega, \mathcal{B})$. As such it has a finite variation $|\langle x^*, \tau\rangle|(\Omega)$. In particular $\sup\{|\langle x^*, \tau(A)\rangle| : A \in \mathcal{B}\}$ is finite for every x^*. By the principle of uniform boundedness $\sup\{\|\tau(A)\| : A \in \mathcal{B}\}$ is finite. The quantity $\sup\{|\langle x^*, \tau\rangle|(A) : \|x^*\| \leq 1\}$ is called the semivariation of τ on A, and it is easy to check that it is bounded below and above by $\sup\{\|\tau(B)\| : B \subset A\}$ and $4\sup\{\|\tau(B)\| : B \subset A\}$ respectively. The semivariation of τ on Ω is denoted by $\|\tau\|$. With this norm the space of E-valued measures on $(\Omega, \mathcal{B})$ becomes a Banach space. Some of the remarks we made concerning measures of finite variation extend to the setting of general vector measures. These extensions are less trivial but are still quite simple. Their proof usually involves some form of the uniform boundedness principle. We mention only two such results. The first, due to Pettis [**457**], states that if τ is a vector measure and if μ is a nonnegative measure on $(\Omega, \mathcal{B})$, and if $\mu(A) = 0$ implies that $\tau(A) = 0$, then for every $\varepsilon > 0$ there is a $\delta > 0$ such that $\mu(A) < \delta$ implies that $\|\tau(A)\| < \varepsilon$. The second result, due to Bartle, Dunford and Schwartz [**37**], says that if τ is an E-valued measure, then $\{\tau(A) : A \in \mathcal{B}\}$ is relatively weakly compact, and there is a nonnegative finite measure μ on $(\Omega, \mathcal{B})$ which is equivalent to the semivariation of τ in the sense that $\mu(A_n) \to 0$ implies that the semivariation of τ on A_n also tends to 0 and, on the other hand, that $\mu(A) \leq \sup\{|\langle x^*, \tau\rangle|(A) : \|x^*\| \leq 1\}$.

We consider here only σ-additive measures on σ-algebras. There are interesting results also on finitely additive vector measures, or on σ-additive measures on algebras, but we do not use them in this book.

Proposition 5.1 is due to Pettis ([**456**]). The Bochner integral was introduced by Bochner in [**69**]. There are several other notions of integral in the literature which are of interest in several contexts. For example, a function $f : \Omega \to E$ is called Pettis integrable (and $x \in E$ is its Pettis integral) if for every $x^* \in E^*$ the function

$\langle x^*, f\rangle$ belongs to $L_1(\Omega, \mu)$ and

$$\langle x^*, x\rangle = \int \langle x^*, f(w)\rangle d\mu(w) \qquad x^* \in E^*.$$

A detailed discussion of vector measures and of integration of vector-valued functions and many additional references can be found in chapters I and II of the book [**146**] by Diestel and Uhl (see also their survey article [**147**]) and in Dunford and Schwartz [**159**], Chapter III and Section IV.10.

The theory of RNP was first developed in the context of the study of Banach spaces with the RNP and was later localized to convex sets with the RNP. Some of these generalizations were routine, but others were not obvious.

The implication (i) $\Rightarrow$ (ii) of Theorem 5.8 is due to Rieffel [**497**]. The main step in the implication (ii) $\Rightarrow$ (i) is due to Maynard [**404**], who used instead of the notion of dentability a certain variant of it. The proof of (ii) $\Rightarrow$ (i) was completed independently by Davis and Phelps [**134**] and by Huff [**257**]. The equivalence of (ii) and (iii) is due to Chatterji [**118**]. Our presentation of the proof of Theorem 5.8 follows that of Bourgain [**83**].

Because of the many equivalent forms of the Radon-Nikodým Property (forms that where not known to be equivalent when the original work was done), and in view of the difference between the unit ball and general sets, it is not easy to pinpoint to whom Theorem 5.11 and Corollary 5.12 are due. Gelfand [**199**] proved Corollary 5.12 with the RNP defined in terms of differentiability of Lipschitz functions (Theorem 5.21). Independently, Dunford and Pettis [**158**] proved the corollary with the RNP defined as in Definition 5.4. Theorem 5.11(i) is due to Lindenstrauss [**354**], where the stronger Corollary 5.18 (in the separable case) is also proved; the nonseparable case of the corollary is due to Troyanski [**559**]. The proof of Theorem 5.11(i) that we presented here is due to Bourgain [**82**]. Theorem 5.11(ii) seems to be due to Bourgin (see [**91**], p. 71]), and we reproduced his proof.

Proposition 5.13 is due to Lindenstrauss. It was first proved in Lindenstrauss [**358**] for closed convex bounded sets in l_1. Bessaga and Pełczyński [**61**] generalized this to closed convex bounded sets in a separable dual space. The general result was first published in Phelps [**459**].

A closed bounded and convex set C is said to have the Krein-Milman property (KMP) if every closed convex subset of C is the closed convex hull of its extreme points. A Banach space is said to have the KMP if its unit ball does. In this terminology Proposition 5.13 states that the RNP implies the KMP. As mentioned in the text, it is unknown whether, conversely, the KMP implies the RNP. Here are some of the many partial results in this direction. Huff and Morris [**258**] proved that dual Banach spaces with the KMP have the RNP. Bourgain and Talagrand [**90**] proved that a Banach lattice with the KMP has the RNP. Schachermayer [**511**] proved that if $E \oplus E$ is isomorphic to E and if E has the KMP, then E has the RNP. The strongest result concerning convex sets C for which the KMP implies the RNP is also due to Schachermayer [**512**]. The two memoirs by Ghoussoub and Maurey [**206**] and by Ghoussoub, Godefroy, Maurey and Schachermayer [**204**] contain several results on related questions.

Theorem 5.16 is due to Stegall [**541**]. The proof here is due to Bourgain [**83**]. Stegall actually proved a stronger result: Let C be a closed bounded and convex set with the RNP in a Banach space E, and let f be a continuous bounded function from C to a Banach space F. Then for every $\varepsilon > 0$ there is a rank-one linear operator $T : E \to F$ with $\|T\| \leq \varepsilon$ such that $f + T$ attains its maximal norm on C.

Theorem 5.17 is due to Phelps [**459**] in the setting of RNP spaces. The localized version, i.e., for sets, was proved first by Bourgain. Kunen and Rosenthal [**341**] gave a martingale proof of the theorem.

A strengthening of Proposition 5.13 (in the separable case) in a different direction from that of Theorem 5.17 was given by Edgar [**164**]. He generalized Choquet's representation theorem to the RNP setting as follows: Let C be a bounded closed convex and separable set with the RNP, and let $x_0 \in C$. Then there is a probability measure μ on the (universally measurable) set $\text{ext}(C)$ such that $x_0 = \int x d\mu(x)$. The uniqueness of μ (for every $x_0 \in C$) is equivalent to C being a simplex. (As in the compact case, a set $C \subset E$ is called a simplex if the cone $\{(\lambda, \lambda x) : \lambda \geq 0,\ x \in C\}$ induces a lattice ordering on the subspace it spans in $R \oplus E$.) The uniqueness theorem is due to Bourgin and Edgar. (See the book [**91**] by Bourgin for a detailed discussion of the generalization of Choquet's theorem and for many additional references.)

We now mention a few results in the context of the RNP for spaces.

A Banach space E has the RNP if and only if $L_p([0,1], E)^*$ is naturally isometric to $L_q([0,1], E^*)$ for some (or for all) $1 \leq p < \infty$, where q is the conjugate exponent to p. The space $L_q([0,1], E^*)$ always embeds isometrically in the obvious way into $L_p([0,1], E)^*$, and the question is whether this embedding is onto. If E has the RNP, then an argument similar to the usual proof of the duality of L_p and L_q shows that the embedding is indeed onto. This result is due (in another equivalent formulation) to Bochner and Taylor [**70**]. A detailed discussion of this result can be found in the book [**146**] by Diestel and Uhl.

The following two quantitative results are due to Schachermayer, Sersouri and Werner [**513**]. They say that if a Banach space fails to have the RNP, then it does so in the strongest possible quantitative way.

(i) If E fails to have the RNP, then for every $\varepsilon > 0$ there is a separable closed convex subset C of E with $\text{diam}(C) = 1$ and $\text{diam}(S) > 1 - \varepsilon$ for any slice S of C.

(ii) If a separable space E fails to have the RNP, then for every $\varepsilon > 0$ there is a symmetric closed convex set C in the unit ball of E so that $d(x^{**}, E) \geq 1 - \varepsilon$ for every extreme point x^{**} of $\tilde{C}$, the w^*-closure of C in E^{**}.

The equivalence of (i) and (ii) in Theorem 5.21 is essentially obvious in view of the classical theory of absolutely continuous functions. That (iii) (with differentiability instead of just ε-differentiability) is equivalent to (i) was a "folklore" result. The first published proof seems to be that of Qian [**481**]. The proof of Qian also establishes (iii) $\Rightarrow$ (i) as stated in Theorem 5.21 here.

Proposition 5.22 is taken from Bates, Johnson, Lindenstrauss, Preiss and Schechtman [**42**]. Theorem 5.23 is due to Stegall [**540**]. The proof we gave is based on Stegall's idea, but it underwent several simplifications over the years (see e.g. Bourgain [**83**]).

There is a neat generalization of Corollary 5.24 to nonseparable spaces. A Banach space is an Asplund space if and only if E^* has the RNP. As mentioned in the Notes and Remarks section of Chapter 4, E is Asplund if and only if every separable subspace of E has a separable dual. A slight modification of the proof of Theorem 5.23 shows that if E is not Asplund, then the unit ball of E^* contains an ε-dyadic tree for some $\varepsilon > 0$. Conversely, Namioka and Phelps [**434**] proved that if E is Asplund, then every closed bounded and convex subset of E^* has slices (which are even determined by elements of E) of arbitrarily small diameter. This is a stronger version of Corollary 5.12(i) even for separable E. For details and for further references see Phelps [**461**].

The first examples of separable Banach spaces with the RNP which fail to be isomorphic to subspaces of separable conjugate spaces were found independently and about the same time by McCartney and O'Brien [**409**] and by Bourgain and Delbaen [**88**]. Example 5.25 that we presented here is that of McCartney and O'Brien. Bourgain and Delbaen constructed a separable $\mathcal{L}_\infty$ space E with the RNP. That E fails to be a subspace of a separable conjugate is an immediate consequence of the fact that every operator from l_∞ into a separable space is weakly compact (compare with the Notes and Remarks section in Chapter 1). Indeed, the identity map from E to itself can be extended to a bounded linear operator from $l_\infty \supset E$ into any conjugate space containing E (by the definition of an $\mathcal{L}_\infty$ space and by the w^*-compactness of the unit ball in a conjugate space.) The example of Bourgain and Delbaen has several additional remarkable properties. In particular, it was the first example of an $\mathcal{L}_\infty$ space that does not contain a subspace isomorphic to c_0.

Johnson and Lindenstrauss [**291**] have used the McCartney-O'Brien approach to construct separable $\mathcal{L}_1$ spaces with the RNP which fail to be isomorphic to subspaces of separable conjugate spaces.

Talagrand [**553**] proved that such examples cannot be Banach lattices. A separable Banach lattice has the RNP (if and) only if it is isometric to a dual space (see also Ghoussoub and Maurey [**206**], page 75).

The first examples of separable spaces which fail to have the RNP but do not contain subspaces isomorphic to c_0 and $L_1(0,1)$ were obtained independently and about the same time by Talagrand [**552**] (who even constructed such a Banach lattice) and by Bourgain and Rosenthal [**89**]. We presented here the example of Bourgain and Rosenthal (and followed the presentation in Bourgain [**83**]). Proposition 5.28 and Example 5.30 (as well as all the lemmas in Section 4) are due to them. It should be noted, however, that Lemma 5.26 is very close to some earlier work of Roberts, who constructed examples of convex compact sets which fail to have extreme points in the nonlocally convex spaces L_p, $0 < p < 1$. The example of Bourgain and Rosenthal has, as shown in their paper, several additional interesting properties (e.g., it has the so-called strong Schur property).

In [**554**] Talagrand combines the features of his earlier example with that of Bourgain and Rosenthal and constructs a separable Banach lattice which does not contain any ε-dyadic tree and which does not have the RNP.

The following interesting structure theorem for general Banach spaces with the RNP was proved by Ghoussoub and Maurey [**205**]: Every infinite-dimensional Banach space with the RNP contains an infinite-dimensional separable subspace which is isometric to a dual space. A closely related result was proved independently by Fonf [**195**].

Further results, examples, and references on the RNP can be found in the books by Diestel and Uhl [**146**], Bourgin [**91**], and Phelps [**461**].

CHAPTER 6

Negligible Sets and Gâteaux Differentiability

1. Haar Null Sets

One of the purposes of this chapter is to generalize Lebesgue's differentiability theorem to infinite-dimensional Banach spaces. More precisely, we shall prove that every Lipschitz function f from a separable Banach space E into $\mathbb{R}$ (or more generally into a Banach space having the Radon-Nikodým Property) is Gâteaux differentiable almost everywhere. To even formulate such a result we first need to explain what "almost everywhere" means. When E is infinite-dimensional it is impossible to find one measure μ such that the class of sets with μ-measure zero is a useful class of negligible sets, and we need to define classes of negligible sets by other means. Since Banach spaces are complete metric spaces, a natural class of negligible sets in E is that of sets of first category. However, as is apparent already on the real line, this notion is not suitable in the context of differentiability theorems for Lipschitz functions. There are several nonequivalent ways to define appropriate notions of negligible sets. These notions are of interest in themselves, and we present here a rather detailed study of them.

In this section we define Haar null sets and study their properties. We first study this notion in the more general context of separable complete metric Abelian groups. In the second part of this section we turn to results on Haar null sets which are specific to Banach spaces. In particular, we show that the class of Haar null sets is not preserved under Lipschitz equivalences between Banach spaces.

Section 2 presents the basic properties of Gaussian measures on Hilbert and Banach spaces. These are used in Section 3 to introduce Gauss null sets, i.e., the class of Borel sets on which all nondegenerate Gaussian measures vanish. We also introduce the class of Aronszajn null sets, defined by direct requirements on their structure. One of the main results of this section is that the two notions coincide. The class of Gauss null sets is properly contained in the class of Haar null sets. It will be proved that the class of Gauss null sets is also not preserved under bi-Lipschitz maps between infinite-dimensional Banach spaces.

The real line can be decomposed as the union of a set of measure 0 and a set of first category. We also show in Section 3 that an analogous decomposition, in a form which is even more striking than the possible decompositions in finite-dimensional spaces, is possible in infinite-dimensional Banach spaces.

In Section 4 we prove the differentiability theorem stated above and some related results. The proof of the differentiability theorem is quite simple and uses only the definitions and immediate properties of either Haar or Aronszajn null sets. It does not require the more delicate results proved in the first sections.

Section 5 is mainly devoted to examples that show that the results of Section 4 do not usually hold when Gâteaux differentiability is replaced by Fréchet differentiability. In the Notes and Remarks section we define and discuss briefly another notion of negligible sets which is relevant to Gâteaux differentiability.

We conclude this chapter with a section which gives an overview and summarizes the various classes of negligible sets which are defined and discussed in Chapter 4 and in the present chapter.

We consider in this chapter only functions and measures defined on separable Abelian groups and separable Banach spaces. Nonseparability introduces many new issues (e.g., measurability questions) of a different nature.

Let G be an Abelian group having a translation invariant metric d, with respect to which G is complete and separable. We denote the unit element in G by e. If G is locally compact, then, as is well known, there is a unique (up to multiplication by a scalar) translation invariant measure μ on G, called the Haar measure. Although no invariant (or even "quasi-invariant" as defined below) measure exists if G fails to be locally compact, it is still possible to define the notion of Haar null sets without the assumption of local compactness. We refer to Appendix D for basic facts on measures on complete separable metric spaces. We shall consider only Borel measures even when "Borel" is omitted.

Definition 6.1. *Let G be as above. A Borel set A is called a Haar null set if there is a Borel probability measure μ on G such that $\mu(x+A) = 0$ for every $x \in G$.*

The next proposition shows that this is, indeed, a generalization of the usual notion of sets of Haar measure 0.

Proposition 6.2. *Let G be locally compact. Then a Borel set A in G is Haar null if and only if its Haar measure is 0.*

Proof. Let μ_0 be the Haar measure on G. If $\mu_0(A) = 0$, then $\mu_0(A + x) = 0$ for all $x \in G$ and thus also $\mu(A + x) = 0$ for all x whenever μ is a probability measure which is absolutely continuous with respect to μ_0. Conversely, assume that $\mu(A + x) = 0$ for all x and for some probability measure μ. By Fubini's theorem and the translation invariance of μ_0

$$0 = \int\int \chi_A(x+y)d\mu(x)d\mu_0(y) = \mu_0(A).$$

□

Proposition 6.3. *A countable union of Haar null sets is a Haar null set.*

Proof. We show first that the union of two Haar null sets is Haar null. Recall that the convolution $\mu * \nu$ of two Borel probability measures μ and ν on G is defined by the relation

$$\int_G \int_G f(x+y)d\mu(x)d\nu(y) = \int_G f(x)d(\mu * \nu)(x) \tag{6.1}$$

for every bounded continuous real-valued function f on G, or alternatively, by

$$\mu * \nu(A) = \int_G \nu(A - x)d\mu(x)$$

for every Borel set A in G. Then $\mu*\nu = \nu*\mu$; and if μ and ν are probability measures such that μ vanishes on every translate of A and ν vanishes on every translate of B, then $\mu * \nu$ is a probability measure which vanishes on every translate of $A \cup B$.

To handle infinite unions of Haar null sets, we use infinite convolutions. The set of all probability measures on G in the topology of weak convergence of measures is a complete metric space (with metric ρ, say); see Appendix D. It follows from (6.1) and from Lebesgue's dominated convergence theorem that if $\mu_n \to \mu$, then $\mu_n * \nu \to \mu * \nu$. We note also that if a measure μ vanishes on some set A and on all its translates, then the same is true for every translate of μ and for every measure which is absolutely continuous with respect to μ. In particular, this holds for measures of the form $\mu_B(A) = \mu(A \cap B)/\mu(B)$ (where B is a Borel set with $\mu(B) > 0$). Hence, for every $\varepsilon > 0$ there is a measure ν so that $\nu(A+x) = 0$ for all x and $\rho(\nu, \delta_e) < \varepsilon$ (where δ_e is the Dirac measure at the unit element e of G).

Let $\{A_n\}_{n=1}^\infty$ be a sequence of Haar null sets in G. By the previous observations we can choose inductively probability measures μ_n on G such that

$$\mu_n(A_n + x) = 0 \quad \text{for all} \quad x \in G \text{ and } \rho(\alpha * \mu_n, \alpha) \leq 2^{-n}$$

where α is any convolution of different μ_k's with $1 \leq k \leq n-1$. By the completeness of ρ the measure $\mu = \prod_{k=1}^\infty *\mu_k$ is well defined. Since $\mu = \mu_n * \nu_n$ for every n, where $\nu_n = \prod_{k \neq n} *\mu_k$, it follows that $\mu_n(A_n + x) = 0$ for all n and x, and hence $\mu(x + \bigcup_{n=1}^\infty A_n) = 0$ for every x. $\square$

It follows from Proposition 6.3 that Haar null sets have empty interiors. Indeed, since G is separable, it can be covered by a countable union of translates of any set with nonempty interior, and clearly G itself is not Haar null.

A simple example of a Haar null set in a Banach space E is any Borel set which intersects all the translates of one fixed line L in sets of one-dimensional Lebesgue measure 0. (These sets will be called later (Definition 6.33) directionally null sets.) Indeed, any probability measure on the line L which is equivalent to the Lebesgue measure on L can be used as μ of Definition 6.1. A proper Borel linear subspace is an important example of a set as above. More generally, if there are n independent vectors $\{x_i\}_{i=1}^n$ such that $\{(\alpha_1, \dots, \alpha_n) : x + \sum_{i=1}^n \alpha_i x_i \in A\}$ is a set of Lebesgue measure 0 in $\mathbb{R}^n$ for every $x \in E$, then A is Haar null.

If A is a Borel set of positive Haar measure in a locally compact group G, then $A - A$ is a neighborhood of the identity in G. To see this, let λ be the Haar measure on G, and assume that $0 < \lambda(A) < \infty$. The convolution $\chi_{-A} * \chi_A$, where χ_A is the characteristic function of A, is a continuous function which is supported in $A - A$. The result follows from the fact that its value at the identity e is nonzero. Indeed, $\chi_{-A} * \chi_A(e) = \int \chi_{-A}(-x)\chi_A(x) d\lambda(x) = \lambda(A)$. The next result generalizes this phenomenon.

Proposition 6.4. *Let A be a Borel set which is not Haar null. Then $A - A$ is a neighborhood of the identity e in G.*

Proof. We shall actually prove the stronger statement that the set

$$K(A) = \{x : A \cap (A + x) \text{ is not Haar null}\}$$

is a neighborhood of e.

Assume that this is false, and choose a sequence $\{x_n\}$ with $d(x_n, e) < 2^{-n}$ so that $A \cap (A + x_n)$ is Haar null. By Proposition 6.3 the set $\bigcup_{n=1}^\infty ((A + x_n) \cap A)$ is

Haar null, hence its complement in A, i.e., $\tilde{A} = A \setminus \bigcup_{n=1}^{\infty} \big((A + x_n) \cap A\big)$ is a Borel set which is not Haar null. Clearly $(\tilde{A} + x_n) \cap \tilde{A} = \emptyset$ for all n.

Let $C = \{0, 1\}^{\aleph_0}$ be the Cantor group. We denote its elements by $\theta = (\theta_n)$, where $\theta_n \in \{0, 1\}$ for every n. The element of C all of whose coordinates are zero except for a 1 in the nth coordinate is denoted by $\theta^{(n)}$. Define $\varphi : C \to G$ by

$$\varphi(\theta) = \sum_{n=1}^{\infty} \theta_n x_n.$$

By the choice of $\{x_n\}$ and by the completeness of the metric d (recall that d is invariant), the map φ is continuous. Note that $\varphi(\theta^{(n)}) = x_n$.

Since $\tilde{A}$ is not Haar null, it follows that the image under φ of the Haar measure on C does not vanish on some translate of $\tilde{A}$; i.e., there is some $y \in G$ such that $\varphi^{-1}(\tilde{A} + y)$ is a set of positive Haar measure in C. Hence $\varphi^{-1}(\tilde{A} + y) - \varphi^{-1}(\tilde{A} + y)$ is a neighborhood of 0 in the compact group C, and it contains the points $\theta^{(n)}$ for large enough n. It follows that there are $\tau, \eta \in \varphi^{-1}(\tilde{A} + y)$ that differ only in their nth coordinate, and hence $x_n \in \tilde{A} - \tilde{A}$ for such n, contradicting the fact that $(\tilde{A} + x_n) \cap \tilde{A} = \emptyset$ for all n. □

Corollary 6.5. (i) *A compact set in a nonlocally compact group is Haar null.*

(ii) *A weakly compact set in a nonreflexive separable Banach space is Haar null.*

Haar null sets in nonlocally compact groups seem to behave like sets of Haar measure 0, but there are important aspects in which they behave differently.

We first give an example which shows that there is no Fubini type theorem for Haar null sets. Let $\mathbb{T}$ be the circle group with Haar measure μ, and let H be the Hilbert space $L_2(\mu)$. Let G be the product group $H \times \mathbb{T}$, and set

$$A = \Big\{(f, t) : \text{ the Fourier series of } f \text{ converges at the point } t\Big\}.$$

It is easy to check that A is a Borel set in G. For every fixed $t_0 \in T$ the set $A_{t_0} = \{f : (f, t_0) \in A\}$ is a Haar null set. Indeed, fix $g_0 \in H$ whose Fourier series does not converge at t_0. Then for every $h \in H$ the set $\{\lambda : \lambda g_0 + h \in A_{t_0}\}$ consists of at most one point.

On the other hand, the famous theorem of Carleson implies that for every fixed $f_0 \in H$ the set $A^{f_0} = \{t : (f_0, t) \in A\}$ is of full Haar measure. Note that the complement of A in G is actually a Haar null set. Indeed, consider the probability measure $\delta_0 \times \mu$ on $G = H \times \mathbb{T}$ (where δ_0 is the Dirac measure at the origin of H), and use Carleson's theorem again.

A more elementary example that uses Lebesgue's theorem on the differentiation of the integral rather than Carleson's theorem is given by

$$A = \Big\{(f, t) \in L_1(\mathbb{T}) \times \mathbb{T} : \lim_{h \to 0} h^{-1} \int_t^{t+h} f(s)ds \text{ exists}\Big\}.$$

In a separable locally compact group the Haar measure is σ-finite; hence any collection of disjoint sets of positive Haar measure is at most countable. This also fails badly in the nonlocally compact case.

Theorem 6.6. *Let G be a nonlocally compact complete metric and separable Abelian group. Then there is an uncountable collection $\{F_\theta\}$ of disjoint closed subsets of G such that each F_θ contains a translate of every compact subset in G. In particular the F_θ's are not Haar null.*

Proof. We start by proving that the sets F_θ are not Haar null. Indeed, fix θ and assume that μ is any probability measure on G. Since μ is tight (see Proposition D.1), there is a compact set K such that $\mu(K) > 0$. Since F_θ contains a translate of K, it follows that μ does not vanish on some translate of F_θ.

Let $\{Q_k\}_{k=1}^\infty$ be an increasing sequence of finite subsets of G whose union is dense in G. Since G is not locally compact, it is possible to find for every $\varepsilon > 0$ a $\delta > 0$ such that the closed ball $B(\varepsilon)$ (with center at e and radius ε) contains an infinite sequence of 2δ-separated points. It is thus possible to construct inductively points $\{y_k\}_{k=1}^\infty$ in $B(\varepsilon)$ such that

$$d\Big(y_k + Q_k, \bigcup_{i<k}(y_i + Q_i)\Big) \geq \delta$$

for $k \geq 2$. (Recall that $\bigcup_{i<k}(y_i + Q_i) - Q_k$ is a finite set.)

Using this observation repeatedly, define inductively positive numbers $\{\varepsilon_n\}_{n=1}^\infty$, $\{\delta_n\}_{n=1}^\infty$ and points $y_{k,n} \in G$ such that

$$y_{k,n} \in B(\varepsilon_n) \quad \text{for every} \quad k, n = 1, 2, \ldots \tag{6.2}$$

$$\sum_{m>n} \varepsilon_m < \delta_n/2 \quad \text{for every} \quad n = 1, 2, \ldots \tag{6.3}$$

$$d\Big(y_{k,n} + Q_k, \bigcup_{i<k}(y_{i,n} + Q_i)\Big) \geq 3\delta_n \quad \text{for every} \quad k, n = 1, 2, \ldots. \tag{6.4}$$

It follows from (6.4) that for each fixed n, the closed sets $\big(y_{k,n} + Q_k + B(\delta_n)\big)$ are δ_n separated. Hence, denoting the set of even numbers by I_0 and the set of odd numbers by I_1, it follows that the two sets $\bigcup_{k \in I_0}\big(y_{k,n} + Q_k + B(\delta_n)\big)$ and $\bigcup_{k \in I_1}\big(y_{k,n} + Q_k + B(\delta_n)\big)$ are closed and disjoint.

Fix $\theta = (\theta_n)$ in $\{0,1\}^{\aleph_0}$, and let I_θ^n be I_0 or I_1 according to the parity of θ_n. Then put

$$F_\theta = \bigcap_{n=1}^\infty \bigcup_{k \in I_\theta^n} \big(y_{k,n} + Q_k + B(\delta_n)\big).$$

By the observation above, the different F_θ's are closed and mutually disjoint. To conclude the proof it remains to show that for every compact $K \subset G$ and for every $\theta \in \{0,1\}^{\aleph_0}$ there is a $z \in G$ such that $z + K \subset F_\theta$.

To this end, define inductively the sequence $\{z_n\}_{n=1}^\infty$ as follows: Having defined $\{z_i\}_{i<n}$, consider the compact set $K + \sum_{i<n} z_i$. Since $\{Q_k\}_{k=1}^\infty$ is an increasing sequence of sets with dense union, it follows that for large enough k

$$K + \sum_{i<n} z_i \subset Q_k + B(\delta_n/2).$$

Fix such a k with the same parity as θ_n (i.e., $k \in I_\theta^n$), and let $z_n = y_{k,n}$. In view of (6.2), (6.3) and the completeness of d, the series $z = \sum_{i=1}^\infty z_i$ converges, and $d(z, \sum_{i \leq n} z_i) \leq \delta_n/2$. Hence for every n

$$K + z \subset y_{k,n} + Q_k + B(\delta_n/2) + B(\delta_n/2) \subset y_{k,n} + Q_k + B(\delta_n).$$

Since n was arbitrary, and since k was chosen to have the same parity as θ_n, it follows that $K + z \subset F_\theta$. $\square$

A measure μ is called quasi-invariant if it is mutually absolutely continuous with any of its translates.

Corollary 6.7. *Let G be as in Theorem 6.6. Then there is no nonnegative σ-finite quasi-invariant Borel measure μ on G except $\mu \equiv 0$.*

Proof. Assume that μ is quasi-invariant. Since μ is tight, there is a compact set K such that $\mu(K) > 0$. It follows from the quasi-invariance that $\mu(F_\theta) > 0$ for every θ. Since the F_θ's are disjoint, this contradicts the σ-finiteness of μ. □

The assumption that μ is nonnegative can be removed from the corollary by passing to the Jordan decomposition. While non-σ-finite invariant measures do exist (e.g. $\mu(A) =$ the number of points in A), they are not useful in our context.

We pass now to Banach spaces. A typical example of a relatively "large" Haar null set is the positive cone in l_p, $1 \le p < \infty$. Note that when A is such a cone, $A - A$ is the entire space. More generally, we have

Proposition 6.8. *Let E be a Banach space with a normalized basis $\{x_n\}_{n=1}^\infty$. Assume that there is a $q < \infty$ such that $\sum |\beta_n|^q < \infty$ whenever $\sum \beta_n x_n$ converges. Then the positive cone $\{x = \sum \beta_n x_n \in E : \beta_n \ge 0 \text{ for all } n\}$ of E is Haar null.*

Proof. Denote the Lebesgue measure on $[0, 1]$ by λ, and consider independent random variables $\{f_n\}_{n=1}^\infty$ on $[0, 1]$ such that

$$\lambda\{t : f_n(t) = 0\} = (n-1)/n, \qquad \lambda\{t : f_n(t) = n^{-1/q}\} = 1/n$$

for every $n \ge 1$. Since $\sum \|f_n\|_{L_1} = \sum n^{-(q+1)/q} < \infty$, it follows that the sequence $F(t) = (f_1(t), f_2(t), \dots)$ is in l_1 for almost all t.

Let μ be the distribution of F; i.e., μ is the probability measure on l_1 defined by $\mu(B) = \lambda(F^{-1}(B))$. We claim that if $(\beta_1, \beta_2, \dots) \in l^q$, then

$$\mu\{(\alpha_1, \alpha_2, \dots) \in l_1 : \alpha_n \le \beta_n \text{ for all } n\} = 0.$$

Indeed, this is evident if some $\beta_n < 0$. If all $\beta_n \ge 0$, then $\sum \beta_n^q < \infty$ implies that $\sum_{n^{-1} \le \beta_n^q} 1/n < \infty$, and it follows that the measure of the set in question is $\prod_{n^{-1} > \beta_n^q}(1 - \frac{1}{n}) = 0$, because

$$\sum_{n^{-1} > \beta_n^q} 1/n = \sum 1/n - \sum_{n^{-1} \le \beta_n^q} 1/n = \infty.$$

Define $T : l_1 \to E$ by $T(\alpha_1, \alpha_2, \dots) = \sum \alpha_n x_n$, and let τ be the probability measure on E defined by $\tau(B) = \mu(T^{-1}B)$.

Let A be the positive cone in E, and let $x = \sum \beta_n x_n \in E$. By our assumption on the basis, $\tau(x - A) = \mu\{(\alpha_1, \alpha_2, \dots) \in l_1 : \alpha_n \le \beta_n \text{ for all } n\} = 0$. It follows that $-A$ is a Haar null set, and so also is A. □

Theorem 6.9. *Every closed convex set with empty interior in a superreflexive space is Haar null.*

We shall need the following technical lemma.

Lemma 6.10. *Let B be a bounded closed convex set in a Banach space E such that B has empty interior and is not contained in a hyperplane in E. Then for every bounded projection P from E onto a closed subspace of finite codimension, the set PB has empty interior in PE.*

Lemma 6.10 follows from the next lemma, which is of some independent interest.

Lemma 6.11. *Let U be an open set in a Banach space E, and let f be a nonaffine real-valued continuous function on U. Then the convex hull of the graph of f has nonempty interior in $E \times \mathbb{R}$.*

Proof. Replacing f by $-f$ if necessary, we may assume that there are points $x, y \in U$ such that $\big(f(x)+f(y)\big)/2 > f\big((x+y)/2\big)$. Let $v \in E \times \mathbb{R}$ be the point whose E coordinate is 0 and whose $\mathbb{R}$ coordinate is 1, and denote the convex hull of the graph of f by C. By rescaling, translating, and adding a constant, we may thus assume that $v \in C$; that U contains $2B(E)$, the ball of radius 2 in E; and that $-1 \leq f(z) \leq 0$ for every $z \in 2B(E)$. We claim that C contains the convex hull of v and of $B(E)$ (hence its interior is certainly nonempty). To show this, we shall show that for every $z \in B(E)$ the line through v and z intersects the graph of f at some point $\big(tz, f(tz)\big)$ with $1 \leq t \leq 2$. Indeed, the function $g(t) = 1 - t - f(tz)$ satisfies $g(1) \geq 0$ and $g(2) \leq 0$. Hence there is a $1 \leq t \leq 2$ such that $g(t) = 0$; i.e., $\big(tz, f(tz)\big)$ is the required point of intersection. □

Proof of Lemma 6.10. It is enough to consider a projection P on a hyperplane H, so identify E with $H \oplus \mathbb{R}$ and let the projection be $P(x,t) = x$. Assume that PB has a nonempty interior U in H. For $z \in U$ put $f(z) = \inf\{\lambda : (z, \lambda) \in B\}$. The function f is bounded and convex on U, hence continuous. The convex hull of the graph of f is contained in B. Since B has empty interior, it follows from Lemma 6.11 that f is affine on U. Consequently the graph of f on U, hence also B, contains an open set V in a hyperplane $\tilde{H}$ of E. Since B is not contained in $\tilde{H}$, it contains the convex hull of some $x_0 \notin \tilde{H}$ and the set V. It follows that B has nonempty interior in E, a contradiction. ◇

Proof of Theorem 6.9. Let B be a closed convex set with empty interior in the superreflexive space E. Since the union of countably many Haar null sets is Haar null, we may assume that B is bounded. We can also assume that B is not contained in a hyperplane in E, since otherwise the result follows from Proposition 6.4.

We shall construct a quotient map T from E onto a Banach space F with a (normalized) basis $\{v_n\}$, and a vector $z \in E$ such that $T(B+z)$ is contained in the positive cone with respect to $\{v_n\}$. Since F is superreflexive, the basis satisfies the assumption of Proposition 6.8 for some $q < \infty$ (see Theorem E.3(ii)); hence there is a Borel probability measure τ on F such that

$$\tau\big(T(B+x)\big) = \tau\big(T(B+z) + T(x-z)\big) = 0$$

for every $x \in E$. By Proposition 1.19 there is a continuous selection φ for T^{-1}. Thus φ is a homeomorphism from F into E, and since F is complete, $\varphi(F)$ is a Borel (actually G_δ) set in E. The measure $\nu(D) = \tau(\varphi^{-1}D)$ on $\varphi(F)$ can be regarded as a Borel probability measure on E (vanishing outside $\varphi(F)$). Since $\varphi^{-1}(D) = TD$ for every D, it follows that $\nu(B+x) = 0$ for all x, and B is Haar null.

It remains to construct the quotient map and the vector z. Let $\{E_n\}_{n=1}^{\infty}$ be an increasing sequence of finite-dimensional subspaces of E whose union is dense in E. We shall construct inductively a sequence of integers $\{k_n\}$, functionals $\{x_n^*\} \in E^*$, and elements $\{z_n\} \in E$ as follows: Let $k_0 = 1$. Let P_n be a bounded linear projection in E with kernel $E_{k_{n-1}}$, and set $H_n = P_n E$. The operator $P_n^* : H_n^* \to E^*$ extends functionals on H_n to functionals on E, hence $\|P_n^* u^*\| \geq \|u^*\|$ for every

$u^* \in H_n^*$. The convex set $P_n B$ is closed (since P_n has a finite-dimensional kernel). By Lemma 6.10 it has empty interior in H_n, and we can thus separate it from points that are arbitrarily close to 0. Thus there is a $y_n^* \in H_n^*$ such that $\|y_n^*\| = 1$ and such that $\langle y_n^*, u\rangle \geq -4^{-(n+1)}$ for every $u \in P_n B$. Put $x_n^* = P_n^* y_n^*/\|P_n^* y_n^*\|$. Then $\|x_n^*\| = 1$ and $\langle x_n^*, x\rangle = 0$ for every $x \in E_{k_{n-1}}$. If $u \in P_n B$, then

$$\langle x_n^*, u\rangle = \langle y_n^*, u\rangle/\|P_n^* y_n^*\| \geq -4^{-(n+1)}/\|P_n^* y_n^*\| \geq -4^{-(n+1)}.$$

Having chosen x_n^*, choose next $k_n > k_{n-1}$ and a vector $z_n \in E_{k_n}$ such that $\langle x_n^*, z_n\rangle \geq 1$ and $\|z_n\| \leq 3/2$.

Now take $z = \sum 4^{-i} z_i$. For any $x \in B$ and for every integer n

$$\begin{aligned}\langle x_n^*, x+z\rangle &= \langle x_n^*, P_n x\rangle + 4^{-n}\langle x_n^*, z_n\rangle + \sum_{i>n} 4^{-i}\langle x_n^*, z_i\rangle \\ &\geq -4^{-(n+1)} + 4^{-n} - 3/2\sum_{i>n} 4^{-i} > 0.\end{aligned}$$

Since the sequence $\{x_n^*\}$ tends weakly to 0, it follows from Theorem E.3(i) that it has a subsequence which is a basic sequence. We may thus assume that $\{x_n^*\}_{n=1}^\infty$ is a basis of a subspace of E^*. By passing to duals (recall that E is reflexive), there is a quotient map $T : E \to F$ such that F has a basis $\{v_n\}$ with biorthogonal functionals $\{v_n^*\}$ and $T^* v_n^* = x_n^*$ for all n, and the computation above shows that T maps $B + z$ into the positive cone corresponding to the basis $\{v_n\}$. ♢

We now state without proof the following counterpart to the preceding result.

Theorem 6.12. *A separable Banach space E is nonreflexive if and only if there is a closed convex set with empty interior B in E which contains a translate of any compact set in E.*

Such a set B cannot, of course, be Haar null. The main part of the theorem is the "only if" part, but we give here examples only in c_0 and in l_1.

In c_0, the natural positive cone is closed and has empty interior. If K is a compact subset of c_0, then the coordinates of the points in K tend to 0 uniformly. Thus K can be translated by an element $y \in c_0$ so that all the coordinates of the elements in $K + y$ are nonnegative.

The natural positive cone in l_1 does not work. In fact, it is Haar null by Proposition 6.8. To give an example in l_1, we use the fact that there is a quotient map from l_1 onto c_0 (as on any separable space). The inverse image of the positive cone of c_0 is a closed convex set in l_1 with empty interior, and, by the same argument as in c_0, it contains a translate of every compact set in l_1.

Sets of measure zero in finite-dimensional spaces are preserved not only by linear automorphisms but also by bi-Lipschitz maps on the space. Theorem 6.14 will show that the situation is completely different in infinite-dimensional spaces for Haar null sets. We start with a simple lemma.

Lemma 6.13. *Let $\{x_n\}_{n=1}^\infty$ be a normalized 1-unconditional basis in a Banach space E, and let $\varepsilon > 0$. Let $\varphi_n : \mathbb{R} \to \mathbb{R}$ be surjective functions such that $\varphi_n(0) = 0$ and $(1+\varepsilon)^{-1}|s-t| \leq |\varphi_n(s) - \varphi_n(t)| \leq (1+\varepsilon)|s-t|$ for every n and $s, t \in \mathbb{R}$. Then*

(i) *The map $f : E \to E$ defined by $f\left(\sum \lambda_n x_n\right) = \sum \varphi_n(\lambda_n) x_n$ is a surjective Lipschitz equivalence with $\|f\|_{\mathrm{Lip}} \leq 1+\varepsilon$ and $\|f^{-1}\|_{\mathrm{Lip}} \leq 1+\varepsilon$.*

(ii) *If also $\sum |\varphi_n(t) - t| \leq \varepsilon$ for all $t \in \mathbb{R}$, then $\|f(x) - x\| \leq \varepsilon$ for all $x \in E$.*

(iii) *If the φ_n's are also differentiable, then f is Gâteaux differentiable.*

Proof. Parts (i) and (ii) are evident. To check (iii), fix $x = \sum \lambda_n x_n \in E$. Define an operator $D : E \to E$ by $D(\sum \tau_n x_n) = \sum \tau_n \varphi'_n(\lambda_n) x_n$. Then D is bounded (because $|\varphi'_n(s)| \le 1 + \varepsilon$ for all n and s), and it is the Gâteaux derivative of f at x. To see this it suffices to check the directional derivatives in the directions of finitely supported vectors in E, but this is obvious for such vectors. □

The basic idea in the construction in the next theorem is as follows. Fix n basis vectors in E, and denote their span by E^n. Consider the cube Q in E^n, centered at the origin with edge size one (i.e., with vertices all of whose coordinates are $\pm 1/2$). We shall use the Lebesgue measure on E^n normalized so that the volume of Q is 1. Remove from Q a concentric cube of volume $1/n$ (i.e., of edge size $1/\sqrt[n]{n}$), and call the remaining set D_n. Also remove from Q a slightly smaller concentric cube of volume $1/n^2$ (i.e., of edge size $1/\sqrt[n]{n^2}$), and call the remaining set A_n. Map A_n onto D_n in the obvious way, keeping the boundary of the original cube fixed. This is a Lipschitz equivalence with uniformly bounded Lipschitz constants (because $(1-1/\sqrt[n]{n})/(1-1/\sqrt[n]{n^2}) \to 1/2$). On the other hand, there is a substantial difference between the volumes of the two cubes we took out, a difference that stems from the fact that $\sum n^{-2} < \infty$ while $\sum n^{-1} = \infty$. This difference will be used in the construction to obtain Haar null sets and sets whose complement are Haar null. In order to get maps which are not only bi-Lipschitz but also arbitrarily small perturbations of the identity, we just replace n^2 above by n^{γ_n}, where $\gamma_n \searrow 1$ but $\sum n^{-\gamma_n} < \infty$.

Theorem 6.14. *Let E be a Banach space with a normalized 1-unconditional basis $\{x_n\}_{n=1}^{\infty}$. Then there are Borel sets $A, D \subset E$ such that D and $E \setminus A$ are Haar null sets with the following property: For every $\varepsilon > 0$ there is a map f_ε from E onto itself with $f_\varepsilon(A) = D$; both f_ε and f_ε^{-1} are C^1-maps; $\|f_\varepsilon(x) - x\| \le \varepsilon$ for all $x \in E$; and for all $x, y \in E$*

$$(1+\varepsilon)^{-1}\|x-y\| \le \|f_\varepsilon(x) - f_\varepsilon(y)\| \le (1+\varepsilon)\|x-y\|.$$

Proof. Write $E = E^1 \oplus E^2 \oplus E^3 + \dots$, where $E_1 = \operatorname{span}\{x_1\}$, $E_2 = \operatorname{span}\{x_2, x_3\}$ and so on. Denote the natural projections from E onto E^n by P_n.

Choose $\gamma_n \searrow 1$ with $\sum n^{-\gamma_n} < \infty$, and put

$$A_n = E^n \setminus \left(\mathbb{Z}^n + [-r_n^{\gamma_n}/2, r_n^{\gamma_n}/2]^n\right); \quad D_n = E^n \setminus \left(\mathbb{Z}^n + [-r_n/2, r_n/2]^n\right)$$

where $r_n = 1/\sqrt[n]{n}$ and $\mathbb{Z}^n$ is the integer lattice in E^n (i.e., the vectors in E^n with integer coordinates).

Let $\alpha_n > 0$ satisfy $\sum n\alpha_n < \infty$, and put

$$\tilde{A}_n = \{x \in E : P_n x \in \alpha_n A_n\} \text{ and } A = \bigcup_{k=1}^{\infty} \bigcap_{n=k}^{\infty} \tilde{A}_n$$

and

$$\tilde{D}_n = \{x \in E : P_n x \in \alpha_n D_n\} \text{ and } D = \bigcup_{k=1}^{\infty} \bigcap_{n=k}^{\infty} \tilde{D}_n.$$

To see that D and $E \setminus A$ are Haar null sets, let μ_n be the restriction of the Lebesgue measure to $\alpha_n[0,1]^n$, normalized so that this set has measure 1. By the periodicity of the sets A_n and D_n, it follows that

$$\mu_n(\alpha_n A_n + x) = 1 - n^{-\gamma_n} \text{ and } \mu_n(\alpha_n D_n + x) = 1 - n^{-1}$$

for every $x \in E^n$. Let μ be the product measure $\prod_{n=1}^{\infty} \mu_n$. It is a well-defined probability measure on E because $\sum n\alpha_n < \infty$.

Since $\tilde{D}_n + z = \{x \in E : P_n x \in P_n z + \alpha_n D_n\}$ for every n and $z \in E$, it follows that $\mu(z + \bigcap_{n \geq k} \tilde{D}_n) = \prod_{n \geq k}(1 - n^{-1}) = 0$ for every $z \in E$ and integer k. Thus $\mu(z + D) = 0$ for every $z \in E$, and hence D is Haar null. Similarly,

$$\begin{aligned} \mu(z + A) &\geq \sup_k \mu\Big(z + \bigcap_{n \geq k} \tilde{A}_n\Big) = \sup_k \prod_{n \geq k} \mu_n(P_n z + \alpha_n A_n) \\ &= \sup_k \prod_{n \geq k} (1 - n^{-\gamma_n}) = 1 \end{aligned}$$

for every $z \in E$; hence $E \setminus A$ is Haar null.

We now define the maps $f_\varepsilon : E \to E$, and we first ignore the differentiability requirement.

Fix j, and assume that $x_j \in E^n$, i.e., that $n(n-1)/2 < j \leq n(n+1)/2$. Define $\varphi_j : \mathbb{R} \to \mathbb{R}$ by

$$\varphi_j(k + r_n^{\gamma_n}/2) = k + r_n/2; \quad \varphi_j(k - r_n^{\gamma_n}/2) = k - r_n/2 \tag{6.5}$$

for every integer k, where φ_j is linear in all the intervals determined by the points $k \pm r_n^{\gamma_n}/2$. There are numbers $\varepsilon_n \searrow 0$ (which depend on n and not on the specific j) such that $|\varphi_j(t) - t| \leq \varepsilon_n$ and $(1+\varepsilon_n)^{-1}|s-t| \leq |\varphi_j(t) - \varphi_j(s)| \leq (1+\varepsilon_n)|t-s|$ for all $s, t \in \mathbb{R}$ and j. The first assertion is evident, and for the second we need only to check the slopes in the two types of segments appearing in the definition of φ_j, which we leave for the reader.

Now put $\psi_j(t) = \alpha_n \varphi_j(t/\alpha_n)$. Then $\psi_j(0) = 0$, and it satisfies $|\psi_j(t) - t| \leq \alpha_n \varepsilon_n$ and $(1+\varepsilon_n)^{-1}|t-s| \leq |\psi_j(t) - \psi_j(s)| \leq (1+\varepsilon_n)|t-s|$. Given $\varepsilon > 0$, find an integer k such that $\varepsilon_k < \varepsilon$ and $\sum_{n \geq k} n\alpha_n\varepsilon_n < \varepsilon$, and define $f_\varepsilon : E \to E$ by

$$f_\varepsilon\Big(\sum_{j=1}^{\infty} \lambda_j x_j\Big) = \sum_{j<k} \lambda_j x_j + \sum_{j \geq k} \psi_j(\lambda_j) x_j.$$

Then $f_\varepsilon(\tilde{A}_n) = \tilde{D}_n$ for every $n \geq k$, and since the sets $\bigcap_{n \geq h} \tilde{A}_n$ and $\bigcap_{n \geq h} \tilde{D}_n$ increase with h, it follows that $f_\varepsilon(A) = D$. Lemma 6.13 implies that f_ε satisfies the other required conditions.

We now claim that if the φ_j's are chosen to be continuously differentiable, strictly monotone, satisfying (6.5) as well as $\varphi_j(0) = 0$ for all j, and such that their Lipschitz constants are about the same as those of the piecewise linear functions we chose above, then f_ε and f_ε^{-1} will be C^1 functions. Lemma 6.13 gives that f_ε is Gâteaux differentiable. We shall check that its derivative is continuous. This will also show that f_ε^{-1} is a C^1 function either by the inverse mapping theorem or by the fact that f_ε^{-1} has the same form as f_ε.

Since $D_{f_\varepsilon}(x)y = \sum_{j<k} \tau_j x_j + \sum_{j \geq k} \tau_j \psi_j'(\lambda_j) x_j$ for every $x = \sum \lambda_j x_j$ and $y = \sum \tau_j x_j$, it follows that if $z = \sum \mu_j x_j$, then

$$\|D_{f_\varepsilon}(x) - D_{f_\varepsilon}(z)\| \leq \sup\big\{|\psi_j'(\lambda_j) - \psi_j'(\mu_j)| : j \geq k\big\}.$$

Recall that if j is such that $x_j \in E_n$, then $|\psi_j'(t) - 1| \leq \varepsilon_n$ for all t, and $\varepsilon_n \to 0$. Thus, given $\delta > 0$, we can find $m \geq k$ such that $|\psi_j'(t) - 1| \leq \delta/2$ for all t and $j \geq m$. It follows that if $|\psi_j'(\lambda_j) - \psi_j'(\mu_j)| \leq \delta$ for $k \leq j \leq m$, then $\|D_{f_\varepsilon}(x) - D_{f_\varepsilon}(z)\| \leq \delta$. □

In Chapters 14 and 15 we discuss perturbations of isometries (and in particular of the identity). The functions f_ε of Theorem 6.14 are small perturbations of the identity in the sense of both chapters.

2. Gaussian Measures

In this section we introduce Gaussian measures on Banach spaces. The theory of Gaussian measures is a well-developed area with many applications. Its history goes back to Wiener's measure on $C(0,1)$, and it is closely connected to the theory of Brownian motion. We restrict ourselves, however, to the basic facts which are needed for the study of Gaussian null sets and their applications.

By their very nature Gaussian measures are closely related to some Hilbertian structure. We shall therefore start this section by considering measures on Hilbert spaces (although some of the results can be stated in a more general context).

Let H be a separable Hilbert space. For every finite-dimensional subspace F of H let P_F be the orthogonal projection onto F, and denote the σ-algebra of Borel subsets of H that depend only on F by $\mathcal{A}_F$, i.e.,

$$\mathcal{A}_F = \{P_F^{-1}(A) : A \text{ is a Borel subset of } F\}.$$

Assume that for every F there is a probability measure μ_F defined on $\mathcal{A}_F$ and that the measures $\{\mu_F\}$ are compatible in the sense that μ_{F_1} and μ_{F_2} agree on sets in $\mathcal{A}_{F_1}$ whenever $F_1 \subset F_2$. The σ-algebra generated by $\mathcal{A} = \bigcup \mathcal{A}_F$ is the σ-algebra of all Borel subsets of H. By the compatibility of the measures, the finitely additive measure on $\mathcal{A}$ given by $\mu(A) = \mu_F(A)$ for $A \in \mathcal{A}_F$ is well defined.

Our first result is a criterion for this measure μ to extend to a σ-additive measure on all the Borel sets of H. We denote the ball $\{x : \|x\| \le r\}$ in H by $B(r)$.

Proposition 6.15. *The compatible family $\{\mu_F\}$ extends to a σ-additive measure on all the Borel subsets of H if and only if for every $\varepsilon > 0$ there is an r such that $\mu_F\Big(P_F^{-1}(B(r) \cap F)\Big) > 1 - \varepsilon$ for all F.*

Proof. Assume that for every $\varepsilon > 0$ there is an r as required. By the general criterion for the existence of such an extension (see Appendix D), we need to check that if B_n is a decreasing sequence of sets in $\mathcal{A}$ with empty intersection, then $\mu(B_n) \to 0$ as $n \to \infty$.

Write $B_n = P_{F_n}^{-1} A_n$, where $\{F_n\}$ are finite-dimensional subspaces of H and $A_n \subset F_n$, and assume, as we may, that the sequence F_n is increasing.

We may assume that each A_n is compact. Indeed, for each $\delta_n > 0$ there is a compact subset $C_n \subset A_n$ such that $\mu_{F_n}(P_{F_n}^{-1} C_n) \ge \mu_{F_n}(P_{F_n}^{-1} A_n) - \delta_n$. The sets $A'_n = P_{F_n}\big(\bigcap_{m \le n} P_{F_m^{-1}} C_m\big) \subset A_n$ are compact, and the sequence $\{P_{F_n}^{-1} A'_n\}$ is also decreasing and satisfies $\mu_{F_n}(P_{F_n}^{-1} A'_n) \ge \mu_{F_n}(P_{F_n}^{-1} A_n) - \delta$ where $\delta = \sum \delta_n$ is arbitrarily small. It is thus enough to prove that $\mu_{F_n}(P_{F_n}^{-1} A'_n) \to 0$; i.e., we may assume that each A_n is compact.

Given $\varepsilon > 0$, choose a suitable r. Since $B(r)$ is weakly compact and since the A_n's are compact, the sets $B_n \cap B(r)$ are weakly compact, and by assumption they have empty intersection. Thus there is an n_0 such that $B_n \cap B(r) = \emptyset$ for $n > n_0$, i.e., $\|x\| > r$ for every $x \in A_n$, and hence $\mu(B_n) = \mu_{F_n}(P_{F_n}^{-1} A_n) \le \varepsilon$ for $n > n_0$.

The other direction is obvious. □

It will be convenient to reformulate Proposition 6.15 in another form. Let $\tilde{\mu}_F$ be the projection of μ_F on F, i.e., $\tilde{\mu}_F(A) = \mu_F(P_F^{-1} A)$ for Borel subsets A of F.

Considering each measure $\tilde{\mu}_F$ as a measure on H concentrated in F, we see that $\mu_F\Big(P_F^{-1}\big(B(r)\cap F\big)\Big) \to 1$ uniformly as $r\to\infty$ if and only if $\tilde{\mu}_F\big(B(r)\cap F\big)$ does. Thus μ has a σ-additive extension if and only if for every $\varepsilon>0$ there is a $\delta>0$ with

$$\int_F e^{-\delta\|x\|^2}\,d\tilde{\mu}_F(x)\ge 1-\varepsilon \tag{6.6}$$

for every finite-dimensional subspace $F\subset H$.

To every compatible family of measures $\{\mu_F\}$ on H we can associate a positive definite function φ on H as follows. Put

$$\varphi_F(y)=\int_F e^{i\langle x,y\rangle}\,d\tilde{\mu}_F(x),\qquad F\subset H,\ y\in F.$$

Since the family $\{\mu_F\}$ is compatible, it follows that $\varphi_{F_1}(y)=\varphi_{F_2}(y)$ whenever $F_1\subset F_2$ and $y\in F_1$, i.e., the φ_F's are the restrictions of a well-defined function φ on H. The function φ is called the characteristic function of the family $\{\mu_F\}$. It follows from its definition that φ is positive definite with $\varphi(0)=1$ and that its restriction to any finite-dimensional subspace of H is continuous. Conversely, by Bochner's theorem (see Theorem D.5(ii)), every function φ with these properties is the characteristic function of some compatible family of measures. The next theorem gives a characterization of those positive definite functions φ on H with $\varphi(0)=1$ which are the characteristic functions of a single probability measure on H. It can be viewed as an infinite-dimensional version of Bochner's theorem.

Theorem 6.16. *A function φ on H is of the form*

$$\varphi(y)=\int_H e^{i\langle x,y\rangle}\,d\mu(x) \tag{6.7}$$

for some probability measure μ on H if and only if φ is positive definite, $\varphi(0)=1$ and for every $\varepsilon>0$ there is a nonnegative trace class operator T_ε on H such that $\langle T_\varepsilon y,y\rangle\le 1$ implies $\operatorname{Re}\big(\varphi(0)-\varphi(y)\big)<\varepsilon$.

Proof. Assume that φ has the form (6.7). Given $\varepsilon>0$, choose r such that $\mu\{x:\|x\|>r\}<\varepsilon/4$, and define the operator T_ε on H by

$$\langle T_\varepsilon u,v\rangle=\varepsilon^{-1}\int_{\|x\|\le r}\langle x,u\rangle\langle x,v\rangle\,d\mu(x).$$

Then T_ε is nonnegative, and, fixing an orthonormal basis $\{e_n\}$ of H, its trace is

$$\operatorname{tr}T_\varepsilon=\sum\langle T_\varepsilon e_n,e_n\rangle=\varepsilon^{-1}\int_{\|x\|\le r}\sum\langle x,e_n\rangle^2 d\mu(x)\le\varepsilon^{-1}r^2.$$

Assume now that $\langle T_\varepsilon y,y\rangle\le 1$. Using the inequality $1-\cos t\le t^2/2$, we obtain

$$\begin{aligned}\operatorname{Re}\big(\varphi(0)-\varphi(y)\big) &= \int_H\big(1-\cos\langle y,x\rangle\big)\,d\mu(x)\\ &\le 2\mu\{x:\ \|x\|>r\}+\int_{\|x\|\le r}\langle y,x\rangle^2/2\,d\mu(x)\\ &< \varepsilon/2+\varepsilon\langle T_\varepsilon y,y\rangle/2\le\varepsilon.\end{aligned}$$

Conversely, let φ be given. By the remarks preceding the theorem, φ is the characteristic function of a compatible family of measures $\{\mu_F\}$, and we need to show that for every $\varepsilon>0$ there is a $\delta>0$ such that (6.6) holds for all F.

Fix a finite-dimensional subspace F of H, and denote its dimension by n and the Lebesgue measure on F by dz. (Since we are dealing with subspaces of one fixed Hilbert space H, the normalization is evident.) As before, we denote the projection of μ_F on F by $\tilde{\mu}_F$. Since

$$e^{-\delta\|x\|^2} = (4\pi\delta)^{-n/2} \int_F e^{i\langle z,x\rangle - \|z\|^2/4\delta} dz$$

for every $x \in F$, Fubini's theorem gives

$$\int_F e^{-\delta\|x\|^2} d\tilde{\mu}_F(x) = (4\pi\delta)^{-n/2} \int_F \varphi(z) e^{-\|z\|^2/4\delta} dz.$$

Consequently,

$$\begin{aligned} 1 - \int_F e^{-\delta\|x\|^2} d\tilde{\mu}_F(x) &= (4\pi\delta)^{-n/2} \int_F e^{-\|z\|^2/4\delta} \big(1 - \varphi(z)\big) dz \\ &= (4\pi\delta)^{-n/2} \int_F e^{-\|z\|^2/4\delta} \mathrm{Re}\big(1 - \varphi(z)\big) dz. \end{aligned}$$

Fix $\varepsilon > 0$, and let T_ε be a positive trace class operator such that $\langle T_\varepsilon z, z\rangle \leq 1$ implies that $\mathrm{Re}\big(1 - \varphi(z)\big) \leq \varepsilon$. It follows that $\mathrm{Re}\big(1 - \varphi(z)\big) \leq \varepsilon + 2\langle T_\varepsilon z, z\rangle$ for all z. If $\{e_k\}_{k=1}^n$ is an orthonormal basis for F, then

$$(4\pi\delta)^{-n/2} \int_F e^{-\|z\|^2/4\delta} \langle z, e_k\rangle \langle z, e_j\rangle dz = 0$$

for $j \neq k$, and

$$(4\pi\delta)^{-n/2} \int_F e^{-\|z\|^2/4\delta} \langle z, e_j\rangle^2 dz = 2\delta.$$

Thus

$$\begin{aligned} 1 - \int_F e^{-\delta\|x\|^2} d\tilde{\mu}_F(x) &\leq (4\pi\delta)^{-n/2} \int_F e^{-\|z\|^2/4\delta} (\varepsilon + 2\langle T_\varepsilon z, z\rangle) dz \\ &= \varepsilon + 2\delta \sum_{k=1}^{n} 2\langle T_\varepsilon e_k, e_k\rangle \leq \varepsilon + 4\delta \operatorname{tr} T_\varepsilon \end{aligned}$$

and (6.6) holds (with ε replaced by 2ε) provided δ satisfies $4\delta \operatorname{tr} T_\varepsilon \leq \varepsilon$. □

We turn now to Gaussian measures.

Definition 6.17. *A probability measure μ on a Banach space E is called a Gaussian measure if for every $x^* \in E^*$ the measure μ_{x^*} on the real line, defined by $\mu_{x^*}(A) = \mu\{y : \langle x^*, y\rangle \in A\}$, has a Gaussian distribution. The Gaussian measure μ is called nondegenerate if for every $x^* \neq 0$ the distribution of μ_{x^*} is nondegenerate.*

We first analyze the structure of Gaussian measures on separable Hilbert spaces.

Proposition 6.18. *A probability measure μ on H is Gaussian if and only if there is an $x_0 \in H$ and a nonnegative trace class operator S on H such that*

$$\int_H e^{i\langle x,y\rangle} d\mu(y) = e^{i\langle x,x_0\rangle - \langle Sx,x\rangle/2}, \quad x \in H. \tag{6.8}$$

The measure μ is nondegenerate if and only if $\ker(S) = \{0\}$.

Proof. Assume that μ satisfies (6.8). For each fixed x let μ_x denote the distribution of $\langle x, \cdot\rangle$; then its characteristic function is equal to

$$\int_{\mathbb{R}} e^{its} d\mu_x(t) = \int_H e^{is\langle x,y\rangle} d\mu(y) = e^{is\langle x,x_0\rangle - s^2\langle Sx,x\rangle/2}$$

which is the characteristic function of a Gaussian distribution on $\mathbb{R}$.

Assume conversely that μ is a Gaussian measure on H. Then every μ_x has a Gaussian distribution; hence

$$\int_H e^{i\langle x,y\rangle} d\mu(y) = e^{im_x - \sigma_x^2/2}$$

where

$$m_x = \int_H \langle x,y\rangle d\mu(y) \qquad \sigma_x^2 = \int_H \big(\langle x,y\rangle - m_x\big)^2 d\mu(y)$$

(and in particular these integrals converge). Since

$$m_x = \lim_{r\to\infty} \int_{\|y\|\le r} \langle x,y\rangle d\mu(y)$$

it follows from the uniform boundedness principle that the linear functional $x \to m_x$ is continuous on H, i.e., $m_x = \langle x, x_0\rangle$ for some x_0. Translating μ by x_0, we may assume that $x_0 = 0$. Hence

$$\sigma_x^2 = \int_H \langle x,y\rangle^2 d\mu(y) = \langle Sx, x\rangle$$

for some positive operator S on H. Thus the characteristic function φ of μ satisfies $\varphi(x) = e^{-\langle Sx,x\rangle/2}$, and we need only to check that S is a trace class operator.

By Theorem 6.16 there is a nonnegative trace class operator T on H such that $\langle Tx, x\rangle \le 1$ implies that $1 - e^{-\langle Sx,x\rangle/2} \le 1/2$. Thus $e^{-\langle Sx,x\rangle/2} \ge 1/2$ and hence $\langle Sx, x\rangle \le \log 4$, i.e., $S \le \log 4 \cdot T$. Since T is a trace class operator, so is S.

The measure μ is nondegenerate if and only if $\langle Sx, x\rangle > 0$ for every $x \ne 0$, i.e., when $\ker(S) = \{0\}$. □

Remarks. (i) For every $x_0 \in H$ and for every nonnegative trace class operator S there is a (unique) probability measure μ on H so that (6.8) holds. Indeed, the right-hand side of (6.8) satisfies the assumptions of Theorem 6.16. (As an operator T in Theorem 6.16, take $T = \varepsilon^{-1}(S + \|x_0\|^2 P)$, where P is the orthogonal projection on the one-dimensional subspace spanned by x_0.)

(ii) For μ as in (6.8) put $H_1 = \ker(S)$ and $H_0 = H_1^\perp$. Then μ is a product of a nondegenerate Gaussian measure μ_0 on H_0 and a dirac measure on H_1.

If S is a nonnegative trace class operator on a Hilbert space H, then H has an orthonormal basis $\{e_n\}_{n=1}^\infty$ consisting of eigenvectors of S. Thus $Se_n = \lambda_n e_n$ with $\lambda_n \ge 0$ for all n and $\operatorname{tr} S = \sum \lambda_n < \infty$. The characteristic function of the measure μ given by (6.8) is the infinite product $\varphi(x) = \prod e^{i\alpha_n a_n - \lambda_n a_n^2/2}$, where $a_n = \langle x, e_n\rangle$ and $\alpha_n = \langle x_0, e_n\rangle$. It follows that the coordinates of x with respect to the basis $\{e_n\}$ are independent Gaussian variables with mean α_n and standard deviation $\sigma_n = \lambda_n^{1/2}$ respectively.

We can thus represent the Gaussian measure μ as a product measure as follows. Let U be the linear map from H to $\mathbb{R}^{\aleph_0}$ defined by

$$Ux = \Big(\langle x, e_1\rangle, \langle x, e_2\rangle, \dots, \langle x, e_n\rangle, \dots\Big)$$

and let $\tilde{\mu}$ be the product measure on $\mathbb{R}^{\aleph_0}$, where the measure on the n-th copy of $\mathbb{R}$ has a Gaussian density with mean α_n and standard deviation $\lambda_n^{1/2}$. The set UH is a Borel set (actually F_σ) in $\mathbb{R}^{\aleph_0}$. Since the expectation (with respect to $\tilde{\mu}$) of $(a_n - \alpha_n)^2$ is λ_n, and since $\sum \lambda_n < \infty$, it follows that $\sum(a_n - \alpha_n)^2 < \infty$ a.e. with respect to $\tilde{\mu}$, i.e., $\tilde{\mu}$ vanishes outside UH. The restriction of $\tilde{\mu}$ to UH is the measure μ we started with (or, more precisely, the image of μ via the 1-1 map U). For future reference we record here that similarly, if $\beta_n > 0$ for every n and if they satisfy $\sum \lambda_n/\beta_n < \infty$, then the measure μ is supported in the Borel set

$$\left\{x = \sum a_n e_n : \sum(a_n - \alpha_n)^2/\beta_n < \infty\right\}. \tag{6.9}$$

To illustrate a computation with Gaussian measures on a Hilbert space, let us calculate the expectation of $e^{\beta\|x\|^2}$. Assume for simplicity that the mean of μ is 0, and pass to the product representation of μ. Notice that the expectation of $e^{\beta t^2}$ with respect to a symmetric Gaussian variable on the line with variance $\sigma^2 = \lambda$ is finite if and only if $2\beta\lambda < 1$. When this is the case, this expectation is equal to $(2\pi\lambda)^{-1/2}\int_{\mathbb{R}} e^{\beta t^2 - t^2/2\lambda}dt = (1-2\beta\lambda)^{-1/2}$. Hence the expectation of $e^{\beta\|x\|^2}$ is finite if and only if $2\beta\sup_n \lambda_n = 2\beta\|S\| < 1$. If this holds, the expectation is $\prod(1-2\beta\lambda_n)^{-1/2}$ (which is finite because $\sum \lambda_n < \infty$).

We now consider the effect of a translation in the Hilbert space on Gaussian measures. Assume that μ is a Gaussian measure on H and that $\{e_n\}$ is an orthogonal basis for H consisting of eigenvectors of the trace class operator S associated with μ. Write $Se_n = \lambda_n e_n$.

We start with the very simple case when the translation is in the direction of some e_n. If n is such that $\lambda_n > 0$, then the translated measure $\nu(A) = \mu(A - \gamma e_n)$ is equivalent to μ (i.e., they have the same sets of measure 0). Indeed, in the product measure representation of μ we shift just one coordinate, and a nondegenerate Gaussian measure on the line is equivalent to all its translates. On the other hand, when $\lambda_n = 0$, then the translated measure ν is clearly singular with respect to μ.

It turns out that this dichotomy holds in general, and a translate of μ is either equivalent to μ or is singular with respect to it.

Proposition 6.19. *With the notation as above, a translate ν of μ by a vector $x \in H$ is either equivalent to μ or is singular with respect to it. It is equivalent to μ if and only if $x \in S^{1/2}H$, i.e., when $\langle x, e_n\rangle = \lambda_n^{1/2} a_n$ with $\sum a_n^2 < \infty$.*

Proof. This is a direct consequence of Kakutani's theorem D.4. We can assume the μ has mean zero. Representing μ as a product measure and normalizing, the assertion becomes: Let all the μ_n's in the product representation of μ be standard Gaussian measures, and let ν_n be the translation of μ_n by a_n. Then $\mu \sim \nu$ if and only if $\sum a_n^2 < \infty$. Indeed (with the notation of Appendix D)

$$H(\mu_n, \nu_n) = \frac{1}{\sqrt{2\pi}} \int e^{-x^2/4} e^{-(x-a_n)^2/4} dx = e^{-a_n^2/8}$$

and the result follows from Kakutani's theorem. It also follows that if the measures are not equivalent, they are mutually singular. □

We pass now to Gaussian measures on separable Banach spaces. An easy way to construct such measures is as images of Gaussian measures on Hilbert space.

Proposition 6.20. *Let E be a Banach space, and let T be a continuous linear operator from a Hilbert space H into E. If μ is a Gaussian probability measure* *on H, then $\nu(A) = \mu(T^{-1}A)$ defines a Gaussian probability measure on E. If μ is nondegenerate and T has a dense range, then ν is also nondegenerate.*

Proof. The only thing to check is that TH is a Borel subset of E (and hence ν is, indeed, a Borel measure). This follows from the fact that closed balls in H are weakly compact. Thus their images in E are closed, hence certainly Borel. □

If E is separable, there is a continuous operator with dense range from H to E. Indeed, fix a sequence $\{x_n\}$ in E with a dense linear span and $\sum \|x_n\|^2 < \infty$, and define $T : l_2 \to E$ by $Te_n = x_n$. It follows that T carries nondegenerate Gaussian measures on H to nondegenerate Gaussian measures on E. By what we have proved for Gaussian measures on H, the Gaussian measures ν on E that we get in this way satisfy $\int_E e^{\beta\|y\|^2} d\nu(y) < \infty$ for small enough positive β.

Now let μ be a Gaussian measure on a Banach space E. Since we have detailed information on Gaussian measures on Hilbert spaces, it is convenient to study μ in terms of its image under an appropriate operator from E to a Hilbert space. Such an operator is constructed in the following lemma.

Lemma 6.21. *Let E be a separable Banach space. Then there is an injective operator U from E to some Hilbert space H with a dense range such that UA is a Borel subset of H for every Borel set A in E.*

Proof. We shall show that the formal identity operator from $C[0,1]$ to $L_2(0,1)$ takes Borel sets to Borel sets. Since every separable Banach space E can be embedded as a subspace of $C[0,1]$, we can take U to be the identity from $E \subset C[0,1]$ to L_2 and H to be the closure of E in L_2.

The identity is, of course, injective, and $C[0,1]$ is separable; thus it is enough to show that the unit ball of $C[0,1]$ is a Borel subset of L_2. Let $\mathcal{J}$ be the countable collection of intervals with rational endpoints in $[0,1]$, and let $\mathcal{J}_m$ be the set of pairs (I,J) from $\mathcal{J}$ so that their lengths $|I|$ and $|J|$ as well as their mutual distance are all smaller than $1/m$. Then the unit ball of $C[0,1]$ is represented in L_2 as the intersection of the closed set $\bigcap_{I\in\mathcal{J}}\{f \in L_2 : \int_I |f| \le |I|\}$ and the Borel set

$$\bigcap_n \bigcup_m \bigcap_{(I,J)\in\mathcal{J}_m} \Big\{f \in L_2 : \Big||I|^{-1}\int_I f - |J|^{-1}\int_J f\Big| < 1/n\Big\}.$$

Indeed, it follows from Lebesgue's theorem on the differentiation of the integral that there is a subset A of $[0,1]$ of full Lebesgue measure such that for every $x \in A$ there is a sequence of intervals $I_j \in \mathcal{J}$ which decreases to x and such that $f(x) = \lim_{j\to\infty} |I_j|^{-1}\int_{I_j} f$. It then follows from the second condition that the restriction of f to A is uniformly continuous on A. Hence the equivalence class of f contains the continuous extension of this restriction to all of $[0,1]$. The first condition then implies that this function is, indeed, in the unit ball of $C[0,1]$. □

Theorem 6.22. *Let μ be a Gaussian measure on E, and let $U : E \to H$ be as in the previous lemma. Let $\tilde{\mu}$ be the Gaussian measure on H, defined for Borel subsets $A \subset H$ by $\tilde{\mu}(A) = \mu(U^{-1}A)$. Denote the trace class operator on H associated to $\tilde{\mu}$ by S, and let $\{e_n\}$ be an orthonormal basis for H, where each e_n is an eigenvector of S with eigenvalue $\lambda_n \ge 0$.*

(i) *Denote the mean of $\tilde{\mu}$ by $\tilde{m}$. Then the range UE contains all the vectors of the form $x = \tilde{m} + \sum \lambda_n^{1/2} a_n e_n$ with $\sum a_n^2 < \infty$. In particular, $e_n \in UE$ whenever n is such that λ_n is strictly positive.*

(ii) *If μ is nondegenerate, so is $\tilde{\mu}$, and the vectors $z_n = U^{-1}e_n$ span a dense subspace of E.*

Proof. Without loss of generality, we can assume that μ, hence also $\tilde{\mu}$, are centered, i.e., $\tilde{m} = 0$.

(i) Assume that $x = \sum \lambda_n^{1/2} a_n e_n$ with $\sum a_n^2 < \infty$. If $x \notin UE$, then the translate of $\tilde{\mu}$ by x is singular with respect to $\tilde{\mu}$ (because $\tilde{\mu}$ is supported in UE). But this contradicts Proposition 6.19.

(ii) It is clear that $\tilde{\mu}$ is nondegenerate. Assume that $\{z_n\}$ span a proper closed subspace of E, and find $0 \neq x^* \in E^*$ such that $\langle x^*, z_n\rangle = 0$ for all n. Since the Gaussian measure μ on E is nondegenerate, the distribution of $\langle x^*, x\rangle$ is nondegenerate. Representing it by the (measurable but not necessarily continuous!) linear functional $\psi = x^* \circ U^{-1}$ on UE, we shall obtain a contradiction by showing that ψ is a constant $\tilde{\mu}$-almost everywhere.

The functional ψ satisfies $\psi(e_n) = \langle x^*, z_n\rangle = 0$ for all n, and we shall construct a sequence $\{y_m\}$ in H such that

(i) $\langle y_m, e_n\rangle = 0$ for every $n \leq m$

(ii) each y_m is a finite linear combination of the e_j's

(iii) $\langle y_m, x\rangle \to \psi(x)$ for $\tilde{\mu}$-almost every $x \in UE$.

This will finish the proof. Indeed, by passing to a subsequence we can assume that the y_m's are disjoint linear combinations of the e_j's. Thus they are independent Gaussian variables on $(UE, \tilde{\mu})$, and by the zero-one law their limit, namely ψ, must be constant, contradicting its choice.

To construct the y_m's, use the tightness of μ to find a compact convex symmetric set $K \subset E$ with $\mu(E \setminus K) < 1/m$, and put $\tilde{K} = UK$. Then $\tilde{\mu}(UE \setminus \tilde{K}) < 1/m$, and since the restriction of U to K is a homeomorphism, ψ is continuous on $\tilde{K}$. It follows that the set $B = \{x \in \tilde{K} : \psi(x) \geq 1/2m\}$ is compact.

Since B is convex and disjoint from the closed subspace spanned by $\{e_n\}_{n\leq m}$ (because $\psi(e_i) = 0$ for every i), there is a $z \in H$ such that $\langle z, e_n\rangle = 0$ for every $n \leq m$ and $\inf_{x\in B}\langle z, x\rangle > 0$. Truncating a tail of the expansion of z with respect to the e_i's, we can assume that $z = \sum_{i=m+1}^k a_i e_i$ is a finite linear combination, and we normalize it so that $\max_{x\in\tilde{K}}\langle z, x\rangle = 1$. Choose $x_0 \in \tilde{K}$ so that $\langle z, x_0\rangle = 1$, and put $y_m = \psi(x_0)z$.

Let $x \in \tilde{K}$; then also $u = \big(x - \langle z, x\rangle x_0\big)/2 \in \tilde{K}$ because $\tilde{K}$ is convex and symmetric. Since u belongs to the kernel of z, it does not belong to B nor to $-B$. Hence $\big|\psi(x) - \langle y_m, x\rangle\big| = \big|\psi\big(x - \langle z, x\rangle x_0\big)\big| = \big|2\psi(u)\big| \leq 1/m$.

Denote the set constructed in the mth step by $\tilde{K}_m$. Assuming, as we may, that $\tilde{K}_m \supset \tilde{K}_{m-1}$, the result follows. □

3. Gauss Null Sets

We begin this section with the definition of three natural classes of negligible sets in a Banach space: Gauss null sets, Aronszajn null sets, and cube null sets. We then show that these three classes actually coincide (see Propositions 6.25 and 6.27 and Theorem 6.32). We first introduce some notation. For each $0 \neq y \in E$ let $\mathcal{A}(y)$ be the family of all Borel sets A in E which intersect each line parallel to y in a set of one-dimensional Lebesgue measure 0. If $\{x_n\}$ is a finite or infinite sequence of nonzero points in E, we denote by $\mathcal{A}(\{x_n\})$ the collection of all Borel sets A which can be decomposed as $A = \bigcup A_n$, where $A_n \in \mathcal{A}(x_n)$ for every n.

Denote the Hilbert cube $[0,1]^{\aleph_0}$ by Q, and let τ be the standard product measure on Q. A subset of E is called a cube if it is the image of Q under a one-one affine map T of the form $T(q) = x + \sum t_n x_n$ for $q = (t_1, t_2, \dots) \in Q$, where $x \in E$ and the linearly independent vectors $x_n \in E$ have a dense linear span and satisfy $\sum \|x_n\| < \infty$. The image of τ under T is denoted by τ_T, and such measures on E are called cube measures.

Definition 6.23. *Let A be a Borel set in a separable Banach space E.*

(i) *The set A is called a Gauss null set if $\mu(A) = 0$ for every nondegenerate Gaussian measure on E.*

(ii) *The set A is called Aronszajn null if it belongs to $\bigcap \mathcal{A}(\{x_n\})$, where the intersection is taken over all sequences $\{x_n\}$ whose linear span is dense in E; i.e., for every sequence $\{x_n\}$ of nonzero vectors with dense linear span, A can be decomposed into a union of Borel sets $\{A_n\}$ so that $A_n \in \mathcal{A}(x_n)$ for every n.*

(iii) *The set A is called cube null if $\tau_T(A) = 0$ for each cube measure on E.*

It follows from the definition that all these families are closed under countable unions, translations and passing to (Borel) subsets.

Gauss null sets are Haar null. Indeed, let A be Gauss null. If μ is any nondegenerate Gaussian measure on E, then so are its translates; hence $\mu(A + x) = 0$ for every $x \in E$. The converse is, however, false. By Corollary 6.5 every compact subset of E is Haar null. On the other hand,

Proposition 6.24. *Let $\{x_n\}_{n=1}^{\infty}$ be a sequence with dense linear span in E such that $\|x_n\| \to 0$. Then the compact set $K = \overline{\mathrm{co}}\{\pm x_n\}$ is not Gauss null.*

Proof. Let H be a Hilbert space with an orthonormal basis $\{e_n\}$. Consider a one-dimensional symmetric Gaussian measure, normalized so that the measure of the interval $[-2^{-n}, 2^{-n}]$ is at least $1 - 2^{-n}$. Denote its variance by α_n. Then $\sum \alpha_n < \infty$, and the nonnegative operator S on H defined by $Se_n = \alpha_n e_n$ for all n is a trace class operator. Let μ be the nondegenerate Gaussian measure on H with mean 0 corresponding to this operator. Let B be the unit ball in H; then the representation of μ as a product measure shows that $\mu(B) \geq \prod(1 - 2^{-n}) > 0$. (Actually $\mu(B) > 0$ for any nondegenerate Gaussian measure on H, but we do not need this.) Define $T : H \to E$ by $Te_n = 2^{-n}x_n$, and let $\tilde{\mu}$ be the nondegenerate Gaussian measure on E defined by $\tilde{\mu}(A) = \mu(T^{-1}A)$. Then $TB \subset K$, and hence $\tilde{\mu}(K) \geq \tilde{\mu}(TB) = \mu(B) > 0$. □

Proposition 6.25. *Aronszajn null sets are Gauss null.*

Proof. Let A be an Aronszajn null set, and let μ be a nondegenerate Gaussian measure on E. Let $H, U, \tilde{\mu}$ and $\{z_n\}$ be as in Theorem 6.22. By (ii) of that theorem $\{z_n\}$ span a dense subspace of E. Since A is Aronszajn null it can be represented as $\bigcup A_n$, where each A_n is a Borel set which intersects every line in the direction of z_n in a set of linear measure 0. We shall prove that $\mu(A_n) = 0$ for all n.

Let μ_n be the n-th component in the product representation of $\tilde{\mu}$. Since UA_n intersects every line in H in the direction of e_n in a set of linear measure 0, and since μ_n is nondegenerate, it follows that $\mu_n(x + UA_n) = 0$ for every $x \in E$. By Fubini's theorem $\tilde{\mu}(UA_n) = 0$ and, of course, $\mu(A_n) = \tilde{\mu}(UA_n)$. □

If $A \in \mathcal{A}(\{y_m\})$, then it certainly also belongs to $\mathcal{A}(\{x_m\})$ for some sequence $\{x_n\}$ with a dense linear span. Indeed, just extend $\{y_m\}$ to a sequence with a dense span. It is important to note that the definition of an Aronszajn null set requires consideration of all sequences with dense span in E. The next proposition shows that one sequence cannot suffice.

Proposition 6.26. *Let μ be a σ-finite measure on a Banach space E. Then there is a sequence $\{y_n\}$ in E and a Borel set $A \subset E$ such that $A \in \mathcal{A}(\{y_n\})$ and $\mu(E \setminus A) = 0$. In particular, starting with a nondegenerate Gaussian measure μ, the set A is not Gauss null (because $\mu(A) = 1$), hence also not Aronszajn null. Yet $A \in \mathcal{A}(\{y_n\})$.*

Proof. Since every finite measure is tight, there are compact sets D_n such that $\mu(E \setminus \bigcup D_n) = 0$. Thus the theorem follows from the fact that for every compact set D in E there is a point $y \neq 0$ in E such that any line parallel to y intersects D by at most one point. Indeed, the symmetric convex hull K of D is also compact, and $\bigcup nK$ is not all of E, so pick $y \notin \bigcup nK$. □

Remark. The argument above gives a very simple proof to the fact that there is no quasi-invariant σ-finite measure on a separable Banach space (see Corollary 6.7). Indeed, assume that μ is a nonnegative σ-finite measure, and choose a compact set D such that $\mu(D) > 0$. Choose $y \in E$ such that the closed symmetric convex hull of D intersects the line in the direction of y only at 0. Then the translations $\{D + \alpha y : \alpha \in \mathbb{R}\}$ are mutually disjoint, and μ, being σ-finite, cannot give positive measure to all of them.

Proposition 6.27. *Gauss null sets are cube null.*

Proof. Let A be a Borel subset of the separable Banach space E, and assume that it is not cube null. Choose $x \in E$ and $\{x_n\} \subset E$ such that $\tau_T(A) > 0$ for the induced operator $T : Q \to E$. By translation we can assume that $x = 0$, and we put $\|x_n\| = \alpha_n^2$. Identify Q with the set $\{q = (t_n) : |t_n| \leq \alpha_n\}$ in l_2. The operator T is then identified with the restriction to Q of the bounded linear operator $Tq = \sum t_n \alpha_n^{-1} x_n$ from l_2 to E.

Choose $\beta_n < \alpha_n$ such that $\prod(\beta_n/\alpha_n) > 1 - \tau_T(A)$. It follows that $\tau(F) > 0$, where $F = \{|t_n| \leq \beta_n\} \cap T^{-1}(A)$. We shall construct a Gaussian measure ν on l_2 such that $\nu(F) > 0$. Its image under T will be a Gaussian measure on E which gives positive measure to A.

Consider the symmetric subset $W = \{q = (t_n) : |t_n| \leq \alpha_n - \beta_n\}$ of l_2. Then $q + F \subset Q$ for every $q \in W$, and thus $\tau(q + F) = \tau(F) > 0$ for such q. Let μ_n be a centered Gaussian measure on $\mathbb{R}$ with variance δ_n, and let $\mu = \prod \mu_n$ be the product measure on $\mathbb{R}^{\mathbb{N}}$. If the δ_n's satisfy $\sum \delta_n < \infty$, then μ is supported in l_2. We also choose $\delta_n \to 0$ so fast that $\mu(W) > 0$. We use an averaging argument to show that there is a $q \in Q$ such that $\mu(-q + F) > 0$; i.e., the q-translate of μ is the required Gaussian measure ν. Indeed, by Fubini's theorem and the symmetry of W

$$\begin{aligned} \int_Q \mu(-q + F) d\tau(q) &= \int_Q \int_{l_2} \chi_F(q' + q) d\mu(q') d\tau(q) \\ &\geq \int_W \int_Q \chi_F(q' + q) d\tau(q) d\mu(q') \geq \mu(W)\tau(F) > 0. \end{aligned}$$

It follows that there is a $q \in Q$ for which $\mu(-q + F) > 0$. □

The definition of $\mathcal{A}(\{x_n\})$ requires that the decomposition be made with Borel sets. The next lemma will be useful for checking that certain sets are indeed Borel.

Lemma 6.28. *Let F be a k-dimensional subspace of E, and let λ_k denote the Lebesgue measure on F. Then the function $f_B(x) = \lambda_k\big(F \cap (B+x)\big)$ is a Borel function for every Borel subset B of E.*

Proof. One checks directly that $\{x : f_B(x) \geq \alpha\}$ is closed for every closed and bounded set B and for every α. When B_j is a sequence of Borel sets with finite measure that either decreases or increases to the set B, then $f_{B_j}(x) \to f_B(x)$. Since every Borel set is obtained from bounded closed sets by countable iteration of such intersections and unions, the result holds for every Borel set. □

Proposition 6.29. *Let F be an n-dimensional subspace of E, and let $\{y_k\}_{k=1}^n$ be a basis for F. Let λ_n be the Lebesgue measure on F, and let A be a Borel subset of E such that $\lambda_n\big(F \cap (A+x)\big) = 0$ for every $x \in E$. Then $A \in \mathcal{A}(\{y_k\})$.*

Proof. We prove the proposition by induction on the dimension n of F. For $n = 1$ this is the definition of $\mathcal{A}(y_1)$. For the induction step, let G be the span of $\{y_k\}_{k=1}^{n-1}$, and assume that λ_n is normalized as the product of the Lebesgue measure λ_{n-1} in G and the measure λ_1 on the line determined by y_n. We shall also denote by the same symbol λ_1 the translates of this measure to lines parallel to y_n.

Let $A_n = \{x \in A : \lambda_1\big(A \cap (x + \mathbb{R}y_n)\big) = 0\}$. By Lemma 6.28 the set A_n is Borel; hence so is $A' = A \setminus A_n$, and it has the property that for each line L parallel to y_n either $L \cap A' = \emptyset$ or $L \cap A' = L \cap A$ and $\lambda_1\big(L \cap A'\big) > 0$.

Fix any $x \in E$. The result will follow from the induction hypothesis once we show that $\lambda_{n-1}\big(G \cap (A'+x)\big) = 0$.

The set $B = G \cap (A'+x)$ is contained in the projection C of $F \cap (A'+x)$ onto G in the direction of y_n. By the observation above, for every $u \in B$ the line L parallel to y_n that passes through u satisfies $\lambda_1\big(L \cap (A'+x)\big) > 0$. Thus

$$\begin{aligned} 0 &= \lambda_n\big((A+x) \cap F\big) \geq \lambda_n\big((A'+x) \cap F\big) \\ &= \int_C \lambda_1\Big((u + \mathbb{R}y_n) \cap (A'+x)\Big) d\lambda_{n-1} \\ &\geq \int_B \lambda_1\Big((u + \mathbb{R}y_n) \cap (A'+x)\Big) d\lambda_{n-1} \end{aligned}$$

by Fubini's theorem. It follows that $\lambda_{n-1}(B) = 0$, because the integrand in the last integral is strictly positive. □

Corollary 6.30. *Let $\{x_n\}$ and $\{y_n\}$ be two sequences with the same linear span. Then $\mathcal{A}(\{x_n\}) = \mathcal{A}(\{y_n\})$.*

Proof. Fix $y \in \operatorname{span}\{x_n\}_{n=1}^m$, and let $A \in \mathcal{A}(y)$. By Fubini's theorem $A + x$ intersects $F = \operatorname{span}\{x_n\}_{n=1}^m$ in a set of measure 0 for every x. It follows from Proposition 6.29 that $\mathcal{A}(y) \subset \mathcal{A}(\{x_n\}_{n=1}^m)$. □

Corollary 6.31. *Let $\{y_n\}$ be a sequence in E with a dense linear span. Then there are sequences $\{x_n\}$ in E and $\{x_n^*\}$ in E^* with $\langle x_n^*, x_m\rangle = \delta_{n,m}$ such that the linear span of $\{x_n\}$ is dense in E, the sequence $\{x_n^*\}$ separates the points of E, and $\mathcal{A}(\{x_n\}) = \mathcal{A}(\{y_n\})$.*

Proof. Let $\{z_n^*\}$ be a sequence in E^* that separates the points of E. The construction alternates between choices in E and E^* to ensure that the span of the x_n's (respectively x_n^*'s) is the same as that of the y_n's (respectively z_n^*'s). Choose $x_1 = y_1$ and take x_1^* in the linear span of $\{z_n^*\}$ so that $\langle x_1^*, x_1\rangle = 1$. Next pick $x_2^* \in \operatorname{span}\{z_n^*\}$ so that $z_1^* \in \operatorname{span}\{x_1^*, x_2^*\}$ and $\langle x_2^*, x_1\rangle = 0$. Then choose $x_2 \in \operatorname{span}\{y_n\}$ so that $\langle x_1^*, x_2\rangle = 0$ and $\langle x_2^*, x_2\rangle = 1$. In the next step we continue in the obvious way, ensuring that $y_2 \in \operatorname{span}\{x_1, x_2, x_3\}$, and so on. □

The next theorem completes (and is the main part of) the proof that Gauss, Aronszajn and cube null sets coincide. We shall nevertheless continue to use these different terms, and adopt in each situation the term which fits the discussion best.

Theorem 6.32. *Cube null sets are Aronszajn null.*

Proof. Let A be a Borel set in a separable Banach space E, and assume that A is not Aronszajn null. Choose a sequence $\{x_n\} \subset E$ whose linear span is dense in E and such that $A \notin \mathcal{A}(\{x_n\})$. We assume that $\|x_n\| = 1$ for all n, and in view of Corollary 6.31 we may also assume that there is a sequence $\{x_n^*\}_{n=1}^\infty \subset E^*$ with $\langle x_n^*, x_m\rangle = \delta_{n,m}$ for all n and m which separates the points of E.

We introduce some notation. The Lebesgue measure of a subset $C \subset \mathbb{R}$ is denoted by $|C|$. Let λ_k be the Lebesgue measure on the k-dimensional subspace E_k spanned by $\{x_n\}_{n=1}^k$, normalized as the image of the standard Lebesgue measure on $\mathbb{R}^k$ under the isomorphism $x \leftrightarrow \big(\langle x_1^*, x\rangle, \dots, \langle x_k^*, x\rangle\big)$. We shall also denote by the same symbol λ_k the translates of this measure to $x + E_k$ for every $x \in E$.

We shall find a closed set $F \subset A$ and closed intervals $\{I_n\}_{n=1}^\infty$ such that

(i) $0 < |I_n| \le 2^{-n}$ for all n,

(ii) $\langle x_n^*, x\rangle \in I_n$ for all n and for all $x \in F$,

(iii) there is a point $x_0 \in F$ with $\lambda_k\big(F \cap (x_0 + E_k)\big) \ge \frac{1}{2}\prod_{n=1}^k |I_n|$ for all k.

This will show that A is not cube null. Indeed, identify the Hilbert cube with $Q = \prod_{n=1}^\infty I_n$, and let $P_k : Q \to \prod_{n=1}^k I_n$ be the natural projections. Let τ_k and τ be the normalized product measures on $\prod_{n=1}^k I_n$ and on Q respectively. By translation we may assume that $x_0 = 0$. In particular $F \cap E_1 \neq \emptyset$ and thus $0 \in I_n$ for $n \ge 2$ by (ii). The affine map $T(t_1, t_2, \dots) = \sum t_n x_n$ from Q to E is one-one (since $\langle x_n^*, x_m\rangle = \delta_{n,m}$), $\sum \|t_n x_n\| < \infty$ for each $(t_1, t_2, \dots) \in Q$ by (i), and $F \subset TQ$ by (ii) and since $\{x_n^*\}$ separate the points of E.

Put $\tilde{F} = T^{-1}(F)$. We show that $\tau(\tilde{F}) \ge 1/2$, and hence $\tau_T(F) = \tau(\tilde{F}) > 0$ as required. Indeed, $\tilde{F}$ is a compact subset of Q (because F is closed), and hence $P_k^{-1}(P_k\tilde{F}) \searrow \tilde{F}$. Since $\tau\big(P_k^{-1}(P_k\tilde{F})\big) = \tau_k(P_k\tilde{F}) \ge 1/2$ for every k by (iii), it follows that $\tau(\tilde{F}) = \lim \tau\big(P_k^{-1}(P_k\tilde{F})\big) \ge 1/2$.

We need some more notation. We shall use Δ to denote a closed nondegenerate interval in $\mathbb{R}$ with rational endpoints. By W we shall denote finite sequences $W = (\Delta_1, \dots, \Delta_k)$ of such intervals, and for such a W and $j \le k$ we put $W_j = (\Delta_1, \dots, \Delta_j)$. Note that since we consider only intervals with rational endpoints, there are countably many Δ's and W's.

Each $W = (\Delta_1, \dots, \Delta_k)$ determines a cylinder C_W given by

$$C_W = \big\{x \in E : \langle x_i^*, x\rangle \in \Delta_i \text{ for every } 1 \le i \le k\big\}.$$

To every Borel set B, every sequence of intervals W as above and $0 < \gamma < 1$ we now assign a Borel set $B(W, \gamma)$. These sets are defined by induction on the length of the sequence W as follows: If W is empty, we put $B(W, \gamma) = B$. Assume now that $B(W, \gamma)$ is already defined for all sequences W of length $k-1$, and fix a sequence $W = (\Delta_1, \dots, \Delta_k)$ of length k. Consider the $(k-1)$ initial segment $W_{k-1} = (\Delta_1, \dots, \Delta_{k-1})$ of W, and put

$$B(W, \gamma) = \bigcup \left\{ B(W_{k-1}, \gamma) \cap C_W \cap F_k \right\}$$

where the union is taken over all the translates $F_k = x + E_k$ of E_k for which $\lambda_k\big(B(W_{k-1}, \gamma) \cap C_W \cap F_k\big) \geq \gamma \prod_{n=1}^{k} |\Delta_n|$.

It follows from Lemma 6.28 that each $B(W, \gamma)$ is a Borel set. The sets $B(W, \gamma)$ have the following properties:

(1) If $B \subseteq B_1$, $\gamma_1 \leq \gamma$ and W^* extends W, then $B(W^*, \gamma) \subseteq B_1(W, \gamma_1)$.

(2) If $\{B_i\}_{i=1}^{\infty}$ is a sequence of Borel sets which increases to the set B and if $\gamma_1 < \gamma$, then

$$B(W, \gamma) \subset \bigcup B_i(W, \gamma_1).$$

(3) Let $x \in B(W, \gamma)$, where W has length k, and let $j \leq k$. Then

$$\lambda_j\big(B(W, \gamma) \cap (x + E_j)\big) \geq \gamma \prod_{n=1}^{j} |\Delta_n|.$$

(4) For every Borel set B, every W, every $\gamma_1 < \gamma$ and every $\delta > 0$,

$$D = B(W, \gamma) \setminus \bigcup_{W^*} B(W^*, \gamma_1) \in \mathcal{A}(x_{k+1})$$

where the union is taken over all extensions of the k-tuple $W = (\Delta_1, \dots, \Delta_k)$ to a $(k+1)$-tuple $W^* = (\Delta_1, \dots, \Delta_{k+1})$ for which $|\Delta_{k+1}| < \delta$.

Property (1) is obvious, and (2) follows easily by induction on the length of W.

To prove (3), fix W of length k and $x \in B(W, \gamma)$. By the inductive definition of $B(W, \gamma)$, the whole k-dimensional section $B(W_{k-1}, \gamma) \cap C_W \cap (x + E_k)$ is included in $B(W, \gamma)$. In particular, $B(W, \gamma)$ contains the $(k-1)$-dimensional section $B(W_{k-1}, \gamma) \cap (x + E_{k-1})$; hence $B(W, \gamma) \cap (x + E_{k-1}) = B(W_{k-1}, \gamma) \cap (x + E_{k-1})$. More generally, $B(W, \gamma) \cap (x + E_j) = B(W_j, \gamma) \cap (x + E_j)$ for every $j < k$, and this proves (3).

The proof of (4) requires some more work. Note that D is indeed a Borel set, and we need to show only that its intersection with every line in the direction of x_{k+1} is a set of one-dimensional Lebesgue measure zero.

The image C of $B(W, \gamma)$ under x_{k+1}^* is a Lebesgue measurable subset of the real line. By Lebesgue's theorem almost every point of C belongs to the set C' of density points of C. Let $\mathcal{D}$ be the collection of closed intervals Δ with rational endpoints such that $|\Delta| < \delta$ and such that $|\Delta \cap C| \geq \gamma_1 |\Delta| / \gamma$. By the definition of density points, $\mathcal{D}$ covers C', and hence for every line L in the direction of x_{k+1} the one-dimensional measure of the set

$$L \cap B(W, \gamma) \setminus \bigcup_{\Delta \in \mathcal{D}} \{x \in L : \langle x_{k+1}^*, x \rangle \in \Delta\}$$

is zero. It remains to check that if $\Delta \in \mathcal{D}$, and if W^* denotes the extension of W by the interval Δ, then $x \in B(W^*, \gamma_1)$ whenever $x \in B(W, \gamma)$ and $\langle x_{k+1}^*, x \rangle \in \Delta$.

That $x \in C_{W^*}$ is clear. Denote the translate of E_{k+1} that passes through x by F_{k+1}, fix $t \in \Delta \cap C$, and let $F_k(t)$ be the translate of E_k that passes through

$x + tx_{k+1}$. Since (as in the proof of (3)) $B(W^*, \gamma) \cap F_k(t) = B(W, \gamma) \cap F_k(t)$, we see that

$$\lambda_k\big(B(W^*, \gamma) \cap C_{W^*} \cap F_k(t)\big) \geq \gamma \prod_{n=1}^{k} |\Delta_n|.$$

The measure of such t's in Δ is at least $\gamma_1|\Delta|/\gamma$ by the definition of $\mathcal{D}$. Hence, by Fubini's theorem

$$\begin{aligned}\lambda_{k+1}&\big(B(W^*, \gamma) \cap C_{W^*} \cap F_{k+1}\big)\\ &= \int_C \lambda_k\big(B(W^*, \gamma) \cap C_{W^*} \cap F_k(t)\big)dt\\ &\geq (\gamma_1|\Delta|/\gamma)\Big(\gamma \prod_{n=1}^{k} |\Delta_n|\Big) = \gamma_1 \prod_{n=1}^{k+1} |\Delta_n|.\end{aligned}$$

We now turn to the construction of the set F and of the intervals I_n. We refer to Appendix D for information on the Souslin operation. (Note that our description of the Souslin operation is somewhat unusual. The sets $A_{m_1,\dots,m_k}$ are not required to be closed, but are required instead to be contained in A.)

Since A is Borel, it can be represented via the Souslin operation as

$$A = \bigcup_{\overline{m}} \bigcap_k A_{m_1,\dots,m_k}$$

so that the intersection $\bigcap_k \overline{A_{m_1,\dots,m_k}}$ of the closures of the $A_{m_1,\dots,m_k}$'s is contained in A for every sequence of integers $\{m_k\}_{k=1}^{\infty}$.

Fix a sequence γ_k that strictly decreases to $1/2$. By property (4) above the sets $A_{m_1,\dots,m_k}(W, \gamma) \setminus \bigcup_{W*} A_{m_1,\dots,m_k}(W^*, \gamma_k)$ belong to $\mathcal{A}(x_{k+1})$ for every choice of k, integers $m_1, \dots, m_k$, a rational $\gamma > \gamma_k$, and a k-tuple of intervals W, and where the union is taken with respect to all extensions of W by intervals of length smaller than 2^{-k-1}. Since this is a countable collection of Borel subsets of A, and since $A \notin \mathcal{A}(\{x_n\})$, it follows that there is an $x_0 \in A$ such that

$$x_0 \notin A_{m_1,\dots,m_k}(W, \gamma) \setminus \bigcup_{W^*} A_{m_1,\dots,m_k}(W^*, \gamma_k) \tag{6.10}$$

for every choice of k, integers $m_1, \dots, m_k$, a rational $\gamma > \gamma_k$, and a k-tuple W, and where the union is taken with respect to extensions of W by intervals of length smaller than 2^{-k-1}.

We next choose inductively integers $\{m_n\}_{n=1}^{\infty}$ and intervals $\{I_n\}_{n=1}^{\infty}$ by using the representation of A via the Souslin operation and repeated use of (6.10).

Choose m_1 such that $x_0 \in A_{m_1}$ (this is possible because $A = \bigcup_{m_1} A_{m_1}$). By using (6.10) for A_{m_1} and the empty sequence as W, we obtain an interval, which we denote by I_1, such that $|I_1| < 2^{-1}$ and $x_0 \in A_{m_1}(I_1, \gamma_1)$.

Fix a rational $\gamma_2 < \gamma_1' < \gamma_1$. Since $A_{m_1,m_2} \nearrow A_{m_1}$ as $m_2 \nearrow \infty$, property (2) gives us an m_2 for which $x_0 \in A_{m_1,m_2}(I_1, \gamma_1')$. We then use (6.10) again to find I_2 with $|I_2| < 2^{-2}$ such that $x_0 \in A_{m_1,m_2}(W_2, \gamma_2)$, where $W_2 = (I_1, I_2)$.

The integer m_3 is chosen so that $x_0 \in A_{m_1,m_2,m_3}(W_2, \gamma_2')$ for some rational $\gamma_3 < \gamma_2' < \gamma_2$, and we continue in an obvious way. Since $\gamma_k > 1/2$ for all k, property (1) gives

$$x_0 \in A_{m_1,m_2,\dots,m_k}(W_k, 1/2)$$

for every k, where $W_k = (I_1, \dots, I_k)$.

Let $F_k = \overline{A_{m_1,\dots,m_k}(W_k, 1/2)}$, and put $F = \bigcap_{k=1}^{\infty} F_k$. Then F is a closed nonempty set (it contains x_0) contained in A (because $\bigcap_k \overline{A_{m_1,\dots,m_k}} \subset A$).

Condition (i) holds by the construction, and (ii) holds because $F \subset F_k \subset C_{W_k}$ for every k. To check (iii) fix k, and let $p \geq k$. Since $x_0 \in A_{m_1,\dots,m_p}(W_p, 1/2)$, it follows from property (3) that

$$\lambda_k\big(A_{m_1,\dots,m_p}(W_p, 1/2) \cap (x_0 + E_k)\big) \geq \frac{1}{2} \prod_{n=1}^{k} |I_n|.$$

Thus certainly

$$\lambda_k\big(F_p \cap (x_0 + E_k)\big) \geq \frac{1}{2} \prod_{n=1}^{k} |I_n|$$

and since $F_p \searrow F$, it follows that $\lambda_k\big(F \cap (x_0 + E_k)\big) \geq \frac{1}{2} \prod_{n=1}^{k} |I_n|$. □

Definition 6.33. *A Borel set A in a separable Banach space is called a directionally null set if there is an $x \in E$ such that $A \in \mathcal{A}(x)$. It is called σ-directionally null if it is a countable union of directionally null sets; i.e., $A \in \mathcal{A}(\{x_n\})$ for some sequence $\{x_n\}_{n=1}^{\infty}$.*

The σ-directionally null sets form a σ-ideal of Haar null sets. A set $A \subset E$ all of whose translates intersect a fixed finite-dimensional subspace of E by sets of measure zero is σ-directionally null by Proposition 6.29. By Corollary 6.31 we may assume that if $A \subset E$ is σ-directionally null, then it belongs to $A \in \mathcal{A}(\{x_n\})$ where $\{x_n\}$ span a dense subspace of E and there are biorthogonal functionals $\{x_n^*\} \subset E^*$.

Proposition 6.34. *A Borel set A is σ-directionally null if and only if there are $x_n \in E$ whose linear span is dense in E and biorthogonal vectors $x_n^* \in E^*$ such that the following statement holds: Let τ be the canonical product measure on $[0,1]^{\aleph_0}$. Then*

$$\tau\Big\{(t_1, t_2, \dots) \in [0,1]^{\aleph_0} : \, y + \sum_{n=1}^{\infty} t_n \lambda_n x_n \in A\Big\} = 0$$

for every $y \in E$ and positive scalars λ_n such that $\sum \lambda_n \|x_n\| < \infty$.

Proof. The "only if" part follows directly from Fubini's theorem. The "if" part follows from the proof of Theorem 6.32. □

The positive cone $C = \{\sum t_n e_n : t_n \geq 0\}$ in l_p $(1 \leq p < \infty)$ is a Haar null set (see Proposition 6.8) which is not σ-directionally null. Indeed, given any sequence $\{x_n\}$ in l_p, fix $\lambda_n > 0$ such that $\sum \lambda_n \|x_n\| < \infty$. Then $y = \sum \lambda_n |x_n| \in l_p$ and $y + \sum t_n \lambda_n x_n \in C$ for every $(t_1, t_2, \dots) \in [0,1]^{\aleph_0}$. By (the trivial part of) Proposition 6.34 C is not σ-directionally null.

There is a natural family of negligible sets which is properly contained in the class of Aronszajn null sets, namely the Borel sets which meet every line in E in a set of measure 0. Like the previous classes of negligible sets this class is closed under countable unions, translations and linear automorphisms of E. However, the sets in this class are too small to be appropriate from the point of view of differentiation theory. Already on $\mathbb{R}^2$ the set of points of nondifferentiability of the function $f(x, y) = |x|$ does not belong to this class. There are some instances, however, in which this class enters the discussion. This is the case in Example 6.35 and Theorems 6.36 and 6.39 below.

We now give an example of a Gauss null set which is mapped by a Lipschitz equivalence onto a set which is not Gauss null. It should be pointed out that the set D and $E \setminus A$ which appear in Theorem 6.14 are not Gauss null. Each of them contains translates of "cubes" of the form $\{\sum t_n \lambda_n x_n : 0 \le t_n \le 1\}$, where $\{x_n\}$ is a basis of E, and λ_n are some positive scalars.

Example 6.35. *There is a bi-Lipschitz map f from a ball B in a Hilbert space H onto another open subset of H, and a Borel subset $A \subset B$ which is not Gauss null and such that $f(A)$ meets every line in H in a set of measure 0 (actually in at most three points).*

Let S_1 and S_2 be two positive trace class operators on H with dense ranges, so that $S_1 H \cap S_2 H = \{0\}$. One way to construct such operators is as follows: Take H to be $L_2(0, 2\pi)$, and define S_1 by $S_1 e^{int} = 2^{-|n|} e^{int}$, $n \in \mathbb{Z}$. Then S_1 is a positive trace class operator with a dense range, and its range contains only real analytic functions. Let $V : H \to H$ be defined by $Vf = hf$ where $h = 1$ on $[0, \pi]$ and $h = -1$ on $(\pi, 2\pi]$. Then V is an isometry of H onto itself and the ranges of S_1 and of $S_2 = VS_1$ have only the origin in common.

Let $\{e_n\}$ be an orthonormal basis for H with $S_1 e_n = \lambda_n e_n$ for every n. We shall construct a nondegenerate Gaussian measure μ on H which is supported on $S_1 H$. Indeed, let μ be the mean zero Gaussian measure on H associated with the operator $S = S_1^3$. Its eigenvalues are $\{\lambda_n^3\}$. Taking $\beta_n = \lambda_n^2$ in (6.9), we conclude that μ is supported in $\{x = \sum a_n e_n \in H : \sum (a_n/\lambda_n)^2 < \infty\}$, which is exactly the range of S_1. It follows that $\mu(B \cap S_1 H) > 0$ for every ball B centered at 0. (We checked this in the proof of Proposition 6.24 if $\lambda_n \to 0$ sufficiently fast, which we can assume here also.) Thus $B \cap S_1 H$ is not a Gauss null set for any such ball.

The map $f(x) = x + \|x\|^2 S_2 x$ is a C^1 (actually C^∞) map which is Lipschitz on every bounded set and whose derivative at 0 is the identity. By the inverse mapping theorem it maps some open ball B around 0 in a bi-Lipschitz manner onto some open set in H.

To conclude the proof we check that $f(S_1 H)$ intersects any line in H in at most three points. Assume that $x, y, z \in S_1 H$ and that $f(x) - f(y) = t\big(f(y) - f(z)\big)$ for some $t \ne 0$. Since $S_1 H \cap S_2 H = \{0\}$, it follows that

$$x - y = t(y - z); \quad \|x\|^2 S_2 x - \|y\|^2 S_2 y = t(\|y\|^2 S_2 y - \|z\|^2 S_2 z).$$

Consequently, using the injectivity of S_2 and eliminating z, we obtain

$$t^2(\|x\|^2 x - \|y\|^2 y) = t^3 \|y\|^2 y - \|(t+1)y - x\|^2 \big((t+1)y - x\big).$$

This is a nontrivial quadratic equation in t (the term $t^3\|y\|^2 y$ cancels out when we expand $\|(t+1)y - x\|^2$, and $t = 0$ is not a solution since $y \ne x$). Thus it has at most one solution besides $t = -1$.

Note that the set A constructed above is contained in a proper (non-closed) linear subspace of H; hence it is Haar null. The next result requires a more elaborate construction and shows that sets which are not Haar null can also be mapped by Lipschitz equivalences onto Gauss null sets.

Theorem 6.36. *Let E be a separable infinite-dimensional Banach space. Then there is a Lipschitz equivalence h of E onto itself and a closed set A in E which is* *not Haar null such that $h(A)$ meets every line in E in a set of measure zero. In particular, $h(A)$ is Gauss null.*

The set A will be similar to the sets F_θ constructed in Theorem 6.6. It will contain a translate of every compact set in the unit ball of E. The inverse map h^{-1} maps the Gauss null set $h(A)$ onto a set which is not Haar null.

Fix $\alpha > 0$, $v \in E$ and $x^* \in E^*$ with $\langle x^*, v\rangle = 0$. The maps of the form

$$\psi_{\alpha,v,x^*}(x) = x + \alpha\langle x^*, x\rangle^2 v \tag{6.11}$$

will serve as building blocks in the construction. Each such map is a homeomorphism of E onto itself (its inverse is $\psi^{-1}_{\alpha,v,x^*}(y) = y - \alpha\langle x^*, y\rangle^2 v$, because if $y = \psi_{\alpha,v,x^*}(x)$, then $\langle x^*, y\rangle = \langle x^*, x\rangle$), and it is bi-Lipschitz on bounded sets.

We start with two lemmas. The first shows that when a map of the form (6.11) is applied to a thin board whose base is the unit ball of a hyperplane which does not contain v, then the image meets every line in a direction close to the hyperplane in a set of small measure. We denote the one-dimensional Lebesgue measure on lines in E, normalized so that the measure of an interval of length 1 is 1, by λ. The closed ball of radius r in E centered at the origin is denoted by $B(r)$.

Lemma 6.37. *For every $0 < \varepsilon, \alpha < 1$ there is a $\delta = \delta(\varepsilon, \alpha) > 0$ with the following property. Let E be a Banach space, and let $u, v \in E$ and $x^*, y^* \in E^*$ satisfy $\|u\| = \|v\| = \|x^*\| = \|y^*\| = 1$ and*

$$\langle x^*, v\rangle = \langle y^*, u\rangle = 0\,;\ \langle x^*, u\rangle = 1\,;\ \langle y^*, v\rangle \geq 1/4.$$

Let $\psi = \psi_{\alpha,v,x^*}$ be as in (6.11), and let $w \in E$ satisfy $\|u - w\| \leq 1/2$. Then $\psi\big(B(1)\bigcap \ker(y^*)\big) + B(\delta)$ intersects every line in the direction of w in a set of measure smaller than ε.

Proof. We start with a simple observation on quadratic polynomials: For each $\varepsilon > 0$ and $c_0 > 0$ there is an $\eta = \eta(\varepsilon, c_0) > 0$ such that whenever $q(t) = a + bt + ct^2$ is a quadratic polynomial with $c \geq c_0$, then the measure of $\{t : |q(t)| \leq \eta\}$ is smaller than ε. Indeed, by adding a constant to q and by a translation, it suffices to show that the measure of $\{t : ct^2 \in I\}$ is smaller than ε for every interval I of length 2η, and this is obvious.

The sets $\psi\big(B(1)\big) + B(\delta)$ are uniformly bounded for all $0 < \alpha, \delta < 1$. It follows that there is an absolute constant β such that $\psi^{-1}(x) = x - \alpha\langle x^*, x\rangle^2 v$ is β-Lipschitz on these sets for all such α and δ.

Pick any $z \in E$. We need to estimate the measure of the set of t's so that $z + tw \in \psi\big(B(1)\bigcap \ker(y^*)\big) + B(\delta)$. Such t's satisfy $\big|\big\langle y^*, \psi^{-1}(z + tw)\big\rangle\big| \leq \beta\delta$, i.e.,

$$\big| - \langle y^*, z + tw\rangle + \alpha\langle x^*, z + tw\rangle^2 \langle y^*, v\rangle\big| \leq \beta\delta.$$

Since $\langle x^*, w\rangle \geq \langle x^*, u\rangle - \|u - w\| \geq 1/2$, the coefficient of t^2 in the quadratic polynomial on the left-hand side, namely $\alpha\langle x^*, w\rangle^2\langle y^*, v\rangle$, is bounded below by $\alpha/16$, and the result follows from the observation in the beginning of the proof. □

Lemma 6.38. *Let E be a separable infinite-dimensional Banach space, and let $\{C_n\}_{n=1}^\infty$ be a family of finite subsets of $B(1)$ such that $\operatorname{dist}(C_n, C_m) \geq \eta$ for some $0 < \eta < 1$ and for all $n \neq m$. Then for every $u \in E$ with $\|u\| = 1$, $0 < \alpha < 1/2$ and $\varepsilon > 0$ there is a $\delta > 0$ and a Lipschitz equivalence f of E onto itself such that*

(i) $(1 - \alpha)\|x - y\| \leq \|f(x) - f(y)\| \leq (1 + \alpha)\|x - y\|$ *for all $x, y \in E$,*

(ii) $\|f(x) - x\| \leq \alpha$ *for all $x \in E$,*

(iii) $\lambda\Big(I \cap \bigcup_{n=1}^{\infty} \big(f(C_n) + B(\delta)\big)\Big) < \varepsilon$ *for every segment I of length one in a direction $w \in E$ which satisfies $\|w - u\| \le 1/2$.*

Proof. Fix $0 < \varepsilon_0 < \varepsilon\eta/2$ and $\alpha_0 = \eta\alpha/6$, and let $\delta = \min\{\delta(\varepsilon_0, \alpha_0), \eta/12\}$, where $\delta(\varepsilon_0, \alpha_0)$ is given by Lemma 6.37.

Choose $x^* \in E^*$ with $\|x^*\| = \langle x^*, u\rangle = 1$. For each n let $y_n^* \in E^*$ be a norm-one functional with $\langle y_n^*, u\rangle = 0$ such that $\langle y_n^*, x\rangle = 0$ for every $x \in C_n$. Choose also a norm-one $v_n \in E$ with $\langle x^*, v_n\rangle = 0$ and $\langle y_n^*, v_n\rangle \ge 1/4$.

Define a function $\varphi_n : C_n \cup \{x : \operatorname{dist}(x, C_n) \ge \eta/3\} \to \mathbb{R}$ by

$$\varphi_n(x) = \begin{cases} \alpha_0\langle x^*, x\rangle^2 & \text{if } x \in C_n \\ 0 & \text{if } \operatorname{dist}(x, C_n) \ge \eta/3. \end{cases}$$

Then φ_n is Lipschitz with constant $\alpha/2$ by the choice of α_0. By Lemma 1.1 it can be extended to an $\alpha/2$-Lipschitz function on all of E, which is bounded by α_0.

Let $f(x) = x + \sum_{n=1}^{\infty} \varphi_n(x)v_n$. The assumption on the distance between the different C_n's implies that f satisfies (i) and that (ii) holds with α even replaced by the smaller number α_0. We now check (iii).

By Lemma 6.37 any line in the direction w intersects $f(C_n) + B(\delta)$ (for any n) in a set of measure smaller than ε_0. By (ii) (with α_0) the distance between $f(C_n) + B(\delta)$ and $f(C_m) + B(\delta)$ with $n \ne m$ is at least $\eta - 2\alpha_0 - 2\delta \ge \eta/2$; hence any segment of length 1 can meet at most $2/\eta$ of these sets. It follows that if I is a segment of length 1 in the direction of w, then

$$\lambda\Big(I \cap \bigcup_{n=1}^{\infty} \big(f(C_n) + B(\delta)\big)\Big) \le 2\varepsilon_0/\eta < \varepsilon.$$

It remains to show that f is surjective. Since it is the identity outside the set $\bigcup_{n=1}^{\infty} \big(C_n + B(\eta/3)\big)$, it suffices to show that $f\big(C_n + B(\eta/3)\big) \supset C_n + B(\eta/3)$ for each n. Let L be any line in the direction of v_n. It follows from the definition of f that $f\big(L \cap \big(C_n + B(\eta/3)\big)\big) \subset L$. The surjectivity now follows because L was arbitrary and since $L \cap \big(C_n + B(\eta/3)\big)$ consists of a finite number of segments so that f is the identity on their endpoints. □

Proof of Theorem 6.36. Let $Q_1 \subset Q_2 \subset \ldots$ be an increasing sequence of finite sets in $B(1)$ whose union is dense in $B(1)$, and let $\{u_j\}_{j=1}^{\infty}$ be a dense sequence in the unit sphere of E. Put $\delta_0 = 1$. We shall define inductively positive numbers $\{\delta_n\}_{n=1}^{\infty}$, $\{\alpha_n\}_{n=1}^{\infty}$, $\{\eta_n\}_{n=1}^{\infty}$, vectors $\{y_{k,n}\}_{k,n=1}^{\infty}$ in E, and Lipschitz equivalences $\{f_n\}_{n=1}^{\infty}$ of E onto itself such that the following conditions hold for all relevant indices.

(1) $1/2 < \prod_{n=1}^{\infty}(1 - \alpha_n) < \prod_{n=1}^{\infty}(1 + \alpha_n) < 2$.

(2) $\sum_{m \ge n} \alpha_m \le \delta_{n-1}$.

(3) $y_{k,n} \in B(\delta_{n-1}/4)$.

(4) $\operatorname{dist}(Q_k + y_{k,n}, Q_l + y_{l,n}) > \eta_n$ whenever $l \ne k$.

(5) $\delta_n \le \min\{\eta_n/4, \alpha_n\}$.

(6) $(1 - \alpha_n)\|x - y\| \le \|f_n(x) - f_n(y)\| \le (1 + \alpha_n)\|x - y\|$ for all $x, y \in E$ and $\|f_n(x) - x\| \le \alpha_n$ for all $x \in E$.

(7) Let $h_n = f_n \circ f_{n-1} \circ \ldots \circ f_1$, and let I be a segment of length 1 in a direction $w \in E$ with $\|w - u_n\| \le 1/2$. Then $\lambda\Big(I \cap \bigcup_{k=1}^{\infty} \big(h_n(Q_k + y_{k,n}) + B(3\delta_n)\big)\Big) < 1/n$.

In the nth step we first choose α_n such that $\alpha_n < \min\{2^{-n+2}, \delta_{n-1}/2\}$. This ensures that (1) holds and since the δ_n's will be chosen to satisfy (5), (2) will hold too. Next, since E is infinite-dimensional, it is possible to choose $\{y_{k,n}\}_{k=1}^{\infty}$ and η_n such that (3) and (4) hold (see the proof of Theorem 6.6). In view of (1) and the assumption that (6) holds for indices smaller than n, it follows from (4) that

$$\operatorname{dist}\Big(h_{n-1}(Q_k + y_{k,n}), h_{n-1}(Q_l + y_{l,n})\Big) > \eta_n/2 \text{ whenever } l \neq k.$$

(Here we take h_0 to be the identity map.) We can now apply Lemma 6.38 for the sets $C_k = h_{n-1}(Q_k + y_{k,n})$ to find f_n and a δ_n such that (5), (6) and (7) hold for n.

The limit $h(x) = \lim h_n(x)$ exists by (1) and (6) and defines a surjective map of E onto itself such that $\|x - y\|/2 \leq \|h(x) - h(y)\| \leq 2\|x - y\|$ for all $x, y \in E$. As the set A we take

$$A = \bigcap_{n=1}^{\infty} \bigcup_{k=1}^{\infty} \Big(Q_k + y_{k,n} + B(\delta_n)\Big)$$

which is closed by (4) and (5). The argument at the end of the proof of Theorem 6.6 shows that A contains a translate of every compact set in $B(1/2)$.

By (2) and (6), $\|h(x) - h_n(x)\| \leq \delta_n$ for every n and $x \in E$. Since h_n is 2-Lipschitz, it follows that for all n and k

$$h\big(Q_k + y_{k,n} + B(\delta_n)\big) \subset h_n\big(Q_k + y_{k,n}\big) + B(3\delta_n)$$

hence, if $\|w - u_n\| \leq 1/2$ and if I is an interval of length 1 in the direction of w, then (7) yields

$$\lambda\big(h(A) \cap I\big) \leq \lambda\Big(I \cap \bigcup_{k=1}^{\infty} h\big(Q_k + y_{k,n} + B(\delta_n)\big)\Big) \leq 1/n.$$

Since $\{u_j\}$ is dense, we can choose n to be arbitrarily large for any given w, and thus $\lambda\big(h(A) \cap I\big) = 0$ for any interval of length 1 in E; i.e., $h(A)$ meets every line in a set of measure zero. ◇

We now pass to the decomposition result.

Theorem 6.39. *Every separable infinite-dimensional Banach space E can be decomposed into two Borel subsets $E = A \cup B$ so that the intersection of A with any line in E has measure 0 and B is a countable union of closed porous sets.*

Porous sets are defined in Definition 4.18. The set B is σ-porous, hence of the first category. The porous sets defining B are actually closed, and this will be used in Section 5. Note that such a decomposition is impossible for a finite-dimensional E. Indeed, by the Lebesgue density theorem porous (hence also σ-porous) sets in finite-dimensional spaces have Lebesgue measure 0.

Proof. We normalize the Lebesgue measure on a line in E according to the distance induced by the norm in E. All balls in this proof are open.

We first construct a sequence $\{z_n\}$ in E such that

$$\bigcup B(z_n, 6) = E \quad \text{and} \quad d\big(z_n, \operatorname{span}\{z_k : k < n\}\big) > 5.$$

To this end fix a dense sequence $\{x_n\} \subset E$, and then choose $\{z_n\}$ inductively such that $\|x_n - z_n\| \leq 11/2$ and $d\big(z_n, \operatorname{span}\{z_k : k < n\}\big) > 5$ for all $n > 1$. Indeed, having chosen z_k for $k < n$, let $x_n^* \in E^*$ satisfy $\|x_n^*\| = 1$, $\langle x_n^*, x_n\rangle \geq 0$ and $\langle x_n^*, z_k\rangle = 0$

for $k < n$, and choose $\|u_n\| = 1$ such that $\langle x_n^*, u_n\rangle > 10/11$. The required vector is $z_n = x_n + 11u_n/2$. It is clear that $\bigcup B(z_n, 6) = E$, and we put $G = \bigcup B(z_n, 1)$.

Let L be any line in E. If $d(z_n, L) < 1$, choose a point $w_n \in L$ such that $\|w_n - z_n\| < 1$. The distance between any two different w_n's is at least 3. Also if $i < j < k$ and the points w_i, w_j and w_k are defined, then the distance of w_k from the segment $[w_i, w_j]$ is at least $\|w_j - w_i\|$. Indeed, assume this were false, and write $w_k = \alpha w_i + \beta w_j$ for some α and β with $|\alpha|, |\beta| \leq 2$. By the construction $\|z_k - (\alpha z_i + \beta z_j)\| > 5$, and thus

$$0 = \big\|w_k - (\alpha w_i + \beta w_j)\big\| > 5 - \|w_k - z_k\| - 2\big(\|w_i - z_i\| + \|w_j - z_j\|\big) > 0$$

a contradiction. Easy induction shows that the diameter of a set of w_i's of cardinality $n \geq 2$ is larger than 2^{n-1}.

Let x, y be points in L with $\|x - y\| > 1$, and choose an integer m such that $2^{m-1} < \|y - x\| \leq 2^m$. It follows from the argument above that the segment $[x, y]$ intersects at most $m + 1$ of the balls defining G. Hence the Lebesgue measure of the intersection of $[x, y]$ with G is at most $2(m + 1) \leq (m + 1)2^{-m+2}\|y - x\|$. In other words, for every $k \geq 1$ there is an $s_k > 0$ such that every segment of length $t \geq s_k$ in E intersects G in a set whose Lebesgue measure is at most $2^{-k}t$.

Rescaling and putting $G_k = 2^{-k}s_k^{-1}G$, we find that every segment of length $t \geq 2^{-k}$ intersects G_k in a set of measure at most $2^{-k}t$. Put

$$A = \bigcap_{j=1}^{\infty} \bigcup_{k=j}^{\infty} G_k.$$

Given a segment I of length $t > 0$, choose j such that $t \geq 2^{-j}$. Then I intersects $\bigcup_{k=j}^{\infty} G_k$ (hence also A) in a set of measure at most $\sum_{k\geq j} 2^{-k}t = 2^{-j+1}t$. Since j can be chosen to be arbitrarily large, any line intersects A in a set of measure 0.

The (closed!) complement B_j of the set $\bigcup_{k=j}^{\infty} G_k$ is porous for every j. Indeed, fix $x \in B_j$ and $\delta > 0$, and choose $k > j$ such that $r_k = 2^{-k}s_k^{-1} < \delta/6$. Each G_k is a union of open balls of a common radius r_k, and the union of the balls with the same centers and radius $6r_k$ covers E. Choose one of these balls such that $x \in B(y, 6r_k)$. Then $\|y - x\| < \delta$ and $B(y, r_k) \subset G_k$; i.e., it is disjoint from B_j. Hence each B_j is porous, and the complement $B = \bigcup B_j$ of A has the desired form. □

4. Gâteaux Differentiability of Lipschitz Functions

The theorem of Lebesgue that every Lipschitz function from $\mathbb{R}$ to $\mathbb{R}$ is differentiable a.e. was generalized in Chapter 5 to mappings into a space with the Radon-Nikodým Property (RNP). By Rademacher's classical generalization of Lebesgue's theorem to functions of several variables, a Lipschitz map from $\mathbb{R}^n$ to $\mathbb{R}^m$ is differentiable almost everywhere. The first proposition in this section combines these two generalizations. Our next step will be to generalize the result further by allowing the domain space to be infinite-dimensional. Here we use the notions of negligible sets from Sections 1 and 3. This will be the main result of this section. We start with a simple technical lemma.

Lemma 6.40. *Let E and F be Banach spaces, and let f be a Lipschitz function from an open set in E into F. Let $G \subset E$ be a dense additive subgroup, and assume that for some $x_0 \in E$ and for all $u \in G$ the directional derivatives at x_0*

$$\lim_{t\to 0} \big(f(x_0 + tu) - f(x_0)\big)/t$$

exist and are additive as a function of u. *Then* f *is Gâteaux differentiable at* x_0.

Proof. The functions $h_t(u) = \big(f(x_0+tu) - f(x_0)\big)/t$ for $t \neq 0$ are equicontinuous. (They are Lipschitz in u with the same constant as f.) Hence the existence of $\lim_{t\to 0} h_t(u)$ for $u \in G$ implies that the limit actually exists for all $u \in \overline{G} = E$. Also the additivity of the limit on G implies its additivity on E, and since by definition $\lim_{t\to 0} h_t(\alpha u) = \alpha \lim_{t\to 0} h(u)$, the limit is a linear operator. This linear operator is bounded by the Lipschitz constant of f. □

Proposition 6.41. *Let* F *be a Banach space with the RNP. Then every Lipschitz function* f *from an open subset* U *of* $\mathbb{R}^n$ *into* F *is Gâteaux differentiable a.e. in* U.

Proof. It clearly suffices to prove the proposition for f which is defined on all of $\mathbb{R}^n$. Let G be a countable dense additive subgroup of $\mathbb{R}^n$. By Theorem 5.21 the directional derivatives of f in the directions of the vectors in G exist almost everywhere. In view of Lemma 6.40 it suffices to show that these directional derivatives are almost everywhere additive functions on G. This will be done by using convolutions. Let $\varphi(x)$ be a C^1 function on $\mathbb{R}^n$ with compact support such that $\int_{\mathbb{R}^n} \varphi(x)dx = 1$. (For example $\varphi(x) = Ce^{-1/(1-\|x\|^2)}$ for $\|x\| < 1$ and $\varphi(x) = 0$ for $\|x\| \geq 1$ for an appropriate C, where the norm in $\mathbb{R}^n$ is the Euclidean norm.)

The function $g(x) = f * \varphi(x) = \int_{\mathbb{R}^n} f(y)\varphi(x-y)dy$ is a C^1 function; hence $D_g(x)u = \big(f * D_\varphi(x)\big)u$ is a linear function of $u \in \mathbb{R}^n$ for every x.

On the other hand, for every x and for all $u \in G$, Lebesgue's dominated convergence theorem gives

$$\begin{aligned} D_g(x)u &= \lim_{t\to 0} \big(g(x+tu) - g(x)\big)/t \\ &= \int_{\mathbb{R}^n} \varphi(y) \lim_{t\to 0} \big(f(x+tu-y) - f(x-y)/t\big)dy. \end{aligned}$$

In other words, $D_g(x)u = \varphi * h_u(x)$ where h_u is the bounded measurable function $\lim_{t\to 0}\big(f(x+tu) - f(x)\big)/t$. Consequently, $\varphi * (h_{u+v} - h_u - h_v) = 0$ for every $u, v \in G$. The same equation will hold if we replace $\varphi(x)$ by $\varphi_k(x) = k^n\varphi(kx)$. Since $\lim_{k\to\infty} \varphi_k * \tilde{h} = \tilde{h}$ pointwise a.e. (at every Lebesgue point of $\tilde{h}$) for every bounded measurable function $\tilde{h}$ from $\mathbb{R}^n$ to $\mathbb{R}$, the same is valid for functions from $\mathbb{R}^n$ to F. (Apply linear functionals on F, and note that we may assume F to be separable since f has a separable image.) We conclude that for every u and v in G the identity $h_{u+v}(x) = h_u(x) + h_v(x)$ holds a.e. on $\mathbb{R}^n$. Since G is countable the exceptional set of x's can be chosen independently of u and v.

Second proof. We now give a second proof of the linearity a.e. of the directional derivatives. Actually, we prove a stronger and more general statement:

Let E *and* F *be Banach spaces with* F *separable, and let* $f : E \to F$ *be a Lipschitz function. Fix* $u, v \in E \setminus \{0\}$, *and let* A *be the set of all* $x \in E$ *such that both* $D_f(x)u$ *and* $D_f(x)v$ *exist, but* $D_f(x)(u+v)$ *either fails to exist or is different from* $D_f(x)u + D_f(x)v$. *Then* A *is* σ*-porous.*

This will give the result, because σ-porous sets in finite-dimensional spaces have measure zero. The statement will follow, using the separability of F, once we show that for all integers k and l and for all $y, z \in F$, the set $A(k,l,y,z)$ of all $x \in E$ satisfying the three conditions

(i) $\|f(x+tu) - f(x) - ty\| \leq |t|/k$ for every $|t| \leq 1/l$,

(ii) $\|f(x+tv)-f(x)-tz\| \leq |t|/k$ for every $|t| \leq 1/l$,

(iii) $\limsup_{s\to 0} \|f\big(x+s(u+v)\big)-f(x)-s(y+z)\|/|s| > 3/k$

is porous.

Denote the Lipschitz constant of f by L. Given $x \in A(k,l,y,z)$, choose arbitrarily small s such that $\|f\big(x+s(u+v)\big)-f(x)-s(y+z)\| > 3|s|/k$, and consider an arbitrary w with $\|w\| < |s|/2kL$. Then (assuming, as we may, that $|s| < 1/l$)

$$\begin{aligned}&\|f\big((x+su+w)+sv\big)-f(x+su+w)-sz\| \\ &\quad\geq \|f(x+su+sv)-f(x+su)-sz\|-2L\|w\| \\ &\quad\geq \|f(x+su+sv)-f(x)-s(y+z)\|-2L\|w\|-|s|/k > |s|/k.\end{aligned}$$

Hence $x+su+w \notin A(k,l,y,z)$; i.e., $B(x+su, |s|/2kL)\bigcap A(k,l,y,z)=\emptyset$. □

We now come to the main result of this section.

Theorem 6.42. *Let E be a separable Banach space, and let F be a space with RNP. Let f be a Lipschitz function from an open set U in E into F. Then the set of points in U in which f is not Gâteaux differentiable is Aronszajn null.*

Proof. We note first that the set of points where f is Gâteaux differentiable is a Borel set. This follows from Lemma 6.40 and from the fact that the derivative of f in the direction u exists at x if and only if for every integer n there is an integer m such that $\big|\big(f(x+tu)-f(x)\big)/t-\big(f(x+su)-f(x)\big)/s\big| < 1/n$ whenever s, t are rationals with $|s|, |t| \leq 1/m$. The other sets appearing in the proof below are Borel for a similar reason. It is clearly enough to consider the case where f is defined on all of E.

Let $\{x_n\}$ be any sequence of linearly independent vectors whose linear span is dense in E, and put $V_n = \operatorname{span}\{x_k : k \leq n\}$. Let D_n be the set of those $x \in E$ such that the directional derivative of f exists at x in all the directions $u \in V_n$ and is linear in $u \in V_n$. For every $y \in E$ the set $\big((E\setminus D_n)+y\big)\cap V_n$ is the set where the restriction of the function $f_y(x) = f(x-y)$ to V_n fails to be differentiable. By Proposition 6.41 this set has measure 0 in V_n; hence by Proposition 6.29 $E\setminus D_n$ belongs to $\mathcal{A}\big(\{x_k : k \leq n\}\big)$. It follows that $E\setminus\big(\bigcap_n D_n\big) \in \mathcal{A}\big(\{x_k\}\big)$. The function f is Gâteaux differentiable on $\bigcap_n D_n$ in view of Lemma 6.40. □

Remark. For the weaker statement that the set of nondifferentiability is Haar null, we do not need to introduce the notion of Aronszajn null sets. In this case, we do not need Proposition 6.29 and use instead the fact that a set A, all of whose translates meet a fixed finite-dimensional space in a set of Lebesgue measure 0, is Haar null. In this case, the V_n's could be any sequence of finite-dimensional subspaces whose union is dense in E.

Corollary 6.43. *Let E be separable, and let F have the RNP. Then any Lipschitz function $f : E \to F$ can be approximated uniformly by Lipschitz functions which are everywhere Gâteaux differentiable.*

Proof. Fix a nondegenerate Gaussian measure μ on E with $\int_E \|y\|d\mu(y) < \infty$ (see the discussion following Proposition 6.20), and let M be the Lipschitz constant of f. Put $g_n(x) = \int_E f(x+n^{-1}y)d\mu(y)$. All the g_n's are Lipschitz with constant M and are everywhere Gâteaux differentiable by Theorem 6.42 and by Lebesgue's dominated convergence theorem. Also

$$\|f(x) - g_n(x)\| = \Big\| \int_E \big(f(x) - f(x + n^{-1}y)\big) d\mu(y) \Big\| \le n^{-1} M \int_E \|y\| d\mu(y).$$

□

In view of the results of Chapter 5, the assumption that F has the RNP is essential in all the results above. However, when we consider weaker notions than Gâteaux differentiability, there are versions of Theorem 6.42 for more general F.

Corollary 6.44. *Let E and F be separable Banach spaces, and let $f : E \to F^*$ be a Lipschitz function. Then f is w^*-Gâteaux differentiable outside an Aronszajn null set; i.e., off this set $\lim_{t\to 0} \big(f(x+tu) - f(x)\big)/t$ exists in the w^* sense for every $u \in E$, and the limit is a bounded linear operator in u.*

Proof. We follow the proofs of Proposition 6.41 and Theorem 6.42, but replace the application of Theorem 5.21 by the classical theorem of Lebesgue applied to the scalar functions $\langle f(x), w_j \rangle$, where $\{w_j\}$ is any dense set in F. □

Our next result explains why we restrict ourselves to Lipschitz functions.

Proposition 6.45. *Let f be a continuous function from an open set V in a Banach space E into a Banach space F. Assume that*

$$\limsup_{t\to 0} \big\| \big(f(x + tu) - f(x)\big)/t \big\| < \infty$$

for a set (x, u) of second category in $V \times E$. Then the restriction of f to some ball inside V is Lipschitz.

Proof. Fix an integer n. The set A_n of all pairs (x, u) such that $x \in V$ and such that $\|\big(f(x + tu) - f(x)\big)/t\| \le n\|u\|$ whenever $|t| \le 1/n$ and $x + tu \in V$ is a relatively closed set in $V \times E$ (since f is continuous). Since $\bigcup A_n$ is of second category, there are $(x_0, u_0) \in V \times E$, a natural number N, and $r > 0$ such that

$$\|f(x + tu) - f(x)\| \le N|t|\|u\|$$

whenever $|t| \le 1/N$, the points $x, x + tu$ are in $B(x_0, r)$, and $u \in B(u_0, r)$. We claim that the restriction of f to $B(x_0, r/2)$ is a Lipschitz function.

Let $x, y \in B(x_0, r/2)$ with $\|x - y\| \le \min\big(r/N, r^2/2(\|u_0\| + r)\big)$, and put $t = \|x - y\|/r$ and $v = r(y - x)/\|x - y\|$. Then $|t| \le 1/N$, $\|v\| = r$, $y = x + tv$ and $y + tu_0 \in B(x_0, r)$. Taking $K = N + 2r^{-1}N\|u_0\|$, we obtain

$$\begin{aligned}
\|f(x) - f(y)\| &\le \|f(x) - f(y + tu_0)\| + \|f(y + tu_0) - f(y)\| \\
&= \|f\big(x + t(v + u_0)\big) - f(x)\| + \|f(y + tu_0) - f(y)\| \\
&\le Nt\big(\|v + u_0\| + \|u_0\|\big) \le Nt\big(\|v\| + 2\|u_0\|\big) \le K\|x - y\|.
\end{aligned}$$

□

5. Examples Related to Fréchet Differentiability

In this section we show that the analog of Theorem 6.42 fails completely for Fréchet differentiability.

If E^* is nonseparable, then we have already proved that E has an equivalent norm which is nowhere Fréchet differentiable (Proposition 4.12). If E^* is separable, then continuous convex functions on E are Fréchet differentiable outside a σ-porous set (Theorem 4.19). But are they also Fréchet differentiable outside a null set? That the answer is negative in nonreflexive spaces follows from Theorem 6.12. Indeed, for every convex set B in E the function $f(x) = d(x, B)$ is a convex function which fails to be Fréchet differentiable on ∂B; and if E is a nonreflexive (separable) space, then by Theorem 6.12 it contains a closed convex set B with empty interior which is not Haar null. This approach fails if E is reflexive, but convex functions need not be a.e. Fréchet differentiable even in reflexive spaces.

Example 6.46. *There is a continuous convex function $f : l_2 \to \mathbb{R}$ such that the set in which f is Fréchet differentiable is an Aronszajn null set.*

For the construction we need some notation and a geometric lemma. A 5-dimensional cube in l_2 with edges of length K is a set of the form

$$T = \Big\{x + \sum_{i=1}^{5} \alpha_i u_i : 0 \le \alpha_i \le K\Big\}$$

for some $x \in l_2$ and orthonormal vectors $\{u_i\}_{i=1}^5$. The normalized Lebesgue measure on T is denoted by μ_T.

Lemma 6.47. *For every $\varepsilon > 0$ there is an $r > 0$ and a countable set A in l_2 such that*

$$\bigcup_{x \in A} B(x, r+\delta) = l_2 \quad \text{and} \quad \mu_T\Big(T \cap \bigcup_{x\in A} B(x,r)\Big) \le \varepsilon$$

for every $\delta > 0$ and 5-dimensional cube T with edges of length 1 in l_2.

Proof. Denote the unit vector basis of l_2 by $\{e_k\}$, and let $\{x_n\}$ be a dense sequence in l_2 consisting of finitely supported vectors. Choose a strictly increasing sequence of integers $\{k_n\}$ such that x_n is supported on $\{e_k\}_{k<k_n}$ for every n, and put $z_n = x_n + e_{k_n}$. Let $A' = \{z_n\}$. Since $\{x_n\}$ is dense in l_2 it follows that

$$\bigcup B(z_n, 1+\delta) = l_2$$

for every $\delta > 0$. We shall verify that there is a constant $K = K(\varepsilon)$ such that for any 5-dimensional cube T with edges of size K

$$\mu_T\Big(T \cap \bigcup_n B(z_n, 1)\Big) \le \varepsilon. \tag{6.12}$$

The desired result then follows by taking $A = K^{-1}A'$ and $r = K^{-1}$.

Fix $\varepsilon > 0$ and $K > 6$ (to be specified later), and let x and $\{u_i\}_{i=1}^5$ define a cube T with edge size K. To prove (6.12) we have to consider only those indices n for which the orthogonal projection w_n of z_n on the affine span of T satisfies

$$\|z_n - w_n\| < 1 \quad \text{and} \quad d(w_n, T) \le 1.$$

Denote the set of these indices by I, and put for $j = 0, 1, \dots$

$$I_j = \big\{n \in I : 1 - 2^{-j} \le \|z_n - w_n\| \le 1 - 2^{-j-1}\big\}.$$

If $n \in I_j$, then $T \cap B(z_n, 1)$ is contained in $T \cap B\Big(w_n, \big(1-(1-2^{-j})^2\big)^{1/2}\Big)$. It follows that there is a constant c_1 such that

$$\mu_T\big(T \cap B(z_n, 1)\big) \le c_1 K^{-5} 2^{-5j/2}$$

for every $n \in I_j$. To estimate the cardinality of I_j, decompose it as $I'_j \cup I''_j$, where

$$I'_j = \left\{n \in I_j : |\langle u_i, e_{k_n}\rangle| \leq \left(5(K+1)2^{j+2}\right)^{-1} \text{ for all } i \leq 5\right\}$$

and $I''_j = I_j \setminus I'_j$. Since $\|u_i\| = 1$ for each i, it follows that $|I''_j| \leq c_2 K^2 2^{2j}$ for every j. To estimate $|I'_j|$, note that for every $n \in I$ there are $-1 \leq \alpha_{i,n} \leq K+1$ such that

$$w_n = x + \sum_{i=1}^{5} \alpha_{i,n} u_i.$$

Hence $\left|\langle e_{k_n}, \sum_{i=1}^5 \alpha_{i,n} u_i\rangle\right| \leq 2^{-(j+2)}$ for $n \in I'_j$.

Write $x = z_n + (w_n - z_n) - \sum_{i=1}^5 \alpha_{i,n} u_i$. Since $\langle e_{k_n}, z_n\rangle = 1$ by the definition of k_n, it follows that

$$\begin{aligned}\langle e_{k_n}, x\rangle &\geq \langle e_{k_n}, z_n\rangle - \|w_n - z_n\| - 2^{-j-2} \\ &\geq 1 - (1 - 2^{-j-1}) - 2^{-j-2} = 2^{-j-2}.\end{aligned}$$

Denote the orthogonal projection in l_2 onto $\operatorname{span}\{e_k : k > m\}$ by Q_m. Since $Q_{k_n} z_n = 0$ for every n, it follows that if $n \in I$, then

$$\begin{aligned}\|Q_{k_n} x\| &\leq \|x - z_n\| \leq \|w_n - z_n\| + \Big\|\sum_{i=1}^{5} \alpha_{i,n} u_i\Big\| \\ &\leq 1 + \sqrt{5}(K+1) < 3K.\end{aligned}$$

Let m be the smallest integer so that $\|Q_m x\| \leq 3K$. The argument above shows that $m \leq k_n$ for all $n \in I$. Hence for every j

$$9K^2 \geq \|Q_m x\|^2 \geq \sum_{n \in I, k_n \neq m} |\langle e_{k_n}, x\rangle|^2 \geq \sum_{n \in I_j, k_n \neq m} |\langle e_{k_n}, x\rangle|^2 \geq 2^{-2j-4}(|I'_j| - 1).$$

Thus $|I'_j| \leq c_3 K^2 2^{2j}$, and hence $|I_j| \leq (c_2 + c_3) K^2 2^{2j}$ for all j. Consequently,

$$\mu_T\Big(T \cap \bigcup_n B(z_n, 1)\Big) \leq \sum_{j=0}^{\infty} |I_j| c_1 K^{-5} 2^{-5j/2} \leq c_4 K^{-3}$$

and we only have to choose K such that $c_4 K^{-3} < \varepsilon$. □

Construction of the example. Fix $0 < \varepsilon < 1$, and consider the set A and the r corresponding to this ε by the lemma. Given $M > 0$, consider the two sets

$$D = \Big\{y \in l_2 : \|y\| \leq M \text{ and } y \notin \bigcup_{x \in A} B(x, r)\Big\}$$

$$K = \Big\{(y, t) : t \geq \|y\|^2, \text{ and either } y \in D \text{ or } \|y\| \geq M + 1\Big\}$$

in l_2 and $l_2 \oplus \mathbb{R}$ respectively. Let C be the closed convex hull of K, and consider the function $g : l_2 \to \mathbb{R}$ defined by $g(y) = \inf\{t : (y, t) \in C\}$.

It follows from the convexity of the function $y \to \|y\|^2$ on l_2 that $g(y) = \|y\|^2$ for $y \in D$ and for $\|y\| \geq M + 1$. Also $g(y) \geq \|y\|^2$ for all y, and $g(y) \leq (M+1)^2$ if $\|y\| \leq M + 1$. It is also clear that g is a convex continuous function on l_2. We claim that g is not Fréchet differentiable on the set D.

Fix $y_0 \in D$. Since $D_{\|y\|^2}(y_0) u = 2\langle y_0, u\rangle$, it follows that the functional $\langle 2y_0, \cdot\rangle$ belongs to the subdifferential of g at y_0. We shall show that $\langle 2y_0, \cdot\rangle$ cannot be the Fréchet derivative of g at y_0, and this will prove our claim.

Let $0 < \delta < 1/6$. Since $\bigcup_A B(x, r(1+\delta)^{1/2})$ covers l_2, there is an $x \in A$ such that $\|y_0 - x\|^2 \le r^2(1+\delta)$. The ball $B(x,r)$ is disjoint from D as well as from the set $\{y : \|y\| \ge M+1\}$ (recall that by the construction $r \le 1/6$, and by our choice $\delta < 1/6$ too). It follows that whenever $(z,t) \in K$, then $z \notin B(x,r)$; hence

$$\begin{aligned}\|z\|^2 &= \|y_0\|^2 + 2\langle y_0, z - y_0\rangle + \|z - y_0\|^2 \\ &= \|y_0\|^2 + 2\langle y_0, z - y_0\rangle - \|y_0 - x\|^2 + \|z - x\|^2 + 2\langle x - y_0, z - y_0\rangle \\ &\ge \|y_0\|^2 + 2\langle y_0, z - y_0\rangle - \delta r^2 + 2\langle x - y_0, z - y_0\rangle\end{aligned}$$

and thus

$$t \ge \|y_0\|^2 + 2\langle y_0, z - y_0\rangle - \delta r^2 + 2\langle x - y_0, z - y_0\rangle.$$

Since this is a linear inequality, it also holds for every $(z,t) \in C$. Since also $g(y_0) = \|y_0\|^2$, it follows that

$$g(z) \ge g(y_0) + 2\langle y_0, z - y_0\rangle - \delta r^2 + 2\langle x - y_0, z - y_0\rangle$$

for every $z \in l_2$. Applying this inequality to $z = y_0 + \delta(x - y_0)$ (and using twice $\|x - y_0\| \ge r$, which holds because $y_0 \in D$), we obtain

$$g(z) - g(y_0) - 2\langle y_0, z - y_0\rangle \ge 2\delta\|x - y_0\|^2 - \delta r^2 \ge \delta\|x - y_0\|^2 \ge \|z - y_0\| r.$$

This shows that $\langle 2y_0, \cdot\rangle$ is not the Fréchet derivative of g at y_0. Indeed, for any small $\delta > 0$ we found a point z with $\|z - y_0\| \le 2r\delta$, for which the left-hand side of the inequality above (i.e., the approximation of $g(z)$ by the value of the linear function determined by $\langle 2y_0, \cdot\rangle$) is not $o(\|z - y_0\|)$.

Now let $\{\varepsilon_n\}$ be a sequence of positive numbers decreasing to 0, and let A_n and r_n correspond to ε_n by the lemma. By the above, there are convex and continuous functions $g_n : l_2 \to \mathbb{R}$ such that $|g_n(y)| \le (n+1)^2$ if $\|y\| \le n+1$ and such that g_n fails to be Fréchet differentiable at the points of the set

$$\Big\{y : \|y\| \le n \text{ and } y \in l_2 \setminus \bigcup_{x \in A_n} B(x, r_n)\Big\}.$$

The function $f(y) = \sum n^{-4} g_n(y)$, which is continuous and convex on l_2, is the required function. Indeed, by the discussion preceding Proposition 4.7, if the sum of convex functions is Fréchet differentiable at some point, then each summand must be Fréchet differentiable at this point. Hence the set of points of Fréchet differentiability of f in $\{y : \|y\| \le n\}$ is contained in $\bigcap_{k=n}^{\infty} \bigcup_{x \in A_k} B(x, r_k)$. However, this set meets every 5-dimensional cube with edge length 1 by a set of measure 0, and therefore the set of points of Fréchet differentiability of f meets every 5-dimensional affine subspace of l_2 by a set of measure 0. Such a set is Aronszajn null by Proposition 6.29, and this finishes the proof that f has the required properties.

We show next, using Theorem 6.39 and the comments following its statement, that on every separable Banach space E there is a Lipschitz (not necessarily convex) function $f : E \to \mathbb{R}$ such that the set on which f is Fréchet differentiable meets every line in E in a set of measure 0.

Theorem 6.48. *Let E be a separable Banach space, and let A be a subset of E which is a countable union of closed porous sets. Then there is a Lipschitz function from E to $\mathbb{R}$ which is not Fréchet differentiable at any point of A.*

We shall need the following lemma.

Lemma 6.49. *Let A be a closed porous set in a Banach space E. Then there is a disjoint family $\mathcal{B}$ of closed balls of radii at most 1 in E which are disjoint from A, and there is a constant $M > 1$ with the following property: Denote by $\mathcal{B}(\delta)$ the collection of all balls of the form $B(z, Mr)$, where $B(z,r) \in \mathcal{B}$ and $r < \delta$. Then the union of $\mathcal{B}$ and $\mathcal{B}(\delta)$ covers E for every $\delta > 0$.*

Proof. For each $i \geq 0$ let D_i be a maximal subset of E such that $B(x, 2^{-i}) \cap A = \emptyset$ for $x \in D_i$ and $\|x-y\| > 2^{-i+2}$ for $x \neq y$ in D_i. Define families $\mathcal{B}_i$ inductively as follows. $\mathcal{B}_0 = \{B(z,1) : z \in D_0\}$, and for $i > 0$ the family $\mathcal{B}_i$ consists of those balls $B(z, 2^{-i})$, $z \in D_i$ which are disjoint from all the balls of the preceding families. Put $\mathcal{B} = \bigcup_i \mathcal{B}_i$. Let $\delta > 0$, and fix $x \notin \bigcup_{B \in \mathcal{B}} B$. We have to show that there is an $r < \delta$ and a ball $B(z,r)$ in some $\mathcal{B}_k$ such that $x \in B(z, Mr)$ for a suitably chosen M. Since $\bigcup_{B \in \mathcal{B}_i} B$ is closed for every i (by the separation requirement on D_i), there is a $\sigma > 0$ such that any ball in $\mathcal{B}$ meeting $B(x,\sigma)$ has radius smaller than δ. Let $0 < \lambda < 1$ be the constant associated to A by the definition of porosity, and choose $y \in E$ such that $0 < \|y - x\| < \sigma/4$ and $A \cap B(y, \lambda\|y-x\|) = \emptyset$. (For $x \in A$ such a y exists by the definition of porosity. For $x \notin A$ this follows from the assumption that A is closed.) Let j be the least natural number such that $2^{-j+2} < \lambda\|y-x\|$. By the maximality of D_j there is a $z \in D_j \cap B(y, \lambda\|x-y\|) \subset B(x, \sigma/2)$. The inductive construction of the $\mathcal{B}_i$'s implies that there is a $0 \leq k \leq j$ and a ball $B(u, 2^{-k})$ in $\mathcal{B}_k$ which intersects $B(z, 2^{-j}) \subset B(x,\sigma)$. Hence $B(u, 2^{-k})$ intersects $B(x,\sigma)$ and thus $2^{-k} < \delta$. Also

$$\begin{aligned} \|x-u\| &\leq \|x-y\| + \|y-z\| + \|z-u\| \leq (1+\lambda)\|x-y\| + \|z-u\| \\ &\leq (1+\lambda)2^{-j+3}/\lambda + 2^{-j} + 2^{-k} \leq 18 \cdot 2^{-k}/\lambda \end{aligned}$$

and the result follows by taking $M = 18/\lambda$. □

Proof of Theorem 6.48. We first check that for every porous (not necessarily closed) set A, the Lipschitz function $f(x) = d(x, A)$ is not Fréchet differentiable at any point of A. Indeed, let λ be the porosity constant of A, and fix $x \in A$. For every $\delta > 0$ find $u \in E$ with $\|u\| < \delta$ such that $A \cap B(x+u, \lambda\|u\|) = \emptyset$. Then $f(x+u) \geq \lambda\|u\|$, $f(x) = 0$ and $f(x-u) \geq 0$. Thus

$$\limsup_{u \to 0} \big(f(x+u) + f(x-u) - 2f(x)\big)/\|u\| \geq \lambda$$

and hence f is not Fréchet differentiable at x.

For a union of porous sets the situation is less simple. We first remark that by Proposition 4.12 we may assume that E^* is separable, and thus, by Theorem 4.13, it has a Fréchet differentiable norm away from 0. Assume that $A = \bigcup_n A_n$ where each A_n is closed and porous. To each A_n we associate a family of balls $\mathcal{B}_n$ and a constant M_n as in the statement of Lemma 6.49.

For each n define a function $f_n : E \to \mathbb{R}$ as follows. If x belongs to a ball $B(z,r) \in \mathcal{B}_n$, put $f_n(x) = (r^2 - \|z-x\|^2)/r$ and if x is outside the union of the balls in $\mathcal{B}_n$, put $f_n(x) = 0$. Then $|f_n(x)| \leq 1$ for every x, the Lipschitz constant of f_n is at most 2 and, by our assumption on the norm, it is Fréchet differentiable in the interior of the balls. If x is in the complement of the union of the interior of the balls, then for every $\delta > 0$ there is a ball $B(z,r) \in \mathcal{B}_n$ with $M_n r \leq \delta$ and such that if we put $u = z - x$, then $\|u\| \leq M_n r$. By the same argument as above

$$\limsup_{u \to 0} \big(f_n(x+u) + f_n(x-u) - 2f_n(x)\big)/\|u\| \geq 1/M_n.$$

Choose $0 < \alpha_n < 2^{-n}$ such that $\alpha_n/M_n > 4\sum_{j>n}\alpha_j$, and put $f = \sum \alpha_n f_n$. The function f is a well-defined Lipschitz function.

Let $x \in A_n$ for some n. Then x does not belong to the union of the balls in $\mathcal{B}_n$. Let m be the smallest integer such that x does not belong to the interior of the balls in $\mathcal{B}_m$. Then f_j is Fréchet differentiable at x for $j \le m-1$; hence

$$\begin{aligned}
&\limsup_{u\to 0} \big(f(x-u) + f(x+u) - 2f(x)\big)/\|u\| \\
&\quad = \limsup_{u\to 0} \sum_{j\ge m} \Big(\alpha_j f_j(x-u) + \alpha_j f_j(x+u) - 2\alpha_j f_j(x)\Big)/\|u\| \\
&\quad \ge \alpha_m/M_m - 4\sum_{j>m}\alpha_j > 0
\end{aligned}$$

and hence f is not Fréchet differentiable at x. $\diamondsuit$

To complement Theorem 6.48 we remark that when E^* is separable, Lipschitz functions from E to $\mathbb{R}$ do have points of Fréchet differentiability. This deep result, as well as the study of Fréchet differentiability of Lipschitz maps between two infinite-dimensional spaces, will be discussed in the second volume.

6. Notes and Remarks

Haar null sets were introduced by Christensen [**120**], and all the results in Section 1 preceding Theorem 6.6 are due to him. The paper of Hunt, Sauer and Yorke [**259**] contains some examples and results on dynamical systems related to Haar null sets. (They were unaware of the previous work on this subject (see their addendum [**260**]) and also used a different term, shy sets, for Haar null sets.)

That many nonlocally compact, separable, complete metric groups (in particular all separable Banach spaces) contain an uncountable collection of disjoint Borel sets, none of them Haar null, was first proved by Dougerty [**154**]. Theorem 6.6 in its full generality, and the proof presented here, are due to Solecki [**539**]. Actually, the results presented in Section 1 on complete metric separable groups are valid (with obvious minor modifications in the statements and proofs) for nonabelian groups.

The fact that there are no quasi-invariant measures on an infinite-dimensional Banach space is due to Girsanov and Mityagin [**209**] and also Sudakov [**548**]. A characterization of topological (including noncommutative) groups which admit a σ-finite translation invariant measure can be found in the book of Yamasaki [**586**]. The method of proof there goes back to the work of Weil [**580**]. The book of Yamasaki also contains a treatment of quasi-invariant measures and, in particular, a proof that no infinite-dimensional locally convex space admits such a σ-finite measure.

In view of results like Corollary 6.7, I. M. Gelfand suggested to use the term "quasi-invariant" in a different meaning. He suggested calling a measure μ on E quasi-invariant if there is a dense linear subspace $E_0 \subset E$ such that every translate of μ by a vector in E_0 is absolutely continuous with respect to μ. Such measures do exist in abundance (we shall discuss this below in connection with Gaussian measures). This notion of "quasi-invariant" is the one most commonly used today (see e.g. Chapter 4 of Skorohod [**537**]), and we shall call it Gelfand quasi-invariant.

Proposition 6.8 and its proof are due essentially to Borwein and Noll [**80**]. They considered the case of the positive cone in l_q. In its present form the proposition appears in Matoušková [**389**]. Lemmas 6.10 and 6.11 appear in Matoušková and Stegall [**392**]. Theorem 6.9 is due to Matoušková [**389**], and Theorem 6.12 is due to Matoušková and Stegall [**392**].

Theorems 6.9 and 6.12 combined leave open the question whether a reflexive (but not superreflexive) space E can contain a closed convex set with empty interior which is not Haar null. Matoušková [**390**] proved a strengthening of Theorem 6.9: If E^* has the so-called Banach-Saks property, then E does not have such subsets.

Theorem 6.14 is due to Lindenstrauss, Matoušková and Preiss [**363**].

Theorem 6.16 was proved independently by Milnos [**423**] and Sazonov [**509**]. We followed the presentation in Skorohod [**537**].

Proposition 6.19 shows in particular that a nondegenerate Gaussian measure on a Hilbert space is Gelfand quasi-invariant. The same holds for nondegenerate Gaussian measures on a general Banach space by Theorem 6.22. It was proved independently by Feldman [**184**] and by Hajek [**244**] that any two Gaussian measures on a Hilbert space are either equivalent or orthogonal. They found criteria (in terms of the corresponding means and trace class operators) when two such measures are equivalent. A proof of this result (based on Kakutani's theorem) and related material can be found, e.g., in Chapter II of Kuo [**342**] or in Bogachev [**73**].

The earliest and most important Gaussian measure on Banach spaces is the Wiener measure ρ on $C[0,1]$. This measure is characterized by the property that under ρ for each $0 \le t_1 < \ldots < t_n \le 1$ the random variables $\{x(t_{k+1}) - x(t_k)\}_{k=1}^{n-1}$ are independent symmetric Gaussian variables with standard deviation $(t_{k+1} - t_k)^{1/2}$ respectively. (One uses the normalization $x(0) = 0$, and then ρ is nondegenerate on the subspace of functions vanishing at 0.)

The simplest explicit construction of ρ is, perhaps, as follows: Let $\{\xi_i\}$ be a sequence of independent standard Gaussian variables on some probability space Ω. Put $S_k = \sum_{i=1}^k \xi_i$, and let $[\,]$ denote integer part. For $t \in [0,1]$ and $\omega \in \Omega$ define

$$x_n(t,\omega) = n^{-1/2}\big(S_{[nt]}(\omega) + (nt - [nt])\xi_{[nt+1]}(\omega)\big)$$

and let ρ_n be the distribution of $x_n(t,\omega)$ on $C(0,1)$. A direct computation shows that for every $0 \le t_1 < t_2 < \ldots < t_m \le 1$ the distribution of $x(t_1), x(t_2), \ldots, x(t_m)$ with respect to ρ_n tends to that required for ρ. In addition, one proves that the sequence $\{\rho_n\}$ is tight, and thus by Prohorov's theorem (Theorem D.2) it has a subsequence converging weakly to some measure on $C[0,1]$ which has to be Wiener's measure (see e.g. Billingsley [**63**] pp. 61–64 for details).

If $x(t)$ has a derivative in $L^2[0,1]$ (in the sense of distributions), then the translation of ρ by x is equivalent to ρ. Otherwise it is orthogonal to it. For a discussion of this and for references to the original papers, see the lecture notes of Kuo mentioned above.

For $\alpha < 1/2$ the measure ρ is concentrated on functions $x(t)$ which are Hölder of order α, while the ρ measure of the Hölder functions of order $1/2$ or of those of bounded variation on $[0,1]$ is 0. The random process $x(t)$ (i.e., the family of random variables $\{x(t)\}_{0 \le t \le 1}$) with respect to the measure ρ is called the Brownian motion on $[0,1]$ (normalized by the requirements that $x(0) = 0$ and that the variance of $x(1)$ is 1). Thus, we can rephrase the results mentioned above as follows: Almost all Brownian paths are Hölder of order α for every $\alpha < 1/2$, but fail to be Hölder $1/2$ or of bounded variation.

All these results are just a tip of the iceberg of the theories of Gaussian measures or Brownian motion. The books of Billingsley and Kuo mentioned above, as well as Ledoux and Talagrand [**349**], contain introductions to these theories; see also Bogachev [**73**].

Theorem 6.22(i) is due to Kuelbs [**340**]. The proof of Theorem 6.22(ii) is based on an argument in Gihman and Skorohod [**207**] (see the appendix there).

Fernique [**186**] proved that for every Gaussian measure μ on a Banach space there is a $\beta > 0$ such that $\int e^{\beta\|x\|^2} d\mu(x) < \infty$. A proof, as well as an investigation of the tail distribution of Gaussian measures, can be found in the book of Ledoux and Talagrand mentioned above.

Aronszajn null sets were introduced by Aronszajn [**23**], Gauss null sets by Phelps [**460**], and cube null sets were first used (without explicit definition) by Mankiewicz [**383**]. Propositions 6.26 and 6.29 are due to Aronszajn. Propositions 6.24 and 6.25 were noted by Phelps.

The remarkable result that Gauss null sets are Aronszajn null (Proposition 6.27 and Theorem 6.32) is due to M. Csörnyei [**127**]. We emphasize again that the main point is to ensure that the decomposition of A into sets in $\mathcal{A}(x_n)$ is made with Borel sets. If we relax this requirement, a representation becomes almost trivial. Let $\{x_n\}_{n=1}^{\infty}$ be any sequence of linearly independent vectors in E whose linear span is dense in E. Put $F_n = \operatorname{span}\{x_k\}_{k=1}^{n}$ and $F = \bigcup_n F_n$. Let $D \subset E$ be a (highly nonmeasurable, chosen by the axiom of choice) set which contains exactly one representative from any coset of F in E, and put $D_n = D + F_{n-1}$. Then $\bigcup_n D_n = E$, and it is trivial to verify that each D_n intersects every line parallel to x_n by at most a single point. Assume now that A is a Gauss null set. Then $A \cap D_n$ is measurable with respect to the completion of all nondegenerate Gaussian measures on E, and $A = \bigcup_n (A \cap D_n)$ is a representation of the form required by Aronszajn (but with Borel replaced by measurable). This observation is due to Bogachev [**71**].

Proposition 6.34 and the other comments on σ-directionally null sets were observed during a conversation with E. Matoušková.

A probability measure μ on a Banach space E is called differentiable in the direction x if for every Borel set A in E

$$\nu(A) = \lim_{t \to 0} \big(\mu(A + tx) - \mu(A)\big)/t$$

exists. A measure is called densely differentiable if it is differentiable in directions which form a dense linear subspace of E. Nondegenerate Gaussian measures are densely differentiable, and every densely differentiable measure vanishes on all Gauss null sets. For these results, as well as other related results and notions, and also for further references, see Bogachev [**71, 72, 73**].

Example 6.35 is due to Bogachev [**72**]. The trick of constructing T_1 and T_2 is due to Dixmier [**151**]. Dixmier proved the following general result. If T is an operator on H such that TH is not closed, then there is a unitary operator V on H such that $TH \cap VTH = \{0\}$. See also Fillmore and Williams [**194**].

Theorem 6.36 (as well as lemmas 6.37 and 6.38) is due to Matoušková [**388**]. This article also contains a result in the opposite direction. Let K be a closed convex set in a Hilbert space H with an empty interior (hence Haar null by Theorem 6.9), and let f be a Lipschitz equivalence of H onto itself which is also Fréchet differentiable. Then $f(K)$ is Haar null.

Theorem 6.39 is due to Preiss and Tišer [**472**].

Aronszajn [**23**], Christensen [**121**] and Mankiewicz [**383**] proved Theorem 6.42 independently. They introduced and used different notions of "almost everywhere" and thus proved somewhat different versions of the theorem. Also, Aronszajn and Mankiewicz worked in the framework of Fréchet spaces.

The first proof of Proposition 6.41 is the same as the currently common proof of Rademacher's theorem. It appears in several books, and we do not know whose proof it is originally. The second proof is due to Preiss and Zajíček [**475**].

Corollary 6.43 and Propositions 6.45 are due to Bogachev and Shkarin [**74**]. Shkarin [**533**] showed that Propositions 6.45 also holds if f is only assumed to be measurable in the sense that $f^{-1}(B)$ is Baire for every Borel set B.

Example 6.46 is due to Matoušek and Matoušková [**386**]. In their original approach they proved Lemma 6.47 (with 5 replaced by 12) by an interesting and quite involved finite-dimensional covering argument. The proof of the lemma presented here is due to Preiss. (Both proofs appear in [**386**].) The construction of the example itself is based on a construction due to Preiss and Zajíček [**477**]. Actually, Matoušek and Matoušková even construct in [**386**] an example of an equivalent norm on l_2 which is Fréchet differentiable only on an Aronszajn null set. In a subsequent paper [**391**], Matoušková generalized the construction to any superreflexive space.

Theorem 6.48 is due to Preiss and Tišer [**472**].

We now describe another class of negligible sets, which is relevant to Gâteaux differentiability of Lipschitz functions.

Definition 6.50. *A set A in a Banach space E is called directionally porous if there is a number $0 < \lambda < 1$ with the following property. For every $x \in A$ there is a vector $u \in E$ with $\|u\| = 1$ such that for every $\delta > 0$ there is a $0 < \mu < \delta$ for which $A \cap B(x + \mu u, \lambda\mu) = \emptyset$. A countable union of directionally porous sets is called σ-directionally porous.*

This class was introduced by Zajíček [**589**] and was also studied by Preiss and Zajíček in [**474, 475**]. A directionally porous set is porous; the additional requirement here is that the vectors y in Definition 4.18 are restricted to be of the form $y = x + \mu u$ for a fixed u (depending on x). For finite-dimensional spaces the two notions coincide.

If A is a directionally porous set, then the Lipschitz function $g(x) = \operatorname{dist}(x, A)$ fails to be Gâteaux differentiable at any point $x \in A$. Indeed, the directional derivative of g at x in the direction u (given by the definition) fails to exist. By Theorem 6.42 (and recall that we consider only separable spaces in this chapter) directionally porous sets (hence also σ-directionally porous sets) are Gauss null. It follows from Theorem 6.39 that every infinite-dimensional Banach space contains a porous set which is not σ-directionally porous.

An argument due to Kirchheim (and presented in Preiss and Zajíček [**474**]) shows, moreover, that if A is any σ-directionally porous set, then there is a real-valued Lipschitz function on E which is not Gâteaux differentiable at any point $x \in A$. Incidentally, the same argument shows that Theorem 6.48 holds for every σ-porous set (and not only for countable union of closed porous sets).

Note that the proof of the statement which was used in the second proof of Proposition 6.41 shows that the set A there is actually σ-directionally porous.

Let f be a Lipschitz equivalence from a Banach space E onto a Banach space F. The image under f of a porous (resp. σ-porous) set is also porous (resp. σ-porous). If f is assumed, in addition, to be everywhere Gâteaux differentiable, then it maps directionally porous (resp. σ-directionally porous) sets to directionally porous (resp. σ-directionally porous) sets. (Indeed, if $\{y_n - x\}$ satisfy $\|y_n - x\| \to 0$ and are in a fixed direction, then the vectors $f(y_n) - f(x)$ are almost in a fixed direction as $n \to \infty$.) This assertion fails without the differentiability assumption, as will follow from Example 7.17.

Theorem 4.20 characterizes sets A where some real-valued continuous convex function is not Gâteaux differentiable at any point in A. The structure of the sets of non Gâteaux differentiability of general real-valued Lipschitz functions on a Banach space E is at present unclear except for $E = \mathbb{R}$. This case is classic: For any set $A \subset \mathbb{R}$ with $m(A) = 0$ there is a Lipschitz function $f : \mathbb{R} \to \mathbb{R}$ which is nondifferentiable on A (see Zahorski [**587**] for a more precise result). Indeed, choose a decreasing sequence of open sets $\{G_n\}$ containg A such that $m(G_n) \le 2^{-n}$ and such that $m\big((a,b) \cap G_{n+1}\big) \le (b-a)/3$ for every component (a,b) of G_n. Put $f = \sum (-1)^{n+1} f_n$ where $f_n(x) = m\big((-\infty, x) \cap G_n\big)$. The series converges uniformly and f is Lipschitz. If $x \in \bigcap G_n$, then f is not differentiable at x. Indeed, let (α_n, β_n) be the component of G_n containing x, and put $\lambda_j = \big(f_j(\beta_n) - f_j(\alpha_n)\big)/(\beta_n - \alpha_n)$. The sequence $\{\lambda_j\}$ is monotone decreasing, $\lambda_j = 1$ for $1 \le j \le n$ and $\lambda_{n+1} \le 1/3$. Thus $\big(f(\beta_n) - f(\alpha_n)\big)/(\beta_n - \alpha_n) = 1 - 1 + \ldots - 1 + \lambda_{n+1} - \lambda_{n+2} \ldots \le 1/3$ for even n and is equal to $1 - 1 + \ldots + 1 - \lambda_{n+1} + \lambda_{n+2} \ldots \ge 2/3$ for odd n.

The situation is much more involved for functions $f : \mathbb{R}^n \to \mathbb{R}$ when $n \ge 2$. Preiss [**471**] proved that if $A \subset \mathbb{R}^2$ is a G_δ-set containing all lines through any two rational points, then any Lipschitz function $f : \mathbb{R}^2 \to \mathbb{R}$ must be differentiable at some point of A. Moreover, Preiss showed (personal communication) that the family of Borel sets $A \subset \mathbb{R}^2$ for which there is a Lipschitz function $f : \mathbb{R}^2 \to \mathbb{R}$ which is nowhere differentiable on A is not even an ideal of sets. It follows that the same is true when $\mathbb{R}^2$ is replaced by any Banach space of dimension larger than 1.

On the other hand, if E is any Banach space, then the class of Borel sets $A \subset E$ for which there is a Lipschitz function $f : E \to l_2$ which is nowhere Gâteaux differentiable in A is even a σ-ideal. Indeed, if $f_i : E \to l_2$ are Lipschitz with constant smaller than 2^{-i} and $f_i(0) = 0$ for all i, then $f : E \to (\sum \oplus l_2)_2 = l_2$ defined by $f(x) = \big(f_1(x), f_2(x), \ldots\big)$ is Lipschitz and is not Gâteaux differentiable at any point where one of the f_i's is not differentiable.

Similarly, the Borel sets where some Lipschitz function to a space with the RNP is nowhere Gâteaux differentiable is also a σ-ideal. It seems to be unknown whether these two σ-ideals coincide, and whether they are equal to the σ-ideal generated by the set of non Gâteaux differentiability of real-valued functions. It seems to be reasonable to conjecture that if E is finite-dimensional, then these three σ-ideals coincide with the class of sets of Lebesgue measure zero.

We now show that the σ-ideal of sets of non Gâteaux differentiability of Lipschitz functions from l_2 to a space F with RNP is properly contained in the class of Gauss null sets. Let A and f be as in Example 6.35; i.e., there is an open ball B in l_2 such that f is a Lipschitz equivalence from B onto an open subset of l_2; both f and f^{-1} are C^∞-functions; and $A \subset B$ is not Gauss null, but $f(A)$ meets every line in at most three points. The set $f(A)$ is thus Gauss null (in a very strong sense), but every Lipschitz function $g : l_2 \to F$ must be Gâteaux differentiable at some point of $f(A)$. Indeed, $h = g \circ f : B \to F$ is Lipschitz, and since A is not Gauss null, h must be Gâteaux differentiable at some point $x_0 \in A$. But then, by the chain rule, $g = h \circ f^{-1}$ is Gâteaux differentiable at $f(x_0)$.

7. Summary

This section is a short overview and a summary of the various σ-ideals of negligible sets which were considered in this chapter and in Chapter 4. All the sets below are assumed to be Borel sets in a separable Banach space E.

We first recall the various classes.

(1) Haar null sets: A is Haar null if there is a probability measure μ on E such that $\mu(A + x) = 0$ for every $x \in E$.

(2) σ-directionally null sets: A is σ-directionally null if it is a countable union of directionally null sets. A set A is directionally null if there is an $x \neq 0$ in E such that any line in the direction of x meets A in a set of measure 0.

(3) Gauss null sets: A is Gauss null if $\mu(A) = 0$ for every nondegenerate Gaussian measure μ on E.

(4) Aronszajn null sets: A is Aronszajn null if for every sequence $\{x_n\}_{n=1}^{\infty} \subset E$ with a dense linear span, A can be represented as a countable union $A = \bigcup A_n$ such that every line in the direction of x_n meets A_n in a set of measure 0.

(5) Cube null sets: A is cube null if $\tau_T(A) = 0$ for each cube measure τ_T on E. A measure on E is called a cube measure if it is the image of the standard product measure τ on the Hilbert cube $Q = [0,1]^{\aleph_0}$ under a one-one affine map T of the form $T(q) = x + \sum t_n x_n$ for $q = (t_1, t_2, \dots) \in Q$, where $x \in E$ and the linearly independent vectors $x_n \in E$ have a dense linear span and satisfy $\sum \|x_n\| < \infty$.

(6) Sets A which meet any line in E in a set of measure 0.

(7) Sets of the first category.

(8) σ-porous sets: A is σ-porous if it is a countable union of porous sets. A set A is porous if there is a $0 < \lambda < 1$ such that for every $x \in A$ and $\varepsilon > 0$ there is a $y \in E$ with $0 < \|y - x\| < \varepsilon$ such that the ball $B(y, \lambda\|y - x\|)$ is disjoint from A.

(9) Sets of non-Gâteaux differentiability of Lipschitz functions into l_2: A belongs to this class if there is a Lipschitz function $f : E \to l_2$ which does not have a Gâteaux derivative at any point of A.

(10) σ-directionally porous sets: A is σ-directional porous if it is a countable union of directionally porous sets. A set A is directionally porous if there is a $0 < \lambda < 1$ such that for every $x \in A$ there is a direction $u \in E$ and points $y_n = x + \varepsilon_n u$ with $\varepsilon_n \to 0$ such that $A \bigcap B(y_n, \lambda\|y_n - x\|) = \emptyset$.

(11) Sets of non-Gâteaux differentiability of continuous convex functions: A belongs to this class if there is a continuous convex $f : E \to \mathbb{R}$ which does not have a Gâteaux derivative at any point of A.

(12) Countable union of subsets of closed hyperplanes.

The diagram below summarizes all the inclusion relations between these classes (for a general separable Banach space E). We use $(a) \Rightarrow (b)$ to mean that any set of class (a) is also of class (b).

$$\begin{array}{ccccccccccccc}
 & & & & & & & & (3) & & & & \\
 & & & & & & & & \Updownarrow & & & & \\
(12) & \Rightarrow & (11) & \Rightarrow & (10) & \Rightarrow & (9) & \Rightarrow & (4) & \Rightarrow & (2) & \Rightarrow & (1) \\
 & & & & \Downarrow & & & & \Updownarrow & & & & \\
 & & & & (8) & & & & (5) & & & & \\
 & & & & \Downarrow & & & & \Uparrow & & & & \\
 & & & & (7) & & & & (6) & & & &
\end{array}$$

(2) $\Rightarrow$ (1) follows from the definitions. See the discussion after Proposition 6.3.

(4) $\Rightarrow$ (2) is evident from the definitions.

(3) $\Leftrightarrow$ (4) $\Leftrightarrow$ (5) is proved in Propositions 6.25 and 6.27 and Theorem 6.32.

(6) $\Rightarrow$ (4) is evident from the definitions.

(9) $\Rightarrow$ (4) follows from Theorem 6.42.

(10) $\Rightarrow$ (9) is explained in the Notes and Remarks section.

(10) $\Rightarrow$ (8) $\Rightarrow$ (7) is evident from the definitions.

Theorem 4.20 characterizes sets of class (11): A is in class (11) if and only if it is contained in a countable union of hypersurfaces, each of which is the graph of a difference of two convex Lipschitz functions defined on a hyperplane in E.

Since the graph of a Lipschitz function defined on a hyperplane is evidently directionally porous, it follows that (11) $\Rightarrow$ (10). It is also clear that (12) $\Rightarrow$ (11).

We now show that for a general space E there are no other valid inclusion relations between the classes.

The positive cone in l_p $(1 < p < \infty)$ is an example of a Haar null set which is not σ-directionally null (see the discussion following Proposition 6.34).

Any compact set in a Banach space is directionally null; see the proof of Proposition 6.26. But in general it is not Gauss null (e.g., the Hilbert cube).

In the end of the Notes and Remarks section we pointed out that there is a set of class (6) (in a Hilbert space) which is not of class (9).

Evidently (12) $\not\Leftarrow$ (6) and (6) $\not\Leftarrow$ (7).

If $E = \mathbb{R}$, then the class (9) coincides with the sets of measure 0. Such sets do not have to belong to the class (10) or even only to (7).

If $E = \mathbb{R}$, then a set is of class (11) if and only if it is countable; hence (10) $\not\Leftarrow$ (11).

If $\dim E > 1$, it is evident that (11) $\not\Leftarrow$ (12).

By Theorem 6.39 every infinite-dimensional separable space E can be decomposed as a union of two sets, A which belongs to class (6) and B which belongs to class (8). In particular a set of class (8) may have a complement which belongs to class (1), and therefore (8) and (1) are incomparable.

For finite-dimensional spaces the classes (1)–(5) are just the sets of Lebesgue measure 0. For such spaces it is also easy to see that classes (8) and (10) coincide. The structure of sets of the class (9) is at present unclear even for finite-dimensional spaces (when $\dim E > 1$). For such spaces the class (8) is a proper subclass of $(1) \cap (7)$; see Zajíček [**590**].

The following criterion was used several times in this chapter to prove that a given Borel set A is *not* Haar null: A set A is not Haar null if there is a ball B in E such that A contains a translate of every compact subset of B. This criterion does not, however, characterize the sets which are not Haar null. For example, let D be a first category subset of $\mathbb{R}$ such that $\mathbb{R} \setminus D$ has measure 0. Then for any Banach space E the set $E \oplus D$ is not Haar null in $E \oplus \mathbb{R}$ (its complement is actually a directionally null set). But $E \oplus D$ contains no interval in the direction of $\mathbb{R}$.

We now summarize the stability properties of the classes considered here under Lipschitz equivalences.

Any homeomorphism of E onto itself preserves the class (7).

Any Lipschitz equivalence of E onto itself preserves the class (8).

It is also obvious that a Lipschitz equivalence of E onto itself which is everywhere Gâteaux differentiable preserves classes (9) and (10).

All the classes considered here, except (7) and (8), are not preserved in general by Lipschitz equivalences from E onto itself. A strong result proving this is the following result from Lindenstrauss, Matoušková and Preiss [**363**], whose proof is based on the proof of Theorem 6.36 and on methods described in the next chapter: There is a subset A of l_2 which contains a translate of every compact subset of the unit ball of l_2, and there is a Lipschitz equivalence f of l_2 onto itself such that $f(A)$ is contained in a closed hyperplane. In other words, f^{-1} maps a set of class (12) into a set which is not even of class (1).

Example 6.35 shows that the class (6) is not preserved by Lipschitz equivalences.

Theorem 6.14 says that if E has an unconditional basis, then there is a Haar null set in E which can be mapped by a Lipschitz equivalence of E onto a set whose complement is Haar null. This equivalence can be accomplished by a C^1-map which is an arbitrarily small perturbation of the identity.

It is unknown if there is a Lipschitz equivalence of a Banach space which maps a Gauss null set onto a set whose complement is Gauss null.

CHAPTER 7

Lipschitz Classification of Banach Spaces

1. Linearization of Lipschitz Maps

In this chapter we show that under quite general conditions Lipschitz equivalent Banach spaces are, in fact, linearly isomorphic. This is not always the case (see Example 7.12), but the results of this chapter (especially Corollary 7.7) and results on Fréchet derivatives to be presented in the second volume indicate that it is quite possible that the Lipschitz and linear classifications coincide for all separable (or at least for separable and reflexive) Banach spaces. The spaces in Example 7.12 are nonseparable and nonreflexive.

The reduction of the Lipschitz to the linear classification is obtained by linearization of Lipschitz maps, and in this section we present two methods to do so: differentiation, which uses the local behavior of the function near a point, and invariant means, which use its behavior at infinity. The method of invariant means applies to general uniformly continuous functions and not only to Lipschitz functions and is very efficient in the study of retraction or extensions. However, averaging procedures, such as invariant means, do not preserve the property of being one-one and are thus usually not useful in the study of embeddings.

We now exhibit the most direct application of differentiation. Let f be a Lipschitz embedding of E into F such that $\|x-y\|/K \leq \|f(x)-f(y)\| \leq K\|x-y\|$ for every $x, y \in E$. If f is Gâteaux differentiable at a point $x_0 \in E$, then the derivative D of f at x_0 is a linear operator from E to F and $\|x\|/K \leq \|Dx\| \leq K\|x\|$ for all $x \in E$; hence E is isomorphic to a subspace of F. If E is separable and F has the Radon-Nikodým property, then Theorem 6.42 implies that f has points of Gâteaux differentiability. Thus E is Lipschitz embeddable into F under these circumstances if and only if it is isomorphic to a linear subspace of F. Linear theory (see the Notes and Remarks section for references) then yields:

(i) If E is Lipschitz equivalent to a subset of F, and if F is reflexive (or has the RNP, or is isomorphic to a Hilbert space), then E has the same property. Indeed, if E is separable, this follows directly from Theorem 6.42. The general case then follows because a Banach space is reflexive, or has the RNP, or is isomorphic to a Hilbert space whenever all its separable subspaces are.

(ii) If $1 \leq p, q < \infty$ with $p \neq q$, then l_p (and c_0) are not Lipschitz equivalent to a subset of l_q.

(iii) $L_p[0,1]$ is not Lipschitz equivalent to a subset of l_q unless $p = q = 2$ or $q = \infty$.

(iv) If $q < \infty$, then $L_p[0,1]$ is not Lipschitz equivalent to a subset of $L_q[0,1]$ unless $2 \geq p \geq q \geq 1$ or $p = q$. (The case $p > 2$ and $q = 1$ does not follow from Theorem 6.42 because L_1 does not have the RNP. But, as we note after Corollary 7.10, the result still holds in this case.)

In trying to use direct methods of differentiation to surjective Lipschitz equivalences, we encounter the following basic problem. (We pose the problem for superreflexive spaces although a reasonable broader context is that of spaces with the Radon-Nikodým property.)

Problem 7.1. *Let E and F be separable superreflexive Banach spaces, and let $f : E \to F$ be a Lipschitz equivalence of E onto F. Does there exist a point $x_0 \in E$ such that the Gâteaux derivative $D_f(x_0)$ of f at x_0 is surjective?*

Several natural approaches suggest themselves in an attempt to give a positive answer to the problem.

(i) If f is Fréchet differentiable at some point $x_0 \in E$, then a direct computation shows that $D_f(x_0)$ is surjective. Easy examples show, however, that such a point may fail to exist. For example, let $g(t) = 2t$ for $t > 0$ and $g(t) = t$ for $t \le 0$, and define $f : l_2 \to l_2$ by $f(x_1, x_2 \dots) = (g(x_1), g(x_2) \dots)$. Then f is a Lipschitz equivalence of l_2 onto itself which is nowhere Fréchet differentiable.

(ii) If there is a point $x_0 \in E$ such that f is Gâteaux differentiable at x_0 and f^{-1} is Gâteaux differentiable at $f(x_0)$, then $D_f(x_0)$ is a surjective isomorphism, and its inverse is $D_{f^{-1}}\big(f(x_0)\big)$. We saw in Chapter 6 that both f and f^{-1} are Gâteaux differentiable outside "small" exceptional sets, but further results and examples in that chapter (and in particular Theorem 6.14) show that in general Lipschitz equivalences may map exceptional sets onto "large" sets. At present it is unknown whether such a point x_0 necessarily exists.

(iii) If f is Gâteaux differentiable at some point $x_0 \in E$, and if the scalar-valued functions $y_i^* \circ f$, where $\{y_i^*\}$ is a dense sequence in the unit ball of F^*, are all Fréchet differentiable at x_0, then it is easy to verify that $D_f(x_0)$ is surjective. It even suffices that all the $y_i^* \circ f$'s are ε-Fréchet differentiable at x_0 for some $\varepsilon > 0$ which is small enough, depending on the Lipschitz constant of f^{-1}. (See the Notes and Remarks section of Chapter 11 for the notion of ε-Fréchet differentiability.) For every $\varepsilon > 0$ and for every *finite* number of Lipschitz functions $\{g_i\}_{i \le n}$ from E to $\mathbb{R}$ there is a point $x_0 \in E$ where f is Gâteaux differentiable and where the g_i's are ε-Fréchet differentiable; see Theorem 10.36. The proof of Theorem 10.36 does not seem, however, to extend to an infinite family of scalar-valued Lipschitz functions.

Thus all these natural approaches to a positive answer to Problem 7.1 do not seem to work (at least with the current methods). On the other hand, we present in this chapter some examples that seem to indicate that Problem 7.1 may, perhaps, have a negative answer. In Example 7.17 (see also the Notes and Remarks section) we show that there is a Lipschitz equivalence f of l_2 onto itself such that the Gâteaux derivatives $\{D_f(x)\}$ fail to be surjective in a rather "large" subset of l_2.

In the remainder of this first section we prove some simple linearization theorems for Lipschitz maps. These results allow us to circumvent the difficulties related to Problem 7.1 and to obtain surjective isomorphisms under certain rather weak additional conditions. They are also of independent interest.

As usual, the modulus of continuity of the function f is denoted by ω_f. If f is a Lipschitz function, its Lipschitz constant is denoted by $\|f\|_{Lip}$.

Theorem 7.2. *Let E and F be Banach spaces, and let E_0 be a subspace of E. Let $S : E_0 \to F$ be a bounded linear operator, and assume that $f : E \to F$ is a uniformly continuous function such that $f|_{E_0} = S$. Then there is a linear operator*

$T : E \to F^{**}$ such that $T|_{E_0} = S$ and such that $\|T\| \leq \omega_f(1)$. In particular, if f is Lipschitz, then $\|T\| \leq \|f\|_{Lip}$.

Proof. We give two proofs of the theorem. The first uses invariant means. For information on invariant means, see Appendix C.

Consider F as a subspace of the dual space F^{**}, and let M and N be invariant means on the spaces of bounded functions from E_0 and E respectively into F^{**}. (E_0 and E are regarded here as additive groups.)

We first average in the direction of E_0, i.e., use M. Since f is uniformly continuous, $f(z+\cdot) - f(\cdot)$ is a bounded function for each fixed $z \in E$. It follows that $g(z) = M_x\big(f(z+x) - f(x)\big)$ is well defined, and it satisfies $\omega_g(t) \leq \omega_f(t)$ for every $t > 0$ because $\|g(z) - g(w)\| = \big\|M_x\big(f(z+x) - f(w+x)\big)\big\| \leq \omega_f\big(\|z - w\|\big)$ for every $z, w \in E$.

Assume now that $y \in E_0$. Since M is E_0-invariant, and since the variable x is taken in E_0, we obtain

$$\begin{aligned} g(y+z) &= M_x\Big(f(z+y+x) - f(y+x)\Big) + M_x\Big(f(y+x) - f(x)\Big) \\ &= M_x\Big(f(z+x) - f(x)\Big) + M_x\Big(Sy\Big) = g(z) + Sy. \end{aligned}$$

In particular $g(y) = Sy$ for every $y \in E_0$.

Now put $Tz = N_x\Big(g(z+x) - g(x)\Big)$. The invariance of N gives

$$T(z+w) = N_x\Big(g(z+w+x) - g(w+x)\Big) + N_x\Big(g(w+x) - g(x)\Big) = Tz + Tw$$

i.e., T is linear. That $\|T\| \leq \omega_g(1)$ follows directly from the definition of T. Finally, if $y \in E_0$, then $g(y+x) - g(x) = Sy$ implies that $Ty = Sy$.

A second proof of Theorem 7.2. The second proof uses "smoothing" and differentiation and thus applies to Lipschitz functions only. Assume first that E is finite-dimensional, and write (algebraically) $E = E_0 \oplus E_1$.

We first smooth f in the direction of E_0. Let $\varphi \geq 0$ be a C^∞-function with compact support on E_0 such that $\int_{E_0} \varphi = 1$ and $\varphi(x) = \varphi(-x)$ for all $x \in E_0$. Put

$$g(z) = \int_{E_0} f(z+x)\varphi(x)\,dx.$$

Then $\|g\|_{Lip} \leq \|f\|_{Lip}$. If $z \in E$ and $y \in E_0$, then

$$g(z+ty) = \int_{E_0} f(z+ty+x)\varphi(x)\,dx = \int_{E_0} f(z+u)\varphi(u-ty)\,du$$

and hence g is differentiable in the directions of E_0 and

$$D_g(z)y = -\int_{E_0} f(z+x)D_\varphi(x)y\,dx.$$

(Note the slight abuse of notation: $D_g(z)y$ is defined only for $y \in E_0$, and it is linear as a function of such y's only. Our usual notation requires $D_g(z)$ to be a linear operator defined on all of E. However, we use this notation only for the values $D_g(z)y$ at points $y \in E_0$, and there should be no confusion.)

It follows from this formula that $D_g(z)y$ is a continuous function of $z \in E$ and that it is linear in $y \in E_0$. Moreover, for every $y \in E_0$

$$
\begin{aligned}
(7.1) \qquad g(y) &= \int_{E_0} f(y+x)\varphi(x)\,dx = \int_{E_0} S(y+x)\varphi(x)\,dx \\
&= \int_{E_0} (Sy+Sx)\varphi(x)\,dx = Sy
\end{aligned}
$$

because $\int_{E_0} \varphi = 1$ and because $\varphi(x) = \varphi(-x)$ implies that $\int_{E_0} Sx\,\varphi(x)dx = 0$.

We now "smooth" in the E_1 direction. Let $\psi \geq 0$ be a C^∞-function with compact support on E_1 such that $\int_{E_1} \psi = 1$, and define

$$f_n(z) = \int_{E_1} g(z + \frac{u}{n})\psi(u)\,du.$$

Since g is already smooth in the E_0 directions, the f_n's are differentiable. The differentials T_n of f_n at 0 are linear operators from E to F with $\|T_n\| \leq \|g\|_{Lip}$, and we consider them as operators into F^{**}. Let $\mathcal{U}$ be a free ultrafilter on $\mathbb{N}$ (see Appendix F), and put $Tz = w^* - \lim_{\mathcal{U}} T_n z$. Then T is a bounded linear operator with $\|T\| \leq \|g\|_{Lip} \leq \|f\|_{Lip}$. To check that $T|_{E_0} = S$, fix $x \in E_0$. For every $u \in E_1$ the directional derivative $D_g(u)x$ of g at u in the direction x exists, and

$$T_n x = D_{f_n}(0)x = \int_{E_1} D_g(u/n)x\psi(u)\,du.$$

Since $g = S$ on E_0, it follows also that $D_g(0)x = Sx$, and hence

$$(7.2) \qquad Tx - Sx = \lim_{\mathcal{U}} \int_{E_1} \Big(D_g(u/n)x - D_g(0)x\Big)\psi(u)\,du = 0$$

because ψ has a compact support and $D_g(\cdot)x$ is continuous.

This completes the proof when E is finite dimensional. The general case follows by a standard compactness argument. For each finite-dimensional subspace G of E put $G_0 = G \cap E_0$, and find a linear operator $T_G : G \to F^{**}$ such that $\|T_G\| \leq \|f\|_{Lip}$ and such that $T_G|_{G_0} = S|_{G_0}$. We then use the w^*-compactness of the unit ball of F^{**} and take T to be the limit of any convergent subnet of the net $\{T_G\}$. □

Remarks. (i) If we do not require $f|_{E_0}$ to be linear, and if S is an arbitrary bounded linear operator from E_0 to F, then the operator T that results from the proofs would satisfy $\|T - S\| \leq \omega_{f-S}^{E_0}(1)$, where $\omega_h^{E_0}$ denotes the modulus of continuity of the function h on E_0. We check this using the second proof (hence under the assumption that f is Lipschitz).

As in (7.1) we obtain $g(y) - Sy = \int_{E_0} \big(f(y+x) - S(y+x)\big)\varphi(x)\,dx$ for every $y \in E_0$; hence

$$\big\|g|_{E_0} - S\big\|_{Lip(E_0)} \leq \int_{E_0} \big\|f(\cdot + x) - S(\cdot + x)\big\|_{Lip(E_0)}\varphi(x)\,dx = \big\|f|_{E_0} - S\big\|_{Lip(E_0)}.$$

Since $\lim_{\mathcal{U}} \int_{E_1} \big(D_g(y/n)x - D_g(0)x\big)\psi(y)\,dy = 0$ as in (7.2), it remains to check that $\|D_g(0) - S\|_{Lip(E_0)} \leq \|g|_{E_0} - S\|_{Lip(E_0)}$. The expression on the left is the operator norm of the linear operator $D_g(0) - S$ on E_0. Assuming, as we may, that $g(0) = 0$, it follows that $D_g(0)x = \lim g(tx)/t$, and since S is linear we obtain

$$\big\|D_g(0)x - Sx\big\| = \lim_{t\to 0} \big\|\big(g(tx) - S(tx)\big)/t\big\| \leq \big\|g|_{E_0} - S\big\|_{Lip(E_0)}$$

for every $x \in E_0$ with $\|x\| = 1$.

(ii) Note that E_0 is a negligible subset of E. Thus even when f is a Lipschitz function, it does not have to have a point of Gâteaux differentiability in E_0, and we cannot deduce Theorem 7.2 directly from the results of Chapter 6.

Corollary 7.3. *Let E_0 be a subspace of E such that there is a uniformly continuous retraction r from E onto E_0. Then there is a linear operator $T : E \to E_0^{**}$ satisfying $Tx = x$ for every $x \in E_0$ and $\|T\| \leq \omega_r(1)$. Equivalently, there is a linear operator $W : E_0^* \to E^*$ with $\|W\| \leq \omega_r(1)$ such that $\langle Wx^*, x\rangle = \langle x^*, x\rangle$ for every $x \in E_0$ and $x^* \in E_0^*$; i.e., Wx^* extends x^*. In particular if E_0 is reflexive or is complemented in E_0^{**}, then it is complemented in E.*

Proof. This follows directly from Theorem 7.2 when we take $F = E_0$, the identity on E_0 as the operator S, and $f = r$. The operator W is the restriction of T^* to E_0^*.

If $P : E_0^{**} \to E_0$ is a projection, then PT is a projection from E onto E_0. □

Corollary 7.4. *Let $q : F \to G$ be a linear quotient map. Let E be the kernel of q, and assume that E is complemented in E^{**}. Assume also that q admits a uniformly continuous lifting $\varphi : G \to F$; i.e., $q\varphi$ is the identity on G. Then E is complemented in F and there is a linear lifting $T : G \to F$.*

Proof. Without loss of generality we may assume that $\varphi(0) = 0$. The map $r(x) = x - \varphi(qx)$ is then a uniformly continuous retraction from F onto E, and the result follows from Corollary 7.3. □

Let $E^\#$ denote the Lipschitz dual of E, namely, the space of Lipschitz functions $f : E \to \mathbb{R}$ with $f(0) = 0$ under the Lipschitz norm. The dual E^* of E is naturally isometric to a subspace of $E^\#$. If E_0 is a subspace of E, we denote by $q : E^* \to E_0^*$ and by $q^\# : E^\# \to E_0^\#$ the operators of the restriction of the linear functionals on E (respectively, the Lipschitz functions on E) to the subspace E_0.

The proof of the next proposition follows the same lines as that of Theorem 7.2. We choose the method of invariant means. The same proof also shows that an analogous result holds when the spaces of Lipschitz functions are replaced by the spaces of uniformly continuous functions with the norm $\omega_f(1)$.

Proposition 7.5. *For every Banach space E there is a norm-one projection from $E^\#$ onto E^*. Moreover, if $E_0 \subset E$, then the projections $P : E^\# \to E^*$ and $P_0 : E_0^\# \to E_0^*$ can be chosen so that $qP = P_0 q^\#$.*

Proof. Let M be an invariant mean on the space of bounded real-valued functions on E. The projection $P : E^\# \to E^*$ is defined by the explicit formula

$$Pf(z) = M_x\Big(f(z+x) - f(x)\Big)$$

for $f \in E^\#$ and $z \in E$.

For the second part, let M and N be invariant means on the bounded real-valued functions on E_0 and on E respectively. Then put

$$P_0 f(z) = M_y\Big(f(z+y) - f(y)\Big)$$

for $f \in E_0^\#$ and $z \in E_0$, and

$$Pf(z) = N_x\Big(M_y\Big(f(z+x+y) - f(x+y)\Big)\Big)$$

for $f \in E^{\#}$ and $z \in E$. Fix $z \in E_0$, and write $f(z+x+y) - f(x+y)$ as the sum of three terms $\big(f(z+x+y) - f(z+y)\big) - \big(f(x+y) - f(y)\big) + \big(f(z+y) - f(y)\big)$. The first term is obtained from the second by translation by z. Since $z \in E_0$, the E_0-invariant mean M gives the same value to these two terms. The third term is independent of x. It follows that

$$N_x\Big(M_y\Big(f(z+x+y) - f(x+y)\Big)\Big) = M_y\Big(f(z+y) - f(y)\Big) = P_0 q^{\#} f(z).$$

This proves that $qP = P_0 q^{\#}$, and we leave the other details to the reader. □

2. Applications and Examples

Our first result combines differentiation with the results of the first section to obtain linear embeddings with a complemented range.

Theorem 7.6. *Let E and F be Banach spaces, and let f be a Lipschitz embedding of E as a subset of F. Assume that there is a Lipschitz retraction from F onto the image $f(E)$ and that f is Gâteaux differentiable at a point $x_0 \in E$. If E is a Lipschitz retract of E^{**}, then there is a Lipschitz retraction from F onto its linear subspace $D_f(x_0)(E)$. If E is linearly complemented in E^{**}, then there is a linear projection from F onto $D_f(x_0)(E)$.*

Proof. Assume, as we may, that $x_0 = 0$ and that $f(0) = 0$. The derivative $D_f(0)$ (which we denote by D) is a linear embedding of E into F. Let $r : F \to f(E)$ be a Lipschitz retraction, and define mappings g and g_n from F into E by $g = f^{-1} \circ r$ and by $g_n(y) = ng(y/n)$. The functions $g_n : F \to E$ are uniformly Lipschitz. Let $\mathcal{U}$ be a free ultrafilter on $\mathbb{N}$, and consider E as a subspace of E^{**}. Since the sequence $\{g_n(y)\}$ is bounded for each $y \in F$, it follows that $h(y) = w^* - \lim_{\mathcal{U}} g_n(y)$ is a well-defined function from F to E^{**}. It is Lipschitz with the same constant as g by the w^*-lower semi-continuity of the norm.

We claim that $h(Dx) = x$ for all $x \in E$. Indeed, put $y = Dx = \lim_n nf(x/n)$. Then $g_n\big(nf(x/n)\big) = ng\big(f(x/n)\big) = x$ for every $x \in E$. Since the g_n's are uniformly Lipschitz, we deduce that $\|g_n(y) - x\| = \|g_n(y) - g_n\big(nf(x/n)\big)\| \to 0$, i.e., that $h(y) = w^* - \lim_{\mathcal{U}} g_n(y) = x$.

The desired retraction is now $D \circ \rho \circ h$, where ρ is the retraction from E^{**} onto E. Since $D(E)$ is isomorphic to E, the last part follows from the first part and from Corollary 7.3. □

The various tools developed so far, together with the decomposition method (see Proposition F.9), give many situations in which it is possible to deduce linear isomorphism from Lipschitz equivalence.

Corollary 7.7. *Let E be a separable Banach space which is complemented in E^{**}, and assume that E and F are Lipschitz equivalent, or more generally, that E is Lipschitz equivalent to a subset of F which is the range of a Lipschitz retraction from F. If F has the RNP, then E is isomorphic to a complemented subspace of F. It follows that if E and F are Lipschitz equivalent separable dual spaces which satisfy the decomposition scheme, then they are linearly isomorphic.*

Proof. The first part follows from Theorems 6.42 and 7.6. For the second part, note first that a dual space G^* is always complemented in its second dual G^{***}. The projection is just the restriction of elements $x^{***} \in G^{***}$ considered as functions on G^{**}, to the subspace G of G^{**}. Thus by the first part E and F are linearly isomorphic to complemented subspaces of each other. □

Corollary 7.8. *Let $1 < p < \infty$. If E is Lipschitz equivalent to l_p or to $L_p[0,1]$, then it is linearly isomorphic to it. The same is true for l_1 if E is also assumed to be a dual space.*

Proof. Since E is isomorphic to a subspace of l_p or L_p, it follows that if $1 < p < \infty$, then E is reflexive. Also, if $F = l_p$ or L_p for $1 \leq p < \infty$, then it is isomorphic to $(\sum \oplus F)_p$. Thus E and F satisfy (ii) of Proposition F.9, and Corollary 7.7 applies. □

Remark. If the spaces E and F are not known to satisfy the decomposition scheme, then Corollary 7.7 is not sufficient to show that they are actually isomorphic and more information is needed. Corollary 10.38 is one instance where such additional information can be obtained.

Gâteaux derivatives can be applied only when the range space of the Lipschitz embedding has the RNP. When the range space is a dual space, we can obtain similar results by using w^*-Gâteaux derivatives without assuming that it has the RNP. Recall that $f : E \to F^*$ is said to be w^*-Gâteaux differentiable at $x_0 \in E$ if the w^*-limit

$$D_f^*(x_0)x = w^* - \lim_{t \to 0} \big(f(x_0 + tx) - f(x_0)\big)/t$$

exists for all $x \in E$ and is a bounded linear operator in x. The linear operator $D_f^*(x_0)$ is called the w^*-Gâteaux derivative of f at x_0.

If f satisfies a Lipschitz condition, then the w^*-lower semi-continuity of the norm implies that $\|D_f^*(x_0)\| \leq \|f\|_{Lip}$.

Theorem 7.9. *Let E and F be separable Banach spaces, and let $f : E \to F^*$ be a Lipschitz map. Then f is w^*-Gâteaux differentiable outside a Gauss null set. Moreover, if f is a Lipschitz embedding, then $D_f^*(x_0)$ is an into-isomorphism outside a Gauss null set.*

Proof. The first part is Corollary 6.44. The new ingredient here is the second part. This requires a new argument because lower estimates on the norm are not preserved by w^*-limits.

Normalize f so that $\|x - y\| \leq \|f(x) - f(y)\| \leq K\|x - y\|$, and denote the set where D_f^* exists by W. We shall show that if $\delta < 1$, then the set

$$W_{z,\delta} = \big\{x \in W : \ \|D_f^*(x)z\| \leq \delta\|z\|\big\}$$

is Gauss null for every $z \in E$. Taking z in a countable dense subset of E and a sequence $\delta_j \to 1$ and using the fact that the $D_f^*(x)$'s are bounded linear operators, we obtain $\|D_f^*(x)z\| \geq \|z\|$ for every $z \in E$ outside a set of x's which is Gauss null.

To see that $W_{z,\delta}$ is Gauss null, fix $z \in E$ with $\|z\| = 1$, and check that for each line L in the direction of z the intersection $A = L \cap W_{z,\delta}$ has one-dimensional Lebesgue measure zero. If this is not so, choose a density point z_0 of A in L, and denote the Lebesgue measure on L by m. Fix $\varepsilon > 0$ so small that

$$m\big([z_0, z_0 + \varepsilon z] \cap A\big) \geq \varepsilon\Big(1 - \frac{1-\delta}{3K}\Big)$$

and let $B = [z_0, z_0 + \varepsilon z] \setminus A$. Since $\|f(z_0 + \varepsilon z) - f(z_0)\| \geq \varepsilon$ and $\delta < 1$, we can find $y \in F$ with $\|y\| = 1$ such that $\langle f(z_0 + \varepsilon z) - f(z_0), y \rangle \geq \varepsilon(1 + \delta)/2$. Put $\varphi(t) = \langle f(z_0 + tz), y \rangle$. The scalar function φ is Lipschitz, hence the integral of its derivative, and

$$|\varphi'(t)| = \left|\langle D_f^*(z_0 + tz)z, y\rangle\right| \leq \left\|D_f^*(z_0 + tz)z\right\| \leq K$$

for almost all t. Thus,

$$\begin{aligned} \varepsilon(1+\delta)/2 &\leq \varphi(\varepsilon) - \varphi(0) = \int_0^\varepsilon \varphi'(t)dt \leq \int_{A\cap[z_0, z_0+\varepsilon z]} \delta + \int_B K \\ &\leq \varepsilon\delta + Km(B) \leq \varepsilon\big(\delta + (1-\delta)/3\big) = \varepsilon(1+2\delta)/3 \end{aligned}$$

a contradiction. □

Theorem 7.9 applies to duals of separable spaces only, since the differentiation theorem does not hold when F is not separable. For the purposes of reducing the Lipschitz to the linear classification, this separability assumption is, however, not needed.

Corollary 7.10. *Let E be a separable Banach space, and let F be any Banach space. If E is Lipschitz equivalent to a subset of F^*, then it is isomorphic to a subspace of F^*. In particular, if E is Lipschitz equivalent to a subset of an arbitrary Banach space G, then it is isomorphic to a subspace of G^{**}.*

Proof. Assume that $f : E \to F^*$ is a Lipschitz embedding. Since E is separable, there is a separable subspace H of F which norms $f(E)$. If G is any separable subspace of F containing H, then f can be regarded as a map from E to G^*, and by Theorem 7.9 E is isomorphic to a subspace of G^*. By Proposition F.8 G can be chosen so that G^* is isometric to a (complemented) subspace of F^*; hence E is also isomorphic to a subspace of F^*. □

Remark. We can now show that L_p is not Lipschitz equivalent to a subset L_1 for $p > 2$. Indeed, by the corollary this would have implied that L_p is isomorphic to a subspace of L_1^{**}. But L_1^{**} has cotype 2 and hence cannot contain such a subspace, because the cotype of L_p for $p > 2$ is only p.

The essential tool used in this chapter was differentiation, and we had to assume that the range space has the RNP or is a dual space. The next result shows that these are not just technical assumptions and that without them the Lipschitz structure of the spaces is really different from the linear one: The space c_0, which is "small" in the linear category, is "very large" in the Lipschitz category.

Theorem 7.11. *Every separable metric space (X, d) is Lipschitz equivalent to a subset of c_0. Moreover, for each $K > 3$ there is a map $f : X \to c_0$ such that*

$$d(x, y) \leq \|f(x) - f(y)\| \leq Kd(x, y)$$

for every $x, y \in X$.

Proof. Fix a point $e \in X$, and denote the open ball with center e and radius r by $B(r)$. Fix $t < 1$ so that $Kt^2 > 3$.

The heart of the proof is the construction of a doubly infinite sequence of subsets $\{M_{n,i}\}_{n\in\mathbb{Z}, i\geq 0}$ of X with the following properties:

(i) $E \setminus B(3t^n/2) = \bigcup\{M_{n,i} : i \geq 0\}$ for each n,

(ii) $\operatorname{diam}(M_{n,i}) \leq 2t^n$ for all n and i,

(iii) the set $\{i : d(x, M_{n,i}) < t^{n+1}\}$ is finite for each $x \in X$ and n.

We first complete the proof under the assumption that the sets $M_{n,i}$ have already been constructed. Put

$$f_{n,i}(x) = K\left(t^{n+1} - d(x, M_{n,i})\right)^+.$$

These functions have the following properties:

(iv) $f_{n,i}$ is Lipschitz with constant K,

(v) $f_{n,i}$ is supported in the complement of $B(t^n/2)$,

(vi) $0 \leq f_{n,i} \leq Kt^{n+1}$,

(vii) for each $x \in X$ and for each $\delta > 0$ the set $\{(n,i) : f_{n,i}(x) > \delta\}$ is finite.

Properties (iv)-(vi) follow from the definition of $f_{n,i}$ and from (i). To check (vii) note that $\|f_{n,i}\|_\infty \leq Kt^{n+1}$ by (vi); hence there is only a finite number of positive n's for which there is any i with $f_{n,i}(x) > \delta$. By (v) there are also finitely many such negative n's. For each of these n's the definition of $f_{n,i}$ implies that $f_{n,i}(x) > 0$ only if i belongs to the set $\{i : d(x, M_{n,i}) < t^{n+1}\}$, and this set is finite by (iii).

Let $(e_{n,i})_{n \in \mathbb{Z}, i \geq 0}$ be an enumeration of the standard basis of c_0. The embedding of X into c_0 is defined by

$$f(x) = \sum f_{n,i}(x) e_{n,i}.$$

Condition (vii) ensures that f is a well-defined mapping from X into c_0, and by (iv) it satisfies $\|f(x) - f(y)\| \leq Kd(x,y)$ for every x, y in X.

To obtain a lower bound, fix $x \neq y$, and choose n such that

$$3t^n < d(x,y) \leq 3t^{n-1}.$$

Interchanging x and y if necessary, we can assume that $d(x,e) \geq 3t^n/2$. By (i) there is an i such that $x \in M_{n,i}$. Hence

$$f_{n,i}(x) = Kt^{n+1}.$$

On the other hand, by (ii)

$$d(y, M_{n,i}) \geq d(y,x) - \operatorname{diam}(M_{n,i}) \geq 3t^n - 2t^n = t^n > t^{n+1}$$

and it follows that $f_{n,i}(y) = 0$. By the choice of n and t

$$\|f(x) - f(y)\| \geq |f_{n,i}(x) - f_{n,i}(y)| = Kt^{n+1} \geq (Kt^2/3)d(x,y) > d(x,y).$$

It remains to construct the sets $M_{n,i}$. The construction is done for each n separately, and by rescaling it suffices to consider only $n = 0$, i.e., when $t^n = 1$. We simplify the notation and write M_i instead of $M_{0,i}$.

Fix a dense sequence $\{y_i : i \geq 0\}$ in X, and define $M_0 = B(y_0, 1) \setminus B(3/2)$ and

$$M_i = B(y_i, 1) \setminus \left(B(3/2) \cup \bigcup \{M_j : j < i\}\right)$$

for $i \geq 1$.

Conditions (i) and (ii) are immediate. To check (iii), fix x in X, and choose i_0 such that $d(x, y_{i_0}) < 1 - t$. We shall show that if $d(x, M_i) < t$, then $i \leq i_0$, and hence that the cardinality of $\{i : d(x, M_i) < t\}$ is at most i_0. Indeed, for each such i

$$d(y_{i_0}, M_i) \leq d(y_{i_0}, x) + d(x, M_i) < 1 - t + t = 1.$$

Thus there is a point $z \in M_i$ which is a "possible candidate" for membership in M_{i_0}. If i_0 were smaller than i, the point z would have been chosen in the i_0th step (or even earlier). Since M_i is constructed to be disjoint from all previously chosen M's, this z would not belong to M_i. Thus $i \leq i_0$ as claimed. □

Our next example shows that the Lipschitz and the linear classifications of general Banach spaces are different. The spaces in the example are not separable and do not have the RNP.

Example 7.12. *There is a Banach space E which is Lipschitz equivalent to $c_0(\Gamma)$ for some uncountable set Γ and which is not linearly isomorphic to it. In fact, $c_0(\Gamma)$ is not even isomorphic to a subspace of E.*

Let $\{N_\gamma : \gamma \in \Gamma\}$ be an uncountable collection of infinite sets of natural numbers such that $N_\gamma \cap N_\beta$ is finite for each $\gamma \neq \beta$. (For example, replace the natural numbers by the rationals, and take Γ to be the reals. Then assign to each $\gamma \in \Gamma$ a sequence of rationals converging to it.) Let E be the subspace of l_∞ spanned by c_0 and by the indicator functions I_{N_γ} of the sets N_γ. The coordinate functionals on l_∞ form a countable family in E^* which separates the points of E. Since no countable family in the dual of $c_0(\Gamma)$ separates its points, it cannot be linearly embedded into E.

The quotient space E/c_0 is isometric to $c_0(\Gamma)$. Indeed, given $\gamma_1, \dots, \gamma_n \in \Gamma$, let N be the finite set $\bigcup_{i \neq j} N_{\gamma_i} \cap N_{\gamma_j}$. Then for every $a_1, \dots, a_n$,

$$\Big\| \sum_1^n a_j I_{N_{\gamma_j}} \Big\|_{E/c_0} = \Big\| \sum_1^n a_j I_{N_{\gamma_j} \setminus N} \Big\|_\infty = \max |a_j|$$

i.e., the images of the I_{N_γ}'s in the quotient are isometrically equivalent to the unit vectors of $c_0(\Gamma)$. Let $q : E \to E/c_0$ be the quotient map. The heart of the construction is to show that q admits a Lipschitz lifting; i.e., that there is a Lipschitz map $f : E/c_0 \to E$ such that $q \circ f$ is the identity on E/c_0. It then follows that E is Lipschitz equivalent to $c_0 \oplus E/c_0$ via the map $x \to \big(x - f\big(q(x)\big), q(x)\big)$ whose inverse is $(y, z) \to y + f(z)$. But $c_0 \oplus E/c_0 = c_0 \oplus c_0(\Gamma)$ is isometric to $c_0(\Gamma)$, which is thus Lipschitz equivalent to E.

Identifying E/c_0 with $c_0(\Gamma)$, we shall first define $f(y)$ for $y \in c_0(\Gamma)^+$. Write $y = \sum_{j=1}^\infty a_j e_{\gamma_j}$ with $a_j \searrow 0$, and define inductively a sequence of sets by $M_1 = N_{\gamma_1}$ and $M_n = N_{\gamma_n} \setminus \bigcup_{j<n} N_{\gamma_j}$ for $n > 1$. We then define

$$f(y) = \sum a_j I_{M_j}.$$

Note that $I_{M_j} \in E$ because it differs from $I_{N_{\gamma_j}}$ by the characteristic function of a finite set; i.e., by an element of c_0. Also $q(I_{M_j}) = q(I_{N_{\gamma_j}}) = e_{\gamma_j} \in c_0(\Gamma)$; hence $q\big(f(y)\big) = y$. To see that f satisfies a Lipschitz condition, we give another formula for $f(y)$. Let A_n be the closed subspace of $c_0(\Gamma)$ spanned by $\{e_\gamma : n \notin N_\gamma\}$. Since the nth coordinate $\big(f(y)\big)_n$ of $f(y)$ is equal to a_i if and only if $n \in N_{\gamma_i}$ and $n \notin N_{\gamma_1} \cup \dots \cup N_{\gamma_{i-1}}$, it follows from the monotonicity of the a_i's that

$$\big(f(y)\big)_n = \operatorname{dist}\big(y, A_n\big).$$

It follows that if $y, z \in c_0(\Gamma)^+$, then

$$\big|\big(f(y)\big)_n - \big(f(z)\big)_n\big| = \big|\operatorname{dist}\big(y, A_n\big) - \operatorname{dist}\big(z, A_n\big)\big| \leq \|y - z\|$$

i.e., $\|f(y) - f(z)\| \leq \|y - z\|$.

A general element $y \in c_0(\Gamma)$ admits a representation $y = y^+ - y^-$ as a difference of two disjointly supported nonnegative vectors. Since the definition of $f(x)$ for $x \in c_0(\Gamma)^+$ depends only on the restriction of x to its support, it follows that $f(y) = f(y^+) - f(y^-)$ is well defined and it satisfies

$$\begin{aligned} \|f(y) - f(z)\| &\leq \|f(y^+) - f(z^+)\| + \|f(z^-) - f(y^-)\| \\ &\leq \|y^+ - z^+\| + \|y^- - z^-\| \leq 2\|y - z\|. \end{aligned}$$

This extends f to a 2-Lipschitz map on all of $c_0(\Gamma)$ and finishes the construction.

The next result characterizes the $C(K)$ spaces which are Lipschitz equivalent to $c_0(\Gamma)$ for some set Γ. (Note that the space E in the example above, being a subalgebra of the $C(K)$ space l_∞, is indeed isomorphic to a $C(K)$ space.) We first need some notation and simple facts concerning scattered topological spaces.

Let K be a compact Hausdorff space. Its derived set K' is the (closed) set of all nonisolated points of K. For any ordinal α we define the αth derived set inductively by $K^{(\alpha+1)} = (K^{(\alpha)})'$ and by $K^{(\alpha)} = \bigcap_{\beta<\alpha} K^{(\beta)}$ when α is a limit ordinal. The space K is said to be scattered if there is an ordinal α such that $K^{(\alpha)} = \emptyset$. The least ordinal α with $K^{(\alpha)} = \emptyset$ is a successor, $\alpha = \beta + 1$, and $K^{(\beta)}$ is a finite set. If it contains N points, then K can be represented as the disjoint union $K = \bigcup_{j \leq N} K_j$, where each K_j is a closed and open subset of K such that $K_j^{(\beta)}$ is a singleton. In this case $C(K)$ is isometric to $\left(\sum_{j \leq N} \oplus C(K_j)\right)_\infty$.

Let K be scattered, and fix $k_0 \in K$. Then $C_0(K, k_0) = \{h \in C(K) : h(k_0) = 0\}$ is isomorphic to $C(K)$. Indeed, all hyperplanes of a Banach space are isomorphic to each other, and we can thus assume that k_0 is isolated. For such a choice of k_0 the space K is homeomorphic to $K \setminus \{k_0\}$, and hence $C_0(K, k_0)$ is even isometric to $C(K)$. Since we shall be interested only in the isomorphic type of the space, we shall simply write $C_0(K)$ and will not specify k_0.

We shall say that the scattered space K is of finite order if $K^{(\omega)} = \emptyset$, where ω denotes the first infinite ordinal. In this case there is an integer n such that $K^{(n)}$ is a finite set and K is said to be of order n. If $n = 1$, then K is a finite union of the one-point compactifications of some discrete sets Γ_i, and $C_0(K)$ is isomorphic to $c_0(\bigcup \Gamma_i)$. If $n > 1$, then the space $C_0(K, K^{(n-1)}) = \{h \in C(K) : h|_{K^{(n-1)}} \equiv 0\}$ is isometric to $C_0(L)$, where the space L is obtained from K by identifying the closed set $K^{(n-1)}$ to a single point. In particular, L is of order $n - 1$.

Theorem 7.13. *The space $C(K)$ is Lipschitz equivalent to $c_0(\Gamma)$ for some set Γ if and only if K is a scattered space of finite order.*

Proof. We prove here only one direction. We shall see later (see the Notes and Remarks section in Chapter 10) that if $K^{(\omega)} \neq \emptyset$, then $C(K)$ is not even uniformly homeomorphic to $c_0(\Gamma)$.

We prove the theorem by induction on the order n of K. As noted above, $C(K)$ is isomorphic to $c_0(\Gamma)$ for $n = 1$, so assume that $n > 1$. Fix $\bar{k} \in K^{(n-1)}$, and let $Q : C_0(K) \to C_0(K^{(n-1)})$ be the restriction operator (where both spaces consist of the functions vanishing at $\bar{k}$). The kernel of Q is $C_0(K, K^{(n-1)})$, and by Tietze's extension theorem Q is surjective. By the induction hypothesis and the remarks above, there are sets Γ_1 and Γ_2 such that $C_0(K, K^{(n-1)})$ is Lipschitz equivalent to $c_0(\Gamma_1)$ and $C_0(K^{(n-1)})$ is isomorphic to $c_0(\Gamma_2)$. As in Example 7.12 we finish the proof by constructing a Lipschitz lifting $f : C_0(K^{(n-1)}) \to C_0(K)$ of Q. The construction is very similar to that of the example, and we only sketch it.

Without loss of generality we can assume that $K^{(n)}$ is the singleton $\bar{k}$. Each $k \in K^{(n-1)}$ which is different from $\bar{k}$ is isolated in $K^{(n-1)}$, and we choose a closed and open subset G_k of K such that $G_k \cap K^{(n-1)} = \{k\}$. The indicator function I_k of G_k is continuous. Note that if $k \neq l$, then $G_k \cap G_l$ is disjoint from $K^{(n-1)}$, and therefore its indicator function belongs to $C_0(K, K^{(n-1)})$, and we can identify $C_0(K^{(n-1)})$ with the image under Q of the span of $\{I_k : \bar{k} \neq k \in K^{(n-1)}\}$. As in the example, it is enough to define $f(y)$ for nonnegative elements $y \in C_0(K^{(n-1)})$.

Write $y = Q\big(\sum a_j I_{k_j}\big)$, where $a_j \searrow 0$ and $k_j \in K^{(n-1)} \setminus \{\bar{k}\}$. Define inductively $M_1 = G_{k_1}$ and $M_n = G_{k_n} \setminus \bigcup_{j<n} G_{k_j}$ for $n > 1$, and put

$$f(y) = \sum a_n I_{M_n}.$$

Since each M_n is closed and open, $f(y)$ is a continuous function on K. It is also clear that $f(y)(\bar{k}) = 0$ and that $Q\big(f(y)\big) = y$. To see that f is Lipschitz, we use again an alternative formula for f. For each $p \in K$ let A_p be the closed linear span of $\big\{I_k : \bar{k} \neq k \in K^{(n-1)} \text{ and } p \notin G_k\big\}$. Then

$$f(y)(p) = \operatorname{dist}\big(y, A_p\big).$$

We leave the details to the reader. □

We conclude this chapter with examples which show that the Gâteaux derivative of an onto Lipschitz equivalence can indeed be a properly into map. The following lemma may also be useful for other constructions of Lipschitz equivalences of Banach spaces.

Lemma 7.14. *Let E be a Banach space, and let $\{A_k\}_{k=1}^\infty$ be subsets of E such that $A_k \subset A_{k-1}$. Let $\{g_k\}_{k=1}^\infty$ be homeomorphisms of E onto itself such that all the g_k's and their inverses are L-Lipschitz for some constant L. Assume also that*

(i) *the restriction of g_k to A_k is an isometry of A_k onto itself,*

(ii) *$g_k(x) = x$ for every $x \notin A_{k-1}$,*

(iii) *$\alpha_k = \sup\big\{\operatorname{dist}(x, E \setminus A_k) : x \in A_k\big\} \to 0$.*

Then the sequence $\varphi_n = g_n \circ \cdots \circ g_1$ converges pointwise to a Lipschitz equivalence φ of E onto itself, and φ^{-1} is the pointwise limit of the φ_n^{-1}'s.

Proof. We first show that φ_n and φ_n^{-1} are also L-Lipschitz for every n. Fix n and x. It follows from (i) and from the monotonicity of the A_k's that if $\varphi_i(x) \notin A_{i+1}$ for some i, then $\varphi_k(x) \notin A_{k+1}$ for every $k \geq i$. Thus the sets

$$B_j = \big\{x : \ \varphi_k(x) \in A_{k+1} \text{ for } 1 \leq k < j \ \text{ and } \ \varphi_k(x) \notin A_{k+1} \text{ for } j \leq k < n\ \big\}$$

for $1 < j < n$, together with the three sets $B_n = \{x : \varphi_k(x) \in A_{k+1} \text{ for } 1 \leq k < n\}$, $B_1 = \{x \in A_1 : g_1(x) \notin A_2\}$, and $B_0 = E \setminus A_1$ form a cover of E.

Since g_k is an isometry on A_k (by (i)) and on the complement of A_{k-1} (by (ii)), it follows that when we restrict φ_n to B_j, then all the factors in its representation as a composition, except possibly g_{j+1}, are isometries. Since g_{j+1} is L-Lipschitz, it follows that φ_n is L-Lipschitz on each B_j and hence also on $\overline{B_j}$. It also follows that φ_n^{-1} is L-Lipschitz on each $\varphi_n\big(\overline{B_j}\big)$.

To see that φ_n and φ_n^{-1} are actually L-Lipschitz on all of E, it suffices to show that if the interval $[a, b]$ is covered by closed sets $\{M_j\}_{j=1}^n$, and if a continuous function $f : [a, b] \to E$ is L-Lipschitz on each M_j, then $\|f(t) - f(a)\| \leq L|t - a|$ for every t in the interval. Let t_j be the minimal point in M_j, and assume that

$a = t_1 \leq \ldots \leq t_n$. We shall prove the claim by induction. If $s < t_k$, then $[a, s]$ is covered by $\{M_j\}_{j=1}^{k-1}$; hence $\|f(s) - f(a)\| \leq L|s - a|$, and by continuity this also holds for $s = t_k$. Fix $a \leq t \leq b$, and assume that $t \in M_k$. Then $t \geq t_k$, and since f is L-Lipschitz on M_k it follows that

$$\begin{aligned} \|f(t) - f(a)\| &\leq \|f(t) - f(t_k)\| + \|f(t_k) - f(a)\| \\ &\leq L\big(|t - t_k| + |t_k - a|\big) = L|t - a|. \end{aligned}$$

Now fix k. If $j \geq k + 1$, then g_j acts as the identity on $E \setminus A_k$ by (ii). Thus $\varphi_n(x) = \varphi_k(x)$ for $n \geq k$ and for every $x \in \varphi_k^{-1}(E \setminus A_k) = E \setminus \varphi_k^{-1}(A_k)$. It follows that $\varphi(x) = \lim \varphi_n(x) = \varphi_k(x)$ for $x \in E \setminus \varphi_k^{-1}(A_k)$. Thus $\varphi = \lim \varphi_n$ is a L-Lipschitz equivalence of $E \setminus \bigcap \varphi_n^{-1}(A_n)$ onto $E \setminus \bigcap A_n$ which is equal to φ_k on $E \setminus \varphi_k^{-1}(A_k)$, takes it onto $E \setminus A_k$, and with $\varphi^{-1} = \varphi_k^{-1}$ on $E \setminus A_k$.

Given ε, choose n such that $L\alpha_n \leq \varepsilon$. For every $x \in \bigcap \varphi_k^{-1}(A_k)$ there is a $z \notin \varphi_n^{-1}(A_n)$ with $\|z - x\| < \varepsilon$ (since φ_n is an L-Lipschitz equivalence, (iii) holds for $\varphi_n^{-1}(A_n)$ with $L\alpha_n$ replacing α_n). Hence $\varphi_m(z) = \varphi_k(z)$ for $m, k \geq n$, and thus

$$\|\varphi_m(x) - \varphi_k(x)\| \leq \|\varphi_m(x) - \varphi_m(z)\| + \|\varphi_k(z) - \varphi_k(x)\| \leq 2L\varepsilon.$$

It follows that $\varphi(x) = \lim \varphi_m(x)$ exists. It is clear that φ is L-Lipschitz, and the same argument shows that $\lim \varphi_n^{-1}$ exists and is an inverse of φ. □

The examples below will follow the scheme of the lemma. For $0 \leq t \leq 1$ let T_t be the clockwise rotation of the plane in an angle of $(1 - t)\pi/2$, and put $T_t = I$ for $t \geq 1$. Each T_t is an isometry (in the usual Euclidean norm), and $\|T_t - T_s\| \leq \pi|t - s|/2$.

For each $k = 2, 3, \ldots$ let $E_k = \operatorname{span}\{e_1, e_k\}$, let P_k be the orthogonal projection of l_2 onto E_k, and let $Q_k = I - P_k$. Identifying E_k with $\mathbb{R}^2$, define linear operators $T_{k,t}$ on l_2 by $T_{k,t}x = T_t P_k x + Q_k x$.

In Examples 7.15 and 7.16 below we shall define a decreasing sequence of sets $A_k \subset l_2$ with the following properties:

(1) $T_{k,t}A_k = A_k$ for every k and every $t \geq 0$,

(2) $\operatorname{dist}\big(x, A_k\big) \leq 2^{-k}$ implies that $\|P_k x\| \leq 2^{-k+4}$.

Put

$$g_k(x) = T_{k, 2^k \operatorname{dist}(x, A_k)}(x).$$

Then g_k is the isometric rotation $g_k(a_1, a_2, \ldots) = (a_k, a_2, \ldots, a_{k-1}, -a_1, a_{k+1}, \ldots)$ on A_k and is the identity when $\operatorname{dist}(x, A_k) \geq 2^{-k}$. Moreover, it follows from (2) that if $\|P_k x\| \geq 2^{-k+4}$, then $g_k(x) = x$; and thus to check the Lipschitz constant of g_k, we may assume that $\|P_k x\| \leq 2^{-k+4}$. Putting $s = 2^k \operatorname{dist}(x, A_k)$ and $t = 2^k \operatorname{dist}(y, A_k)$, we then obtain

$$\begin{aligned} \|g_k(x) - g_k(y)\| &\leq \|T_{k,s}x - T_{k,t}x\| + \|T_{k,t}x - T_{k,t}y\| \\ &= \|T_s P_k x - T_t P_k x\| + \|T_{k,t}x - T_{k,t}y\| \\ &\leq \pi|s - t|\big\|P_k x\big\|/2 + \|x - y\| \\ &\leq \pi 2^k \|x - y\| 2^{-k+3} + \|x - y\| \leq 30\|x - y\|. \end{aligned}$$

By (1) g_k^{-1} has a representation analogous to that of g_k, and a similar argument shows that it is 30-Lipschitz.

The following two examples are obtained by suitable choices of the sets A_k and by verifying that together with the g_k's as above they satisfy the conditions of the lemma. The first example is a weaker version of the second. We denote the right shift on l_2 by S; i.e., $S(a_1, a_2, \dots) = (0, a_1, a_2, \dots)$.

Example 7.15. *There is a Lipschitz equivalence $\varphi : l_2 \to l_2$ which is Gâteaux differentiable at 0, and $D_\varphi(0)$ is not surjective.*

Let $A_k = 2^{-k+3}B(l_2)$. Then (1) and (2), as well as the conditions of the lemma (starting the composition with g_2 instead of g_1), are obviously satisfied. To see that $D_\varphi(0) = -S$, take first a finitely supported vector $x = (a_1, \dots, a_N, 0, 0 \dots)$ with $\|x\| = 1$. If $t < 2^{-N}$, then a direct computation gives $\varphi(tx) = -S(tx)$. It follows that $\lim_{t\to 0} \varphi(tx)/t = -Sx$; and since φ is Lipschitz, this holds for every x.

Example 7.16. *There is a Lipschitz equivalence $\varphi : l_2 \to l_2$ such that the restriction of φ to the Hilbert cube $C = \{x \in l_2 : |a_j| \le 2^{-j}\}$ is $-S$. In particular φ is Gâteaux differentiable on $\{x \in l_2 : |a_j| < 2^{-j}\}$ and $D_\varphi = -S$ there.*

For $k \ge 2$ put

$$A_k = \Big\{x \in l_2 : \ \|P_k x\| \le 2^{-k+3} \text{ and } |a_j| \le 2^{-j+1} + 2^{-k+1} \text{ for } 2 \le j < k\Big\}.$$

Then (1) and (2) obviously hold. To see that this is a decreasing sequence of sets and that (ii) of the lemma holds, we check that actually every y with $\operatorname{dist}(y, A_k) \le 2^{-k}$ belongs to A_{k-1}. Indeed, if $x = (a_1, \dots) \in A_k$, then

$$\|P_{k-1}x\|^2 = |a_1|^2 + |a_{k-1}|^2 \le 2^{-2k+6} + (2^{-k+2} + 2^{-k+1})^2 = 25 \cdot 2^{-2(k-1)}.$$

It follows that if $y = (b_1, \dots)$ satisfies $\|x - y\| \le 2^{-k}$, then

$$\|P_{k-1}y\| \le 2^{-k} + \|P_{k-1}x\| \le 2^{-k} + 5 \cdot 2^{-(k-1)} < 2^{-(k-1)+3}$$

and if $2 \le j < k-1$, then

$$|b_j| \le 2^{-k} + |a_j| \le 2^{-k} + 2^{-j+1} + 2^{-k+1} \le 2^{-j+1} + 2^{-(k-1)+1}.$$

Condition (iii) of the lemma holds with $\alpha_k = 2^{-k+3}$ because $P_k A_k = 2^{-k+3}B(E_k)$.

To compute φ on C, put

$$C_k = \Big\{x \in l_2 : \ |a_1| \le 2^{-k+1}; \ |a_j| \le 2^{-j+1} \text{ for } 2 \le j < k; \ |a_j| \le 2^{-j} \text{ for } j \ge k\Big\}$$

and note that $C = C_2$ and that $C_k \subset A_k$.

The explicit representation $g_k(x) = (a_k, a_2, \dots, a_{k-1}, -a_1, a_{k+1}, \dots)$ whenever $x \in A_k$ implies that g_k maps C_k into C_{k+1}. It follows that if $y = (b_1, \dots) \in C$, then

$$g_k \circ \dots \circ g_2(y) = (b_k, -b_1, \dots, -b_{k-1}, b_{k+1}, \dots) \to -Sy.$$

The final claim follows by considering finitely supported vectors first, as in the previous example.

The next example uses the above for general compact sets in l_2.

Example 7.17. *Let K be a compact set in l_2. Then there is a Lipschitz equivalence $\psi : l_2 \to l_2$ whose restriction to K is an affine map that takes K into a closed hyperplane in l_2.*

Choose a sequence $\beta_n \to 0$ such that $|a_n| \le \beta_n$ for every $x = (a_1, \dots) \in K$ and for all n. Fix a subsequence $\{n_j\}$ such that $\beta_{n_j} \le 2^{-j}$, and consider the orthogonal decomposition $l_2 = H_1 \oplus H_2$ with $H_1 = \operatorname{span}\{e_{n_j}\}$. Let C be the Hilbert cube $\{|a_{n_j}| \le 2^{-j}\}$ in H_1, and let $\varphi : H_1 \to H_1$ be the map of Example 7.16. The required map is $\psi(x, y) = \big(\varphi(x), y\big)$ for $x \in H_1$ and $y \in H_2$.

Remark. Closed hyperplanes are Gauss null, and this is not necessarily so for compact sets. Thus Examples 7.16 and 7.17 give other proofs (see also Chapter 6) that being a Gauss null set is not preserved under a Lipschitz equivalence.

3. Notes and Remarks

The introduction of null sets and the versions of Theorem 6.42 on the existence of points of Gâteaux differentiability for Lipschitz functions were proved by Christensen [**121**], Mankiewicz [**383**] and Aronszajn [**23**]. The first two were motivated by the possible applications to the Lipschitz classification of Banach (and Fréchet) spaces, as explained in the beginning of this chapter. Some of the results (i)-(iv) were proved earlier by Lindenstrauss [**357**] and by Enflo [**169, 171**].

For a proof that l_p and c_0 are not isomorphic to a subspace of l_q (for $p \neq q$), see Lindenstrauss and Tzafriri [**371**], page 54. Since L_p contains a subspace isomorphic to l_2, it cannot be isomorphic to a subspace of l_q. The type and cotype of L_p are $\min(p, 2)$ and $\max(p, 2)$ respectively (see Lindenstrauss and Tzafriri [**372**], page 73). Since a space with a given type or cotype cannot be isomorphic to a subspace with a bigger type (or, respectively, smaller cotype), it follows that L_p cannot be isomorphic to a subspace of L_q unless, possibly, when $2 \leq p \leq q$ or $2 \geq p > q$. The first case is still impossible (for $p \neq 2$) by Kadec and Pełczyński [**304**], and it is explained in Appendix D how to embed L_p in L_q when $2 \geq p > q$. (Another construction is given in Corollary 8.8.)

The results of Section 1 are due to Lindenstrauss [**357**]. He used smoothing and differentiation. Pełczyński [**451**], page 61, applied invariant means in this context.

The recurring assumption that some space is complemented in its second dual is essential. We saw in Example 1.5 that c_0 is a Lipschitz retract of l_∞, but it is not linearly complemented in it. This follows from the fact (mentioned in the Notes and Remarks section in Chapter 1) that there are no separable $\mathcal{P}_\lambda$ spaces. A direct proof follows from the observation, made in the proof of Example 7.12, that l_∞/c_0 is not isomorphic to a subspace of l_∞ (or of any space E such that E^* contains a countable family that separates the points of E), and therefore c_0 is not complemented in l_∞.

The results in Section 2 up to (and not including) Theorem 7.11 are due to Heinrich and Mankiewicz [**252**].

In connection with Theorem 7.6 we mention that it is unknown whether every Banach space E is a Lipschitz retract of E^{**}. Another open problem, related to Corollary 7.8, is whether a Banach space which is Lipschitz equivalent to a dual space is actually linearly isomorphic to one. We also note that observation (i) in the beginning of this chapter implies that the RNP is invariant under Lipschitz equivalence. It follows from Theorem 11.21 in Chapter 11 that for a separable space E the fact that the dual E^* has the RNP is also invariant under Lipschitz equivalence. (We use here that E^* has the RNP if and only if it is separable; see Corollary 5.24.)

Note also that if E does not contain a proper subspace isomorphic to itself (such as the spaces constructed by Gowers and Maurey [**226**]), and if it has the RNP, then E is uniquely determined by its Lipschitz structure. Indeed, if F is Lipschitz equivalent to E, then it also has the RNP, and by differentiation we find that E and F are isomorphic to subspaces of each other and these isomorphisms must be surjective by the condition on E.

Theorem 7.11 is due to Aharoni [**1**]. The proof here is due to Assouad [**28**]. The optimal value of K is not known, but it is proved in [**1**] that $K \geq 2$ for any embedding of l_1 into c_0. Note that the proof actually gives an embedding into c_0^+, and Aharoni [**2**] showed that $K \geq 3$ for any Lipschitz embedding of c into c_0^+.

It is unknown if a Banach space E which is Lipschitz universal for separable metric spaces (i.e., such that every separable metric space is Lipschitz equivalent to a subset of E) must contain a linear subspace which is isomorphic to c_0.

The space c_0 in Theorem 7.11 cannot, of course, be replaced by a reflexive space. It is unknown if this is possible if we require only that the embedding be uniform rather than Lipschitz. The next proposition shows that compact spaces are uniformly homeomorphic (with homeomorphisms which are "almost" Lipschitz) to subsets of the reflexive space $\left(\sum \oplus l_\infty^n\right)_2$.

Proposition 7.18. *Let (X,d) be a compact metric space. Then there is an embedding $f : X \to \left(\sum \oplus l_\infty^n\right)_2$ such that*

$$d(x,y)/K \leq \|f(x) - f(y)\| \leq K d(x,y)\big|\log d(x,y)\big|^{1/2}$$

for some constant K and for all $x, y \in X$ with $d(x,y) \leq 1/2$.

Proof. Assume, as we may, that the diameter of X is smaller than $1/2$, and for each n let $\{x_{n,j} : j \leq j_n\}$ be a maximal 2^{-n}-separated set in X. For each n and $j \leq j_n$ put $f_{n,j}(x) = \operatorname{dist}\bigl(x, X \setminus B(x_{n,j}, 2^{-n+1})\bigr)$, and let $\{e_{n,j}\}$ be the natural basis for $\left(\sum \oplus l_\infty^{j_n}\right)_2$. The map $f(x) = \sum_{n,j} f_{n,j}(x) e_{n,j}$ is well defined because $\|\sum_j f_{n,j}(x) e_{n,j}\| \leq \max_j \|f_{n,j}\|_\infty = 2^{-n+1}$ for every n, and it follows that $\|f(x)\| \leq \left(\sum 2^{2(-n+1)}\right)^{1/2} < \infty$.

Fix $x, y \in X$, and choose N such that $2^{-N-1} < d(x,y) \leq 2^{-N}$. Using the estimates $\|f_{n,j}\|_{Lip} \leq 1$ for $n \leq N$ and $\|f_{n,j}\|_\infty \leq 2^{-n+1}$ for $n > N$, we obtain

$$\|f(x) - f(y)\|^2 \leq N d(x,y)^2 + 4 \sum_{n>N} 2^{2(-n+1)} \leq K d(x,y)^2 \,\big|\log d(x,y)\,\big|.$$

Conversely, choose j with $d(x, x_{N+3,j}) < 2^{-(N+3)}$. Then $f_{N+3,j}(x) \geq 2^{-(N+3)}$, while $f_{N+3,j}(y) = 0$. Hence $\|f(x) - f(y)\| \geq 2^{-(N+3)} \geq d(x,y)/8$. □

Example 7.12 is due to Aharoni and Lindenstrauss [**3**]. The part of Theorem 7.13 presented in this chapter is due to Deville, Godefroy and Zizler [**142**].

Lemma 7.14 and the examples following it are due to Ives and Preiss [**269**]. They also constructed a Lipschitz equivalence of l_2 which is not Gâteaux differentiable at 0 but such that $t^{-1}\bigl(\varphi(tx) - \varphi(0)\bigr)$ converges weakly to zero as $t \to 0$ for every x.

As we mentioned in Chapter 6, Lindenstrauss, Matoušková and Preiss [**363**] constructed a Lipschitz equivalence f of l_2 onto itself and a Borel set A which contains a translate of every compact set in the unit ball (thus A is not Haar null) such that $f(A)$ is contained in a closed hyperplane. There is also a Borel set $\tilde{A}$ which contains a translate of every compact set in the unit ball of l_2 such that f has a nonsurjective Gâteaux derivative at every point of $\tilde{A}$. The proof is based on Lemma 7.14 and the setup of the examples that follow it, as well as on the method of proof of Theorem 6.36.

CHAPTER 8

Uniform Embeddings into Hilbert Space

1. Positive Definite and Negative Definite Functions

In this chapter we study uniform embeddings of metric spaces, and in particular normed and quasi-normed spaces, into a Hilbert space. This is done in the second section by a systematic application of positive definite kernels on metric spaces. In the first section we present the basic definitions and results on positive definite and on negative definite kernels and functions.

Definition 8.1. *An Hermitian kernel on a set X (i.e., a complex-valued function $K(x,y)$ on $X \times X$ such that $K(x,y) = \overline{K(y,x)}$) is said to be positive definite if $\sum K(x_i,x_j)c_i\overline{c_j} \geq 0$ for all $x_1,\dots,x_n$ in X and for all complex scalars $c_1,\dots,c_n$. A function f on an abelian group X is positive definite if the kernel $K(x,y) = f(x-y)$ it induces on X is positive definite.*

The fact that K is Hermitian actually follows from the positivity of the quadratic form (take $c_1 = c_2 = 1$ and then $c_1 = 1, c_2 = i$).

To say that a kernel K is positive definite means that every $n \times n$ matrix $K(x_i,x_j)$, where $x_1,\dots,x_n \in X$, is positive definite. By basic linear algebra every Hermitian matrix A is diagonalizable; i.e., there is a unitary matrix U such that $D = U^*AU$ is a diagonal matrix. The matrix A is positive definite when the entries of D are nonnegative. Another basic result is that an Hermitian matrix is positive definite if and only if the determinant of every $k \times k$ submatrix obtained by removing the same $n-k$ rows and columns is nonnegative.

It follows directly from the definition that the set of positive definite kernels is a convex cone which is closed under pointwise limits, complex conjugation, and taking real parts, and that $K(x,y) = g(x)\overline{g(y)}$ is positive definite for every function $g : X \to \mathbb{C}$. It turns out that this set is also closed under multiplication.

Proposition 8.2. (i) *If K_1 and K_2 are positive definite kernels on the same set X, then so is $K = K_1K_2$.*

(ii) *Let K be a positive definite kernel, and let $f(z) = \sum a_n z^n$ be an analytic function in a disk containing the range of K with $a_n \geq 0$ for every n. Then $f \circ K$ is also positive definite. In particular, e^K is positive definite for every positive definite kernel K.*

Proof. It is enough to consider matrices. For every vector $u = (a_1,\dots,a_n)$ the rank-one matrix $(a_i\overline{a_j})$ is positive definite, and by diagonalization every positive definite matrix is a sum of n such matrices (the u's are just the eigenvectors of the matrix). We can thus assume that $K_1 = (a_i\overline{a_j})$ and $K_2 = (b_i\overline{b_j})$. But then $K = \big((a_ib_i)(\overline{a_jb_j})\big)$ is also of the same form, hence positive definite. This proves (i), and (ii) follows immediately. □

Closely related to positive definite kernels are the negative definite ones.

Definition 8.3. *An Hermitian kernel N on a set X is called negative definite if $\sum N(x_i, x_j)c_i\overline{c_j} \leq 0$ for all $x_1, \ldots, x_n$ in X and for all complex scalars $c_1, \ldots, c_n$ satisfying $\sum c_j = 0$. Negative definite functions are defined similarly.*

The kernel $-K$ is clearly negative definite whenever K is positive definite, but the exact relation between these two notions is somewhat more involved.

Proposition 8.4. *A kernel N is negative definite if and only if e^{-tN} is positive definite for every $t > 0$.*

Proof. If e^{-tN} is positive definite for every $t > 0$, then the kernels $(1-e^{-tN})/t$ are negative definite, hence so is their limit as $t \to 0$, namely N.

Conversely, assume that N is negative definite, and fix $x_1, \ldots, x_n$ in X and $c_1, \ldots, c_n \in \mathbb{C}$. Fix $x_0 \in X$, and let $c_0 = -\sum_{i=1}^n c_j$; then

$$\sum_{i,j=0}^{n} N(x_i, x_j)c_i\overline{c_j} = \sum_{i,j=1}^{n} \Big(N(x_i, x_j) - N(x_i, x_0) - N(x_0, x_j) + N(x_0, x_0)\Big)c_i\overline{c_j}$$

and the left-hand side is nonpositive because N is negative definite. It follows that the kernel

$$K(x, y) = -N(x, y) + N(x, x_0) + N(x_0, y) - N(x_0, x_0)$$

is positive definite. Since $N(x_0, y) = \overline{N(y, x_0)}$ and since $g(x)\overline{g(y)}$ is positive definite for any function g, the result follows from

$$e^{-tN(x,y)} = e^{tK(x,y)}\Big(e^{-tN(x,x_0)}\overline{e^{-tN(y,x_0)}}\Big)e^{tN(x_0,x_0)}$$

and from Proposition 8.2. □

Examples. (i) Let H be a Hilbert space. Then the kernel $K(x, y) = \langle x, y\rangle$ is positive definite. Indeed, given $x_1, \ldots, x_n \in H$ and scalars $c_1, \ldots, c_n$, one sees that

$$\sum \langle x_i, x_j\rangle c_i\overline{c_j} = \left\|\sum c_j x_j\right\|^2 \geq 0.$$

A similar computation shows that the function $\|x\|^2$ is negative definite. We shall see below that these simple examples generate the most general positive, respectively negative, definite kernels on arbitrary sets.

(ii) The function $x \to e^{i\langle t,x\rangle}$ is positive definite on $\mathbb{R}^n$ for every $t \in \mathbb{R}^n$. It follows that the Fourier transform $\hat{\mu}(x) = \int e^{i\langle t,x\rangle}d\mu(t)$ of a finite positive measure is a continuous positive definite function. Bochner's theorem (see Theorem D.5(ii)) says that every continuous positive definite function on $\mathbb{R}^n$ is the Fourier transform of such a positive measure.

(iii) Let N be a negative definite kernel such that $N(x, y) \geq 0$ for all x, y. Then so is $1 - e^{-tN}$ for every $t > 0$; hence also $\int(1 - e^{-tN})d\mu(t)$ for every finite positive measure on $\mathbb{R}^+$. Since $x^\alpha = c_\alpha \int_0^\infty(1 - e^{-tx})t^{-\alpha-1}dt$ for every $x > 0$ and for every $0 < \alpha < 1$, it follows that N^α is negative definite for each such α.

By (i) (or by direct computation) the function x^2 is negative definite on the real line. It follows that if $0 < p \leq 2$, then $|x|^p$ is a negative definite function and that $\exp(-|x|^p)$ is a positive definite function on $\mathbb{R}$. (See also Theorem D.8(i) and (ii) for another proof that $\exp(-|x|^p)$ is positive definite if and only if $0 < p \leq 2$.)

By integrating the pointwise inequalities that follow from the above, we deduce that $\|x\|^p$ is a negative definite function and that $\exp(-\|x\|^p)$ is a positive definite function on $L_p(\mu)$ for every measure μ and for every $0 < p \leq 2$. For $p = 0$ we consider $L_0(\mu)$ with the metric $d(x, y) = \int [1 - \exp(-|x - y|)] d\mu$ which induces the topology of convergence in probability. Similar arguments show that the function $\exp\big(-d(x, 0)\big)$ is positive definite on $L_0(\mu)$.

The next theorem relates positive and negative definite kernels to mappings into a Hilbert space. It is the key for our study of uniform embeddings.

Proposition 8.5. (i) *The kernel $K(x, y)$ on X is positive definite if and only if there is a Hilbert space H and a map T from X into H such that*

$$K(x, y) = \langle T(x), T(y) \rangle \text{ for all } x, y \in X.$$

(ii) *Let $N(x, y)$ be a real-valued kernel on X satisfying $N(x, x) = 0$ for every $x \in X$. Then N is negative definite if and only if there is a Hilbert space H and a map T from X into H such that*

$$N(x, y) = \|T(x) - T(y)\|^2 \text{ for all } x, y \in X.$$

Proof. One direction was noted in Example (i) above.

Conversely, assume that K is a positive definite kernel on X. Define a semi-inner product on the space E of finitely supported complex-valued functions on X by $\langle g, h \rangle = \sum K(x, y) g(x) \overline{h(y)}$, and put $F = \{g \in E : \langle g, g \rangle = 0\}$. The required Hilbert space is the completion of the quotient $H = E/F$, and the map T is defined by $T(x) = [\delta_x]$. (Here $\delta_x(y) = 1$ if $y = x$ and 0 otherwise, and $[g]$ denotes the image of g under the quotient map.) We leave the easy details to the reader.

To prove (ii) fix some $x_0 \in X$. As in the proof of Proposition 8.4, the kernel $K(x, y) = \big(-N(x, y) + N(x, x_0) + N(x_0, y)\big)/2$ is positive definite. If H and T are the Hilbert space and map which are obtained by applying (i) to K, then

$$\begin{aligned} \|T(x) - T(y)\|^2 &= \langle T(x), T(x) \rangle + \langle T(y), T(y) \rangle - 2\langle T(x), T(y) \rangle \\ &= K(x, x) + K(y, y) - 2K(x, y) = N(x, y). \end{aligned}$$

□

From now on we shall assume that the positive definite kernels K are real and that $K(x, x)$ is a constant independent of x. (These conditions are automatically fulfilled when the kernel is induced by a real positive definite function.) Note that to check the positive definiteness of a real kernel it is enough to use real scalars, but we need to assume that $K(x, y) = K(y, x)$, since this does not follow from the positivity of the quadratic form when we use real scalars only. We normalize so that $K(x, x)$ is identically 1. In this case the embedding T of Proposition 8.5 satisfies $\|T(x)\| = 1$ for all $x \in X$; i.e., it takes X into the unit sphere of H. Under these conditions

$$\|Tx - Ty\|^2 = \|Tx\|^2 + \|Ty\|^2 - 2\langle Tx, Ty \rangle = 2\big(1 - K(x, y)\big) \tag{8.1}$$

for all $x, y \in X$, and also, by the Cauchy-Schwartz inequality,

$$|K(x, y)| = |\langle T(x), T(y) \rangle| \leq \|T(x)\| \, \|T(y)\| = 1.$$

Corollary 8.6. *Let f be a real positive definite function on an abelian group X normalized so that $f(0) = 1$. Then*

(i) *$|f(x)| \leq f(0)$ for every $x \in X$.*

(ii) $|f(x) - f(y)|^2 \leq 2\big(1 - f(x-y)\big)$ *for every* $x, y \in X$*. In particular, if* X *is equipped with a translation invariant metric and if* f *is continuous at* 0*, then it is uniformly continuous on all of* X*.*

(iii) $1 - f(nx) \leq n^2\big(1 - f(x)\big)$ *for every* $x \in X$ *and for every positive integer* n*.*

Proof. All these properties can be proved directly from the definition, but the representation in Hilbert space makes them obvious.

Part (i) is a special case of $|K(x,y)| \leq 1$ above. Fix a Hilbert space H and a map $T : X \to H$ such that $f(x-y) = \langle T(x), T(y)\rangle$. Part (ii) follows from the identity (8.1) above because

$$|f(x) - f(y)|^2 = \big|\langle T(x) - T(y), T(0)\rangle\big|^2 \leq \|T(x) - T(y)\|^2 = 2\big(1 - f(x-y)\big)$$

and part (iii) follows from the triangle inequality in H. Indeed,

$$\begin{aligned} 1 - f(nx) &= \|T(0) - T(nx)\|^2/2 \\ &\leq \Big(\sum \|T(mx) - T\big((m-1)x\big)\|\Big)^2/2 = n^2\big(1 - f(x)\big). \end{aligned}$$

□

Proposition 8.5(ii) gives a characterization of metric spaces (X, d) which are isometric to subsets of a Hilbert space. A necessary and sufficient condition is that $d(x,y)^2$ is a negative definite kernel on X or, equivalently, that $\exp\big(-td(x,y)^2\big)$ is positive definite on X for every $t > 0$. In the next section we shall give conditions, in terms of the existence of certain positive or negative definite kernels on X, so that it embeds uniformly in a Hilbert space. We shall be mainly interested in normed and quasi-normed spaces, and the next result will be used to enable us to reformulate the problem as a problem on linear embeddings in L_0. We refer to Appendix D for notions and results that we use from probability theory.

Proposition 8.7. *Let* f *be a continuous positive definite function on a real linear metric space* E *satisfying* $f(0) = 1$*. Then there is a probability space* $(\Omega, \mathcal{B}, \mu)$ *and a continuous linear operator* $U : E \to L_0(\mu)$ *such that the characteristic function* $\mathbb{E}\exp(itUx)$ *of* Ux *is equal to* $f(tx)$ *for every* $x \in E$ *and* $t \in \mathbb{R}$*.*

Proof. For each finite subset $A = \{a_1 \dots a_n\} \subset E$ identify $\mathbb{R}^n$ with $\mathbb{R}^A$, and consider the positive definite function $f_A(t_1, \dots, t_n) = f\big(\sum t_j a_j\big)$. By Bochner's theorem it is the Fourier transform of a probability measure μ_A on $\mathbb{R}^A$. It is easy to check that if $A \subset B$, then the measure μ_A is the image of μ_B under the canonical projection from $\mathbb{R}^B$ onto $\mathbb{R}^A$. By Kolmogorov's consistency theorem (see Appendix D) there is a probability space $(\Omega, \mathcal{B}, \mu)$ and a family $\{U_x : x \in E\}$ of random variables on it such that for each finite subset A of E the joint distribution of $\{U_x : x \in A\}$ is μ_A. The map U is defined by $U(x) = U_x$.

To show that U is linear, fix $x, y \in E$ and scalars a, b. We shall show that $U(ax + by) - (aUx + bUy) = 0$ a.e. by checking that its characteristic function is identically equal to 1. Indeed, take $A = \{(ax+by), ax, by\}$. The joint distribution of $\big(U(ax+by), aUx, bUy\big)$ is the same as that of $\big(U(ax+by), U(ax), U(by)\big)$ and is given by μ_A. The definition of μ_A then gives that for each t

$$\begin{aligned} &\mathbb{E}\exp\Big(it\big(U(ax+by) - aU(x) - bU(y)\big)\Big) \\ &= f\big(t(ax+by) - tax - tby\big) = f(0) = 1. \end{aligned}$$

Finally, U is continuous because if $x_n \to 0$ and if $t \in \mathbb{R}$ is arbitrary, then $\mathbb{E}\exp(itUx_n) = f(tx_n) \to f(0) = 1$ by the continuity of f. Hence $Ux_n \to 0$ in probability. $\square$

The next results are not directly connected with uniform embeddings into Hilbert space, but they use the same methods. See Appendix H for information on quasi-normed spaces.

Corollary 8.8. *If $0 < r < p \leq 2$, then L_p is isometric to a subspace of L_r.*

Proof. Fix $r < p$ and use the positive definite function $f(x) = \exp(-\|x\|^p)$ on L_p to find a measure space $(\Omega, \mathcal{B}, \mu)$ and an operator $U : L_p \to L_0(\mu)$ as in the proposition. Then U is actually a multiple of an isometry into $L_r(\mu)$. Indeed, if $\|x\| = 1$, then Ux has a symmetric stable distribution with characteristic function $\exp(-|t|^p)$. By Theorem D.8(iii) there is a constant c, depending on r and p, such that $\|Ux\|_r = \big(\mathbb{E}|Ux|^r\big)^{1/r} = c$. $\square$

Notice that for $p = 2$ the space L_2 is actually isometric to a subspace of L_r for every r and not just for $r < 2$.

Theorem 8.9. *Let E be a real quasi-normed space, and let $0 < p \leq 2$. Then E is isometric to a subspace of $L_p(\mu)$ for some measure μ if and only if $\|x\|^p$ is a negative definite function on E or, equivalently, $\exp(-\|x\|^p)$ is positive definite.*

Proof. By Corollary F.4 we may assume that E is finite-dimensional. (The definition and basic results concerning ultraproducts of quasi-Banach spaces are analogous to those of Banach spaces. In particular, Corollary F.4 also holds for quasi-Banach spaces and for $0 < p < 1$.) Identifying E as $\mathbb{R}^n$, we shall find a measure ν on the Euclidean unit sphere S in $\mathbb{R}^n$ such that $\|x\| = \big(\int |\langle x, y\rangle|^p d\nu(y)\big)^{1/p}$; i.e., E is isometric to a subspace of $L_p(\nu)$.

Fix a basis $\{e_k\}_{k=1}^n$ for E, and consider E as $\mathbb{R}^n$ by identifying $x = \sum a_k e_k$ with $(a_1, \dots, a_n)$. By Proposition 8.7 there is a probability space $(\Omega, \mathcal{B}, \mu)$ and an operator $U : E \to L_0(\mu)$ such that $\mathbb{E}e^{itUx} = \exp(-|t|^p\|x\|^p)$ for every $x \in E$. We first replace Ω and μ by $\mathbb{R}^n$ with a measure η and identify Ux with the inner product $\langle x, \cdot\rangle$. Indeed, let η be the measure on $\mathbb{R}^n$ which is the image of μ under the map $\omega \to \big(Ue_1(\omega), \dots, Ue_1(\omega)\big)$. Then Ux and $\langle x, \cdot\rangle$ have the same distribution and the same characteristic function. To see this write $x = \sum a_k e_k$; hence

$$\mathbb{E}e^{itUx} = \int e^{it\sum a_k Ue_k(\omega)} d\mu(\omega) = \int e^{it\langle x,y\rangle} d\eta(y).$$

Fix an arbitrary $r < p$. As in the corollary above, there is a multiple η_r of η such that $\|x\|^r = \int |\langle x, y\rangle|^r d\eta_r(y)$ for every $x \in E$. Since $|\langle x, \cdot\rangle|^r$ is r-homogeneous, it follows that there is a measure ν_r on the Euclidean unit sphere S in $\mathbb{R}^n$ such that

$$\|x\|^r = \int_{\mathbb{R}^n} |\langle x, y\rangle|^r d\eta_r(y) = \int_S |\langle x, y\rangle|^r d\nu_r(y).$$

We estimate the total mass of ν_r. Since $r < 2$, it follows from Hölder's inequality that $1 = \big(\sum |\langle e_k, y\rangle|^2\big)^{r/2} \leq \sum |\langle e_k, y\rangle|^r$ for every $y \in S$. Hence

$$\int d\nu_r(y) \leq \sum \int |\langle e_k, y\rangle|^r d\nu_r(y) = \sum_{k=1}^n \|e_k\|^r \leq C$$

where the constant C depends on n but not on r. It follows that the measures ν_r are uniformly bounded, and we can thus find a sequence $r_j \nearrow p$ such that ν_{r_j} converge weakly to some measure ν. Clearly $\|x\|^p = \int_S |\langle x, y\rangle|^p d\nu(y)$. □

Remark. The embedding in Proposition 8.7 is only real-linear even for complex vector spaces. Indeed, if X is a complex-random variable, then $\mathbb{E}\exp\{itX\}$ is not even necessarily defined for all t. However, consequences such as Theorem 8.9 still hold in the complex case if we replace isometry by a linear isomorphism. Indeed, let E_R denote the space E as a vector space over the reals, and let $U : E_R \to L_0(\mu)$ be the isometry of Theorem 8.9. The space $E_R \oplus E_R$ is made into a complex space when multiplication by i is defined by $i(x, y) = (-y, x)$. The map $x \to (ix, x)$ is then a complex-linear isomorphism of E into $E_R \oplus E_R$. The operator $U \oplus U$ is a complex-linear isomorphism of $E_R \oplus E_R$ into $L_0(\mu) \oplus L_0(\mu)$ (which is a complex vector space under an analogous definition of multiplication by i). This latter space with its complex structure is the same as the complex $L_0(\mu)$.

2. Uniform Embeddings into Hilbert Space

In this section we study the uniform embeddability of metric spaces into a Hilbert space. Since Hilbert spaces over the reals suffice for this purpose, we shall assume that all the spaces are, indeed, real. Let X be a metric space, and let K be a positive definite kernel on X. We denote the map from X to a Hilbert space H, induced by K as in Proposition 8.5, by T_K. If K is induced by the positive definite function f, we write simply T_f.

We shall assume that K is real and that $K(x, x) \equiv 1$. The identity (8.1) reduces the study of the continuity of T and of T^{-1} to that of the behavior of K near the diagonal. It is natural to introduce for each $\varepsilon > 0$ the following parameters:

$$g_K(\varepsilon) = \inf\{1 - K(x, y) :\ d(x, y) \geq \varepsilon\},$$
$$h_K(\varepsilon) = \sup\{1 - K(x, y) :\ d(x, y) \leq \varepsilon\}.$$

If X is an abelian group and if K is induced by a real positive definite function f on X satisfying $f(0) = 1$, formula (8.1) takes the form

$$\|Tx - Ty\|^2 = 2\big(1 - f(x - y)\big) \tag{8.2}$$

and we define similarly

$$g_f(\varepsilon) = \inf\{1 - f(x) : d(x, 0) \geq \varepsilon\},$$
$$h_f(\varepsilon) = \sup\{1 - f(x) : d(x, 0) \leq \varepsilon\}.$$

We assume here and later on, without further mention, that the metric on the abelian group is invariant, i.e., that $d(x, y) = d(x - y, 0)$ for all x and y in X.

With the aid of these parameters we obtain the following criterion:

Proposition 8.10. *Let K be a real positive definite kernel on a metric space X such that $K(x, x) \equiv 1$. Then the embedding $T = T_K$ is a uniform homeomorphism if and only if $g_K(\varepsilon) > 0$ for every $\varepsilon > 0$ and $\lim_{\varepsilon \to 0} h_K(\varepsilon) = 0$.*

Proof. The explicit formula (8.1) for $\|Tx - Ty\|$ shows that T is uniformly continuous if and only if $\lim_{\varepsilon \to 0} h_K(\varepsilon) = 0$ and that T^{-1} is uniformly continuous if and only if $g_K(\varepsilon) > 0$ for all $\varepsilon > 0$. □

Example. The spaces $L_p(\mu)$ for $0 \le p \le 2$ embed uniformly into a Hilbert space. Indeed, consider $L_0(\mu)$ with the metric $d(x,y) = \int[1 - \exp(-|x-y|)]d\mu$. Then the kernels induced by the functions $f(x) = \exp\big(-d(x,0)\big)$ on $L_0(\mu)$ and $f(x) = \exp(-\|x\|^p)$ on L_p for $0 < p \le 2$ satisfy the conditions of Proposition 8.10.

Corollary 8.11. *An infinite-dimensional Hilbert space is uniformly homeomorphic to a subset of its unit sphere. In particular, if a metric space embeds uniformly in a Hilbert space, then it also embeds uniformly in its unit sphere.*

Proof. Let H be a Hilbert space. Then $f(x) = \exp(-\|x\|^2)$ is a real continuous positive definite function on H with $f(0) = 1$. Applying Proposition 8.10 to the kernel $f(x-y)$ gives that H is uniformly homeomorphic to a subset of the unit sphere of some Hilbert space. □

The question whether a metric space is uniformly homeomorphic to a subset of a Hilbert space is already determined by its finite subsets.

Proposition 8.12. *A metric space X is uniformly homeomorphic to a subset of a Hilbert space H if (and clearly only if) all its finite subsets are embeddable into H with maps which have uniform estimates on the moduli of continuity of the embeddings and of their inverses.*

Proof. We can restrict ourselves to mappings into the unit sphere of H. For every finite subset A of X find a kernel K_A as above so that we have uniform estimates on $g_{K_A}(\varepsilon)$ and $h_{K_A}(\varepsilon)$. Extend each K_A to a function $\tilde{K}_A$ defined on all of $X \times X$ by defining it to be zero off $A \times A$. By Tychonoff's theorem the closure of the set $\{\tilde{K}_A : A \text{ finite}\}$ is compact in the topology of pointwise convergence, so let K be the limit of a pointwise convergent subnet of $\tilde{K}_A$. Then K is a positive definite kernel on all of X which induces a uniform homeomorphism of X into H.

Note that if H is finite-dimensional, a direct compactness argument shows that X actually embeds uniformly into the same space H. □

The fact that we used positive definite functions in Corollary 8.11 and in the example preceeding it instead of general kernels is not accidental. This can always be done in metric abelian groups.

Proposition 8.13. *Let X be a metric abelian group which embeds uniformly into a Hilbert space. Then there is a real continuous positive definite function on X with $f(0) = 1$ such that $g_f(\varepsilon) = \inf\{1 - f(x) : d(x,0) \ge \varepsilon\} > 0$ for every $\varepsilon > 0$.*

Proof. We use an averaging argument. Let $M_x(\cdot)$ be an invariant mean on the bounded real-valued functions on the group X (see Appendix C). Assume that T is a uniform homeomorphism of X into a real Hilbert space with inner product $\langle \cdot, \cdot \rangle$. By Corollary 8.11 we can assume that T maps into the unit sphere; hence the kernel $K(x,y) = \langle Tx, Ty \rangle$ satisfies the conditions in Proposition 8.10. Define a function f on X by the formula

$$f(x) = M_y\Big(K(y+x, y)\Big)$$

for $x \in X$. Then $f(0) = 1$ and f is positive definite. Indeed, if $x_j \in X$ and $c_j \in \mathbb{R}$, then the translation invariance of M_y and the fact that K is positive definite give

$$\begin{aligned} \sum f(x_i - x_j)c_ic_j &= M_y\Big(\sum K(y + x_i - x_j, y)c_ic_j\Big) \\ &= M_y\Big(\sum K(y + x_i, y + x_j)c_ic_j\Big) \geq 0. \end{aligned}$$

By the definition of g_K and by the invariance of the metric on X it follows that $1 - K(y + x, y) \geq g_K(\varepsilon)$ whenever $d(x, 0) \geq \varepsilon$. Thus

$$1 - f(x) = M_y\Big(1 - K(y + x, y)\Big) \geq g_K(\varepsilon)$$

i.e., $g_f(\varepsilon) \geq g_K(\varepsilon) > 0$. By a similar argument $h_f(\varepsilon) \leq h_K(\varepsilon) \to 0$ as $\varepsilon \to 0$. □

The image of a Banach space under a uniform homeomorphism into the unit sphere of H is usually a complicated geometric object. For example, since it is bounded it is very far from being metrically convex.

An interesting property of an embedding T induced by a positive definite function is that T^{-1} satisfies a Lipschitz condition for *small* distances; i.e., there is a constant C such that $\|z - y\| \leq C\|Tz - Ty\|$ whenever $\|z - y\| \leq 1$. Indeed, put $C^{-2} = 2g_f(1/2)$, and fix z and y such that $\|z - y\| \leq 1$. Applying Corollary 8.6(iii) with $x = z - y$ and with n chosen so that $1 \geq \|nx\| \geq 1/2$, we obtain

$$\begin{aligned} \|Tz - Ty\|^2 &= 2\big(1 - f(x)\big) \geq 2n^{-2}\big(1 - f(nx)\big) \\ &\geq 2n^{-2}g_f(1/2) \geq C^{-2}\|x\|^2 = C^{-2}\|z - y\|^2. \end{aligned}$$

The following theorem gives a complete linear characterization of the linear metric spaces that embed uniformly into a Hilbert space.

Theorem 8.14. *A real linear metric space E embeds uniformly into a Hilbert space if and only if there is a probability space $(\Omega, \mathcal{B}, \mu)$ such that E is linearly isomorphic to a subspace of $L_0(\mu)$.*

Proof. We have seen that $L_0(\mu)$ embeds uniformly into a Hilbert space, and this also holds, of course, for its subspaces.

Conversely, if E embeds uniformly into a Hilbert space let f be a continuous positive definite function on E with $f(0) = 1$ and $g_f(\varepsilon) > 0$ for every $\varepsilon > 0$.

Let $U : E \to L_0(\mu)$ be the operator of Proposition 8.7. Let $\{x_n\}$ be a sequence in E such that $Ux_n \to 0$ in $L_0(\mu)$, i.e., in probability. Then $\mathbb{E}\exp(itUx_n) \to 1$ for each fixed t in $\mathbb{R}$. Taking $t = 1$ we see that $f(x_n) = \mathbb{E}\exp(iUx_n) \to 1 = f(0)$. As $g_f(\varepsilon) > 0$ for every $\varepsilon > 0$, it follows that $x_n \to 0$; hence U is one-one and its inverse is continuous. □

To apply the theorem we need to understand which linear metric spaces are isomorphic to a subspace of $L_0(\mu)$. From now on we shall consider only quasi-Banach spaces (see Appendix H) and, in particular, Banach spaces.

A measurable function f is said to be in weak L_p if there is a constant K such that $t^p\mu\{|f| > t\} \leq K^p$ for every $t > 0$. The minimal possible constant K defines a quasi-norm on the space $L_{p,\infty}(\mu)$ of all equivalence classes of such functions. If $f \in L_{p,\infty}(\mu)$ and if $r < p$, then there is a constant C (depending only on p and r) such that $\|f\|_r \leq C\|f\|_{p,\infty}$.

Theorem 8.15. *Let $0 < p \leq 2$, and assume that E is a quasi-Banach subspace of $L_0(\mu)$ with type p. Then there is an everywhere positive function ϕ such that the* *operator* $M_\phi f = \phi \cdot f$ *is an isomorphism from* E *into* $L_{p,\infty}(\mu)$ *as well as into* $L_r(\mu)$ *for every* $r < p$. *In particular, if* E *is isomorphic to a Banach space (so has type 1), it is isomorphic to a subspace of* $L_{1,\infty}$ *and to a subspace of* $L_r(\mu)$ *for every* $r < 1$.

The theorem will follow from

Proposition 8.16. *Let* $0 < p \le 2$, *and let* T *be a continuous linear operator from a quasi-Banach space* E *of type* p *into* $L_0(\mu)$. *Then for every* $\varepsilon > 0$ *there is a set* A *with* $\mu(A) \ge 1 - \varepsilon$ *such that* $P_A T$ *is a continuous operator from* E *into* $L_{p,\infty}(\mu)$, *where* $P_A f(s) = \chi_A(s) f(s)$ *for* $f \in L_0(\mu)$.

Proof. We first establish the following claim.

For every $\varepsilon > 0$ *there is a constant* K *such that given any* $\{y_i\} \subset E$, *the set*

$$B = \left\{ s : \left(\sum |Ty_i(s)|^2 \right)^{1/2} \le K \left(\sum \|y_i\|^p \right)^{1/p} \right\}$$

satisfies $\mu(B) \ge 1 - \varepsilon$.

We may assume that $\sum \|y_i\|^p = 1$. Let m be the Lebesgue measure on $[0,1]$, and let C be the type-p constant of E. Thus $(\int \| \sum r_i(t) y_i\|^p dt)^{1/p} \le C^{1/p}$ and

$$m\left\{ t : \left\| \sum r_i(t) y_i \right\| > (16C/\varepsilon)^{1/p} \right\} \le \varepsilon/16.$$

For every $\lambda > 0$ and $\delta > 0$ the continuity of T yields that there is a constant K such that if $\|y\| \le \lambda$, then $\mu\{s : |Ty(s)| > K/4\} < \delta$. Taking $\lambda = (16C/\varepsilon)^{1/p}$ and $\delta = \varepsilon/16$, we see that there is a constant K (depending on C and on ε) so that if $\| \sum r_i(t) y_i \| \le (16C/\varepsilon)^{1/p}$; i.e., if t does not belong to the set above, then

$$\mu\left\{ s : \left| T\left(\sum r_i(t) y_i \right)(s) \right| > K/4 \right\} < \varepsilon/16.$$

It follows that the measure of the set

$$Q = \left\{ (s,t) : \left| T\left(\sum r_i(t) y_i \right)(s) \right| \le K/4 \right\}$$

satisfies $\mu \otimes m(Q) \ge 1 - \varepsilon/8$.

Put $Q_s = \{t : (s,t) \in Q\}$, and let $\tilde{B} = \{s : m(Q_s) \ge 7/8\}$. By Fubini's Theorem $\mu(\tilde{B}) \ge 1 - \varepsilon$.

By Proposition D.6(ii) the inequality $m\{t : |\sum r_i(t) a_i| \le K/4\} \ge 7/8$ implies that $(\sum a_i^2)^{1/2} \le K$. Applying this estimate with $a_i = Ty_i(s)$ for $s \in \tilde{B}$ shows that $\tilde{B} \subset B$, and the claim follows.

Fix K by the claim, and define inductively sets A_i, vectors $x_i \in E$, and numbers ξ_i. Starting with $A_0 = \Omega$ we distinguish two cases.

If $\|P_{A_{i-1}} T\|_{p,\infty} > K$, choose x_i with $\|x_i\| = 1$ and numbers ξ_i such that $\mu(A_{i-1} \cap \{|Tx_i| > K\xi_i\}) > \xi_i^{-p}$ and such that $\xi > \xi_i/2$ for every other admissible choice of x and ξ. We then take $A_i = A_{i-1} \cap \{|Tx_i| \le K\xi_i\}$.

If $\|P_{A_{i-1}} T\|_{p,\infty} \le K$, put $x_i = 0$, $\xi_i = \infty$, and $A_i = A_{i-1}$.

Note that $\mu(A_i) \le \mu(A_{i-1}) - \xi_i^{-p}$, and thus $\sum \xi_i^{-p} \le 1$. In particular $\xi_i \to \infty$.

Put $A = \bigcap A_i$. We first show that $\|P_A T\|_{p,\infty} \le K$. Indeed, assume that $\|x\| \le 1$ and $\xi \ge 1$ satisfy $\mu(A \cap \{|Tx| > K\xi\}) > \xi^{-p}$. Then for every i we obtain

$$\xi^{-p} < \mu\Big(A \cap \{|Tx| > K\xi\} \Big) \le \mu\Big(A_{i-1} \cap \{|Tx| > K\xi\} \Big).$$

Hence $\xi \ge \xi_i/2$, contradicting $\xi_i \to \infty$.

It remains to estimate $\mu(A)$. If $s \notin A$, then there is an i such that $s \in A_{i-1} \setminus A_i$ and hence $\xi_i^{-1}|Tx_i(s)| > K$; thus

$$\Big(\sum \xi_i^{-2}|Tx_i(s)|^2\Big)^{1/2} > K.$$

Applying the claim for $y_i = \xi_i^{-1}x_i$, we deduce that if $s \notin A$, then s is not in the set B of the claim. Hence $B \subset A$, and consequently $\mu(A) \geq 1 - \varepsilon$. □

Proof of Theorem 8.15. We apply Proposition 8.16 to the identity operator on E. Iterative application of the proposition gives a disjoint decomposition $\Omega = \bigcup A_n$ such that $\mu(A_n) = 2^{-n}$ and such that $\|P_{A_n} f\|_{p,\infty} \leq \alpha_n \|f\|_E$, say.

The required function ϕ is defined by $\phi(t) = 2^{-n}\alpha_n^{-1}$ on A_n. The operator M_ϕ is a continuous linear operator from E to $L_{p,\infty}(\mu)$ by the choice of ϕ, and it is an isomorphism on all of $L_0(\mu)$ because ϕ is everywhere positive. Thus the $L_{p,\infty}$ and all the L_r (for $0 \leq r < p$) metrics are equivalent to each other on $M_\phi(E)$. ◇

An immediate consequence of the next corollary is that the spaces c_0 and l_p for $2 < p < \infty$ do not embed uniformly into Hilbert space.

Corollary 8.17. *Every quasi-Banach subspace of $L_0(\mu)$ has cotype 2.*

Proof. By Proposition H.2 there is a $0 < p \leq 1$ so that $\|x+y\|^p \leq \|x\|^p + \|y\|^p$ for all x, y in the given subspace E. Thus E has type p, and fixing $0 < r < p$ we can assume by Theorem 8.15 that E is a subspace of $L_r(\mu)$.

Fix $\{f_i\}_{i=1}^n \subset L_r(\mu)$. By Kahane's inequality (see Appendix G), the expressions $\big(\int \|\sum r_i(t) f_i\|_r^2 dt\big)^{1/2}$ and $\big(\int \|\sum r_i(t) f_i\|_r^r dt\big)^{1/r}$ are equivalent. We thus need to show that $\int \|\sum r_i(t) f_i\|_r^r dt \geq C(\sum \|f_i\|_r^2)^{r/2}$. By Khintchine's inequality the left-hand side is

$$\begin{aligned}\iint \Big|\sum r_i(t) f_i(s)\Big|^r d\mu(s)dt &= \iint \Big|\sum r_i(t) f_i(s)\Big|^r dt d\mu(s)\\ &\sim \int \Big(\sum |f_i(s)|^2\Big)^{r/2} d\mu(s).\end{aligned}$$

Applying the triangle inequality $\int \|F(s)\|_{2/r} d\mu(s) \geq \|\int F(s) d\mu(s)\|_{2/r}$ to the function $F(s) = \big(|f_1(s)|^r, \ldots, |f_n(s)|^r\big)$ finishes the proof. □

Remark. It is not true that every Banach space with cotype 2 embeds isomorphically into $L_0(\mu)$. For example, when $1 \leq p < 2$, the space C_p of Schatten class operators on a Hilbert space has cotype 2 (see Proposition J.1(ii)), but it is not isomorphic to a subspace of $L_0(\mu)$ (and hence it does not embed uniformly into a Hilbert space). If $p > 1$, then Theorem 8.15 implies that if C_p were isomorphic to a subspace of $L_0(\mu)$, then it would also embed into $L_r(\mu)$ for some $r > 1$, but by Proposition J.1(iv) it does not. For $p = 1$ see the Notes and Remarks section.

We finish this section with a simple averaging argument in spaces with a symmetric basis. A function f on a Banach space with a symmetric basis $\{e_n\}$ will be called symmetric if $f\big(\sum \theta_n a_n e_{\pi(n)}\big) = f\big(\sum a_n e_n\big)$ for all choices of signs $\{\theta_n\}$ and for all permutations π of the natural numbers.

Lemma 8.18. *Let E be a Banach space with a symmetric basis, and let f be a positive definite function on E. Then there is a symmetric positive definite function F on E such that for every $\varepsilon > 0$*

$$g_F(\varepsilon) \geq g_f(\varepsilon) \text{ and } h_F(\varepsilon) \leq h_f(\varepsilon).$$

Proof. For each m let $F_m\left(\sum a_n e_n\right)$ be the average of $f\left(\sum \theta_n a_n e_{\pi(n)}\right)$ taken over all choices of signs $\{\theta_n\}$ with $\theta_n = 1$ for $n > m$ and permutations π with $\pi(n) = n$ for $n > m$. Any limit point F of the F_m's is symmetric and satisfies the required inequalities. □

It follows from the lemma that if a space with a symmetric basis is uniformly homeomorphic to a subset of a Hilbert space, then the homeomorphism can be chosen to be induced by a symmetric positive definite function.

3. Notes and Remarks

Proposition 8.2 is due to Schur [**525**]. Proposition 8.5(i) was probably proved first by Moore [**427**].

The notion of a negative definite kernel, Proposition 8.4, part (ii) of Proposition 8.5, and the idea to study isometric embeddings of metric spaces by the analysis of positive and negative definite kernels on them are all due to Schoenberg [**522**]. For more information and references on positive definite functions see the books by Berg, Christensen and Ressel [**54**] and by Sasvari [**508**].

Theorem 8.9 is due to Bretagnolle, Dacunha-Castelle and Krivine [**92**]. They considered only $1 \leq p \leq 2$, but the proof applies to $0 < p < 1$ as well. They also used another variant of Proposition 8.7. The variant presented here is taken from Aharoni, Maurey and Mityagin [**5**]. The survey article by Koldobsky [**333**] contains further information on the connections between positive definite functions and the isometric theory of Banach spaces.

The systematic application of the theory of positive definite functions to the study of uniform embeddings into Hilbert space is due to Aharoni, Maurey and Mityagin [**5**]. The main results of Section 2 follow their article.

An analog of Proposition 8.12 holds for uniform embedding into $L_p(\mu)$ spaces for $p \leq 2$ (and for embeddings into uniformly bounded subsets of $L_p(\mu)$ for all p). Assume that for each finite subset A of X the map T_A is a uniform embedding of A into an L_p space, and that the embeddings and their inverses are equicontinuous. Since for $p \leq 2$ the space $L_p(\mu)$ is uniformly homeomorphic to a bounded subset of itself, and we can assume that all the T_A's map into a some ball of a fixed radius. But then X is naturally uniformly embedded into the ultraproduct of the $L_p(\mu)$'s. Such an ultraproduct is itself an L_p space; see Appendix F.

Theorem 8.15 and Proposition 8.16 are due to Nikišin [**437**]. The proof here is taken from Kalton [**311**]. See also Maurey [**395**].

In connection with Theorem 8.15, it is an open problem (raised by Kwapien) whether every Banach subspace of $L_0(\mu)$ is isomorphic to a subspace of $L_1(\mu)$. Kalton [**313**] contains more information on this problem and the strongest known result: A Banach space E is isomorphic to a subspace of $L_1(\mu)$ if and only if $l_1(E)$ is linearly isomorphic to a subspace of $L_0(\mu)$. The answer is positive for a reflexive Banach subspace E of $L_0(\mu)$. By Rosenthal [**503**] and Maurey [**396**] E has type p for some $p > 1$. Since it embeds uniformly into a Hilbert space, it is isomorphic to a subspace $L_r(\mu)$ for some $r > 1$, hence also to a subspace of $L_1(\mu)$. The isometric version of the Kwapien problem has a negative answer. Koldobsky [**332**] constructed a subspace of $L_{1/2}$ which is not isometric to a subspace of L_1.

We can now complete the remark after Corollary 8.17. Since C_1 is isomorphic to $l_1(C_1)$, and since by Proposition J.1(iv) it does not embed in $L_1(\mu)$, it cannot be isomorphic to a subspace of $L_0(\mu)$.

Enflo [**168**] was the first to prove that c_0 does not embed uniformly into a Hilbert space. His proof will be reproduced in the second volume. Another proof, due to Raynaud, will be presented in Chapter 9. Enflo's proof gives the stronger result that the subsets $U_n \subset l_\infty^n$, consisting of the vectors all of whose coordinates are 0 or 1, do not not embed uniformly in n into a Hilbert space.

We saw that a Hilbert space is uniformly homeomorphic to a subset of its unit sphere. It follows from Theorem 8.14 that every linear metric subspace of $L_0(\mu)$ which contains an isomorphic copy of a Hilbert space (such as $L_p(\mu)$ for $p \leq 2$) is uniformly homeomorphic to a bounded subset of itself. On the other hand, for $2 < p < \infty$ infinite-dimensional $L_p(\mu)$ spaces are not uniformly homeomorphic to any bounded subset of themselves. Indeed, by Mazur's theorem the unit balls of $L_p(\mu)$ and $L_2(\mu)$ are uniformly homeomorphic; thus every bounded subset of $L_p(\mu)$ embeds uniformly into a Hilbert space, while, by Corollary 8.17, the whole space does not. The same reasoning shows that C_p is not uniformly homeomorphic to a bounded subset of itself even for $p < 2$.

The identification of the Banach spaces which are uniformly homeomorphic to a bounded subset of themselves seems to be a difficult problem. It is not clear, for example, whether every subspace of $L_0(\mu)$ (which may contain no copy of a Hilbert space) has this property. On the other hand, it is proved in Aharoni [**1**] that c_0 is uniformly homeomorphic to a bounded subset of itself. By Theorem 7.11 this is also true for any separable Banach space containing a subspace isomorphic to c_0.

To say that a map $f : X \to Y$ is a uniform embedding imposes a uniformity on how f and f^{-1} change distances locally. Sometimes, e.g., when X or Y is metrically convex, this also implies a uniformity on the changes that f (or f^{-1}) makes for distant points (see Proposition 1.11), but usually this is not so: the embeddings in the chapter map unbounded sets into the unit sphere of a Hilbert space.

In some situation one is only interested in the behavior of f and f^{-1} for large distances, for example, when X is a discrete metric space the local condition is meaningless. In order to treat such questions Gromov [**229**] introduced the following notion: A map f from an unbounded metric space X into a metric space Y is called a uniform embedding at infinity if there are two functions $\rho_1, \rho_2 : (0, \infty) \to (0, \infty)$ with $\lim_{t\to\infty} \rho_1(t) = \lim_{t\to\infty} \rho_2(t) = \infty$ such that

$$\rho_1\big(d_X(u, v)\big) \leq d_Y\big(f(u), f(v)\big) \leq \rho_2\big(d_X(u, v)\big)$$

for all $u, v \in X$. The embeddings in this chapter, being maps into the unit sphere, are not uniform embeddings at infinity.

The study of uniform embeddings at infinity of discrete metric spaces into a Hilbert space is currently a very active research area, connected to several questions in topology and differential geometry. Of particular interest is the case when X is a finitely generated group with the word distance.

CHAPTER 9

Uniform Classification of Spheres

1. The Mazur Map

In this chapter we study the uniform classification of the unit spheres of Banach spaces.

The unit spheres of two Banach spaces E and F are uniformly homeomorphic if and only if there is a positive homogeneous bijective map φ from E onto F so that both $\varphi_{|B(E)}$ and $\varphi^{-1}_{|B(F)}$ are uniformly continuous (and the map can then be chosen so that $\|\varphi(x)\| = \|x\|$ for all $x \in E$). Indeed, assume that such a homogeneous φ exists. Since φ and φ^{-1} are both Lipschitz for large distances, there are $0 < c < C < \infty$ such that $\|x\| = 1$ implies that $c \leq \|\varphi(x)\| \leq C$. It follows that $\psi(x) = \varphi(x)/\|\varphi(x)\|$ is a uniform homeomorphism from $S(E)$ onto $S(F)$. Conversely, if ψ is a uniform homeomorphism of the spheres, then the map $\varphi(x) = \|x\|\psi(x/\|x\|)$ is a homogeneous homeomorphism as required.

In particular, if E is obtained from F by renorming, then $S(E)$ is even Lipschitz equivalent to $S(F)$. Also, if the unit spheres of two Banach spaces are uniformly homeomorphic, then so are their unit balls. Conversely, we shall see in Section 4 that when the unit balls of two Banach spaces are uniformly homeomorphic, then (under mild and perhaps superfluous assumptions) so are their spheres. Thus the uniform classification of spheres is essentially identical to that of balls.

The uniform classification of spaces is much finer than that of spheres. We shall see in Section 4 that under the same, perhaps superfluous, assumptions the unit spheres of two uniformly homeomorphic spaces are themselves uniformly homeomorphic. On the other hand, the results of this chapter give, for example, a large class of spaces whose unit spheres are uniformly homeomorphic to that of a Hilbert space. We shall see in the next chapter that the only Banach space uniformly homeomorphic to a Hilbert space is the Hilbert space itself. (In the Lipschitz category the situation is different. If the unit spheres of two Banach spaces are Lipschitz equivalent, then the radial extension of the homeomorphism is a Lipschitz equivalence of the spaces.)

In this connection it is worthwhile to recall that all compact convex and infinite-dimensional sets in a Banach space are homeomorphic to the Hilbert cube $[0, 1]^{\aleph_0}$, hence by compactness, uniformly homeomorphic to it.

Our first result, due to Mazur, is one of the earliest results concerning the topology of infinite-dimensional Banach spaces. It gives a simple explicit uniform homeomorphism (called the Mazur map) between the unit spheres of the different L_p spaces.

Theorem 9.1. *Let μ and ν be two measures, and fix $1 \leq p, q < \infty$. If $L_p(\mu)$ and $L_q(\nu)$ are infinite-dimensional and have the same density character, then their unit spheres are uniformly homeomorphic.*

Proof. We show that the unit spheres of $L_p(\mu)$ and $L_q(\mu)$ are uniformly homeomorphic. Since $L_2(\mu)$ and $L_2(\nu)$ are isometric when they have the same density character, this suffices to prove the theorem. We give the proof in the complex case; the details in the real case are simpler. For a complex number $z = re^{i\theta}$ we put $\operatorname{sign}(z) = e^{i\theta}$ if $z \neq 0$ and $\operatorname{sign}(0) = 0$. We shall assume that $p > q$.

For $f \in S(L_p)$ define $\phi(f) = |f|^{p/q}\operatorname{sign}(f)$. Then ϕ maps L_p to L_q and takes the unit sphere of L_p onto the unit sphere of L_q.

We first show that ϕ is Gâteaux differentiable everywhere (as a map between the complex spaces L_p and L_q considered as real spaces) and that the norm of the derivative $D_\phi(f)$ of ϕ at a point f with $\|f\|_p \le 1$ is bounded by p/q. Hence ϕ is a Lipschitz function on $S(L_p)$ with Lipschitz constant p/q.

The function $F(z) = |z|^{p/q}\operatorname{sign}(z)$ is a C^1 function from the complex plane, which we identify with $\mathbb{R}^2$, to itself. Easy computation shows that its derivative at the point z is the real-linear operator $\xi \to \frac{p}{q}|z|^{\frac{p}{q}-1}\operatorname{sign}(z)\Re\big(\xi \operatorname{sign}(\overline{z})\big)$. It follows that for every $f, h \in L_p$ the quotient $\big(\phi(f+th) - \phi(f)\big)/t$ converges pointwise to $\frac{p}{q}|f|^{\frac{p}{q}-1}\operatorname{sign}(f)\Re\big(h \operatorname{sign}(\overline{f})\big)$.

This formula defines an operator $D_\phi(f)$ from L_p to L_q. We estimate its norm by Hölder's inequality

$$\|D_\phi(f)h\|_q^q \le (p/q)^q \int |f|^{p-q}|h|^q \le (p/q)^q\Big(\int |f|^p\Big)^{(p-q)/p}\Big(\int |h|^p\Big)^{q/p}.$$

Thus $\|D_\phi(f)\| \le (p/q)\|f\|_p^{(p-q)/q} \le p/q$ whenever $\|f\|_p \le 1$.

We leave it to the reader to check that the limit in the definition of $D_\phi(f)h$ actually exists in the L_q norm, i.e., that $D_\phi(f)$ is indeed the Gâteaux derivative of ϕ at f.

We now estimate $\|\phi(f)-\phi(g)\|_q^q = \int \big||f|^{p/q}\operatorname{sign}(f) - |g|^{p/q}\operatorname{sign}(g)\big|^q$ from below. We first check that there is a constant $c > 0$ (which depends on p/q) such that $|a^{p/q} - \theta b^{p/q}| \ge c|a - \theta b|^{p/q}$ for every $a \ge b \ge 0$ and for every complex number θ with $|\theta| = 1$. Indeed, we may assume that $b = 1$, and we distinguish two cases. If $|a - \theta| \ge 1$, it is clear that such a constant exists because both terms have the same asymptotics as $a \to \infty$. It is geometrically clear that since $a \ge 1$ and $p/q \ge 1$, then $|a^{p/q} - \theta| \ge |a - \theta|$, which is larger than $|a - \theta|^{p/q}$ when $|a - \theta| \le 1$. Integrating, we obtain that $\|\phi(f) - \phi(g)\|_q^p \ge c^q \int |f - g|^p = c^q\|f - g\|_p^p$; i.e., ϕ^{-1} is Hölder with exponent q/p. □

The map above between the spheres extends to the whole space by homogeneity. The extension

$$\psi(f) = \psi_{p,q}(f) = \|f\|_p\phi(f/\|f\|_p) = \|f\|_p^{1-p/q}|f|^{p/q}\operatorname{sign}(f)$$

satisfies $\psi_{p,q}^{-1} = \psi_{q,p}$ and is a uniform homeomorphism between spheres (and balls) of any radius. The modulus of continuity of this homeomorphism deteriorates, however, when the radius of the ball tends to infinity. The next proposition records the quantitative relations between the size of the ball, the exponents p and q, and the modulus of continuity of ψ.

Proposition 9.2. *The modulus of continuity of $\psi_{p,q}$ on the ball of radius R in L_p (or l_p) is bounded by $p\delta/q$ when $p > q$, and by $cR^{1-p/q}\delta^{p/q}$ if $p < q$ (where c depends only on p/q).*

In particular, for every M the family $\{\psi_{p,q}(f) : 1 \leq p, q \leq M\}$ is a family of equi-uniform homeomorphisms between the R-balls in L_p and the R-balls in L_q provided $R \leq \exp(1/|p-q|)$.

Proof. Since the ψ's are homogeneous, it suffices to check them on the sphere of radius R. But this follows from inspection of the estimates in the proof of Theorem 9.1:

For $p > q$ the map ϕ is Lipschitz with constant p/q, and this is preserved by the extension.

When $p < q$ the modulus of continuity of ϕ is bounded by $c\delta^{p/q}$; hence the modulus of $\psi = R\phi(\cdot/R)$ is bounded by $cR^{1-p/q}\delta^{p/q}$. □

Let E be a Banach lattice (see Appendix E for information on lattices and on the p-convexification $E^{(p)}$ of the lattice E). The map $\phi_p : S(E^{(p)}) \to S(E)$ defined by $\phi_p(f) = |f|^p \text{sign}(f)$ is the analogue of the Mazur map from L_p to L_1. The proof of the next proposition is a variation of the proof of Theorem 9.1.

Proposition 9.3. *Let E be a Banach lattice, and let $1 < p < \infty$. Then the map ϕ_p is a uniform homeomorphism between $S(E^{(p)})$ and $S(E)$.*

Proof. To simplify the notation we consider the real case only. As in the proof of Theorem 9.1, the derivative of ϕ_p at f is the operator of multiplication by $p|f|^{p-1}\text{sign}(f)$. To estimate its norm, choose $\|f\|_{E^{(p)}} = \|h\|_{E^{(p)}} = 1$ and a norm-one functional $g \in \left(E^{(p)}\right)^*$ so that $\left\||f|^{(p-1)/p}|h|^{1/p}\right\|_{E^{(p)}} = \int |f|^{(p-1)/p}|h|^{1/p}g$. Then

$$\begin{aligned}
\left\||f|^{p-1}\text{sign}(f) \cdot h\right\|_E^{1/p} &= \left\||f|^{p-1}|h|\right\|_E^{1/p} = \left\||f|^{(p-1)/p}|h|^{1/p}\right\|_{E^{(p)}} \\
&= \int |f|^{(p-1)/p}|h|^{1/p}g = \int \left(|f|g\right)^{(p-1)/p}\left(|h|g\right)^{1/p} \\
&\leq \left(\int |f|g\right)^{(p-1)/p}\left(\int |h|g\right)^{1/p} \\
&\leq \|f\|_{E^{(p)}}^{(p-1)/p}\|h\|_{E^{(p)}}^{1/p}\|g\|_{(E^{(p)})^*} = 1.
\end{aligned}$$

Hence ϕ_p is Lipschitz with constant p.

Conversely, if $\|f\|_{E^{(p)}} = \|g\|_{E^{(p)}} = 1$, choose a norm-one functional $h \in E^*$ so that $\int |f-g|^p h = \left\||f-g|^p\right\|_E$. Then, since there is a constant $c_p > 0$ such that $|a^p - \theta b^p| \geq c_p|a - \theta b|^p$ for every $a \geq b \geq 0$ and $\theta = \pm 1$, we obtain that

$$\begin{aligned}
\|\phi_p(f) - \phi_p(g)\|_E &\geq \int \left||f|^p\text{sign}(f) - |g|^p\text{sign}(g)\right|h \\
&\geq c_p \int |f-g|^p h = c_p\left\||f-g|^p\right\|_E = c_p\|f-g\|_{E^{(p)}}^p
\end{aligned}$$

i.e., ϕ_p^{-1} is Hölder with exponent $1/p$. □

2. Unit Spheres of Banach Lattices

In this section we give necessary and sufficient conditions on a Banach lattice so that its unit sphere is uniformly homeomorphic to that of a Hilbert space. Example 9.23 will show that our criterion (namely, that E does not contain l_∞^n's uniformly) is not sufficient for general Banach spaces and that some structural assumptions on the space (such as the lattice structure we assume in this section) are also needed.

Theorem 9.4. *Let E be a Banach space with an unconditional basis. Then the unit sphere of E is uniformly homeomorphic to that of a Hilbert space if and only if E does not contain l_∞^n's uniformly.*

We first need some notation and preparations. If E is smooth we denote the unique supporting functional at $x \in S(E)$ by $j(x)$, i.e., $\|j(x)\| = \langle j(x), x\rangle = 1$. If E is uniformly smooth, then j is uniformly continuous; see Proposition A.5.

Remark. If E is both uniformly convex and uniformly smooth, then j is a uniform homeomorphism between the unit spheres of E and of E^*. By Proposition A.8 superreflexive spaces admit an equivalent norm with these properties; hence $S(E)$ and $S(E^*)$ are uniformly homeomorphic for such spaces.

Let E be a Banach space with a 1-unconditional basis. We shall use the coordinates to consider its elements (and those of E^*) as functions on the integers, and use notations such as $|x|$, x/y or x^*x to denote the absolute value, the coordinatewise division of x by y or the coordinatewise product of x^* and x, respectively, as such functions. If A is a subset of the integers, we denote by Ax the vector whose coordinates off A are zero and on A they agree with those of x.

If $x \in E$ and $z^* \in E^*$, then $\|xz^*\|_1 = \langle |x|, |z^*|\rangle \le \|x\|\,\|z^*\|$, where $\|\cdot\|_1$ denotes the norm in l_1. Thus, by the definition of $|j(x)|$, the coordinatewise product $F_E(x) = x|j(x)|$ is an element in the unit sphere $S(l_1)$, and its coordinates have the same signs as those of x. Note also that $j(|x|) = |j(x)|$.

We shall consider the map F_E defined above only when E is also uniformly convex and uniformly smooth. The next lemma lists the properties of F_E when E is, in addition, finite-dimensional.

Lemma 9.5. *Let E be an n-dimensional Banach space. Assume that E is uniformly convex and uniformly smooth and that it has a 1-unconditional basis. Let $F_E : S(E) \to S(l_1^n)$ be the map described above. Then*

(i) *F_E is uniformly continuous with modulus of continuity that depends on the modulus of smoothness of E.*

(ii) *F_E is one-one and the inverse map F_E^{-1} is uniformly continuous with modulus of continuity that depends on the modulus of convexity of E.*

(iii) *F_E is onto, hence a uniform homeomorphism between $S(E)$ and $S(l_1^n)$.*

Proof. (i) By the definition of F_E

$$\begin{aligned} \big\|F_E(x) - F_E(y)\big\|_1 &= \big\|x|j(x)| - y|j(y)|\big\|_1 \\ &\le \big\|x \cdot \big(|j(x)| - |j(y)|\big)\big\|_1 + \big\|(x-y)\cdot |j(y)|\big\|_1 \\ &\le \|x\|\big\||j(x)| - |j(y)|\big\| + \|x-y\|\big\||j(y)|\big\| \\ &\le \big\||j(x)| - |j(y)|\big\| + \big\|x-y\big\|. \end{aligned}$$

Since $\big\||j(x)| - |j(y)|\big\| \le \big\|j(x) - j(y)\big\|$, the result follows from the uniform continuity of j in uniformly smooth spaces, namely from Proposition A.5.

(ii) We shall use the convention that $0 \cdot t/0 = 0$.

Fix $x, y \in S(E)$, and let $f = x|j(x)|$ and $h = y|j(y)|$ be their images in $S(l_1^n)$. Put $\|f - h\|_1 = \varepsilon$ and estimate $\|x - y\|$ in terms of ε.

Let $\Lambda = \{i : \text{sign} f = \text{sign} h\}$, and assume, as we may, that f and h are nonnegative on Λ. It follows that x and y are also nonnegative on Λ and that the same is true for $j(x)$ and $j(y)$.

Define $g \in l_1^n$ to be 0 off Λ and $g = f \wedge h$, the minimum of f and h, on Λ. Then $\|g\|_1 = 1 - \varepsilon/2$, and its support is contained in the intersection of Λ with the supports of both x and y.

It follows from $\|gy/x\|_1 \le \|fy/x\|_1 = \langle |y|, |j(x)| \rangle \le 1$ and from the analogous inequality $\|gx/y\|_1 \le 1$ that $\|g \cdot (y/x + x/y)\|_1 \le 2$. Fix $0 < \delta < 1$ and put

$$A = \{j \in \Lambda : \min(y/x, x/y) < 1 - \delta\}, \quad B = \{j \in \Lambda : y/x + x/y > 2 + \delta^2\}.$$

The inequality $t + t^{-1} > 2 + \delta^2$ for $0 < t < 1 - \delta$ implies that $A \subseteq B$. Let $C = \Lambda \setminus B$. Using $t + t^{-1} > 2$ for $0 < t < 1$, we obtain

$$\begin{aligned} 2 &\ge \|g \cdot (y/x + x/y)\|_1 = \|Bg \cdot (y/x + x/y)\|_1 + \|Cg \cdot (y/x + x/y)\|_1 \\ &\ge (2 + \delta^2)\|Bg\|_1 + 2\|Cg\|_1 = 2\|g\|_1 + \delta^2\|Bg\|_1 = 2 - \varepsilon + \delta^2\|Bg\|_1 \end{aligned}$$

i.e., $\|Bg\|_1 \le \varepsilon/\delta^2$. Denoting the complement of A by D, it then follows that $\|Dg\|_1 = \|g\|_1 - \|Ag\|_1 \ge \|g\|_1 - \|Bg\|_1 \ge 1 - \varepsilon/2 - \varepsilon/\delta^2$. Since $y/x \ge 1 - \delta$ on $D \cap \Lambda$ we obtain that

$$\|Dgy/x\|_1 \ge (1 - \delta)\|Dg\|_1 \ge (1 - \delta)(1 - \varepsilon/2 - \varepsilon/\delta^2).$$

Taking $\delta = \varepsilon^{1/3}$ we obtain that

$$\begin{aligned} \langle \Lambda y, j(x) \rangle &= \|\Lambda fy/x\|_1 \ge \|gy/x\|_1 \ge \|Dgy/x\|_1 \\ &\ge (1 - \delta)(1 - \varepsilon/2 - \varepsilon/\delta^2) \ge 1 - 2\varepsilon^{1/3}. \end{aligned}$$

Since also $\langle |y|, |j(x)| \rangle \le 1$, it follows that $\langle y, j(x) \rangle \ge 1 - 4\varepsilon^{1/3}$, and hence that $\|x + y\| \ge \langle x + y, j(x) \rangle \ge 2 - 4\varepsilon^{1/3}$. By the definition of the modulus of convexity $\|x - y\| \le \delta_E^{-1}(2\varepsilon^{1/3})$.

(iii) By (ii) F_E is a homeomorphism of $S(E)$ into $S(l_1^n)$. But both sets are homeomorphic to the $(n-1)$-sphere, and any homeomorphism of a sphere into itself is necessarily onto. Indeed, by Brouwer's theorem the sphere is not contractible, but any proper closed subset of S^{n-1} can be contracted to a point. □

Corollary 9.6. *Let E be a uniformly convex and uniformly smooth Banach space with a 1-unconditional basis. Then its unit sphere $S(E)$ is uniformly homeomorphic to the unit sphere $S(l_1)$ of l_1.*

Proof. Denote the unconditional basis for E by $\{x_j\}$ and the unit vector basis in l_1 by $\{e_j\}$.

For each finite subset A of the integers, Lemma 9.5 gives a uniform homeomorphism between the unit spheres of $\operatorname{span}\{x_j : j \in A\}$ and of $\operatorname{span}\{e_j : j \in A\}$. These homeomorphisms, as well as their inverses, are equicontinuous, and they are consistent: If $x \in \operatorname{span}\{x_j : j \in A \cap B\}$, then the homeomorphisms associated with the sets A and B agree on x.

We thus obtain a uniform homeomorphism between the set of all finitely supported vectors in $S(E)$ and the set of finitely supported vectors in $S(l_1)$. This homeomorphism extends by continuity to a uniform homeomorphism between the whole spheres. □

Proof of Theorem 9.4: Enflo [**168**] proved that if a Banach space contains l_∞^n's uniformly, then its unit ball cannot be uniformly embedded into a Hilbert space (see also Corollary 9.21 below).

Conversely, let $E^{(2)}$ be the 2-convexification of E. Then $E^{(2)}$ is 2-convex, and if E does not contain l_∞^n's uniformly, it is also q-concave for some $q < \infty$. By

Proposition E.10 it can be renormed to be uniformly convex and uniformly smooth. By Corollary 9.6 the unit sphere of $E^{(2)}$ is uniformly homeomorphic to $S(l_1)$, hence by Mazur's theorem to the unit sphere of l_2. But $S(E)$ and $S(E^{(2)})$ are uniformly homeomorphic by Proposition 9.3. $\diamondsuit$

The next theorem is a more general version of Theorem 9.4.

Theorem 9.7. *Let E be an infinite-dimensional Banach lattice with a weak unit. Then the unit sphere of E is uniformly homeomorphic to the unit sphere of a Hilbert space if and only if E does not contain l_∞^n's uniformly.*

Proof. Since E does not contain l_∞^n's uniformly, it is order continuous. By Proposition E.9 we can assume that it is a lattice of functions on some measure space (Ω, Σ, μ) and that $\|f\|_1 \leq \|f\|_E \leq 2\|f\|_\infty$ for every $f \in E$. Using the fact that E does not contain l_∞^n's uniformly again, and by applying 2-convexification as in the proof of Theorem 9.4, we reduce the general case to the case where E is uniformly convex and uniformly smooth.

For each finite measurable decomposition $\mathcal{F} = \{A_1, \ldots, A_n\}$ of Ω let $F_{\mathcal{F}}$ be the map of Lemma 9.5 from the unit sphere of the span of $\{\chi_A : A \in \mathcal{F}\}$ in E to the unit sphere of their span in $L_1(\mu)$.

As in the proof of Corollary 9.6, the family $\{F_{\mathcal{F}}\}$ is a consistent net of equicontinuous homeomorphisms. Thus their limit is a uniform homeomorphism between the simple functions in the unit sphere of E and the simple functions in the unit sphere of $L_1(\mu)$. Since these are dense sets, the homeomorphism extends to the whole spheres. $\square$

Remark. Recall that every separable lattice has a weak unit. Also, the assumption that E has a weak unit was used only to represent the lattice as a lattice of functions. If E is already given as a lattice of functions, the assumption is no longer necessary and the assertion holds under the sole assumption that E does not contain l_∞^n's uniformly.

The next theorem applies the very useful "decomposition method" to extend the scope of the previous theorems. In what follows, it is convenient to introduce the notation $E \sim F$ to mean that $S(E)$ is uniformly homeomorphic to $S(F)$.

Theorem 9.8. *Assume that $E \supset F \supset G$ are superreflexive Banach spaces so that $E \sim G \sim H$ (where H is a Hilbert space). Then also $F \sim H$.*

The proof requires two lemmas.

Lemma 9.9. (i) *If $E_1 \sim F_1$ and $E_2 \sim F_2$, then $E_1 \oplus E_2 \sim F_1 \oplus F_2$.*

(ii) *If $E \sim F$, then $l_2(E) \sim l_2(F)$.*

Proof. We prove only part (ii). Let $\varphi : E \to F$ be the homogeneous extension of the uniform homeomorphism between the respective unit spheres. We define $\hat{\varphi} : l_2(E) \to l_2(F)$ by $\hat{\varphi}(x_1, x_2, \ldots) = \big(\varphi(x_1), \varphi(x_2), \ldots\big)$. Then $\hat{\varphi}$ is a well-defined positive homogeneous bijective map from $l_2(E)$ onto $l_2(F)$, and we check that the restriction of $\hat{\varphi}$ to the unit ball of $l_2(E)$ is uniformly continuous.

Let $\varepsilon > 0$ and choose $\delta < \varepsilon$ such that if $x, y \in E$ satisfy $\|x\|, \|y\| \leq 1$ and if $\|x - y\| \leq \sqrt{\delta}$, then $\|\varphi(x) - \varphi(y)\| \leq \varepsilon$. By the homogeneity of φ we obtain that if $\|x\|, \|y\| \leq \eta \leq 1$ and if $\|x - y\| \leq \eta\sqrt{\delta}$, then $\|\varphi(x) - \varphi(y)\| \leq \varepsilon\eta$.

Assume now that $\sum_{n=1}^{\infty}\|x_n\|^2 \le 1$, $\sum_{n=1}^{\infty}\|y_n\|^2 \le 1$ and $\sum_{n=1}^{\infty}\|x_n - y_n\|^2 \le \delta^2$, and let $I = \{n : \|x_n - y_n\| \le \sqrt{\delta}\max(\|x_n\|, \|y_n\|)\}$. Then

$$\sum_{n\in I}\|\varphi(x_n) - \varphi(y_n)\|^2 \le \varepsilon^2 \sum_{n\in I}\max\left(\|x_n\|, \|y_n\|\right)^2 \le 2\varepsilon^2$$

and

$$\begin{aligned}\sum_{n\notin I}\|\varphi(x_n) - \varphi(y_n)\|^2 &\le \sum_{n\notin I}\left(\|x_n\| + \|y_n\|\right)^2 \le 4\sum_{n\notin I}\max\left(\|x_n\|, \|y_n\|\right)^2 \\ &\le 4\delta^{-1}\sum_{n\notin I}\|x_n - y_n\|^2 \le 4\delta \le 4\varepsilon.\end{aligned}$$

Hence

$$\sum_{n=1}^{\infty}\|\varphi(x_n) - \varphi(y_n)\|^2 \le 2\varepsilon^2 + 4\varepsilon$$

and $\hat{\varphi}$ is uniformly continuous on the unit ball of $l_2(E)$. By the same reasoning $\hat{\varphi}^{-1}$ is uniformly continuous on the unit ball of $l_2(F)$. □

Lemma 9.10. *Assume that E is a superreflexive Banach space and that F is a closed subspace of E. Then $E \sim E/F \oplus F$.*

Proof. Let $q : E \to E/F$ be the quotient map. Since E can be renormed to be uniformly convex, and since uniformly convex spaces have uniformly normal structure (see Proposition A.9), it follows from Corollary 1.25 that there is a homogeneous map $\tau : E/F \to E$ which is uniformly continuous on bounded sets and such that $q\tau$ is the identity on E/F. The map $\varphi : E \to F \oplus E/F$ defined by

$$\varphi(x) = (x - \tau q x, qx)$$

is bijective, homogeneous, and onto. Hence it induces a uniform homeomorphism between the unit spheres. □

Proof of Theorem 9.8. Combining Lemmas 9.9 and 9.10, we obtain

$$\begin{aligned} H &= l_2(H) \sim l_2(E) \sim l_2(E/F) \oplus l_2(F) \\ &\sim l_2(E/F) \oplus l_2(F) \oplus F \sim l_2(E) \oplus F \sim H \oplus F.\end{aligned}$$

Hence

$$\begin{aligned} F &\sim F/G \oplus G \sim F/G \oplus H \sim F/G \oplus H \oplus H \\ &\sim F \oplus H \sim H.\end{aligned}$$

◊

Corollary 9.11. *Let F be an infinite-dimensional subspace of a superreflexive Banach lattice. Then the unit sphere of F is uniformly homeomorphic to the unit sphere of a Hilbert space.*

Moreover, if E is a superreflexive Banach space with an unconditional basis, then the unit sphere of any infinite-dimensional subspace of any quotient of E is uniformly homeomorphic to the unit sphere of a Hilbert space.

Proof. Let E be the Banach lattice. It follows from Proposition E.8 that every subspace of E contains a subspace G with an unconditional basis. Thus both $S(E)$ and $S(G)$ are uniformly homeomorphic to $S(H)$, and the first part follows from Theorem 9.8.

Assume now that E has an unconditional basis, and let F be a quotient of E. By duality F^* is a subspace of E^*; hence $S(F^*)$ is uniformly homeomorphic to $S(H)$ by the first part. By the remark after the formulation of Theorem 9.4 the spheres $S(F^*)$ and $S(F)$ are uniformly homeomorphic, and thus $S(F)$ is uniformly homeomorphic to $S(H)$.

Now let G be an infinite-dimensional subspace of F. By Theorem E.3(iii) it has a further subspace with an unconditional basis, and the result now follows from Theorem 9.8. □

3. Applications of the Complex Interpolation Method

In this section we apply the complex interpolation method to construct uniform homeomorphisms between the unit spheres of certain uniformly convex Banach spaces. We refer to Appendix I for unexplained notation, terminology and results concerning interpolation spaces. In the case of Banach lattices, the homeomorphism we construct here is the same as the one constructed in the previous section. All the spaces in this section will be complex, but, as we point out after the proof of Theorem 9.12, the results apply also to real Banach lattices.

Theorem 9.12. *Let (E_0, E_1) be an interpolation pair so that E_0 and E_1 are uniformly convex. Then for every $0 < \sigma, \tau < 1$ the unit spheres of E_σ and E_τ are uniformly homeomorphic.*

Proof. Fix $x \in E_\sigma$ with $\|x\|_\sigma = 1$. By Proposition I.3 there is a unique function $f_x \in \mathcal{F}_2(\sigma)$ satisfying $\|f_x\|_{\mathcal{F}_2(\sigma)} = 1$ and $f_x(\sigma) = x$. We shall show that the map $\phi : S(E_\sigma) \to S(E_\tau)$ defined by $\phi(x) = f_x(\tau)$ is the required homeomorphism.

It follows from Proposition I.3(ii) that $\phi(S(E_\sigma)) \subset S(E_\tau)$. Put $w = \phi(x)$ and denote by f_w^τ the unique function in $\mathcal{F}_2(\tau)$ such that $\|f_w^\tau\|_{\mathcal{F}_2(\tau)} = 1$ and $f_w^\tau(\tau) = w$. By Proposition I.3(i) and (ii) f_x also satisfies these two conditions. The uniqueness of f_w^τ implies that $f_x = f_w^\tau$, and hence, that if we denote by ψ the corresponding map from $S(E_\tau)$ to $S(E_\sigma)$, then $x = \psi(w)$. In other words $\psi = \phi^{-1}$. We show next that ϕ is uniformly continuous. A similar argument shows that ψ is uniformly continuous, and hence ϕ is a uniform homeomorphism.

The map ϕ is the composition of three maps: $x \to f_x$ from E_σ to $\mathcal{F}_2(\sigma)$, the identity map from $\mathcal{F}_2(\sigma)$ to $\mathcal{F}_2(\tau)$, and the evaluation map $f \to f(\tau)$ from $\mathcal{F}_2(\tau)$ to E_τ. The last two maps are bounded linear maps. (That the identity map from $\mathcal{F}_2(\sigma)$ to $\mathcal{F}_2(\tau)$ is an isomorphism follows from the boundedness of the Radon-Nikodým derivatives $d\mu_\sigma/d\mu_\tau$ and $d\mu_\tau/d\mu_\sigma$.) It remains to show that the map $x \to f_x$ is a uniformly continuous map from E_σ to $\mathcal{F}_2(\sigma)$, and this follows from the uniform convexity of $\mathcal{F}_2(\sigma)$: Let $\varepsilon > 0$ and fix x, y in $S(E_\sigma)$ with $\|x-y\|_\sigma = \varepsilon$. Let f_x and f_y be the corresponding optimal analytic functions in $\mathcal{F}_2(\sigma)$. By the triangle inequality $\|x+y\|_\sigma \geq 2 - \varepsilon$, and hence $\|f_x + f_y\|_{\mathcal{F}_2(\sigma)} \geq \|(f_x + f_y)(\sigma)\|_\sigma = \|x+y\|_\sigma \geq 2 - \varepsilon$. Since also $\|f_x\|_{\mathcal{F}_2(\sigma)} = \|f_y\|_{\mathcal{F}_2(\sigma)} = 1$, it follows that $\|f_x - f_y\|_{\mathcal{F}_2(\sigma)} \leq \delta^{-1}(\varepsilon)$, where $\delta(\cdot)$ is the modulus of uniform convexity of $\mathcal{F}_2(\sigma)$. □

Remarks. (i) The method of this section applies to complex spaces. It turns out, however, that for Banach lattices the map actually sends real elements to real elements: If $x \in E_\sigma$ is real, then $g(z) = \overline{f_x(\overline{z})}$ is an analytic function that satisfies $\|g\|_{\mathcal{F}_2(\sigma)} = \|f_x\|_{\mathcal{F}_2(\sigma)}$ and $g(\sigma) = x$. Since f_x is the unique function with these properties it follows that $g = f_x$ and thus $f_x(\tau) = g(\tau) = \overline{f_x(\tau)}$; i.e., $f_x(\tau)$ is real. Thus Theorem 9.12 also holds for real Banach lattices.

(ii) In the case of Banach lattices the map $x \to f_x(\tau)$ from $S(E_\sigma)$ to $S(E_\tau)$ is exactly the homeomorphism of the previous section. It is enough to check this in the finite-dimensional Lemma 9.5, so let (E_0, E_1) be an interpolation pair of uniformly convex and uniformly smooth n-dimensional spaces with a 1-unconditional basis, and let $0 < \theta < 1$. For each $x \in S(E_\theta)$ let f_x be as above. Similarly, for every $x^* \in S(E_\theta^*)$ we denote by f_{x^*} the unique function in $\mathcal{F}_2(\theta)^\#$ so that $f_{x^*}(\theta) = x^*$ and $\|f_{x^*}\|_{\mathcal{F}_2(\theta)^\#} = \|x^*\|_\theta$, where $\mathcal{F}_2(\theta)^\#$ is the space analogous to $\mathcal{F}_2(\theta)$, with E_0 and E_1 replaced by E_0^* and E_1^*.

The value of the map $F_{E_\theta}: S(E_\theta) \longrightarrow S(l_1^n)$ of Lemma 9.5 at x is given by the coordinatewise product $x|j(x)| = f_x(\theta) f_{|j(x)|}(\theta)$; i.e., it is the value of the l_1^n-valued analytic function $f_x(z) f_{|j(x)|}(z) = \sum_{j=1}^n a_j(z) e_j$ at the point $z = \theta$. The claim will follow once we show that this function is actually constant. By the unconditionality of the basis in E_θ we can assume that the vector x is nonnegative, and so $|j(x)| = j(x)$.

It follows from $\|f_x\|_{\mathcal{F}_2(\theta)} = \|f_{j(x)}\|_{\mathcal{F}_2(\theta)^\#} = 1$ that

$$\sum |a_j(z)| = \Big\| \sum a_j(z) e_j \Big\|_{l_1^n} = \big|\langle f_x(z), f_{j(x)}(z) \rangle\big| \le 1. \tag{9.1}$$

On the other hand, $\sum a_j(\theta) = \langle f_x(\theta), f_{j(x)}(\theta)\rangle = \langle x, j(x)\rangle = 1$. By the maximum principle

$$\sum a_j(z) \equiv 1. \tag{9.2}$$

Comparing with (9.1) we see that the functions $a_j(z)$ must all be nonnegative; hence, being analytic they are constant.

We now give some applications of the theorem.

(i) An alternative proof of Theorem 9.7: As in the first proof we may assume that E is both uniformly convex and uniformly smooth, hence also r-convex and s-concave for some $1 < r < s < \infty$. By Corollary I.5, there is a Banach lattice F so that both E and L_2 can be represented as interpolation spaces between F and F^*. By the reiteration theorem (Proposition I.1(iii)) we can replace F by $[F, F^*]_\theta$ for small θ, and since L_2 is an interpolation space of the original spaces we may assume (by Proposition I.1(iv)) that both F and F^* are uniformly convex. Thus Theorem 9.12 applies, and $S(E)$ and $S(L_2)$ are uniformly homeomorphic.

(ii) The unit spheres of the various $L_r(\mu)$ spaces (for $1 < r < \infty$) are uniformly homeomorphic to each other because by Proposition I.1(v) if $1/p = (1-\theta)/p_0 + \theta/p_1$, then $L_p(\mu) = [L_{p_0}(\mu), L_{p_1}(\mu)]_\theta$. Moreover, it is easy to verify by the formulas relating p, p_0, p_1 and θ that ϕ is exactly the Mazur map of Theorem 9.1.

(iii) Proposition I.1(vi) implies similarly that the unit spheres of the Schatten class spaces C_p are uniformly homeomorphic to each other for $1 < p < \infty$.

Unfortunately, the proofs in (ii) and (iii) do not extend directly to $p = 1$.

The next proposition gives some information on the modulus of continuity of the homeomorphism ϕ of Theorem 9.12 and of its extension $\phi_R(x) = R\phi(x/R)$ to spheres of radius R. This information will be used in Chapter 10.

Proposition 9.13. *Let E_0 and E_1 be as in Theorem 9.12. Then there is a function γ satisfying $\lim_{\varepsilon\to 0}\gamma(\varepsilon)=0$, which depends only on the moduli of convexity of E_0 and of E_1, with the following property: For every $R>0$ there is an $\eta>0$ so that if $1/4\le\sigma,\tau\le 3/4$ satisfy $|\sigma-\tau|<\eta$, then the modulus of continuity of ϕ_R as a map from the sphere of radius R in E_σ into E_τ is bounded by γ.*

Proof. The numbers 1/4 and 3/4 are, of course, arbitrary, and we need only σ and τ to be bounded away from $\partial\mathcal{S}$.

Fix σ and τ and denote the moduli of continuity of ϕ on $S(E_\sigma)$ and of ϕ_R on $R\cdot S(E_\sigma)$ by ω and ω_R respectively. We shall combine two estimates on ω_R.

Since both E_0 and E_1 are uniformly convex, the spaces $\mathcal{F}_2(\theta)$ are uniformly convex for all $0<\theta<1$, and there is a function $\delta(\varepsilon)>0$ which uniformly bounds their moduli of convexity from below. The proof of Theorem 9.12 shows that there is a constant $K(\sigma,\tau)$, depending on σ and τ, such that $\omega(\varepsilon)\le K(\sigma,\tau)\delta^{-1}(\varepsilon)$. Since σ and τ are bounded away from 0 and 1, the constants $K(\sigma,\tau)$ are uniformly bounded by some constant K_1, and we obtain the first estimate

$$\omega_R(\varepsilon)\le K_1R\delta^{-1}(\varepsilon/R). \tag{9.3}$$

For the second estimate choose $x,y\in S(E_\sigma)$ with $\|x-y\|_\sigma=\varepsilon$. Fix $v^*\in E_\tau^*$ with $\|v^*\|_\tau=1$ so that $\langle v^*,\phi(x)-\phi(y)\rangle=\|\phi(x)-\phi(y)\|_\tau$. Since $E_\tau^*=[E_0^*,E_1^*]_\tau$ there is a function g in $\mathcal{F}_2(\tau)^\#$ satisfying $g(\tau)=v^*$ and $\|g\|_{\mathcal{F}_2(\tau)^\#}=1$. (Recall that $\mathcal{F}_2(\tau)^\#$ is the analog of $\mathcal{F}_2(\tau)$ with the E's replaced by their duals.)

The scalar analytic function $\psi(z)=\langle g(z),f_x(z)-f_y(z)\rangle$ is bounded by 2. By Cauchy's formula for the derivative, and since σ and τ are bounded away from $\partial\mathcal{S}$, there is a constant K_2, independent of σ and τ, such that $|\psi(\sigma)-\psi(\tau)|\le K_2|\sigma-\tau|$. Since $|\psi(\sigma)|=|\langle g(\sigma),x-y\rangle|\le\|x-y\|_\sigma=\varepsilon$ we obtain

$$\|\phi(x)-\phi(y)\|_\tau=\psi(\tau)\le|\psi(\sigma)|+|\psi(\sigma)-\psi(\tau)|\le\varepsilon+K_2|\sigma-\tau|.$$

Thus $\omega(\varepsilon)\le\varepsilon+K_2|\sigma-\tau|$, and we obtain the second estimate

$$\omega_R(\varepsilon)\le\varepsilon+K_2R|\sigma-\tau|. \tag{9.4}$$

To simplify the final computation we assume (by renorming E_0 and E_1 if necessary; see Theorem A.6(iii')) that there is an $0<\alpha\le 1/2$ such that $\delta^{-1}(\varepsilon)\le C\varepsilon^\alpha$. The function $\gamma(\varepsilon)=\max\{K_1C\varepsilon^{\alpha^2},(1+K_2)\varepsilon\}$ satisfies the requirements.

Indeed, take $\eta=R^{-1-1/\alpha}$ and distinguish two cases. If $\varepsilon\le R^{-1/\alpha}$, then substituting $\varepsilon^\alpha=\varepsilon^{\alpha(1-\alpha)}\varepsilon^{\alpha^2}\le R^{\alpha-1}\varepsilon^{\alpha^2}$ in (9.3) gives that $\omega_R(\varepsilon)\le K_1C\varepsilon^{\alpha^2}$. If $\varepsilon\ge R^{-1/\alpha}$, then $R|\sigma-\tau|\le R\eta\le R^{-1/\alpha}\le\varepsilon$ and $\omega_R(\varepsilon)\le(1+K_2)\varepsilon$ by (9.4). □

Remark. The same function γ bounds, of course, the modulus of continuity of ϕ_R on the whole ball of radius R and not only on the sphere.

4. Spheres and Balls

A well-known result in infinite-dimensional topology is that the unit ball $B(E)$ and the unit sphere $S(E)$ of any infinite-dimensional Banach space E are homeomorphic. The analogous problem in the uniform category is open:

Problem 9.14. *Let E be an infinite-dimensional Banach space. Is the unit ball $B(E)$ of E uniformly homeomorphic to its unit sphere $S(E)$?*

By Corollary 3.5, $S(E)$ is Lipschitz contractible; hence there is no obvious "uniform topological" obstruction to the existence of such a uniform homeomorphism.

It would follow from a positive answer to this problem that the uniform classifications of spheres and of balls coincide. Theorem 9.15 below shows that these two classifications "almost" coincide regardless of the solution to the problem.

In this section we denote the direct sum $(G \oplus \mathbb{R})_\infty$ for a Banach space G by $\tilde{G}$.

Theorem 9.15. *Let E and F be two Banach spaces. If their unit spheres are uniformly (respectively Lipschitz) homeomorphic, then so are their unit balls. Conversely, if $B(E)$ and $B(F)$ are uniformly (respectively Lipschitz) homeomorphic, then so are $S(\tilde{E})$ and $S(\tilde{F})$.*

The main tool for the proof will be Proposition 9.16 below. We start with a few remarks. For most Banach spaces $S(E)$ and $S(\tilde{E})$ are, in fact, Lipschitz equivalent, e.g., when E and $E \oplus \mathbb{R}$ are isomorphic. It follows from the theorem that for such spaces the uniform classifications of balls and of spheres coincide.

There are examples of spaces E such that $S(E)$ and $S(\tilde{E})$ are not Lipschitz equivalent. Let F be a separable infinite-dimensional reflexive space which is not isomorphic to any of its proper subspaces (see Gowers and Maurey [**226**]), and take E to be a codimension-one subspace of F. If $S(E)$ and $S(F)$ were Lipschitz equivalent, so would be their unit balls. Thus assume that $f : B(F) \to B(E)$ is a Lipschitz homeomorphism. By Theorem 6.42 f is Gâteaux differentiable at some interior point. Its derivative at this point is a linear isomorphism of F into its proper subspace E, which is impossible.

This example also shows that there are counterexamples to Problem 9.14 in the Lipschitz category. It is unknown, however, whether $S(E)$ and $S(\tilde{E})$ are uniformly homeomorphic for all Banach spaces. It is actually unknown whether there is any Banach space for which the answer to the problem is positive.

A metric space X is said to be uniformly (or Lipschitz) homogeneous if for every $x \neq y$ in X there is a uniform (respectively Lipschitz) homeomorphism of X onto itself which maps x to y. Since the unit sphere $S(E)$ of any Banach space is Lipschitz homogeneous, it follows that if $B(E)$ is uniformly homeomorphic to $S(E)$ (or to $S(F)$ for any Banach space F), then it must also be uniformly homogeneous.

Proposition 9.16. *Let E be a Banach space so that $B(E)$ is uniformly homogeneous. Then $B(E)$ is uniformly homeomorphic to $S(\tilde{E})$. If $B(E)$ is Lipschitz homogeneous, then $B(E)$ is Lipschitz equivalent to $S(\tilde{E})$.*

Proof. Before embarking on the proof, we first introduce some notation and terminology and describe some useful constructions of Lipschitz homeomorphisms.

Let C be a subset of E, and let Φ be a homeomorphism of C onto itself. We shall say that Φ is supported in a subset A of C if $\Phi(z) = z$ for every $z \notin A$. Some of the homeomorphisms that we define below will also be naturally defined on all of E but supported in $B(E)$ (or some other set). They can thus be considered as homeomorphisms of any set containing the support.

Fix any $x, y \in E$, and let $\varepsilon > 0$. Then there is a Lipschitz homeomorphism $\Phi : E \to E$ which is supported in the 4ε-neighborhood of the interval $[x, y]$ and so that Φ acts as the translation $\Phi(z) = z + (y - x)$ of $B(x, \varepsilon)$ onto $B(y, \varepsilon)$. We shall call such a Φ a Lipschitz translation. The existence of Lipschitz translations is intuitively clear, and we also give an explicit construction. Assume, without loss of generality, that $\|y - x\| = 1$. Fix $z^* \in E^*$ so that $\|z^*\| = \langle z^*, y - x \rangle = 1$, and put $\langle z^*, x\rangle = a$ and $\langle z^*, y\rangle = b$ (hence $b - a = 1$). Let f be a Lipschitz homeomorphism of $\mathbb{R}$ which is supported in the interval $[a-2\varepsilon, b+2\varepsilon]$ and which acts as the translation $f(t) = 1+t$ when $|t-a| \le \varepsilon$ (so, in particular, $f(a) = b$). Denote the distance of z from the line passing through x and y by $d(z)$, and let $0 \le g \le 1$ be a Lipschitz function on E such that $g(z) = 1$ when $d(z) \le \varepsilon$ and $g(z) = 0$ when $d(z) \ge 2\varepsilon$. The map Φ given by

$$\Phi(z) = z + g(z)\Big(f\big(\langle z^*, z\rangle\big) - \langle z^*, z\rangle\Big)(y - x)$$

has the desired properties. Indeed, Φ is a Lipschitz map which acts as a Lipschitz homeomorphism on each line parallel to $y - x$. (It acts as $t \to t + \lambda\big(f(t) - t\big)$ for some fixed $0 \le \lambda \le 1$ on each such line.) Hence Φ is one-one, and using its special form, one checks easily that it is a Lipschitz homeomorphism. If $d(z) \ge 2\varepsilon$, then $\Phi(z) = z$. A direct computation shows that if $d(z) \le 2\varepsilon$ for a point z whose distance from $[x, y]$ is larger than 4ε, then $\langle z^*, z\rangle$ does not belong to $[a - 2\varepsilon, b + 2\varepsilon]$. Hence $f\big(\langle z^*, z\rangle\big) = \langle z^*, z\rangle$ and $\Phi(z) = z$ again. The existence of Lipschitz translations shows in particular that every open ball is Lipschitz homogeneous.

Let $B_1 \subset B_2 \subset B_3$ be three concentric balls in E. Then there is a Lipschitz homeomorphism Φ of E onto itself which is supported in B_3 and which acts on B_1 as the dilation $x \to x_0 + \frac{r_2}{r_1}(x - x_0)$, where r_1 and r_2 are the radii of B_1 and B_2 respectively and x_0 is the common center of the balls. We shall call such a Φ a Lipschitz dilation.

By using a Lipschitz dilation followed by a Lipschitz translation we see that for any two balls B_1 and B_2 in E and for any $\varepsilon > 0$ there is a Lipschitz homeomorphism of E onto itself which maps B_1 onto B_2 and which is supported in an ε-neighborhood of the convex hull of B_1 and B_2.

Let F be a codimension-one subspace of E. By renorming we may assume that $E = (F \oplus \mathbb{R})_\infty = \tilde{F}$. Consider the truncated cone

$$C = \big\{(u, t) \in E : \ u \in F \text{ and } 0 \le \|u\| \le t \le 1\big\}.$$

Since C is a bounded closed and convex subset of E with a nonempty interior, it is Lipschitz equivalent to $B(E)$ under a mapping that takes its boundary ∂C onto $S(E)$. We shall also consider the subset

$$\tilde{C} = \big\{(x, t) : \ x \in E \text{ and } 0 \le \|x\| \le t \le 1\big\}$$

of $\tilde{E}$. Its boundary in $\tilde{E}$ will be used as a model for $S(\tilde{E})$.

Fix $0 \le \alpha \le \beta \le 1$. We denote the set $\{(u, t) \in C : t \ge \alpha\}$ by C_α^+, the set $\{(u, t) \in C : t \le \alpha\}$ by C_α^-, and the set $\{(u, t) \in C : \alpha \le t \le \beta\}$ by $C_{\alpha,\beta}$. Analogous notation will be used for $\tilde{C}$. Note that $\tilde{C}_1^+$ is a translate of $B(E)$. The point $(0, 1)$ in C will be denoted by e. We shall need two types of natural Lipschitz homeomorphisms of C onto itself, which we now describe without giving explicit formulas.

Fix $0 < \alpha < \beta < 1$. The first type is supported in C_β^- and acts as the dilation $(u, t) \to (u/2, t/2)$ on C_α^-. For the second type fix also $0 < \delta < 1 - \beta$. Then there is a Lipschitz homeomorphism Φ of C onto itself, supported in $C_\alpha^+ \setminus B(e, \delta/2)$, such that $C_\beta^+ \subset \Phi\big(C \cap B(e, \delta)\big) \subset C_\alpha^+$. Note that the image of $C \cap \partial B(e, \delta)$ is contained in $C_{\alpha,\beta}$.

We are now ready for the proof of Proposition 9.16. We shall construct a uniform homeomorphism φ from $2B(E)$ into C so that

(i) $C^+_{1/3} \subset \varphi\big(B(E)\big) \subset C^+_{1/4}$.

(ii) The sets $\varphi\big(2B(E)\big)$ and $\frac{1}{2}\varphi\big(2B(E) \setminus B(E)\big)$ are disjoint.

(iii) $\varphi(2y) = \varphi(y)/2$ for every $y \in S(E)$.

(iv) $C \setminus \varphi\big(B(E)\big) = 2\left(C \setminus \varphi\big(2B(E)\big)\right)$.

Once such a φ is constructed, we define a uniform homeomorphism f from $\partial\tilde{C}$ onto C as follows:

For $n \geq 0$, put $S_n = \{(x,t) \in \tilde{C} : 2^{-n-1} \leq \|x\| = t \leq 2^{-n}\}$. Then $S_{n+1} = \frac{1}{2}S_n$, and the boundary of $\tilde{C}$ is the union of $\tilde{C}^+_1$, $\{0\}$, and the S_n's. Identify the ball $2B(E)$ with $\tilde{C}^+_1 \cup S_0$ in such a way that $B(E)$ is identified with the "top" $\tilde{C}^+_1$ of $\tilde{C}$, and so that if $x \in S(E)$ and $1 < \lambda \leq 2$ (hence $\lambda x \in 2B(E) \setminus B(E)$), then λx is identified with $(3-\lambda)/2 \cdot (x,1)$. (This identification is certainly a Lipschitz homeomorphism.) We shall thus consider φ as being defined on $\tilde{C}^+_1 \cup S_0$.

With this identification the homeomorphism $f : \partial\tilde{C} \to C$ is defined as follows:

$$f(x,t) = \begin{cases} \varphi(x,t) & \text{if } (x,t) \in \tilde{C}^+_1 \cup S_0 \\ \varphi(2^n x, 2^n t)/2^n & \text{if } (x,t) \in S_n \text{ and } n \geq 1 \\ 0 & \text{at } (0,0). \end{cases}$$

Condition (iii) ensures that f is well-defined, and it is clearly a uniformly continuous function from $\partial\tilde{C}$ into C. By (iv) it is actually onto. The restriction of f to each S_n separately is a uniform homeomorphism, and the moduli of continuity of f and of f^{-1} on the different S_n's are bounded by those of φ. Thus, to finish the proof, we need only to find lower estimates on $\|f(x,t) - f(y,s)\|$ when $(x,t) \in S_n$ and $(y,s) \in S_m$ with $n > m$.

The case $n \geq m+2$ is easy. In this case $\|(x,t) - (y,s)\| \leq 2^{-m+1}$, and since $\varphi\big(2B(E) \setminus B(E)\big)$ is contained in $C_{1/8,1/3}$ (by (i) and (iv)) it follows that $\|f(x,t) - f(y,s)\| \geq 2^{-m}/24$. We thus need to check only the case $n = m+1$, and rescaling by 2^m we may assume that $m = 0$ and that $n = 1$.

By (ii) there is a z with $\|z\| = 1/2$ so that $f(z,1/2)$ lies on the segment connecting $f(x,t)$ and $f(y,s)$. Let ω be the modulus of continuity of $f^{-1} = \varphi^{-1}$ on S_0. Then ω also bounds the modulus of continuity of f^{-1} on S_1, and we obtain that

$$\begin{aligned} \|(x,t) - (y,s)\| &\leq \|(x,t) - (z,1/2)\| + \|(z,1/2) - (y,s)\| \\ &\leq \omega\big(\|f(x,t) - f(z,1/2)\|\big) + \omega\big(\|f(z,1/2) - f(y,s)\|\big) \\ &\leq 2\omega\big(\|f(x,t) - f(y,s)\|\big) \end{aligned}$$

which is the required lower estimate.

We now turn to the construction of φ.

Let $B = B(z,r)$ be a ball so that $B(z,2r) \subset C_{2/3,3/4}$. We shall find a concentric ball $B_1 = B(z,r_1)$ contained in B and construct a uniform homeomorphism φ_1 of C onto itself so that the following conditions hold:

(v) $C^+_{1/3} \subset \varphi_1(B_1) \subset C^+_{1/4}$.

(vi) $\varphi_1(B)$ is disjoint from $\frac{1}{2}\varphi_1\big(B \setminus B_1\big)$.

(vii) $\varphi_1(z + ry) = \varphi_1(z + r_1 y)/2$ for every $\|y\| = 1$.

(viii) $C \setminus \varphi_1(B_1) = 2\big(C \setminus \varphi_1(B)\big)$.

Identify $2B(E)$ with B by a suitable dilation in such a way that $B(E)$ is identified with B_1 and $2B(E) \setminus B(E)$ is thus identified with $B \setminus B_1$. The restriction of φ_1 to B is a map φ satisfying (i)–(iv) as required.

The map φ_1 will be constructed in several steps.

Recall that the point $(0,1) \in C$ is denoted by e, and observe that for every $\varepsilon > 0$ there is a uniform homeomorphism ψ of C onto itself which is supported in $C \cap B(e, \varepsilon)$ and such that $\psi(e)$ is interior to C. Indeed, by the homogeneity of C there is a uniformly continuous homeomorphism ψ_1 of C taking e to $e/2$. Fix $\delta > 0$ so that $\psi_1(C \cap B(e, \varepsilon)) \supset B(e/2, 2\delta)$, and choose x, y with $\|y - e/2\| < \delta$ and so that $x = \psi_1^{-1}(y) \in B(e, \varepsilon)$ is in the interior of C. Let Ψ be a Lipschitz translation of C which maps $e/2$ to y and which is supported in $B(e/2, 2\delta)$. Then $\psi = \psi_1^{-1}\Psi\psi_1$ takes e to x and is supported in $B(e, \varepsilon)$.

Composing the map ψ above with appropriate Lipschitz dilations and translations, we obtain that there is a uniform homeomorphism φ_2 of C onto itself, supported in $C^+_{1/2}$, which maps the point z to e.

Choose $\delta > 0$ so that the interior of $\varphi_2(B)$ contains $C \cap B(e, 2\delta)$, and then choose the ball $B_1 = B(z, r_1)$ so that $\varphi_2(B(z, r_1)) \subset B(e, \delta)$. Let Φ be a Lipschitz homeomorphism of C onto itself which is supported in $C^+_{1/4} \setminus B(e, \delta)$ so that $\Phi(B(e, 2\delta)) \supset C^+_{1/3}$. Since the support of φ_2 is contained in $C^+_{1/2}$, it follows that $\varphi_3 = \varphi_2^{-1}\Phi\varphi_2$ satisfies:

(ix) φ_3 is supported in $C^+_{1/4}$.

(x) $C^+_{1/3} \subset \varphi_3(B) \subset C^+_{1/4}$.

(xi) φ_3 is the identity on B_1.

Finally, let Φ_1 be a Lipschitz homeomorphism of C which is supported in $(C_{2/3,3/4}) \cup (C^-_{1/2})$ and which acts in $C_{2/3,3/4}$ as the Lipschitz dilation taking B_1 to B, and as the multiplication $x \to x/2$ on $C^-_{1/3}$. We shall see that the map $\varphi_1 = \varphi_3\Phi_1\varphi_3$ satisfies (v)–(viii).

Note that φ_1 acts on B_1 just as φ_3 acts on B (up to a dilation), and that if $x \in \partial B$ (or more generally, if $\varphi_3(x) \in C^-_{1/3}$), then $\varphi_1(x) = \frac{1}{2}\varphi_3(x)$.

Condition (v) is immediate, and drawing a picture may help the reader verify (vi)–(viii). We check only (vi). Assume that $(x,t) \in \varphi_1(B) \cap \frac{1}{2}\varphi_1(B \setminus B_1)$, and write $(x,t) = \frac{1}{2}\varphi_1(z)$ for some $z \in B \setminus B_1$. Since $\varphi_1(B_1) = \varphi_3(B) \supset C^+_{1/3}$, it follows that $\varphi_1(B \setminus B_1) \subset C^-_{1/3}$. Hence $(x,t) \in \frac{1}{2}\varphi_1(B \setminus B_1) \subset C^-_{1/6}$. On the other hand, write $(x,t) = \varphi_1(y)$ with $y \in B$. Since φ_3 is the identity on $C^-_{1/4}$ and since Φ_1 acts on $C^-_{1/3}$ as multiplication by $1/2$, it follows from $(x,t) \in C^-_{1/6}$ that

$$(x,t) = \varphi_1(y) = \varphi_3(\Phi(\varphi_3(y))) = \frac{1}{2}\varphi_3(y).$$

Using $\varphi_1(B_1) = \varphi_3(B)$ again, we see that there is a point $w \in B_1$ such that $(x,t) = \frac{1}{2}\varphi_1(w)$. Thus $\varphi_1(w) = \varphi_1(z)$, and this is impossible since φ_1 is one-one.

The same proof establishes the proposition in the Lipschitz category. □

Proof of Theorem 9.15. The first part is obvious, so assume that f is a uniform homeomorphism of $B(E)$ onto $B(F)$ and distinguish two cases. If $S(E)$ is mapped by f onto $S(F)$, then they are uniformly homeomorphic, and hence so are $S(\tilde{E})$ and $S(\tilde{F})$. If this is not the case, then either f maps a point in $S(E)$

into the interior of $B(F)$, or f^{-1} maps a point in $S(F)$ into the interior of $B(E)$. Assume, for example, that $x \in E$ with $\|x\| = 1$ and $f(x) = y$ with $\|y\| < 1$. Since f maps neighborhoods of x onto neighborhoods of y, it follows that there are $x_1 \in E$ with $\|x_1\| < 1$ and $y_1 \in F$ with $\|y_1\| < 1$ so that $f(x_1) = y_1$. Let Φ be a Lipschitz translation of F supported in $B(F)$ so that $\Phi(y) = y_1$. Then $f^{-1}\Phi f$ is a uniform homeomorphism of $B(E)$ taking x to the interior point x_1. It follows that $B(E)$ is uniformly homogeneous. By Proposition 9.16 the sphere $S(\tilde{E})$ is uniformly homeomorphic to $B(E)$. Since $B(F)$ is uniformly homeomorphic to $B(E)$, it is also homogeneous, hence uniformly homeomorphic to $S(\tilde{F})$. It follows that the balls in E and F and the spheres in $\tilde{E}$ and $\tilde{F}$ are all mutually uniformly homeomorphic.

The same proof shows that if f is assumed to be a Lipschitz homeomorphism, then $S(\tilde{E})$ is Lipschitz equivalent to $S(\tilde{F})$. ◊

The next result shows that the uniform classification of Banach spaces is (essentially) strictly finer than that of their unit spheres. (The proof also shows that if E and F are Lipschitz equivalent, then so are $S(\tilde{E})$ and $S(\tilde{F})$.)

Proposition 9.17. *If the two Banach spaces E and F are uniformly homeomorphic, then $S(\tilde{E})$ is uniformly homeomorphic to $S(\tilde{F})$.*

Proof. Let $\varphi : E \to F$ be the uniform homeomorphism, and assume that $\varphi(0) = 0$. Since uniformly continuous functions are Lipschitz for large distances, there is a constant K such that $\|x - y\|/K \le \|\varphi(x) - \varphi(y)\| \le K\|x - y\|$ whenever $\|x - y\| \ge 1$ or $\|\varphi(x) - \varphi(y)\| \ge 1$.

Using the same notation as in the proof of Proposition 9.16, we identify $S(\tilde{E})$ with the boundary of the truncated cone $\tilde{C}$ in $\tilde{E}$. The map $f_E : \partial\tilde{C} \setminus \{(0,0)\} \to E$ defined by

$$f_E(x,t) = \begin{cases} x & \text{if } t = 1 \\ x/\|x\|^2 & \text{if } t = \|x\| < 1 \end{cases}$$

is one-one and surjective, and its inverse is

$$f_E^{-1}(x) = \begin{cases} (x,1) & \text{if } \|x\| \le 1 \\ \|x\|^{-2}(x, \|x\|) & \text{if } \|x\| > 1. \end{cases}$$

Similarly, we consider the truncated cone $\tilde{D}$ in $\tilde{F}$ and the analogous map $f_F : \partial\tilde{D} \setminus \{(0,0)\} \to F$.

The map $\psi = f_F^{-1} \circ \varphi \circ f_E$ (and $\psi(0,0) = (0,0)$) is continuous at $(0,0)$, and for each $0 < \lambda \le 1$ it is uniformly continuous on $\{(x,t) : t \ge \lambda\}$. We shall check that it is uniformly continuous on all of $\partial\tilde{C}$.

Fix $\|x\| < 1$. Then $\|f_E(x, \|x\|)\| = 1/\|x\| > 1$. By the choice of K and since $\varphi(0) = 0$, it follows that

$$K^{-1}\|x\|^{-1} \le \|\varphi\big(f_E(x, \|x\|)\big)\| \le K\|x\|^{-1}. \tag{9.5}$$

If moreover $\|x\| < 1/K$, then it follows that $\|\varphi\big(f_E(x, \|x\|)\big)\| \ge 1$ and hence that $\psi(x, \|x\|) = (z, \|z\|)$ where $K^{-1}\|x\| \le \|z\| \le K\|x\|$.

As mentioned above, we have good a estimate for $\|\psi(x,t) - \psi(y,s)\|$ when $\min(s,t) \ge K^{-4}$. It is also easy to estimate it when $s \ge 1/K$ and $t \le K^{-4}$ and to check that if $\|x\| \le \|y\|/K^4$, then $\|\psi(x,t) - \psi(y,s)\| \le K\|(x,t) - (y,s)\|$.

So assume that $\|(x,t)-(y,s)\| = \delta$ with $\|y\|/K^4 \le \|x\| \le \|y\| \le 1/K$, and put $u = \varphi\big(f_E(x,t)\big)$ and $v = \varphi\big(f_E(y,s)\big)$. Then $K\|x\|^{-1} \ge \|u\| \ge K^{-1}\|x\|^{-1} \ge 1$ and $K\|y\|^{-1} \ge \|v\| \ge K^{-1}\|y\|^{-1} \ge 1$ by (9.5). In particular, $\|u\| \le K^6\|v\|$.

Let ω be the modulus of continuity of φ. Since

$$\begin{aligned} \big\|x/\|x\|^2 - y/\|y\|^2\big\| &\le \|x-y\|/\|x\|^2 + \big|\|x\|^2 - \|y\|^2\big|/\|x\|^2\|y\| \\ &\le 3\|x-y\|/\|x\|^2 = 3\delta/\|x\|^2 \end{aligned}$$

it follows that $\|u-v\| \le \omega(3\delta/\|x\|^2)$. Being monotone and subadditive, ω satisfies $\omega(ab) \le ([a]+1)\omega(b)$ for every $a,b>0$. Thus $\|u-v\| \le 6\omega(\delta)/\|x\|^2$. Hence

$$\begin{aligned} \|\psi(x,t)-\psi(y,s)\| &= \|f_F^{-1}(u) - f_F^{-1}(v)\| = \big\|u/\|u\|^2 - v/\|v\|^2\big\| \\ &\le \|u-v\|/\|u\|^2 + \big|\|u\|^2 - \|v\|^2\big|/\|u\|^2\|v\| \\ &\le (2+K^6)\|u-v\|/\|u\|^2 \le (2+K^6)6\omega(\delta)/\|x\|^2\|u\|^2 \\ &\le 6K^2(2+K^6)\omega(\delta) \end{aligned}$$

because $\|x\|^{-1}\|u\|^{-1} \le K$. The uniform continuity of ψ^{-1} is proved in a similar way. □

5. Stable Metrics

Example 9.23 at the end of this section will show that some structural conditions on the space E, besides not containing l_∞^n's uniformly, are necessary for its unit sphere to be uniformly homeomorphic to $S(L_2)$. Thus the problem of uniform classification of the unit spheres of general Banach spaces seems to be very difficult.

A Banach space E is called stable if for any two bounded sequences $\{x_n\}$ and $\{y_n\}$ in E and for any two free ultrafilters $\mathcal{U}$ and $\mathcal{V}$ on $\mathbb{N}$

$$\lim_{n,\mathcal{U}} \lim_{m,\mathcal{V}} \|x_n + y_m\| = \lim_{m,\mathcal{V}} \lim_{n,\mathcal{U}} \|x_n + y_m\|.$$

Examples. (i) Hilbert space is stable. Indeed, if the weak limits of $\{x_n\}$ and $\{y_n\}$ along the ultrafilters are x and y respectively, then the iterated limits are both equal to $\big(\lim_{n,\mathcal{U}}\|x_n\|^2 + \lim_{m,\mathcal{V}}\|y_m\|^2 + 2\langle x,y\rangle\big)^{1/2}$.

(ii) $L_p(\mu)$ spaces are stable for $0<p<\infty$. A proof that holds for all $p>0$ is given in Krivine and Maurey [**339**]. For $0<p\le 2$ a shorter proof is possible. In this range the kernel $\exp(-\|x-y\|_p^p)$ is positive definite (see example (iii) after Proposition 8.4 or Theorem D.8(i)). By Theorem 8.5 it has the form $\langle Tx, Ty\rangle$ where T is a map into some Hilbert space. The argument above for Hilbert space now gives the result.

(iii) c_0 is not stable. Take $x_n = e_n$ and $y_m = \sum_{i\le m} e_i$.

When $\phi : X \to E$ is a uniform embedding of a metric space X into a stable Banach space E, the metric $d(x,y) = \|\phi(x) - \phi(y)\|$ on X is uniformly equivalent to the original one, and it is stable in the following sense.

Definition 9.18. *A metric space (X,d) is called stable if for any two bounded sequences $\{x_n\}$ and $\{y_n\}$ in X and for any two free ultrafilters $\mathcal{U}$ and $\mathcal{V}$ on $\mathbb{N}$*

$$\lim_{n,\mathcal{U}} \lim_{m,\mathcal{V}} d(x_n, y_m) = \lim_{m,\mathcal{V}} \lim_{n,\mathcal{U}} d(x_n, y_m).$$

Lemma 9.19. *Let (X, d) be a stable metric space, and fix integers $n > l \geq 1$. Let $\{x_{i_1 \ldots i_l}\}$ and $\{y_{i_{l+1} \ldots i_n}\}$ be two bounded multi-indexed sequences in X. Fix n free ultrafilters $\mathcal{U}_1, \ldots, \mathcal{U}_n$ on $\mathbb{N}$, and let π be a permutation of $\{1, \ldots, n\}$ which preserves the order on each of the sets $A = \{1, \ldots, l\}$ and $B = \{l+1, \ldots, n\}$. Then*

$$\lim_{i_1, \mathcal{U}_1} \ldots \lim_{i_n, \mathcal{U}_n} d(x_{i_1 \ldots i_l}, y_{i_{l+1} \ldots i_n}) = \lim_{i_1, \mathcal{U}_1} \ldots \lim_{i_n, \mathcal{U}_n} d(x_{i_{\pi(1)} \ldots i_{\pi(l)}}, y_{i_{\pi(l+1)} \ldots i_{\pi(n)}}).$$

Proof. For $n = 2$ this is just the definition of stability. We proceed by induction on n and distinguish two cases:

(i) There are two consecutive integers, say k and $k+1$, in the same set A or B so that $\pi(k)$ and $\pi(k+1)$ are also consecutive integers.

In this case, identifying $\mathbb{N}^2$ with $\mathbb{N}$, we replace (i_k, i_{k+1}) by a single index and the filters $\mathcal{U}_k$ and $\mathcal{U}_{k+1}$ by the product ultrafilter. This reduces the number of indices from n to $n-1$ and induction applies.

(ii) If case (i) does not hold, then $n-2$ and n belong to the image of one set and $n-1$ to the image of the other. Since the order in the sets A and B is preserved by π, we shall assume that $\pi(n) = n$, $\pi(l) = n-1$ and $\pi(n-1) = n-2$. (The other possibility is treated in a similar way.) For each fixed $i_1, \ldots, i_{n-2}$, stability gives that

$$\begin{aligned} &\lim_{i_{n-1}, \mathcal{U}_{n-1}} \lim_{i_n, \mathcal{U}_n} d(x_{i_{\pi(1)} \ldots i_{\pi(l)}}, y_{i_{\pi(l+1)} \ldots i_{\pi(n)}}) \\ &\quad = \lim_{i_{n-1}, \mathcal{U}_{n-1}} \lim_{i_n, \mathcal{U}_n} d(x_{i_{\pi(1)} \ldots i_{n-1}}, y_{i_{\pi(l+1)} \ldots i_n}) \\ &\quad = \lim_{i_n, \mathcal{U}_n} \lim_{i_{n-1}, \mathcal{U}_{n-1}} d(x_{i_{\pi(1)} \ldots i_{n-1}}, y_{i_{\pi(l+1)} \ldots i_n}) \\ &\quad = \lim_{i_{n-1}, \mathcal{U}_{n-1}} \lim_{i_n, \mathcal{U}_n} d(x_{i_{\sigma(1)} \ldots i_{\sigma(l)}}, y_{i_{\sigma(l+1)} \ldots i_{\sigma(n)}}) \end{aligned}$$

where the permutation σ agrees with π except on l and n and is defined on these points by $\sigma(l) = n$ and $\sigma(n) = n-1$. It follows that we can replace π by σ in the statement of the lemma. But case (i) applies to σ because $\sigma(n-1) = n-2$ and $\sigma(n) = n-1$ are consecutive. □

For information on spreading bases see Appendix E.

Theorem 9.20. *Let E be a Banach space with a spreading basis $\{e_n\}$ whose unit ball $B(E)$ admits a stable metric d, uniformly equivalent to the norm. Then $\{e_n\}$ is unconditional.*

Proof. Renorming, we may assume that $\{e_n\}$ is 1-spreading and bi-monotone; i.e., that $\| \sum_k^l a_i e_{m_i} \| \leq \| \sum_1^n a_i e_i \|$ for all scalars $\{a_i\}$, integers $m_1 < m_2 < \ldots$, and $1 \leq k \leq l \leq n$.

Using the uniform equivalence of d and the norm, choose $\alpha > 0$ and $r > 0$ so that

$$d(x, y) < \alpha \text{ implies that } \|x - y\| < 1/2 \tag{9.6}$$

and

$$\|x - y\| < r \text{ implies that } d(x, y) < \alpha/2. \tag{9.7}$$

Fix n, scalars $a_1, \ldots, a_n$, and signs $\theta_1, \ldots, \theta_n = \pm 1$. We shall show that if $1 = \| \sum a_i e_i \| \geq \| \sum \theta_i a_i e_i \| = B$, then $B \geq r$. This will show that the unconditional constant of $\{e_n\}$ is at most $1/r$.

Let l denote the number of $+$ signs among the θ_i's. Since $\{e_n\}$ is spreading and bi-monotone, we obtain that

$$1 = \left\|\sum a_i e_i\right\| = \left\|\sum a_i e_{m_i}\right\| \le 2\left\|\sum_{i\le l} a_i e_{m_i} - \sum_{i>l} a_i e_{m_i}\right\| \tag{9.8}$$

for every $m_1 < \ldots < m_n$.

Using (9.6) with $x = \sum_{i\le l} a_i e_{m_i}$ and $y = \sum_{i>l} a_i e_{m_i}$, inequality (9.8) gives

$$\lim_{m_1,\mathcal{U}_1} \cdots \lim_{m_n,\mathcal{U}_n} d\Big(\sum_{i\le l} a_i e_{m_i}, \sum_{i>l} a_i e_{m_i}\Big) \ge \alpha. \tag{9.9}$$

On the other hand, since

$$\Big\|\sum_{\theta_i=+1} a_i e_{m_i} - \sum_{\theta_i=-1} a_i e_{m_i}\Big\| = B$$

for every $m_1 < \ldots < m_n$, it follows that

$$\lim_{m_1,\mathcal{U}_1} \cdots \lim_{m_n,\mathcal{U}_n} d\Big(\sum_{\theta_i=+1} a_i e_{m_i}, \sum_{\theta_i=-1} a_i e_{m_i}\Big) \le \sup\{d(x,y) : \|x-y\| \le B\}. \tag{9.10}$$

Let π be the permutation that maps $\{1,\ldots,l\}$ onto $\{i : \theta_i = +1\}$ and $\{l+1,\ldots,n\}$ onto $\{i : \theta_i = -1\}$, preserving the order in each of these sets. Applying Lemma 9.19 with this π, we see that the left-hand sides of (9.9) and (9.10) are equal. It follows that $\alpha \le \sup\{d(x,y) : \|x-y\| \le B\}$, and thus $B \ge r$ by (9.7). $\square$

As an application of Theorem 9.20 we obtain

Corollary 9.21. *The unit ball of c_0 does not embed uniformly into a Hilbert space. Consequently, if a Banach space contains l_∞^n's uniformly, its unit ball does not embed uniformly into a Hilbert space.*

Proof. Since Hilbert space is stable, such an embedding would have induced a uniformly equivalent stable metric on the unit ball of c_0. But this would contradict Theorem 9.20 because the summing basis $s_n = \sum_{i\le n} e_i$ of c_0 is spreading, yet conditional.

By Proposition 8.12, if the unit balls of l_∞^n were equi-uniformly homeomorphic to subsets of a Hilbert space, then the unit ball of c_0 would embed uniformly into a Hilbert space. Thus the second part of the corollary follows from the first. $\square$

Corollary 9.22. *Let E be a nonreflexive Banach space, and assume that $B(E)$ admits a stable metric, uniformly equivalent to the norm. Then E contains l_1^n's uniformly.*

Proof. Since E is nonreflexive, Proposition E.7 implies that it contains a normalized sequence $\{e_i\}$ so that

$$\Big\|\sum_{j=1}^{l} a_j e_{i_j} - \sum_{j=l+1}^{n} a_j e_{i_j}\Big\| \ge \frac{1}{2}\sum_{i=1}^{n} a_j$$

for every $1 \le l \le n$ and for all nonnegative coefficients $\{a_j\}$.

Let $\mathcal{U}$ be a free ultrafilter on $\mathbb{N}$. Then for every $1 \le l \le n$

$$\lim_{i_1,\mathcal{U}} \ldots \lim_{i_n,\mathcal{U}} \Big\|\sum_{j=1}^{l} a_j e_{i_j} - \sum_{j=l+1}^{n} a_j e_{i_j}\Big\| \ge \frac{1}{2}\sum_{j=1}^{n} a_j.$$

Inspection of the proof of Theorem 9.20 shows that there is a constant $\lambda > 0$ such that

$$\lim_{i_1,\mathcal{U}} \dots \lim_{i_n,\mathcal{U}} \Big\| \sum_{j=1}^{n} \theta_j a_j e_{i_j} \Big\| \geq \lambda \sum_{j=1}^{n} a_j$$

for all n, for all nonnegative scalars $\{a_j\}$ and for all choices of signs $\{\theta_j\}$.

Using the compactness of the unit ball of l_1^n, we find that for each n, there are $i_1 \dots i_n$ so that

$$\Big\| \sum_{j=1}^{n} \theta_j a_j e_{i_j} \Big\| \geq \frac{\lambda}{2} \sum_{j=1}^{n} a_j$$

for all choices of nonnegative scalars $\{a_j\}$ and for all choices of signs $\{\theta_j\}$. These $\{e_{i_j}\}$'s are then $2/\lambda$ equivalent to the unit vectors of l_1^n. □

Example 9.23. *There is a Banach space F which does not contain l_∞^n's uniformly and whose unit sphere is not uniformly homeomorphic to the unit sphere of a Hilbert space.*

James' "nonoctahedral" space (see [**276**]) is a nonreflexive space F which does not contain l_1^n's uniformly. It follows from Corollary 9.22 that $B(F)$ does not embed uniformly into a stable Banach space; hence, in particular, it does not embed uniformly into a Hilbert space. However, since l_1^n is isometric to a subspace of $l_\infty^{2^n}$, the space F also does not contain l_∞^n's uniformly.

Remark. Since F above does not contain l_∞^n's uniformly, it has finite cotype (see Corollary G.4). We shall see in the second volume that if E contains l_∞^n's uniformly, then $B(E)$ does not embed uniformly into a space with a finite cotype. Thus $B(F)$ is also not uniformly homeomorphic to the unit balls of c_0 and of $\left(\sum \oplus l_{p_k}\right)_2$ whenever $p_k \to \infty$.

Note that the unit balls of these two spaces, which are not uniformly homeomorphic to the unit ball of a Hilbert space, are also not uniformly homeomorphic to each other: $B(c_0)$ is an absolute Lipschitz retract (Example 1.5), while the unit ball of $\left(\sum \oplus l_{p_k}\right)_2$ is not even an absolute neighborhood uniform retract (Example 1.30).

6. Notes and Remarks

Keller [**324**] proved that all infinite-dimensional compact convex subsets of l_2 are homeomorphic to the Hilbert cube. Klee observed that this holds in all Banach spaces, because for every separable Banach space E there is a continuous injective linear map from E into l_2. For a proof of Keller's theorem, as well as for proofs of all the results in infinite-dimensional topology that we quote in this chapter without specific reference, see Bessaga and Pełczyński [**62**], or van Mill [**576**].

Mazur's theorem appears in [**405**].

Endow L_q for $q < 1$ with the metric $d_q(f,g) = \int |f-g|^q$. Weston [**582**] proved that Mazur's theorem holds for all $0 < p < \infty$: There are constants K_1 and K_2 so that $K_1 d_q(f,g)^{1/q} \leq \|\phi(f) - \phi(g)\|_p \leq K_2 d_q(f,g)^{1/p}$ for $0 < q < 1 \leq p < \infty$, and so that $K_1 d_q(f,g)^{p/q} \leq d_p\big(\phi(f),\phi(g)\big) \leq K_2 d_q(f,g)$ for $q < p < 1$.

The estimates of the moduli of continuity in the proof of Theorem 9.1 are best possible (up to constants). For $1 \leq q \leq p \leq 2$ it was proved by Lövblom [**376**] that for any homeomorphism $\psi : S(L_p) \to S(L_q)$ the moduli of continuity of ψ and ψ^{-1}

satisfy $\omega_\psi(\omega_{\psi^{-1}}(\varepsilon)) \geq K\varepsilon^{p/q}$. The same result for all p and q (with the exception of the case $q = 1$ and $1 \leq p \leq 2$) follows also from Theorem 2.3.

Proposition 9.3 appears as Proposition 2.8 in Odell and Schlumprecht [**443**] (for spaces with unconditional basis) and as Proposition 2.4 in Chaatit [**116**] (for general Banach lattices).

The "if" part of Theorem 9.4 was proved by Odell and Schlumprecht [**443**]. The "only if" part is due to Enflo [**168**]. The original proof of Enflo will be reproduced in the second volume.

Theorem 9.7 is due to Chaatit [**116**].

It follows from Lemma 9.5 that if E is an n-dimensional Banach space with a 1-unconditional basis, then every $h \in S(l_1^n)$ can be represented as the coordinatewise product $h = x \cdot x^*$ for some $x \in S(E)$ and $x^* \in S(E^*)$. Indeed, Lemma 9.5 gives such a representation when E is uniformly convex and uniformly smooth, and the general case follows by taking a sequence of uniformly convex and uniformly smooth renormings of E that converge to the original norm.

Several authors proved, independently, the existence of such a factorization (and various generalizations), and it is usually referred to as "the Lozanovskii factorization". See Lozanovskii [**378, 379**], Jamison and Ruckle [**278**] and Gillespie [**208**]. These authors used a variational argument to prove the result, and this is also the approach in Odell and Schlumprecht [**443**] and in Chaatit [**116**]. In Part (iii) of Lemma 9.5 the variational argument is replaced by a topological one, the Brouwer theorem. This approach was first noted by Bollobás, Leader and Radcliffe [**76**].

Theorem 9.8 and its corollaries are due to N. Kalton (oral communication). We thank him for allowing us to present them here. It also holds, with the same proof, when E is assumed to have only uniformly normal structure.

The application of the complex interpolation method to the uniform classification of spheres is due to Daher [**130**], and was also discovered independently by Kalton (unpublished).

Raynaud [**484**] gives another approach to the proof of Theorem 9.7. He uses the Lozanovskii factorization above for lattices of functions. Assuming the lattice is uniformly convex and uniformly smooth, he defines $F_E : S(E) \to S(L_1(\mu))$ by $F_E(x) = x|j(x)|$ as we did in Lemma 9.5. All the parts of the proof are as in our proofs, except that he gives an abstract nonquantitative argument that F_E^{-1} is uniformly continuous: If this were not so, we could find $u_n, v_n \in L_1(\mu)$ and $\varepsilon > 0$ so that $\|u_n - v_n\| \to 0$, yet $\|x_n - y_n\| \geq \varepsilon$ where $F_E(x_n) = u_n$ and $F_E(y_n) = v_n$. Let $\mathcal{U}$ be a free ultrafilter on $\mathbb{N}$. The ultrapower of $L_1(\mu)$ is an abstract L space and can be represented as L_1 of some measure space $(\Omega', \mathcal{A}', \mu')$. The ultrapower $\tilde{E}$ of E is a uniformly convex and uniformly smooth function space on some measure space. The main result in [**484**] implies that $\tilde{E}$ is actually representable on $(\Omega', \mathcal{A}', \mu')$. Using this result, we see that the function $F_{\tilde{E}} : S(\tilde{E}) \to S(L_1(\mu'))$ is also well-defined and one-one. But if we put $\tilde{x} = (x_n)_{\mathcal{U}}$, and define similarly $\tilde{y}, \tilde{u}$ and $\tilde{v}$, then our choice shows that $\|\tilde{x} - \tilde{y}\| \geq \varepsilon$, yet $F_{\tilde{E}}(\tilde{x}) = \tilde{u} = \tilde{v} = F_{\tilde{E}}(\tilde{y})$.

The results of section 4 are due to R. Nahum [**431, 432**].

We mention two problems related to Problem 9.14. It is unknown whether a uniformly continuous involution on the unit ball of a Banach space must have an approximate fixed point. (A function f on $B(E)$ is an involution if f^2 is the identity on $B(E)$.) See Goebel [**214**] for more information on this problem. Since the map $x \to -x$ is a Lipschitz involution on $S(E)$ without an approximate fixed point, a positive solution to Problem 9.14 for some Banach space E will give a counterexample on E.

The second problem was posed by Nirenberg [**438**] in the context of existence problems for solutions of nonlinear equations. Suppose that f is a continuous function from a Banach space E into itself so that there is a constant $k > 0$ with $\|f(x)-f(y)\| \geq k\|x-y\|$ for every $x, y \in E$. Assume further that the image of f has a nonempty interior. Does it follow that f is onto? Morel and Steinlein [**428**] gave a counterexample in l_1. It is unknown whether such examples exist in other classical spaces, and in particular in Hilbert space. Chang and Shujie [**117**] observed that if f is also assumed to be Fréchet differentiable with $\limsup_{x\to x_0} \|D_f(x)-D_f(x_0)\| < k$ for every $x_0 \in E$ (and in particular when f is C^1), then the answer is yes.

We note that if E is such that $B(E)$ is Lipschitz equivalent to $S(E)$, then there is even a nonsurjective Lipschitz embedding of E into itself so that its image has a nonempty interior. Indeed, assume that $\phi : S(E) \to B(E)$ is the homeomorphism, and define a Lipschitz homeomorphism f from $B(E)$ onto $\tilde{C}$ (using the notation of section 4) by

$$f(x) = \begin{cases} \|x\|(\phi(x/\|x\|), 1) & \text{if } x \neq 0 \\ (0,0) & \text{if } x = 0. \end{cases}$$

Let g be a Lipschitz homeomorphism of $\tilde{C}$ onto $\tilde{C}^+_{1/2}$ so that $g(y) = y$ for every $y \in \tilde{C}^+_1$. Then the map $h = f^{-1}gf$ is a Lipschitz homeomorphism of $B(E)$ onto the "annulus" $\{x \in E : 1/2 \leq \|x\| \leq 1\}$, and $h(x) = x$ when $\|x\| = 1$. Extending h to be the identity off $B(E)$, we obtain a Lipschitz homeomorphism of E onto $\{x \in E : \|x\| \geq 1/2\}$.

It follows from the proof of Proposition 9.16 that the homogeneity of $B(E)$ with respect to locally uniformly continuous (or locally Lipschitz) homeomorphisms implies that it is actually homogeneous with respect to uniformly continuous (or Lipschitz) homeomorphisms. (A map φ is locally uniformly continuous if for every x there is a neighborhood U of x such that φ is uniformly continuous on U.) Indeed, in the construction of ψ from ψ_1 in the proof of Proposition 9.16 we need only the uniform continuity of ψ_1 and of its inverse on appropriate small neighborhoods of e and $e/2$.

It should be noted, however, that the study of the locally uniform (or locally Lipschitz) category is in many respects closer in spirit to the topological category than to the uniform category. For example, completeness is not preserved under such maps. Nahum [**430, 432**] proved that if E is a nonreflexive space, then the ball $B(E)$ is locally Lipschitz equivalent to $B(E) \setminus \{x\}$ for any $x \in B(E)$. If x is an interior point, then the map can be chosen to be the identity on $S(E)$ and can then be extended to yield a locally Lipschitz homeomorphism of E onto $E \setminus \{x\}$. This result clearly holds also if $B(E)$ is uniformly homeomorphic (or even only locally uniformly homeomorphic) to the unit ball $B(F)$ of a nonreflexive space F. Conceivably $B(E)$ is uniformly homeomorphic to such $B(F)$ for every Banach space

E. (For Hilbert space, or more generally for $L_p(\mu)$, the Mazur map gives such a uniform homeomorphism onto the unit ball of $L_1(\mu)$.)

In this connection we mention that Väisälä [**572**] proved that a normed space E is Lipschitz equivalent to $E \setminus \{0\}$ if and only if it is not complete.

It is interesting to study the Lipschitz (or uniform) homogeneity of other subsets of Banach spaces. Väisälä [**569**] and Hohti [**255**] studied the Lipschitz homogeneity of the Hilbert cube $Q = [0,1]^{\aleph_0}$ under metrics of the form $d_\lambda(x,y) = \sup \lambda_j |x_j - y_j|$, where $\{\lambda_j\}$ is a sequence of positive numbers which decreases to zero. Väisälä showed that if (Q, d_λ) is Lipschitz homogeneous, then $R(\lambda) = \sup \lambda_j/\lambda_{j+1} < \infty$. Hohti proved conversely that if $R(\lambda) < \infty$, then the cube is Lipschitz homogeneous.

Hohti also proved that the space of all closed subsets of the unit interval with the Hausdorff metric (which is known to be homeomorphic to the Hilbert cube) is not Lipschitz homogeneous.

The results of section 5 are due to Raynaud [**483**].

Stable Banach spaces were first introduced by Krivine and Maurey [**339**]. See the book [**239**] by Guerre-Delabriére for their properties and for their role in the theory of general Banach spaces.

The first construction of a nonreflexive "nonoctahedral" space as in Example 9.23 is due to R.C. James in [**276**]. For a simplification of this construction see James and Lindenstrauss [**277**]. For a different approach see Pisier and Xu [**468**].

The following two problems are naturally related to the results of this chapter.

(i) Assume that the unit sphere of the Banach space E is uniformly homeomorphic to a subset of a Hilbert space. Does it follow that $S(E)$ is, in fact, uniformly homeomorphic to the unit sphere of a Hilbert space?

(ii) Assume that E is superreflexive. Is $S(E)$ uniformly homeomorphic to the unit sphere of a Hilbert space?

While a positive answer to problem (ii) above will identify the uniform type of the unit spheres of all superreflexive spaces, the examples mentioned at the end of section 5 seem to indicate that one cannot expect to obtain a reasonable uniform classification of the spheres $S(E)$ for (separable) spaces E which are not superreflexive. We have not checked whether even some simple constructions such as $(\sum \oplus l_{p_k}^k)_2$, $(\sum \oplus l_{p_k})_0$ or $(\sum \oplus l_\infty^k)_2$ already yield new uniform types of spheres.

CHAPTER 10

Uniform Classification of Banach Spaces

1. Reduction to Lipschitz and Linear Maps

The uniform classification of Banach spaces uses a combination of metric, geometric, and topological arguments. The first section uses "limits at infinity" to pass from uniformly continuous to Lipschitz maps. This is done by passing from the original spaces to their ultrapowers. The results for the uniform category are then obtained by a combination of linear methods and the results for the Lipschitz category in Chapter 7.

The second section presents a geometric tool, the set of (approximate) midpoints. It turns out that the metric or metric-topological structure of these sets remains controlled under uniform homeomorphisms. It follows that spaces for which the structures of these sets are different cannot be uniformly homeomorphic. In particular we exhibit spaces, such as l_p for $1 < p < \infty$, whose linear structure is determined by their uniform structure; i.e., a Banach space which is uniformly homeomorphic to such a space must be linearly isomorphic to it.

Discrete nets approximate the geometry of the whole space, and a uniform homeomorphism between Banach spaces yields Lipschitz equivalent nets in them. In the third section we present results, analogous to those of the first two sections, where the assumption on the existence of uniform homeomorphism is replaced by the existence of Lipschitz equivalent nets. We show that in infinite-dimensional spaces all nets are Lipschitz equivalent to each other. Surprisingly, this is no longer true for nets in $\mathbb{R}^n$ for $2 \leq n < \infty$.

In the fourth section we construct explicit uniform homeomorphisms between spaces with certain linear structure. Appropriate choice of the spaces yields uniformly homeomorphic spaces which are not isomorphic to each other. In the last section we give examples of spaces whose uniform structure determines exactly two linear structures, i.e., two uniformly homeomorphic spaces E and F which are not isomorphic to each other, and such that any Banach space which is uniformly homeomorphic to them is either isomorphic to E or to F.

A uniformly continuous map $f : E \to F$ between Banach spaces satisfies a Lipschitz condition for large distances (Proposition 1.11); i.e., for each $d > 0$ there is a constant K_d such that $\|f(x)-f(y)\| \leq K_d\|x-y\|$ for all $x, y \in E$ with $\|x-y\| \geq d$. This allows "passage to limit at infinity" (or, formally, to ultrapowers) and the reduction of some questions on uniform classification to the Lipschitz category. For information on ultrapowers of Banach spaces see Appendix F.

Theorem 10.1. (i) *Let E and F be uniformly homeomorphic Banach spaces, and let $\mathcal{U}$ be a free ultrafilter on the set of natural numbers. Then $(E)_{\mathcal{U}}$ and $(F)_{\mathcal{U}}$ are Lipschitz equivalent.*

(ii) *More generally, assume that E is uniformly homeomorphic to a subset of F on which there is a uniform retraction from F, and that $\mathcal{U}$ is a free ultrafilter on $\mathbb{N}$. Then there is a Lipschitz embedding of $(E)_{\mathcal{U}}$ into $(F)_{\mathcal{U}}$ such that there is a Lipschitz retraction from $(F)_{\mathcal{U}}$ onto the range of the embedding.*

Proof. (i) Let $\varphi : E \to F$ be an onto uniform homeomorphism. Since φ and φ^{-1} satisfy a Lipschitz condition for large distances, there is a constant K such that

$$\|x - y\|/K \leq \|\varphi(x) - \varphi(y)\| \leq K\|x - y\| \tag{10.1}$$

for all $x, y \in E$ with $\|x - y\| \geq 1$.

Assume that $\varphi(0) = 0$, and put $\varphi_n(x) = \frac{1}{n}\varphi(nx)$. Then $\varphi_n^{-1}(y) = \frac{1}{n}\varphi^{-1}(ny)$, the modulus of continuity of φ_n and of φ_n^{-1} are bounded by those of φ and φ^{-1}, and φ_n already satisfies (10.1) whenever $\|x - y\| \geq 1/n$.

If $\{x_n\}$ is a bounded sequence in E, then $\{\varphi_n(x_n)\}$ is bounded in F. Indeed, let ω be the modulus of continuity of φ; then $\|\varphi_n(x_n)\| \leq \frac{1}{n}\omega(n\|x_n\|) \leq \omega(\|x_n\|)$ because ω is subadditive. Let $\mathcal{U}$ be a free ultrafilter on $\mathbb{N}$. Then the natural map $\Phi = (\varphi_n) : (E)_{\mathcal{U}} \to (F)_{\mathcal{U}}$ between the ultrapowers of E and F is well-defined, and we check that it is a Lipschitz equivalence. Indeed, fix two different points $x = (x_1, \dots)$ and $y = (y_1, \dots)$ in $(E)_{\mathcal{U}}$. Then $\{n : \|x_n - y_n\| \leq 1/n\} \notin \mathcal{U}$; hence Φ satisfies (10.1) for every $x, y \in (E)_{\mathcal{U}}$.

The map Φ^{-1} from $(F)_{\mathcal{U}}$ to $(E)_{\mathcal{U}}$ is defined similarly by using the φ_n^{-1}'s.

(ii) The proof is similar to that of (i). Let $r : F \to \varphi(E)$ be the uniform retraction, and, assuming $\varphi(0) = 0$, define φ_n as before and define $\psi_n : F \to E$ by $\psi_n(x) = \frac{1}{n}\varphi^{-1}\big(r(nx)\big)$. The same argument as in part (i) shows that the maps $\Phi = (\varphi_n) : (E)_{\mathcal{U}} \to (F)_{\mathcal{U}}$ and $\Psi = (\psi_n) : (F)_{\mathcal{U}} \to (E)_{\mathcal{U}}$ are well-defined Lipschitz maps, and that $\Psi \circ \Phi$ is the identity on $(E)_{\mathcal{U}}$. □

The theorem does not extend to general *into* uniform embeddings. For example, if $p < 2$, then L_p is uniformly homeomorphic to a subset of a Hilbert space (see Chapter 8) but is not Lipschitz equivalent to such a subset. If $\varphi : E \to F$ is an into map such that both φ and φ^{-1} satisfy a Lipschitz condition for large distances, then the proof shows that $(E)_{\mathcal{U}}$ is, indeed, Lipschitz equivalent to a subset of $(F)_{\mathcal{U}}$.

Corollary 10.2. *Let E be a separable Banach space, and assume that E is uniformly homeomorphic to F or, more generally, that there is a map φ from E into F such that both φ and φ^{-1} satisfy a Lipschitz condition for large distances. Let $\mathcal{U}$ be a free ultrafilter on $\mathbb{N}$. Then E is linearly isomorphic to a subspace of $(F)_{\mathcal{U}}^{**}$.*

Proof. This follows directly from Corollary 7.10 and from Theorem 10.1. □

Corollary 10.3. *Let $1 \leq p \leq \infty$, and let E be a Banach space which is uniformly homeomorphic to a linear subspace of an $L_p(\mu)$-space. Then E is isomorphic to a subspace of an $L_p(\mu)$-space.*

Proof. This follows from the previous corollary and from the following three properties of $L_p(\mu)$-spaces: A Banach space is isomorphic to a subspace of an $L_p(\mu)$-space if and only if each of its separable subspaces is (see Corollary F.4). Ultraproducts of $L_p(\mu)$-spaces are $L_p(\mu)$-spaces (see Proposition F.3). The second dual of an $L_p(\mu)$-space is also an $L_p(\mu)$-space. □

Remarks. (i) If $p \neq q$, then $L_p(\mu_1)$ and $L_q(\mu_2)$ are not uniformly homeomorphic to each other. Indeed, their respective ultrapowers are $L_p(\mu)$ and $L_q(\mu)$-spaces, and at least one of these spaces is not isomorphic to a subspace of the other.

(ii) For $p = 1$ we obtain that if a Banach space E is uniformly homeomorphic to a linear subspace F of L_1, then E is reflexive if and only if F is. Indeed, E is isomorphic to a subspace of an $L_1(\mu)$-space by Corollary 10.3. Hence, by a theorem of Rosenthal [**503**], it is reflexive if and only if it is isomorphic to a subspace of L_p for some $p > 1$. By Corollary 10.3 and Rosenthal's theorem again, this happens if and only if F is isomorphic to a subspace of L_p; i.e., it is reflexive.

Proposition 10.4. *Let F be a superreflexive Banach space.*

(i) *Let E be uniformly homeomorphic to a subset of F on which there is a uniformly continuous retraction. Then E is superreflexive and there is an ultrafilter $\mathcal{U}$ such that E is isomorphic to a complemented subspace of $(F)_{\mathcal{U}}$.*

(ii) *Let E and F be uniformly homeomorphic Banach spaces. Then there are ultrafilters $\mathcal{U}_1$ and $\mathcal{U}_2$ such that E is isomorphic to a complemented subspace of $(F)_{\mathcal{U}_1}$ and F is isomorphic to a complemented subspace of $(E)_{\mathcal{U}_2}$. In particular, if E and F satisfy one of the "decomposition schemes" of Proposition F.9, then there is an ultrafilter $\mathcal{V}$ such that $(E)_{\mathcal{V}}$ and $(F)_{\mathcal{V}}$ are isomorphic.*

Proof. By Theorem 10.1 there is an ultrafilter $\mathcal{U}$ and there are Lipschitz maps $\Phi : (E)_{\mathcal{U}} \to (F)_{\mathcal{U}}$ and $\Psi : (F)_{\mathcal{U}} \to (E)_{\mathcal{U}}$ such that $\Psi \circ \Phi$ is the identity on $(E)_{\mathcal{U}}$.

We first show that E is superreflexive. Indeed, it is Lipschitz equivalent to a subset of the superreflexive space $(F)_{\mathcal{U}}$; hence by the argument in the first few paragraphs of Chapter 7 each of its separable subspaces is isomorphic to a subspace of $(F)_{\mathcal{U}}$, hence superreflexive. The result follows because a Banach space is superreflexive if and only if all its separable subspaces are.

Consider E as a subspace of $(E)_{\mathcal{U}}$. Since E is reflexive, there is a norm-one projection $P : (E)_{\mathcal{U}} \to E$.

Let $\{G_i\}_{i \in I}$ be the set of all separable norm-one complemented subspaces of E, and let $P_i : E \to G_i$ be the projections. For each i the map $\Phi \circ P_i \circ P \circ \Psi$ is a Lipschitz retraction from $(F)_{\mathcal{U}}$ onto $\Phi(G_i)$. Hence, by Corollary 7.7, G_i is isomorphic to a complemented subspace of $(F)_{\mathcal{U}}$. Let $T_i : G_i \to (F)_{\mathcal{U}}$ be the isomorphism, and let $Q_i : (F)_{\mathcal{U}} \to T_i G_i$ be the projection. The norms of T_i and Q_i are bounded by a constant that depend only on the Lipschitz constants of Φ and Ψ, and not on G_i.

By Proposition F.8 every separable subspace of E is contained in one of the G_i's; hence, by Proposition F.5, there is an ultrafilter $\mathcal{V}$ such that E is isometric to a subspace of $(G_i)_{\mathcal{V}}$. Being reflexive, E is norm-one complemented in $(G_i)_{\mathcal{V}}$. But $(T_i)_{\mathcal{V}}$ is an isomorphism of $(G_i)_{\mathcal{V}}$ into $(F)_{\mathcal{U}\times\mathcal{V}}$, and $(Q_i)_{\mathcal{V}}$ is a bounded projection from $(F)_{\mathcal{U}\times\mathcal{V}}$ onto the image of $(G_i)_{\mathcal{V}}$ under $(T_i)_{\mathcal{V}}$. Hence E is isomorphic to a complemented subspace of $(F)_{\mathcal{U}\times\mathcal{V}}$.

Part (ii) follows from part (i) and from Proposition F.10. □

Theorem 10.5. *Let F be an $\mathcal{L}_p$-space ($1 < p \leq \infty$), and let $\varphi : E \to F$ be an into uniform homeomorphism such that there is a uniform retraction from F onto $\varphi(E)$. If $p < \infty$, then E is either an $\mathcal{L}_p$-space or is isomorphic to a Hilbert space. If φ is onto, or if $p = \infty$, then E is an $\mathcal{L}_p$-space.*

Proof. We first consider the range $1 < p < \infty$. Thus F is superreflexive, and by Proposition 10.4 there is an ultrafilter $\mathcal{U}$ such that E is isomorphic to a complemented subspace of $(F)_{\mathcal{U}}$. But ultrapowers of $\mathcal{L}_p$-spaces are $\mathcal{L}_p$-spaces (see Proposition F.3), and a complemented subspace of an $\mathcal{L}_p$-space for $1 < p < \infty$ is either an $\mathcal{L}_p$-space or is isomorphic to a Hilbert space (see Theorem F.2). If φ is onto, then E cannot, of course, be a Hilbert space unless $p = 2$.

For $p = \infty$ we shall use the characterization in Theorem F.2(v), namely, that a space G is an $\mathcal{L}_\infty$-space if and only if G^{**} is injective.

Let $f : (E)_{\mathcal{U}} \to (F)_{\mathcal{U}}$ be a Lipschitz embedding such that there is a Lipschitz retraction from the $\mathcal{L}_\infty$-space $G = (F)_{\mathcal{U}}$ onto $f\big((E)_{\mathcal{U}}\big)$. By Proposition F.6 E^{**} is naturally identified with a norm-one complemented subspace of $(E)_{\mathcal{U}}$, and thus there are Lipschitz maps $f : E^{**} \to G$ and $g : G \to E^{**}$ such that $g \circ f$ is the identity on E^{**}. The map g extends to a Lipschitz map $h : G^{**} \to E^{**}$. Indeed, G^{**} is a subspace of $(G)_{\mathcal{U}}$, and h is defined on all of $(G)_{\mathcal{U}}$ by the formula

$$h\big((x_i)_{\mathcal{U}}\big) = w^* - \lim_{\mathcal{U}} g(x_i).$$

It follows that E^{**} is Lipschitz equivalent to a Lipschitz retract of the injective space G^{**}, hence an absolute Lipschitz retract. To finish the proof, consider E^{**} as a linear subspace of the injective space $l_\infty(\Gamma)$ for some Γ. Since it is an absolute Lipschitz retract there is a Lipschitz retract from $l_\infty(\Gamma)$ onto E^{**}. By Corollary 7.3 E^{**} is linearly complemented in $l_\infty(\Gamma)$, hence injective too. □

Remark. It is unknown whether an analogous result holds for $p = 1$.

The next theorem shows that uniformly homeomorphic spaces have "the same" finite-dimensional subspaces. It follows that they must share every local property, i.e., a property that is determined by their finite-dimensional subspaces.

Theorem 10.6. *Let E and F be uniformly homeomorphic. Then there is a positive constant K such that for each finite-dimensional subspace E_1 of E, there is a subspace F_1 of F with $d(E_1, F_1) \le K$ and vice versa. In particular E has type-p for some $p > 1$ (or cotype-q for some $q < \infty$) if and only if F does.*

Proof. We give two proofs. The first is an easy deduction from Corollary 10.2, and the second uses averaging and instructive combinatorial and geometric arguments. The second proof is self-contained and, in particular, does not use differentiation or ultraproducts. It is thus naturally considerably longer.

We can assume that E and F are separable. Indeed, given a finite-dimensional subspace $E_1 \subset E$, a standard back and forth argument yields a separable space G with $E_1 \subset G \subset E$ whose image under the homeomorphism is a linear subspace of F. By Corollary 10.2 E is isomorphic to a subspace of $(F)_{\mathcal{U}}^{**}$, and the result follows from the principle of local reflexivity (Proposition F.1) and from the fact that F and $(F)_{\mathcal{U}}$ have the same finite-dimensional subspaces (Proposition F.6).

An alternative proof. Let $\varphi : E \to F$ be the uniform homeomorphism. Since both φ and φ^{-1} are Lipschitz for large distances, φ can be normalized so that there is a constant K for which $\|x - y\| \le \|\varphi(x) - \varphi(y)\| \le K\|x - y\|$ for every $x, y \in E$ with $\|x - y\| \ge 1$.

Let E_1 be an n-dimensional subspace of E, and fix $\{x_i\}_{i=1}^n$ in E_1 such that $\|\sum k_i x_i\| \ge 1$ for every choice of integers $\{k_i\}_{i=1}^n \subset \mathbb{Z}$, not all equal to 0 (for example, take $\{x_i\}$ to be an Auerbach basis; see Lemma E.5). Denote the lattice generated by $\{x_i\}$ by $\mathcal{M}$. The finite lattice of points $x = \sum k_i x_i \in \mathcal{M}$ with $|k_i| \le m$ for every i is denoted by $\mathcal{M}_m$. A set of the form $u + s\mathcal{M}_m$ with $u \in \mathcal{M}$ and $s \in \mathbb{N}$ will be called a finite lattice of size $(2m+1)^n$ and step s.

For each finite lattice $\mathcal{L}$ with step s, define a function $\varphi_{\mathcal{L}}$ on $s\mathcal{M}$ by

$$\varphi_{\mathcal{L}}(y) = |\mathcal{L}|^{-1} \sum_{x \in \mathcal{L}} \big(\varphi(x+y) - \varphi(x)\big).$$

Then $\varphi_{\mathcal{L}}$ is K-Lipschitz, and it is "nearly additive" for large $\mathcal{L}$. Indeed

$$\begin{aligned}\varphi_{\mathcal{L}}(y+z) &= |\mathcal{L}|^{-1} \sum_{x \in \mathcal{L}} \big(\varphi(x+y+z) - \varphi(x)\big)\\ &= |\mathcal{L}|^{-1} \sum_{x \in \mathcal{L}} \big(\varphi(x+y+z) - \varphi(x+y)\big) + |\mathcal{L}|^{-1} \sum_{x \in \mathcal{L}} \big(\varphi(x+y) - \varphi(x)\big).\end{aligned}$$

The second term is $\varphi_{\mathcal{L}}(y)$, and the first is $\varphi_{y+\mathcal{L}}(z)$. For fixed y and z, if $\mathcal{L}$ is large enough, then most of the terms in $\varphi_{y+\mathcal{L}}(z)$ and $\varphi_{\mathcal{L}}(z)$ coincide. Since each term in these averages is bounded by $K\|z\|$, it follows that for each $\varepsilon > 0$ there is an m (which depends only on ε, K and $\|y\|$) such that if $|\mathcal{L}| > (2m+1)^n$, then

$$\|\varphi_{\mathcal{L}}(y+z) - \varphi_{\mathcal{L}}(y) - \varphi_{\mathcal{L}}(z)\| \le \varepsilon \|z\|.$$

More generally, for every m_0 and $\varepsilon > 0$ there is an m (which depends only on ε, K, m_0 and s) such that if $|\mathcal{L}| > (2m+1)^n$, then

$$\Big\|\varphi_{\mathcal{L}}(x) - \sum k_i \varphi_{\mathcal{L}}(sx_i)\Big\| \le \varepsilon \|x\|$$

for every $x = s\sum k_i x_i \in s\mathcal{M}_{m_0}$. Let $T_{\mathcal{L}} : E_1 \to F$ be the linear operator defined by $T_{\mathcal{L}}(sx_i) = \varphi_{\mathcal{L}}(sx_i)$ for $1 \le i \le n$. The inequality above means that

$$\|\varphi_{\mathcal{L}}(x) - T_{\mathcal{L}} x\| \le \varepsilon \|x\| \tag{10.2}$$

for every $x \in s\mathcal{M}_{m_0}$. Since $\varphi_{\mathcal{L}}$ is K-Lipschitz, a simple argument shows that if ε is small enough, and if m_0 is large enough so that $\{x/\|x\| : 0 \ne x \in s\mathcal{M}_{m_0}\}$ forms an ε-dense set in the unit sphere of E_1, then $\|T_{\mathcal{L}}\| \le 2K$.

The upper bounds for the Lipschitz constant of $\varphi_{\mathcal{L}}$ and for $\|T_{\mathcal{L}}\|$ followed directly from the triangle inequality. Lower bounds are usually lost by averaging, and generally $\varphi_{\mathcal{L}}$ need not be one-one and $T_{\mathcal{L}}$ need not be invertible. To obtain an invertible operator (with a bound on $\|T_{\mathcal{L}}^{-1}\|$) we need to make a special choice of $\mathcal{L}$, and this will be done by using the following claim:

Let l be a positive integer, and let $y \in \mathcal{M}$. Then there is an $M = M(l, y)$ such that for every finite lattice $\mathcal{L}^1$ of size $(2m+1)^n$ with $m > M$ there is a sublattice $\mathcal{L}^2 \subset \mathcal{L}^1$ of size $(2l+1)^n$, and a norm-one functional $u^* \in F^*$ such that

$$\big\langle u^*, \varphi(x+sy) - \varphi(x)\big\rangle \ge s\|y\|/2$$

for every $x \in \mathcal{L}^2$, where s is the step of $\mathcal{L}^2$.

To see that the claim proves the theorem, fix a small $\varepsilon > 0$ and m_0 as above, and enumerate the nonzero points in $\mathcal{M}_{m_0}$ as $\{y_i\}_{i=1}^p$ where $p = (2m_0+1)^n - 1$.

Starting with a very large finite lattice $\mathcal{M}_{m_1} = \mathcal{L}^1$ (where m_1 depends on ε, m_0, p via the estimates in the claim), use the claim successively (with $y = y_i$ at the ith step) to find (still very large) sublattices $\mathcal{L}^1 \supset \mathcal{L}^2 \supset \ldots \supset \mathcal{L}^{p+1} = \mathcal{L}$ and unit vectors $u_i^* \in F^*$ such that the final sublattice $\mathcal{L}$ will be large enough to ensure that both (10.2) and $\|T_{\mathcal{L}}\| \le 2K$ hold, and such that

$$\big\langle u_i^*, \varphi(x+sy_i) - \varphi(x)\big\rangle \ge s\|y_i\|/2$$

for every $x \in \mathcal{L}$, where s is the step of $\mathcal{L}$. Thus

$$\begin{aligned} \|\varphi_{\mathcal{L}}(sy_i)\| &= \Big\| |\mathcal{L}|^{-1} \sum_{x\in\mathcal{L}} \big(\varphi(x+sy_i) - \varphi(x)\big) \Big\| \\ &\geq |\mathcal{L}|^{-1} \sum_{x\in\mathcal{L}} \Big\langle u_i^*, \big(\varphi(x+sy_i) - \varphi(x)\big) \Big\rangle \geq s\|y_i\|/2 \end{aligned}$$

for every i. Assuming that $\varepsilon < 1/6$, it follows from (10.2) that $\|T_{\mathcal{L}} y_i\| \geq \|y_i\|/3$ for every i. If $z \in E_1$ and $\|z\| = 1$, choose i such that $\big\|z - y_i/\|y_i\|\big\| < \varepsilon$. Then

$$\|T_{\mathcal{L}} z\| \geq \big\|T_{\mathcal{L}}\big(y_i/\|y_i\|\big)\big\| - \big\|T_{\mathcal{L}}\big(z - y_i/\|y_i\|\big)\big\| \geq 1/3 - 2K\varepsilon \geq 1/4$$

provided $\varepsilon < 1/24K$.

It remains to prove the claim. To simplify the notation we shall assume that $\mathcal{L}^1$ has step 1. (This can always be attained by changing the scale.) Fix an integer N which will be specified later (and is large compared to l), and put

$$\alpha_j = \sup\Big\{2^{-j}\|\varphi(x+2^j y) - \varphi(x)\|/\|y\| : x \in \mathcal{L}^1 \text{ such that } x + 2^j y \in \mathcal{L}^1\Big\}.$$

Then $1 \leq \alpha_j \leq K$ for every j and $\alpha_j \leq \alpha_{j-1}$ by the triangle inequality. We choose M such that if $m > M$, then the sets defining α_j are nonempty for a large enough number of j's to ensure that there is a j_0 for which $\alpha_{j_0} \leq \alpha_{j_0+N} + 2^{-N}$.

Fix $z \in \mathcal{L}^1$ for which the supremum in the definition of α_{j_0+N} is attained. Translating $\mathcal{L}^1$ by z, we may assume that $z = 0$. By passing to the sublattice $\mathcal{L}^1 \bigcap 2^{j_0}\mathcal{L}^1$ and then changing the scale, we may also assume that $j_0 = 0$, i.e.,

$$\alpha_0 - 2^{-N} \leq \alpha_N = 2^{-N}\|\varphi(2^N y) - \varphi(0)\|/\|y\|.$$

Let $u^* \in F^*$ be a norm-one functional such that $\langle u^*, \varphi(2^N y) - \varphi(0)\rangle = 2^N \alpha_N \|y\|$.

Now let $v \in \mathcal{M}_m$, and put $C = 1 + 2K\max\{\|v\|/\|y\| : v \in \mathcal{M}_m\}$. Then

$$\begin{aligned} &\langle u^*, \varphi(v + 2^N y) - \varphi(v)\rangle \\ &\quad\geq \langle u^*, \varphi(2^N y) - \varphi(0)\rangle - \|\varphi(v+2^N y) - \varphi(2^N y)\| - \|\varphi(v) - \varphi(0)\| \\ &\quad\geq 2^N \alpha_N\|y\| - 2K\|v\| \geq 2^N(\alpha_0 - 2^{-N})\|y\| - 2K\|v\| \geq (2^N\alpha_0 - C)\|y\|. \end{aligned}$$

The left-hand side is a sum $\sum \langle u^*, \varphi(v + (k+1)y) - \varphi(v+ky)\rangle$, where each of the 2^N terms is bounded from above by $\alpha_0\|y\|$. Since $\alpha_0 \geq 1$, it follows that for each such v, the number of k's for which $\langle u^*, \varphi(v+(k+1)y) - \varphi(v+ky)\rangle \leq \|y\|/2$ is at most $2C$. By the pigeonhole principle, for large enough N there is a k_0 such that $\mathcal{L}^2 = k_0 y + \mathcal{M}_m$ is contained in $\mathcal{L}^1$, and $\langle u^*, \varphi(x+y) - \varphi(x)\rangle \geq \|y\|/2$ for every $x \in \mathcal{L}^2$. □

We shall show in the next section that for each $1 \leq p < \infty$, $p \neq 2$, the spaces l_p and L_p, which have the same finite-dimensional subspaces, are not uniformly homeomorphic. Thus the converse to Theorem 10.6 does not hold.

The next result is a stronger version of Theorem 10.6 for superreflexive spaces.

Theorem 10.7. *Assume that the superreflexive spaces E and F are uniformly homeomorphic. Then there is a positive constant K such that for each finite-dimensional subspace G of E and for each integer $n \geq \dim(G)$ there is a linear embedding $S : G \to F$ with the following property: Let H be an arbitrary n-dimensional subspace of F containing $S(G)$. Then there is a linear operator $T : H \to E$ such that TS is the identity on G and $\|S\|\,\|T\| \leq K$.*

Proof. By Proposition 10.4 there is a set I and an ultrafilter $\mathcal{U}$ on I such that E is isomorphic to a complemented subspace of $(F)_{\mathcal{U}}$. Thus there are bounded linear operators $A : E \to (F)_{\mathcal{U}}$ and $B : (F)_{\mathcal{U}} \to E$ such that $BA = id_E$ (the identity on E). We put $K = \|A\|\,\|B\| + 1$.

We shall use the information on finite-dimensional subspaces of ultraproducts and on operators to and from such subspaces given in Proposition F.7.

Represent the restriction of A to G by a family of operators $A_i : G \to F$. We claim that one of the operators A_i can be taken as the required operator S. If this is not the case, choose, for each i, an n-dimensional subspace $H_i \supset A_i(G)$ such that $\|A_i\|\,\|T\| > K$ for every operator $T : H_i \to E$ with $TA_{i\,|G} = id_G$. Then $(H_i)_{\mathcal{U}}$ is an n-dimensional subspace of $(F)_{\mathcal{U}}$ containing $A(G)$ and clearly $BA_{|G} = id_G$. Choose bases $\{x_{i,j}\}_{j=1}^n$ for H_i and $\{x_j\}_{j=1}^n$ for $(H_i)_{\mathcal{U}}$, and let $B_i : H_i \to E$ be as in Proposition F.7(ii) and (iii), i.e., such that $Bx_j = B_i x_{i,j}$ for every i and j. Thus $B_i A_{i\,|G} = id_G$ for every $i \in \mathcal{U}$. It follows from $\|A_{|G}\| = \lim_{\mathcal{U}} \|A_i\|$ and from $\|B_{|(H_i)_{\mathcal{U}}}\| = \lim_{\mathcal{U}} \|B_i\|$ that there is an $i \in \mathcal{U}$ with $\|B_i\|\,\|A_i\| < K + 1/2$, contradicting the choice of H_i. □

As mentioned earlier, Theorem 10.6 by itself is not sufficient to determine whether specific Banach spaces are uniformly homeomorphic to each other. The next theorem is an example of a situation where other considerations are necessary. It should be compared with the fact that $C(S^n)$ (where S^n is the n-dimensional unit sphere in $\mathbb{R}^{n+1}$) is an absolute Lipschitz retract; see Theorem 1.6. In particular, it will follow that spaces of smooth functions on higher-dimensional manifolds are not uniformly homeomorphic to spaces of continuous functions. (Note that both $C(S^n)$ and $C^1(S^n)$ contain an isometric copy of every separable Banach space; hence the non-uniform homeomorphism cannot be deduced from results like Theorem 10.6.) The proof uses only results from Chapter 7, but the theorem belongs naturally to this chapter.

Let $C^1(S^n)$ be the space of continuously differentiable functions on S^n, and let $C_0^1(S^n)$ be its codimension-one subspace of functions satisfying $\int_{S^n} f(x)dx = 0$ (where dx is the normalized Lebesgue measure on S^n). Let V denote the space of all continuous tangent vector fields on S^n, i.e., continuous functions $F : S^n \to \mathbb{R}^{n+1}$ such that $\langle F(x), x\rangle = 0$ for every $x \in S^n$. V is equipped with the supremum norm. The gradient ∇ of $f \in C^1(S^n)$ is defined by extending f to be constant on rays emanating from the origin and then taking the usual gradient in $\mathbb{R}^{n+1}$. It is clear that $\nabla f(x)$ is tangent to S^n at x; hence ∇ is an injection from $C_0^1(S^n)$ into V, and we use it to identify $C_0^1(S^n)$ as a closed subspace of V with the norm induced from V (which is equivalent to the usual norm on $C_0^1(S^n)$).

The orthogonal group O_{n+1} acts on S^n, and this action induces actions of O_{n+1} on $C_0^1(S^n)$ and on V by the formulas $f_u(x) = f(ux)$ and $F_u(x) = u^{-1}F(ux)$ for $x \in S^n$, $u \in O_{n+1}$, $f \in C_0^1(S^n)$, and $F \in V$. These actions are continuous in the strong operator topology, and they satisfy $(f_u)_v = f_{uv}$ and $(F_u)_v = F_{uv}$ for every $u, v \in O_{n+1}$. They also commute with the embedding ∇ of $C_0^1(S^n)$ into V. Indeed, since u is orthogonal

$$\nabla(f_u) = \nabla(f \circ u) = u^*\Big((\nabla f) \circ u\Big) = u^{-1}\Big((\nabla f) \circ u\Big) = (\nabla f)_u$$

by the chain rule.

Theorem 10.8. *If $n \geq 2$, then $C_0^1(S^n)$ is not an absolute uniform retract.*

Proof. It is enough to prove the theorem for $n = 2$. We shall show that there is no uniformly continuous retraction from V onto $C_0^1(S^2)$.

For $x \in S^2$ denote by δ_x the linear functional $\delta_x(f) = f(x)$ on $C_0^1(S^2)$. The theorem follows from the following claim.

There is no Lipschitz mapping $\phi : S^2 \to V^*$ such that for each $x \in S^2$ the functional $\phi(x)$ is an extension of the functional δ_x to a functional on V.

To deduce the theorem, assume that $C_0^1(S^2)$ is a uniform retract of V. By Corollary 7.3 there is a bounded linear operator $W : C_0^1(S^2)^* \to V^*$ such that $\langle W\nu, f\rangle = \langle \nu, f\rangle$ for every $f \in C_0^1(S^2)$ and $\nu \in C_0^1(S^2)^*$; i.e., $W\nu$ is an extension of ν. Hence the mapping $\phi(x) = W\delta_x$, which is Lipschitz because $x \to \delta_x$ is Lipschitz and W is a bounded linear operator, contradicts the claim.

To prove the claim, assume that ϕ is a Lipschitz extension map. We shall show

(i) There is a Lipschitz extension map ψ which commutes with the action of O_3; i.e., $\psi(x)(F_u) = \psi(ux)(F)$ for every $x \in S^2$, $u \in O_3$, and $F \in V$.

(ii) There is a unique extension map (Lipschitz or not) which commutes with the action of O_3, and there is an explicit formula for it.

(iii) The map in (ii) is not Lipschitz.

Combining (i)–(iii) gives a contradiction.

(i) We use an averaging argument. The map $\psi(x)(F) = \int_{O_3} \phi(vx)(F_{v^{-1}})dv$ (where dv denotes the Haar measure on O_3) is well-defined, because the integrand is a continuous function of v. To see that $\psi(x)$ extends δ_x, choose $f \in C_0^1(S^2)$. Then $\phi(vx)(f_{v^{-1}}) = f_{v^{-1}}(vx) = f(v^{-1}vx) = f(x)$ for every $v \in O_3$; hence the same also holds for the integral. The map ψ commutes with the action of O_3 because

$$\begin{aligned}\psi(ux)(F) &= \int_{O_3} \phi(vux)(F_{v^{-1}})dv = \int_{O_3} \phi(wx)(F_{uw^{-1}})dw \\ &= \int_{O_3} \phi(wx)\big((F_u)_{w^{-1}}\big)dw = \psi(x)(F_u).\end{aligned}$$

(ii) Put $x_0 = (0,0,1)$, and let G be the subgroup of O_3 which fixes x_0. We start with the observation that if $H \in V$ is invariant under G, then $H = \nabla h$ for some function $h \in C_0^1(S^2)$. Indeed, since H is invariant under G and since G contains the rotations around the z-axis as well as the reflections with respect to planes containing the z-axis, it follows that it has the form $H(x) = s(\alpha)\tau_x$, where $0 \le \alpha \le \pi$ is the angle between x and $-x_0$, the function s is a scalar continuous function, and τ_x is the unit vector tangent to S^2 at x in the direction of the short arc on the great circle joining x and x_0. (Note that $\tau(x_0)$ and $\tau(-x_0)$ are not defined, but this causes no problem because $H(x_0) = H(-x_0) = 0$, hence also $s(0) = s(\pi) = 0$.) Now define $h(x) = c + \int_0^\alpha s(\beta)d\beta$, where the constant c is chosen so that $\int_{S^2} h(x)dx = 0$. Then $h \in C_0^1(S^2)$, and clearly $\nabla h = H$.

We can now show the uniqueness of the extension map. Let ψ be such a map, and fix any $F \in V$ and $x \in S^2$. Choose the coordinate system so that $x = (0,0,1)$, let G be as above, and put $H = \int_G F_u d\mu(u)$ where μ is the normalized Haar measure on G. Then H is invariant under G, hence of the form $H = \nabla h$ for some function $h \in C_0^1(S^2)$. Since ψ commutes with the action of G, and since G fixes x, it follows that $\psi(x)(F) = \psi(x)(F_u)$ for every $u \in G$; hence

$$\psi(x)(F) = \int_G \psi(x)(F_u)d\mu(u) = \psi(x)\left(\int_G F_u d\mu(u)\right) = \psi(x)(H) = h(x).$$

Thus $\psi(x)$ is completely determined by its values on $C_0^1(S^2)$, hence unique.

We now give an explicit formula for the unique extension map ψ. Denote the geodesic length between two points $x, y \in S^2$ by $\rho(x,y)$, and let $\tau_{y,x}$ be the unit tangent vector at y in the direction of the short arc joining x and y. We show that

$$\psi(x)(F) = \int_{S^2} \operatorname{ctg}\Big(\frac{1}{2}\rho(x,y)\Big)\langle F(y), \tau_{y,x}\rangle dy$$

gives the required extension. (The singularity of $\operatorname{ctg} t$ at $t = 0$ is like $1/t$, hence integrable in two dimensions.) To check that ψ commutes with the action of O_3, fix any $F \in V$ and $u \in O_3$. Then

$$\begin{aligned}
\psi(ux)(F) &= \int_{S^2} \operatorname{ctg}\Big(\frac{1}{2}\rho(ux,y)\Big)\langle F(y), \tau_{y,ux}\rangle dy \\
&= \int_{S^2} \operatorname{ctg}\Big(\frac{1}{2}\rho(ux,uy)\Big)\langle F(uy), \tau_{uy,ux}\rangle dy \\
&= \int_{S^2} \operatorname{ctg}\Big(\frac{1}{2}\rho(x,y)\Big)\langle F(uy), u\tau_{y,x}\rangle dy \\
&= \int_{S^2} \operatorname{ctg}\Big(\frac{1}{2}\rho(x,y)\Big)\langle u^{-1}F(uy), \tau_{y,x}\rangle dy = \psi(x)(F_u).
\end{aligned}$$

To see that ψ is an extension map, fix x and choose spherical coordinates (θ, α) on S^2 with $0 \le \theta \le 2\pi$ and $0 \le \alpha \le \pi$ in such a way that x corresponds to $\alpha = \pi$. In these coordinates $\langle \nabla f(y), \tau_{y,x}\rangle = \frac{\partial f}{\partial \alpha}(\theta,\alpha)$ for $f \in C_0^1(S^2)$ and $y = (\theta, \alpha) \in S^2$.

The identity $\operatorname{ctg}\frac{1}{2}(\pi - \alpha)\sin\alpha = 1 - \cos\alpha$ and integration by parts show that for every $h \in C^1[0,\pi]$

$$\int_0^\pi \operatorname{ctg}\Big(\frac{1}{2}(\pi-\alpha)\Big)h'(\alpha)\sin\alpha\, d\alpha = 2h(\pi) - \int_0^\pi h(\alpha)\sin\alpha\, d\alpha.$$

Let $f \in C_0^1(S^2)$, and write the integral for $\psi(x)(\nabla f)$ in spherical coordinates. Since $f(\theta,\pi) = f(x)$ for all θ, and since $\int_{S^2} f(y)dy = 0$, we obtain that

$$\begin{aligned}
\psi(x)(\nabla f) &= \frac{1}{4\pi}\int_0^{2\pi} d\theta \int_0^\pi \operatorname{ctg}\Big(\frac{1}{2}(\pi-\alpha)\Big)\frac{\partial f}{\partial\alpha}(\theta,\alpha)\sin\alpha\, d\alpha \\
&= \frac{1}{4\pi}\int_0^{2\pi} d\theta\Big(2f(x) - \int_0^\pi f(\theta,\alpha)\sin\alpha\, d\alpha\Big) \\
&= f(x) - \frac{1}{4\pi}\int_0^{2\pi}\int_0^\pi f(\theta,\alpha)\sin\alpha\, d\alpha\, d\theta \\
&= f(x) - \int_{S^2} f(y)dy = f(x).
\end{aligned}$$

(iii) We use spherical coordinates as above and put $x_\xi = (\xi, \pi/2)$ for $0 \le \xi \le 2\pi$. To show that ψ is not Lipschitz we shall construct, for each $0 < \delta < \pi/20$, a vector field $F = F_\delta$ with $\|F\| \le 1$ such that

$$\psi(x_\delta)(F) - \psi(x_0)(F) \ge C\delta|\log\delta|$$

where C is a constant independent of δ.

Let $G \in V$ be such that if $y = (\theta,\alpha)$ satisfies $|\alpha - \pi/2| \le \pi/4$, then $G(y)$ is the unit vector tangent to S^2 at y, parallel to the plane $\alpha \equiv \pi/2$, and directed clockwise. The values of G when $|\alpha - \pi/2| > \pi/4$ are immaterial, and we require only that $\|G\| = 1$. For $A > 0$ put

$$W_{A,\delta} = \Big\{ y = (\theta, \alpha) : \; 2\delta \le \theta \le \pi/2 \; \text{ and } \; |\alpha - \pi/2| \le A\theta \Big\}.$$

Note that if $A < 1/2$ and $y = (\theta, \alpha) \in W_{A,\delta}$, then $|\alpha - \pi/2| \le \pi/4$; hence G is given by the description above. We first show that it is possible to find A (independent of δ) such that

$$\operatorname{ctg}\Big(\frac{1}{2}\rho(x_\delta, y)\Big)\langle G(y), \tau_{y,x_\delta}\rangle - \operatorname{ctg}\Big(\frac{1}{2}\rho(x_0, y)\Big)\langle G(y), \tau_{y,x_0}\rangle \ge \frac{\delta}{10(\theta-\delta)^2} \tag{10.3}$$

for every $y = (\theta, \alpha) \in W_{A,\delta}$.

Indeed, write the left-hand side as the difference of two terms

$$\Big(\operatorname{ctg}\Big(\frac{1}{2}\rho(x_\delta, y)\Big) - \operatorname{ctg}\Big(\frac{1}{2}\rho(x_0, y)\Big)\Big)\langle G(y), \tau_{y,x_\delta}\rangle$$

and

$$\Big(\langle G(y), \tau_{y,x_0}\rangle - \langle G(y), \tau_{y,x_\delta}\rangle\Big) \operatorname{ctg}\Big(\frac{1}{2}\rho(x_0, y)\Big).$$

To estimate the first term note that if A is small enough (the choice is independent of δ), then $\rho(x_0, y) \ge \rho(x_\delta, y) + \delta/2$. It follows from the monotonicity and concavity of ctg that $\operatorname{ctg}\beta - \operatorname{ctg}\gamma \ge (\beta-\gamma)\operatorname{ctg}'\beta = (\gamma-\beta)/\sin^2\beta$ whenever $0 < \beta < \gamma < \pi/2$. Since $\theta \ge 2\delta$, it follows that $\rho^2(x_\delta, y) \le (\theta - \delta + A\theta)^2 \le 2(\theta-\delta)^2$ provided A is small enough. Thus

$$\begin{aligned}\operatorname{ctg}&\Big(\frac{1}{2}\rho(x_\delta, y)\Big) - \operatorname{ctg}\Big(\frac{1}{2}\rho(x_0, y)\Big)\\ &\ge \big(\rho(x_0, y) - \rho(x_\delta, y)\big)/2\sin^2\Big(\frac{1}{2}\rho(x_\delta, y)\Big) \ge \delta/2(\theta-\delta)^2\end{aligned}$$

and of course $\langle G(y), \tau_{y,x_\delta}\rangle \ge 1/2$ for small enough A. Thus the first term is bounded from below by $\delta/4(\theta-\delta)^2$.

To estimate the second term, put $y_\delta = (2\delta, A2\delta)$. Then

$$\|\tau_{y,x_\delta} - \tau_{y,x_0}\| \le \|\tau_{y_\delta,x_\delta} - \tau_{y_\delta,x_0}\| < \delta/100$$

provided A is small enough. Since also

$$\operatorname{ctg}\Big(\frac{1}{2}\rho(x_0, y)\Big) \le 1/\sin(\theta/2) \le 4/\theta \le 4/(\theta-\delta)$$

the second term is bounded from above by $4\delta/100(\theta-\delta)$. Combining the two estimates gives (10.3).

Now fix $A < 1/2$ so that (10.3) holds in $W_{A,\delta}$, and put

$$V_{A,\delta} = \Big\{ y = (\theta, \alpha) : \; 3\delta \le \theta \le \pi/4 \; \text{ and } \; |\alpha - \pi/2| \le A\theta/2 \Big\}.$$

Let $0 \le f \le 1$ be a continuous function on S^2 which is supported in $W_{A,\delta}$ and such that $f \equiv 1$ on $V_{A,\delta}$. The desired vector field is $F(y) = f(y)G(y)$. Indeed, since $\langle F(y), \tau_{y,x}\rangle = f(y)\langle G(y), \tau_{y,x}\rangle$, it follows that

$$\begin{aligned}&\psi(x_0)(F) - \psi(x_\delta)(F)\\ &= \int_{S^2}\Big[\operatorname{ctg}\Big(\frac{1}{2}\rho(x_0, y)\Big)\langle G(y), \tau_{y,x_0}\rangle - \operatorname{ctg}\Big(\frac{1}{2}\rho(x_\delta, y)\Big)\langle G(y), \tau_{y,x_\delta}\rangle\Big]\, dy\\ &\ge \int_{V_{A,\delta}} \frac{\delta\, dy}{10(\theta-\delta)^2}\end{aligned}$$

by (10.3), since the integrand is nonnegative and $f \equiv 1$ on $V_{A,\delta}$. Finally, using spherical coordinates, we obtain that

$$\begin{aligned}\int_{V_{A,\delta}} \frac{dy}{(\theta-\delta)^2} &= \int_{3\delta}^{\pi/4} \int_{\pi/2-A\theta/2}^{\pi/2+A\theta/2} \frac{\sin\alpha \, d\alpha \, d\theta}{(\theta-\delta)^2} \\ &\geq C_1 \int_{3\delta}^{\pi/4} \frac{\theta \, d\theta}{(\theta-\delta)^2} \geq C_2 |\log\delta|.\end{aligned}$$

□

2. Approximate Midpoints

Definition 10.9. *Let x and y be two points in the Banach space E, and let $0<\delta<1$. The set of δ-approximate midpoints between x and y is the set*

$$\mathrm{Mid}(x,y,\delta) = \{z \in E : \ \|x-z\|, \|y-z\| \leq (1+\delta)\|x-y\|/2\}.$$

If $z \in \mathrm{Mid}(x,y,\delta)$, then $\|x-z\| \geq (1-\delta)\|x-y\|/2$, and similarly for $\|y-z\|$. When $\delta=0$ we obtain the set of metric midpoints, a basic tool in the study of isometries (see, e.g., Theorem 14.1). The approximate midpoints will play a similar role in the study of uniform homeomorphisms.

We start by identifying the structure of the approximate midpoints for some concrete spaces. By translation it is enough to consider the case $y=-x$.

Proposition 10.10. (i) *Let $1 \leq p < \infty$, and let $0 \neq x \in l_p$. Then there is a finite-dimensional subspace $E_0 \subset l_p$ and a finite-codimensional subspace $E_1 \subset l_p$ such that*

$$\frac{1}{2}\delta^{1/p}\|x\|B(E_1) \subset \mathrm{Mid}(x,-x,\delta) \subset E_0 + (3p\delta)^{1/p}\|x\|B(l_p).$$

(ii) *More generally, assume that E has a 1-unconditional basis $\{x_n\}$. If the basis satisfies an upper p-estimate with constant one* (see Appendix E), *then for every $0 \neq x \in E$ there is a subspace $E_1 \subset E$ of finite codimension such that*

$$\frac{1}{2}\delta^{1/p}\|x\|B(E_1) \subset \mathrm{Mid}(x,-x,\delta).$$

If the basis satisfies a lower p-estimate with constant one, then for every $0 \neq x \in E$ there is a finite-dimensional subspace $E_0 \subset E$ such that

$$\mathrm{Mid}(x,-x,\delta) \subset E_0 + (3p\delta)^{1/p}\|x\|B(E).$$

(iii) *Let $0 \neq x \in c_0$. Then for every $\varepsilon < \delta$ there is a subspace $E_1 \subset c_0$ of finite codimension such that*

$$(1+\varepsilon)\|x\|B(E_1) \subset \mathrm{Mid}(x,-x,\delta).$$

(iv) *Let $0 \neq x \in L_1$. Then there is an infinite sequence $\{y_j\}$ of exact metric midpoints between x and $-x$ such that $\|y_i - y_j\| = \|x\|$ for every $i \neq j$.*

(v) *Let $2 < p < \infty$, and let $0 \neq x \in L_p$. Then there is a constant C, depending only on p, and a subspace F_1 of L_p of infinite codimension such that*

$$\mathrm{Mid}(x,-x,\delta) \subset F_1 + C\delta^{1/2}\|x\|B(L_p).$$

Proof. By normalization we may assume that $\|x\| = 1$.

(i) Assume first that x is finitely supported, and put $E_0 = \operatorname{span}\{e_j : j \leq N\}$ and $E_1 = \operatorname{span}\{e_j : j > N\}$, where N is chosen so that $x \in E_0$.

If $y \in \delta^{1/p}B(E_1)$, then $\|x \pm y\| = (\|x\|^p + \|y\|^p)^{1/p} \leq (1+\delta)^{1/p} \leq 1+\delta$, i.e., $y \in \operatorname{Mid}(x, -x, \delta)$. Conversely, assume that $\|x \pm y\| \leq 1+\delta$, and write $y = z_0 + z_1$ where $z_i \in E_i$. For a suitable choice of a sign θ we obtain that $\|x + \theta z_0\| \geq 1$; hence $(1+\delta)^p \geq \|x + \theta z_0\|^p + \|z_1\|^p \geq 1 + \|z_1\|^p$ and $\|z_1\| \leq \big((1+\delta)^p - 1\big)^{1/p} \leq (2p\delta)^{1/p}$. If the support of x is not finite, we use the argument above for a finitely supported vector which approximates x (the degree of approximation depends only on δ).

The proofs of (ii) and (iii) are similar to that of (i).

(iv) Inductively divide the interval $[0,1]$ to 2^j disjoint subsets $\{A_{i,j} : i \leq 2^j\}$ such that $A_{i,j} = A_{2i-1,j+1} \cup A_{2i,j+1}$ and such that $\int_{A_{i,j}} |x| = 2^{-j}$. Then define $y_j(t) = (-1)^i x(t)$ for $t \in A_{i,j}$ and $i = 1, \dots, 2^j$.

(v) Assume first that $x \equiv 1$. Let F_2 be the span of the Rademacher functions. By Khintchine's inequality there is a constant A_p such that $A_p\|y\|_p \leq \|y\|_2$ for every $y \in F_2$. Take F_1 to be the closed subspace of L_p orthogonal to $\operatorname{span}\{1, F_2\}$, i.e.,

$$F_1 = \Big\{ z \in L_p : \int z(t)\big(1 + y(t)\big)dt = 0 \quad \text{for every } y \in F_2 \Big\}.$$

Fix $z \in \operatorname{Mid}(x, -x, \delta)$, and write it as $z = \alpha + x_1 + x_2$ where $x_i \in F_i$. Assume, as we may, that $\alpha \geq 0$. Then

$$1 + \delta \geq \|1 + z\|_p = \|(1+\alpha) + x_1 + x_2\|_p \geq \|(1+\alpha) + x_1 + x_2\|_2 \geq 1 + \alpha$$

i.e., $\alpha \leq \delta$. Also, since $\|x_2\|_2 \leq 1 + \delta < 2$

$$\begin{aligned} 1+\delta &\geq \|(1+\alpha) + x_1 + x_2\|_2 \geq \|1 + x_2\|_2 \\ &= \big(1 + \|x_2\|_2^2\big)^{1/2} \geq 1 + \|x_2\|_2^2/8 \geq 1 + A_p^2\|x_2\|_p^2/8 \end{aligned}$$

i.e., $\|x_2\|_p \leq (8\delta/A_p^2)^{1/2}$. Thus $\operatorname{dist}(z, F_1) \leq \|\alpha + x_2\| \leq \delta + (8\delta/A_p^2)^{1/2}$.

For a general x we may assume by passing to $|x|$ and by an arbitrarily small perturbation that x is continuous and strictly positive. Then $f(t) = \int_0^t x(s)^p ds$ is strictly monotone and $f' = x^p$. The operator $Ty(t) = y\big(f(t)\big)x(t)$ preserves the norm on L_p because

$$\|Ty(t)\|_p^p = \int \big|y\big(f(t)\big)x(t)\big|^p dt = \int \big|y\big(f(t)\big)\big|^p f'(t)dt = \|y\|_p^p.$$

Since $\min x(t) > 0$, the inverse $T^{-1}y(t) = y\big(f^{-1}(t)\big)/x\big(f^{-1}(t)\big)$ is well-defined on all of L_p; hence T is an isometry of L_p onto itself. Finally $T1 = x$, and the result follows with F_1 of the case $x \equiv 1$ replaced by $T^{-1}F_1$. □

Assume that a map f between Banach spaces satisfies a Lipschitz condition for large distances, and denote its Lipschitz constant for distances larger than d by K_d. If x and y satisfy $\|x - y\| \geq 2d$ and $\|f(x) - f(y)\| = K_d\|x - y\|$, then the triangle inequality shows that f takes metric midpoints between x and y to metric midpoints between $f(x)$ and $f(y)$. The next lemma is a quantitative version of this observation for approximate midpoints when K_d is only nearly attained. Note that K_d is a nonincreasing function of d, and put $K_\infty = \lim_{d\to\infty} K_d$.

Lemma 10.11. *Assume that $f : E \to F$ satisfies a Lipschitz condition for large distances and that $K_\infty > 0$. Then for every $0 < \delta < 1/2$ there are points $x, y \in E$ with $\|x - y\|$ as large as we please so that*

$$f\big(\mathrm{Mid}(x, y, \delta)\big) \subset \mathrm{Mid}\big(f(x), f(y), 5\delta\big).$$

Proof. Let $0 < \delta < 1/2$, and assume that d is large enough to ensure that $K_d \leq (1+\delta)K_\infty$. Choose $x, y \in E$ with $\|x - y\| > 4d$ and

$$\|f(x) - f(y)\| \geq K_{4d}\|x - y\|/(1+\delta) \geq K_\infty \|x - y\|/(1+\delta).$$

If $z \in \mathrm{Mid}(x, y, \delta)$, then $\|x - z\| \geq (1-\delta)\|x - y\|/2 > d$. Thus

$$\begin{aligned} \|f(x) - f(z)\| &\leq K_d\|x - z\| \leq K_\infty(1+\delta)^2\|x - y\|/2 \\ &\leq (1+\delta)^3\|f(x) - f(y)\|/2 \leq (1+5\delta)\|f(x) - f(y)\|/2 \end{aligned}$$

because $\delta < 1/2$. A similar estimate holds for $\|f(y) - f(z)\|$. □

Since uniform homeomorphisms between Banach spaces satisfy a Lipschitz condition for large distances, it follows from the lemma that if two Banach spaces are uniformly homeomorphic, then it is impossible for the approximate midpoint sets to be "large" in one space and "small" in the other. In applying this principle, the "size" of sets will be measured either by purely metric means or by a combination of metric and topological considerations. The next theorem gives a quantitative measure that results from such a combination. Heuristically it says that *a uniform homeomorphism between two Banach spaces cannot take a large ball in a subspace of finite codimension in one space into a small neighborhood of a subspace of infinite codimension in the other.* Note that we shall actually only assume that the inverse of the homeomorphism is uniformly continuous.

Theorem 10.12. *Let E and F be Banach spaces, and let $\varphi : E \to F$ be a homeomorphism with a uniformly continuous inverse φ^{-1}. Assume that there are a subspace $E_1 \subset E$ of finite codimension, a subspace $F_1 \subset F$ of infinite codimension, and positive constants α and β such that*

$$\varphi\big(\alpha B(E_1)\big) \subset F_1 + \beta B(F) = \big\{x \in F : \mathrm{dist}(x, F_1) \leq \beta\big\}. \tag{10.4}$$

Then the modulus of continuity of φ^{-1} satisfies $\omega_{\varphi^{-1}}(2\beta) \geq \alpha/4$.

Proof. The proof will follow from the following two facts:

(i) *There is a compact set $A \subset \frac{\alpha}{2}B(E)$ such that whenever a continuous function $f : A \to E$ satisfies $\|f(x) - x\| \leq \alpha/4$ for every $x \in A$, then $f(A) \cap E_1 \neq \emptyset$.*

Indeed, by Proposition 1.19(ii) it is possible to select a continuous right-inverse $\psi : E/E_1 \to E$ to the quotient map $\pi : E \to E/E_1$ so that ψ maps $\frac{3\alpha}{8}B(E/E_1)$ into $\frac{\alpha}{2}B(E)$. We then put $A = \psi\big(\frac{3\alpha}{8}B(E/E_1)\big)$ (which is a compact set because E/E_1 is finite-dimensional). If $f : A \to E$ satisfies $\|f(x) - x\| \leq \alpha/4$ for every $x \in A$, then the map $g(y) = y - \pi f\psi(y)$ maps $\frac{3\alpha}{8}B(E/E_1)$ into itself because

$$\|g(y)\| = \Big\|\pi\Big(\psi(y) - f\big(\psi(y)\big)\Big)\Big\| \leq \alpha/4 < 3\alpha/8.$$

By Brouwer's fixed point theorem g has a fixed point y_0, and it follows that $f\big(\psi(y_0)\big) \in f(A) \cap E_1$.

(ii) *Let B be a compact subset of F. Then there is a point $y \in F$ with $\|y\| < 2\beta$ and $\mathrm{dist}(B + y, F_1) > \beta$.*

Indeed, let B_1 be a finite $\beta/2$-dense set in B. The subspace G spanned by F_1 and B_1 is a proper subspace of F; hence there is a point $y \in F$ with $\|y\| < 2\beta$ and $\operatorname{dist}(y, G) > 3\beta/2$. If $x \in F_1$ and $z \in B$, choose $z_1 \in B_1$ with $\|z - z_1\| < \beta/2$. Then $z_1 - x \in G$; hence $\|z + y - x\| \geq \operatorname{dist}(y, G) - \|z - z_1\| > \beta$.

To prove the theorem, find a compact set $A \subset \frac{\alpha}{2}B(E)$ as in (i). Then use (ii) for the compact set $B = \varphi(A)$, and find $y \in F_1$ with $\|y\| < 2\beta$ for which $\operatorname{dist}\big(\varphi(A) + y, F_1\big) > \beta$.

The map $f(x) = \varphi^{-1}\big(\varphi(x) + y\big)$ maps A into E, and if we had $\omega_{\varphi^{-1}}(2\beta) < \alpha/4$, then f would satisfy

$$\begin{aligned}\|f(x) - x\| &= \|\varphi^{-1}\big(\varphi(x) + y\big) - \varphi^{-1}\big(\varphi(x)\big)\| \\ &\leq \omega_{\varphi^{-1}}(\|y\|) \leq \omega_{\varphi^{-1}}(2\beta) < \alpha/4\end{aligned}$$

for every $x \in A$. By (i) there is an $x_0 \in A$ with $f(x_0) \in E_1$. Also $f(x_0) \in \alpha B(E)$ because $\|f(x_0)\| \leq \|f(x_0) - x_0\| + \|x_0\| < 3\alpha/4 < \alpha$. Condition (10.4) then implies that $\varphi\big(f(x_0)\big) = \varphi(x_0) + y \in F_1 + \beta B(F)$, contradicting $\operatorname{dist}\big(\varphi(A) + y, F_1\big) > \beta$. $\square$

Remark. A useful variation of the theorem is the following: If φ, E, E_1 and F are as above, and if $\omega_{\varphi^{-1}}(2\beta) < \alpha/4$, then there is a compact set $K \subset F$ such that $2\beta B(F) \subset K + \varphi\big(\alpha B(E_1)\big)$. Indeed, let $A \subset \frac{\alpha}{2}B(E)$ be as in (i), and put $K = -\varphi(A)$. Fix an arbitrary $y \in F$ with $\|y\| < 2\beta$, and put $f(x) = \varphi^{-1}\big(\varphi(x) + y\big)$. As in the proof of the theorem $\|f(x) - x\| < \alpha/4$ for all $x \in A$; hence there is an $x_0 \in A$ such that $f(x_0) \in \alpha B(E_1)$, i.e., $y \in -\varphi(x_0) + \varphi\big(\alpha B(E_1)\big) \subset K + \varphi\big(\alpha B(E_1)\big)$.

We are now ready to apply approximate midpoints to problems of uniform classifications. Theorem 10.15 below will give a stronger version of the next result (but only for $p > 1$).

Theorem 10.13. *Let $1 \leq p < \infty$, $p \neq 2$. Then l_p and L_p are not uniformly homeomorphic.*

Proof. We start by observing that if $\varphi : E \to F$ is an onto homeomorphism such that φ^{-1} is uniformly continuous, then there is a positive constant L such that

$$\|\varphi(x) - \varphi(y)\| \geq L\|x - y\| \quad \text{whenever } \|x - y\| \geq 1. \tag{10.5}$$

We now distinguish three cases according to the value of p.

(i) $p = 1$. Although this is actually covered by the next case, we give a separate proof. Assume that $\varphi : L_1 \to l_1$ is a uniform homeomorphism, and let L be as in (10.5). By Lemma 10.11 we can find for every $0 < \delta < 1/2$ points $x, y \in L_1$ with $\|x - y\| \geq 2$ and $\varphi\big(\operatorname{Mid}(x, y, \delta)\big) \subset \operatorname{Mid}\big(\varphi(x), \varphi(y), 5\delta\big)$.

By translations we can assume that $y = -x$ and that $\varphi(-x) = -\varphi(x)$. In particular, since $\|x - (-x)\| \geq 2$, it follows that $\|\varphi(x)\| \leq K_2\|x\|$, where K_2 is the Lipschitz constant of φ for distances larger than 2. By Proposition 10.10(iv) there is an infinite sequence $\{y_j\} \subset \operatorname{Mid}(x, -x, \delta)$ such that $\|y_i - y_j\| = \|x\| \geq 1$. Combining with the right-hand side in part (i) of Proposition 10.10 for $p = 1$, we obtain that there is a finite-dimensional subspace $E_0 \subset l_1$ such that

$$\varphi(y_j) \in E_0 + 3 \cdot 5\delta\|\varphi(x)\|B(l_1) \subset E_0 + 15\delta K_2\|x\|B(l_1).$$

Write $\varphi(y_j) = z_j' + z_j''$ where $z_j' \in E_0$ and $\|z_j''\| \leq 15\delta K_2\|x\|$. Since E_0 is finite-dimensional, and since $\{z_j'\}$ is bounded, there are $i \neq j$ with $\|z_i' - z_j'\| < \delta K_2\|x\|$. By (10.5) we obtain that

$$L\|x\| = L\|y_i - y_j\| \leq \|\varphi(y_i) - \varphi(y_j)\| \leq 31\,\delta K_2\|x\|$$

which is impossible for $\delta < L/31K_2$.

(ii) $1 \le p < 2$. Let $\varphi : L_p \to l_p$ be a uniform homeomorphism satisfying (10.5). Since L_p contains an isometric copy of l_2 (e.g., the span of independent standard Gaussian variables; see Appendix D), we obtain that there is a (nonsurjective) map $\psi : l_2 \to l_p$ which satisfies (10.5) and a Lipschitz condition for large distances.

Fix $\delta > 0$ and choose d such that $\delta^{1/2}d > 2$. As in the case $p = 1$, there is (after translations) a point $x \in l_2$ with $\|x\| \ge d$ for which

$$\psi\big(\mathrm{Mid}(x, -x, \delta)\big) \subset \mathrm{Mid}\big(\psi(x), -\psi(x), 5\delta\big).$$

By the left-hand side in Proposition 10.10(i) for l_2 we can find an infinite sequence $\{y_j\} \subset \mathrm{Mid}(x, -x, \delta)$ such that $\|y_i - y_j\| \ge \frac{1}{2}\delta^{1/2}\|x\|$. Thus $\|y_i - y_j\| \ge 1$ by the choice of d, and $\|\psi(y_i) - \psi(y_j)\| \ge L\|y_i - y_j\| \ge L\delta^{1/2}\|x\|/2$. The right-hand side in Proposition 10.10(i) for l_p and the finite-dimensionality of E_0 give, as for $p = 1$, that there are $i \ne j$ with $\|\psi(y_i) - \psi(y_j)\| \le 3(15p\delta)^{1/p}\|\psi(x)\| \le 3(15p\delta)^{1/p}K_2\|x\|$. Thus $3(15p\delta)^{1/p}K_2 \ge L\delta^{1/2}/2$, which is impossible for small enough δ because $p < 2$.

(iii) $p > 2$. Let $\varphi : l_p \to L_p$ be a uniform homeomorphism satisfying (10.5). For each $\delta > 0$ and $d > 1$ find, as before, a point $x \in l_p$ for which $\|x\| > d$ and $\varphi\big(\mathrm{Mid}(x, -x, \delta)\big) \subset \mathrm{Mid}\big(\varphi(x), -\varphi(x), 5\delta\big)$.

By parts (i) and (v) of Proposition 10.10 there are a subspace $E_1 \subset l_p$ of finite codimension such that

$$\mathrm{Mid}(x, -x, \delta) \supset \frac{1}{2}\delta^{1/p}\|x\|B(E_1)$$

and a subspace F_1 of L_p of infinite codimension such that

$$\mathrm{Mid}(\varphi(x), -\varphi(x), 5\delta) \subset F_1 + C(5\delta)^{1/2}\|\varphi(x)\|B(L_p).$$

Thus $\varphi\big(\frac{1}{2}\delta^{1/p}\|x\|B(E_1)\big) \subset F_1 + K\delta^{1/2}\|x\|B(L_p)$ (where the constant K depends only on p). By Theorem 10.12 $\omega_{\varphi^{-1}}\big(2K\delta^{1/2}\|x\|\big) \ge \delta^{1/p}\|x\|/4$. Since (10.5) implies that $\omega_{\varphi^{-1}}(t) \le t/L$ whenever $t > 1$, it follows that if d is chosen in such a way that $2K\delta^{1/2}\|x\| \ge 2K\delta^{1/2}d > 1$, then $\omega_{\varphi^{-1}}\big(2K\delta^{1/2}\|x\|\big) \le 2K\delta^{1/2}\|x\|/L$. Hence $2K\delta^{1/2}/L \ge \delta^{1/p}/8$, which is impossible for small enough δ because $p > 2$. □

The proofs in the next proposition use the same ideas.

Proposition 10.14. *Let E and F be uniformly homeomorphic Banach spaces.*

(i) *If E has an unconditional basis with an upper p-estimate, then no quotient of F can have an unconditional basis which has a lower r-estimate with $r < p$.*

(ii) *If E has an unconditional basis with a lower r-estimate for $r < 2$, then F does not contain a subspace isomorphic to a Hilbert space.*

(iii) *If E is superreflexive and has an unconditional basis with an upper p-estimate for some $p > 2$, then F does not contain a subspace isomorphic to a Hilbert space.*

Proof. (i) Let $\varphi : E \to F$ be a uniform homeomorphism satisfying (10.5), and assume that $Q : F \to G$ is a quotient map, where G has an unconditional basis with a lower r-estimate for some $r < p$. By renorming we may assume that the unconditionality constants and the upper or the lower estimate constants of the bases are equal to one.

By (10.5), and since Q is linear, the limit of the Lipschitz constants for large distances of the uniformly continuous function $Q \circ \varphi$ satisfies $K_\infty > 0$. Using translations as before, Lemma 10.11 gives for each $\delta > 0$ and $d > 1$ a point $x \in E$ with $\|x\| > d$ and $Q\varphi\big(\mathrm{Mid}(x, -x, \delta)\big) \subset \mathrm{Mid}\big(Q\varphi(x), -Q\varphi(x), 5\delta\big)$.

By Proposition 10.10(ii), there is a subspace E_1 of E of finite codimension such that $\mathrm{Mid}(x, -x, \delta) \supset \frac{1}{2}\delta^{1/p}\|x\|B(E_1)$, and a subspace G_1 of G of infinite codimension such that there is an absolute constant K for which

$$\begin{aligned}\mathrm{Mid}(Q\varphi(x), -Q\varphi(x), 5\delta) &\subset G_1 + (3r5\delta)^{1/r}\|Q\varphi(x)\|B(G)\\ &\subset G_1 + K\delta^{1/r}\|x\|B(G).\end{aligned}$$

Putting $F_1 = Q^{-1}G_1$, we obtain that $\varphi\big(\frac{1}{2}\delta^{1/p}\|x\|B(E_1)\big) \subset F_1 + K\delta^{1/r}\|x\|B(F)$, and hence $\omega_{\varphi^{-1}}\left(2K\delta^{1/r}\|x\|\right) \geq \delta^{1/p}\|x\|/8$ by Theorem 10.12. Since $\omega_{\varphi^{-1}}(t) \leq t/L$ whenever $t > 1$ by (10.5), it follows that if d satisfies $2K\delta^{1/r}\|x\| \geq 2K\delta^{1/2}d > 1$, then $2K\delta^{1/r}/L \geq \delta^{1/p}/8$, which is impossible for small enough δ because $p > r$.

(ii) The proof is the same as in the case $1 \leq p < 2$ of Theorem 10.13.

(iii) E has type 2 by Theorem G.2(iii); hence by Theorem 10.6 so has F. By Theorem G.2(ii) every subspace of F which is isomorphic to a Hilbert space would necessarily be complemented; i.e., l_2 (which, of course, satisfies a lower 2-estimate) would be a quotient of F, contradicting part (i). □

We now show that the uniform structure of l_p spaces ($1 < p < \infty$) and of some of their direct sums determines their linear structure.

Theorem 10.15. *Let $1 < p < \infty$, and let F be uniformly homeomorphic to l_p. Then F is isomorphic to l_p. More generally, if $1 < p_1 < \ldots < p_n < 2$ or $2 < p_1 < \ldots < p_n < \infty$, and if F is uniformly homeomorphic to $E = \sum_{i=1}^{n} \oplus l_{p_i}$, then F is isomorphic to E.*

Proof. Assume first that F is uniformly homeomorphic to l_p. By Theorem 10.5 F is an $\mathcal{L}_p$-space. If $p = 2$, then F is isomorphic to a Hilbert space by Theorem F.2(ii). Otherwise, Proposition 10.14(ii) (for $p < 2$) and (iii) (for $p > 2$) give that F does not contain a subspace isomorphic to l_2; hence F is isomorphic to l_p by Theorem F.2(vi).

For the second part take for simplicity $n = 2$, and assume that F is uniformly homeomorphic to $E = l_p \oplus l_q$. By Proposition 10.4 F is superreflexive, and there is an ultrapower $E_{\mathcal{U}}$ of E such that F is isomorphic to a complemented subspace of $E_{\mathcal{U}}$. Write $E_{\mathcal{U}}$ as a direct sum $E_{\mathcal{U}} = L_p(\mu) \oplus L_q(\mu)$ for some measure μ. Then every separable subspace W of $L_p(\mu) \oplus L_q(\mu)$ is contained in a separable space which is isometric to $L_p(\nu) \oplus L_q(\nu)$ for some measure ν. (Indeed, ν may be taken to be the restriction of μ to the σ-algebra generated by the elements of W.) Since separable $L_r(\nu)$-spaces are isometric to complemented subspaces of L_r, it follows that there is a linear isomorphic embedding $J : F \to L_p \oplus L_q$ such that the range $J(F)$ is complemented.

Assume now that $p, q > 2$, and let P and Q be the projections of $L_p \oplus L_q$ on L_p and L_q respectively. By Proposition 10.14(iii) F does not contain an isomorphic copy of l_2; hence by Theorem F.2(vii) the operators PJ and QJ factor through l_p and l_q respectively. Thus J factors through $l_p \oplus l_q$, and since $J(F)$ is complemented, it follows that F is isomorphic to a complemented subspace of $l_p \oplus l_q$. By Theorem 10.41 (in the Notes and Remarks section) and by the remarks following it, F is isomorphic to l_p, l_q or $l_p \oplus l_q$. But the first two possibilities are excluded by the first part of the theorem.

If $p, q < 2$, we use duality. The space F^* is a complemented subspace of $L_{p'} \oplus L_{q'}$ (where $1/p + 1/p' = 1/q + 1/q' = 1$), and it cannot contain an isomorphic copy of l_2. Indeed, it has type 2, and if it contained such a subspace, it would have been complemented by Theorem G.2(ii). But then F would also contain a copy of l_2, contradicting Proposition 10.14(ii). It follows, as above, that F^* is isomorphic to $l_{p'} \oplus l_{q'}$, and hence F is isomorphic to $l_p \oplus l_q$. □

We now turn to spaces uniformly homeomorphic to c_0, and we first give some background from the linear theory. References, more details, and further notation are given in the Notes and Remarks section.

Every countable compact metric space is homeomorphic to the set of ordinals $[1, \alpha] = \{\beta : 1 \leq \beta \leq \alpha\}$ for some countable ordinal α, equipped with the order topology. The space of continuous functions on $[1, \alpha]$ is denoted by $C(\alpha)$. Denote the first infinite ordinal by ω. Then c_0 is identified with the codimension-one subspace of $C(\omega)$ consisting of the functions vanishing at the point ω, and it is isomorphic to $C(\omega)$. Fix $\alpha < \beta$; then $C(\alpha)$ and $C(\beta)$ are linearly isomorphic if and only if $\beta \leq \alpha^\omega$. Thus $C(\omega^\omega)$ is the "next simplest" such space after c_0.

Any separable complemented subspace E of a $C(K)$ space contains a complemented subspace isomorphic to c_0. If such an E is not actually isomorphic to c_0, then it has a quotient space isomorphic to $C(\omega^\omega)$.

Theorem 10.16. *Let E be uniformly homeomorphic to c_0. Then E does not have a quotient space isomorphic to $C(\omega^\omega)$. In particular, if E is isomorphic to a complemented subspace of a $C(K)$ space, then E is isomorphic to c_0.*

Proof. We start by computing approximate midpoints sets in $C(\omega^\omega)$. Assume that $f \in C(\omega^\omega)$ is such that there is a point $t_0 \in [1, \omega^\omega]'$, the first derived set of $[1, \omega^\omega]$, such that $|f(t_0)| > (1 - \eta)\|f\|$ for some $\eta > 0$. Since t_0 is not isolated, there is an infinite set $B \subset [1, \omega^\omega]$ such that $|f(t)| > (1 - \eta)\|f\|$ for every $t \in B$. It follows that if $g \in \operatorname{Mid}(f, -f, \eta)$, then its restriction to B satisfies $\|g_{|B}\| \leq 2\eta\|f\|$. Put $G_1 = \{g \in C(\omega^\omega) : g_{|B} \equiv 0\}$. Then $\operatorname{Mid}(f, -f, \eta) \subset G_1 + 2\eta\|f\|B(C(\omega^\omega))$, and G_1 has infinite codimension because B is infinite.

Now let $\varphi : c_0 \to E$ be an onto uniform homeomorphism, and assume that there is a quotient map $Q : E \to C(\omega^\omega)$. The nth derived set $K_n = [1, \omega^\omega]^{(n)}$ is homeomorphic to $[1, \omega^\omega]$ for every n, and we put $R_n : C(\omega^\omega) \to C(K_n)$ for the restriction map.

Fix $\delta > 0$. The Lipschitz constants $K_d(n)$ of $R_n Q\varphi$ for distances larger than d are bounded away from zero and are nonincreasing in both d and n. We normalize φ so that their greatest lower bound is 1, and by replacing $C(\omega^\omega)$ by $C(K_N)$ for large enough N, we may assume that there is a $d > 0$ such that the Lipschitz constants for distances larger than d of both $Q\varphi$ and $R_1 Q\varphi$ are smaller than $1 + \delta$.

Let $x, y \in c_0$ satisfy $\|x - y\| > 2d$ and $\|R_1 Q\varphi(x) - R_1 Q\varphi(y)\| > (1 - \delta)\|x - y\|$. The choice of d implies that we also have $\|Q\varphi(x) - Q\varphi(y)\| \leq (1 + \delta)\|x - y\|$. By translations we may assume that $y = -x$ and that $\varphi(-x) = -\varphi(x)$. Hence $\|R_1 Q\varphi(x)\| > \frac{1-\delta}{1+\delta}\|Q\varphi(x)\| > (1 - 5\delta)\|Q\varphi(x)\|$; i.e., there is a point $t_0 \in [1, \omega^\omega]'$ with $|Q\varphi(x)(t_0)| > (1 - 5\delta)\|Q\varphi(x)\|$. By Lemma 10.11 and by the first part of the proof (with $\eta = 5\delta$) there is a $G_1 \subset C(\omega^\omega)$ of infinite codimension such that

$$Q\varphi(\operatorname{Mid}(x, -x, \delta)) \subset \operatorname{Mid}(Q\varphi(x), -Q\varphi(x), 5\delta) \subset G_1 + 10\delta\|Q\varphi(x)\|B(C(\omega^\omega)).$$

The subspace $F_1 = Q^{-1}G_1$ has infinite codimension in E. Since $\|Q\| = 1$, we get from Proposition 10.10(iii) that there is a subspace E_1 of c_0 of finite codimension such that $\varphi\big(\|x\|B(E_1)\big) \subset F_1 + 10\delta\|\varphi(x)\|B\big(C(\omega^\omega)\big)$. Since δ was arbitrary, this contradicts the uniform continuity of φ^{-1} by Theorem 10.12. □

It is unknown whether the uniform structure of c_0 determines its linear structure. The next result shows that this is true for the Lipschitz structure.

Theorem 10.17. *If a Banach space F is Lipschitz equivalent to a subspace of c_0, then it is actually isomorphic to a subspace of c_0. Moreover, if F is Lipschitz equivalent to c_0, then it is isomorphic to it.*

Proof. Let E be a subspace of c_0, and let $\varphi : E \to F$ be a Lipschitz equivalence from E onto F, normalized so that $\|x-y\| \le \|\varphi(x)-\varphi(y)\| \le C\|x-y\|$. Define a new norm on F^* by

$$|\!|\!| x^* |\!|\!| = \sup\Big\{\frac{\langle x^*, \varphi(u)-\varphi(v)\rangle}{\|u-v\|} : u, v \in F \text{ and } u \neq v\Big\}.$$

Then $\|x^*\| \le |\!|\!| x^* |\!|\!| \le C\|x^*\|$ for every $x^* \in F^*$ and $|\!|\!|\cdot|\!|\!|$ is w^*-lower semi-continuous, hence a dual of an equivalent norm on F, which we also denote by $|\!|\!|\cdot|\!|\!|$.

The first part of the theorem will follow from Theorem E.4(ii) once we show that $\lim |\!|\!| x_n^* + x^* |\!|\!| \ge 1 + \varepsilon/10C$ for every $x^* \in F^*$ with $|\!|\!| x^* |\!|\!| = 1$ and $x_n^* \in F^*$ such that $|\!|\!| x_n^* |\!|\!| \ge \varepsilon$ and w^*-$\lim x_n^* = 0$.

Let $\delta > 0$ (to be specified later), and choose vectors $u, v \in E$ such that $\langle x^*, \varphi(u)-\varphi(v)\rangle \ge (1-\delta)\|u-v\|$. By translations and rescaling we may assume that $v = -u$, $\varphi(v) = -\varphi(u)$ and $\|u\| = 1$. Thus $\langle x^*, \varphi(u)\rangle \ge (1-\delta)$.

Since E is a subspace of c_0, there is a subspace $E_1 \subset E$ of finite codimension so that $\|u+w\| \le 1+\delta$ for every $w \in B(E_1)$. Since φ^{-1} is 1-Lipschitz, it follows from the remark after Theorem 10.12 (with $\alpha = 1$ and $\beta = 1/10$) that there is a compact set $K \subset F$ such that $B(F) \subset K + 5\varphi\big(B(E_1)\big)$.

Since $x_n^* \to 0$ uniformly on the compact set K, we may assume by discarding finitely many x_n^*'s that for each n there is a $w_n \in B(E_1)$ such that

$$\langle x_n^*, -\varphi(w_n)\rangle \ge 5^{-1}\|x_n^*\| - \delta \ge (5C)^{-1}|\!|\!| x_n^* |\!|\!| - \delta \ge (5C)^{-1}\varepsilon - \delta.$$

Note that $\langle x^*, \varphi(w_n)\rangle \le 2\delta$. Indeed, $-\varphi(-u) = \varphi(u)$ and $\|u+w_n\| \le 1+\delta$ imply that $1 = |\!|\!| x^* |\!|\!| \ge \langle x^*, \varphi(w_n)+\varphi(u)\rangle/\|w_n+u\| \ge (1+\delta)^{-1}\big(\langle x^*, \varphi(w_n)\rangle + 1 - \delta\big)$.

Finally, $\|u-w_n\| \le 1+\delta$ and $\big|\langle x_n^*, \varphi(u)\rangle\big| \le \delta$ for large enough n give

$$\begin{aligned}
|\!|\!| x^* + x_n^* |\!|\!| &\ge \langle x^* + x_n^*, \varphi(u) - \varphi(w_n)\rangle/\|u-w_n\| \\
&\ge (1+\delta)^{-1}\Big(\langle x^*, \varphi(u)\rangle + \langle x_n^*, -\varphi(w_n)\rangle + \langle x_n^*, \varphi(u)\rangle - \langle x^*, \varphi(w_n)\rangle\Big) \\
&\ge (1+\delta)^{-1}\big(1 - \delta + (5C)^{-1}\varepsilon - \delta - \delta - 2\delta\big) \ge 1 + \varepsilon/10C
\end{aligned}$$

provided δ is small enough.

For the second part of the theorem, observe that F is an $\mathcal{L}_\infty$-space by Theorem 10.5 and that it is isomorphic to a subspace of c_0 by the first part of the theorem. Thus F is isomorphic to c_0 by Theorem F.2(viii). □

3. Discrete Nets

Definition 10.18. *An (a, b)-net in a metric space X is a subset $\mathcal{N}$ such that $d(z, z') \ge a$ for every $z \neq z'$ in $\mathcal{N}$ and $d(x, \mathcal{N}) < b$ for every $x \in X$. When $a = 1$ and $b \le \lambda$ we say that $\mathcal{N}$ is a λ-net.*

Nets give a discrete approximation to the geometry of X, and it is thus natural to investigate the relations between the metric structure of the whole space X and the structure of the nets it contains. If E and F are Banach spaces, then a uniform homeomorphism between them is Lipschitz for large distances; hence it carries nets in E to nets in F, and its restriction to such a net is a Lipschitz equivalence. The following two questions are therefore very natural:

(1) Do nets determine the structure of the spaces? More precisely, if E and F have Lipschitz equivalent nets, are they uniformly homeomorphic?

(2) Conversely, does the structure of the spaces determine that of nets in them? In particular, given two nets in the same Banach space E, are they necessarily Lipschitz equivalent?

In this section we give partial positive answers to the first question (the general question seems to be open). We also give a positive answer to (2) for infinite-dimensional spaces. We show, however, that the answer to (2) is negative in finite-dimensional spaces.

The next result is analogous to Theorem 10.1.

Proposition 10.19. (i) *Let $\mathcal{N}$ and $\mathcal{M}$ be Lipschitz equivalent nets in the two Banach spaces E and F respectively, and let $\mathcal{U}$ be a free ultrafilter on $\mathbb{N}$. Then the ultrapowers $(E)_{\mathcal{U}}$ and $(F)_{\mathcal{U}}$ are Lipschitz equivalent.*

(ii) *More generally, assume that $\mathcal{N}$ is Lipschitz equivalent to a subset of $\mathcal{M}$ on which there is a Lipschitz retraction from all of $\mathcal{M}$. Then there is a Lipschitz embedding of $(E)_{\mathcal{U}}$ into $(F)_{\mathcal{U}}$ for which there is a Lipschitz retraction from $(F)_{\mathcal{U}}$ onto the range of the embedding.*

Proof. We prove only (i). Let $\varphi : \mathcal{N} \to \mathcal{M}$ be the Lipschitz equivalence. The maps $\varphi_n : n^{-1}\mathcal{N} \to n^{-1}\mathcal{M}$ defined by $\varphi_n(x) = n^{-1}\varphi(nx)$ are Lipschitz equivalences, and their Lipschitz constants (and those of their inverses) are uniformly bounded. Thus $(\varphi_n)_{\mathcal{U}}$ is a Lipschitz equivalence between the dense sets $\left(n^{-1}\mathcal{N}\right)_{\mathcal{U}} \subset (E)_{\mathcal{U}}$ and $\left(n^{-1}\mathcal{M}\right)_{\mathcal{U}} \subset (F)_{\mathcal{U}}$ and extends by continuity to a Lipschitz equivalence between the whole spaces. □

By Proposition 10.19, or by arguments similar to its proof, the results of Section 1 remain true when the assumption of the existence of homeomorphisms which are uniform or satisfy a Lipschitz condition for large distances is replaced by the existence of Lipschitz equivalence between nets. The same is true for most of the results of Section 2. Some care is needed however when Theorem 10.12 is used. Rather than formulating a general discrete version of Theorem 10.12, we just consider the spaces l_p for $p < \infty$. We denote the span of the first N unit vectors in l_p by E_N, the natural basis projections on E_N by P_N, and $(I - P_N)l_p$ by E^N.

Proposition 10.20. *Let $\lambda \geq 1$, and let $\mathcal{N}$ and $\mathcal{M}$ be two λ-nets in the Banach spaces l_p and F respectively. Let $K \geq 1$, and let $\varphi : \mathcal{N} \to \mathcal{M}$ be a Lipschitz equivalence so that $\|x - y\| \leq \|\varphi(x) - \varphi(y)\| \leq K\|x - y\|$ for every $x, y \in \mathcal{N}$. Then there is a constant α_0 such that if F_1 is a subspace of F of infinite codimension, then for every $\alpha > \alpha_0$ and for every integer N there is a point $z_0 \in \mathcal{N}$ such that* $\operatorname{dist}\left(z_0, \alpha B(E^N)\right) \leq \alpha/10$ *and* $\operatorname{dist}\left(\varphi(z_0), F_1\right) \geq \alpha/10$.

Proof. We consider only $p > 2$. The proof for $p < 2$ is similar. By Proposition 9.2 there are constants K_1 and K_2 such that the Mazur map $\psi : l_p \to l_2$; i.e.,

$\big(\psi(x)\big)_n = \|x\|^{1-p/2}|x_n|^{p/2}\text{sign}x_n$ is Lipschitz with constant K_1, and such that the restriction of ψ^{-1} to $\alpha B(l_2)$ is $2/p$-Hölder with constant $K_2\alpha^{1-2/p}$.

We shall first prove the following claim.

$(*)$ *There is an* $\alpha_0 = \alpha_0(p, K, \lambda)$ *such that if* $\alpha > \alpha_0$ *and if* N *is a positive integer, then there is a finite set* $A \subset \mathcal{N} \bigcap \frac{3\alpha}{4} B(l_p)$ *with the following property: Whenever* $f : A \to l_p$ *is a Lipschitz function with constant* $2\lambda + K$ *satisfying* $\|f(x) - x\| \le \alpha/4$ *for every* $x \in A$*, then* $\text{dist}\big(f(A), E^N\big) \le \alpha/10$.

Since E_N is finite-dimensional there is a finite set $A \subset \mathcal{N} \bigcap \frac{3\alpha}{4} B(l_p)$ (provided $\alpha > 4\lambda$) such that for every $x \in \frac{\alpha}{2} B(E_N)$ there is a $y \in A$ with $\|x - y\| < \lambda$.

Given $f : A \to E$, we first use ψ to extend f to $\frac{3\alpha}{4} B(E_N)$. Indeed, $\psi \circ f \circ \psi^{-1}$ is uniformly continuous on $\psi(A)$. By Theorem 1.12 it extends to a function $\tilde{f}$ on all of l_2 with the same modulus of continuity as $\psi \circ f \circ \psi^{-1}$. The extension of f is then taken as $\psi^{-1} \circ \tilde{f} \circ \psi$ on $\frac{3\alpha}{4} B(E_N)$, and it is $(2/p)^2$-Hölder there with constant $C\alpha^{1-(2/p)^2}$, where C depends on p, K and λ.

We now choose $\alpha_0 \ge 40\lambda$ so that $C\alpha_0^{1-(2/p)^2}\lambda^{(2/p)^2} \le \alpha_0/10$. From now on we shall assume that $\alpha > \alpha_0$, and thus in particular $\|f(x) - f(y)\| < \alpha/10$ whenever $\|x\|, \|y\| \le 3\alpha/4$ and $\|x - y\| < \lambda$.

Note that the extended function f satisfies $\|f(x) - x\| \le \alpha/2$ for every x in $\frac{\alpha}{2} B(E_N)$. Indeed, choose $y \in A$ with $\|x - y\| < \lambda$. Then, by the choice of α_0

$$\begin{aligned} \|f(x) - x\| &\le \|x - y\| + \|y - f(y)\| + \|f(y) - f(x)\| \\ &\le \lambda + \alpha/4 + \alpha/10 < \alpha/2. \end{aligned}$$

Since $\|P_N\| = 1$, it follows that $g(x) = P_N\big(x - f(x)\big)$ maps $\frac{\alpha}{2} B(E_N)$ into itself, and thus has a fixed point x_0 there. In particular $P_N f(x_0) = 0$, i.e., $f(x_0) \in E^N$. Let $y_0 \in A$ be such that $\|x_0 - y_0\| < \lambda$. Then the choice of α_0 gives that

$$\text{dist}\big(f(A), E^N\big) \le \|f(y_0) - f(x_0)\| < \alpha/10.$$

We now proceed with the proof of the proposition. Let N be an integer, α_0 as in $(*)$, $\alpha > \alpha_0$ and A the set appearing in $(*)$. As in Theorem 10.12, and since $\lambda < \alpha/8$, there is a point $y \in F$ with $\|y\| + \lambda < \alpha/4$ and $\text{dist}\big(\varphi(A) + y, F_1\big) > \alpha/8$. For each $x \in A$ choose a point $y(x) \in \mathcal{M}$ such that $\|\varphi(x) + y - y(x)\| < \lambda$, and put $f(x) = \varphi^{-1}\big(y(x)\big)$. Then f is $(2\lambda + K)$-Lipschitz. Indeed, $\|x_1 - x_2\| \ge 1$ for $x_1 \ne x_2 \in A$, and φ^{-1} is 1-Lipschitz; hence

$$\begin{aligned} \|f(x_1) - f(x_2)\| &\le \|y(x_1) - y(x_2)\| \le 2\lambda + \|\varphi(x_1) - \varphi(x_2)\| \\ &\le (2\lambda + K)\|x_1 - x_2\|. \end{aligned}$$

Also

$$\|f(x) - x\| = \|\varphi^{-1}\big(y(x)\big) - \varphi^{-1}\big(\varphi(x)\big)\| \le \|y(x) - \varphi(x)\| = \|y\| + \lambda < \alpha/4.$$

By $(*)$ there is a point $x_0 \in A$ such that $\text{dist}\big(f(x_0), E^N\big) \le \alpha/10$, and we take $z_0 = f(x_0)$. Then $\|z_0\| \le \|x_0\| + \|x_0 - f(x_0)\| \le 3\alpha/4 + \alpha/4 = \alpha$, and thus also $(I - P_N)z_0 \in \alpha B(E^N)$. Hence $\text{dist}\big(z_0, \alpha B(E^N)\big) = \text{dist}(z_0, E^N) \le \alpha/10$. On the other hand, $\varphi(z_0) = \varphi\big(f(x_0)\big) = y(x_0)$; hence $\alpha_0 \ge 40\lambda$ gives that

$$\begin{aligned} \text{dist}\big(\varphi(z_0), F_1\big) &= \text{dist}\big(y(x_0), F_1\big) \ge \text{dist}\big(\varphi(x_0) + y, F_1\big) - \lambda \\ &\ge \alpha/8 - \lambda \ge \alpha/10. \end{aligned}$$

□

The following theorem is one example of the various possible classification results analogous to those of Section 2. We only sketch its proof.

Theorem 10.21. *Let $1 < p < \infty$. If a Banach space F contains a net which is Lipschitz equivalent to a net in l_p, then F is isomorphic to l_p.*

Proof. For $p \le 2$ Theorem 10.12 was not used in the proof of the first part of Theorem 10.15; hence the proof here is completely analogous and will not be given. For $p > 2$ we only indicate the changes in the proof of Proposition 10.14(i) (with E there taken as l_p). The rest of the proof is again completely analogous to the proof of Theorem 10.15.

Let $\mathcal{N}$ and $\mathcal{M}$ be λ-nets in l_p and F respectively, and let $\varphi : \mathcal{N} \to \mathcal{M}$ be a Lipschitz equivalence normalized so that $\|x - y\| \le \|\varphi(x) - \varphi(y)\| \le K\|x - y\|$ for every $x, y \in \mathcal{N}$. Let α_0 be as in Proposition 10.20 above. Choose d and δ such that $\delta^{1/p}d > \alpha_0$. As in Proposition 10.14(i), there are a subspace $F_1 \subset F$ of infinite codimension, a vector x with $\|x\| \ge d$, and an integer N such that $\varphi(\mathcal{N} \bigcap \delta^{1/p}\|x\|B(E^N)) \subset F_1 + C\delta^{1/r}\|x\|B(F)$ for some constant C which depends only on p and r. (We leave it to the reader to supply the easy details needed to overcome the fact that the net $\mathcal{N}$ is not invariant under translation and is not symmetric with respect to the origin.) But this contradicts Proposition 10.20 for $\alpha = \delta^{1/p}\|x\|$ if δ is chosen to be small enough so that $\delta^{1/p-1/r} > 100C$. □

We now discuss question (2) posed in the beginning of this section, and we first consider nets in infinite-dimensional spaces.

Proposition 10.22. *Let E be an infinite-dimensional Banach space. Then any two nets in E are Lipschitz equivalent.*

We shall need the following simple lemma.

Lemma 10.23. *Let X be a metric space, and let $\mathcal{N}_1$ and $\mathcal{N}_2$ be two nets in X. Let $T : \mathcal{N}_1 \to \mathcal{N}_2$ be a bijective map such that $\sup\{d(x, Tx) : x \in \mathcal{N}_1\} < \infty$. Then T is a Lipschitz equivalence between the two nets.*

Proof. By symmetry it is enough to check that T^{-1} is Lipschitz. Assume that $\mathcal{N}_2$ is an (a, b)-net in X, and put $c = \sup\{d(x, Tx) : x \in \mathcal{N}_1\}$. Fix two points $x \neq y$ in $\mathcal{N}_1$, and we distinguish two cases:
If $d(x, y) \le 4c$, then

$$d(Tx, Ty) \ge a \ge \frac{a}{4c} d(x, y)$$

and if $d(x, y) \ge 4c$, then

$$d(Tx, Ty) \ge d(x, y) - d(x, Tx) - d(y, Ty) \ge d(x, y) - 2c \ge d(x, y)/2.$$

Thus T^{-1} is Lipschitz with constant $\max\{2, 4c/a\}$. □

Proof of Proposition 10.22. Assume that $\mathcal{N}_1$ and $\mathcal{N}_2$ are two (a, b)-nets in E, and note that if B is a ball in E of radius $2b$, then $B \cap \mathcal{N}_1$ and $B \cap \mathcal{N}_2$ are infinite sets of the same cardinality as the density character of E.

Let $\{y_\alpha\}$ be a maximal $5b$-separated set in E, and let $\{C_\alpha\}$ be a partition of E into disjoint sets, with $\operatorname{diam}(C_\alpha) \le 10b$ and $C_\alpha \supset B(y_\alpha, 2b)$ for each α. (Any x which does not belong to $\bigcup B(y_\alpha, 2b)$ is assigned to some C_α such that $\|x - y_\alpha\| \le 5b$.)

Hence, for each α the two intersections $C_\alpha \cap \mathcal{N}_1$ and $C_\alpha \cap \mathcal{N}_2$ are infinite and have the same cardinality. It follows that there are bijections T_α of $C_\alpha \cap \mathcal{N}_1$ onto $C_\alpha \cap \mathcal{N}_2$. We then define $T : \mathcal{N}_1 \to \mathcal{N}_2$ by $Tx = T_\alpha x$ whenever $x \in C_\alpha$. Since $\operatorname{diam}(C_\alpha) \leq 10b$ for each α, it follows that $d(x, Tx) \leq 10b$ for all $x \in \mathcal{N}_1$; hence T is a Lipschitz equivalence by the lemma. ♢

Surprisingly the situation is completely different when E is finite-dimensional with $\dim(E) > 1$. We first introduce the notion of homogeneous α-Hölder maps, which is the appropriate notion in this context for the study of α-Hölder equivalence. It will turn out that nets in $\mathbb{R}^n$ are not necessarily Lipschitz equivalent, but that they are α-Hölder equivalent for every $\alpha < 1$.

Definition 10.24. *Let E and F be Banach spaces. A map φ from a subset $A \subset E$ to F is said to be homogeneous α-Hölder with constant C if*

$$\|\varphi(x) - \varphi(y)\| \leq CR^{1-\alpha}\|x - y\|^\alpha$$

for every $R > 0$ and for every $x, y \in A$ satisfying $\|x\|, \|y\| \leq R$.

For $\alpha = 1$, this is just the Lipschitz condition. If φ is a homogeneous Hölder map, then so is $r\varphi(\cdot/r)$ for every $r > 0$ (and with the same exponent and constant). If both φ and its inverse are homogeneous Hölder, then $\|\varphi(x)\| \sim \|x\|$.

To simplify the notation we shall consider only nets in the plane. Only simple modifications are needed for $\mathbb{R}^n$ with general n. We shall also assume that all the nets contain 0 and that all the mappings map 0 to 0.

Denote the normalized Lebesgue measure on $\mathbb{R}^2$ by m and the unit disk in $\mathbb{R}^2$ by B. We first show that the problem is equivalent to that of finding Lipschitz (or Hölder) homeomorphisms which map any prescribed measure of the form $f dm$ (where f is bounded from above and below) to the Lebesgue measure on the image.

Proposition 10.25. *Fix $0 < \alpha \leq 1$. The following two statements are equivalent:*

(i) *For every $\lambda \geq 1$ there is a $C = C(\alpha, \lambda)$ with the following property: If f is any measurable function satisfying $1 \leq f(x) \leq \lambda$ for all $x \in B$, then there is a homeomorphism φ of B into $\mathbb{R}^2$ such that both φ and φ^{-1} are α-homogeneous with constant C and*

$$m\Big(\varphi(A)\Big) = \int_A f\,dm \tag{10.6}$$

for every measurable set $A \subset B$.

(ii) *For every $\lambda \geq 1$ there is a $K = K(\alpha, \lambda)$ with the following property: If $\mathcal{N}$ is any λ-net in $\mathbb{R}^2$, then there is a one-one map ψ of $\mathcal{N}$ onto $\mathbb{Z}^2$ such that both ψ and ψ^{-1} are α-homogeneous with constant K.*

Note that for $\alpha = 1$ the map φ in (i) is differentiable a.e., and condition (10.6) is then equivalent to the existence of a Lipschitz homeomorphism which solves the differential equation $|J(\varphi)| = f$ a.e. (where $J(\varphi)$ is the Jacobian of φ).

Proof. (i)⇒(ii). Note first that if (i) holds, then for every measurable function satisfying $1 \leq f \leq \lambda$ on all of $\mathbb{R}^2$ there is a homogeneous α-Hölder homeomorphism of $\mathbb{R}^2$ onto itself with constant C for which (10.6) holds on all of $\mathbb{R}^2$. Indeed, find for each k a $\varphi_k : B \to \mathbb{R}^2$ satisfying (10.6) for $f_k(x) = f(kx)$. The functions $k\varphi_k(x/k)$ on $\{x : \|x\| \leq k\}$, respectively, and their inverses are homogeneous α-Hölder maps with the same constant C. By the Arzelà-Ascoli theorem they have a subsequence which converges uniformly on compact sets. The limit of the subsequence is the desired homeomorphism, it is onto because its image is both open (by invariance of domain) and closed (as the image under a uniform homeomorphism).

Now let $\mathcal{N}$ be a λ-net, and let $C_z = \{x \in \mathbb{R}^2 : d(x, \mathcal{N}) = d(x, z)\}$ for $z \in \mathcal{N}$ be its Voronoi cells. Define a function f by $f(x) = 1/m(C_z)$ for $x \in C_z$. Then f is bounded from below and above by constants that depend only on λ. Let φ be a homogeneous α-Hölder homeomorphism of $\mathbb{R}^2$ satisfying (10.6) for this f. Then the sets $D_z = \varphi(C_z)$ have measure 1, and their diameter is bounded by $C_1(1+\|z\|^{1-\alpha})$, where C_1 depends only on α and λ.

Let E_y be the square of unit area centered at y. We claim that there is a one-one map ψ from $\mathcal{N}$ onto $\mathbb{Z}^2$ such that $D_z \cap E_{\psi(z)} \neq \emptyset$ for every $z \in \mathcal{N}$. Indeed, consider the subset $\mathcal{R} \subset \mathcal{N} \times \mathbb{Z}^2$ of all pairs (z, y) such that $D_z \cap E_y \neq \emptyset$. For each $z \in \mathcal{N}$ (resp. $y \in \mathbb{Z}^2$) there are only finitely many y's (resp. z's) with $D_z \cap E_y \neq \emptyset$ (i.e., $(z, y) \in \mathcal{R}$). Moreover, let $A \subset \mathcal{N}$ be a finite set, and let B be the set $\{y : (z, y) \in \mathcal{R} \text{ for some } z \in A\}$. Then $|B| \geq |A|$ because $\bigcup_{y \in B} E_y$ covers $\bigcup_{z \in A} D_z$ and the area of each of the D's and the E's is 1. Similarly, if $B \subset \mathbb{Z}^2$ is a finite set and if $A = \{z : (z, y) \in \mathcal{R} \text{ for some } y \in B\}$, then $|A| \geq |B|$.

By Hall's marriage theorem (see [**425**], Theorem 4.2.1), $\mathcal{R}$ contains the graphs of two one-one mappings, ψ_1 from $\mathcal{N}$ into $\mathbb{Z}^2$ and ψ_2 from $\mathbb{Z}^2$ into $\mathcal{N}$. By the proof of the Schröder-Bernstein theorem (see [**247**], Section 22) $\mathcal{R}$ contains the graph of a one-one mapping ψ from $\mathcal{N}$ onto $\mathbb{Z}^2$.

The condition $D_z \cap E_{\psi(z)} \neq \emptyset$ implies that

$$\|\varphi(z) - \psi(z)\| \leq 1 + \operatorname{diam}(D_z) \leq C_2(1 + \|z\|^{1-\alpha}).$$

It follows that if $x, z \in \mathcal{N}$ with $\|x\|, \|z\| \leq R$ and $x \neq z$, then

$$\begin{aligned}
\|\psi(x) - \psi(z)\| &\leq \|\psi(x) - \varphi(x)\| + \|\varphi(x) - \varphi(z)\| + \|\varphi(z) - \psi(z)\| \\
&\leq 2C_2(1 + R^{1-\alpha}) + CR^{1-\alpha}\|x - z\|^{\alpha} \\
&\leq KR^{1-\alpha}\|x - z\|^{\alpha}
\end{aligned}$$

for some $K = K(\alpha, \lambda)$, because $\|x - z\| \geq 1$. A similar argument shows that ψ^{-1} is homogeneous α-Hölder.

(ii)$\Rightarrow$(i). For each $k > 1$ tile the plane with the squares generated by the lattice $\big((k-1)!\big)^{-1}\mathbb{Z}^2$. For each such square S which is contained in B, pick the points it contains from the net $(k!\lambda_S)^{-1}\mathbb{Z}^2$, where $\lambda_S = |S|^{-1}\int_S f\,dm$. Since $1 \leq \lambda_S \leq \lambda$, we can add points in the squares which are not fully contained in B, and we may also have to add or delete points near the boundaries of the squares, and in this way we obtain a set $\mathcal{N}_k$ which is a $\big(1/(k!\lambda), 1/k!\big)$-net in B. These nets have the property that the normalized counting measures on $\mathcal{N}_k$ converge weakly to $f\,dm$ as $k \to \infty$.

Let $\mathcal{N}$ be the λ-net in $\mathbb{R}^2$ which is obtained by using the points of $k!\lambda\mathcal{N}_k$ in the annulus $\{(k-1)!\lambda < \|x\| \leq k!\lambda\}$ and by adding or deleting points in the annuli $\{k!\lambda - \lambda \leq \|x\| \leq k!\lambda + \lambda\}$ of width 2λ to ensure that the resulting set is, indeed, a λ-net. Let ψ be a homogeneous α-Hölder homeomorphism of $\mathcal{N}$ onto $\mathbb{Z}^2$. Then the restriction $\psi_k(x)$ of $\psi(k!\lambda x)/k!\lambda$ to $B \cap (k!\lambda)^{-1}\mathcal{N}$ maps it into $(k!\lambda)^{-1}\mathbb{Z}^2$, and these maps, as well as their inverses, are homogeneous α-Hölder with a uniform constant. By the Arzelà-Ascoli theorem they have a subsequence which converges uniformly to a homogeneous α-Hölder homeomorphism φ of B into $\mathbb{R}^2$.

For each k the mapping ψ_k transfers the normalized counting measure μ_k on $B \cap (k!\lambda)^{-1}\mathcal{N}$ to the normalized counting measure m_k on its image in $(k!\lambda)^{-1}\mathbb{Z}^2$. Hence the limit φ maps the weak limit of the μ_k's, namely fdm, to the limit of the measures m_k, namely, to the Lebesgue measure m. □

Theorem 10.26. (i) *For every net in* $\mathbb{R}^2$ *there is an* $0 < \alpha \leq 1$ *such that the net is homogeneous* α*-Hölder equivalent to* $\mathbb{Z}^2$*.*

(ii) *There is a net in* $\mathbb{R}^2$ *which is not Lipschitz equivalent to* $\mathbb{Z}^2$*.*

Both parts are proved by considering the existence of appropriately smooth homeomorphisms which map a prescribed measure to the Lebesgue measure. We shall not prove part (i) (see the Notes and Remarks section for references). Part (ii) will follow from the next lemma.

Lemma 10.27. *For every* $\lambda > 1$ *and* $L > 1$ *there is a* $\mu > 1$ *with the following property: Let* $x, y \in \mathbb{R}^2$*, and let* U *be an open neighborhood of the segment* $[x, y]$*. Then there are a function* f *on* U *taking only the values 1 and* λ*, a finite set of disjoint segments* $I_j = [x_j, y_j]$ *in* U*, and an* $\varepsilon > 0$ *such that if* $\varphi : U \to \mathbb{R}^2$ *is a Lipschitz homeomorphism for which the Lipschitz constants of* φ *and of* φ^{-1} *are bounded by* L *and* $m\{u \in U : J(\varphi) \neq f\} < \varepsilon$*, then there is a* j *such that*

$$\|\varphi(x_j) - \varphi(y_j)\| / \|x_j - y_j\| \geq \mu \|\varphi(x) - \varphi(y)\| / \|x - y\|.$$

Proof. Choose a positive integer M and an $\eta > 0$ such that

$$8L(\eta L + L/M) + \pi(\eta L + L/M)^2 < (\lambda - 1)/2.$$

The required $\mu > 1$ is chosen so that it satisfies the conditions $\mu^2 < 1 + \eta^2/5$ and $(1 - \eta^2/5) + \big((M+1)^2 - 1\big)\mu < (M+1)^2$.

By scaling and appropriate choice of coordinates, assume that U is a neighborhood of the segment joining the points $x = (0,0)$ and $y = (1,0)$. If N is large enough (we shall add below further conditions on the necessary size of N), then U contains the rectangle R whose base is the segment $[x, y]$ and with height $1/N$. For $1 \leq i \leq N$ put $S_i = \{(s,t) : (i-1)/N \leq s \leq i/N \ , \ 0 \leq t \leq 1/N\}$, and define the function f on R to be 1 on the square S_i when i is even, and λ when i is odd.

Let $z_{k,l}(i) = \big((i-1)/N + k/MN, l/MN\big)$ be the points of the $(MN)^{-1}$-grid in R, where $1 \leq i \leq N$, $0 \leq k < M$, and $0 \leq l \leq M$.

Now let $\varphi : U \to \mathbb{R}^2$ be a homeomorphism such that the Lipschitz constant of φ and of its inverse are bounded by L. Choosing the coordinates appropriately, assume that $\varphi(x) = (0,0)$ and that $\varphi(y) = (a, 0)$ for some $a > 0$. For $i < N$ put $w_{k,l}(i) = \varphi\big(z_{k,l}(i+1)\big) - \varphi\big(z_{k,l}(i)\big)$.

We shall show that if ε is small enough and if $m\{x \in U : J(\varphi) \neq f\} < \varepsilon$, then $\|w_{k,l}(i)\| > \mu a/N$ for some k, l, i. This will prove the lemma. Indeed, the finitely many intervals $J_{k,l,i} = [z_{k,l}(i), z_{k,l}(i+1)]$ are all parallel to the x-axis. Since φ is Lipschitz, they can be moved to be disjoint with very small displacement in such a way that the inequality is preserved.

Assume that $\|w_{k,l}(i)\| \leq \mu a/N$ for all k, l, i. We first show that there is an i such that $\|w_{k,l}(i) - w\| < \eta a/N$ for all k, l where $w = (a/N, 0)$. To this end, let $\rho_{k,l}(i)$ be the length of the projection of $w_{k,l}(i)$ on the x-axis. Since by the assumption $\|w_{k,l}(i)\| \leq \mu a/N$, and since $\mu^2 < 1 + \eta^2/5$, it is enough to show that there is an i such that $\rho_{k,l}(i) \geq (1 - \eta^2/5)a/N$ for every k, l. So assume this is false.

By the pigeonhole principle there are indices $0 \leq r < M$, $0 \leq s \leq M$ and a subset $I \subset \{0, \ldots, N-1\}$ with $|I| \geq N/(M+1)^2$ such that $\rho_{r,s}(i) < (1 - \eta^2/5)a/N$

for every $i \in I$. The image under φ of the polygonal path connecting the points $(0,0), (0, s/MN), z_{r,s}(1), \dots, z_{r,s}(N), (1, s/MN)$, and $(1,0)$ is a path connecting $(0,0)$ and $(a,0)$; hence the length of its projection on the x-axis is at least a. Using the estimates $\rho_{r,s}(i) < (1-\eta^2/5)a/N$ for $i \in I$ and $\rho_{r,s}(i) \le \mu a/N$ for $i \notin I$, and the bound L/N for the lengths of the images of the two vertical segments and of the first and last horizontal segments, we obtain

$$a \le \frac{N}{(M+1)^2}(1-\eta^2/5)a/N + \big(N - \frac{N}{(M+1)^2}\big)\mu a/N + 4L/N$$

which contradicts the choice of μ for large N (and this is how we chose N).

Thus fix i so that $\|w_{k,l}(i) - w\| < \eta a/N$ for all k, l, and compare the area of $\varphi(S_i)$ (which is the same as the area of its translate $\varphi(S_i) + w$) with that of $\varphi(S_{i+1})$. Since $\|\varphi\big(z_{k,l}(i)\big) + w - \varphi\big(z_{k,l}(i+1)\big)\| < \eta a/N$ (this is just a rewriting of the inequality $\|w_{k,l}(i) - w\| < \eta a/N$), it follows that the Hausdorff distance between the boundaries of $\varphi(S_i) + w$ and of $\varphi(S_{i+1})$ is at most $\eta a/N + L/MN$. Since the length of these boundaries is at most $4L/N$, the upper estimate $2\delta l + \pi\delta^2$ for the area of a δ-neighborhood of a curve of length l gives that

$$\begin{aligned}
\big|m\big(\varphi(S_i)\big) - m\big(\varphi(S_{i+1})\big)\big| &= \big|m\big(\varphi(S_i) + w\big) - m\big(\varphi(S_{i+1})\big)\big| \\
&\le 8(L/N)(\eta a/N + L/MN) + \pi(\eta a/N + L/MN)^2 \\
&< (\lambda - 1)/2N^2
\end{aligned}$$

since $a \le L$, and by the choice of M and η.

On the other hand, since $J(\varphi) = f$ off a set of measure ε, then assuming i is even, say, we obtain that the Jacobian is 1 on S_i and λ on S_{i+1} except for sets of measure at most ε. It follows that $m\big(\varphi(S_i)\big) \le 1/N^2 + \varepsilon L^2$ and that $m\big(\varphi(S_{i+1})\big) \ge \lambda/N^2 - \varepsilon L^2$. Thus $(\lambda-1)/2N^2 > (\lambda-1)/N^2 - 2\varepsilon L^2$, which is impossible if $\varepsilon < (\lambda-1)/4N^2L^2$. □

Proof of Theorem 10.26(ii). Fix $\lambda > 1$, and let $\{U_L : L = 1, \dots\}$ be disjoint open sets in B. We shall find functions $f_L : U_L \to \{1, \lambda\}$ with the property that there is no Lipschitz homeomorphism $\varphi : U_L \to \mathbb{R}^2$ such that the Lipschitz constants of φ and of φ^{-1} are bounded by L and such that $J(\varphi) = f_L$ on U_L. The desired f is defined to be f_L on U_L and 1 off $\bigcup_L U_L$.

Thus fix L, and let $\mu = \mu(\lambda, L)$ be as in the lemma. Choose k so that $\mu^k/L > L$, and using the lemma k times, define inductively for every $n \le k$ disjoint finite collections of segments $\{I_{n,j}\}_j$, numbers $\varepsilon_n > 0$, functions g_n on U_L, and disjoint open sets $U_{n,j} \supset I_{n,j}$ as follows:

Fix two points x, y in the interior of U_L, and use Lemma 10.27 to find a function g_1, disjoint segments $\{I_{1,j}\}$ in U_L, and $\varepsilon_1 > 0$ as in the lemma. Then choose a collection of disjoint open sets $U_{1,j} \supset I_{1,j}$ in U_L with total area smaller than $\varepsilon_1/2$. This completes the first step.

In the second step we work in each $U_{1,j}$ separately, and g_2 will be defined as g_1 off $\bigcup_j U_{1,j}$. Thus fix j, and apply the lemma with the endpoints of $I_{1,j}$ to find a function $g_{1,j}$ on $U_{1,j}$, a number $\varepsilon_{1,j} > 0$, and a disjoint collection of segments $\{I_{1,j,k}\}$ in $U_{1,j}$ as in the lemma. Then define g_2 to be $g_{1,j}$ on $U_{1,j}$, take $\varepsilon_2 = \min_j \varepsilon_{1,j}$, and choose for each $I_{1,j,k}$ an open set $U_{1,j,k} \subset U_{1,j}$ containing it and such that the total area of $\bigcup_{j,k} U_{1,j,k}$ is smaller than $\varepsilon_2/2$. The second step is completed when we renumber the segments $I_{1,j,k}$ and the open sets $U_{1,j,k}$ as $I_{2,j}$ and $U_{2,j}$ respectively.

This procedure is iterated k times, and we finally take $f_L = g_k$.

Assume that $\varphi : U_L \to \mathbb{R}^2$ is a Lipschitz homeomorphism, that the Lipschitz constants of φ and of φ^{-1} are bounded by L, and that $J(\varphi) = f_L$ on U_L. Then $\|\varphi(x) - \varphi(y)\|/\|x - y\| \geq 1/L$, and hence by the construction in the first step there is an interval $I_j^1 = [x_j^1, y_j^1]$ such that $\|\varphi(x_j^1) - \varphi(y_j^1)\|/\|x_j^1 - y_j^1\| \geq \mu/L$. By the construction in the second step there is an interval $I_i^2 = [x_i^2, y_i^2]$ such that $\|\varphi(x_i^2) - \varphi(y_i^2)\|/\|x_i^2 - y_i^2\| \geq \mu^2/L$. After k iterations, we find points u and v with $\|\varphi(u) - \varphi(v)\|/\|u - v\| \geq \mu^k/L > L$, contradicting the fact that φ is L-Lipschitz. $\diamondsuit$

4. Nonisomorphic Uniformly Homeomorphic Spaces

The next theorem gives a method for the construction of pairs of uniformly homeomorphic Banach spaces. When E is chosen appropriately, the two spaces are not mutually isomorphic.

We shall use the following normalization procedure repeatedly. Assume that $\varphi : E \to F$ is a positively homogeneous homeomorphism from E onto F so that there is a constant C with $C^{-1}\|x\| \leq \|\varphi(x)\| \leq C\|x\|$ for every $x \in E$. Put $\psi(x) = \|x\|\varphi(x)/\|\varphi(x)\|$. Then $\psi^{-1}(y) = \|y\|\varphi^{-1}(y)/\|\varphi^{-1}(y)\|$ and ψ is a positively homogeneous and norm-preserving homeomorphism from E onto F. If φ is uniformly continuous on E (resp. uniformly continuous on bounded sets in E), then the same is true for ψ. We only check the uniform continuity of ψ. Write $\|\psi(x) - \psi(z)\| \leq \big\|\|x\|\,\|\varphi(z)\|\varphi(x) - \varphi(z)\|z\|\,\|\varphi(x)\|\big\|/\|\varphi(x)\|\,\|\varphi(z)\|$. Then the numerator is bounded by

$$\|\varphi(x) - \varphi(z)\|\,\|x\|\,\|\varphi(z)\| + \|\varphi(z)\|\,\|x\|\,\|\varphi(z) - \varphi(x)\| + \|\varphi(z)\|\,\|\varphi(x)\|\,\|z - x\|$$

hence $\|\psi(x) - \psi(z)\| \leq 2C\|\varphi(x) - \varphi(z)\| + \|x - z\|$.

Theorem 10.28. *Let $1 \leq p < \infty$, and let $\{E_n\}_{n=1}^{\infty}$ be Banach spaces with the following property: For each $k \geq 0$ there is an $n(k)$ and a homogeneous norm-preserving homeomorphism $\varphi_k : E_{n(k)} \to l_p$ such that the family of restrictions of φ_k to $2^k B(E_{n(k)})$ and of φ_k^{-1} to $2^k B(l_p)$ is equi-uniformly continuous.*

Assume that the Banach space E has a Schauder decomposition $E = \sum \oplus E_n$. Then E and $l_p \oplus E$ are uniformly homeomorphic.

Proof. Recall that $E = \sum \oplus E_n$ is a Schauder decomposition if each E_n is a subspace of E and if every $x \in E$ has a unique representation as a convergent series $x = \sum x_n$ with $x_n \in E_n$. We call x_n the nth coordinate of x, and by a renorming which does not change the norms on the individual E_n's we may assume that $\|x_n\| \leq \|x\|$ for every n and for every $x \in E$. We shall also assume that $l_p \oplus E$ is an l_1 sum, and to simplify the notation, we shall assume that $n(k) = k$ for every $k \geq 0$. (See also the last paragraph of the proof.)

The mappings φ_n induce homogeneous (but not norm-preserving) homeomorphisms ψ_n from $E_n \oplus E_{n+1}$ onto $(l_p \oplus l_p)_p$ by $\psi_n(x_n, x_{n+1}) = \big(\varphi_n(x_n), \varphi_{n+1}(x_{n+1})\big)$, and we denote by $\omega(\varepsilon)$ a common bound for the moduli of continuity of φ_n and ψ_n and of their inverses on the appropriate balls in the appropriate spaces.

The proof will be accomplished in several steps.

Write vectors $x \in (l_p \oplus l_p \oplus l_p)_p$ as (u, v, w), and similarly write (u, v) for points in $(l_p \oplus l_p)_p$. Let T be a linear isometry from $(l_p \oplus l_p)_p$ onto l_p. The two linear operators $S_0(u, v, w) = \big(T(u, v), w\big)$ and $S_1(u, v, w) = \big(v, T(u, w)\big)$ are isometries of the space $(l_p \oplus l_p \oplus l_p)_p$ onto $(l_p \oplus l_p)_p$.

The first step is the construction, for $0 \le \tau \le 1$, of homogeneous norm-preserving homeomorphisms $h_\tau : (l_p \oplus l_p \oplus l_p)_p \to (l_p \oplus l_p)_p$ such that $h_0 = S_0$ and $h_1 = S_1$, and such that there is a constant K for which

$$\|h_\tau(x) - h_\eta(y)\| \le K\Big(\|x - y\| + |\tau - \eta| \max\left(\|x\|, \|y\|\right)\Big) \tag{10.7}$$

and similarly for their inverses.

To this end, we first construct a continuous path $\tau \to S_\tau$ of invertible operators S_τ from $(l_p \oplus l_p \oplus l_p)_p$ onto $(l_p \oplus l_p)_p$ so that S_0 and S_1 are the two given operators. Let $V_\tau(u, v, w) = (u_\tau, v_\tau, w_\tau)$ for $0 \le \tau \le 1/2$ be a path of invertible operators of $(l_p \oplus l_p \oplus l_p)_p$ onto itself such that V_0 is the identity and $V_{1/2}(u, v, w) = (u, w, v)$. (Such V_τ are obtained by considering the formulas for the appropriate rotations in $\mathbb{R}^3$ and by working with the coordinates u, v, w as if they were scalars.) We then define $S_\tau(u, v, w) = \big(T(u_\tau, v_\tau), w_\tau\big)$ for $0 \le \tau \le 1/2$. For $1/2 \le \tau \le 1$ put $S_\tau = U_\tau S_{1/2}$ where U_τ is a path of invertible operators on $(l_p \oplus l_p)_p$ connecting the identity to the operator $(u, v) \to (v, u)$.

Using the continuity of $\tau \to S_\tau$ and the compactness of $[0, 1]$, we can approximate, and then replace, the S_τ's by a map $\tau \to S_\tau$ which is Lipschitz in τ (even piecewise linear). We can thus also assume that there is a constant K_1 such that

$$\|S_\tau - S_\eta\|,\ \|S_\tau^{-1} - S_\eta^{-1}\| \le K_1 |\tau - \eta|.$$

The maps $h_\tau(x) = \|x\| S_\tau(x)/\|S_\tau(x)\|$ satisfy the required conditions.

For $n = 0, 1, \dots$ put $\alpha_n = 2^n - 1$. The second step of the proof is to construct for each $t > 0$ a homogeneous norm-preserving homeomorphism g_t of $l_p \oplus E$ onto E such that there are constants L_1 and L_2 for which

$$\|g_t(x) - g_s(y)\| \le L_1 \omega\Big(L_2\Big(\omega\big(\|x - y\|\big) + |s - t|\Big)\Big) \tag{10.8}$$

whenever $\|x - y\| < 1$ and there is an n for which $\alpha_n \le s, t \le \alpha_{n+1}$ and also $\|x\|, \|y\| \le \alpha_{n+1}$. A similar estimate should hold for their inverses.

The g_t's are obtained by "transplanting" the h_τ's. For $x = u + \sum x_j$ and $\alpha_n \le t \le \alpha_{n+1}$ put

$$\tilde{g}_t(x) = \psi_n^{-1}\Big(h_{2^{-n}(t-\alpha_n)}\big(u, \psi_n(x_n, x_{n+1})\big)\Big) + \sum_{j \ne n, n+1} x_j.$$

Note that the two definitions of $\tilde{g}_{\alpha_n}(x)$, the one using $h_1\big(u, \psi_{n-1}(x_{n-1}, x_n)\big)$ and the one using $h_0\big(u, \psi_n(x_n, x_{n+1})\big)$, coincide; hence $\tilde{g}_{\alpha_n}$ is well-defined. Each $\tilde{g}_t$ is homogeneous; and if $y = v + \sum y_j$ and $\alpha_n \le s \le \alpha_{n+1}$ are also given and if $\|x\|, \|y\| \le \alpha_{n+1}$ and $\|x - y\| < 1$, then $\|\tilde{g}_t(x) - \tilde{g}_s(y)\|$ is bounded by

$$\Big\| \sum_{j \ne n, n+1} (x_j - y_j) \Big\|$$
$$+ \Big\| \psi_n^{-1}\Big(h_{2^{-n}(t-\alpha_n)}\big(u, \psi_n(x_n, x_{n+1})\big)\Big) - \psi_n^{-1}\Big(h_{2^{-n}(s-\alpha_n)}\big(v, \psi_n(y_n, y_{n+1})\big)\Big) \Big\|.$$

The first term is bounded by $3\|x - y\|$, and by (10.7) the second is bounded by

$$\omega\left(K\Big\{\|u - v\| + \omega\big(\|x_n - y_n\| + \|x_{n+1} - y_{n+1}\|\big)\Big\} + \alpha_{n+1}|2^{-n}s - 2^{-n}t|\right).$$

Since there is a constant C such that $\varepsilon \le C\omega(\varepsilon)$ for $\varepsilon \le 1$, it follows that

$$\|\tilde{g}_t(x) - \tilde{g}_s(y)\| \le L_1 \omega\Big(L_2\Big(\omega\big(\|x - y\|\big) + |s - t|\Big)\Big).$$

Thus the family $\{\tilde{g}_t\}$ satisfies (10.8). It follows that this is also true for the renormalized maps $g_t(x) = \|x\|\tilde{g}_t(x)/\|\tilde{g}_t(x)\|$.

Finally the uniform homeomorphism $\phi : l_p \oplus E \to E$ is defined by

$$\phi(x) = g_{\|x\|}(x).$$

We remark at the end that it was possible to work with a subsequence of the E_n's even though the decomposition is not assumed to be unconditional because all the maps in the proof are compositions of maps which either act on at most three coordinates in the decomposition or are renormalizations of vectors. □

Corollary 10.29. *Let $1 \le p, q, p_n < \infty$ satisfy $p_n \to p$, and assume that p is different from q and from all the p_n's. Then $(\sum \oplus l_{p_n})_q$ is uniformly homeomorphic to $l_p \oplus (\sum \oplus l_{p_n})_q$, but is not isomorphic to it.*

Proof. By Proposition 9.2 $E_n = l_{p_n}$ satisfy the conditions of Theorem 10.28 (with the Mazur maps ψ_{p,p_n} as the maps φ_n in the statement of the theorem). Hence the two spaces are uniformly homeomorphic. They are not isomorphic because l_p is not even isomorphic to a subspace of $(\sum \oplus l_{p_n})_q$. Indeed, since $p \ne q$, a standard "gliding hump" procedure shows that if l_p were isomorphic to a subspace of $(\sum \oplus l_{p_n})_q$, then it would have been isomorphic to a subspace of a finite sum of the l_{p_n}'s. But this is impossible because p is different from all the p_n's. □

Remark. Note that when $p = 1$ and $q, p_n > 1$, then one of the spaces is reflexive and the other is not. Also, if $p, q, p_n > 1$, then both spaces are uniformly convex.

The following corollary will be used in the next section.

Corollary 10.30. *Let $1 < r < \infty$, and let $C < \infty$. Assume that the Banach space E has a Schauder decomposition $E = \sum E_n$, where all the E_n's have C-unconditional bases which are r-convex and $(r + \varepsilon_n)$-concave with constant C. If there is a subsequence with $\varepsilon_{n_k} \to 0$, then $l_r \oplus E$ and E are uniformly homeomorphic.*

Proof. By renorming we can assume that $C = 1$ (see Appendix E). Combining Corollary I.5 and Proposition 9.13, we obtain the l_r and the E_n's satisfy the assumptions of Theorem 10.28. □

5. Uniform Types that Determine a Finite Number of Linear Structures

The examples in this section are based on the Tsirelson space, which we denote by T. We start with the construction of T and of some of its modifications and present their basic properties. We then show that the uniform type of the modified space $\mathcal{T}^{(p)}$ determines exactly two isomorphism classes of Banach spaces.

Let E_0 be the space of all finitely supported sequences. We denote the unit vector basis of E_0 by $\{e_n\}$. If A is a subset of $\mathbb{N}$, we also denote by the same letter A the natural associated projection of E_0 onto span$\{e_n : n \in A\}$; i.e., if $x = \sum a_n e_n$, then $Ax = \sum_{n \in A} a_n e_n$. We shall consider 1-unconditional norms on E_0. For such norms $\|Ax\| \le \|x\|$ for every $x \in E_0$.

Let A and B be two finite subsets of $\mathbb{N}$. We shall write $A < B$ if $i < j$ for every $i \in A$ and $j \in B$. If $i \in \mathbb{N}$, we write $i \le B$ if $i \le j$ for all $j \in B$. A collection $\{A_j : j \le n\}$ of finite subsets of $\mathbb{N}$ is called admissible if $n \le A_1 < \ldots < A_n$.

Define inductively a sequence of norms on E_0 as follows: $\|x\|_0 = \|x\|_{c_0}$ and

$$\|x\|_{m+1} = \max\Big(\|x\|_m, \frac{1}{2}\sup\big\{\sum \|A_j x\|_m\big\}\Big)$$

where the supremum is taken over all admissible collections $\{A_j\}$.

These norms are 1-unconditional, and the sequence $\{\|x\|_m\}$ is monotone non-decreasing for each x (and bounded above by the norm of x in l_1); hence the limit $\lim \|x\|_m = \|x\|$ exists and is a 1-unconditional norm on E_0. The Tsirelson space T is the completion of E_0 with respect to this norm.

It is clear that the norm on T satisfies the identity

$$\|x\| = \max\Big(\|x\|_{c_0}, \frac{1}{2}\sup\big\{\sum \|A_j x\|\big\}\Big) \tag{10.9}$$

where the supremum is taken over all admissible collections $\{A_j\}$.

Proposition 10.31. (i) *Assume that $\{x_j\}_{j=1}^n$ is a normalized block basic sequence of length n in T and that all the x_j's are supported in $[n, \infty)$. Then*

$$\big\|\sum a_j x_j\big\| \geq \frac{1}{2}\sum |a_j|$$

for every choice of scalars $\{a_j\}$; i.e., the x_j's are 2-equivalent to the unit vector basis of l_1^n.

(ii) *T and T^* do not contain a superreflexive subspace or a subsymmetric basic sequence. In particular they do not contain a subspace isomorphic to c_0 or to l_p for any $1 \leq p < \infty$, and they are reflexive.*

(iii) *T is isomorphic to its finite-codimensional subspaces.*

Proof. (i) This follows directly from (10.9).

(ii) It follows from (i) that T does not contain a subspace which is isomorphic to a uniformly convex space, i.e., a superreflexive subspace. It also follows from (i) that the only possible subsymmetric basic sequences in T are equivalent to the unit vector basis in l_1. Once we also rule out this possibility, T is reflexive by Theorem E.3(iv). (The statements on T^* hold because by duality it satisfies a property analogous to (i) with l_∞^n's replacing the l_1^n's.)

Thus assume that T contains a subspace isomorphic to l_1. By Proposition 13.6 there is a normalized block sequence $\{v_i\}$ in T such that

$$\sum |a_i| \geq \big\|\sum a_i v_i\big\| \geq \frac{8}{9}\sum |a_i|$$

for every choice of scalars $\{a_i\}$. In particular

$$\Big\|v_0 + r^{-1}\sum_{i=1}^r v_i\Big\| \geq 16/9$$

for every $r \geq 1$. Let $\{A_j\}_{j=1}^k$ be an admissible collection such that

$$\Big\|v_0 + r^{-1}\sum_{i=1}^r v_i\Big\| = \frac{1}{2}\sum_{j=1}^k \Big\|A_j\Big(v_0 + r^{-1}\sum_{i=1}^r v_i\Big)\Big\|$$

and let m be the largest integer in the support of v_0. We first note that necessarily $k \leq m$. Indeed, otherwise $A_j v_0 = 0$ for all j, and the right-hand side would be bounded by 1.

Split the set $1 \le i \le r$ into two sets:

$$B = \{i : A_j v_i \neq 0 \text{ for at least two values of } j\},$$
$$C = \{i : A_j v_i \neq 0 \text{ for at most one value of } j\}.$$

Then the cardinality of B is at most $k-1$, and if $i \in C$, then $\sum_{j=1}^k \|A_j v_i\| \le \|v_i\|$. (If $i \in B$, we obtain from (10.9) the weaker estimate $\sum_{j=1}^k \|A_j v_i\| \le 2\|v_i\|$.) Hence

$$\begin{aligned}
16/9 &\le \frac{1}{2}\sum_{j=1}^{k} \Big\| A_j\Big(v_0 + r^{-1}\sum_{i=1}^{r} v_i\Big)\Big\| \\
&\le \frac{1}{2}\sum_{j=1}^{k}\|A_j v_0\| + \frac{1}{2}r^{-1}\Big(\sum_{i\in B}\sum_{j=1}^{k}\|A_j v_i\| + \sum_{i\in C}\sum_{j=1}^{k}\|A_j v_i\|\Big) \\
&\le \|v_0\| + \frac{1}{2}r^{-1}\Big(2\sum_{i\in B}\|v_i\| + \sum_{i\in C}\|v_i\|\Big) \\
&\le 1 + \frac{1}{2}r^{-1}\big(2(k-1)+r\big) \le 3/2 + k/r \le 3/2 + m/r
\end{aligned}$$

which is impossible when $r > 4m$.

(iii) The basis $\{e_n\}_{n\ge 1}$ is equivalent to its subsequence $\{e_n\}_{n\ge 2}$. We do not prove this here, and refer to Casazza and Shura [**112**] for a detailed study of subsequences of T. □

We now present a modification $\mathcal{T}$ of the Tsirelson space. It differs from T in the collections of finite sets that are used to define the norm. The first difference, made for convenience and not really essential, is that we allow $(n+1)^n$ sets in $[n,\infty)$ (and not only n). The second, more important difference is that we allow collections of any disjoint sets, and not just successive blocks. This gives control on general disjointly supported vectors in $\mathcal{T}$ and not only on block basic sequences.

A collection $\{A_j : j \le k\}$ of finite subsets of $\mathbb{N}$ is called allowable if they are disjoint and if there is an index n such that $k \le (n+1)^n$ and $A_j \subset [n,\infty)$ for every j. The sequence of norms $\|x\|_m$ is defined as before, replacing the condition that the collections of sets be admissible by the requirement that they are allowable. The completion of E_0 under the resulting limit norm is denoted by $\mathcal{T}$, and its norm satisfies (10.9), where the supremum is now taken over all allowable collections $\{A_j\}$.

Proposition 10.32. (i) *Let $\{x_j\}_{j=1}^k$ be normalized disjointly supported vectors in $\mathcal{T}$. Assume that there is an n with $k \le (n+1)^n$ such that all the x_j's are supported in $[n,\infty)$. Then*

$$\Big\|\sum a_j x_j\Big\| \ge \frac{1}{2}\sum |a_j|$$

for every choice of scalars $\{a_j\}$; i.e., the x_j's are 2-equivalent to the unit vector basis of l_1^k.

(ii) *$\mathcal{T}$ and $\mathcal{T}^*$ do not contain a superreflexive subspace or a subsymmetric basic sequence. In particular, they do not contain a subspace isomorphic to c_0 or to l_p for any $1 \le p < \infty$, and they are reflexive.*

(iii) *$\mathcal{T}$ is isomorphic to its finite-codimensional subspaces.*

(iv) *There is a constant* M *such that for every* $q > 1$ *there is an* $N = N(q)$ *for which the "tail"* $\{e_n : n \geq N\}$ *is a* q*-concave basis with* q*-concavity constant smaller than* M.

Proof. The proof of (i) is the same as in Proposition 10.31, and for (iii) we refer again to Casazza and Shura [**112**]. To prove that $\mathcal{T}$ does not contain a subspace isomorphic to l_1, and hence to prove (ii), is a little more complicated than for T (it is no longer true that the cardinality of the set B is bounded by $k-1$, and a more delicate combinatorial argument is necessary). The Notes and Remarks section contains a reference for the proof and further comments.

We now prove (iv). By Proposition E.11(i) it is enough to show the analogous statement with q-concavity replaced by lower q-estimate. (We also use here the last statement in Proposition E.11(i), i.e., that it is possible to choose the parameters in such a way that the constants remain uniformly bounded.)

Fix N, and let $\{x_i\}_{i=1}^l$ be disjointly supported vectors supported in $[N+1,\infty)$. Put $L = \|\sum_{i=1}^l x_i\|$. Assume that $\|x_1\| \geq \|x_2\| \geq \dots$ and that the vectors have finite supports. Put $m_0 = 0$, and define inductively $m_k = m_{k-1} + N^k$ for $k \geq 1$. We may assume (by adding some zero vectors) that $l = m_s$ for some s.

Fix m, and let F be a set with $|F| \leq (m+1)^m$. Let $\{x_j\}_{j\in F}$ be disjointly supported vectors, all supported in $[m,\infty)$. Then their supports form an allowable collection; hence

$$L \geq \frac{1}{2}\sum_{i\in F}\|x_i\| \geq \frac{1}{2}|F|\min\bigl\{\|x_i\| : i \in F\bigr\}.$$

Now fix $1 < k \leq s$. Then the supports of at most m_{k-1} of the first m_k vectors $\{x_i\}_{i=1}^{m_k}$ intersect $[1, m_{k-1}]$, and hence there is a subset $F_k \subset \{1,\dots,m_k\}$ of cardinality $m_k - m_{k-1} = N^k \leq (m_{k-1}+1)^{m_{k-1}}$ such that $\{x_j\}_{j\in F}$ are supported in $[m_{k-1},\infty)$. By the estimate above and by the monotonicity of the $\|x_j\|$'s it follows that $\sum_{i=m_{k-1}+1}^{m_k}\|x_i\| \leq \sum_{i\in F_k}\|x_i\| \leq 2L$ and $\|x_{m_k+1}\| \leq \|x_{m_k}\| \leq 2N^{-k}L$.

Since all the x_i's are supported in $[m_1,\infty]$, a similar, and even simpler, argument shows that the first estimate holds also for $k=1$, and clearly $\|x_j\| \leq L$ for $j = 1$ or $j = m_1 + 1$. Hence

$$\begin{aligned}\sum_{i=1}^{l}\|x_i\|^q &\leq \sum_{k=1}^{s}\|x_{m_{k-1}+1}\|^{q-1}\sum_{i=m_{k-1}+1}^{m_k}\|x_i\| \leq 2L\sum_{k=1}^{s}\|x_{m_{k-1}+1}\|^{q-1}\\ &\leq 2L\Bigl(2L^{q-1} + \sum_{k=3}^{s}\bigl(2N^{-(k-1)}L\bigr)^{q-1}\Bigr) \leq 5L^q\end{aligned}$$

provided N is large enough (depending on q). □

Let $1 < p < \infty$, and denote the p-convexification of $\mathcal{T}$ by $\mathcal{T}^{(p)}$. Thus $\mathcal{T}^{(p)}$ has an unconditional basis, and the norm is given by

$$\bigl\|\sum a_n e_n\bigr\|_{\mathcal{T}^{(p)}} = \bigl\|\sum |a_n|^p e_n\bigr\|_{\mathcal{T}}^{1/p}.$$

Proposition 10.33. (i) *Let* $\{x_j\}_{j=1}^k$ *be normalized disjointly supported vectors in* $\mathcal{T}^{(p)}$. *Assume that there is an* n *with* $k \leq (n+1)^n$ *such that all the* x_j*'s are supported in* $[n,\infty)$. *Then the* x_j*'s are equivalent to the unit vector basis of* l_p^k *(with a constant that depends only on* p*).*

(ii) $\mathcal{T}^{(p)}$ *and* $\mathcal{T}^{(p)*}$ *are reflexive and do not contain a subspace isomorphic to* l_r *for any* $1 < r < \infty$.

(iii) $\mathcal{T}^{(p)}$ *is isomorphic to its finite-codimensional subspaces.*

(iv) *There is a constant* M *such that for every* $q > p$ *there is an* $N = N(q)$ *such that* $\{e_n : n \geq N\}$ *is* q*-concave with* q*-concavity constant smaller than* M.

Proof. The proof follows immediately from Proposition 10.32 and from the explicit formula for the norm. We recall that the p-convexification of an s-concave lattice is ps-concave (see Appendix E). □

Proposition 10.34. *Let* $1 < p < \infty$. *Then* $\mathcal{T}^{(p)}$ *is uniformly homeomorphic to* $\mathcal{T}^{(p)} \oplus l_p$.

Proof. Choose $q_n \searrow p$, and split $\mathbb{N}$ as a disjoint union $A_0 \cup \bigcup_{n=1}^{\infty} A_n$, where A_0 is a finite set, and for each $n \geq 1$ the set A_n is an infinite subset of $[N(q_n), \infty)$. By the choice of $N(q_n)$ each $F_n = \operatorname{span}\{e_j : j \in A_n\}$ is q_n-concave with concavity constant at most M. Since $\mathcal{T}^{(p)}$ is isomorphic to its finite-codimensional subspaces, it is isomorphic to the 1-unconditional sum $\sum \oplus F_n$, and Corollary 10.30 applies. □

Our next goal is to show that the only Banach spaces which are uniformly homeomorphic to $\mathcal{T}^{(p)}$ are $\mathcal{T}^{(p)} \oplus l_p$ and $\mathcal{T}^{(p)}$ itself. We first need some preparations.

Recall that when F is reflexive there is a natural norm-one projection Q from $(F)_{\mathcal{U}}$ onto the canonical image of F in $(F)_{\mathcal{U}}$ given by

$$Q\big((x_i)_{\mathcal{U}}\big) = (x)_{\mathcal{U}}$$

where $x = w - \lim_{\mathcal{U}} x_i$. The kernel of Q is denoted by $(F)_{\mathcal{U},0}$.

Lemma 10.35. $(\mathcal{T}^{(p)})_{\mathcal{U},0}$ *is an* $\mathcal{L}_p$*-space.*

Proof. Let F be an n-dimensional subspace of $(\mathcal{T}^{(p)})_{\mathcal{U},0}$, and let $N = N(n)$ be as in Lemma E.6. By Proposition F.7 there are n-dimensional subspaces $F_i \subset \mathcal{T}^{(p)}$ with $F = (F_i)_{\mathcal{U}}$. Since $F \subset \ker(Q)$, it follows that $Q\big((x_i)_{\mathcal{U}}\big) = 0$ for every bounded family $x_i \in F_i$. Assume (by a small perturbation) that $F_i \subset \operatorname{span}\{e_j : j \geq m_i\}$ for every i, where $m_i \to \infty$ and $m_i > N$ for all i. By the choice of N there are N-dimensional subspaces $G_i \supset F_i$ which are 2-isomorphic to the spans of N disjoint vectors supported in $[m_i, \infty)$. By Proposition 10.33(i) the G_i's are C-isomorphic (with C depending only on p) to l_p^N; hence so is $G = (G_i)_{\mathcal{U}}$. Then G contains F, and it is contained in $(\mathcal{T}^{(p)})_{\mathcal{U},0}$ because $m_i \to \infty$. □

Corollary 10.38 below is a variation of Corollary 7.7 for superreflexive spaces. It only gives an embedding of E as a complemented subspace of an ultrapower $(F)_{\mathcal{U}}$ of F (rather than in F itself), but it also gives information on the complement. This more precise information is needed because direct application of Corollary 7.7 gives only (by the same methods as in the proof of Theorem 10.39) that $E \oplus l_p$ and $\mathcal{T}^{(p)} \oplus l_p$ are isomorphic to complemented subspaces of each other. It is unknown, however, if this alone suffices to show that they are actually isomorphic.

We first need some preparations. The first theorem (quoted without proof; see the Notes and Remarks section for reference) belongs to the circle of ideas, to be discussed in detail in Volume 2, related to Fréchet differentiability of Lipschitz maps. It gives a point where two functions are simultaneously Gâteaux differentiable, and where one of the functions is actually "almost" Fréchet differentiable.

Theorem 10.36. *Let E and F be superreflexive separable Banach spaces, and let G be a finite-dimensional space. Let $f : E \to F$ and $g : E \to G$ be two Lipschitz functions. Then for every $\varepsilon > 0$ there is a point $x_0 \in E$ such that both f and g are Gâteaux differentiable at x_0, and moreover, there is a $\delta > 0$ such that*

$$\|g(x_0 + u) - g(x_0) - D_g(x_0)u\| \leq \varepsilon\|u\|$$

for every $u \in E$ with $\|u\| \leq \delta$.

Lemma 10.37. *Let $f : E \to F$ be a Lipschitz equivalence between two separable superreflexive spaces E and F, normalized so that*

$$c\|x - y\| \leq \|f(x) - f(y)\| \leq \|x - y\|$$

for every $x, y \in E$. Let G be a finite-dimensional subspace of F^*, and let $\varepsilon > 0$. Then there is a linear isomorphism T from E into F such that

(i) $c\|x\| \leq \|Tx\| \leq \|x\|$ *for every $x \in E$.*

(ii) *TE is complemented in F with a projection P satisfying $\|P\| \leq c^{-1}$.*

(iii) $\|P^*y^*\| \geq (1 - \varepsilon)c\|y^*\|$ *for every $y^* \in G$.*

Proof. The operator T will be the Gâteaux derivative of f at some point x_0 which depends on G. Thus (i) and (ii) follow directly from Theorem 7.6.

Let $Q : F \to G^*$ be the natural quotient map defined by $\langle Qz, y^*\rangle = \langle y^*, z\rangle$ for $z \in F$ and $y^* \in G$. We use Theorem 10.36 for the two functions f and $g = Q \circ f$, with ε replaced by $c\varepsilon$. Let x_0 be as in Theorem 10.36. By translations in E and in F we may assume that $x_0 = 0$ and $f(0) = 0$, and hence also $g(0) = 0$.

Denote the Gâteaux derivative of f at 0 by T. Then the Gâteaux derivative of g at 0 is QT. With this notation Theorem 10.36 gives that there is a $\delta > 0$ such that

$$\|g(x) - QTx\| \leq c\varepsilon\|x\|$$

whenever $\|x\| \leq \delta$.

Let $y^* \in G$ with $\|y^*\| = 1$. Since $T^* = T^*P^*$ and $\|T^*\| \leq 1$ it suffices to show that $\|T^*y^*\| \geq (1 - \varepsilon)c$. Choose $y \in F$ with $\|y\| = \langle y^*, y\rangle = c\delta$ (this is possible because F is reflexive), and then $x = f^{-1}(y)$ satisfies $\|x\| \leq \delta$. Hence

$$\begin{aligned} \delta\|T^*y^*\| &\geq \langle T^*y^*, x\rangle = \langle y^*, Tx\rangle = \langle QTx, y^*\rangle \\ &\geq \langle g(x), y^*\rangle - c\varepsilon\|x\| = \langle y^*, f(x)\rangle - c\varepsilon\|x\| \\ &= \langle y^*, y\rangle - c\varepsilon\|x\| \geq c\delta - c\varepsilon\delta \end{aligned}$$

i.e., $\|T^*y^*\| \geq (1 - \varepsilon)c$ as required. □

Corollary 10.38. *Let E, F and f be as in the lemma, and let $\mathcal{U}$ be a free ultrafilter on $\mathbb{N}$. Then there is a complemented subspace $E_1 \subset (F)_\mathcal{U}$, isomorphic to E, and a projection $P : (F)_\mathcal{U} \to E_1$ with $(I - P)(F)_{\mathcal{U},0} = (I - P)(F)_\mathcal{U}$. In particular, the complement $W = (I - P)(F)_\mathcal{U}$ of E_1 is isomorphic to a quotient of $(F)_{\mathcal{U},0}$.*

Proof. Let G_n be an increasing sequence of finite-dimensional subspaces of F^* such that $\bigcup G_n$ is dense in F^*. Let the operators T_n and the projections P_n be as in the lemma for $G = G_n$ and $\varepsilon = 1/n$.

The ultrapower $(T_n)_\mathcal{U}$ is an isomorphism of $(E)_\mathcal{U}$ onto a subspace of $(F)_\mathcal{U}$ which is complemented by $(P_n)_\mathcal{U}$. Let $Q : (E)_\mathcal{U} \to E$ be the canonical projection, and let T be the restriction of $(T_n)_\mathcal{U}$ to E, i.e., $Tx = (T_n x)_\mathcal{U}$. Then T is an isomorphism onto its image, which we denote by E_1, and E_1 is complemented in $(F)_{\mathcal{U}}$ by the projection $P = TQ(T_n^{-1})_{\mathcal{U}}(P_n)_{\mathcal{U}}$. Also $\|P^*y^*\| \geq c\|y^*\|$ for every $y^* \in F^*$. Indeed, it is enough to check this for $y^* \in \bigcup G_n$, but if $y^* \in G_n$ and if $m \geq n$, then $\|P_m^* y^*\| \geq (1 - 1/m)c\|y^*\|$.

To check that $(I - P)(F)_{\mathcal{U},0} = W$ it is enough, by duality, to show that the restriction to W^* of the canonical projection of $(F)_{\mathcal{U}}^* = (F^*)_{\mathcal{U}}$ onto $(F^*)_{\mathcal{U},0}$ is bounded from below. So let $w^* \in W^*$, and write it as $w^* = y^* + x^*$, where $y^* \in F^*$ and $x^* \in (F^*)_{\mathcal{U},0}$. We shall show that $\|x^*\| \geq c\|w^*\|/(c + \|P\|)$. Indeed, $P^*x^* = -P^*y^*$ (because $P^*w^* = 0$), and the result follows from

$$\|P^*\|\,\|x^*\| \geq \|P^*x^*\| = \|P^*y^*\| \geq c\|y^*\| \geq c\big(\|w^*\| - \|x^*\|\big).$$

□

Theorem 10.39. *A Banach space E is uniformly homeomorphic to $\mathcal{T}^{(p)}$ for some $1 < p < \infty$ if and only if it is either isomorphic to $\mathcal{T}^{(p)}$ or to $\mathcal{T}^{(p)} \oplus l_p$.*

Proof. Since $\mathcal{T}^{(p)}$ and $\mathcal{T}^{(p)} \oplus l_p$ are uniformly homeomorphic, we need to show that if E is uniformly homeomorphic to $\mathcal{T}^{(p)}$, then it is either isomorphic to $\mathcal{T}^{(p)}$ or to $\mathcal{T}^{(p)} \oplus l_p$.

Theorem 10.1 implies that E and $\mathcal{T}^{(p)}$ have Lipschitz equivalent ultrapowers. It follows from Corollary 10.38 and Lemma 10.35 that $(\mathcal{T}^{(p)})_{\mathcal{U}} = \mathcal{T}^{(p)} \oplus (\mathcal{T}^{(p)})_{\mathcal{U},0}$ is isomorphic to $E \oplus W$ where W is a quotient of the $\mathcal{L}_p$-space $(\mathcal{T}^{(p)})_{\mathcal{U},0}$. From this point on the proof uses only results from the linear theory of Banach spaces.

Every subspace of $L_p(\mu)$, hence every $\mathcal{L}_p$-space, contains a subspace isomorphic to l_s for some s (see references in the Notes and Remarks section). Since $\mathcal{T}^{(p)}$ and its dual do not contain such subspaces, it follows that every operator from $\mathcal{T}^{(p)}$ into the $\mathcal{L}_p$-space $(\mathcal{T}^{(p)})_{\mathcal{U},0}$, or between their duals, is strictly singular.

The dual W^* of W is isomorphic to a subspace of the $\mathcal{L}_q$-space $(\mathcal{T}^{(p)})_{\mathcal{U},0}^*$ (where $1/p + 1/q = 1$). Thus each of its infinite-dimensional subspaces contains a subspace isomorphic to l_s for some s. On the other hand, W^* is complemented in ${\mathcal{T}^{(p)}}^* \oplus {(\mathcal{T}^{(p)})_{\mathcal{U},0}}^*$; hence by Theorem 10.41 (in the Notes and Remarks section), it is isomorphic to $V_1 \oplus V_2$, where V_1 is complemented in ${\mathcal{T}^{(p)}}^*$ and V_2 is complemented in ${(\mathcal{T}^{(p)})_{\mathcal{U},0}}^*$. If $V_1 \subset W^*$ were infinite-dimensional, it would have had to contain a subspace isomorphic to l_s for some s. Since this is not so, it follows that W^* is isomorphic to V_2, which is a complemented subspace of an $\mathcal{L}_q$-space. Thus W is not only a quotient of an $\mathcal{L}_p$-space, but is actually an $\mathcal{L}_p$-space.

Since $\mathcal{T}^{(p)}$ and E are separable, and since a separable subspace of an $\mathcal{L}_p$-space is isomorphic to a complemented subspace of L_p, a standard back and forth argument shows that there are complemented subspaces W_1 and W_2 of L_p such that $E \oplus W_1$ is isomorphic to $\mathcal{T}^{(p)} \oplus W_2$. Using the fact that $W_i \oplus L_p$ is isomorphic to L_p, we obtain that there is a surjective isomorphism S from $E \oplus L_p$ onto $\mathcal{T}^{(p)} \oplus L_p$. We now claim that $E \oplus l_p$ is isomorphic to $\mathcal{T}^{(p)} \oplus l_p$.

The proof is similar to that of Theorem 10.15. For $p = 2$ there is nothing to prove. Otherwise Proposition 10.14 implies that E does not contain a subspace isomorphic to l_2, and by duality we can assume that $p > 2$. By Theorem F.2(vii) every operator from E to L_p factors through l_p. Thus the image of E under S is actually isomorphic to a subspace of $\mathcal{T}^{(p)} \oplus l_p$. Since the image SL_p is of the form $V_1 \oplus V_2$ where $V_1 \subset \mathcal{T}^{(p)}$ is finite-dimensional, the claim follows.

To finish the proof we distinguish between two cases:

If E contains a complemented subspace isomorphic to l_p, then $l_p = l_p \oplus l_p$ implies that $E = E \oplus l_p$, and hence it is isomorphic to $\mathcal{T}^{(p)} \oplus l_p$.

If E does not contain a complemented subspace isomorphic to l_p, we analyze the isomorphism between $E \oplus l_p$ and $\mathcal{T}^{(p)} \oplus l_p$. By Theorem 10.41 $E = F_1 \oplus F_2$, where F_1 is isomorphic to a complemented subspace of $\mathcal{T}^{(p)}$ and F_2 to a complemented subspace of l_p. Similarly $l_p = G_1 \oplus G_2$, where G_1 is isomorphic to a complemented subspace of $\mathcal{T}^{(p)}$ and G_2 to a complemented subspace of l_p. But F_2 is finite-dimensional (because E does not contain a complemented copy of l_p and every subspace of l_p does), and G_1 is finite-dimensional (because $\mathcal{T}^{(p)}$ does not contain a subspace isomorphic to l_p). Thus E is isomorphic to a direct sum of a finite-dimensional space and a finite-codimensional subspace of $\mathcal{T}^{(p)}$. Since $\mathcal{T}^{(p)}$ is isomorphic to its finite-codimensional subspaces, E and $\mathcal{T}^{(p)}$ are isomorphic. □

Remark. The linear structure of $\mathcal{T}^{(p)}$ is uniquely determined by its Lipschitz structure. Indeed, otherwise the theorem implies that $\mathcal{T}^{(p)} \oplus l_p$ is Lipschitz equivalent to $\mathcal{T}^{(p)}$. By differentiation, as in the beginning of Chapter 7, this would imply that $\mathcal{T}^{(p)}$ contains a subspace isomorphic to l_p, which is impossible.

The next proposition is stated without proof. Its proof follows the same lines as that of the theorem.

Proposition 10.40. *Let $\{p_i\}_{i=1}^n$ be n different numbers with $1 < p_i < 2$ for all i, or $2 < p_i < \infty$ for all i. Then a Banach space is uniformly homeomorphic to $\left(\sum_{i=1}^n \oplus \mathcal{T}^{(p_i)}\right)$ if and only if it is isomorphic to one of the 2^n isomorphically distinct spaces $\left(\sum_{i=1}^n \oplus \mathcal{T}^{(p_i)}\right) \oplus \left(\sum_{i \in A} l_{p_i}\right)$ for some subset $A \subset \{1, \dots, n\}$.*

6. Notes and Remarks

The systematic application of ultraproducts to uniform classification of Banach spaces is due to Heinrich and Mankiewicz [**252**]. In particular, Theorem 10.1, Corollary 10.2, Proposition 10.4, and the proofs of Corollary 10.3, Theorem 10.5, the first proof of Theorem 10.6, and the proof of Theorem 10.7 are taken from this article.

Corollary 10.3 was observed by Ribe [**492**]. The result in Remark (i) after the corollary is originally due to Lindenstrauss [**357**] when $\max(p, q) > 2$, and to Enflo [**169**] when $\max(p, q) \leq 2$. Remark (ii) was observed by Assaf Naor.

Theorem 10.5 is due to Ribe [**493**] for $1 < p < \infty$, and to Heinrich and Mankiewicz [**252**] for $p = \infty$.

Theorem 10.6 is due to Ribe [**492**], and the second proof follows his original argument. See also the presentation in Enflo [**175**]. Its stronger form for superreflexive spaces, Theorem 10.7, is also due to Ribe [**493**].

Theorem 10.8 was announced by Grothendieck [**234**] with some indications of a proof. The first complete published proof is due to Henkin [**253**], and we reproduced it here. Henkin actually showed that $C^k(S^n)$ is not a uniform absolute retract for any $k \geq 1$ and $n \geq 2$, but we presented only the technically somewhat simpler case $k = 1$. It should be noted (as Henkin does) that $C^k(S^n)$ is isomorphic to $C^k(I^n)$, where I^n is the n-dimensional cube. For $n = 1$ the kth derivative operator $D^k : f \to f^{(k)}$ is an isomorphism of a subspace of codimension k of $C^k(I)$ onto

$C(I)$; hence $C^k(I)$ and $C(I)$ are isomorphic to each other. Kislyakov [**328**] proved that for $k \geq 1$ and $n \geq 2$, the space $C^k(I^n)$ is not isomorphic to an $\mathcal{L}_\infty$-space (not even to a quotient of an $\mathcal{L}_\infty$-space). It follows from Theorem 10.5 that it is not uniformly homeomorphic to an $\mathcal{L}_\infty$-space.

Using the approach of Kislyakov, Pełczyński and Senator [**455**] proved the following generalization of his result. Let W be a finite subset of $(\mathbb{Z}^+)^n$, with the property that if $k = (k_1, \dots, k_n) \in W$ and if $0 \leq h_i \leq k_i$ for every $i \leq n$, then also $h = (h_1, \dots, h_n) \in W$. Denote by C_W the space of all functions on the n-dimensional torus $\mathbb{T}^n$ for which the derivatives $\partial^{|k|} f / \partial^{k_1} x_1 \dots \partial^{k_n} x_n$ exist and are continuous for every $k \in W$. The norm on C_W is the maximum of the sup norms of all these derivatives. Then there are two possibilities. If there is a $\tilde{k} \in W$ such that $k \leq \tilde{k}$ for every $k \in W$, then C_W is isomorphic to $C(\mathbb{T}^n)$. If no such $\tilde{k}$ exists, then C_W is not an $\mathcal{L}_\infty$-space.

The use of midpoints and approximate midpoints and the idea to compare their "size" are due to Enflo in his unpublished proof of the case $p = 1$ of Theorem 10.13.

Theorem 10.12 is a fundamental result of Gorelik [**222**] as formulated by Johnson, Lindenstrauss and Schechtman [**295**]. It is called by them the "Gorelik principle". The remark following the theorem is due to Godefroy, Kalton and Lancien [**212**].

The case $1 < p < 2$ of Theorem 10.13 is due to Bourgain [**87**], and the case $p > 2$ is due to Gorelik [**222**].

Weston [**582**] proved that L_p and l_q are not uniformly homeomorphic also for any $0 < p, q \leq 1$. (The metric in L_p for $p < 1$ is $d_p(x, y) = \int |x - y|^p$, and d_q is defined similarly in l_q for $q < 1$.) The proof is modeled after the proof for $p = q = 1$. The new ingredient is to replace the metric d_q on l_q for $q < 1$ by the metric

$$d(x, y) = \sum \phi(|x_i - y_i|)$$

where $x = (x_i)$, $y = (y_i)$ and where $\phi(t) = t^q$ for $0 \leq t \leq 1$ and $\phi(t) = t$ for $t \geq 1$. The new metric d is uniformly homeomorphic to d_q and is approximately metrically convex for large distances in the sense that if $d(x_1, x_2) > 2$, then there is a point y such that $\frac{1}{2} d(x_1, x_2) - 1 \leq d(x_i, y) \leq \frac{1}{2} d(x_1, x_2) + 1$. It follows that uniformly continuous maps between (l_q, d) and the metrically convex L_p satisfy a Lipschitz condition for large distances. The same argument as in L_1 shows that for every $0 \neq x \in L_p$ there is an infinite sequence $\{y_j\}$ of exact midpoints between x and $-x$ such that $d_p(y_i, y_j) = d_p(x, 0)$ for every $i \neq j$. On the other hand, the set of approximate midpoints (in the d-metric) between $x \in l_q$ and $-x$ is contained in a small perturbation of a compact set.

Lövblom [**376, 377**] used the methods of Theorem 10.13 to give a lower estimate on the modulus of continuity of a uniform homeomorphism φ from the unit ball $B(L_p)$ onto $B(l_p)$ for $1 \leq p < 2$. She showed that there are $\varepsilon_n \to 0$ and a constant $K > 0$ such that $\omega_{\varphi^{-1}}(\omega_\varphi(\varepsilon_n)) \geq K \varepsilon_n |\log \varepsilon_n|^\gamma$ for $\gamma = 1$ when $p = 1$, and for every $\gamma < (2 - p)/2p$ when $1 < p < 2$. It is unknown what the best estimates are.

Proposition 10.14 and Theorems 10.15 and 10.16 are due to Johnson, Lindenstrauss and Schechtman [**295**].

Recall that a bounded linear operator $T : E \to F$ is strictly singular if it is not an isomorphism on any infinite-dimensional subspace of E. The following theorem is due to Edelstein and Wojtaszczyk [**162**]. For a proof and a detailed exposition of the necessary background, see Section 2.c of Lindenstrauss and Tzafriri [**371**].

Theorem 10.41. *Let E and F be Banach spaces such that every operator from E into F is strictly singular, and let G be a complemented subspace of $E \oplus F$. Then there are complemented subspaces E_0 of E and F_0 of F such that G is isomorphic to $E_0 \oplus F_0$.*

Compact operators are, of course, strictly singular. By Proposition 2.c.3 in Lindenstrauss and Tzafriri [**371**] every operator from l_p to l_q $(p > q)$ is compact. The same argument shows that if $p > q_j$ for $j \leq m$, then every operator from l_p to $\sum_{j=1}^{m} \oplus l_{q_j}$ is compact, hence strictly singular. It follows that if E is a complemented subspace of $\sum_{i=1}^{n} \oplus l_{p_i}$ where $p_i \neq p_j$ for $i \neq j$, then E is isomorphic to $\sum_{i=1}^{n} \oplus E_i$, where each E_i is a complemented subspace in l_{p_i}. Since an infinite-dimensional complemented subspace of l_p is isomorphic to l_p, it follows that E is finite-dimensional or is isomorphic to $\sum_{i \in A} \oplus l_{p_i}$ for some subset $A \subset \{1, \dots, n\}$.

Johnson, Lindenstrauss and Schechtman [**295**] proved that Theorem 10.15 also holds in the nonseparable case; i.e., every Banach space which is uniformly homeomorphic to $l_p(\Gamma)$ is isomorphic to it.

Godefroy, Kalton and Lancien [**211**] proved that if E is uniformly homeomorphic to a subspace of l_p (for $1 < p < \infty$), and if E has an unconditional FDD, then E is isomorphic to a subspace of l_p.

Mazurkiewicz and Sierpiński [**408**] proved that countable compact metric spaces are homeomorphic to spaces of ordinals with the order topology. The isomorphic classification of $C(\alpha)$ for countable ordinals is due to Bessaga and Pełczyński [**60**]. It is unknown whether every separable complemented subspace E of a $C(K)$-space is isomorphic to $C(S)$ for some compact metric space S. If E^* is nonseparable, then this was proved to be true by Rosenthal [**502**].

Let F be a Banach space with a separable dual. Szlenk [**551**] associated to F an ordinal-valued function $\eta(F, \varepsilon)$ which is defined for $\varepsilon > 0$ and is now called the Szlenk index of F. The index of c_0 is finite for every $\varepsilon > 0$, and it follows from the proof of Theorem 2 in Benyamini [**48**] that a complemented subspace E of a $C(K)$-space for which $\eta(E, \varepsilon)$ is finite for every $\varepsilon > 0$ is a quotient of c_0. By Johnson and Zippin [**300**] (see also Alspach [**9**]), such E is isomorphic to a subspace of c_0. Being a complemented subspace of a $C(K)$-space, it is actually isomorphic to c_0.

Alspach and Benyamini [**11**] proved that if E is an $\mathcal{L}_\infty$-space and if there is an $\varepsilon > 0$ such that $\eta(E, \varepsilon) \geq \omega$, then E has a quotient isomorphic to $C(\omega^\omega)$. By Theorem 10.16 such a space cannot be uniformly homeomorphic to c_0. If E is uniformly homeomorphic to c_0, then by Theorem 10.5 it is an $\mathcal{L}_\infty$-space. There are, however, $\mathcal{L}_\infty$-spaces E for which $\eta(E, \varepsilon)$ is finite for every $\varepsilon > 0$ but which are not isomorphic to c_0; see Alspach [**8**]. Thus Theorem 10.16 is not sufficient to show that c_0 has a unique uniform structure.

Godefroy, Kalton and Lancien [**211**] study the preservation of the Szlenk index (and of other, related, indices) by uniform homeomorphisms. They use these results to prove that if E is uniformly homeomorphic to a subspace of c_0, then E^* is separable, and that if E is uniformly homeomorphic to c_0 itself, then E^* is isomorphic to l_1.

Theorem 10.17 is due to Godefroy, Kalton and Lancien [**210, 212**]. These articles also contain some more results on spaces which are uniformly homeomorphic to c_0 and on spaces which are Lipschitz equivalent to $c_0(\Gamma)$ for uncountable Γ.

We can now complete the proof of Theorem 7.13. The argument is due to Godefroy, and appears in Johnson, Lindenstrauss and Schechtman [**295**]. Assume that $C(K)$ is uniformly homeomorphic to $c_0(\Gamma)$. By a back and forth argument, every separable subspace of $C(K)$ is contained in a separable subalgebra containing 1 which is uniformly homeomorphic to c_0. But such subalgebras are isometric to $C(S)$-spaces; hence by Theorem 10.16 they are isomorphic to c_0. By Pełczyński and Semadeni [**454**] K must be scattered, and by Baker [**34**] $K^{(\omega)} = \emptyset$. The preceding argument and Theorem 7.13 imply that if E is a $C(K)$ space which is uniformly homeomorphic to $c_0(\Gamma)$, then E is actually Lipschitz equivalent to it.

The techniques which are presently available to show that the uniform structure of certain spaces determines their linear structure are very limited. It is unknown whether this is the case for many natural spaces, for example, l_1, c_0, $L_p(0,1)$ for $1 \le p < \infty$ ($p \ne 2$), and more generally for $\mathcal{L}_p$-spaces ($p \ne 2$) except for l_p, for $l_p \oplus l_q$ when $p < 2 < q$ or when one of p or q is 2, and for separable $C(K)$ spaces.

One can also ask more restricted questions. For example, it is unknown if the uniform and linear classification of separable $C(K)$-spaces coincide.

We mention two results on metric linear spaces. The first, due to Mankiewicz [**382**] (see also further results in [**384**]), is that the uniform structure of $s = \mathbb{R}^{\aleph_0}$ with the standard seminorms uniquely determines its linear structure as a Fréchet space; i.e., if a Fréchet space is uniformly homeomorphic to s, then it is isomorphic to it. The second is that if E is a quasi-Banach space which is uniformly homeomorphic to a metric space X with a homeomorphism f such that both f and f^{-1} satisfy a Lipschitz condition for large distances, then E is, in fact, normable. This was claimed by Ribe [**493**] for general uniform homeomorphisms (and X a Banach space), but his proof uses the Lipschitz condition for large distances, which is not automatic for uniformly continuous maps on quasi-Banach spaces. The following short proof was shown to us by Assaf Naor. Denote the metric on X by d, and choose a constant L such that $\|x-y\| \ge 1$ implies that $\|x-y\|/L \le d\big(f(x), f(y)\big) \le L\|x-y\|$. By the first part of the proof of Proposition H.3 it is enough to show that $\|\sum_{i=1}^n x_i\| \le L^2 n$ whenever $\|x_i\| \le 1$ in E. So choose such x_i's, and put $y_m = \sum_{i=1}^m x_i$ (with $y_0 = 0$) and $\alpha = \min \|x_i\|$. Then

$$\begin{aligned}\Big\|\sum_{i=1}^n x_i\Big\| &= \alpha\Big\|\sum_{i=1}^n \alpha^{-1}x_i\Big\| \le \alpha L d\big(f(\alpha^{-1}y_n), f(0)\big)\\ &\le \alpha L \sum_{m=1}^n d\big(f(\alpha^{-1}y_m), f(\alpha^{-1}y_{m-1})\big) \le L^2 n.\end{aligned}$$

We also mention one result on uniform homeomorphism of nonmetrizable linear spaces. Let K be a compact metric space, and denote by $C_p(K)$ the space of real-valued continuous functions on K in the topology of pointwise convergence. Gulko [**240**] proved that if $C_p(K)$ and $C_p(H)$ are uniformly homeomorphic (with respect to the natural uniform structures), then K and H have the same dimension.

Note that under the norm topology, all $C(K)$-spaces when K is an uncountable compact metric space are even linearly isomorphic; hence no such result is possible for them. This is the famous theorem of Milutin, see for example Pełczyński [**451**].

Proposition 10.19 was observed in Johnson, Lindenstrauss and Schechtman [**295**]. Proposition 10.20 and Theorem 10.21 are new.

We have not checked what exact assumptions are needed for Proposition 10.20 to hold for spaces other than l_p. The same proof works whenever the space has a basis and when its unit sphere is uniformly homeomorphic to the unit sphere of a Hilbert space. It also holds for $E = c_0$, where the existence of the extension of f in the proof of (i) follows from the fact that c_0 is an absolute Lipschitz retract.

Proposition 10.22 was observed by Lindenstrauss, Matoušková and Preiss [**363**].

The question whether nets in $\mathbb{R}^n$ are mutually Lipschitz equivalent was posed independently by Gromov in [**229**] in the context of his studies of manifolds (and more general metric spaces) in terms of the geometry of nets in them, and by Furstenberg, who was studying ways to analyze the dynamics of two-parameter group actions (i.e., actions of $\mathbb{R}^2$) by passing to discrete nets in the groups.

Moser [**429**] proved the existence of C^∞ diffeomorphism of the unit cube in $\mathbb{R}^n$ with a given strictly positive C^∞ Jacobian. We refer to Rivière and Ye [**498**] for the history and for recent results on this subject in other classes of functions.

The proof of part (ii) of Theorem 10.26 (and of (ii)$\Rightarrow$(i) in Proposition 10.25) is due independently to McMullen [**410**] and to Burago and Kleiner [**106**], and we reproduced their proof. This gives a negative solution of the Gromov-Furstenberg problem. McMullen also proved (i)$\Rightarrow$(ii) in Proposition 10.25 and showed how the results of Rivière and Ye [**498**] give part (i) of Theorem 10.26.

The method of proof of Theorem 10.28 is due to Ribe [**495**], who introduced it to prove the case $p = 1$ of Corollary 10.29. The method was later modified by Aharoni and Lindenstrauss [**4**] so that it also applies to $p > 1$ (see also the exposition in Benyamini [**52**]). It is not known whether for p_n, p and q as in Corollary 10.29 the spaces $(\sum \oplus l_{p_n})_q$ and $l_p \oplus (\sum \oplus l_{p_n})_q$ are the only Banach spaces which are uniformly homeomorphic to $(\sum \oplus l_{p_n})_q$.

Corollary10.30 is from Johnson, Lindenstrauss and Schechtman [**295**]. (They formulated it with the additional assumption that the Schauder decomposition is unconditional.)

The only properties of l_p that were used in Theorem 10.28 in an essential way (in the construction of $\{h_\tau\}$) were that it is isomorphic to its square and that its general linear group is arcwise connected. Thus the same proof works, for example, with L_p instead of l_p. The necessary conditions for the construction of $\{h_\tau\}$ can be further reduced and be formulated in terms of genuinely nonlinear maps instead of the construction presented here, where the basic maps S_τ were linear. At the moment we do not have any examples where this is actually useful.

Tsirelson [**563**] constructed the first example of a Banach space which does not contain a subspace isomorphic to c_0 or to l_p for any $1 \leq p < \infty$. The space T described here is the dual of Tsirelson's original space, and we presented its description by Figiel and Johnson [**189**]. They also introduced the idea of convexifying the space. By first convexifying T, and then concavifying the resulting space, they obtained a uniformly convex space which does not contain a subspace isomorphic to l_p for any $1 < p < \infty$.

The modification that uses allowable rather than admissible collections is due to Johnson [**289**], and we refer to this article for the missing part in the proof of Proposition 10.32 (i.e., that $\mathcal{T}$ does not contain a subspace isomorphic to l_1). In fact Casazza and Odell [**111**] proved that T and $\mathcal{T}$ are isomorphic. This result is deeper and we do not use it.

Casazza, Johnson and Tzafriri [**109**] studied the structure of spaces generated by subsequences of the natural basis of T. Changing the definition of admissible (or allowable) collections so as to allow collections of $(n+1)^n$ sets rather than n sets amounts to passing to the subsequence $\{e_{(n+1)^n}\}$, and it follows from their results that this subsequence is equivalent to all of $\{e_n\}$.

We refer to Casazza and Shura [**112**] and its references for more details on Tsirelson-like spaces.

The results starting with Proposition 10.34 are due to Johnson, Lindenstrauss and Schechtman [**295**]. (Except Theorem 10.36, which is due to Lindenstrauss and Preiss [**366**]. Its proof will be given in Volume 2.) Their proof of Corollary 10.30 holds also for $p = 1$; hence $\mathcal{T} \oplus l_1$ is also uniformly homeomorphic to $\mathcal{T}$. At the moment there is no known example of a Banach space whose uniform structure determines exactly n different linear structures when n is not a power of 2.

The results of Johnson, Lindenstrauss and Schechtman [**295**] are formulated for the class of "asymptotic $\mathcal{L}_p$-spaces". A Banach space E is in this class if there is a λ with the following property: for every n there is a finite-codimensional subspace $F \subset E$ such that every n-dimensional subspace of F is contained in some subspace G of F which is λ-isomorphic to l_p^k, where $k = \dim G$. The Tsirelson spaces are the main examples of asymptotic $\mathcal{L}_p$-spaces, and the proofs in the general case are the same as for the Tsirelson spaces; therefore we restricted the exposition to them.

The proof of Theorem 10.39 uses the fact that every subspace of L_p contains a subspace isomorphic to l_s for some s. This was proved by Kadec and Pełczyński [**304**] for $p > 2$ and by Aldous [**7**] for $1 \le p < 2$. The result also follows from the theory of stable Banach spaces developed by Krivine and Maurey [**339**].

It is also interesting to classify Banach spaces with respect to various classes of nonlinear maps (other than uniformly continuous and Lipschitz maps) that appear in the literature. One such class is the class of quasi-symmetric maps, which was introduced and studied in the context of general metric spaces by Tukia and Väisälä [**564**]. For a detailed study of such maps in Banach spaces see Väisälä [**570, 571, 573, 574, 575**]. Let U be an open subset of a Banach space E. A map $f : U \to F$ is called quasi-symmetric if there is a constant $M > 0$ such that whenever $u, v, w \in U$ satisfy $\|u - v\| \le \|u - w\|$, then $\|f(u) - f(v)\| \le M\|f(u) - f(w)\|$. A Lipschitz equivalence is clearly also a quasi-symmetric equivalence. Not much is known on the quasi-symmetric classification of Banach spaces. For example, Väisälä asked whether there is a homeomorphism between l_p and l_q which is bi-quasi-symmetric, i.e., such that both f and f^{-1} are quasi-symmetric. One could ask more generally whether l_p is determined by its "quasi-symmetric structure". Note that if f is quasi-symmetric in an open set U and is Gâteaux differentiable at a point $x_0 \in U$, then its derivative $D_f(x_0)$ is either 0 or an into isomorphism. It follows that if f is a Lipschitz quasi-symmetric map from an open connected set U in a separable Banach space E into a Banach space F which has the Radon-Nikodým property and does not contain a subspace isomorphic to E, then f must be constant (this observation is due to Väisälä and Lindenstrauss). Indeed, by Theorem 6.42 f is Gâteaux differentiable except for a Gauss null set. Since F does not contain a subspace isomorphic to E, this derivative is equal to 0 whenever it exists. Let L be a line in E. The restriction of f to the translate $x+L$ of L has 0 derivative a.e. on $x+L$ except for a Gauss null set of x's. Since f is Lipschitz, it follows that these restrictions are constants, and since the complement of a Gauss null set is dense, it follows that f is actually constant on every line parallel to L. Since L is arbitrary, the result follows.

Another way to classify spaces, finer than the topological classification, is by imposing some standard additional property on the homeomorphism f (such as differentiability, analyticity, etc.), but not on its inverse f^{-1}. The following example, due to Pełczyński [**452**], shows that there are nonisomorphic real Banach spaces E and F and (not uniform) homeomorphisms $f : E \to F$ and $g : F \to E$ such that both f and g are real analytic. (The maps are actually given by third degree polynomials.)

For $k \geq 1$ put $F_k = l_{p_k}$ where $p_k = 3^k$, and let $E = (F_1 \oplus F_3 \oplus F_5 \oplus \ldots)_2$ and $F = (F_1 \oplus F_2 \oplus F_4 \oplus F_6 \oplus \ldots)_2$. The polynomial x^3 is one-one; hence for each $j > 1$ the map $h : l_{p_j} \to l_{p_{j-1}}$ given by $h(a_1, a_2, \ldots) = \big(a_1^3, a_2^3, \ldots\big)$ is a well-defined surjective real analytic (even polynomial) homeomorphism. Define $f : E \to F$ by $f(x_1, x_3, x_5 \ldots) = \big(x_1, h(x_3), h(x_5), \ldots\big)$. To define $g : F \to E$, represent E as $E = (\tilde{F_1} \oplus F_1 \oplus F_3 \oplus \ldots)_2$ where $\tilde{F_1}$ is also isometric to l_{p_1}, and set $g(x_1, x_2, x_4 \ldots) = \big(x_1, h(x_2), h(x_4), \ldots\big)$.

The following proposition of Bonic and Frampton [**77**] leads to a related result (in the opposite direction).

Proposition 10.42. *Let U be a bounded open set in l_1, and let F be a Banach space. Assume that f is a continuous map from the closure $\overline{U}$ of U to F such that f is Fréchet differentiable in U and such that $D_f(u)$ is strictly singular for every $u \in U$. Then $f(\partial U)$ is dense in $f(\overline{U})$.*

Proof. Fix $z_0 \in f(U)$ to be approximated by $f(\partial U)$, and assume that $0 \in U$, that $z_0 = f(0) = 0$ and that U is contained in the unit ball of l_1. Fix $\varepsilon > 0$, and let Ω be the set of all 1-Lipschitz curves $\omega : [0, \alpha] \to \overline{U}$ such that $\omega(0) = 0$, $\|\omega(\alpha)\| \geq \alpha/2$ and $\|f \circ \omega(\alpha)\| \leq \varepsilon\alpha/2$ (α is not fixed and depends on ω). Let $\omega_0 : [0, \alpha_0] \to F$ be maximal with respect to the natural order on Ω, i.e., $\omega_1 \prec \omega_2$ if and only if ω_2 extends ω_1 to a bigger interval. (Such an ω_0 exists by Zorn's lemma.) Note that

$$\|f \circ \omega_0(\alpha_0)\| \leq \varepsilon\alpha_0/2 \leq \varepsilon\|\omega(\alpha_0)\| \leq \varepsilon.$$

Since ε was arbitrary we need to show only that $x_0 = \omega_0(\alpha_0) \in \partial U$. Assume that this is false, i.e., $x_0 \in U$. Since f is Fréchet differentiable at x_0 there is $\delta > 0$ such that $B(x_0, \delta) \subset U$ and

$$\|f(x_0 + y) - f(x_0) - D_f(x_0)y\| \leq \varepsilon\delta/4$$

for every y with $\|y\| \leq \delta$. Let E_n be the subspace of l_1 of all vectors whose first n coordinates vanish, and choose n large enough so that $\|x_0 + y\| \geq \|x_0\| + \delta/2$ for every $y \in E_n$ with $\|y\| = \delta$. Since $D_f(x_0)$ is strictly singular, there is a $y_0 \in E_n$ with $\|y_0\| = \delta$ such that $\|D_f(x_0)y_0\| \leq \varepsilon\delta/4$. We now construct an extension ω_1

of ω_0, contradicting its maximality, by defining $\omega_1(t) = x_0 + \delta^{-1}(t - \alpha_0)y_0$ for $\alpha_0 \le t \le \alpha_0 + \delta$. Then ω_1 is 1-Lipschitz and it satisfies

$$\|\omega_1(\alpha_0 + \delta)\| = \|x_0 + y_0\| \ge \|x_0\| + \delta/2 \ge (\alpha_0 + \delta)/2$$

and

$$\begin{aligned}\|f \circ \omega_1(\alpha_0 + \delta)\| &= \|f(x_0 + y_0)\| \\ &\le \|f(x_0 + y_0) - f(x_0) - D_f(x_0)y_0\| + \|f(x_0)\| + \|D_f(x_0)y_0\| \\ &\le \varepsilon\delta/4 + \varepsilon\alpha_0/2 + \varepsilon\delta/4 = \varepsilon(\alpha_0 + \delta)/2.\end{aligned}$$

□

The same result holds, of course, when l_1 is replaced by any space E containing it. It follows in particular that if E contains a subspace isomorphic to l_1 and if every bounded linear operator from E to F is strictly singular, then there is no homeomorphism from an open set in E into F which is Fréchet differentiable.

The following variant of Proposition 10.42 is proved (with a different proof) in page 99 of Deville, Godefroy and Zizler [**144**]. It applies to a somewhat larger class of Banach spaces, but imposes somewhat stronger conditions on f.

Proposition 10.43. *Let E be a separable Banach space with a nonseparable dual. Let U be a bounded open set in E, and let f be a continuous map from $\overline{U}$ to a Banach space F which is Fréchet differentiable on U such that $D_f(u)$ is compact for every $u \in U$. Then $f(\partial U)$ is dense in $f(\overline{U})$.*

It follows that general Fréchet differentiable maps with E and U as above (but without the compactness assumption on $D_f(u)$) satisfy a "maximum principle".

Corollary 10.44. *Let E, U, F be as above, and let $f : \overline{U} \to F$ be a continuous map which is Fréchet differentiable on U. Then $f(\partial U)$ is weakly dense in $f(\overline{U})$. In particular,* $\sup\{\|f(u)\| : u \in \partial U\} = \sup\{\|f(u)\| : u \in \overline{U}\}$.

Proof. We need to show that for each $u_0 \in U$ and $\{y_i^*\}_{i=1}^n \subset F^*$ there is a point $y \in f(\partial U)$ such that $|\langle y_i^*, y - f(u_0)\rangle| < 1$ for all i. To this end apply Proposition 10.43 to the map $u \to (\langle y_1^*, f(u)\rangle, \dots, \langle y_n^*, f(u)\rangle)$ from $\overline{U}$ to $\mathbb{R}^n$. □

The corollary characterizes spaces with a nonseparable dual among all separable spaces. Indeed, if E^* is separable, then by Theorem 4.13 there is a Fréchet differentiable (even C^1) bump function f on E. This f (with $F = \mathbb{R}$ and U a bounded open set containing the support of f) does not satisfy the conclusion of the corollary.

If f is a Fréchet differentiable uniform homeomorphism from U onto an open set V in F, and if φ is a Fréchet differentiable bump function supported in V, then $\varphi \circ f$ is a Fréchet differentiable bump function supported in U. Hence

Corollary 10.45. *Let E and F be separable Banach space with E^* nonseparable and F^* separable. Then there is no Fréchet differentiable uniform homeomorphism between an open set in E and an open set in F.*

The Mazur map from the open unit ball of L_1 onto the ball in L_2 shows that the corollary fails when the homeomorphism is not assumed to be Fréchet differentiable. Note that the inverse of this map, namely, the Mazur map from the unit ball of L_2 onto the ball of L_1 is a C^1 map.

CHAPTER 11

Nonlinear Quotient Maps

1. Surjective Lipschitz and Smooth Maps

In the first section we shall show that for any two separable Banach spaces E and F there is a surjective C^1 Lipschitz map from E onto F. Thus the existence of such a map gives no information on the structure of these spaces. For many separable spaces E the condition that the derivative is continuous can be strengthened, and it is possible to construct a surjective map whose derivative is Lipschitz on bounded sets (although our construction does not yield a globally Lipschitz map any more). But this is not always possible, e.g., for $E = c_0$.

In the second section we introduce and study a notion of quotient maps which is the proper notion in the uniform (resp. Lipschitz) category. This class of maps properly contains the compositions of linear quotient maps with uniform homeomorphism (Lipschitz equivalences), and if F is the image of E under such a map, then often F is already a linear quotient map of E or at least resembles such a linear quotient space. We show, for example, that a uniform quotient of a Hilbert space is isomorphic to a Hilbert space.

Theorem 11.1. *Let E be a separable Banach space. Then for every separable Banach space F there is a surjective C^1 Lipschitz map φ from E onto F.*

Proof. Using Theorem E.3(v), and replacing E by an appropriate quotient space, we may assume that E has a normalized basis $\{x_n\}$ with biorthogonal functionals $\{x_n^*\}$. Put $Q = \{x \in E : \; |\langle x_n^*, x\rangle| \le 1 \text{ for every } n\}$ and $Q' = 2Q$. For every increasing k-tuple of positive integers $(i_1, \dots, i_k)$, consider the translated dilations

$$Q_{i_1,\dots,i_k} = \sum_{j=1}^{k} 10^{-j} x_{i_j} + 2 \cdot 10^{-(k+1)} Q \; ; \quad Q'_{i_1,\dots,i_k} = \sum_{j=1}^{k} 10^{-j} x_{i_j} + 4 \cdot 10^{-(k+1)} Q.$$

Then

$$Q'_{i_1,\dots,i_k} \bigcap Q'_{j_1,\dots,j_k} = \emptyset \quad \text{whenever} \quad (i_1, \dots, i_k) \ne (j_1, \dots, j_k) \tag{11.1}$$

and the sets decrease in the strong sense that even

$$Q'_{i_1,\dots,i_k,i_{k+1}} \subset Q_{i_1,\dots,i_k}. \tag{11.2}$$

If $i_1, i_2, \dots$ is any increasing infinite sequence of positive integers, then the decreasing chain $\{Q_{i_1,\dots,i_k}\}$ intersects in the single point $\sum_{k=1}^{\infty} 10^{-k} x_{i_k}$.

To prove the theorem, we first construct an auxiliary C^1 Lipschitz function $f : E \to [0,1]$, supported in Q' and such that $f \equiv 1$ on Q. Consider the bounded linear operator $T : E \to c_0$ given by $Tx = \sum \langle x_n^*, x\rangle e_n$. Then T maps Q into the unit ball of c_0 and the complement of Q' into the complement of $2B(c_0)$. It follows that it is enough to find a C^1 Lipschitz function $g : c_0 \to [0,1]$, supported in $2B(c_0)$ and such that $g \equiv 1$ on $B(c_0)$, and then to take $f = g \circ T$. To construct such a function g, we use the fact that c_0^* is separable; hence by Theorem 4.13(ii) there is an equivalent Fréchet differentiable norm $|\!|\!|\cdot|\!|\!|$ on c_0 such that $4\|x\|/5 \le |\!|\!|x|\!|\!| \le 6\|x\|/5$ for every x. We then define $g(x) = h(|\!|\!|x|\!|\!|)$, where $h : \mathbb{R}^+ \to [0,1]$ is a C^1 Lipschitz function such that $h(t) = 1$ for $t \le 6/5$ and $h(t) = 0$ for $t \ge 8/5$.

Fix $k \ge 1$ and $(i_1, \dots, i_k)$, and let $L_{i_1,\dots,i_k}(x) = \frac{1}{2}10^{k+1}\bigl(x - \sum_{j=1}^k 10^{-j}x_{i_j}\bigr)$ be the affine map taking $Q_{i_1,\dots,i_k}$ onto Q and $Q'_{i_1,\dots,i_k}$ onto Q'. Let $\{y_j\}$ be a dense sequence in the unit ball of F, and define for each k

$$\psi_k(x) = \begin{cases} f\bigl(L_{i_1,\dots,i_k}(x)\bigr)y_{i_k} & x \in Q'_{i_1,\dots,i_k} \\ 0 & x \notin \bigcup Q'_{i_1,\dots,i_k}. \end{cases}$$

By (11.1) ψ_k is a well-defined C^1 function, and $\psi_k(x) = y_{i_k}$ for $x \in Q_{i_1,\dots,i_k}$. It is also Lipschitz with constant $10^{k+1}C/2$, where C is the Lipschitz constant of f. It follows that the series $\sum k^{-2}10^{-k}\psi_k$ converges in the Lipschitz norm to a Lipschitz map $\psi : E \to F$. Since the set of C^1 maps from E to F is closed in the space of Lipschitz maps between these spaces, it follows that ψ is also C^1. It is supported in Q', and we check that its image contains the ball of radius 10^{-1} in F. Indeed, if $\|y\| \le 10^{-1}$, then a standard successive approximation procedure, using the density of $\{y_j\}$ in $B(F)$, shows that there is an increasing sequence $\{i_k\}$ such that $y = \sum k^{-2}10^{-k}y_{i_k}$. Put $z = \sum 10^{-k}x_{i_k}$. Then $z \in Q_{i_1,\dots,i_k}$ for every k; hence $\psi_k(z) = y_{i_k}$, and it follows that $y = \sum k^{-2}10^{-k}y_{i_k} = \sum k^{-2}10^{-k}\psi_k(z) = \psi(z)$.

Finally, the sets $Q''_n = nQ' + 5^n x_1$ for $n \ge 1$ are pairwise disjoint. Thus the surjective map $\varphi(x) = \sum n\psi\bigl((x - 5^n x_1)/n\bigr)$ is a well-defined C^1 Lipschitz map. $\square$

The following lemma gives two alternative formulations for the condition which will be used in the next theorem.

Lemma 11.2. *Let $\{x_n^*\}$ be a normalized sequence in the dual E^* of a Banach space E. Then the following two conditions are equivalent.*

(i) *$N^{-1}\|\sum_{j=1}^N x_{n_j}^*\| \to 0$ as $N \to \infty$ for every subsequence $\{x_{n_j}^*\}$ of $\{x_n^*\}$.*

(ii) *For every $\varepsilon > 0$ there is a $k(\varepsilon)$ such that* $\mathrm{Card}\{j : |\langle x_j^*, x\rangle| \ge \varepsilon\} \le k(\varepsilon)$ *whenever $x \in E$ satisfies $\|x\| \le 1$.*

Proof. Assume that there is an $\varepsilon > 0$ for which (ii) is false. Choose an integer $p \ge 2/\varepsilon$, and find unit vectors $\{z_m\}$ in E and an infinite sequence $\{n_j\}$ such that $\langle x_{n_j}^*, z_m\rangle \ge \varepsilon$ whenever $p^{m-1} \le j < p^m$. Then

$$\begin{aligned} p^{-m}\Bigl\|\sum_{j<p^m} x_{n_j}^*\Bigr\| &\ge p^{-m}\Bigl(\sum_{p^{m-1}\le j<p^m}\langle x_{n_j}^*, z_m\rangle - \sum_{j<p^{m-1}}\|x_{n_j}^*\|\Bigr) \\ &\ge p^{-m}\bigl(\varepsilon(p^m - p^{m-1}) - p^{m-1}\bigr) \not\to 0 \end{aligned}$$

as $m \to \infty$ by the choice of p; hence (i) does not hold either.

Conversely, assume that (ii) holds, and let $\varepsilon > 0$. Let $N > k(\varepsilon)/\varepsilon$, and fix $x \in E$ with $\|x\| \le 1$. Since $\mathrm{Card}\{j : |\langle x_j^*, x\rangle| \ge \varepsilon\} \le k(\varepsilon)$, it follows that the estimate $\bigl\langle \sum_{j=1}^N x_{n_j}^*, x\bigr\rangle \le k(\varepsilon) + N\varepsilon \le 2\varepsilon N$ holds for any subsequence $\{x_{n_j}^*\}$. This estimate does not depend on x; hence $N^{-1}\|\sum_{j=1}^N x_{n_j}^*\| \le 2\varepsilon$. $\square$

Theorem 11.3. *Let E be a separable Banach space, and assume that E^* contains a normalized sequence satisfying one of the equivalent conditions of the previous lemma. Then for every separable Banach space F there is a surjective C^∞ map*

φ from E onto F whose derivatives are bounded on bounded sets. In particular φ and D_φ are Lipschitz on bounded sets.

Proof. The proof is similar to that of Theorem 11.1, but there are differences in the construction of the auxiliary function f and in some other details. By Theorem E.3(v) again and by passing to an appropriate subsequence, we may assume that E has a basis $\{x_n\}$ and that the given sequence $\{x_n^*\}$ is its sequence of biorthogonal functionals.

Now let $h : \mathbb{R} \to [0,1]$ be a C^∞ function, supported in $[-2,2]$ and such that $h(t) = 1$ for every $t \in [-1,1]$. The function f defined by

$$f(x) = \prod_{j=1}^{\infty} h(\langle x_j^*, x\rangle)$$

is supported in Q' and is identically 1 in Q. If $\|x\| \leq n$, then $|\langle x_j^*, x\rangle| < 1/2$ for all but at most $k(1/2n)$ indices j. Thus all but at most $k(1/2n)$ of the terms in the product are identically equal to 1 in the ball $B(x, 1/2)$. It follows that f is indeed a C^∞ function and that its derivatives are bounded on bounded sets. Put $\alpha_n = \sup\{\|D_f^{(j)}(x)\| : \|x\| \leq \frac{1}{2}10^{n+1}(n+1) \text{ and } j \leq n\}$.

The ψ_k's are defined as in Theorem 11.1. For each $z \in E$ and for each k there are a $\tilde{z} \in E$ with $\|\tilde{z}\| \leq 1$, a $y \in F$ with $\|y\| \leq 1$, and a neighborhood of z in which ψ_k has the form $\psi_k(x) = f\big(\frac{1}{2}10^{k+1}(x - \tilde{z})\big)y$. It follows that for each $\lambda > 0$ the derivatives of ψ_k satisfy

$$\sup\{\|D_{\psi_k}^{(n)}(x)\| : \|x\| \leq \lambda\} \leq \Big(\frac{1}{2}10^{k+1}\Big)^n \sup\Big\{\|D_f^{(n)}(x)\| : \|x\| \leq \frac{1}{2}10^{k+1}(\lambda+1)\Big\}.$$

In particular $\|D_{\psi_k}^{(j)}(x)\| \leq \frac{1}{2}10^{k(k+1)}\alpha_k$ on $kB(F)$ for all $j \leq k$, and the series $\sum k^{-2}10^{-k(k+1)}\alpha_k^{-1}\psi_k$ converges in the C^m-norm for every m and on every ball in E. Its sum ψ is a C^∞ map whose derivatives are bounded on bounded sets.

The proof that the image of ψ contains some ball in F and the definition of φ are similar to those in Theorem 11.1. (Since bounded sets intersect only a finite number of the sets Q_n'', the derivatives of φ are also bounded on bounded sets.)

That φ and D_φ are Lipschitz on bounded sets follows by the mean value theorem from the boundedness of the derivatives of φ on such sets. □

Remarks. (i) The spaces l_p for $p \geq 1$ satisfy the condition of Theorem 11.3. The same is true for any superreflexive space (see Theorem E.3(ii)). In particular, every separable Banach space is the image, under a surjective C^1 Lipschitz map or under a surjective C^∞ map which is Lipschitz on bounded sets, of a Hilbert space.

(ii) Theorems 11.1 and 11.3 should be compared with the classical Morse-Sard theorem in finite-dimensional spaces, namely, that if $n \geq m$ and if $\varphi : \mathbb{R}^n \to \mathbb{R}^m$ is C^{n-m+1}, then the set of critical values of φ has measure zero in $\mathbb{R}^m$. (A point $y \in \mathbb{R}^m$ is a critical value if $y = \varphi(x)$ where the rank of $D_\varphi(x)$ is smaller than m.) In particular, if φ is surjective, it follows that almost every $y \in \mathbb{R}^m$ is a regular (i.e., noncritical) value.

On the other hand, in both constructions above the derivatives of the surjective maps are such that the rank of $D_\varphi(x)$ is either zero or one for all $x \in E$. Indeed, assume first that $x \in Q'$ and that it is not of the form $\sum 10^{-j}x_{i_j}$. Then x belongs to some finite decreasing chain of the Q''s. By (11.2) there is neighborhood U of x such that the ψ's are constant in U for all but at most one k. Since ψ_k is of the form $g(\cdot)v$ where g is a scalar function and v is a fixed vector in F, it follows that the rank of $D_\varphi(x)$ is at most one. The same argument holds in all the translates $5nx_1 + Q'$, and thus the rank of $D_\varphi(x)$ is at most one on a dense subset of E. Since $D_\varphi(x)$ is continuous, this holds everywhere.

Thus Theorem 11.3 shows that the Morse-Sard theorem fails in a very strong sense for maps defined on infinite-dimensional spaces. Indeed, the map φ in Theorem 11.3 is as singular as a smooth nonconstant map could be.

While Theorem 11.1 holds for every separable Banach space E, Theorem 11.3 does not, and it turns out that the critical smoothness condition which makes the difference is the local uniform continuity of the derivative. (A map is said to be locally uniformly continuous if every $x \in E$ has a neighborhood on which the map is uniformly continuous. This condition is clearly satisfied when the map is uniformly continuous on bounded sets.) The condition of uniform continuity of the derivative will be used through (4.3) of Chapter 4.

In searching for spaces E for which Theorem 11.3 fails, we need to consider only spaces which do not contain a subspace isomorphic to l_1. To see this we first note that if $l_1 \subset E$, then $l_2 \subset E^*$. Indeed, the adjoint of the embedding of l_1 into E is a w^*-continuous quotient map $q : E^* \to l_\infty$. Let $F \subset l_\infty$ be isometric to L_1, and let $j : F \to E^*$ be a norm-one right-inverse for q on F. (To see that such j exists, let $F_n \subset F$ be isometric to l_1^n with $\bigcup F_n$ dense in F. For each n define an inverse on F_n, of norm smaller than $1 + n^{-1}$, by defining it appropriately on the unit vectors of l_1^n and extending it by linearly. The w^*-continuity of q allows passage to a convergent subsequence.) Since L_1 contains a subspace isometric to l_2, so does E^*. It follows that there is a normalized sequence $x_n^* \in E^*$ with $\| \sum_{i=1}^N x_{n_i}^* \| = N^{1/2}$ for each of its subsequences; hence E satisfies the conditions of Theorem 11.3.

That E does not contain a copy of l_1 will be used through Rosenthal's theorem (Theorem E.4(i)), namely, by the property that every bounded sequence in E has a weak Cauchy subsequence.

We now identify a class of Banach spaces E for which Theorem 11.3 fails.

Definition 11.4. *A Banach space E is said to be of class $\mathcal{C}$ if it does not contain l_1, and if there is a nonempty bounded open and convex subset U in E such that every differentiable function $f : U \to \mathbb{R}$ with a uniformly continuous derivative is weakly sequentially continuous; i.e., f takes weak Cauchy sequences in U to Cauchy sequences in $\mathbb{R}$.*

Note that if f is weakly sequentially continuous on U and if $x_n \in U$ converges to $x \in U$ weakly, then $f(x_n) \to f(x)$. Indeed, the sequence $\{y_n\}$ defined by $y_{2n} = x_n$ and $y_{2n+1} = x$ is weak Cauchy, and $\{f(y_n)\}$ necessarily converges to $f(x)$.

Proposition 11.5. *Let $E \in \mathcal{C}$ be a separable Banach space, and let F be superreflexive. Assume that $U \subset E$ is an open set as in Definition 11.4 and that $\varphi : U \to F$ is a differentiable mapping with a uniformly continuous derivative. Then $\varphi(U)$ is a relatively norm-compact subset of F.*

In particular, if F is infinite-dimensional, then there is no surjective differentiable map $\varphi : E \to F$ whose derivative is locally uniformly continuous.

Proof. If $\varphi(U)$ is not relatively compact, find $\delta > 0$ and a sequence $x_n \in U$ such that $\|\varphi(x_n) - \varphi(x_m)\| > 3\delta$ for all $n \neq m$. Since $l_1 \not\subset E$, we may assume by passing to a subsequence that $\{x_n\}$ is a weak Cauchy sequence.

Since F is superreflexive, there is a function $h : F \to \mathbb{R}$ with a uniformly continuous derivative such that $h(0) = 1$ and $h(y) = 0$ when $\|y\| \geq \delta$. Indeed, by Theorem A.6(iv') we may assume that the norm on F is uniformly smooth. Moreover, by Proposition A.5 the map $x \to \|x\|^2$ is differentiable with a uniformly continuous derivative on bounded sets. We then take a C^∞ function $h_0 : \mathbb{R} \to \mathbb{R}$ such that $h_0(0) = 1$ and $h_0(t) = 0$ for $t \geq \delta^2$, and $h(y) = h_0(\|y\|^2)$.

The real-valued function $f(x) = \sum_{n=1}^\infty h\big(\varphi(x) - \varphi(x_{2n})\big)$ has a uniformly continuous derivative on U and should, therefore, be weakly sequentially continuous. However, $f(x_{2n}) = 1$ and $f(x_{2n+1}) = 0$ by the construction.

To prove the second part, assume that such a map φ exists. Then for each $x \in E$ there is a neighborhood V, which is obtained from U by a translation and a dilation, on which φ is uniformly continuous. Since E is separable, we can cover E with a countable collection $\{V_n\}$ of such sets; and since dilations and translations of a set U which satisfies the condition as in Definition 11.4 also satisfy the same condition, it follows from the first part that all the sets $\varphi(V_n)$ are relatively compact. But this is impossible, as they cover the infinite-dimensional space F. □

Remark. The proposition also holds under conditions other than superreflexivity of F, but some conditions on F are necessary. Indeed, if $F = c_0$, then there is even a bounded *linear* noncompact operator from any separable Banach space E into c_0, or equivalently, a bounded sequence $x_n^* \in E^*$ with $w^* - \lim x_n^* = 0$ which is not norm-convergent.

The main tool for exhibiting concrete examples of spaces of class $\mathcal{C}$ is Proposition 11.7 below. We start with a technical preparation; we denote the standard basis in $\mathbb{R}^n$ by $\{e_i\}$.

Lemma 11.6. *For any two vectors $u, v \in \mathbb{R}^n$ there is a vector ξ with coordinates ± 1 such that*

$$\big|\langle \xi, u\rangle\big| + \big|\langle \xi, v\rangle\big| \leq 2\big(\|u\|_\infty + \|v\|_\infty\big).$$

Proof. Consider the operator $T : l_n^\infty \to \mathbb{R}^2$ given by $Ty = \big(\langle y, u\rangle, \langle y, v\rangle\big)$, and let $z = \sum a_i e_i$ be an extreme point in the unit ball of its kernel. Then there are at most two i's with $|a_i| < 1$. Indeed, if there were three such i's, say the first three, then since T is a rank-two operator, we could find a $y \neq 0$ in $\operatorname{span}\{e_1, e_2, e_3\} \cap \ker T$. Then $\|z \pm \varepsilon y\|_\infty \leq 1$ for small enough ε and $T(z \pm \varepsilon y) = 0$, contradicting the extremality of z. Replacing the a_i's for which $|a_i| < 1$ by their signs (and by 1, say, if $a_i = 0$) gives the desired ξ. □

Proposition 11.7. *For each $\varepsilon > 0$, $K > 0$ and modulus of continuity ω there is an $N = N(\varepsilon, K, \omega)$ with the following property: Let $n \geq N$, and let f be a real-valued differentiable function on the unit ball $B(l_n^\infty)$ of l_n^∞ such that $\|D_f(x)\| \leq K$ for every $x \in B(l_n^\infty)$ and such that D_f is uniformly continuous with modulus of continuity bounded by ω. Then $\min_{1\leq i\leq n} |f(e_i) - f(0)| < \varepsilon$.*

The same result also holds if f is defined only on $B(l_n^\infty)^+$, the set of vectors with nonnegative coordinates in $B(l_n^\infty)$.

Proof. We may assume that $f(0) = 0$. By passing to the span of a subset of $\{e_i\}_{i=1}^n$ whose cardinality is at least $n/2$, we may assume that the $f(e_i)$'s have the same sign, say $f(e_i) \geq 0$. Replacing f by the average $\frac{1}{n!}\sum_{\pi\in\Pi_n} f\Big(\sum a_{\pi(i)} e_i\Big)$

(where Π_n is the set of all permutations of $\{1, \dots, n\}$), we may also assume that f is symmetric with respect to Π_n, and we need to show that $f(e_1) < \varepsilon$.

Fix k such that $K/k + \omega(1/k) < \varepsilon/11$, and choose $N > 22K3^k/\varepsilon$. Put $x_0 = 0$ and $y_0 = e_1$, and define inductively x_j and y_j for $1 \le j \le k-1$ such that

(i) $x_j = y_j$ except that their first coordinate is 0 for the x_j's and 1 for the y_j's,

(ii) for each $1 \le i \le j$ the x_j's and y_j's have a coordinate whose value is i/k,

(iii) $|f(x_j) - f(x_{j-1})|, |f(y_j) - f(y_{j-1})| \le 5\varepsilon/11k$.

Once these vectors are defined, we finish the proof as follows. Put $x = x_{k-1}$ and $y = y_{k-1}$, and estimate by the triangle inequality

$$|f(e_1) - f(0)| \le |f(y) - f(x)| + \sum_{j=1}^{k-1} \Big(|f(y_j) - f(y_{j-1})| + |f(x_j) - f(x_{j-1})|\Big).$$

To estimate the first term, we use the symmetry of f. Let A be a set of cardinality $k-1$ where both x and y attain each of the values $1/k, 2/k, \dots, (k-1)/k$, and let $B = A \cup \{1\}$. By (i) the coordinates of $x + k^{-1}\chi_B$ are a permutation of those of y; hence $f(y) = f(x + k^{-1}\chi_B)$ by the symmetry of f. Since f is K-Lipschitz, it follows that $|f(y) - f(x)| \le K/k \le \varepsilon/11$ by the choice of k. The contribution of all the rest of the terms is smaller than $10\varepsilon/11$ by (iii), and this finishes the proof.

For the inductive construction of the x_j's and y_j's we add one more condition.

(iv) The vectors x_j and y_j take the constant value j/k on a set A_j of coordinates with cardinality $|A_j| \ge n/3^j$ (and $|A_0| = n - 1$ for $j = 0$).

Assume that x_{j-1} and y_{j-1} have been chosen, and put

$$B = \Big\{l \in A_{j-1} : \max\Big(\big|\langle D_f(x_{j-1}), e_l\rangle\big|, \big|\langle D_f(y_{j-1}), e_l\rangle\big|\Big) \le \varepsilon/11\Big\}.$$

By the definition of N and since $\|D_f(x_{j-1})\|, \|D_f(y_{j-1})\| \le K$, it follows that $|B| \ge 6n/3^j$. Let B' be a proper subset of B with $|B'| \ge 2n/3^j$.

By Lemma 11.6 there is a ξ which takes the values ± 1 on B' and the value zero outside B' such that $\big|\langle \xi, D_f(x_{j-1})\rangle\big| + \big|\langle \xi, D_f(y_{j-1})\rangle\big| \le 4\varepsilon/11$. Let C be the set where ξ takes the value $+1$, and assume, as we may, that $|C| \ge |B'|/2$. Now take $A_j = C$, $x_j = x_{j-1} + \xi/k$ and $y_j = y_{j-1} + \xi/k$.

Conditions (i), (ii) and (iv) are obvious. To check (iii) use the estimate (4.3) with $\|x_j - x_{j-1}\| = 1/k$ to obtain

$$\big|f(x_j) - f(x_{j-1}) - \langle D_f(x_{j-1}), \xi/k\rangle\big| \le \omega(1/k)/k.$$

But $\big|\langle D_f(x_{j-1}), \xi\rangle\big| \le 4\varepsilon/11$ by the choice of ξ and $\omega(1/k) \le \varepsilon/11$ by the choice of k; hence $|f(x_j) - f(x_{j-1})| \le 5\varepsilon/11k$, and a similar estimate also holds for $|f(y_j) - f(y_{j-1})|$.

For the last part, fix x_1 and y_1 whose first coordinate is 0 and 1 respectively and whose other coordinates are all equal to $1/k$. Then $\|f(x_1) - f(x_0)\| \le K/k$, and $\|f(y_1) - f(y_0)\| \le K/k$. We now repeat the same inductive procedure starting with the vectors x_1 and y_1. The result follows because all the vectors x_i and y_i that are constructed in the proof are now nonnegative. □

Theorem 11.8. *The space c_0 is in the class $\mathcal{C}$.*

Proof. We shall show that if $f : B(c_0) \to \mathbb{R}$ has a uniformly continuous derivative, and if f is not weakly sequentially continuous, then there is actually a weakly convergent sequence $y_n \to y$ such that $f(y_n) \not\to f(y)$ (and not just a weak

Cauchy sequence for which $\{f(y_n)\}$ does not converge). To see that this implies the theorem, assume, by passing to a subsequence and by a small perturbation, that $\inf |f(y_n) - f(y)| > 0$ and that $y_n = y + u_n$ where the supports of the u_n's are mutually disjoint and disjoint from the support of y. The function $g = f \circ \varphi$, where $\varphi(\sum a_n e_n) = y + \sum a_n u_n$, is also a C^1 function on $B(c_0)$ with a uniformly continuous derivative; hence $\inf |g(e_n) - g(0)| = 0$ by Proposition 11.7. But this is impossible, because $g(e_n) = f(y_n)$ and $g(0) = f(y)$.

To prove that such a sequence $\{y_n\}$ exists, we need the following claim.

Let $\{z_n\}$ be a weak Cauchy sequence in $B(c_0)$, and let $\varepsilon > 0$. Then there are an infinite set $M \subset \mathbb{N}$ and an integer k with the property that the set

$$\Big\{n \in M : \ \|z_n - v\| \leq 1 \text{ and } |f(z_n - v) - f(z_n)| \geq \varepsilon\Big\}$$

is finite for every vector $v \in B(c_0)$ whose first k coordinates vanish.

Assume that the claim is false. Since it does not hold for $k = 0$ and $M = \mathbb{N}$, there is a vector $v \in B(c_0)$ such that the set

$$M_1 = \Big\{n : \ \|z_n - v\| \leq 1 \text{ and } |f(z_n - v) - f(z_n)| \geq \varepsilon\Big\}$$

is infinite. A small perturbation of v yields k_1 and a vector $v_1 \in B(c_0)$ supported in $[1, k_1]$ such that $\|z_n - v_1\| \leq 1$ and $|f(z_n - v_1) - f(z_n)| > \varepsilon/2$ for every $n \in M_1$.

Since the claim does not hold for $k = k_1$ and $M = M_1$, we can repeat the same argument and then continue inductively to find a decreasing sequence of infinite subsets M_j of $\mathbb{N}$, an increasing sequence of integers $\{k_j\}$, and vectors $\{v_j\}$ in $B(c_0)$ supported in $[k_{j-1}+1, k_j]$ such that if $n \in M_j$, then $z_n - v_j \in B(c_0)$ and $|f(z_n - v_j) - f(z_n)| > \varepsilon/2$.

Let $N = N(\varepsilon/2, K, \omega)$ be as in Proposition 11.7, where ω is the modulus of continuity of D_f and $K = \sup\{\|D_f(x)\| : x \in B(c_0)\}$. Fix $m \in M_N$, and consider the map $g : B(l_N^\infty)^+ \to \mathbb{R}$ defined by $g\big(\sum_{j=1}^N a_j e_j\big) = f\big(z_m - \sum_{j=1}^N a_j v_j\big)$. Then g is well defined, and its derivative has the same modulus of continuity and bound as f; but $|g(e_j) - g(0)| = |f(z_m - v_j) - f(z_m)| > \varepsilon/2$ for every $j \leq N$, contradicting the choice of N.

Let $\{z_n\}$ be an arbitrary weak Cauchy sequence, and fix $\varepsilon > 0$. We now show that there is a subsequence $\{z_{n_i}\}$ and a weakly convergent sequence $\{y_i\}$ such that $|f(z_{n_i}) - f(y_i)| < \varepsilon$ for every i. It will also be important to identify the weak limit y of the y_i's explicitly. Let k and M be as in the claim. We may assume, by passing to a subsequence contained in M and by a small perturbation, that the restrictions of all the z_n's to $[1, k]$ are a fixed vector y and that for each i the restrictions of all the z_n's for $n \geq i$ to $[k+1, k+i]$ are a fixed vector v_i independent of n. Put $y_i = z_i - v_i$. Since $\|z_i - y_i\| \leq 1$, it follows from the choice of k and by passing to a further subsequence that we may also assume that $|f(z_i) - f(y_i)| < \varepsilon$ for every i. The sequence $\{y_i\}$ clearly converges weakly to y.

We now finish the proof. Let $\{x_n\}$ be a weak Cauchy sequence such that $\{f(x_n)\}$ does not converge, and find two infinite subsequences $\{x'_n\}$ and $\{x''_n\}$ and an $\varepsilon > 0$ such that $|f(x'_n) - f(x''_m)| > 3\varepsilon$ for all n and m. Using the argument above for these two sequences, we can find a common k and two weakly convergent sequences $\{y'_i\}$ and $\{y''_i\}$ such that $|f(x'_{n_i}) - f(y'_i)| < \varepsilon$ and $|f(x''_{n_i}) - f(y''_i)| < \varepsilon$. Since k was common to both sequences, the two weakly convergent sequences have the same weak limit y; indeed, $\langle y, e_m\rangle = \lim_n \langle x_n, e_m\rangle$ for $m \leq k$, and $\langle y, e_m\rangle = 0$

for $m > k$. It follows that the sequence $\{y_n\}$ obtained by interlacing them is also weakly convergent, and by the construction $\{f(y_n)\}$ does not converge. □

Although we do not know of examples of reflexive spaces in the class $\mathcal{C}$, we shall point out in the Notes and Remarks section that a class which is very close to $\mathcal{C}$ (and which may actually coincide with it) does contain reflexive spaces. The next proposition holds also for this modified class.

Proposition 11.9. *Let $E \in \mathcal{C}$ be a separable reflexive Banach space, and let $\varphi : E \to F$ be a surjective map with a locally uniformly continuous derivative. Then F is also reflexive.*

Proof. Cover E, as in the last part of Proposition 11.5, by a countable number of sets as in the definition of the class $\mathcal{C}$ and such that D_φ is uniformly continuous on each of these sets. It follows from Baire's theorem that the norm closure of the image of at least one of these sets contains a ball and hence that it is enough to show that for each such set U every sequence $\{z_j\}$ in $\varphi(U)$ has a weakly convergent subsequence. Choose $x_j \in U$ such that $\varphi(x_j) = z_j$. Since E is reflexive, we may assume that the x_j's converge weakly to a point x, and then the weak sequential continuity of φ implies that the z_j's converge weakly to $\varphi(x)$. □

2. Nonlinear Quotient Maps

As we saw in the previous section, there is a surjective Lipschitz map from any separable infinite-dimensional Banach space onto any other separable Banach space. It follows that unlike the linear situation, where the existence of a linear quotient map from one Banach space onto another gives information on the relations between their structures, surjective Lipschitz maps give no such information. By the open mapping theorem linear quotient maps are open, and they even satisfy a stronger "uniform openness" condition: If $\varphi : E \to F$ is such a map, then there is a $\lambda > 0$ such that $\varphi\big(B(x,r)\big) \supset B\big(\varphi(x), \lambda r\big)$ for every $x \in E$ and for every $r > 0$. These conditions do not hold for general surjective Lipschitz maps, and in this section we study what structural relations between E and F follow when we assume that the surjective Lipschitz (or uniformly continuous) map also satisfies an analogous "uniform openness" condition. We first give the formal definitions.

Definition 11.10. *Let X and Y be metric spaces. A map $\varphi : X \to Y$ is called co-uniformly continuous if for every $\varepsilon > 0$ there is a $\delta = \delta(\varepsilon) > 0$ such that $\varphi\big(B(x,\varepsilon)\big) \supset B\big(\varphi(x), \delta\big)$ for every $x \in X$. If there is a constant $C > 0$ such that it is possible to choose $\delta(\varepsilon) \geq \varepsilon/C$ for every $\varepsilon > 0$, then φ is said to be co-Lipschitz. The co-Lipschitz constant of φ is* $\inf\{C\}$ *where C is as above.*

The map φ is called a uniform (or Lipschitz) quotient map if it is both uniformly continuous and co-uniformly continuous (respectively Lipschitz and co-Lipschitz). When there is a surjective uniform (or Lipschitz) quotient map from X onto Y, we say that Y is a uniform (or Lipschitz) quotient of X.

Note that uniform continuity can also be formulated in the language of balls and their images: The map φ is uniformly continuous with modulus of continuity ω if $\varphi\big(B(x,\varepsilon)\big) \subset B\big(\varphi(x), \omega(\varepsilon)\big)$ for every $x \in E$ and $\varepsilon > 0$.

Compositions of linear quotient maps and uniform homeomorphisms between Banach spaces are uniform quotient maps, but the class of uniform quotient maps is larger. Indeed, if φ is such a composition, then all the preimages $\varphi^{-1}(x)$ are connected. However, $\varphi(re^{i\theta}) = re^{i2\theta}$ is a surjective Lipschitz and co-Lipschitz map of $\mathbb{R}^2$ onto itself, and $\varphi^{-1}(x)$ consists of exactly two points for each $x \neq 0$.

Uniformly continuous functions on metrically convex spaces satisfy a Lipschitz condition for large distances (see Proposition 1.11). In analogy, co-uniformly continuous maps onto metrically convex spaces are co-Lipschitz for large distances.

Lemma 11.11. *Let φ be a co-uniformly continuous map from the metric space X onto the metrically convex space Y. Then for every $\varepsilon > 0$ there is a constant $C = C(\varepsilon)$ such that $\varphi(B(x,r)) \supset B(\varphi(x), r/C)$ for every $x \in E$ and $r \geq \varepsilon$.*

Proof. Let $\delta = \delta(\varepsilon)$ be as in the definition of co-uniform continuity. We check that if $r \geq n\varepsilon$, then $\varphi(B(x,r)) \supset B(\varphi(x), n\delta)$. Indeed, if $d(y, \varphi(x)) \leq n\delta$, then the metric convexity of Y yields points $\varphi(x) = y_0, \dots, y_n = y$ with $d(y_i, y_{i+1}) \leq \delta$ for all i. By the choice of δ we can define inductively points $x = x_0, \dots, x_n$ in X such that $d(x_i, x_{i+1}) \leq \varepsilon$ and $\varphi(x_i) = y_i$. Thus $x_n \in B(x,r)$ and $\varphi(x_n) = y$. $\square$

The fact that a map between two Banach spaces is co-Lipschitz is preserved by small Lipschitz perturbations.

Lemma 11.12. *Let E and F be Banach spaces, and let $f : E \to F$ be a continuous co-Lipschitz map with co-Lipschitz constant smaller than C. If $g : E \to F$ is L-Lipschitz with $LC < 1$, then $f + g$ is co-Lipschitz.*

Proof. Fix $x \in E$ and $r > 0$. We shall show that for every $y \in B(f(x)+g(x), r)$ there is a point $z \in B(x, (1-LC)^{-1}r)$ such that $f(z) + g(z) = y$. By translations we may assume that $x = 0$ and that $f(0) = g(0) = 0$.

Let $z_0 = 0$, and choose inductively $z_n \in E$ such that $\|z_n\| \leq (LC)^{n-1}r$ and $f\big(\sum_{k=0}^{n+1} z_k\big) + g\big(\sum_{k=0}^{n} z_k\big) = y$. The choice of z_1 is similar to the induction step, and we present only the latter. Assume that $z_1, \dots, z_n$ have been chosen to satisfy $\|z_k\| \leq (LC)^{k-1}r$ and $f\big(\sum_{k=0}^{n} z_k\big) + g\big(\sum_{k=0}^{n-1} z_k\big) = y$. Then

$$\begin{aligned} \Big\| f\Big(\sum_{k=0}^{n} z_k\Big) + g\Big(\sum_{k=0}^{n} z_k\Big) - y \Big\| &= \Big\| g\Big(\sum_{k=0}^{n} z_k\Big) - g\Big(\sum_{k=0}^{n-1} z_k\Big) \Big\| \\ &\leq L\|z_n\| \leq L^n C^{n-1} r. \end{aligned}$$

Thus $y - g\big(\sum_{k=0}^{n} z_k\big) \in B\Big(f\big(\sum_{k=0}^{n} z_k\big), L^n C^{n-1} r\Big)$. Since f is co-Lipschitz with co-Lipschitz constant smaller than C, it follows that there is a point z_{n+1} with $\|z_{n+1}\| \leq (LC)^n r$ such that $f\big(\sum_{k=0}^{n+1} z_k\big) = y - g\big(\sum_{k=0}^{n} z_k\big)$.

We now set $z = \sum_{k=0}^{\infty} z_k$ and use the continuity of f and g. $\square$

A Banach space which is a Lipschitz quotient of another Banach space does not have to be a linear quotient of this space. In fact, the subspace of l_∞ constructed in Example 7.12, which is Lipschitz equivalent to $c_0(\Gamma)$ for some uncountable set Γ, is not a linear quotient of $c_0(\Gamma)$. Indeed, $K = \{e_\gamma : \gamma \in \Gamma\} \cup \{0\}$ is a weakly compact set that spans a dense subspace of $c_0(\Gamma)$. It follows that every linear quotient of $c_0(\Gamma)$ contains such a weakly compact generating set (the image of K). On the other hand, every weakly compact subset $A \subset l_\infty$ is separable because the countable set of coordinate functionals separates its points, and thus A is metrizable in the weak topology, hence weakly separable, and hence also norm separable.

It is unknown whether a Lipschitz quotient of a *separable* Banach space must actually be a linear quotient of this space. A natural approach to this question is by differentiation of the quotient map $\varphi : E \to F$ in the hope that the derivative could be used to construct linear quotient maps. Unfortunately Gâteaux derivatives, which were so useful in the study of Lipschitz embeddings (see Chapter 7), do not necessarily give any useful information for Lipschitz quotient maps.

Example 11.13. *There is a Lipschitz quotient map of l_1 onto itself which is Gâteaux differentiable at zero and satisfies $D_\varphi(0) = 0$.*

For each positive integer k define a function $g_k : \mathbb{R} \to \mathbb{R}$ by

$$g_k(t) = \begin{cases} t & \text{if } |t| \geq 2^{-k} \\ 0 & \text{if } |t| \leq 2^{-(k+1)} \end{cases}$$

and g_k is linear in the two intervals $[2^{-(k+1)}, 2^{-k}]$ and $[-2^{-k}, -2^{-(k+1)}]$.

We now define a Lipschitz quotient map $f : l_1 \to \mathbb{R}$ by $f(\sum a_k e_k) = \sum g_k(a_k)$. Since the g_k's are 2-Lipschitz for every k, it follows that so is f. To see that f is co-Lipschitz with constant 2, fix any $x = \sum a_k e_k$, and put $f(x) = s$. Given $r > 0$, choose a coordinate l such that $2^{-l} < r/10$. Then $g_l(a_l + 2r) > g_l(a_l) + r$, and it follows that there is a $a_l \leq b \leq a_l + 2r$ such that $g_l(b) = g_l(a_l) + r$. The point $y = \sum b_k e_k$, where $b_k = a_k$ for $k \neq l$ and $b_l = b$, satisfies $\|x - y\| \leq 2r$ and $f(y) = s + r$. A similar construction gives a point y with $\|x - y\| \leq 2r$ and $f(y) = s - r$. To see that f is Gâteaux differentiable at 0 and that $D_f(0) = 0$, it suffices to check the directional derivative in the direction of vectors x with a finite support (because these are dense in l_1 and f is Lipschitz). But if x is such a vector, then actually $f(tx) = 0$ for small enough t.

To construct the required example, write $l_1 = \left(\sum \oplus l_1\right)_1$. If $x = (x_1, x_2, \dots)$ is a vector in this sum, with x_n in the nth copy of l_1, put $\varphi(x) = \sum f(x_n) e_n$. We leave the easy details to the reader.

Remarks. (i) A modification of this construction yields a similar example in l_p for every $1 \leq p < \infty$.

(ii) Let T be any bounded linear operator on l_1. Then there is a Lipschitz quotient map ψ form l_1 onto itself such that $D_\psi(0) = T$. This follows from Lemma 11.12 when we take $\psi = T + \lambda\varphi$ with φ as above and with a large enough λ.

The failure of the Gâteaux derivative at a point to be surjective is due to the different degrees in which it approximates the map in the different directions. It follows from the next lemma that when the approximation of a co-Lipschitz map by a linear operator is in some sense uniform, then the linear map is, indeed, surjective.

Lemma 11.14. *Let E and F be Banach spaces. Fix $0 < \varepsilon < \delta$ and $r > 0$, and let B be a ball of radius r and center x_0 in E. Assume that $\varphi(B) \supset B(\varphi(x_0), \delta r)$. If $T : E \to F$ is a continuous linear map with $\|\varphi(x) - Tx\| < \varepsilon r$ for every $x \in B$, then $T(B)$ contains the ball $B(\varphi(x_0), (\delta - \varepsilon) r)$ in F. In particular, T is surjective.*

Proof. By translation and normalization we may assume that B is the unit ball in E, that $\varphi(0) = 0$, and that $\delta = 1$.

It follows that for every $y \in F$ with $\|y\| \leq 1$ there is a point $x \in B$ such that $\varphi(x) = y$, and hence $\|Tx - y\| = \|Tx - \varphi(x)\| < \varepsilon$. Since T is continuous and linear, a successive approximation argument as in the proof of the open mapping theorem shows that $T(B)$ contains the ball $B(0, 1 - \varepsilon)$ in F. □

Remark. We formulated the lemma for linear maps. It is clear, however, that it also holds (with minor changes in the constants) for general affine maps.

To obtain such a uniform approximation by differentiation, we have to use Fréchet derivatives or one of its variations. Unfortunately, there are no suitable general existence theorems for these derivatives. Good affine approximations may, however, exist without differentiation. Moreover, there are situations in which there are affine approximations which are better than those obtained by differentiation even when the latter do exist. For example, given $\delta > 0$, consider the function $f_\delta : \mathbb{R} \to \mathbb{R}$ that takes the value $(-1)^n\delta$ at $n\delta$ and is linear between these points. Then f_δ is 2-Lipschitz, and it is δ-approximated by the function which is identically zero. However, the derivative of f_δ at any point does not give a global approximation at all. (See the Notes and Remarks section for some comments on these issues.) These considerations lead to the following definition.

Definition 11.15. *Let E and F be Banach spaces. A pair (E, F) is said to satisfy the property of uniform approximation by affine functions (UAAP) if for every $\varepsilon > 0$ there is a constant $c = c(\varepsilon) > 0$ with the following property: Let B be a ball in E, and let r be its radius. Let $f : B \to F$ be a Lipschitz function with constant L. Then there is a ball $B_1 \subset B$ of radius $r_1 > cr$ and a continuous affine function $g : E \to F$ such that $\|f(x) - g(x)\| \le \varepsilon r_1 L$ for every $x \in B_1$.*

(Note that B_1 is not required to have the same center as B.)

Proposition 11.16. *Let E be a superreflexive Banach space, and let F be finite-dimensional. Then (E, F) satisfies the UAAP.*

Proof. The UAAP is preserved when E or F is replaced by isomorphic spaces. By Theorem A.6 we may thus assume that E is uniformly smooth.

Assume first that $F = \mathbb{R}$ is one-dimensional. Let B be a ball in E, and let $f : B \to \mathbb{R}$ be a Lipschitz function. We may assume that the radius of B is 1 and that the Lipschitz constant of f is 1. Fix $\varepsilon > 0$.

Let $L(t) = \sup\{\|f(x) - f(y)\|/\|x - y\| : \|x - y\| \ge t\}$ be the Lipschitz constant of f for distances larger than t.

If $L(1) \le \varepsilon$ and if x_0 is the center of B, then the constant map $f(x_0)$ is a 2ε-approximation of f on all of B.

We may thus assume that $L(1) \ge \varepsilon$. By the uniform smoothness of E there is a $\delta = \delta(\varepsilon)$ with $0 < \delta < 1/2$ such that if $\|x\| = 1$ and $\|y\| \le \delta$, and if x^* is the unique norming functional at x, i.e., $\|x^*\| = \langle x^*, x\rangle = 1$, then $\|x + y\| = 1 + \langle x^*, y\rangle + r(y)$, where the remainder (which depends on x) satisfies $|r(y)| \le \varepsilon\|y\|$.

Let $k = k(\varepsilon)$ be the first integer for which $(1 + \delta\varepsilon)^k\varepsilon > 1$. The choice of k and the inequalities $1 \ge L(4^{-k}) \ge L(4^{-k+1}) \ge \ldots \ge L(1) \ge \varepsilon$ give that there is a $j \le k$ such that $L(4^{-j}) < (1 + \delta\varepsilon)L(4^{-j+1})$. It follows that there is a $d > 0$ (which is bounded below by a constant depending only on ε) and there are two points $y, z \in B$ with $\|z - y\| = 2d$ such that $L(4^{-j}) < (1 + \delta\varepsilon)\big(f(z) - f(y)\big)/2d$. Indeed, choose y, z with $\|y - z\| \ge 4^{-j+1}$ such that $L(4^{-j}) < (1+\delta\varepsilon)\big(f(z) - f(y)\big)/\|z - y\|$, and put $\|z - y\| = 2d$.

Translating and adding a constant to f, we may assume that $y = -z$ and that $f(-z) = -f(z)$. Thus $\|z\| = d$ and $B \bigcap B(0, d\delta)$ contains a ball B_1 of radius $r_1 = \delta d/2$ (centered at $\delta d x_0/2\|x_0\|$, where x_0 is the center of B). Put $\alpha = f(z)$. Replacing f by $-f$ if necessary, we may assume that $\alpha > 0$, and clearly $\alpha \le d$. Since $L(d/2) \le L(4^{-j})$ (because $4^{-j} \le d/2$), it follows that if $\|x - y\| \ge d/2$, then $|f(x) - f(y)| \le (1 + \delta\varepsilon)\alpha\|x/d - y/d\|$.

Let z^* be the norming functional at z/d. We show that the linear function $g = (\alpha/d)z^*$ satisfies $\sup\{|f(x) - g(x)| : x \in B_1\} \leq 4\varepsilon\delta d = 8\varepsilon r_1$. Indeed, fix any $x \in B \bigcap B(0, d\delta)$ (which contains B_1). Since $\delta < 1/2$, we see that $\|x - z\|$ and $\|x + z\|$ are both larger than $d/2$. Since also $|r(x/d)| < \varepsilon\delta$, it follows that

$$\begin{aligned} f(x) &= f(x) - f(-z) - \alpha \leq (1 + \delta\varepsilon)\alpha\|z/d + x/d\| - \alpha \\ &= (1 + \delta\varepsilon)\alpha\Big(1 + \langle z^*, x/d\rangle + r(x/d)\Big) - \alpha \\ &= g(x) + \alpha r(x/d) + \delta\varepsilon\alpha\big(1 + \langle z^*, x/d\rangle + r(x/d)\big) \\ &\leq g(x) + 4\varepsilon\delta d. \end{aligned}$$

Using z instead of $-z$ we obtain similarly that $f(x) \geq g(x) - 4\varepsilon\delta d$ and hence that $|f(x) - g(x)| \leq 4\varepsilon\delta d$ on B_1.

We now generalize the first step above to finite collections of functions. Let $\{f_i\}_{i=1}^n$ be 1-Lipschitz real-valued functions on a ball B of radius r, and find a constant $c_n = c_n(\varepsilon)$, a ball $B_n \subset B$ of radius $r_n \geq c_n r$, and affine functions $g_i : E \to \mathbb{R}$ such that $|f_i(x) - g_i(x)| \leq \varepsilon r_n$ for every $x \in B_n$ and $i \leq n$. The first part of the proof gives this for $n = 1$, and we only check $n = 2$.

By the case $n = 1$ there is a ball $B_1 \subset B$ of radius $r_1 \geq c_1\big(\varepsilon c_1(\varepsilon)\big)r$ and an affine function g_1 such that $|f_1(x) - g_1(x)| \leq \varepsilon c_1(\varepsilon)r_1$ on B_1. Applying the case $n = 1$ again, find a ball $B_2 \subset B_1$ of radius $r_2 \geq c_1(\varepsilon)r_1$ and an affine function g_2 such that $|f_2(x) - g_2(x)| \leq \varepsilon r_2$ on B_2. Then $|f_1(x) - g_1(x)| \leq \varepsilon c_1(\varepsilon)r_1 \leq \varepsilon r_2$ on B_2 because $B_2 \subset B_1$. We can thus take $c_2(\varepsilon) = c_1\big(\varepsilon c_1(\varepsilon)\big)c_1(\varepsilon)$.

Passing to a general finite-dimensional space F, we may assume by renorming that it is isometric to l_1^n. Denote the natural basis of l_1^n by $\{y_i\}_{i=1}^n$, and the coordinate functionals by $\{y_i^*\}_{i=1}^n$.

Fix a ball B of radius r in E, a 1-Lipschitz map $f : B \to F$ and $\varepsilon > 0$. The argument above (with the functions $f_i = y_i^* \circ f$ and with $\varepsilon/8n$ instead of ε) yields a ball $B_n \subset B$ with radius $r_n \geq c_n(\varepsilon)r$ and affine functions g_i such that $|f_i(x) - g_i(x)| \leq r_n\varepsilon/n$ on B_n. If $x \in B_n$, then $f(x) = \sum_{i=1}^n f_i(x)y_i$ gives

$$\Big\|f(x) - \sum_{i=1}^n g_i(x)y_i\Big\| = \sum_{i=1}^n |f_i(x) - g_i(x)| \leq \varepsilon r_n.$$

□

We now consider uniform quotients. As noted above, even Lipschitz quotients do not have to be isomorphic to linear quotients. Also, by Corollary 10.29 and the remark following it, there is a separable reflexive Banach space which is uniformly homeomorphic to a nonreflexive space. Some structural information on uniform quotients can, however, be obtained under certain circumstances, and we follow the program in the first section of Chapter 10 and study uniform quotients by passing to Lipschitz quotients on ultrapowers. Together with Proposition 11.16 this will give control on the local linear structure of uniform quotients of superreflexive Banach spaces.

The image of a Banach space under a uniform quotient map does not have to have a linear structure. For example, a Banach space can be uniformly homeomorphic to a subset, say of the unit sphere, of another Banach space. This is why we made it a point to say explicitly in the next few results that we assume that the image is a Banach space.

Proposition 11.17. *Let E be a Banach space, and assume that the Banach space F is a uniform quotient of E. Let $\mathcal{U}$ be a free ultrafilter on the set of natural numbers. Then the ultrapower $(F)_{\mathcal{U}}$ is a Lipschitz quotient of the ultrapower $(E)_{\mathcal{U}}$.*

Proof. Let $\varphi : E \to F$ be the quotient map. Since φ is both Lipschitz and co-Lipschitz for large distances, there is a constant C such that

$$B\big(\varphi(x), Cr\big) \supset \varphi\big(B(x,r)\big) \supset B\big(\varphi(x), r/C\big)$$

for every $x \in E$ and for every $r \geq 1$. The maps $\varphi_n(x) = \varphi(nx)/n$ satisfy the same condition for $r \geq 1/n$, and thus their ultraproduct $\tilde{\varphi} : (E)_{\mathcal{U}} \to (F)_{\mathcal{U}}$ defined by $\tilde{\varphi}(x_1, x_2, \dots) = \big(\varphi_1(x_1), \varphi_2(x_2), \dots\big)$ satisfies it for every $r > 0$. □

Theorem 11.18. *Let the Banach space F be a uniform quotient of a super-reflexive Banach space E. Then there is a constant K such that for every finite-dimensional subspace G of F^* there is a subspace G_1 of E^* with $d(G, G_1) \leq K$. In particular, F is isomorphic to a linear quotient of some ultrapower of E.*

Proof. Since a space and its ultrapowers have the same finite-dimensional subspaces (see Proposition F.6(i)), we may pass to ultrapowers and assume by Proposition 11.17 that F is a Lipschitz quotient of E. Let $\varphi : E \to F$ be the quotient map, and assume, as we may, that φ is 1-Lipschitz and that its co-Lipschitz constant is smaller than $1/\delta$ for some $\delta > 0$.

Let G be a finite-dimensional subspace of F^*. Denote its annihilator in F by $G_\perp$, and let $Q : F \to F/G_\perp$ be the natural quotient map. It follows that the composition $Q \circ \varphi$ is 1-Lipschitz and that it satisfies

$$Q \circ \varphi\big(B(x,r)\big) \supset B\big(Q \circ \varphi(x), \delta r\big) \tag{11.3}$$

for every $x \in E$ and $r > 0$.

Fix $0 < \varepsilon < \delta$. Using Proposition 11.16 with a large enough r, we can find a ball of radius 1, which we may translate to be the unit ball of E, and an affine map $T : E \to F/G_\perp$ such that

$$\sup\big\{\|Q \circ \varphi(x) - Tx\| \,:\, x \in B\big\} < \varepsilon. \tag{11.4}$$

By adding constants to T and φ we may assume that $T(0) = 0$, i.e., that T is linear. By Lemma 11.14, conditions (11.3) and (11.4) imply that $T(B)$ contains the ball $B(0, \delta - \varepsilon)$ in $F/G_\perp$. Identifying $F/G_\perp$ with G^*, and since $\|T\| \leq 1 + \varepsilon$ by (11.4), we find that T^* is a $(1+\varepsilon)/(\delta - \varepsilon)$-isomorphism of G into E^*, and this proves the first part of the theorem.

As in Proposition F.6(i), it follows from the first part that F^* is isomorphic to a subspace of an ultrapower of E^*. Hence it is reflexive, and F is isomorphic to a linear quotient of an ultrapower of E. □

Remark. The explicit estimate $(1+\varepsilon)/(\delta - \varepsilon)$ for K in the proof above is arbitrarily close to 1 if δ can be taken arbitrarily close to 1 and if ε is then chosen to be small enough. It follows that if F is an isometric Lipschitz quotient of E in the sense that for every $\eta > 0$ there is a quotient map from E onto F whose Lipschitz and co-Lipschitz constants are η-close to 1, then F^* is actually finitely representable in E^* and F is isometric to a linear quotient of an ultrapower of E.

The next two corollaries give some consequences of the theorem.

Corollary 11.19. *Let $1 < p < \infty$. Then every Banach space F which is a uniform quotient of L_p is isomorphic to a linear quotient of L_p. In particular, a uniform quotient of a Hilbert space is isomorphic to a Hilbert space.*

Proof. By the theorem F is isomorphic to a linear quotient of an ultrapower of L_p, i.e., of $L_p(\mu)$ for some measure μ (see Proposition F.3(ii)). Since F is separable, it is actually isomorphic to a quotient of L_p. □

The next corollary is an instance where the Banach spaces which are uniform quotients of a Banach space E can be classified completely even though not all of them are linear quotients of E. For information on the Tsirelson space, see Chapter 10, Section 5.

Corollary 11.20. *A Banach space is a uniform quotient of the 2-convexified Tsirelson space $\mathcal{T}^{(2)}$ if and only if it is isomorphic to a linear quotient of $\mathcal{T}^{(2)} \oplus l_2$.*

Proof. By Proposition 10.34 the spaces $\mathcal{T}^{(2)}$ and $\mathcal{T}^{(2)} \oplus l_2$ are uniformly homeomorphic; thus they have the same uniform quotients. If F is such a uniform quotient, then by the theorem it is isomorphic to a linear quotient of an ultrapower $(\mathcal{T}^{(2)})_{\mathcal{U}}$. By Lemma 10.35 $(\mathcal{T}^{(2)})_{\mathcal{U}}$ is isomorphic to $\mathcal{T}^{(2)} \oplus H$ for some Hilbert space H, and since F is separable, we can take $H = l_2$. □

The corollaries above used the local structure of a uniform quotient. The next result (on Lipschitz quotients) is purely infinite-dimensional and gives new information even for Lipschitz equivalences.

Theorem 11.21. *Let E be a Banach space with a separable dual, and assume that the Banach space F is a Lipschitz quotient of E. Then F^* is separable.*

Proof. Let $\varphi : E \to F$ be the quotient map, and assume that it is 1-Lipschitz with co-Lipschitz constant smaller than C. If F^* is nonseparable, then there is an uncountable set of norm-one functionals $\{y^*_\gamma\}_{\gamma\in\Gamma}$ in F^* such that $\|y^*_\gamma - y^*_\beta\| > 1/2$ for $\gamma \neq \beta$. The functions $f_\gamma = y^*_\gamma \circ \varphi$ are 1-Lipschitz, and since E^* is separable, Preiss's theorem ([**471**]; see also the second volume of this book) gives that for each γ there is a point x_γ in the unit ball of E where f_γ is Fréchet differentiable.

Fix $\varepsilon > 0$, and for each γ find a $\delta_\gamma > 0$ such that

$$\left|f_\gamma(x_\gamma + z) - f_\gamma(x_\gamma) - \langle D_{f_\gamma}(x_\gamma), z\rangle\right| \leq \varepsilon\|z\|$$

whenever $\|z\| \leq \delta_\gamma$. By passing to an uncountable subset we may assume that there is a $\delta > 0$ such that $\delta_\gamma > \delta$ for every γ. Since E, E^* and $\mathbb{R}$ are separable, we may also assume that the inequalities $\|x_\gamma - x_\beta\| < \varepsilon\delta$, $|f_\gamma(x_\gamma) - f_\beta(x_\beta)| < \varepsilon\delta$, and $\|D_{f_\gamma}(x_\gamma) - D_{f_\beta}(x_\beta)\| < \varepsilon$ hold for every $\beta, \gamma \in \Gamma$. It follows that if $\|z\| < \delta$, then

$$\begin{aligned}\left|f_\gamma(x_\gamma + z) - f_\beta(x_\gamma + z)\right| &\leq \left|f_\gamma(x_\gamma + z) - f_\gamma(x_\gamma) - \langle D_{f_\gamma}(x_\gamma), z\rangle\right| \\ &+ \left|f_\beta(x_\beta + z) - f_\beta(x_\beta) - \langle D_{f_\beta}(x_\beta), z\rangle\right| + \left|f_\gamma(x_\gamma) - f_\beta(x_\beta)\right| \\ &+ \left|\langle D_{f_\gamma}(x_\gamma) - D_{f_\beta}(x_\beta), z\rangle\right| + \left|f_\beta(x_\gamma + z) - f_\beta(x_\beta + z)\right| < 5\varepsilon\delta.\end{aligned}$$

On the other hand, $\varphi\big(B(x_\gamma, \delta)\big) \supset B\big(\varphi(x_\gamma), \delta/C\big)$ implies that

$$\begin{aligned}&\sup\left\{\left|f_\gamma(x_\gamma + z) - f_\beta(x_\gamma + z)\right| : \|z\| < \delta\right\} \\ &= \sup\left\{\left|\langle y^*_\gamma - y^*_\beta, \varphi(x_\gamma + z)\rangle\right| : \|z\| < \delta\right\} \\ &\geq \sup\left\{\left|\langle y^*_\gamma - y^*_\beta, \varphi(x_\gamma) + y\rangle\right| : \|y\| < \delta/C\right\} \\ &\geq \|y^*_\gamma - y^*_\beta\|\,(\delta/C) > \delta/2C\end{aligned}$$

and this contradicts the previous inequality if $\varepsilon < 1/10C$. □

Remark. There is no analogous result in the uniform category even for uniform homeomorphisms. Indeed, as noted before, Corollary 10.29 and the remark that follows it show that there is a separable reflexive space which is uniformly homeomorphic to a space containing a copy of l_1.

Recall that a Banach space E is an Asplund space if every separable subspace of E has a separable dual.

Corollary 11.22. *A Banach space which is a Lipschitz quotient of an Asplund space is also Asplund.*

Proof. To deduce the corollary from the theorem, we need to show only that if $\varphi : E \to F$ is a Lipschitz quotient map, and if F_0 is a separable subspace of F, then there is a separable subspace $\tilde{E}$ of E such that φ maps $\tilde{E}$ onto a closed linear subspace $\tilde{F} \supset F_0$ of F and such that the restriction of φ to $\tilde{E}$ is co-Lipschitz as a map from $\tilde{E}$ onto $\tilde{F}$. To see this assume that the co-Lipschitz constant of φ is smaller than 1, and define inductively closed linear separable subspaces $F_0 \subset F_1 \subset \dots$ of F and countable sets $W_0 \subset W_1 \subset \dots$ in E which are closed under rational linear combinations such that $\varphi(W_n)$ spans F_n and such that

(11.5) $\quad \varphi\big(B(x,r) \cap W_{n+1}\big)$ contains a dense set in the ball $B\big(\varphi(x),r\big)$ in F_n

for every $x \in W_n$ and for every rational number $r > 0$. (Note that once (11.5) holds for rational numbers r, it actually holds for every real $r > 0$.)

The set W_0 is chosen so that $\varphi(W_0)$ is dense in F_0. If W_n and F_n have been chosen, then W_{n+1} exists by the co-Lipschitz condition on φ, and we then take F_{n+1} to be the closed linear span of $\varphi(W_{n+1})$.

Taking $\tilde{E}$ as the closure of $\bigcup W_n$ and $\tilde{F}$ as the closure of $\bigcup F_n$, it is clear that these are separable subspaces and that $\varphi(\tilde{E})$ is dense in $\tilde{F}$. We shall show that $\varphi(\tilde{E}) = \tilde{F}$ and that the co-Lipschitz constant of φ as a map from $\tilde{E}$ to $\tilde{F}$ is at most 2. From now on balls will be taken in $\tilde{F}$ and in $\tilde{E}$.

Fix $x_0 \in \tilde{E}$ and $z \in \tilde{F}$ with $\|z - \varphi(x_0)\| < r$. We may assume that $x_0 \in \bigcup W_n$. Approximate z by a point $z_1 \in \bigcup F_n$ such that $\|z_1 - z\| < r/2$ and $z_1 \in B\big(\varphi(x_0), r\big)$. By (11.5) there is an $x_1 \in \bigcup W_n$ such that $\|\varphi(x_1) - z_1\| < 1/2$ and $\|x_1 - x_0\| \leq r$.

Inductively, having chosen $x_i \in \bigcup W_n$ and $z_i \in \bigcup F_n$ for $i \leq n$ in such a way that $\|\varphi(x_i) - z_i\| < 2^{-i}$, $\|z_i - z\| < r2^{-i}$, and $\|x_i - x_{i-1}\| \leq r2^{-(i-1)}$, approximate z by $z_{n+1} \in \bigcup F_n$ so that $\|z_{n+1} - z\| < r2^{-(n+1)}$ and $z_{n+1} \in B\big(z_n, r2^{-n}\big)$. By (11.5) there is an $x_{n+1} \in \bigcup W_n$ with $\|\varphi(x_{n+1}) - z_{n+1}\| < 2^{-(n+1)}$ and $\|x_{n+1} - x_n\| \leq r2^{-n}$.

It follows that $x_n \to x$ with $\|x - x_0\| \leq 2r$ and $\varphi(x) = z$. □

The "Gorelik principle" (Theorem 10.12) is an important tool in the study of uniform homeomorphisms. In an imprecise formulation, it says that a uniform homeomorphism between two Banach spaces E and F cannot take a "large" ball in a finite-codimensional subspace of E into a "small" neighborhood of an infinite-codimensional subspace of F. By the open mapping theorem this property also holds for compositions of uniform homeomorphisms and linear quotient maps. The following two examples show, however, that this important tool is not available for the study of general uniform or even Lipschitz quotient maps.

Example 11.23. *For every Banach space F there is a uniform quotient map φ from $E = F \oplus F \oplus \mathbb{R}$ onto F which is Lipschitz and maps $B(F \oplus F)$ to 0.*

We use the norm $\|(x_1, x_2, \lambda)\| = \max\{\|x_1\|, \|x_2\|, |\lambda|\}$ on E, and φ will have the form $\varphi(x_1, x_2, \lambda) = f(x_1, \lambda) + g(x_2, \lambda)$ for two Lipschitz maps $f, g : F \oplus \mathbb{R} \to F$.

For $x \in F$ put $f_0(x) = x$ if $\|x\| \le 1$ and $f_0(x) = \big(2 - \|x\|\big)^+ x$ if $\|x\| \ge 1$. Let $\{u_j\}$ be a maximal 4-separated set in F, and put $f(x, \lambda) = |\lambda| \sum f_0\big(|\lambda|^{-1}x - u_j\big)$ for $\lambda \ne 0$ and $f(x, 0) = 0$. The different terms in the sum are disjointly supported, and hence f is a well-defined Lipschitz map with $\|f(x, \lambda)\| \le |\lambda|$ for every $x \in F$. For each fixed λ the map f is supported in $\bigcup B\big(|\lambda|u_j, 2|\lambda|\big)$, and for each j it translates the ball $B\big(|\lambda|u_j, |\lambda|\big)$ onto the ball $B\big(0, |\lambda|\big)$.

We now define g. For $t, \lambda \ge 0$ let $\beta(t, \lambda) = t$ if $\lambda \ge 1$; and if $\lambda \le 1$, put $\beta(t, \lambda) = t$ for $t \ge 2$, $\beta(t, \lambda) = \lambda t$ for $t \le 1 - \lambda$, and $\beta(t, \lambda)$ to be linear in t in the interval $[1 - \lambda, 2]$. Let $g(x, \lambda) = \beta(\|x\|, |\lambda|)x/\|x\|$. Then g is Lipschitz, and the explicit formulas show that $\varphi(x_1, x_2, 0) = 0$ for $x_1, x_2 \in F$ with $\|x_2\| \le 1$.

It remains to show that φ is co-uniform. We shall show that if $0 < r < 1/4$, then the image of $B\big((x_1, x_2, \lambda), r\big)$ under φ contains $B\big(\varphi(x_1, x_2, \lambda), r^2/100\big)$ for every $(x_1, x_2, \lambda) \in E$. We may assume by symmetry that $\lambda \ge 0$, and we distinguish two cases. If $\lambda \le r/10$, we actually show that $f\big(B\big((x_1, \lambda), r\big)\big) \supset B\big(f(x_1, \lambda), r/10\big)$, and hence certainly

$$\begin{aligned}
\varphi\Big(B\big((x_1, x_2, \lambda), r\big)\Big) &\supset g(x_2, \lambda) + f\big(B\big((x_1, \lambda), r\big)\big) \\
&\supset g(x_2, \lambda) + B\big(f(x_1, \lambda), r/10\big) \\
&= B\big(\varphi(x_1, x_2, \lambda), r/10\big) \supset B\big(\varphi(x_1, x_2, \lambda), r^2/100\big).
\end{aligned}$$

If $\lambda \ge r/10$, we shall show that $g\big(B(x_2, r), \lambda\big) \supset B\big(g(x_2, \lambda), r^2/100\big)$ and proceed similarly.

Assume first that $\lambda \le r/10$, and fix j such that $\|5x_1/r - u_j\| \le 4$. It follows that $\|x_1 - ru_j/5\| \le 4r/5$, and of course also $|\lambda - r/5| \le 4r/5$; hence

$$f\Big(B\big((x_1, \lambda), r\big)\Big) \supset f\Big(B\big(\tfrac{r}{5}u_j, \tfrac{r}{5}\big), \tfrac{r}{5}\Big) = B\big(0, \tfrac{r}{5}\big) \supset B\big(f(x_1, \lambda), r/10\big)$$

because $\|f(x_1, \lambda)\| \le \lambda \le r/10$.

Assume now that $\lambda \ge r/10$. If $\lambda \ge 1$, then $g(\cdot, \lambda)$ is the identity on F, and hence $g\big(B(x_2, r), \lambda\big) = B\big(g(x_2, \lambda), r\big)$.

So fix $1 > \lambda \ge r/10$. Then $g(\cdot, \lambda)$ is a Lipschitz equivalence of F. Indeed, it is a piecewise linear mapping of each ray emanating from 0 onto itself, and the slopes depend only on λ and not on the ray. The Lipschitz constant of $g^{-1}(\cdot, \lambda)$ is bounded by $1/\lambda$ on each such ray, and thus an easy calculation shows that it is certainly smaller than $10/\lambda$ on all of F. Hence

$$g\big(B(x_2, r), \lambda\big) \supset B\big(g(x_2, \lambda), r\lambda/10\big) \supset B\big(g(x_2, \lambda), r^2/100\big).$$

Remark. If F is isomorphic to its square and to its hyperplanes (as all the standard classical infinite-dimensional spaces are), then φ can be considered as a map of F onto itself.

Example 11.24. *For every separable Banach space F there is a Lipschitz quotient map φ from l_1 onto F which maps a hyperplane to 0.*

The two functions $g, h : l_1^2 \to \mathbb{R}$ defined by

$$g(s, t) = \begin{cases} s - t & \text{if } s \ge t \ge 0 \\ 0 & \text{if } t \ge s \ge 0 \\ s & \text{otherwise} \end{cases} \qquad h(s, t) = \begin{cases} t - s & \text{if } 2s \ge t \ge s \ge 0 \\ s & \text{if } t \ge 2s \ge 0 \\ 0 & \text{otherwise} \end{cases}$$

are 1-Lipschitz, and since they have disjoint supports, they actually satisfy

$$|g(s,t) - g(s',t')| + |h(s,t) - h(s',t')| \le |s - s'| + |t - t'|. \tag{11.6}$$

Let $\{(x^n, y^n)\}_{n=0}^{\infty}$ be dense in the product $l_1 \times S(F)$, and assume that each x^n is in l_1^n; i.e., in the span of the first n basis vectors $\{e_i\}_{i=1}^n$ in l_1.

We define inductively a sequence of maps $\varphi_n : l_1^n \to F$. For $n = 1$ the space l_1^1 is just the real line, and we put $\varphi_1(x) = xy^0$. Denote the coordinates of x^n by x_j^n and those of $z \in l_1^{n+1}$ by z_j, and define inductively

$$\varphi_{n+1}(z) = \varphi_n\Big(x_1^n + g(z_1 - x_1^n, z_{n+1}), z_2, \dots, z_n\Big) + h(z_1 - x_1^n, z_{n+1})y^n$$

if $x_1^n \ge 0$, and if $x_1^n < 0$, then

$$\varphi_{n+1}(z) = \varphi_n\Big(x_1^n - g(-z_1 + x_1^n, z_{n+1}), z_2, \dots, z_n\Big) + h(-z_1 + x_1^n, z_{n+1})y^n.$$

It follows from (11.6) that the φ_n's are 1-Lipschitz. Also φ_{n+1} extends φ_n; hence, putting $\varphi(x) = \varphi_n(x)$ for $x \in l_1^n$ and then extending by continuity gives a 1-Lipschitz map $\varphi : l_1 \to F$.

It follows easily from the formulas that φ takes the hyperplane consisting of all vectors in l_1 whose first coordinate is 0 to 0, and it remains to check that it is co-Lipschitz.

Fix $x \in l_1$ and $r > 0$, and let $y \in F$ with $\|y\| < r$. Given any $0 < \varepsilon < r$, choose n such that $\|x - x^n\| < \varepsilon$ and $\big\|y - \|y\|y^n\big\| < \varepsilon$. Put $z^n = x^n + \theta\|y\|e_1 + 2\|y\|e_{n+1}$, where $\theta = \operatorname{sign}(x_1^n)$ (and $\operatorname{sign}(0) = 1$). Then

$$\|z^n - x\| \le \|z^n - x^n\| + \|x - x^n\| < 3\|y\| + \varepsilon < 4r.$$

Assume now that $x_1^n \ge 0$. Then $g(z_1^n - x_1^n, z_{n+1}^n) = g\big(\|y\|, 2\|y\|\big) = 0$, and $h(z_1^n - x_1^n, z_{n+1}^n) = h\big(\|y\|, 2\|y\|\big) = \|y\|$. Hence

$$\varphi(z^n) = \varphi_{n+1}(z^n) = \varphi(x^n) + \|y\|y^n$$

and the same formula holds also when $x_1^n \le 0$. Since φ is 1-Lipschitz, we obtain

$$\big\|\varphi(z^n) - \big(\varphi(x) + y\big)\big\| \le \|x - x^n\| + \big\|y - \|y\|y^n\big\| < 2\varepsilon.$$

Since ε was arbitrary, it follows that the image of $B(x, 4r)$ under φ is dense in $B\big(\varphi(x), r\big)$, and a standard successive approximation argument shows that it actually contains it.

3. Notes and Remarks

Theorems 11.1 and 11.3 are due to Bates [**41**]. His original motivation was the failure of the Morse-Sard theorem in infinite dimensions. See also Bates [**39, 40**] and their references for further information in this direction.

For some spaces E the map in Theorem 11.3 can even be taken as a polynomial map. The next (unpublished) proposition is due to P. Hájek, and we thank him for allowing us to reproduce it here.

Proposition 11.25. *Let E be a separable Banach space such that there is a $p \ge 1$ and a noncompact operator from E into l_p. Then for every separable Banach space F there is a surjective polynomial map from E onto F.*

Proof. Fix $r \le 1$ and put $K_r = \{\sum a_j e_j \in l_1 : a_j \ge 0 \text{ and } \sum a_j^r \le 1\}$. We first observe that there is a bounded linear operator $T : l_1 \to F$ such that $T(K_r)$ contains the ball of radius $2^{-1/r}$ in F. Indeed, let $\{x_i\}$ be a dense set in $B(F)$ and define the operator by $Te_i = x_i$. Given $x \in F$ with $\|x\| \le 2^{-1/r}$, choose i_0 such that $\|2^{-1/r}x_{i_0} - x\| \le 2^{-1/r}$, and continue inductively to choose a subsequence such that $\|2^{-(n+1)/r}x_{i_n} - (x - \sum_{j=0}^{n-1} 2^{-(j+1)/r}x_{i_j})\| \le 2^{-(n+2)/r}$. Then the point $y = \sum_{j=0}^{\infty} 2^{-(j+1)/r}e_{i_j}$ is in K_r and $Ty = x$.

The second observation is that if $p \ge 1$ and $k \ge p$ is an integer, then the polynomial $P(\sum a_j e_j) = \sum a_j^k e_j$ maps l_p to l_1 and takes the closed convex hull of $\{e_n\} \subset l_p$ onto $K_{1/k}$. Thus $T \circ P$ is a surjective polynomial from l_p onto F.

Finally, if there is a noncompact operator from E into l_p, then a standard "gliding hump" argument shows that there is also another operator $S : E \to l_p$ such that $S(B(E))$ contains the closed convex hull of the unit vectors in l_p, and thus $T \circ P \circ S$ is a surjective polynomial from E onto F. □

The definition of the class $\mathcal{C}$ and all the results related to it are due to Hájek [**245, 246**]. Hájek actually uses a different definition: $E \in \mathcal{C}$ if there is an open bounded and convex set U in E such that for every $f : U \to \mathbb{R}$ with a uniformly continuous derivative the set $D_f(U)$ is relatively norm-compact in E^*. He shows that such spaces do not contain l_1 and that for spaces not containing l_1 the two definitions coincide. We found it more natural to base the discussion on the alternative definition.

It is unclear whether $E \in \mathcal{C}$ implies that every open bounded and convex subset of E satisfies the conditions of the definition. Passing from one such set to another involves the following interesting extension problem. Let $U \subset V$ be two such sets, and assume that $f : U \to \mathbb{R}$ has a uniformly continuous derivative. Does f extend to a function on V with the same property?

We now introduce a class which is close to $\mathcal{C}$ and which actually agrees with it if the answer to one of the questions above is positive. A Banach space E belongs to the class $\mathcal{C}'$ if it does not contain l_1 and if there is an $0 < r \le 1$ such that the restriction of every differentiable function $f : B(E) \to \mathbb{R}$ with a uniformly continuous derivative to $rB(E)$ is weakly sequentially continuous. The results of Section 1 hold, with obvious modifications, also for the class $\mathcal{C}'$.

The Tsirelson space T (see Section 5 in Chapter 10) is reflexive, and its dual T^* is in the class $\mathcal{C}'$ with $r = 1/2$. (This was noted in a private communication by Hájek. His proof in [**246**] that $T^* \in \mathcal{C}$ is incomplete.) The main point in the proof is that T^* has the property that if $\{u_n : n \le 2N\}$ are disjoint normalized blocks in T^*, then $\{u_n : N+1 \le n \le 2N\}$ are 2-equivalent to the unit vector basis in l_∞^N. The proof is now very similar to that of $c_0 \in \mathcal{C}$ and is actually simpler because T^* is reflexive, and therefore weak Cauchy sequences are already weakly convergent.

In [**246**] Hájek shows that $C(K) \in \mathcal{C}$ for every countable compact metric space K but that this is not true for all their subspaces. He also shows that if F has an unconditional basis and does not contain a copy of c_0, then there is no surjective map from c_0 onto F with a locally uniformly continuous derivative. Combined with the analog of Proposition 11.9 for the class $\mathcal{C}'$, this shows that there are no such maps from either c_0 onto T^* or vice versa.

The results of Section 2 are taken from the article [**42**] by Bates, Johnson, Lindenstrauss, Preiss and Schechtman. Co-uniformly continuous and co-Lipschitz maps were considered earlier by Whyburn [**584**] and by Gromov [**230**] respectively. Co-uniformly continuous maps and uniform quotients are discussed in the context of the general theory of uniform spaces in James [**270**].

To be a Lipschitz quotient map is a strong condition on a surjective Lipschitz map. Even Lipschitz retractions do not have to be co-Lipschitz. It is easy to check, for example, that the Lipschitz retraction from l_∞ onto c_0 in Example 1.5 is not co-Lipschitz. In fact, it is unknown whether c_0 is a Lipschitz quotient of l_∞.

In connection with Example 11.13 and Remark (ii) that follows it, it is of interest to observe that if $\varphi : E \to F$ is a co-Lipschitz map which is Gâteaux differentiable at a point x_0 and if x_0 is isolated in its "level set" $\{x : \varphi(x) = \varphi(x_0)\}$, then the derivative T is an (into) isomorphism. Indeed, we may assume that $x_0 = 0$ and $\varphi(x_0) = 0$ and that the co-Lipschitz constant of φ is smaller than 1. Then $\lim_{t\to 0} \varphi(tx)/t = Tx$ for every $x \in E$. If T is not bounded from below, choose $x \in E$ with $\|x\| = 1$ such that $\|Tx\| < 1/4$. Then $\|\varphi(tx)\| < t/4$ for small enough $t > 0$; i.e., $0 \in B\big(\varphi(tx), t/4\big)$. Since the co-Lipschitz constant of φ is smaller than 1, it follows that there is an $x_t \in B(tx, t/4)$ such that $\varphi(x_t) = 0$. Since $3t/4 \le \|x_t\| \le 5t/4$, it follows that $x_t \neq 0$ and $x_t \to 0$, contradicting the assumption that 0 is isolated in its level set.

A map φ from a Banach space E to a Banach space F is said to be ε-Fréchet differentiable at a point x_0 if there is a bounded linear operator $T : E \to F$ and a $\delta > 0$ such that $\|\varphi(x_0 + u) - \varphi(x_0) - Tu\| \le \varepsilon\|u\|$ whenever $\|u\| < \delta$ (it is Fréchet differentiable if this holds for every $\varepsilon > 0$). ε-Fréchet differentiability at a point gives an affine approximation of φ (although this approximation is not as strong as in the definition of UAAP). Unfortunately there are no known existence theorems for Fréchet and ε-Fréchet derivatives between infinite-dimensional Banach spaces which are useful for our purposes. Lindenstrauss and Preiss [**366**] proved the existence of points of ε-Fréchet differentiability for Lipschitz maps from superreflexive to finite-dimensional spaces, but the proof is quite complicated. Theorem 11.16 is much simpler and gives what we need. In this connection, note also that for the proof of Theorem 11.21 it suffices to know that if E^* is separable and if $f : E \to \mathbb{R}$ is Lipschitz, then f has, for every $\varepsilon > 0$, points of ε-Fréchet differentiability. This result is easier than Preiss's theorem from [**471**], and is proved in Fabian and Preiss [**182**]. Fréchet differentiability and its variations, as well as properties related to UAAP, will be studied in detail in the second volume.

The same proof as that of Corollary 11.20 gives that for $1 < p < \infty$ every uniform quotient of $\mathcal{T}^{(p)}$ is isomorphic to a linear quotient of $\mathcal{T}^{(p)} \oplus L_p$.

Example 11.23 can be strengthened for finite-dimensional spaces. For example, for every $n \ge 2$ there is a Lipschitz map from $\mathbb{R}^{n+1}$ onto $\mathbb{R}^n$ which is co-uniform and maps the unit ball in some hyperplane of $\mathbb{R}^{n+1}$ to 0. For $n = 2$ this is proved in [**42**]. The (non obvious) extension to $n > 2$ is due to Maleva [**380**].

In Example 11.24 the domain is l_1. This is used in an essential way in proving that the functions φ_n are 1-Lipschitz. It is unknown whether a Lipschitz quotient map from a superreflexive space (and in particular, from a Hilbert space) onto another Banach space can map a hyperplane to a single point.

Uniformly and co-uniformly continuous maps from $\mathbb{R}^2$ to itself and to the real line are studied in Johnson, Lindenstrauss, Preiss and Schechtman [**293**]. They show that if $\varphi : \mathbb{R}^2 \to \mathbb{R}^2$ is uniformly continuous with $\omega_\varphi(\varepsilon) = o(\varepsilon^{1/2})$ as $\varepsilon \to 0$ and if φ is co-uniformly continuous, then $\varphi = P \circ \psi$ where ψ is a homeomorphism of the plane and P is a complex polynomial. The proof uses a topological characterization of analytic functions due to Stoilow [**544**]. The authors also show that the same result holds under other conditions involving the moduli of uniform and of co-uniform continuity of φ, and they give an example of a uniform quotient map of the plane onto itself for which the result is false. The methods of this article work only for the plane, and nothing much is known in higher dimensions.

Lipschitz quotient maps from $\mathbb{R}^n$ to itself are related to quasi-regular maps. (The book [**496**] by Rickman is a good reference on this topic.) A map φ from a domain $G \subset \mathbb{R}^n$ to $\mathbb{R}^n$ is quasi-regular if

(i) φ is continuous, its restriction to every line parallel to the axes is absolutely continuous, and its partial derivatives are locally in $L_p(\mathbb{R}^n)$ for $p = n$.

(ii) There is a constant K such that $\|D_\varphi(x)\|^n \leq K J_\varphi(x)$ for almost every $x \in G$, where $J_\varphi(x)$ is the Jacobian of φ at x.

Lipschitz quotient maps certainly satisfy (i). Instead of (ii) they satisfy only the weaker estimate with absolute values, namely, $\|D_\varphi(x)\|^n \leq K|J_\varphi(x)|$ (with $K = (M/m)^{n-1}$, where M and m are the Lipschitz and the co-Lipschitz constants of φ).

A key step in the study of uniform quotients of the plane in [**293**] was to find conditions under which the level sets $\varphi^{-1}(y)$ are discrete (even finite) for every point y. A deep theorem of Reshetnyak (which is carefully presented in Rickman's book) says that the level sets of a quasi-regular map φ are discrete. The positivity of $J_\varphi(x)$ (i.e., the fact that φ preserves orientation) is used in a crucial way in this proof. It is not clear whether every Lipschitz quotient map of $\mathbb{R}^n$ onto itself for $n \geq 3$ must either preserve or reverse orientation (i.e., that $J_\varphi(x)$ has constant sign almost everywhere).

The following result is proved in David [**131**] (see also Jones [**301**]): For every $\eta > 0$ and $c > 0$ there are constants $c_1 = c_1(\eta, c, n)$ and $\alpha = \alpha(\eta, c, n)$ with the following property. Let f be a c-Lipschitz function from any ball $B \subset \mathbb{R}^n$ into $\mathbb{R}^n$ such that $m\big(f(B)\big) \geq \eta m(B)$ (where m is the Lebesgue measure in $\mathbb{R}^n$), then there is a subset $A \subset B$ with $m(A) \geq \alpha m(B)$ such that f satisfies $\|f(x) - f(y)\| \geq c_1\|x - y\|$ for all $x, y \in A$. In particular if f is a Lipschitz quotient map from $\mathbb{R}^n$ onto itself, then f is biLipschitz on a fixed percentage of any ball in $\mathbb{R}^n$.

Using this fact, Lemma 11.14 and Proposition 11.16, David and Semmes [**132**] proved that if f is a Lipschitz quotient map from $\mathbb{R}^n$ onto $\mathbb{R}^k$ with $k < n$, then there are constants C and α such that for every ball $B \subset \mathbb{R}^n$ there are a linear map $T : \mathbb{R}^n \to \mathbb{R}^{n-k}$ and a subset A of B with $m(A) \geq \alpha m(B)$ such that the restriction of the map $x \to \big(f(x), Tx\big)$ to A is biLipschitz with constant C. The constants C and α depend only on n, k and the Lipschitz and co-Lipschitz constants of f. (The result in [**132**] is actually proved in the more general context of (n, k)-regular maps which we shall not define.)

CHAPTER 12

Oscillation of Uniformly Continuous Functions on Unit Spheres of Finite-Dimensional Subspaces

1. Dvoretzky's Theorem

Let f be a real-valued Lipschitz function (or more generally, a uniformly continuous function) defined on the unit sphere of an infinite-dimensional Banach space E. In this and the next chapter we discuss the question whether there is an infinite-dimensional subspace F, or at least finite-dimensional subspaces F of arbitrarily high dimension, such that the restriction of f to the unit sphere of F is almost constant, i.e., whether for every $\varepsilon > 0$ there is a value λ and a subspace F such that $|f(x) - \lambda| < \varepsilon$ for every x in the unit sphere of F. It turns out that such λ and finite-dimensional subspaces of arbitrarily high dimension can be found (and this is the subject of this chapter), but it is generally impossible to find an infinite-dimensional subspace (and this is discussed in the next chapter).

We start this section with a proof of the fundamental theorem of Dvoretzky, namely, that every infinite-dimensional Banach space E contains almost Euclidean subspaces of arbitrarily high finite dimension. Actually the theorem does not require E to be infinite-dimensional. For every k and $\varepsilon > 0$ there is a finite n such that any n-dimensional Banach space has a k-dimensional subspace whose distance to an Euclidean space is smaller than $1 + \varepsilon$. To find further subspaces (again of arbitrarily high dimension) where a given Lipschitz function is almost constant, we apply again an argument similar to the proof of Dvoretzky's theorem (in a simplified version).

The procedure described in this section proves the existence of "good" finite-dimensional subspaces by a probabilistic (or measure theoretic) argument. Although it gives good quantitative control of their dimension, it gives no information on their form. In Section 2 we present a different approach, due to Krivine, which yields subspaces of a more specified form (subspaces spanned by "block bases"; see Appendix E for the definition). Again the proof is in two steps which essentially repeat the same argument. In the first step we find a $1 \leq p \leq \infty$ and produce for each k a subspace which is almost isometric to l_p^k (instead of subspaces which are almost Euclidean as in Dvoretzky's theorem). We then apply a similar argument again in order to obtain the desired blocks so that the given function f is almost constant on their span. The proofs in Section 2 are combinatorial in nature and (at least in the way we present them here) do not give quantitative estimates on the dimension.

The problem of getting good estimates (mainly in the proof of Dvoretzky's theorem, but also in Krivine's theorem and in other related results) is of fundamental significance in the local theory of Banach spaces. We shall not, however, study this topic in any detail here.

Let σ_n be the normalized rotation invariant measure on the unit sphere S^{n-1} in $\mathbb{R}^n$ (with its usual Euclidean structure), and let $d(x, y)$ be the natural geodesic distance on S^{n-1} (arc length on great circles). We shall use the term *cap* for a ball in this metric. For a Borel set $A \subset S^{n-1}$ and $\varepsilon > 0$ put $A_\varepsilon = \{x \in S^{n-1} : d(x, A) \leq \varepsilon\}$.

We shall base the proofs in this section on the following "concentration of measure" phenomenon

$$\text{(12.1)} \qquad \textit{If } \sigma_n(A) \geq 1/2, \textit{ then } \sigma_n(A_\varepsilon) \geq 1 - 2e^{-n\varepsilon^2/2}.$$

If $\sigma_n(A) = 1/2$, then this estimate applies to both A and its complement. It follows that if $\varepsilon > 0$ is fixed and if n is large enough, then "most" of S^{n-1} consists of points whose distances to both A and its complement are smaller than ε (i.e., σ_n is "concentrated" in the ε-neighborhood of the boundary of A).

We shall give a proof of (12.1) in Proposition 12.1 below, but we first sketch a different natural approach, based on the following isoperimetric inequality on S^{n-1}:

Let A be a subset of S^{n-1}, and let B be a cap with the same measure as A, i.e., $\sigma_n(B) = \sigma_n(A)$. Then $\sigma_n(A_\varepsilon) \geq \sigma_n(B_\varepsilon)$ for every $\varepsilon > 0$.

To see that this is indeed an isoperimetric inequality, note that the $(n-2)$-dimensional measure of the boundary of $A \subset S^{n-1}$ is $\lim_{\varepsilon \to 0} \varepsilon^{-1}\big(\sigma_n(A_\varepsilon) - \sigma_n(A)\big)$ whenever the limit exists. Thus the inequality says that the boundaries of caps have the smallest measure among all sets of a given measure.

Inequality (12.1) follows directly from the isoperimetric inequality. Indeed, if $\sigma_n(A) = 1/2$, then the corresponding cap B is just the half sphere, and then $\sigma_n(B_\varepsilon)$ is given explicitly by

$$\sigma_n(B_\varepsilon) = \Big(\int_{-\pi/2}^{\varepsilon} \cos^{n-2} t dt\Big) \Big/ \Big(\int_{-\pi/2}^{\pi/2} \cos^{n-2} t dt\Big)$$

which can be easily estimated. We give no further details, since Proposition 12.1 proves (12.1) in a more general context. (The exponent we obtain there is $n\varepsilon^2/4$ instead of $n\varepsilon^2/2$, but this is insignificant for our applications.)

Remarks. (i) Inequality (12.1) gives an estimate on the growth of $\sigma_n(A_\varepsilon)$ when $\sigma_n(A) = \alpha$ for arbitrary $\alpha > 0$ (and not only for $\alpha = 1/2$). To see this, choose δ such that $\alpha > 2e^{-n\delta^2/2}$. Then $\sigma_n(A_\delta) \geq 1/2$ (for otherwise $\sigma_n(S^{n-1} \setminus A_\delta) > 1/2$; hence $1 - \alpha = \sigma_n(S^{n-1} \setminus A) > \sigma_n\big((S^{n-1} \setminus A_\delta)_\delta\big) \geq 1 - 2e^{-n\delta^2/2} > 1 - \alpha$ by (12.1) and by the choice of δ). It follows from (12.1) that $\sigma_n(A_{\delta+\varepsilon}) \geq 1 - 2e^{-n\varepsilon^2/2}$.

(ii) The geodesic metric d and the norm distance are close to each other in the sense that

$$d(x, y) \geq \|x - y\| = \big(2 - 2\cos d(x, y)\big)^{1/2} = d(x, y) + O\big(d(x, y)\big)^2.$$

It follows that the only effect of replacing d by the norm distance in the definition of A_ε is a minor change in the constants in (12.1).

Let E be an n-dimensional Banach space, and let K be its unit ball. We denote the normalized Lebesgue measure on K by m. If $A \subset \partial K$, put $\tilde{A} = \bigcup_{0 \leq t \leq 1} tA$. We introduce a probability measure μ on the unit sphere ∂K of E by defining $\mu(A) = m(\tilde{A})$. Note that if K is the Euclidean ball, then $\mu = \sigma_n$.

Proposition 12.1. *Let E be an n-dimensional uniformly convex Banach space with modulus of convexity δ_E. Then*

(i) *If A is a Borel subset of K with $m(A) \geq 1/2$ and $\varepsilon > 0$, then*

$$m(A_\varepsilon) \geq 1 - 2e^{-2n\delta_E(\varepsilon)}$$

where $A_\varepsilon = \{x \in K : d_K(x, A) \leq \varepsilon\}$, and $d_K(x, A)$ denotes the K-distance from x to A.

(ii) *An analogous formula holds for subsets of ∂K (with m replaced by μ) for every $\varepsilon > 0$ with $\delta_E(\varepsilon) \leq 1/2$.*

Proof. (i) Let $B = K \backslash A_\varepsilon$, and let $x \in A$ and $y \in B$. By definition $\|x - y\| \geq \varepsilon$, and hence $\|(x + y)/2\| \leq 1 - \delta_E(\varepsilon)$; i.e., $(A + B)/2 \subset \big(1 - \delta_E(\varepsilon)\big)K$. By the multiplicative form of the Brunn-Minkowski inequality (see Theorem D.12)

$$\big(1 - \delta_E(\varepsilon)\big)^n \geq m\big((A + B)/2\big) \geq \big(m(A)m(B)\big)^{1/2} \geq \big(m(B)/2\big)^{1/2}$$

and thus

$$m(B) \leq 2\big(1 - \delta_E(\varepsilon)\big)^{2n} \leq 2e^{-2n\delta_E(\varepsilon)}.$$

(ii) We use the same argument as in (i). We just observe that if $A, B \subset \partial K$ with $\|x - y\| \geq \varepsilon$ for all $x \in A$ and $y \in B$, then $\|(u + v)/2\| \leq 1 - \delta_E(\varepsilon)$ whenever $u \in \tilde{A}$ and $v \in \tilde{B}$. To check this write $u = tx$ and $v = sy$ with $0 \leq s \leq t \leq 1$. It suffices to treat the case $t = 1$, and then

$$\begin{aligned} \|(u + v)/2\| &= \|(x + sy)/2\| \leq s\|(x + y)/2\| + (1 - s)\|x/2\| \\ &\leq s(1 - \delta_E(\varepsilon)) + (1 - s)/2 \leq 1 - \delta_E(\varepsilon) \end{aligned}$$

by the condition on ε. □

Remark. If E is a Hilbert space, then $\delta_E(\varepsilon) = 1 - \big(1 - \frac{\varepsilon^2}{4}\big)^{1/2} \geq \varepsilon^2/8$, and (12.1) follows with $\varepsilon^2/2$ replaced by $\varepsilon^2/4$ (and with the norm rather than the geodesic distance).

We shall need the following simple lemma.

Lemma 12.2. *Let E be a k-dimensional Banach space, let $0 < \varepsilon < 1$, and let $\{x_j\}_{j=1}^N$ be points in the unit ball of E such that $\|x_j - x_{j'}\| \geq \varepsilon$ for $j \neq j'$. Then $N \leq (1 + 2/\varepsilon)^k$.*

Proof. The balls with centers x_j and radius $\varepsilon/2$ have mutually disjoint interiors and are all contained in the ball with radius $1 + \varepsilon/2$ centered at the origin. Comparison of volumes gives

$$(\varepsilon/2)^k N \leq (1 + \varepsilon/2)^k.$$

□

Let $O(n)$ be the group of orthogonal transformation on $\mathbb{R}^n$, and denote the Haar measure on $O(n)$ by ν. The uniqueness of the (normalized) rotation invariant measure on S^{n-1} implies that ν and σ_n are related by the formula

$$\sigma_n(A) = \nu\big\{U \in O(n) :\ Ux \in A\big\}$$

for every set $A \subset S^{n-1}$ and for every fixed $x \in S^{n-1}$.

Proposition 12.3. *Let f be a real-valued function on S^{n-1} which is τ-Lipschitz with respect to the Euclidean norm, and let $0 < \varepsilon < \tau/2$. Then there is a k-dimensional subspace F of $\mathbb{R}^n$ with $k \geq Cn\varepsilon^2/|\log(\varepsilon/\tau)|\tau^2$ (where C is a universal constant), and there is a number λ_0 such that*

$$\lambda_0 - \varepsilon \leq f(x) \leq \lambda_0 + \varepsilon$$

for every $x \in F \cap S^{n-1}$.

Proof. Let λ_0 be the median of f; i.e., λ_0 is the (unique) number for which

$$\sigma_n\{x : f(x) \geq \lambda_0\} \geq 1/2 \quad \text{and} \quad \sigma_n\{x : f(x) \leq \lambda_0\} \geq 1/2$$

and put $A = \{x : f(x) = \lambda_0\}$. Since f is τ-Lipschitz, it follows that $|\lambda_0 - f(x)| \leq \varepsilon/2$ on $A_{\varepsilon/2\tau}$.

By (12.1), $\nu\{U \in O(n) : Ux \in A_{\varepsilon/2\tau}\} = \sigma_n(A_{\varepsilon/2\tau}) \geq 1 - 4e^{-n\varepsilon^2/8\tau^2}$ for every fixed $x \in S^{n-1}$. Hence, if $\{x_j\}_{j=1}^N$ are any N points in S^{n-1} and if $4N < e^{n\varepsilon^2/8\tau^2}$, then there is a $U_0 \in O(n)$ such that $U_0 x_j \in A_{\varepsilon/2\tau}$ for all j.

Let $F_0 \subset \mathbb{R}^n$ be the subspace spanned by the first k unit vectors of $\mathbb{R}^n$, and let $\{x_j\}_{j=1}^N$ be a maximal $\varepsilon/2\tau$-separated set in $F_0 \cap S^{n-1}$. It follows from Lemma 12.2 that $N \leq (1 + 4\tau/\varepsilon)^k$, and hence if k satisfies

$$4(1 + 4\tau/\varepsilon)^k \leq e^{n\varepsilon^2/8\tau^2} \tag{12.2}$$

then there is a $U_0 \in O(n)$ such that $U_0 x_j \in A_{\varepsilon/2\tau}$ for every $1 \leq j \leq N$; i.e., $|f(U_0 x_j) - \lambda_0| \leq \varepsilon/2$ for all j. If x is any point in $F_0 \cap S^{n-1}$, then there is a j such that $\|x_j - x\| \leq \varepsilon/2\tau$, and hence

$$|f(U_0 x) - \lambda_0| \leq |f(U_0 x) - f(U_0 x_j)| + |f(U_0 x_j) - \lambda_0| \leq \varepsilon/2 + \varepsilon/2 = \varepsilon.$$

Thus f is ε-close to the constant λ_0 on the unit sphere of $F = U_0 F_0$. To estimate the dimension of F, we just need to note that there is a universal constant C such that $k = \left[Cn\varepsilon^2/|\log(\varepsilon/\tau)|\tau^2\right] + 1$ satisfies (12.2). □

Remark. The main part of the proof was to find a maximal well-separated set Λ in the unit sphere of some k-dimensional subspace F of $\mathbb{R}^n$ with a good estimate on $|f(x) - \lambda_0|$ for every $x \in \Lambda$.

This part will be used below several times (with different choices of parameters), and we formulate it separately: Let f be a τ-Lipschitz function on S^{n-1}. Fix $\alpha, \delta > 0$, and assume that k satisfies

$$4(1 + 2/\delta)^k \leq e^{n\alpha^2/8\tau^2}.$$

Then there is a k-dimensional subspace F of $\mathbb{R}^n$ and a maximal δ-separated set Λ in $F \bigcap S^{n-1}$ such that $|f(x) - \lambda_0| < \alpha$ for every $x \in \Lambda$.

The number λ_0 for which we verified Proposition 12.3 is the median of f. In general it is hard to compute the median of a function. However, because of the concentration of the measure, the median of f is close to its average which is much easier to calculate.

Lemma 12.4. *There is a universal constant C_1 such that if f is a real-valued τ-Lipschitz function on S^{n-1} and if λ_0 is the median of f, then*

$$\left|\lambda_0 - \int_{S^{n-1}} f(x)\,d\sigma_n(x)\right| \leq C_1\tau/\sqrt{n}.$$

Proof. Put $\alpha_m = \tau\big((4\log m)/n\big)^{1/2}$ for $m \geq 2$ and $\alpha_1 = 0$. By (12.1)

$$\sigma_n\big\{x : \alpha_m \leq |f(x) - \lambda_0| \leq \alpha_{m+1}\big\} \leq 4e^{-n\alpha_m^2/2\tau^2} = 4/m^2$$

for $m \geq 2$. Hence

$$\Big|\lambda_0 - \int_{S^{n-1}} f(x)d\sigma_n(x)\Big| \leq \alpha_2 + \sum_{m=2}^{\infty} 4\alpha_{m+1}/m^2 \leq C_1\tau/\sqrt{n}.$$

□

We now turn to functions of the form $f(x) = |||x|||$, where $|||\cdot|||$ is a norm on $\mathbb{R}^n$. We start with a simple lemma, which enables us to obtain for these special functions a little sharper estimate on k. We say that a set Λ in a metric space A is θ-dense if for every $y \in A$ there is an $x \in \Lambda$ with $d(x,y) \leq \theta$. A maximal θ-separated set is clearly θ-dense.

Lemma 12.5. *Let E be a Banach space endowed with two norms $\|\cdot\|$ and $|||\cdot|||$, and let S denote the unit sphere of E with respect to the norm $\|\cdot\|$. Assume that Λ is θ-dense in S (with respect to the distance induced by $\|\cdot\|$) for some $0 < \theta < 1$ and that there are $0 < a < b$ such that $a \leq |||x||| \leq b$ for every $x \in \Lambda$. Then*

$$\big(a - b\theta(1-\theta)^{-1}\big)\|y\| \leq |||y||| \leq b(1-\theta)^{-1}\|y\|$$

for every $y \in E$.

Proof. It is enough to prove the lemma for y with $\|y\| = 1$. By the θ-density of Λ and a standard iterative argument $y = x_1 + \lambda_1 x_2 + \lambda_2 x_3 + \cdots$, where $x_j \in \Lambda$ and $0 \leq \lambda_j \leq \theta^j$ for all j. Hence $|||y||| \leq b(1 + \sum_{j=1}^{\infty} \lambda_j) \leq b(1-\theta)^{-1}$.

To prove the other direction, choose $x \in \Lambda$ with $\|y - x\| \leq \theta$. By the first part $|||y - x||| \leq \theta b(1-\theta)^{-1}$; hence $|||y||| \geq |||x||| - |||x - y||| \geq a - b\theta(1-\theta)^{-1}$. □

We are now ready to prove the first result on the existence of almost Euclidean subspaces of a Banach space.

Theorem 12.6. *There is a universal constant $C > 0$ such that for every n-dimensional Banach space E and for every $0 < \varepsilon < 1/2$ there is a k-dimensional subspace F of E with $k \geq Cn\varepsilon^2/d(E, l_2^n)^2|\log\varepsilon|$ and $d(F, l_2^k) \leq 1 + \varepsilon$.*

Proof. Consider E as $\mathbb{R}^n$ with a norm $|||\cdot|||$ satisfying $\|x\| \leq |||x||| \leq d\|x\|$ for all x, where $\|\cdot\|$ is the Euclidean norm and $d = d(E, l_2^n)$. Put $f(x) = |||x|||$ on S^{n-1}. Then f is Lipschitz with constant d. By the remark following Proposition 12.3 it follows that if

$$4(1 + 2/\varepsilon)^k < e^{n\varepsilon^2/8d^2} \tag{12.3}$$

then there is a subspace F of $E = \mathbb{R}^n$ of dimension k and an ε-dense set Λ (with respect to $\|\cdot\|$) in $F \cap S^{n-1}$ such that for some λ_0

$$\lambda_0 - \varepsilon \leq |||x||| \leq \lambda_0 + \varepsilon$$

for every $x \in \Lambda$. By the normalization of the norms $\lambda_0 \geq 1$; hence (assuming, as we may, that $\varepsilon < 1/100$), if $x \in F$ is an arbitrary point with $\|x\| = 1$, then Lemma 12.5 gives

$$\begin{aligned}\lambda_0(1 - 4\varepsilon) &\leq (\lambda_0 - \varepsilon) - \varepsilon(\lambda_0 + \varepsilon)(1-\varepsilon)^{-1} \leq |||x|||\\ &\leq (\lambda_0 + \varepsilon)/(1-\varepsilon) \leq (1 + 3\varepsilon)\lambda_0.\end{aligned}$$

Thus $d\big((F, |||\cdot|||), l_2^k\big) \leq (1 + 3\varepsilon)/(1 - 4\varepsilon) \leq 1 + 25\varepsilon$.

For a suitable absolute constant C, the k in the statement of the theorem satisfies (12.3), and it clearly does not matter that we replaced ε by 25ε. □

If E is any n-dimensional Banach space, then $d(E, l_2^n) \leq \sqrt{n}$ by Proposition G.1. If $d(E, l_2^n)$ is actually of the order of magnitude $\sqrt{n}$, then Theorem 12.6 gives no useful information. In order to obtain a meaningful estimate and find almost Euclidean subspaces in general Banach spaces, we need two additional lemmas. The first is purely computational, while the second is of a geometric nature.

Lemma 12.7. *Denote the coordinates of a vector $x \in \mathbb{R}^n$ by $x_1, \ldots, x_n$. Then there is an absolute positive constant c such that*

$$\int_{S^{n-1}} \max_{1 \leq j \leq n} |x_j| \, d\sigma_n(x) \geq c(\log n/n)^{1/2}$$

for every n.

Proof. Let γ be the probability measure on $\mathbb{R}^n$ whose density with respect to Lebesgue measure is $e^{-\pi \sum_{j=1}^n x_j^2}$. By using polar coordinates in $\mathbb{R}^n$ it follows that for every positive homogeneous function f on $\mathbb{R}^n$

$$\int_{R^n} f(x) d\gamma(x) = \lambda_n \int_{S^{n-1}} f(x) d\sigma_n(x)$$

for a suitable constant λ_n. To estimate λ_n, take $f(x) = \|x\|$. Then the integral on the right is 1; hence the symmetry of γ under permutation of the coordinates gives

$$\begin{aligned} \lambda_n &= \int_{\mathbb{R}^n} \|x\| d\gamma(x) \leq \Big(\int_{\mathbb{R}^n} \|x\|^2 d\gamma(x)\Big)^{1/2} = \Big(n \int_{\mathbb{R}^n} x_1^2 d\gamma(x)\Big)^{1/2} \\ &= \Big(n \int_{-\infty}^{\infty} t^2 e^{-\pi t^2} dt\Big)^{1/2} = (n/2\pi)^{1/2}. \end{aligned}$$

Put $\alpha = \big((2\pi)^{-1} \log n\big)^{1/2}$. Then

$$\int_{\alpha}^{\infty} e^{-\pi t^2} dt = \int_{\pi\alpha^2}^{\infty} e^{-u} \frac{du}{2\sqrt{\pi u}} \geq \frac{1}{2\pi\alpha} \int_{\pi\alpha^2}^{\infty} e^{-u} du = e^{-\pi\alpha^2}/2\pi\alpha \geq 1/2n$$

for large enough n, and hence

$$\gamma\Big\{x \in \mathbb{R}^n : \max_{1 \leq j \leq n} |x_j| \leq \alpha\Big\} = \Big(\int_{-\alpha}^{\alpha} e^{-\pi t^2} dt\Big)^n \leq (1 - 1/n)^n \leq 1/2.$$

Passing to the complement gives

$$\begin{aligned} \int_{S^{n-1}} \max |x_j| \, d\sigma_n(x) &= \lambda_n^{-1} \int_{\mathbb{R}^n} \max |x_j| \, d\gamma(x) \\ &\geq \lambda_n^{-1} \alpha \, \gamma\{x \in \mathbb{R}^n : \max |x_j| \geq \alpha\} \\ &\geq \lambda_n^{-1} \alpha/2 \geq c(\log n/n)^{1/2}. \end{aligned}$$

□

Remark. In probabilistic terms Lemma 12.7 gives an estimate (which is standard in probability theory) for the expectation of the supremum of n independent suitably normalized Gaussian variables.

Lemma 12.8. *Let $\|\cdot\|$ be a norm in $\mathbb{R}^n$ such that the Euclidean unit ball is the ellipsoid of maximum volume in the $\|\cdot\|$ unit ball. Then there is an orthonormal basis $\{x_j\}_{j=1}^n$ in $\mathbb{R}^n$ such that $\|x_j\| \geq 2^{-n/2(n-j+1)}$ for all j.*

Proof. Note that by the assumption $|||x||| \le \|x\|$ for all $x \in \mathbb{R}^n$. We choose the x_j's inductively. Choose $x_1 \in S^{n-1}$ with $|||x_1||| = 1$. (Such an x_1 exists because the maximality of the Euclidean ball implies that S^{n-1} intersects the boundary of the $|||\cdot|||$ unit ball). Having chosen x_k for $k \le j-1$, take $x_j \in S^{n-1}$ to be of maximal $|||\cdot|||$-norm in the orthogonal complement to $\{x_k\}_{k=1}^{j-1}$.

To estimate $a = |||x_j|||$, complete the vectors chosen already to an orthonormal basis $x_1, \dots, x_j, y_{j+1}, \dots, y_n$ of $\mathbb{R}^n$, and consider the ellipsoid

$$\mathcal{E} = \Big\{ \sum_{k=1}^{j} t_k x_k + \sum_{k=j+1}^{n} t_k y_k : 2\sum_{k=1}^{j-1} t_k^2 + 2a^2 \sum_{k=j}^{n} t_k^2 \le 1 \Big\}.$$

This ellipsoid is contained in the $|||\cdot|||$ unit ball. Indeed, for any $\{t_k\}_{k=1}^n$ the choice of x_j gives

$$\Big|\Big|\Big| t_j x_j + \sum_{k=j+1}^{n} t_k y_k \Big|\Big|\Big| \le a \Big\| t_j x_j + \sum_{k=j+1}^{n} t_k y_k \Big\| = a\Big(\sum_{k=j}^{n} t_k^2\Big)^{1/2}.$$

Since also $|||\sum_{k=1}^{j-1} t_k x_k||| \le \|\sum_{k=1}^{j-1} t_k x_k\| = (\sum_{k=1}^{j-1} t_k^2)^{1/2}$, we see that the points in $\mathcal{E}$ satisfy

$$\begin{aligned} \Big|\Big|\Big| \sum_{k=1}^{j} t_k x_k + \sum_{k=j+1}^{n} t_k y_k \Big|\Big|\Big| &\le \Big(\sum_{k=1}^{j-1} t_k^2\Big)^{1/2} + a\Big(\sum_{k=j}^{n} t_k^2\Big)^{1/2} \\ &\le \sqrt{2}\Big(\sum_{k=1}^{j-1} t_k^2 + a^2 \sum_{k=j}^{n} t_k^2\Big)^{1/2} \le 1. \end{aligned}$$

The volume of $\mathcal{E}$ is $2^{-n/2} a^{-(n-j+1)} \alpha_n$, where α_n denotes the volume of the Euclidean unit ball. By the maximality of the Euclidean ball $2^{-n/2} a^{-(n-j+1)} \le 1$; i.e., $a \ge 2^{-n/2(n-j+1)}$. □

Corollary 12.9. *With the notation of Lemma* 12.8, *there are* $[n/2]$ *orthonormal vectors* x_j *such that* $|||x_j||| \ge 1/2$ *for all* j.

We are now ready to prove Dvoretzky's theorem.

Theorem 12.10. *There is a universal constant* C_0 *such that for every Banach space* E *of dimension* n *and for every* $0 < \varepsilon < 1/2$ *there is a* k*-dimensional subspace* F *of* E *with* $k \ge C_0 \varepsilon^2 \log n / |\log \varepsilon|$ *and* $d(F, l_2^k) \le 1 + \varepsilon$.

Proof. In view of Corollary 12.9 we may assume, replacing n by $[n/2]$ if necessary, that E is $\mathbb{R}^n$ with a norm $|||\cdot|||$ and that $\|\cdot\|$ is an Euclidean norm on $\mathbb{R}^n$ such that $|||x||| \le \sqrt{n}\|x\|$ for all $x \in \mathbb{R}^n$ and such that there is an orthonormal basis $\{x_j\}_{j=1}^n$ in $\mathbb{R}^n$ with $|||x_j||| \ge \sqrt{n}/2$ for all j. (Note that we changed the normalization of Lemma 12.8 and multiplied the Euclidean norm by a factor of $\sqrt{n}$.)

We now estimate the mean of $|||x|||$ on S^{n-1}. Let $\{r_j(s)\}_{j=1}^n$ be the Rademacher functions on $[0,1]$. Since the integral $\int_0^1 |||\sum_{j=1}^n r_j(s) t_j x_j||| ds$ is just the average of $|||\sum_{j=1}^n \pm t_j x_j|||$ over all possible 2^n choices of sign, it is bounded from below by $\max_j |||t_j x_j||| \ge \sqrt{n}/2 \max_j |t_j|$.

By the symmetry of S^{n-1} and by Lemma 12.7

$$\begin{aligned}\int_{S^{n-1}} \left\|\left|\sum t_j x_j\right|\right\| d\sigma_n(t) &= \int_{S^{n-1}} \int_0^1 \left\|\left|\sum r_j(s) t_j x_j\right|\right\| ds\, d\sigma_n(t) \\ &\geq \sqrt{n}/2 \int_{S^{n-1}} \max |t_j|\, d\sigma_n(t) \geq c(\log n)^{1/2}/2.\end{aligned}$$

The Lipschitz constant of $f(x) = \|\|x\|\|$ on S^{n-1} is $\sqrt{n}$; hence by Lemma 12.4 the median λ_0 of $f(x)$ satisfies $\lambda_0 \geq c_1(\log n)^{1/2}$ for some absolute constant c_1. By the remark following Proposition 12.3 we get that if

$$4(1+2/\varepsilon)^k < 2e^{n\varepsilon^2 \log n/2nc_2^2} = n^{\varepsilon^2/2c_2^2} \tag{12.4}$$

for some constant c_2, then there is a subspace F of E of dimension k and an ε-dense set Λ in $F \cap S^{n-1}$ such that

$$\lambda_0 - \varepsilon\sqrt{\log n}/c_2 \leq \|\|x\|\| \leq \lambda_0 + \varepsilon(\log n)^{1/2}/c_2$$

for all $x \in \Lambda$. It follows from Lemma 12.5 and from the estimate we have on λ_0 that if c_2 is suitably chosen, then $d(F, l_2^k) \leq 1 + \varepsilon$. It remains to note that once c_2 is chosen, it is possible to choose C_0 such that if k is as in the statement of the theorem, then it satisfies (12.4). □

Combining Proposition 12.3 with Theorem 12.10 gives a positive answer to our original problem on a general finite-dimensional Banach space.

Corollary 12.11. *Let f be a τ-Lipschitz function defined on the unit sphere of an n-dimensional Banach space $(E, \|\|\cdot\|\|)$, and let $0 < \varepsilon < \tau/2$. Then there is a value λ_0 and a subspace F of E of dimension $k \geq C\varepsilon^2 \log n/|\log(\varepsilon/\tau)|\tau^2$ (where C is a universal constant) such that $|f(x) - \lambda_0| \leq \varepsilon$ for all $x \in F$ with $\|\|x\|\| = 1$.*

Proof. By Theorem 12.10 there is a subspace F_1 of E of dimension $k_1 = C_1 \log n$ such that $d(F_1, l_2^{k_1}) \leq 2$. This means that one can introduce an Euclidean norm $\|\cdot\|$ in F_1 such that $\|x\| \leq \|\|x\|\| < 2\|x\|$ for all $x \in F_1$. The function $\tilde{f}$ on the $\|\cdot\|$ unit sphere S defined by $\tilde{f}(x) = f(x/\|\|x\|\|)$ is Lipschitz with constant smaller than 4τ. We now apply Proposition 12.3 to this function. □

Remark. The corollary holds also if we replace the assumption that f is Lipschitz by the assumption that f is uniformly continuous (with a modulus of continuity independent of n). In this case we still get $k \geq \alpha \log n$ where α is a function of ε and of the modulus of continuity of f. This is evident from the proof of Proposition 12.3.

The estimates obtained above are not sharp as far as the dependence on ε or τ are concerned. On the other hand, the dependence on n cannot be improved in general. This follows from the next proposition.

Proposition 12.12. *There is a constant C such that if F is a k-dimensional subspace of l_∞^n with $d(F, l_2^k) \leq 2$, then $k \leq C \log n$.*

Proof. Let F be as in the statement of the proposition. It follows that there are n linear functionals of norm 1 on l_2^k (i.e., n points $\{x_j\}_{j=1}^n$ in S^{k-1}) such that

$$\|x\| \leq 2 \max_j |\langle x, x_j \rangle|$$

for every $x \in l_2^k$. It follows that for every $x \in S^{k-1}$ there is a j such that either $\|x - x_j\| \leq 1$ or $\|x + x_j\| \leq 1$. (If j is such that $\langle x, x_j \rangle \geq 1/2$, then $\|x - x_j\|^2 =$

$2 - 2\langle x, x_j \rangle \leq 1$, and if $\langle x, x_j \rangle \leq -1/2$, then $\|x + x_j\|^2 \leq 1$.) Thus S^{k-1} is covered by $2n$ spherical caps with (Euclidean) radius 1. The σ_k measure of such a cap can be easily computed exactly. However, for our purpose it suffices to apply (12.1): There is an absolute constant $\rho > 0$ such that a cap of radius 1 is at distance ρ from the opposite half sphere. Hence the surface area of a cap is at most $2e^{-k\rho^2/2}$. Consequently $2n \cdot 2e^{-k\rho^2/2} \geq 1$. □

2. Krivine's Theorem

The results of this section are analogous to those of the previous section, but the methods here give some control on the form of the subspace on which the given Lipschitz function has small oscillation. A weak form of the results of the previous section (without the quantitative estimates on the dimension of the subspace) follows also from the results here. Although the results of this section are finite-dimensional in nature, we do not attempt to give any explicit quantitative formulation. We shall actually work in the infinite-dimensional setting, which is far more convenient here. The finite-dimensional results will then follow by a simple compactness argument. For information on bases, and in particular on spreading and unconditional bases, see Appendix E.

The proof of the main theorem (Theorem 12.19) has two parts. The first, Theorem 12.14, is combinatorial in the spirit of Ramsey theory. It says that every infinite sequence (satisfying some simple nondegeneracy condition) has arbitrary long blocks which are very regular (almost 1-spreading and 1-unconditional). The second step, which is also of a combinatorial nature, uses this regularity to find further blocks which are equivalent to l_p^k.

Since we are only interested in a finite number of blocks, it is very convenient to pass freely from one infinite sequence to another, provided all finite block sequences of the new sequence can be represented as blocks on the old sequence. When this new sequence has some further regularity, it is easier to find the required blocks there. We first formalize this procedure.

Definition 12.13. *A sequence $\{y_j\}_{j=1}^{\infty}$ in a Banach space E_1 is said to be block finitely representable in a sequence $\{x_i\}_{i=1}^{\infty}$ in a Banach space E_2 if for every $\varepsilon > 0$ and for every integer n there are vectors $\{z_j\}_{j=1}^{n} \subset E_2$ of the form*

$$z_j = \sum_{j=m_{j-1}}^{m_j - 1} b_i x_i$$

with $m_0 < m_1 < \cdots < m_n$ such that

$$(1-\varepsilon)\Big\|\sum_{j=1}^{n} a_j z_j\Big\| \leq \Big\|\sum_{j=1}^{n} a_j y_j\Big\| \leq (1+\varepsilon)\Big\|\sum_{j=1}^{n} a_j z_j\Big\|$$

for all choices of scalars $\{a_j\}$.

Note that if $\{x_i\}$ is a basic sequence, then the z_j's are a (finite) block basic sequence with respect to the x_i's. The relation of being block finitely representable is obviously a transitive relation.

Theorem 12.14. *Let $\{x_i\}$ be an infinite sequence in a Banach space E_1 such that the set of numbers $\{\|\sum_{k=1}^{n} a_k x_{i_k}\| : i_1 < \ldots < i_n\}$ is bounded and bounded away from 0 for every fixed n and for every fixed choice of $\{a_k\}_{k=1}^{n}$, not all of them* *zero. Then there is a basic sequence* $\{y_j\}_{j=1}^{\infty}$ *in some Banach space* E_2 *which is block finitely representable in* $\{x_i\}$ *and such that*

(i) $\{y_j\}$ *is 1-spreading; i.e.,* $\|\sum a_j y_j\| = \|\sum a_j y_{k_j}\|$ *for every choice of scalars* $\{a_j\}$ *and increasing sequence of integers* $\{k_j\}$.

(ii) $\{y_j\}$ *is 1-unconditional; i.e.,* $\|\sum a_j y_j\| = \|\sum \varepsilon_j a_j y_j\|$ *for every choice of scalars* $\{a_j\}$ *and signs* $\varepsilon_j = \pm 1$.

We shall actually use the theorem only for normalized basic sequences, and in this case the condition that $\|\sum a_k x_{i_k}\|$ are bounded and bounded away from zero certainly holds.

We need two preliminary results. The first is the well-known (infinite) Ramsey theorem, and the second applies it to sequences of vectors in Banach spaces. If M is a subset of the natural numbers $\mathbb{N}$, we denote by $M^{[n]}$ the set of all subsets of M of cardinality n.

Proposition 12.15. *Let* ψ *be a mapping from* $\mathbb{N}^{[n]}$ *to some finite set* A. *Then there is an infinite subset* M *of* $\mathbb{N}$ *such that* ψ *is constant on* $M^{[n]}$.

Proof. By induction on n. For $n = 1$ this is evident, and we pass from $n-1$ to n. Pick any $j \in \mathbb{N}$, and restrict ψ to the n-element subsets of $\mathbb{N}$ which contain j. By the induction hypothesis there is an infinite subset M_1 of $\mathbb{N}$ such that ψ is constant on all the sets composed of j together with $n-1$ elements out of M_1.

We now iterate this observation and choose a decreasing sequence of infinite sets $\mathbb{N} \supset M_1 \supset M_2 \supset \cdots$, and elements $j_i \in M_i - M_{i+1}$ such that the restriction of ψ to any n-tuple consisting of j_i and $n-1$ elements from M_{i+1} is constant. By passing to a subsequence if necessary, we may assume that this constant does not depend on i. The required set is $M = \{j_i\}_{i=1}^{\infty}$. □

Proposition 12.16. *Let* $\{x_j\}_{j=1}^{\infty}$ *be a sequence in a Banach space* E *such that the set of numbers* $\{\|\sum_{k=1}^{n} a_k x_{i_k}\| : i_1 < \ldots < i_n\}$ *is bounded and bounded away from* 0 *for every fixed* n *and for every fixed choice of* $\{a_k\}_{k=1}^{n}$ *not all of them zero. Let* ε_n *be any sequence of positive numbers. Then there is an infinite subset* $L = \{m_n\}_{n=1}^{\infty} \subset \mathbb{N}$ *such that for every* n *and for all choices of* $j_1 < j_2 \cdots < j_n$ *and* $i_1 < i_2 \cdots < i_n$ *in* L *with* $j_1 \geq m_n$ *and* $i_1 \geq m_n$

$$\|\sum_{k=1}^{n} a_k x_{i_k}\| \leq (1+\varepsilon_n)\|\sum_{k=1}^{n} a_k x_{j_k}\| \tag{12.5}$$

for every choice of $\{a_k\}$.

Proof. We first fix n and scalars $\{a_k\}_{k=1}^{n}$ (not all of them zero), and choose $0 < \alpha < \beta < \infty$ such that $\alpha \leq \|\sum_{k=1}^{n} a_k x_{i_k}\| \leq \beta$ for all $i_1 < i_2 < \cdots < i_n$ in $\mathbb{N}$. Divide the interval $[\alpha, \beta]$ into N subintervals $[\alpha_m, \beta_m]$ such that $\beta_m \leq (1+\varepsilon_n)\alpha_m$ for all m, and define $\psi : \mathbb{N}^{[n]} \to \{1, 2, \ldots, N\}$ by $\psi(\{i_1, \ldots, i_n\}) = m$, where m is such that $\|\sum_{k=1}^{n} a_k x_{i_k}\| \in [\alpha_m, \beta_m]$.

By Ramsey's theorem there is an infinite subset M for which (12.5) holds for the given n and $\{a_k\}_{k=1}^{n}$ and for all $i_1, \ldots, i_n, j_1, \ldots, j_n \in M$.

Next, we still fix n and find an infinite set M_n such that (12.5) holds for this n and for *all* choices of $\{a_k\}_{k=1}^{n}$ and all $i_1, \ldots, i_n, j_1, \ldots, j_n \in M_n$. Fix a finite set of choices of $\{a_k\}_{k=1}^{n}$ in the unit ball of l_1^n which is dense enough in the l_1^n norm, and apply the same procedure to all of them simultaneously with ε_n replaced by $\varepsilon_n/2$. Since Proposition 12.15 holds also for a finite number of mappings ψ (instead of just one), we obtain a set M_n as desired. Applying this construction successively for $n = 1, \dots,$ we obtain a decreasing sequence of infinite subsets of the integers $M_1 \supset M_2 \supset \cdots$ such that (12.5) holds for every $i_1, \dots, i_n, j_1, \dots, j_n \in M_n$ and for all choices of scalars. Choose now a strictly increasing sequence $\{m_n\}_{n=1}^{\infty}$ with $m_n \in M_n$ for every n, and then $L = \{m_n\}_{n=1}^{\infty}$ has the desired properties. □

Let $\{x_j\}$ and $L = \{m_n\}_{n=1}^{\infty}$ be as in Proposition 12.16, and assume that $\varepsilon_n \to 0$. Define a norm on the space of all finite sequences by

$$|||\sum_{k=1}^{n} a_k e_k||| = \lim_{j\to\infty} \|\sum_{k=1}^{n} a_k x_{m_{k+j}}\|.$$

where $\{e_i\}_{i=1}^{\infty}$ denote the unit vectors. By Proposition 12.16 the limit exists and the new norm is 1-spreading; i.e., $|||\sum_{k=1}^{n} a_k e_k||| = |||\sum_{k=1}^{n} a_k e_{j_k}|||$ for all choices of n, scalars $\{a_k\}_{k=1}^{n}$ and indices $j_1 < \cdots < j_n$.

Lemma 12.17. *With the notation above, put $u_k = e_{k+1} - e_k$. Then*

$$|||\sum_{k\in A} a_k u_{2k}||| \le |||\sum_{k\in B} a_k u_{2k}|||$$

for every choice of $\{a_k\}$ and for any finite sets $A \subset B$. In particular $\{u_{2k}\}$ is a 2-unconditional basic sequence.

Proof. Rather than giving the full proof, we only illustrate the method by checking that $|||u_2 + u_6||| \le |||u_2 + u_4 + u_6|||$. The proof of the general case is similar. Since $|||u_2 + u_4 + u_6||| = |||u_2 + u_j + u_n|||$ for every $4 \le j \le n-2$, we see that

$$\begin{aligned} |||u_2 + u_4 + u_6||| &= \frac{1}{n-5}\sum_{j=4}^{n-2} |||u_2 + u_j + u_n||| \ge \Big|\Big|\Big|u_2 + \frac{1}{n-5}\Big(\sum_{j=4}^{n-2} u_j\Big) + u_n\Big|\Big|\Big| \\ &\ge |||u_2 + u_n||| - \frac{1}{n-5}\Big|\Big|\Big|\sum_{j=4}^{n-2} u_j\Big|\Big|\Big| \to |||u_2 + u_6||| \end{aligned}$$

because $|||u_2 + u_n||| = |||u_2 + u_6|||$ and $\sum_{j=4}^{n-2} u_j = e_{n-1} - e_4$. □

Proof of Theorem 12.14. By passing to a subsequence as in Proposition 12.16 and to differences as in Lemma 12.17 and by normalizing, we may assume that the x_i's are already a normalized 1-spreading basic sequence and that they satisfy

$$\|\sum_{k\in A} a_k x_k\| \le \|\sum_{k\in B} a_k x_k\| \tag{12.6}$$

for every $A \subset B$ and scalars $\{a_k\}$.

The sequence $\alpha_m = \|x_1 - x_2 + x_3 - x_4 + \cdots - x_{2m}\|$ is monotone nondecreasing by the lemma, and we shall distinguish two cases according to whether it is bounded or not.

If $\lim \alpha_m = \infty$, put $v_1^m = (x_1 - x_2 + x_3 - x_4 + \cdots - x_{2m})/\alpha_m$, and let v_j^m for $j \ge 2$ be the vector obtained by shifting the coordinates of v_1^m by $2m(j-1)$ indices. (Thus all the v_j^m's have the same distribution with respect to the basis $\{x_i\}$, and they have disjoint supports.)

The vector $v_1^m - v_2^m + (x_{2m+1} - x_{4m+1})/\alpha_m$ has the same coordinates as $v_1^m + v_2^m$, except that the second half of its nonzero coordinates is shifted by one index. Since the basis is 1-spreading, they also have the same norm, and since $\|x_j\| = 1$ for all j, it follows that

$$\Big| \|v_1^m + v_2^m\| - \|v_1^m - v_2^n\| \Big| \leq \|x_{2m+1} - x_{4m+1}\|/\alpha_m \leq 2/\alpha_m. \tag{12.7}$$

A similar computation shows that

$$\Big| \big\|\sum_{j=1}^n a_j v_j^m\big\| - \big\|\sum_{j=1}^n \pm a_j v_j^m\big\| \Big| \leq 2\sum_{j=1}^n |a_j|/\alpha_m \tag{12.8}$$

for all choices of scalars $\{a_j\}$.

Define a new norm on the space of all finite sequences by

$$\Big|\Big|\Big| \sum_{j=1}^n a_j y_j \Big|\Big|\Big| = \lim_{p\to\infty} \Big\|\sum_{j=1}^n a_j v_j^{m_p}\Big\|$$

where $\{m_p\}$ is a subsequence of the integers for which this limit exists for all n and for all choices of $\{a_j\}_{j=1}^n$, and where the y_j's are the standard unit vectors. Then the sequence $\{y_j\}$ is still 1-spreading, and by the argument above it is also 1-unconditional.

If $\sup \alpha_m = K < \infty$, we shall show that the usual unit vector basis of c_0 is block finitely represented in $\{x_i\}$. By (12.6), if $a_i \geq 0$, then

$$\Big\|\sum_{i=1}^n a_i x_i\Big\| \leq \Big\|\sum_{i=1}^n x_i\Big\| \cdot \max_{i\leq n} a_i \leq K \max_{i\leq n} a_i$$

Hence $\|\sum a_i x_i\| \leq 2K \max |a_i|$ for all finite sequences $\{a_i\}$. Thus $\{x_i\}$ is $2K$-equivalent to the unit vector basis of c_0. By Theorem 13.6 in the next chapter, for every $\varepsilon > 0$ the equivalence constant can be improved from $2K$ to $1+\varepsilon$ by passing to appropriate blocks of the x_i's. Taking $\varepsilon_n \to 0$, passing to blocks successively, and applying a standard diagonalization procedure give blocks $\{y_i\}$ such that $\{y_i\}_{i\geq n}$ is $1+\varepsilon_n$-equivalent to the unit vector basis of c_0, and this finishes the proof. ◇

Let T be a bounded linear operator on a complex Banach space E. A sequence $\{x_n\}_{n=1}^\infty$ of norm-one vectors in E is called an approximate eigenvector of T if there is a λ such that $\|Tx_n - \lambda x_n\| \to 0$ (and λ is called an approximate eigenvalue). It is a standard fact in functional analysis that every operator has an approximate eigenvector (in fact, every boundary point λ of the spectrum of T is an approximate eigenvalue). The next proposition shows that finitely many commuting bounded operators have a common approximate eigenvector. Consider the case of two commuting operators, S and T. If T actually has a proper eigenvalue λ, this is easy. In this case the space $\{x : Tx = \lambda x\}$ of the eigenvectors of T corresponding to λ is invariant under S (because T and S commute), and any approximate eigenvector for S in this space is a common approximate eigenvector for T and S. We shall use ultraproducts (see Appendix F) to reduce the general case when the operators do not have proper eigenvalues, to the simple case when they do.

Proposition 12.18. *Let $\{T_j\}_{j=1}^m$ be a finite family of commuting bounded linear operators on a complex Banach space E. Then there are complex numbers $\{\lambda_j\}_{j=1}^m$ and a sequence $\{x_n\}_{n=1}^\infty$ with $\|x_n\| = 1$ for all n such that*

$$\lim_{n\to\infty} \|T_j x_n - \lambda_j x_n\| = 0$$

for every $j \leq m$.

Proof. We shall consider only two operators, T and S, but the same proof holds for any finite family of commuting operators. Let $\tilde{E}$ be an ultrapower of E, and let $\tilde{T}$ and $\tilde{S}$ be the corresponding ultrapowers of T and S; then $\tilde{T}$ and $\tilde{S}$ also commute. Fix an approximate eigenvalue λ of T. By the definition of ultrapowers λ is a proper eigenvalue of $\tilde{T}$; hence $\tilde{S}$ has an approximate eigenvalue μ in the space of eigenvectors of $\tilde{T}$ corresponding to this λ.

Passing once more to an ultrapower and recalling that an ultrapower of an ultrapower is an ultrapower (see Appendix F), we get that some ultrapowers of T and S have a common eigenvector $\tilde{x}$ of norm 1. The vector $\tilde{x}$ has a representative $(x_1, x_2, \dots)$ in $l_\infty(E)$ with $\|x_i\| = 1$ for all i, and then

$$\lim_{\mathcal{U}} \|Tx_i - \lambda x_i\| = \lim_{\mathcal{U}} \|Sx_i - \mu x_i\| = 0$$

for the ultrafilter $\mathcal{U}$ defining the ultrapower. The result follows by passing to an appropriate subsequence of the x_i's. □

We are now ready for Krivine's theorem.

Theorem 12.19. *Let $\{x_i\}_{i=1}^\infty$ be a basic sequence in a Banach space. Then the unit vector basis of either c_0 or some l_p (for $1 \le p < \infty$) is block finitely representable in $\{x_i\}$.*

Proof. By Theorem 12.14 we may assume that the x_i's are 1-unconditional and 1-spreading. It will be convenient to introduce a new complex Banach space with a basis indexed by the set of nonnegative rationals $\mathbb{Q}^+$. Let E be the completion of the space of all finitely supported complex sequences indexed by $\mathbb{Q}^+$, with unit vector basis $\{v_q : q \in \mathbb{Q}^+\}$ and with the norm

$$\Big\| \sum_{j=1}^k a_j v_{q_j} \Big\| = \Big\| \sum_{j=1}^k |a_j| x_j \Big\| \qquad \text{whenever } \; 0 \le q_1 < q_2 < \dots < q_k.$$

The new norm is well defined because $\{x_i\}$ is 1-unconditional and 1-spreading. It also satisfies $\| \sum_{j=1}^k a_j v_{q_j} \| = \| \sum_{j=1}^k |a_j| v_{p_j} \|$ whenever $0 \le q_1 \le q_2 \dots < q_k$ and $0 \le p_1 < p_2 < \dots < p_k$; i.e., the v's are also 1-unconditional and 1-spreading.

Put $E_1 = \overline{\operatorname{span}}\{v_q\}_{0 \le q < 1}$, and define two linear operators T and S on E_1 by

$$Tv_q = v_{q/2} + v_{(q+1)/2} \qquad Sv_q = v_{q/3} + v_{(q+1)/3} + v_{(q+2)/3}.$$

The operator T maps each vector $u \in E_1$ to a sum of two disjointly supported vectors, each having the same distribution as u. Thus $\|u\| \le \|Tu\| \le 2\|u\|$ for all $u \in E_1$. Similarly $\|u\| \le \|Su\| \le 3\|u\|$.

Direct computation shows that $ST = TS$; hence by Proposition 12.18 there are finitely supported norm-one vectors $w_n \in \operatorname{span}\{v_q\}_{0 \le q < 1}$ and complex numbers λ and μ such that

$$\lim_{n\to\infty} \|Tw_n - \lambda w_n\| = \lim_{n\to\infty} \|Sw_n - \mu w_n\| = 0.$$

We may assume that all the coordinates of the w_n's, as well as λ and μ, are nonnegative numbers (otherwise replace them by $|w_n|$, $|\lambda|$ and $|\mu|$ respectively). Clearly $1 \le \lambda \le 2$ and $1 \le \mu \le 3$.

For positive integers n and j, let $w_{n,j}$ be the vector obtained from w_n by translating it j units to the right (i.e., $w_{n,j}$ has the same distribution as w_n, and it is supported in $[j, j+1)$). The $w_{n,j}$'s are now used to define a new Banach space

$(\tilde{E}, \|\!| \cdot \|\!|)$ with a basis $\{e_j\}$ which is block finitely represented in $\{x_i\}$. It is the completion of the space of finitely supported sequences under the norm

$$\Big\|\!\Big| \sum_{j=1}^{m} a_j e_j \Big|\!\Big\| = \lim_n \Big\| \sum_{j=1}^{m} a_j w_{n,j} \Big\|$$

where $\{e_j\}$ denote the unit vectors, and the limit is taken with respect to some ultrafilter on $\mathbb{N}$ (or, equivalently, with respect to a suitable subsequence $\{n_k\}$). The vectors $\{e_j\}$ form a normalized, 1-unconditional and 1-spreading basis of $\tilde{E}$, and we shall finish the proof by showing that they are isometrically equivalent to the unit vector basis of c_0 or of l_p for some $1 \le p < \infty$.

We use the notation $u_1 \prec u$ (or $j \prec u$) to mean that the smallest index in the support of u is larger than the maximal one in the support of u_1 (or larger than j). (We use the same notation when the index set is either $\mathbb{Q}^+$ or $\mathbb{N}$.)

Fix $u_1 \prec u_4$ in $\operatorname{span}\{v_q\}_{0 \le q < 1}$, and choose for every n two vectors, u_2^n and u_3^n, with the same distribution as w_n and supported so that $u_1 \prec u_2^n \prec u_3^n \prec u_4$. It follows from the choice of the w_n's and from the fact that $\{v_q\}$ is 1-spreading that

$$\lim_{n \to \infty} \Big(\|u_1 + u_2^n + u_3^n + u_4\| - \|u_1 + \lambda u_2^n + u_4\| \Big) = 0.$$

Consequently, if y_1 and y_2 are two vectors in $\tilde{E}$ and if j is an integer such that $y_1 \prec j$ and $j+1 \prec y_2$, then

$$\|\!| y_1 + e_j + e_{j+1} + y_2 |\!\| = \|\!| y_1 + \lambda e_j + y_2 |\!\|.$$

Similarly, if $y_1 \prec j$ and $j+2 \prec y_2$, then

$$\|\!| y_1 + e_j + e_{j+1} + e_{j+2} + y_2 |\!\| = \|\!| y_1 + \mu e_j + y_2 |\!\|.$$

Using these two identities repeatedly, we get in particular that for any choice of integers k and l

$$\Big\|\!\Big| \sum_{j=1}^{2^k 3^l} e_j \Big|\!\Big\| = \lambda^k \mu^l. \tag{12.9}$$

If $\lambda = 1$, then $\|\!| \sum_{j=1}^{2^k} e_j |\!\| = 1$ for all k; and since the basis is 1-unconditional, it follows that it is isometrically equivalent to the unit vectors basis of c_0. The same is true if $\mu = 1$, and we assume from now on that $\lambda > 1$ and $\mu > 1$.

Consider the multiplicative function $\psi(2^k 3^{-l}) = \lambda^k \mu^{-l}$. It is monotone nondecreasing, because if $2^{k_1} 3^{-l_1} \le 2^{k_2} 3^{-l_2}$, then $2^{k_1} 3^{l_2} \le 2^{k_2} 3^{l_1}$, and (12.9) and the monotonicity of the basis give $\lambda^{k_1} \mu^{l_2} \le \lambda^{k_2} \mu^{l_1}$. Since the set $\{2^k 3^{-l} : k, l \in \mathbb{N}\}$ is dense in $\mathbb{R}^+$, it follows that ψ extends to a monotone multiplicative function from $\mathbb{R}^+$ into itself which is not identically 1 and is thus of the form $\psi(t) = t^{1/p}$ for some $1 \le p < \infty$. (That $p \ge 1$ follows from $\psi(2) = \lambda \le 2$). In particular $\lambda = 2^{1/p}$ and $\mu = 3^{1/p}$. We shall prove that the basis $\{e_j\}$ is isometrically equivalent to the unit vector basis of l_p for this p.

Note first that the same argument used to derive (12.9) from the two identities that precede it also gives the following generalization. Assume that $s = \sum_{i=1}^{m} 2^{k_i} 3^{l_i}$ for some m and integers k_i and l_i; then

$$\Big\|\!\Big| \sum_{j=1}^{s} e_j \Big|\!\Big\| = \Big\|\!\Big| \sum_{i=1}^{m} \lambda^{k_i} \mu^{l_i} e_i \Big|\!\Big\|. \tag{12.10}$$

In particular, if we put $\alpha(m) = \|\sum_{i=1}^m e_i\|$, then taking $k_i = k$ and $l_i = l$ for all $i \leq m$, we obtain

$$\alpha(m2^k3^l) = \lambda^k\mu^l\alpha(m) = (2^k3^l)^{1/p}\alpha(m).$$

We claim that $\alpha(m) = m^{1/p}$ for all integers m. Fix an integer m. Since the set $\{2^k3^{-l} : k, l \in \mathbb{N}\}$ is dense in $\mathbb{R}^+$, we can find for each integer n two integers k and $l > n$ such that $2^k3^{-l} \leq m \leq 2^k3^{-l} + 3^{-n}$; i.e., $2^k \leq 3^l m \leq 2^k + 3^{l-n}$. Since $\alpha(m)$ is a monotone nondecreasing function of m, and since $\alpha(r+s) \leq \alpha(r) + \alpha(s)$ by the triangle inequality, it follows that

$$2^{k/p} = \alpha(2^k) \leq 3^{l/p}\alpha(m) \leq 2^{k/p} + 3^{(l-n)/p}$$

and the claim follows by dividing by $3^{l/p}$ and letting $n \to \infty$. Using this we get from (12.10) that if a_i are of the form $a_i = \lambda^{k_i}\mu^{l_i} = (2^{k_i}3^{l_i})^{1/p}$ for each $i \leq m$, and if we put $s = \sum_{i=1}^m 2^{k_i}3^{l_i}$, then

$$\Big\|\sum_{i=1}^m a_ie_i\Big\| = \Big\|\sum_{i=1}^s e_i\Big\| = s^{1/p} = \Big(\sum_{i=1}^m a_i^p\Big)^{1/p}$$

and since the points $a/\|a\|_p$ with $a = (a_1, \dots, a_m)$ of this form are dense in the unit sphere of l_p^m, the last identity holds for every choice of scalars $\{a_i\}$. □

Remark. To obtain a weak form of Dvoretzky's theorem (with no quantitative estimates), we need to observe only that l_p^n contains a $(1+\varepsilon)$-Euclidean subspace of dimension k if n is large enough. For $p = \infty$ this holds for every k-dimensional space and not only for l_2^k. One way to see this for $1 \leq p < \infty$ is to embed l_2^k isometrically in $L_p[0,1]$, e.g., by independent Gaussian variables (see Appendix D), and then observe that every finite-dimensional subspace of L_p is "almost" a subspace of l_p^n for large enough n.

Note that since blocks of the unit vector basis of l_p^n are equivalent to l_p^k, the Euclidean subspace is not spanned by blocks anymore.

We now reformulate Theorem 12.19 in a finite-dimensional form.

Theorem 12.20. *For every integer k and constants $\varepsilon > 0$ and $C < \infty$ there is an integer $\overline{n} = n(k, \varepsilon, C)$ with the following property: If $n \geq \overline{n}$, and if E is an n-dimensional Banach space with a normalized basis $\{x_i\}_{i=1}^n$ with basis constant smaller than C, then there is a $1 \leq p \leq \infty$ and a block basis of the x_i's which is $(1+\varepsilon)$-equivalent to the unit vector basis of l_p^k.*

Proof. Assume that there are k_0, ε_0 and C_0 for which no appropriate integer $n(k_0, \varepsilon_0, C_0)$ can be found. Then for every n there are normalized vectors $\{x_{i,n}\}_{i=1}^n$ in some Banach space E_n, with basis constant smaller than C_0, such that no block basis of $\{x_{i,n}\}_{i=1}^n$ is $(1+\varepsilon_0)$-equivalent to the unit vector basis of $l_p^{k_0}$ for any $1 \leq p \leq \infty$. Define a norm on the space of finitely supported sequences, with unit vector basis $\{y_i\}$, by

$$\Big\|\sum a_iy_i\Big\| = \lim_n \Big\|\sum a_ix_{i,n}\Big\|$$

where the limit is taken with respect to some ultrafilter on $\mathbb{N}$. Then $\{y_i\}$ is a normalized basis with constant smaller than C_0. By Theorem 12.19 there is a block basis $\{z_j\}_{j=1}^{k_0}$ of the y_i's which is $(1+\varepsilon_0/2)$-equivalent to the unit vector basis of $l_p^{k_0}$ for some $1 \leq p \leq \infty$. Choosing a finite dense enough set in the unit ball of span$\{z_j\}$ and taking n large enough give a block in E_n which is $(1+\varepsilon_0)$-equivalent to the unit vector basis of $l_p^{k_0}$, which is impossible. □

As in Section 1 we now essentially repeat the arguments used to prove Theorems 12.14 and 12.19 in order to find l_p^k-blocks such that a given uniformly continuous function has small oscillation on the unit sphere of their span.

Theorem 12.21. *Let f be a real-valued uniformly continuous function defined on the unit sphere of a Banach space E with a basis $\{x_i\}_{i=1}^\infty$. Then there is a constant λ_0 such that for every $\varepsilon > 0$ and for every integer k there is a block basic sequence $\{y_j\}_{j=1}^k$ of $\{x_i\}$ with the property that $|f(y) - \lambda_0| < \varepsilon$ for every y in* span$\{y_j\}_{j=1}^k$ *with $\|y\| = 1$.*

Proof. By Theorem 12.19 there is a space E_1 which is isometric to c_0 or l_p and is block finitely represented in E. Denote the unit vector basis of E_1 by $\{e_k\}$. We define a Lipschitz function $\tilde{f}$ on E_1 as follows: First extend f by homogeneity to all of E. Next, choose for each m a block basis $\{z_{k,m}\}_{k=1}^m$ of $\{x_i\}$ which is $(1+m^{-1})$-equivalent to $\{e_k\}_{k=1}^m$, and define $\tilde{f}$ for finitely supported vectors by

$$\tilde{f}\Big(\sum a_k e_k\Big) = \lim_{j\to\infty} f\Big(\sum_{k=1}^{m_j} a_k z_{k,m_j}\Big)$$

where $\{m_j\}$ is a subsequence of the integers for which the limit exists for all choices of scalars $\{a_k\}$. It suffices to prove the theorem for E_1 and the newly defined function $\tilde{f}$, and we may thus assume, without loss of generality, that $\{x_i\}$ is already the unit vector basis of c_0 or l_p.

We may also assume that f is 1-spreading; i.e., that $f(\sum a_i x_i) = f(\sum a_i x_{m_i})$ whenever $m_1 < m_2 < \dots$. To see this we repeat the argument from Proposition 12.16. (We do not need here the assumption of boundedness from below, and condition (12.5) is replaced by $\big|f(\sum_{k=1}^n a_k x_k) - f(\sum_{k=1}^n a_k x_{j_k})\big| \le \varepsilon_n$.) The function $f_1(\sum a_k x_k) = \lim_{j\to\infty} f(\sum a_k x_{k+j})$ is a 1-spreading function on c_0 or l_p, which can replace f.

We show next that we may assume that f is also 1-unconditional; i.e., that f satisfies $f(\sum a_i x_i) = f(\sum |a_i| x_{m_i})$ for every choice of scalars. The modulus of continuity of f is denoted by ω.

For $1 \le p < \infty$ we use the same argument as in the proof of the case $\lim \alpha_m = \infty$ in Theorem 12.14. Since f is 1-spreading, it has the same value at the two points $v_1^m - v_2^m + (x_{2m+1} - x_{4m+1})/\alpha_m$ and $v_1^m + v_2^m$. It follows as in (12.7) that

$$\big|f(v_1^m + v_2^m) - f(v_1^m - v_2^n)\big| \le \omega(2/\alpha_m)$$

and similarly (as in (12.8)) that

$$\Big|f\Big(\sum_{j=1}^n a_j v_j^m\Big) - f\Big(\sum_{j=1}^n \pm a_j v_j^m\Big)\Big| \le \omega\Big(2\sum_{j=1}^n |a_j|/\alpha_m\Big)$$

for all choices of scalars $\{a_j\}$. The norm $|||\cdot|||$ defined there is again isometric to l_p, with unit vectors $\{y_j\}$, and f is naturally defined on this space and is 1-unconditional.

Assume now that $\{x_i\}$ is the unit vector basis of c_0. Fix an integer m, and put $u = \sum_{i=1}^{2m+1}\big(1 - \frac{|m+1-i|}{m}\big)(-1)^i x_i$ and $v = \sum_{i=1}^{2m+1}\big(1 - \frac{|m+1-i|}{m}\big)(-1)^i x_{i+1}$. Then u and v are norm-one vectors with the same distribution (with v a translation of

u to the right by one index), and $\|u+v\| = 1/m$ (i.e., v is very close to $-u$). Let $\{u_k\}_{k=1}^n$ be disjointly supported vectors, all having the same distribution as u, and let $\{v_k\}$ be their translation to the right by one index. We assume that $v_k \prec u_{k+1}$ for all k. Fix $\{a_k\}_{k=1}^n$, and put $I = \{k : a_k \geq 0\}$ and $J = \{k : a_k < 0\}$. By the computation above

$$\Big\| \sum_{k=1}^{n} |a_k| u_k - \Big(\sum_I a_k u_k + \sum_J a_k v_k \Big) \Big\| \leq \max |a_k|/m$$

and since f is 1-spreading, we obtain

$$f\Big(\sum_{k\in I} a_k u_k + \sum_{k \in J} a_k v_k \Big) = f\Big(\sum_{k=1}^{n} a_k u_k \Big).$$

Thus $\big|f\big(\sum |a_k| u_k\big) - f\big(\sum a_k u_k\big)\big| \leq \omega\big(\max |a_k|/m\big)$. By passing to limits as in the l_p case, we then obtain a naturally defined function f (on another copy of c_0), which is 1-spreading and 1-unconditional.

We now pass to the final part of the proof. We shall prove this step for l_p only. For c_0 we shall prove a much stronger result in Section 3 of the next chapter. In the proof λ_0 will depend on ε and k, but since f is bounded on the unit sphere of E, it follows that it can be chosen independently of ε and of k.

It is convenient to introduce a notion of distance between the distributions of two vectors $u, v \in E$. Thus, put $\operatorname{dist}_D(u,v) = \inf\{\|u_1 - v_1\|\}$, where the infimum is taken over all vectors u_1, v_1 having the same distribution as u and v respectively. (Note that what we mean by saying that two vectors have the same distribution is, perhaps, somewhat unconventional. We require that the two vectors have the same nonzero coordinates and *in the same order*; i.e., we do not allow permutations.) If $\operatorname{dist}_D(u,v) \leq \delta$, then $|f(u) - f(v)| \leq \omega(\delta)$ because f is 1-spreading.

We shall construct, for every integer k and for every $\delta > 0$, a normalized block basis $\{y_j\}_{j=1}^k$ of $\{x_i\}$ such that $\operatorname{dist}_D(y, y_1) \leq \delta$ for any $y = \sum_{j=1}^k a_j y_j$ with $\|y\| = 1$ and nonnegative coefficients. Since f is 1-spreading and 1-unconditional, this will conclude the proof of the theorem.

As in the proof of Theorem 12.19 we find it convenient to represent E as a space of (complex valued) functions on the rationals Q in $[0,1)$ with the l_p norm, and we denote the function which is 1 at $t \in Q$ and is zero elsewhere by x_t.

Fix m (much larger than k, its exact value will depend on k and δ, as will be explained below). For each $1 \leq n \leq m$ let T_n be the operator on E defined by

$$T_n x_t = \sum_{i=0}^{n-1} x_{(t+i)/n}.$$

Then $\|T_n x\| = n^{1/p}\|x\|$ for every $x \in E$, and the operators $\{T_n\}_{n=1}^m$ commute. By Proposition 12.18 there is a finitely supported nonnegative vector $u \in E$ with $\|u\| = 1$ such that $\|T_n u - n^{1/p} u\| \leq 1$ for every $1 \leq n \leq m$.

We take $\{y_j\}_{j=1}^k$ to be k vectors in E all having the same distribution as u, and such that $y_j \prec y_{j+1}$ for every $j < k$. We shall show that if m is chosen to be large enough, then $\{y_j\}_{j=1}^k$ has the desired property. In the computations below it is convenient to add vectors $\{y_j\}_{j=k+1}^m$, also with the same distribution as u and such that $y_j \prec y_{j+1}$ for every $1 \leq j < m$.

We start by considering y of the form $y = \sum_{i=1}^{k}(m_i/m)^{1/p}y_i$, where m_i are nonnegative integers with $\sum_{i=1}^{k} m_i = m$. Put $z = m^{-1/p}\sum_{j=1}^{m} y_j$. We shall show that $\operatorname{dist}_D(y_1, z) < \delta/3$ and $\operatorname{dist}_D(z, y) < \delta/3$ provided m is large enough. Thus $\operatorname{dist}_D(y_1, y) < 2\delta/3 < \delta$ as required.

Since y_1 has the same distribution as u, and since z has the same distribution as $m^{-1/p}T_m u$, it follows that $\operatorname{dist}_D(y_1, z) \leq \|u - m^{-1/p}T_m u\| \leq m^{-1/p} < \delta/3$ provided m is large enough. Also, for each $i \leq k$ the vector $(m_i/m)^{1/p}y_i$ has the same distribution as $(m_i/m)^{1/p}u$. Decompose the set of integers $[1, m]$ to k successive intervals $\{J_i\}$ of cardinalities m_i respectively, and write $z = m^{-1/p}\sum_{i=1}^{k} z_i$, where $z_i = \sum_{j\in J_i} y_j$. Since each z_i has the same distribution as $T_{m_i}u$, we see that $\operatorname{dist}_D(y, z) \leq m^{-1/p}\sum_{i=1}^{k}\|m_i^{1/p}u - T_{m_i}u\| \leq m^{-1/p}k < \delta/3$ provided m is large enough.

The result now follows by taking m also so large that the k-tuples of the form $\big((m_1/m)^{1/p}, \ldots, (m_k/m)^{1/p}\big)$ are $\delta/3$-dense in the positive part of the unit sphere of l_p^k. □

A finite-dimensional version of the theorem follows as in the proof of Theorem 12.20.

Theorem 12.22. *For every integer k, constants $\varepsilon > 0$ and $C < \infty$, and modulus of continuity ω there is an integer $\overline{n} = n(k, \varepsilon, C, \omega)$ with the following property: Let $n \geq \overline{n}$, and let E be an n-dimensional Banach space with a normalized basis $\{x_i\}_{i=1}^{n}$ whose basis constant is smaller than C. Let f be a real-valued function f on the unit sphere of E whose modulus of continuity is bounded by ω. Then there is a block basis $\{y_j\}_{j=1}^{k}$ of $\{x_i\}_{i=1}^{n}$, and there is a scalar λ such that $|f(y) - \lambda| < \varepsilon$ whenever y in $\operatorname{span}\{y_j\}_{j=1}^{k}$ satisfies $\|y\| = 1$.*

3. Notes and Remarks

Theorem 12.10 (with the estimate $k \geq c(\varepsilon)\sqrt{\log n/\log\log n}$) is due to Dvoretzky [**160**]. Proposition 12.3, Theorem 12.6 and Theorem 12.10 in the formulation given here are due to V. D. Milman [**415, 417**]. We presented the proof of Milman as streamlined in the paper of Figiel, Lindenstrauss and Milman [**190**]. We followed closely the presentation in this latter paper (except for Proposition 12.1 here and for using a somewhat different version of Lemma 12.8). For the historic background of Dvoretzky's theorem, for different proofs, and for its applications we refer to Lindenstrauss [**361**], which also contains a discussion (with references) of the estimate of k appearing in the theorem. The dependence on ε for fixed n can be improved (see e.g. Gordon [**221**]), but the best possible order of magnitude is not yet known except for some special cases.

The article [**419**] by Milman is a survey of the research directions inspired by the theorem of Dvoretzky. In particular, it contains several references to results in which the small oscillation phenomena holds for functions defined on homogeneous spaces other than the sphere S^{n-1}, for example, on the Grassmannian of all finite-dimensional subspaces of l_2, where the distance between two subspaces is the Hausdorff distance between their unit spheres. A detailed study of the "concentration of measure" ideas in the framework of rather general metric spaces is contained in Chapter $3\frac{1}{2}$ of the book [**230**] by Gromov.

Inequality (12.1) appears in Lévy [**351**]. A short proof of the spherical isoperimetric inequality can be found in Figiel, Lindenstrauss and Milman [**190**]. Another proof, due to W. Beckner, appears in Benyamini [**51**]. Proposition 12.1 was first proved by Gromov and Milman [**231**]. The (considerably) simpler proof presented here is due to Arias-De-Reyna, Ball and Villa [**22**].

Lemma 12.8 is a weak version of the main lemma of Dvoretzky and Rogers [**161**]. The (unique) ellipsoid of maximal volume in the unit ball of E is a natural tool in this context (it was used by F. John [**280**] and in the paper of Dvoretzky and Rogers) and has many other applications. The proof presented here is due to W. B. Johnson (and first appeared in Milman and Schechtman [**420**]). Note that in the proof of Lemma 12.8 we did not use the fact, proved by F. John, that (in the notation of the lemma) $\|x\|/\sqrt{n} \leq |||x|||$ for all x. Actually the fact that $d(E, l_2^n) \leq \sqrt{n}$ for every E of dimension n was not used in the proof of Theorem 12.10. In Figiel, Lindenstrauss and Milman [**190**] there is an even weaker version of Lemma 12.8, whose proof does not involve the maximal volume ellipsoid and which is still sufficient for the proof of Dvoretzky's theorem.

We now mention two more results on the estimate of k (as a function of n) proved in Figiel, Lindenstrauss and Milman [**190**]. First, if E has cotype q, then $k \geq \alpha n^{2/q}$ where α depends only on ε and on the cotype-q constant of E. In particular, if $E = l_1^n$, then $k \geq c(\varepsilon)n$. The second result is that for every Banach space E of dimension n there are subspaces $F_1 \subset E$ and $F_2 \subset E^*$ with $d(F_1, l_2^{k_1}) \leq 2$ and $d(F_2, l_2^{k_2}) \leq 2$ such that $k_1 k_2 \geq cn$ for some positive absolute constant c. Equivalently, we can take F_2 to be a quotient of E. The following result of Milman [**418**] also involves subspaces and duality (i.e., subspaces and quotients) and has striking geometric consequences: For every $\varepsilon > 0$ there is a $c(\varepsilon) > 0$ such that every n-dimensional Banach space has a subspace F, such that F has a k-dimensional quotient G with $k \geq c(\varepsilon)n$ and $d(G, l_2^k) < 1 + \varepsilon$. All these results are sharp.

A discussion of these results and of further results in this direction, as well as of their applications to the geometry of convex bodies, can be found in the books of Milman and Schechtman [**420**] and of Pisier [**467**]. We mention here just one additional result due to Milman and Schechtman [**421**], which was proved after these books appeared: There is a universal constant c such that for every Banach space E of dimension n and for every $k \geq c \log n$ there is a k-dimensional subspace F of E such that

$$d(F, l_2^k) \leq c\sqrt{k/\log(n/k)}.$$

This result is, from several points of view, best possible.

Proposition 12.15 is due to Ramsey [**482**]. Brunel and Sucheston [**104, 105**] were the first to apply Ramsey's theorem in Banach space theory. Proposition 12.16 and Theorem 12.14 are due to them. An exposition of further results related to the Brunel-Sucheston theory can be found in Beauzamy and Laprsté [**45**].

The main result of Section 2, namely Theorem 12.19, is due to Krivine [**338**]. The proof we present here (as well as Proposition 12.18) is due to Lemberg [**350**]. Theorem 12.19 implies, in particular, that a normalized basis $\{x_i\}_{i=1}^{\infty}$ is isometrically equivalent to all its normalized block bases if and only if $\{x_i\}$ is isometrically equivalent to the unit vector basis of l_p for some $1 \leq p < \infty$ or c_0. This fact goes back to the paper of Bohnenblust [**75**] and implicitly even to an earlier result of Kolmogorov and (independently) Nagumo on "general means". An isomorphic version of Bohnenblust's result was proved by Zippin [**594**]. He showed that a normalized basis is equivalent to all its block bases if and only if it is equivalent to the unit vector basis of c_0 or some l_p, $1 \leq p < \infty$. This result is also an immediate consequence of Theorem 12.19 (and a simple uniform boundedness argument).

Theorem 12.21 is due to Odell, Rosenthal and Schlumprecht [**441**].

CHAPTER 13

Oscillation of Uniformly Continuous Functions on Unit Spheres of Infinite-Dimensional Subspaces

1. Preliminary Results

Let f be a uniformly continuous real-valued function on the unit sphere of an infinite-dimensional Banach space E. In this chapter we consider the question whether f is almost constant (i.e., has small oscillation) when restricted to a suitable infinite-dimensional subspace of E. In the preceding chapter we showed that the answer is positive if we consider finite-dimensional subspaces (of arbitrary large dimensions). Here we shall show that once we pass to infinite-dimensional subspaces we encounter a substantially different situation. This genuinely infinite-dimensional phenomenon was clarified only recently, solving a major problem that was open for many years, and opening new directions for research.

In the first section we present the basic notions and some rather easy results concerning them.

Sections 2-5 are technically much more involved. In Section 2 we show that the existence of an equivalent norm on a space E which does not have small oscillation on the unit sphere of any infinite-dimensional subspace of E is related to the existence of subspaces of E which are isomorphic to l_p or c_0. At the end of the section we formulate the main results of this chapter, characterizing the Banach spaces E for which every uniformly continuous function on their unit sphere (respectively every equivalent norm) has a small oscillation on the unit sphere of an appropriate infinite-dimensional subspace of E. These results depend on the results proved in Sections 3 and 5. Section 3 considers c_0, and is a nice application of infinite combinatorics to nonlinear functional analysis. In Section 4 we introduce the Schlumprecht space $\mathcal{S}$ and investigate its properties. In Section 5 we show how to transfer the results of Section 4 on $\mathcal{S}$ to results on the structure of l_p for $1 < p < \infty$ and, in particular, Hilbert space. The results of Section 5 show that a uniformly continuous function on the unit sphere of l_p for $1 \leq p < \infty$ need not be nearly constant on any infinite-dimensional subspace.

Definition 13.1. *Let f be a real-valued function defined on the unit sphere $S(E)$ of an infinite-dimensional Banach space E. The function f is said to be oscillation stable if for every infinite-dimensional subspace F of E and for every $\varepsilon > 0$ there is an infinite-dimensional subspace G of F such that the oscillation of f on $S(G)$ is at most ε.*

Definition 13.2. *A subset A of the unit sphere of an infinite-dimensional Banach space E is said to be an asymptotic set if $d(A, F) = 0$ for every infinite-dimensional subspace F of E.*

Remark. If E has a basis $\{x_n\}_{n=1}^\infty$, then in the definition above is suffices to consider only subspaces F which are spanned by a block basis $\{y_k\}_{k=1}^\infty$ of the x_n's. Indeed, each subspace of E contains arbitrarily small perturbations of blocks by Proposition E.1.

The following simple result relates oscillation stability to asymptotic sets.

Proposition 13.3. *Let E be an infinite-dimensional Banach space. Then the following three assertions are equivalent:*

(i) *Every real-valued uniformly continuous function on $S(E)$ is oscillation stable.*

(ii) *Every real-valued Lipschitz function on $S(E)$ is oscillation stable.*

(iii) *There is no infinite-dimensional subspace F of E whose unit sphere contains two asymptotic sets A and B with $d(A, B) > 0$.*

Proof. Clearly (i) $\Rightarrow$ (ii). Assume that (ii) holds, and that there are F, A and B contradicting (iii). The function $f(x) = d(x, A)$ on the unit sphere of E is a Lipschitz function which vanishes on A and is bigger than $d(A, B)$ on B. By the definition of asymptotic sets the oscillation of the restriction of f to the unit sphere of any infinite-dimensional subspace of F is at least $d(A, B)$.

To prove that (iii) $\Rightarrow$ (i), assume that f is a uniformly continuous function on $S(E)$ which is not oscillation stable, and find an infinite-dimensional $F_0 \subset E$ and an $\varepsilon_0 > 0$ such that the oscillation of f on the unit sphere of any subspace of F_0 is at least $2\varepsilon_0$. Put $a = \sup_G \inf\{f(x) : x \in S(G)\}$, where the supremum is taken over all infinite-dimensional subspaces G of F_0, and choose an infinite-dimensional subspace F of F_0 such that $\inf\{f(x) : x \in S(F)\} > a - \varepsilon_0$.

Put $b = \inf_G \sup\{f(x) : x \in S(G)\}$, where the infimum is taken over all infinite-dimensional subspaces G of F. Then $b - a > \varepsilon_0$ by the choice of F_0 and ε_0. Put

$$A = \{x \in S(F) : \ f(x) \geq (a+b)/2 + \varepsilon_0/2\}$$
$$B = \{x \in S(F) : \ f(x) \leq (a+b)/2 - \varepsilon_0/2\}.$$

Then A and B are asymptotic sets in $S(F)$, and $d(A, B) \geq \omega^{-1}(\varepsilon_0)$ where ω is the modulus of continuity of f. □

Remark. If E has the property that every infinite-dimensional subspace F of E contains a further subspace which is isomorphic to E, we can replace (iii) in Proposition 13.3 by

(iii)′ *The unit sphere of E does not contain two asymptotic sets A and B with $d(A, B) > 0$.*

Example of such spaces are c_0 and l_p for $1 \leq p < \infty$, where the span of any block basis of the unit vector basis is even isometric to the whole space.

Definition 13.4. *Let $\lambda > 1$. A norm $\|\cdot\|$ on an infinite-dimensional Banach space E is said to be λ-distortable if there is an equivalent norm $|||\cdot|||$ on E such that for every infinite-dimensional subspace F of E*

$$\sup\left\{|||x|||/|||y||| \ : \ x, y \in F \ \text{ and } \ \|x\| = \|y\| = 1\right\} \geq \lambda.$$

The given norm is said to be distortable if it is λ-distortable for some $\lambda > 1$. It is said to be arbitrarily distortable if it is λ-distortable for all $\lambda > 1$.

Proposition 13.5. *For uniformly convex spaces E the condition* (i)–(iii) *of Proposition* 13.3 *are also equivalent to*

(iv) *There is an infinite-dimensional subspace of E with a distortable norm.*

Proof. An equivalent distorted norm is a Lipschitz function which is not oscillation stable. Conversely, let F be a uniformly convex space such that $S(F)$ contains two asymptotic sets A and B with $d = d(A, B) > 0$, and assume for contradiction that there is no distortable norm on any infinite-dimensional subspace of F.

Consider the equivalent norm $|\!|\!| \cdot |\!|\!|$ on F whose unit ball is the closed convex hull of $A \cup (-A) \cup \{x : \|x\| \le 1/2\}$, and fix $\eta > 0$ (to be determined later). Since $|\!|\!| \cdot |\!|\!|$ does not distort $\| \cdot \|$ on any infinite-dimensional subspace of F, it follows that for every such subspace there is a further subspace G such that the oscillation of $|\!|\!| \cdot |\!|\!|$ on $S(G)$ is smaller than $\eta/2$. Since A is asymptotic, there is a point $w \in G$ with $\|w\| = 1$ and $d(w, A) < \eta/4$. Hence $1 = \|w\| \le |\!|\!| w |\!|\!| < 1 + \eta/2$, and thus $1 \le |\!|\!| z |\!|\!| < 1 + \eta$ for every $z \in G$ with $\|z\| = 1$.

Fix any such z. Since $1 \le |\!|\!| z |\!|\!| < 1 + \eta$, we can find for each $\delta > 0$ positive numbers α, β with $\alpha + \beta < 1 + \eta$ and points $x \in \mathrm{co}\{A \cup -A\}$ and y with $\|y\| \le 1/2$ such that $\|z - (\alpha x + \beta y)\| < \delta$. Since $\|z\| = 1$ and $\|y\| \le 1/2$, it follows that if δ is small enough, then $\beta < 2\eta$. Let $z^* \in F^*$ satisfy $\|z^*\| = \langle z^*, z\rangle = 1$, and consider the slice $S = \{u \in B(F) : \langle z^*, u\rangle \ge 1 - 2\eta\}$. Then $x \in S$ by the computation above, and since $x \in \mathrm{co}\{A \cup -A\}$ it follows that there is also a point $v \in A \bigcup -A$ which belongs to S. The uniform convexity of F implies that $\mathrm{diam}(S) < d/8$ provided η is small enough, hence $\|z - v\| < d/8$.

Using a similar argument with B instead of A and passing to a subspace G_0 of G gives $d(z, A \cup -A), d(z, B \cup -B) < d/8$ for every $z \in S(G_0)$. Since $d(-A, -B) = d$, it follows that there is a $z_0 \in S(G_0)$ such that $d(z_0, -A), d(z_0, -B) \ge d/3$. Hence $d(z_0, A), d(z_0, B) < d/8$, and thus $d(A, B) < d/4$, a contradiction. □

Remarks. (i) The equivalent statements of Proposition 13.3 are preserved under isomorphism. Thus Proposition 13.5 holds for spaces with an equivalent uniformly convex norm; i.e., for superreflexive spaces.

(ii) As in the remark following Proposition 13.3, if $E = l_p$ for $1 < p < \infty$, then statement (iv) can be replaced by the statement that the norm in E is distortable.

(iii) From Proposition 13.6 below and from the main result of Section 5, it will follow that Proposition 13.5 does not hold for $E = l_1$. Thus the uniform convexity assumption cannot be dropped from its statement.

Proposition 13.6. *The spaces c_0 and l_1 do not have a distortable norm.*

Proof. By the remark following Proposition 13.3, we only have to prove that the usual norms of these spaces are not distortable.

Consider first $E = l_1$ with the usual norm $\| \cdot \|$ and with an equivalent norm $|\!|\!| \cdot |\!|\!|$ satisfying $\alpha |\!|\!| x |\!|\!| \le \|x\| \le |\!|\!| x |\!|\!|$ for some $\alpha > 0$ and all $x \in l_1$. For every integer n let

$$\lambda_n = \sup\big\{\|x\| : \ |\!|\!| x |\!|\!| = 1 \text{ and the first } n \text{ coordinates of } x \text{ are } 0\big\}.$$

Then λ_n is a nonincreasing sequence and $\lambda = \lim \lambda_n$ satisfies $\alpha \le \lambda \le 1$. Given $\varepsilon > 0$, choose n_0 such that $\lambda_{n_0} \le \lambda(1 + \varepsilon)$. It follows from the definition of λ that there is a block basis $\{y_k\}_{k=1}^{\infty}$ of the unit vectors basis of l_1 whose first n_0 coordinates vanish such that $|\!|\!| y_k |\!|\!| = 1$ and $\|y_k\| \ge \lambda/(1 + \varepsilon)$ for all k.

Since $\|\sum a_k y_k\| = \sum \|a_k y_k\| \geq \lambda \sum |a_k|/(1+\varepsilon)$ for every choice of finitely many scalars $\{a_k\}$, it follows that

$$\begin{aligned}\sum |a_k|/(1+\varepsilon)^2 &\leq \|\sum a_k y_k\|/\lambda_{n_0} \leq |\!|\!| \sum a_k y_k |\!|\!| \\ &\leq \sum |\!|\!| a_k y_k |\!|\!| \leq \sum |a_k|\end{aligned}$$

and the oscillation of $|\!|\!| \cdot |\!|\!|$ on the unit sphere of $\overline{\operatorname{span}}\{y_k\}_{k=1}^\infty$ is as small as we please if ε was chosen to be small enough.

The proof for c_0 is very similar. Replace λ_n by

$$\mu_n = \inf\big\{\|x\| : \ |\!|\!| x |\!|\!| = 1 \text{ and the first } n \text{ coordinates of } x \text{ are } 0\big\}.$$

Then $\mu_n \nearrow \mu$. Given $\varepsilon > 0$, fix n_0 such that $\mu_{n_0} \geq \mu/(1+\varepsilon)$, and choose blocks $\{z_k\}$ of the unit vector basis in c_0 whose first n_0 coordinates vanish and such that $|\!|\!| z_k |\!|\!| = 1$ and $\|z_k\| \leq \mu(1+\varepsilon)$ for all k. Since

$$\left\|\sum a_k z_k\right\| = \max |a_k| \|z_k\| \leq \mu(1+\varepsilon) \max |a_k|$$

it follows that

$$|\!|\!| \sum a_k z_k |\!|\!| \leq \left\|\sum a_k z_k\right\|/\mu_{n_0} \leq (1+\varepsilon)^2 \max |a_k|$$

while for every j

$$|\!|\!| \sum a_k z_k |\!|\!| \geq |\!|\!| 2a_j z_j |\!|\!| - |\!|\!| \sum a_k z_k - 2a_j z_j |\!|\!| \geq 2|a_j| - (1+\varepsilon)^2 \max |a_k|.$$

Hence $|\!|\!| \sum a_k z_k |\!|\!| \geq \big(2 - (1+\varepsilon)^2\big) \max |a_k|$. □

Remarks. (i) Using the same proof as above for l_p with $1 < p < \infty$ gives, using the λ_n's (resp. the μ_n's), a block basis on whose span the norm $|\!|\!| \cdot |\!|\!|$ would dominate the usual norm in l_p times $1/(1+\varepsilon)^2$ (resp. be dominated by $(1+\varepsilon)^2$ times the given norm). However, the estimate for the reverse inequality would hold only for the unit vectors and not for their linear combinations. What made the proof work for l_1 and c_0 is that in these spaces the needed reverse estimates are automatically true.

It follows from the results of Section 5 that for $1 < p < \infty$ there is a distorted norm on l_p, and hence natural attempts to push the proof, such as by iterating this procedure and choosing alternatively blocks with almost maximal $|\!|\!| \cdot |\!|\!|$ norms and then blocks with almost minimal $|\!|\!| \cdot |\!|\!|$ norm and so on, must fail.

(ii) The Mazur map (see Section 1 of Chapter 9) from the unit sphere of l_{p_1} to the unit sphere of l_{p_2} maps a block basic sequence to a block basic sequence. It thus maps an asymptotic set in the unit sphere of l_{p_1} to an asymptotic set in unit sphere of l_{p_2}. Since this map is uniformly continuous it follows that the question whether l_p has the equivalent properties stated in Proposition 13.3 has the same answer for all $1 \leq p < \infty$. By Proposition 13.5 the answer to the question whether the norm in l_p can be distorted will also be the same in the range $1 < p < \infty$. (By Proposition 13.6 we must, of course, exclude l_1 here.)

We study now a weaker notion of distortion, in which we shall only consider subspaces isomorphic to the original space. The question whether a given space is distortable in this weaker sense is, of course, meaningful only when the space contains "many" subspaces isomorphic to itself. Two such families of spaces are the spaces $L_p[0,1]$ for $1 \leq p < \infty$ and $C(K)$, which are known to be rich with "small" subspaces isometric to the entire space.

Definition 13.7. *A basis $\{x_n\}$ in a Banach space E is said to be λ-reproducible for some $\lambda > 1$ if whenever E is isometric to a subspace of a Banach space $\tilde{E}$ having a basis $\{y_m\}$, then for every $\varepsilon > 0$ there is a block basis $\{z_n\}$ of $\{y_m\}$ which is $(\lambda + \varepsilon)$-equivalent to $\{x_n\}$. A basis is called precisely reproducible if it is 1-reproducible.*

It is trivial to check that the unit vector basis of c_0 and l_p, $1 \le p < \infty$, are precisely reproducible. The following two propositions are less obvious.

Proposition 13.8. *The Haar basis is a precisely reproducible basis in L_p for every $1 \le p < \infty$.*

Proposition 13.9. *Let K be an uncountable compact metric space. Then every basis in $C(K)$ with basis constant λ is λ-reproducible. In particular every monotone basis in $C(K)$ is precisely reproducible.*

We shall prove here only Proposition 13.8.

Proof of Proposition 13.8. We first introduce some notation and recall the definition of the Haar basis.

Denote the dyadic intervals of order n in $[0,1]$ by $\Delta_{n,j} = [j/2^n, (j+1)/2^n]$ for $0 \le j \le 2^n - 1$. The Haar functions (normalized in the L_∞ norm) are the functions $h_0 \equiv 1$ and $h_{n,j} = \chi_{\Delta_{n,2j}} - \chi_{\Delta_{n,2j+1}}$ for $n = 1, 2 \ldots$ and $0 \le j \le 2^{n-1} - 1$, where χ_A denotes the characteristic function of the set A. They form a basis for $L_p[0,1]$ for every $1 \le p < \infty$, which is unconditional when $p > 1$ (see Appendix E).

If $\{A_{n,j}\}$ is any system of measurable subsets of $[0,1]$ such that $|A_{n,j}| = 2^{-n}$ and $A_{n,j}$ is the disjoint union of $A_{n+1,2j}$ and $A_{n+1,2j+1}$, then the functions $g_0 \equiv 1$ and $g_{n,j} = \chi_{A_{n,2j}} - \chi_{A_{n,2j+1}}$ for $n = 1, 2 \ldots$ and $0 \le j \le 2^{n-1} - 1$ are isometrically equivalent to the Haar basis in all the L_p's.

Assume that L_p is contained in a Banach space $\tilde{E}$ with a basis $\{y_m\}$. We shall show that for any choice of positive numbers ε_0 and $\{\varepsilon_{n,j}\}$ there are sets $A_{n,j}$ as above and disjoint blocks z_0 and $\{z_{n,j}\}$ of the y_m's such that $\|g_0 - z_0\| < \varepsilon_0$ and $\|g_{n,j} - z_{n,j}\| < \varepsilon_{n,j}$. If we chose the ε's to be small enough, then the result follows by Proposition E.1.

The construction is inductive. Thus fix n and j, and assume that $A_{m,i}$ and $z_{m,i}$ have been defined for all $0 \le i \le 2^m - 1$ and $m \le n$, as well as $A_{n+1,i}$ and $z_{n+1,i}$ for $i \le 2j - 1$, and we need to "split" $A_{n,j}$.

In the inductive construction we also assume that there is an N such that each of the sets which have already been chosen is a finite union of dyadic intervals of order N and also that all the corresponding vectors are contained in $\operatorname{span}\{y_m : m \le N\}$. We denote the basis projection from $\tilde{E}$ onto this latter space by P_N.

To find the splitting of $A_{n,j}$, let r_k be the Rademacher functions on $[0,1]$, and note that $r_k f \to 0$ weakly in L_p, hence also in $\tilde{E}$, for any function $f \in L_p$. Since P_N is of finite rank, it follows that if k is large enough, then $\|P_N(r_k \chi_{A_{n,j}})\| < \varepsilon_{n,j}$. We can also choose k to be a multiple of N, and then take as $z_{n,j}$ a linear combination of $\{y_m\}_{m>N}$ such that $\|z_{n,j} - r_k \chi_{A_{n,j}}\| < \varepsilon_{n,j}$ and the sets

$$A_{n+1,2j} = \{t \in A_{n,j} : r_k(t) = 1\} \quad \text{and} \quad A_{n+1,2j+1} = \{t \in A_{n,j} : r_k(t) = -1\}.$$

$\diamondsuit$

Remarks. (i) The proposition and its proof remain valid when we replace L_p by any rearrangement invariant lattice E (see Appendix E), provided $r_k \to 0$ weakly in E.

(ii) Since one can check directly that the Haar basis is not unconditional in L_1, it follows from the proposition that L_1 is not isomorphic to a subspace of any Banach space with an unconditional basis. Since $C(0,1)$ contains a copy of L_1, it follows that the same is true for $C(0,1)$.

Proposition 13.10. *Let E be a Banach space with a precisely reproducible basis $\{x_n\}$.*

(i) *If the unconditional constant K of the basis is larger than 1, then there is an equivalent norm $\|\!|\cdot|\!\|$ on E such that*

$$\mu = \sup\big\{\|\!|x|\!\|/\|\!|y|\!\| \,:\, x, y \in F \quad \textit{and} \quad \|x\| = \|y\| = 1\big\} \geq K^{1/2}$$

for any subspace $F \subset E$ which is isometric to E.

(ii) *If the basis is not unconditional, then an equivalent norm $\|\!|\cdot|\!\|$ as in* (i) *exists for every $K < \infty$.*

Proof. (i) The basis $\{x_n\}$ is 1-unconditional under the equivalent norm

$$\Big\|\!\Big|\sum a_n x_n\Big|\!\Big\| = \sup_{\pm}\Big\|\sum \pm a_n x_n\Big\|.$$

Let $F \subset E$ be isometric to E, and let $\{y_n\}$ be a basis for F isometrically equivalent to $\{x_n\}$. Since F is μ-isomorphic to $(F, \|\!|\cdot|\!\|)$, the basis $\{y_n\}$ is μ-reproducible under $\|\!|\cdot|\!\|$; hence for every $\varepsilon > 0$ it is $(\mu+\varepsilon)$-equivalent to some blocks of $\{x_n\}$ (taken with the $\|\!|\cdot|\!\|$ norm). Since $\{x_n\}$ are 1-unconditional under $\|\!|\cdot|\!\|$, and this also holds for its blocks, it follows that $\{y_n\}$ are $(\mu+\varepsilon)$-unconditional under $\|\!|\cdot|\!\|$, and hence $\mu(\mu+\varepsilon)$-unconditional under $\|\cdot\|$. Thus $\mu(\mu+\varepsilon) \geq K$, and the result follows because ε was arbitrary.

(ii) If the basis is not unconditional, we consider for each $0 \geq \lambda > -1$ the semigroup G_λ consisting of operators T on E of the form

$$T\Big(\sum a_n x_n\Big) = \sum \lambda^{m_n} a_n x_n$$

where $\{m_n\}$ is a nondecreasing sequence of positive integers which are eventually constant. One easily verifies that every $T \in G_\lambda$ satisfies $\|T\| \leq (1-\lambda)M/(1-|\lambda|)$ where M is the basis constant of $\{x_n\}$. Hence

$$\|\!|x|\!\|_\lambda = \sup\big\{\|Tx\| \,:\, T \in G_\lambda\big\}$$

is an equivalent norm on E, and $\|\!|Tx|\!\|_\lambda \leq \|\!|x|\!\|_\lambda$ for every $T \in G_\lambda$ and every $x \in E$. Each $T \in G_\lambda$ maps block bases to block bases, and $\sup\{\|T\| : T \in G_\lambda\} \to \infty$ as $\lambda \to -1$ (because $\{x_n\}$ is not unconditional). An argument similar to that of (i), with the norm $\|\!|x|\!\|_\lambda$ instead of $\|\!|x|\!\|$ (and with λ close enough to -1 depending on the given $K < \infty$) finishes the proof. We omit the details. □

Remark. The Haar basis in L_p is unconditional for every $1 < p < \infty$, but its unconditional constant is 1 if and only if $p = 2$. As we saw above, L_1 and $C(0,1)$ do not have unconditional bases. It follows that the norms on these spaces are distortable in the weak sense of Proposition 13.10. We shall obtain much stronger results in Section 5, but chose to also present Proposition 13.10 because its proof is much simpler.

2. Existence of Subspaces Isomorphic to l_p or c_0

This section is devoted to the proof of the following theorem.

Theorem 13.11. *Assume that the separable infinite-dimensional Banach space E does not have a subspace with a distortable norm. Then there is a $1 \le p \le \infty$ such that for every $\varepsilon > 0$ the space E has a subspace which is $(1+\varepsilon)$-isometric to l_p (or c_0 if $p = \infty$).*

We shall see in Section 5 that the spaces l_p actually admit an equivalent distorted norm for $1 < p < \infty$. See also Theorem 13.16 at the end of this section.

The subspaces of E which are almost isometric to c_0 or l_p will be found by means of the following simple proposition.

Proposition 13.12. *Let $\{x_j\}_{j=1}^{\infty}$ be a sequence of norm-one vectors in a Banach space E, and assume that there is a $1 \le p \le \infty$ such that $\lim_{j\to\infty} \|x + x_j\|$ exists for every $x \in E$ and satisfies*

$$\lim_{j\to\infty} \|x + x_j\| = \lim_{j\to\infty} \lim_{k\to\infty} \|x + \alpha x_j + \beta x_k\| \tag{13.1}$$

for every $x \in E$ and for all $\alpha, \beta \in \mathbb{R}$ with $|\alpha|^p + |\beta|^p = 1$ (or $\max(|\alpha|, |\beta|) = 1$ when $p = \infty$). Then for every $\varepsilon > 0$ there is a subsequence of the x_j's which is $(1+\varepsilon)$-equivalent to the unit vector basis of l_p (or c_0 if $p = \infty$).

Proof. Choose a sequence $\varepsilon_n > 0$ with $\prod_{n=1}^{\infty}(1+\varepsilon_n) \le 1+\varepsilon$. We shall write the proof for $1 \le p < \infty$, the proof for $p = \infty$ is identical. By (13.1) for $x = 0$, and by the compactness of the unit sphere of l_p^2 we can choose j_1 such that

$$(1+\varepsilon_1)^{-1} \le \lim_{j\to\infty} \|\alpha x_{j_1} + \beta x_j\| \le (1+\varepsilon_1)$$

whenever $|\alpha|^p + |\beta|^p = 1$. Using (13.1) for $x = \alpha x_{j_1}$ and the compactness of the unit sphere of l_p^2, we can now find $j_2 > j_1$ such that

$$\begin{aligned}(1+\varepsilon_2)^{-1} \lim_j \|\alpha x_{j_1} + \beta x_j\| &\le \lim_j \|\alpha x_{j_1} + \alpha' \beta x_{j_2} + \beta' \beta x_j\| \\ &\le (1+\varepsilon_2) \lim_j \|\alpha x_{j_1} + \beta x_j\|\end{aligned}$$

whenever $|\alpha|^p + |\beta|^p = 1$ and $|\alpha'|^p + |\beta'|^p = 1$. Combining the two sets of inequalities we get that

$$(1+\varepsilon_1)^{-1}(1+\varepsilon_2)^{-1} \le \lim_j \|\alpha x_{j_1} + \beta x_{j_2} + \gamma x_j\| \le (1+\varepsilon_1)(1+\varepsilon_2)$$

whenever $|\alpha|^p + |\beta|^p + |\gamma|^p = 1$. Continuing inductively, find an increasing sequence $\{j_m\}_{m=1}^{\infty}$ such that for every n and every choice of $\{\alpha_m\}_{m=1}^{n}$ with $\sum_{m=1}^{n} |\alpha_m|^p = 1$

$$\prod_{m=1}^{n-1}(1+\varepsilon_m)^{-1} \le \lim_j \Big\| \sum_{m=1}^{n-1} \alpha_m x_{j_m} + \alpha_n x_j \Big\| \le \prod_{m=1}^{n-1}(1+\varepsilon_m).$$

The subsequence $\{x_{j_m}\}_{m=1}^{\infty}$ has the desired property. □

In the proof of Theorem 13.11 we shall use norms on E with a simple explicit form, and assume that they are not distorted. For every $y \in E$ put

$$\|x\|_y = \frac{1}{2}\Big(\Big\| \|x\| y + x \Big\| + \Big\| \|x\| y - x \Big\|\Big), \quad x \in E.$$

Fix any $x, y \in E$ and $0 \le s \le t$. Writing s as a convex combination of t and $-t$ gives, by the triangle inequality in E

$$\|sy + x\| + \|sy - x\| \le \|ty + x\| + \|ty - x\|.$$

Taking $s = \|x + z\|$ and $t = \|x\| + \|z\|$ gives

$$\begin{aligned} 2\|x+z\|_y &= \Big\| \|x+z\|y + (x+z)\Big\| + \Big\| \|x+z\|y - (x+z)\Big\| \\ &\le \Big\|(\|x\| + \|z\|)y + (x+z)\Big\| + \Big\|(\|x\| + \|z\|)y - (x+z)\Big\| \\ &\le 2\big(\|x\|_y + \|z\|_y\big) \end{aligned}$$

i.e., $\|\cdot\|_y$ satisfies the triangle inequality. Taking $s = 0$ and $t = \|x\|$ we see that $\|x\| \le \|x\|_y$, and since $\|x\|_y \le \|x\|(1 + \|y\|)$ is clear, it follows that $\|\cdot\|_y$ defines an equivalent norm on E for every y.

We shall need the following standard result on general Banach spaces.

Proposition 13.13. *Let $\{u_i\}_{i=1}^m$ be m vectors in a Banach space E, and let F be an $(m+1)$-dimensional subspace of E. Then there is a $y \in F$ with $\|y\| = 1$ and $\|y + u_i\| = \|y - u_i\|$ for $1 \le i \le m$.*

Proof. The map

$$f(y) = \big(\|y + u_1\| - \|y - u_1\|, \|y + u_2\| - \|y - u_2\|, \dots, \|y + u_m\| - \|y - u_m\|\big)$$

is a continuous antipodal (i.e., $f(y) = -f(-y)$ for every y) map from the $(m+1)$-dimensional sphere $S(F)$ to $\mathbb{R}^m$. By Borsuk's theorem there is a point $y \in S(F)$ such that $f(y) = 0$. □

The proof of Theorem 13.11 is based on the methods and results of Section 2 of the previous chapter. The next proposition is very close to Proposition 12.16.

Proposition 13.14. *Let E be a separable Banach space, and let $\{y_i\}_{i=1}^\infty$ be a normalized basic sequence in E. Then there is a normalized block basis $\{w_j\}_{j=1}^\infty$ of the y_i's such that*

$$\lim_{k_1\to\infty} \lim_{k_2\to\infty} \dots \lim_{k_n\to\infty} \Big\|x + \sum_{m=1}^n \alpha_m w_{k_m}\Big\| \tag{13.2}$$

exists and

$$\begin{aligned} &\lim_{k_1\to\infty} \lim_{k_2\to\infty} \dots \lim_{k_n\to\infty} \Big\|x + \sum_{m=1}^n \alpha_m w_{k_m}\Big\| \\ &\quad = \lim_{k_1\to\infty} \lim_{k_2\to\infty} \dots \lim_{k_n\to\infty} \Big\|x - \sum_{m=1}^n \alpha_m w_{k_m}\Big\| \end{aligned} \tag{13.3}$$

for all $x \in E$, all integers n and all scalars $\{\alpha_m\}_{m=1}^n$.

Proof. The existence of the limit (13.2) for $x = 0$ is just a restatement of Proposition 12.16. In this case we do not even need to pass to blocks and just take $\{w_j\}_{j=1}^\infty$ to be a suitable subsequence of the y_i's. The same proof as that of Proposition 12.16 (i.e., a Ramsey type argument) guarantees that (13.2) holds for every $x \in E$.

In order to get (13.3) we have to pass to blocks, and we start with an inductive construction of an auxiliary block sequence $\{v_n\}$. We only describe the inductive step. Let $\{x_k\}_{k=1}^\infty$ be a dense sequence in E, and choose coefficients $\{\alpha_j^t\}_{j=1}^{n-1}$ for

$1 \le t \le r$ such that the set of vectors $\sum_{j=1}^{n-1} \alpha_j^t v_j$ is $1/n$-dense in the ball of radius n in $\text{span}\{v_j : j < n\}$. Now apply Proposition 13.13 for the $m = nr$ vectors $x_k + \sum_{j=1}^{n-1} \alpha_j^t v_j$, where $1 \le k \le n$ and $1 \le t \le r$, and to the span F of $m+1$ vectors y_i whose indices are all bigger than the last i in the support of v_{n-1}. The proposition yields a normalized vector v_n. The sequence $\{v_n\}_{n=1}^{\infty}$ is a block basic sequence of the y_i's such that

$$\Big\|x_k + \sum_{j=1}^{n-1} \alpha_j^t v_j - v_n\Big\| = \Big\|x_k + \sum_{j=1}^{n-1} \alpha_j^t v_j + v_n\Big\|$$

for all $1 \le k \le n$ and $1 \le t \le r$. Having chosen the v_n's, we can now apply the proof of Proposition 12.16 and get that (13.2) and (13.3) hold for a suitable subsequence $\{w_j\}_{j=1}^{\infty}$ of the v_n's. □

Let E be a Banach space, and let $\{w_j\} \subset E$ be as in Proposition 13.14. Define a Banach space containing E as follows: Let $\tilde{E}$ be a vector space with an (algebraic) basis $\{e_m\}_{m=1}^{\infty}$, and define a norm $\|\!| \cdot |\!\|$ on the vector space $E \oplus \tilde{E}$ by

$$(13.4) \qquad \Big|\!\Big\|x + \sum_{m=1}^{n} \alpha_m e_m\Big|\!\Big\| = \lim_{k_1 \to \infty} \lim_{k_2 \to \infty} \dots \lim_{k_n \to \infty} \Big\|x + \sum_{m=1}^{n} \alpha_m w_{k_m}\Big\|.$$

If $x \in E$, then $|\!\|x|\!\| = \|x\|$, and thus $|\!\| \cdot |\!\|$ extends $\| \cdot \|$ from E to $E \oplus \tilde{E}$. Note that the $\{e_m\}_{m=1}^{\infty}$ are a 1-spreading basis for the (completion) of $\tilde{E}$ under $|\!\| \cdot |\!\|$, and by (13.3)

$$|\!\|x + e|\!\| = |\!\|x - e|\!\|$$

for every $x \in E$ and $e \in \tilde{E}$. The next proposition shows that under the distortion assumption of Theorem 13.11 we can ensure that the much stronger assertion

$$(13.5) \qquad |\!\|x + u|\!\| = |\!\|x + v|\!\|$$

holds for every $x \in E$ and $u, v \in \tilde{E}$ with $|\!\|u|\!\| = |\!\|v|\!\|$.

Proposition 13.15. *Let E be a separable infinite-dimensional Banach space such that no norm on E of the form $\| \cdot \|_y$ is distorted on any infinite-dimensional subspace of E. Then there is a normalized basic sequence $\{w_j\}_{j=1}^{\infty}$ in E such that the norm $|\!\| \cdot |\!\|$ is well defined by* (13.4) *and satisfies* (13.5).

Proof. Fix a normalized basic sequence $\{y_i\}_{i=1}^{\infty}$ in E. By our assumption for every $\varepsilon > 0$ and $x \in E$ there is a block basis $\{z_j\}_{j=1}^{\infty}$ of $\{y_i\}_{i=1}^{\infty}$ such that $\big|\|v_1\|_x - \|v_2\|_x\big| \le \varepsilon$ for any $v_1, v_2 \in \overline{\text{span}}\{z_j\}_{j=1}^{\infty}$ with $\|v_1\| = \|v_2\| = 1$.

Let $\{x_k\}_{k=1}^{\infty}$ be a dense sequence in E and $\varepsilon_n \searrow 0$. Using the previous remark repeatedly and then passing to the diagonal sequence, there is a normalized block basis $\{w_j\}_{j=1}^{\infty}$ of $\{y_i\}_{i=1}^{\infty}$ such that

$$\big|\|v_1\|_{x_k} - \|v_2\|_{x_k}\big| < \varepsilon_n; \quad 1 \le k \le n$$

for every n and any $v_1, v_2 \in \overline{\text{span}}\{w_j : j \ge n\}$ with $\|v_1\| = \|v_2\| = 1$. Assuming also that (13.2) and (13.3) hold for the w_j's, it follows that for every normalized block sequence $\{v_l\}_{l=1}^{\infty}$ of $\{w_j\}_{j=1}^{\infty}$ and every $x \in E$ the limit $\lim_l \|x + v_l\| = \lim_l \|v_l\|_x$ exists and depends only on x (and not on the particular sequence $\{v_l\}_{l=1}^{\infty}$).

Define $|\!\| \cdot |\!\|$ by (13.4), and fix $x \in E$ and $v = \sum_{m=1}^{n} a_m e_m \in \tilde{E}$ with $|\!\|v|\!\| = 1$. The iterated limit defining $|\!\|x + v|\!\|$ can be replaced by a simple limit of the form

$\lim_l \|x + v_l\|$ where $v_l = \sum_{m=1}^{n} a_m w_{k_m(l)}$ for appropriate sequences $\{k_m(l)\}_{l=1}^{\infty}$. Consequently, $\|x+v\| = \frac{1}{2}\lim\left(\|x+v_l\|+\|x-v_l\|\right) = \lim \|v_l\|_x$ which is independent of the particular v, and (13.5) holds. □

Relation (13.5) has a useful implication concerning the restriction of the norm $\|\cdot\|$ to $\tilde{E}$: If u, v, w are such that there is an n such that $u \in \operatorname{span}\{e_m\}_{m=1}^{n}$ and $v, w \in \operatorname{span}\{e_m\}_{m>n}$ with $\|v\| = \|w\|$, then

$$\|u + v\| = \|u + w\|. \tag{13.6}$$

To see this, let $\{u_l\}_{l=1}^{\infty} \subset E$ be a block basic sequence of the w_j's such that $\|x + u\| = \lim_l \|x + u_l\|$ for every $x \in E$. In the iterated limit defining $\|x + u + v\|$ we first take the limits over the w_j's with $j > n$, and then over the first n terms. Hence $\|x + u + v\| = \lim_l \|x + u_l + v\|$. Similarly $\|x + u + w\| = \lim_l \|x + u_l + w\|$. But $x + u_l \in E$; hence $\|x + u_l + v\| = \|x + u_l + w\|$ for each fixed l by (13.5). Thus $\|x + u + v\| = \|x + u + w\|$ and the claim follows by taking $x = 0$.

After all these preparations we are now ready for the proof of the theorem.

Proof of Theorem 13.11. Find a normalized basic sequence $\{w_k\}_{k=1}^{\infty}$ in E such that (13.4) defines defines a norm $\|\cdot\|$ on $E \oplus \tilde{E}$ which satisfies (13.5) and (13.6).

By Krivine's theorem 12.19 there is a $1 \le p \le \infty$ and a normalized block basic sequence $a_1, b_1, a_2, b_2, \ldots$ of the e_m's such that

$$\left| \|\alpha a_n + \beta b_n\| - 1 \right| \le n^{-1} \tag{13.7}$$

for every n and $\alpha, \beta \in \mathbb{R}$ with $|\alpha|^p + |\beta|^p = 1$. (We shall assume in the proof that $p < \infty$, the proof for $p = \infty$ is identical.)

By (13.6) $\|\alpha a_n + \beta a_k\| = \|\alpha a_n + \beta b_n\|$ for every $k > n$ and $\alpha, \beta \in \mathbb{R}$. Hence by (13.5) $\|x + v\| = \lim_n \lim_k \|x + \alpha a_n + \beta a_k\|$ for every $v \in \tilde{E}$ with $\|v\| = 1$ and $|\alpha|^p + |\beta|^p = 1$. In particular

$$\lim_n \|x + a_n\| = \|x + a_1\| = \lim_n \lim_k \|x + \alpha a_n + \beta a_k\| \tag{13.8}$$

for every $x \in E$ and $|\alpha|^p + |\beta|^p = 1$. (Note that by (13.5) $\|x + a_n\|$ is actually independent of n.)

We shall transfer (13.8) to a similar relation in E, and find a basic sequence $\{v_l\}$ in E such that (13.8) holds with the v_n's replacing the a_n's. The desired result will then follow from Proposition 13.12.

Note first that for every $x \in E$ and $\varepsilon > 0$ there is a subsequence $\{w'_m\}_{m=1}^{\infty}$ of $\{w_k\}$ such that

$$\left| \left\| x + \sum_{m=1}^{n} \alpha_m w'_m \right\| - \left\| x + \sum_{m=1}^{n} \alpha_m e_m \right\| \right| < \varepsilon \tag{13.9}$$

for every n and $\{\alpha_m\}_{m=1}^{n}$ with $\|\sum_{m=1}^{n} \alpha_m e_m\| \le 1$. We choose the w'_m's inductively in such a way that for every integer i, all $\{\alpha_m\}_{m=1}^{i}$ and $|\beta|, |\gamma| \in [-2, 2]$

$$\left| \left\| x + \sum_{m=1}^{i} \alpha_m w'_m + \beta e_1 + \gamma e_2 \right\| - \left\| x + \sum_{m=1}^{i} \alpha_m w'_m + \beta w'_{i+1} + \gamma e_2 \right\| \right| \le \varepsilon 2^{-(i+1)}.$$

(This can be done by the definition (13.4) and compactness considerations). By (13.5) we obtain for every i

$$\left|\!\left|\!\left| x + \sum_{m=1}^{i} \alpha_m w'_m + \sum_{m=i+1}^{n} \alpha_m e_m \right|\!\right|\!\right| = \left|\!\left|\!\left| x + \sum_{m=1}^{i-1} \alpha_m w'_m + \alpha_i w'_i + \left|\!\left|\!\left| \sum_{m=i+1}^{n} \alpha_m e_m \right|\!\right|\!\right| e_2 \right|\!\right|\!\right|$$

and

$$\left|\!\left|\!\left| x + \sum_{m=1}^{i-1} \alpha_m w'_m + \sum_{m=i}^{n} \alpha_m e_m \right|\!\right|\!\right| = \left|\!\left|\!\left| x + \sum_{m=1}^{i-1} \alpha_m w'_m + \alpha_i e_1 + \sum_{m=i+1}^{n} \alpha_m e_m \right|\!\right|\!\right|.$$

By (13.5) and (13.6) the term $\sum_{m=i+1}^{n} \alpha_m e_m$ in the right-hand side can be replaced by the term $|\!|\!|\sum_{m=i+1}^{n} \alpha_m e_m|\!|\!| e_2$. Since $|\!|\!|\sum_{m=1}^{n} \alpha_m e_m|\!|\!| \leq 1$ implies that $|\alpha_m| \leq 2$ for each m, and since $\|x + \sum_{m=1}^{n} \alpha_m w'_m\| = |\!|\!|x + \sum_{m=1}^{n} \alpha_m w'_m|\!|\!|$ (because $\|\cdot\|$ and $|\!|\!|\cdot|\!|\!|$ coincide on E), we obtain

$$\begin{aligned}
&\left| \left\| x + \sum_{m=1}^{n} \alpha_m w'_m \right\| - \left|\!\left|\!\left| x + \sum_{m=1}^{n} \alpha_m e_m \right|\!\right|\!\right| \right| \\
&\leq \sum_{i=1}^{n} \left| \left|\!\left|\!\left| x + \sum_{m=1}^{i} \alpha_m w'_m + \sum_{m=i+1}^{n} \alpha_m e_m \right|\!\right|\!\right| - \left|\!\left|\!\left| x + \sum_{m=1}^{i-1} \alpha_m w'_m + \sum_{m=i}^{n} \alpha_m e_m \right|\!\right|\!\right| \right| \\
&= \sum_{i=1}^{n} \Bigg| \left|\!\left|\!\left| x + \sum_{m=1}^{i-1} \alpha_m w'_m + \alpha_i w'_i + \left|\!\left|\!\left| \sum_{m=i+1}^{n} \alpha_m e_m \right|\!\right|\!\right| e_2 \right|\!\right|\!\right| \\
&\qquad - \left|\!\left|\!\left| x + \sum_{m=1}^{i-1} \alpha_m w'_m + \alpha_i e_1 + \left|\!\left|\!\left| \sum_{m=i+1}^{n} \alpha_m e_m \right|\!\right|\!\right| e_2 \right|\!\right|\!\right| \Bigg| \\
&\leq \sum_{i=1}^{n} \varepsilon 2^{-i} \leq \varepsilon.
\end{aligned}$$

Observe also that (13.9) is valid for every subsequence of $\{w'_m\}$.

Choose $\varepsilon_k \searrow 0$, and let $\{x_k\}_{k=1}^{\infty}$ be dense in E. Passing repeatedly to subsequences and then, using the diagonal sequence, we deduce from (13.9) that there is a subsequence $\{z_j\}_{j=1}^{\infty}$ of $\{w_m\}$ such that for every k and $k < j_1 < j_2 < \cdots < j_n$

$$\left| \left|\!\left|\!\left| x_i + \sum_{m=1}^{n} \alpha_m e_m \right|\!\right|\!\right| - \left\| x_i + \sum_{m=1}^{n} \alpha_m z_{j_m} \right\| \right| < \varepsilon_k$$

whenever $1 \leq i \leq k$ and $|\!|\!|\sum_{m=1}^{n} \alpha_m e_m|\!|\!| \leq 1$. Together with (13.8) this gives that there is a normalized block basis $\{v_l\}_{l=1}^{\infty}$ of $\{z_j\}_{j=1}^{\infty}$ such that $\lim_l \|x + v_l\|$ exists and is equal to $\lim_l \lim_s \|x + \alpha v_l + \beta v_s\|$ for and every $x \in E$ and every $\alpha, \beta \in \mathbb{R}$ with $|\alpha|^p + |\beta|^p = 1$.

An application of Proposition 13.12 concludes the proof. ◇

The following theorems (which are the main results of this chapter) follow by combining Theorem 13.11 with results in the previous and the following sections. We shall prove in the next section that every uniformly continuous function from the unit sphere of c_0 into $\mathbb{R}$ is oscillation stable. In Section 5 we shall prove that l_2 (as well as l_p for $1 < p < \infty$) has a distortable norm. Combining these results with Propositions 13.3, 13.5, 13.6 and Theorem 13.11 gives immediately.

Theorem 13.16. *Let E be an infinite-dimensional Banach space. Then E fails to have an infinite-dimensional subspace with a distortable norm if and only if every infinite-dimensional subspace of E has a further subspace isomorphic (or equivalently, almost isometric) to c_0 or l_1.*

Theorem 13.17. *Let E be an infinite-dimensional Banach space. Then every uniformly continuous (or equivalently, Lipschitz) real-valued function f on the unit sphere of E is oscillation stable if and only if every infinite-dimensional subspace of E has a further subspace which is isomorphic (or equivalently, almost isometric) to c_0.*

3. Uniformly Continuous Functions on the Unit Sphere of c_0

The purpose of this section is to present the proof of the following theorem.

Theorem 13.18. *Let g be a uniformly continuous function from the unit sphere of c_0 into $\mathbb{R}$. Then for every $\varepsilon > 0$ there is an infinite-dimensional subspace F of c_0 such that the oscillation of g on $S(F) = S(c_0) \cap F$ is at most ε.*

The proof of Theorem 13.18 uses methods of infinite combinatorics. At the heart of the proof is the following well-known theorem of Hindman as well as its higher-dimensional generalizations. Denote the set of positive integers by $\mathbb{N}$, and the set of all finite subsets of $\mathbb{N}$ by $\mathbb{N}^{<\omega}$.

Theorem 13.19. (i) *Assume that $\mathbb{N}$ is colored by a finite number of colors. Then there is an increasing sequence of integers $\{n_i\}_{i=1}^{\infty}$ such that the set of all of its finite sums $\sum_{i \in B} n_i$ is monochromatic.*

(ii) *Assume that $\mathbb{N}^{<\omega}$ is colored by a finite number of colors. Then there is an infinite sequence $\{A_i\}_{i=1}^{\infty}$ of disjoint elements of $\mathbb{N}^{<\omega}$ such that the set of all their finite unions $\bigcup_{i \in B} A_i$ is monochromatic.*

For the proof of Theorem 13.19 we need two lemmas. A semigroup $(S, +)$ is said to have right-continuous addition if $y \to y + x$ is a continuous function of y for every fixed $x \in S$. An idempotent in S is an element y which satisfies $y + y = y$.

Lemma 13.20. *Let $(S, +)$ be a compact Hausdorff semigroup, and assume that the addition in S is right-continuous. Then S contains an idempotent.*

Proof. Consider the family Σ of all compact subsemigroups of S, ordered by inclusion. By Zorn's lemma Σ has a minimal element S_0, and fix $y \in S_0$. Then $S_0 + y$ is a subsemigroup of S_0 which is compact by the right-continuity of the addition. By minimality $S_0 = S_0 + y$. The set $S_1 = \{x \in S_0 : x + y = y\}$ is a compact subsemigroup of S_0 (again by the right-continuity of the addition), and it is nonempty (because $S_0 + y = S_0$). By the minimality $S_1 = S_0$ and $y + y = y$. (Note that the minimality of S_0 actually gives that S_0 is the singleton $\{y\}$.) $\square$

We shall use the following notation. Let α be an ultrafilter on a set X, and let $P(x)$ be a proposition concerning the elements of X. We write $\Lambda_\alpha x[P(x)]$ to denote the statement $\{x \in X : P(x)\} \in \alpha$; i.e., that the set of x for which $P(x)$ holds belongs to the ultrafilter α.

Lemma 13.21. *Let $(S,+)$ be a semigroup, and let $\mathcal{U}(S) \subset 2^{2^S}$ be the set of ultrafilters on S. Then $\mathcal{U}(S)$ is compact and Hausdorff in the product topology, and the operation*

$$\alpha + \beta = \Big\{A \subset S : \ \Lambda_\alpha x\big[\Lambda_\beta y[x+y \in A]\big]\Big\}$$

makes $\mathcal{U}(S)$ into a semigroup with right-continuous addition.

Proof. An ultrafilter on S is a set of subsets of S, hence $\mathcal{U}(S) \subset 2^{2^S}$. It is a closed subset of 2^{2^S} because each of the conditions which define filters and ultrafilters can be expressed as in intersection of conditions which involve at most two subsets of S. For example, the maximality property of an ultrafilter is just the statement that for every $A \subset S$ either A or $S \setminus A$ belong to the ultrafilter.

For the second part of the lemma we need to check that $\alpha+\beta$ is an ultrafilter on S and that the addition is right-continuous and associative. This is straightforward, and we only check here that $\alpha+\beta$ is maximal and that addition is right-continuous. It is convenient to use the notation $\tilde{A}(\beta) = \{x : \Lambda_\beta y[x+y \in A]\}$, and then $A \in \alpha+\beta$ for two ultrafilters α and β if and only if $\tilde{A}(\beta) \in \alpha$.

Since β is an ultrafilter, it follows that for every $A \subset S$ and every $x \in S$ either $\Lambda_\beta y[x+y \in A]$ or $\Lambda_\beta y[x+y \in S \setminus A]$. Hence $S \setminus \tilde{A}(\beta) = (S \setminus A)\tilde{\ }(\beta)$. Consequently, since α is also an ultrafilter, either $A \in \alpha + \beta$ or $(S \setminus A) \in \alpha + \beta$.

To check the right-continuity of the addition, observe that a map into a product space is continuous if each coordinate function is continuous. We thus need to show that for every fixed ultrafilter β and for every $A \in S$ the set of α such that $A \in \alpha+\beta$ is closed. But this is obvious because $A \in \alpha + \beta$ if and only if $\tilde{A}(\beta) \in \alpha$. □

Remark. Note that if α is determined by $s_1 \in S$ (i.e., α consists of all subsets of S containing s_1) and if β is determined by $s_2 \in S$, then $\alpha + \beta$ is the ultrafilter determined by $s_1 + s_2$.

Proof of Theorem 13.19. (i) By Lemmas 13.20 and 13.21 there is an ultrafilter α on the semigroup $\mathbb{N}$ of positive integers such that $\alpha+\alpha = \alpha$. By the remark above this ultrafilter is free. For every set $A \in \alpha$ put $\tilde{A} = \{n : A - n \in \alpha\}$, where $A - n = \{m - n : m \in A \text{ and } m > n\}$. Since $\alpha = \alpha + \alpha$ we obtain that also $\tilde{A} \in \alpha$, and hence $A \cap \tilde{A} \in \alpha$.

Assume now that $\mathbb{N}$ is colored by a finite number of colors. Then there is a unique color, say red, such that the set A_0 of red elements belongs to α.

Pick $n_1 \in A_0 \cap \tilde{A}_0$, and let $A_1 = A_0 \cap (A_0 - n_1)$. Then $A_1 \in \alpha$, and we pick $n_2 \in A_1 \cap \tilde{A}_1$ with $n_2 > n_1$. Continue inductively, and construct a decreasing sequence of subsets A_i of A_0 (all belonging to α) and an increasing sequence $\{n_i\}$ by putting

$$A_{i+1} = A_i \cap (A_i - n_{i+1})$$

and then choosing $n_{i+2} \in A_{i+1} \cap \tilde{A}_{i+1}$ with $n_{i+2} > n_{i+1}$. It follows from the construction that all the finite sums of the elements $\{n_i\}_{i=1}^\infty$ are colored red.

(ii) We shall deduce (ii) from a strengthened version of (i): The sequence $\{n_i\}_{i=1}^\infty$ in (i) can be chosen to have the additional property that n_{k+1} is divisible by 2^{n_k+1}. Indeed, assume that $\{m_j\}_{j=1}^\infty$ satisfies (i), and we shall find mutually disjoint finite sets B_i such that the sequence $n_i = \sum_{j \in B_i} m_j$, which clearly satisfies (i), also satisfies the additional claimed property.

The choice of the n_i's is inductive. Take $n_1 = m_1$, and suppose that n_k has already been chosen. By the pigeonhole principle the set $\{m_j : j \notin \bigcup_{i\leq k} B_i\}$ contains infinitely many m_j's which are congruent to each other modulo 2^{n_k+1}. We now take n_{k+1} to be the sum of 2^{n_k+1} such m_j's.

Let $\mathbb{N}_0 = \{0\} \cup \mathbb{N}$, and for each $n \in \mathbb{N}$ let $\varphi(n) \in \mathbb{N}_0^{<\omega}$ be the set of exponents of 2 which appear in the binary expansion of n. Then the map $\varphi : \mathbb{N} \to \mathbb{N}_0^{<\omega}$ is injective and surjective, and any finite coloring of $\mathbb{N}_0^{<\omega}$ corresponds via φ to a finite coloring of $\mathbb{N}$.

Fix a finite coloring of $\mathbb{N}_0^{<\omega}$, and let $\{n_i\}_{i=1}^{\infty}$ be a sequence in $\mathbb{N}$ which satisfies (i) for the corresponding coloring of $\mathbb{N}$, and which also satisfies the additional property above. This property implies that if $A_i = \varphi(n_i)$, then $\max A_i < \min A_{i+1}$ for every i (i.e., the maximal element in A_i is smaller than the minimal element in A_{i+1}). It also implies that for any $j_1 < j_2, \ldots, < j_k$

$$\varphi(\sum_{i=1}^{k} n_{j_i}) = \bigcup_{i=1}^{k} A_{j_i}.$$

Thus $\{A_i\}_{i=1}^{\infty}$ satisfies (ii). $\diamondsuit$

For the proof of Theorem 13.18 we need non-obvious k-dimensional versions of Theorem 13.19. They will be proved by induction on k, and the case $k = 1$ reduces to Theorem 13.19. The tools used in the induction step are essentially the same as those used in the proof of Theorem 13.19, but we separated the case $k = 1$ from the induction step in order to first familiarize the reader with these tools in the simpler setting.

We shall use the following notation. For $k \geq 1$ let $T : \mathbb{N}_0^k \to \mathbb{N}_0^k$ be the shift

$$T(n_1, n_2 \ldots, n_k) = (0, n_1, n_2, \ldots, n_{k-1}).$$

Put $X_k = \mathbb{N}_0^k \setminus T\mathbb{N}_0^k = \{\mathbf{n} = (n_1, n_2, \ldots, n_k) : n_1 \neq 0\}$. Given a subset $A = \{\mathbf{n}_i\}_{i\in I}$ in X_k we define the subspace generated by A to be the set of all elements of $\mathbb{N}_0^k$ of the form

$$\sum_{j=1}^{k} \sum_{i\in B_j} T^{j-1}\mathbf{n_i}$$

where the B_j's are finite disjoint subsets of I with $B_1 \neq \emptyset$, and the summation is the usual vector addition. The assumption that B_1 is nonempty guarantees that for every $A \subset X_k$ the subspace generated by A is also contained in X_k. The terminology "subspace generated by A" stems from the eventual application of this concept for producing discrete nets in actual subspaces of c_0. Note that $X_1 = \mathbb{N}$, and for every subset A of $\mathbb{N}$, the subspace generated by A is exactly the set of all finite sums of elements of A as in Theorem 13.19(i). Thus this theorem is the case $k = 1$ of the following k-dimensional generalization.

Proposition 13.22. *For any $k \geq 1$ and for any finite coloring of X_k there is an infinite subset of X_k which generates a monochromatic subspace.*

For $k > 1$ we shall need a new result on ultrafilters on X_k, and first some more notations and definitions. The shift map $\sigma : \mathcal{U}(X_k) \to \mathcal{U}(X_{k-1})$ (where $k \geq 2$) is defined by

$$\sigma(\alpha) = \{A \subset X_{k-1} : \Lambda_\alpha \mathbf{n}[T\mathbf{n} \in A]\}$$

for $\alpha \in \mathcal{U}(X_k)$. (We use here the obvious identification between TX_k and X_{k-1}.) In the sequel we shall similarly identify $T^{k-j}X_k$ with X_j for $1 \le j < k$. The map σ is continuous and surjective.

Using this identification of X_j with $T^{k-j}X_k$, we define the addition of an element $\mathbf{m} \in X_j$ with an element $\mathbf{n} \in X_k$ for $j < k$ by the natural formula

$$\mathbf{n} + \mathbf{m} = \mathbf{m} + \mathbf{n} = \big(n_1, n_2, \ldots, n_{k-j}, n_{k-j+1} + m_1, \ldots, n_k + m_j\big) \in X_k.$$

There is a corresponding addition of ultrafilters from $\mathcal{U}(X_k) \times \mathcal{U}(X_j)$ into $\mathcal{U}(X_k)$ and from $\mathcal{U}(X_j) \times \mathcal{U}(X_k)$ into $\mathcal{U}(X_k)$ respectively. If $\alpha \in \mathcal{U}(X_k)$ and $\beta \in \mathcal{U}(X_j)$, we define (with the x's taken in X_k and $y \in X_j$)

$$\alpha + \beta = \big\{A \subset X_k : \ \Lambda_\alpha x\big[\Lambda_\beta y[x + y \in A]\big]\big\}$$
$$\beta + \alpha = \big\{A \subset X_k : \ \Lambda_\beta y\big[\Lambda_\alpha x[x + y \in A]\big]\big\}.$$

It is straightforward to check that these additions of ultrafilters are right-continuous. Also if $\alpha \in \mathcal{U}(X_k)$ and $\beta \in \mathcal{U}(X_j)$, then $\sigma(\alpha + \beta) = \sigma(\alpha) + \sigma(\beta)$ for $2 \le j < k$ and $\sigma(\alpha + \beta) = \sigma(\beta + \alpha) = \sigma(\alpha)$ for $j = 1$ and $k > 1$.

With these notations we have

Lemma 13.23. *For every integer k there is an ultrfilter $\alpha \in \mathcal{U}(X_k)$ such that*

$$\sigma^j(\alpha) + \alpha = \alpha + \sigma^j(\alpha) = \alpha \text{ for all } 0 \le j \le k-1.$$

Proof. For $k = 1$ this follows (as observed already) from Lemmas 13.20 and 13.21. Assume that $k \ge 2$ and that the lemma holds for $k-1$, and fix $\beta \in \mathcal{U}(X_{k-1})$ such that $\sigma^j(\beta) + \beta = \beta + \sigma^j(\beta) = \beta$ for $0 \le j \le k-2$.

The set $S = \{\gamma : \sigma(\gamma) = \beta\}$ is a nonempty compact subset of $\mathcal{U}(X_k)$ because σ is continuous and surjective, and since $\mathcal{U}(X_k)$ is compact. The set $S + \beta \subset \mathcal{U}(X_k)$ is also compact (by the right-continuity of addition), and it is a semigroup because if $\sigma(\gamma_1) = \sigma(\gamma_2) = \beta$, then $\gamma_1 + \beta + \gamma_2 \in S$, indeed

$$\sigma(\gamma_1 + \beta + \gamma_2) = \sigma(\gamma_1) + \sigma(\beta) + \sigma(\gamma_2) = \beta + \sigma(\beta) + \beta = \beta + \beta = \beta.$$

(For $k = 2$ this does not formally make sense since $\sigma(\beta)$ is not defined, but in this case $\sigma(\gamma_1 + \beta + \gamma_2) = \sigma(\gamma_1 + \gamma_2) = \beta + \beta = \beta$.)

By Lemma 13.20 there is a $\gamma \in S$ such that $(\gamma + \beta) + (\gamma + \beta) = \gamma + \beta$, and we claim that $\alpha = \beta + \gamma + \beta$ has the desired property for k. Indeed, if $1 \le j \le k-1$, then $\alpha + \sigma^j(\alpha) = \beta + \gamma + \beta + \sigma^j(\beta) + \sigma^{j-1}(\beta) + \sigma^j(\beta) = \beta + \gamma + \beta = \alpha$. (As above, if $j = k-1$, the undefined terms $\sigma^j(\beta)$ simply do not appear.) Similarly $\sigma^j(\alpha) + \alpha = \alpha$. For $j = 0$ we get

$$\begin{aligned}\alpha + \alpha &= \beta + \gamma + \beta + \beta + \gamma + \beta = \beta + \gamma + \beta + \gamma + \beta \\ &= \beta + (\gamma + \beta) + (\gamma + \beta) = \beta + \gamma + \beta = \alpha.\end{aligned}$$

□

Proof of Proposition 13.22. With Lemma 13.23 at our disposal, we can proceed exactly as in the proof of Theorem 13.19(i). Let $\langle \mathbf{n}_1, \mathbf{n}_2, \ldots, \mathbf{n}_j\rangle$ denote the subspace generated in X_k by the finite sequence $\{\mathbf{n_i}\}_{i=1}^j$ of distinct elements in X_k.

Let α be the ultrafilter on X_k given by Lemma 13.23. If X_k is colored by finitely many colors, then there is a monochromatic subset $A_0 \subset X_k$ which belongs to α. By the property of α

$$\Lambda_\alpha \mathbf{n}\big[\Lambda_\alpha \mathbf{m}[\langle \mathbf{n}, \mathbf{m}\rangle \subset A_0]\big].$$

Since α is an ultrafilter, any two sets in α intersect. We can thus pick $\mathbf{n}_1 \in A_0$ such that $\Lambda_\alpha \mathbf{m}\big[\langle \mathbf{n}_1, \mathbf{m}\rangle \subset A_0\big]\big]$, and then set

$$A_1 = A_0 \cap \big\{\mathbf{m} \neq \mathbf{n}_1 : \langle \mathbf{n}_1, \mathbf{m}\rangle \subset A_0\big\}.$$

Then $A_1 \in \alpha$, and we pick $\mathbf{n}_2 \in A_1$ such that $\Lambda_a \mathbf{m}\big[\langle \mathbf{n}_2, \mathbf{m}\rangle \subset A_1\big]\big]$, which is again possible by the choice of α. Consequently $\Lambda_a \mathbf{m}\big[\langle \mathbf{n}_1, \mathbf{n}_2, \mathbf{m}\rangle \subset A_0\big]\big]$.

Continuing inductively, we obtain a sequence of distinct elements $\{\mathbf{n}_i\}_{i=1}^\infty$ which generates a monochromatic subspace. $\diamondsuit$

To formulate the k-dimensional version of part (ii) of Theorem 13.19 we need some more notations. For $k \geq 1$ let Y_k be the set of all finitely supported functions $f : \mathbb{N}_0 \to \{0, 1, \dots, k\}$ which take the value k at least once. Define an operator $T : Y_k \to Y_{k-1}$ by $Tf(n) = (f(n) - 1) \vee 0$.

Let $\{f_i\}_{i=1}^\infty$ be a sequence in Y_k such that $\max\operatorname{supp} f_i < \min\operatorname{supp} f_{i+1}$ for every i. The set of all functions in Y_k which are finite sums of the form $f = \sum_i T^{r_i} f_i$, where $r_i = 0$ for at least one i (to ensure that f takes the value k at least once) is called the subspace generated by the f_i's. For $k = 1$ the set Y_1 is naturally identified with $\mathbb{N}_0^{<\omega}$, and the subspace generated by $\{f_i\}_{i=1}^\infty$ is then the set of all finite unions of the sets corresponding to the f_i's.

Proposition 13.24. *Let k be an integer, and color Y_k with finitely many colors. Then there is a sequence $\{f_i\}_{i=1}^\infty$ in Y_k which generates a monochromatic subspace.*

Proof. Let $\varphi : \mathbb{N} \to Y_1$ be the map which associates with n the characteristic function of the set of exponents of 2 appearing in the binary expansion of n. (We also put $\varphi(0)$ to be the function on $\mathbb{N}_0$ which vanishes identically, but note that it does not belong to Y_1.)

Define a map $\hat{\varphi} : X_k \to Y_k$ by

$$\hat{\varphi}(n_1, n_2, \dots, n_k)(m) = \max_{0 \leq j \leq k-1} (k - j)\varphi(n_{j+1})(m)$$

for $m \in \mathbb{N}_0$. Observe that since $n_1 \neq 0$, this indeed defines an element of Y_k. Any finite coloring of Y_k induces a finite coloring of X_k via $\hat{\varphi}$. By Proposition 13.22 there is a sequence $\{\mathbf{n}_i\}_{i=1}^\infty$ in X_k which generates a monochromatic subspace. The same argument as in the proof of Theorem 13.19(ii) shows that we may assume in addition that if $\mathbf{n}_i = (n_1^i, \dots, n_k^i)$, then for each i all the k sets $\{\varphi(n_j^i)\}_{j=1}^k$ precede all the k sets $\{\varphi(n_j^{i+1})\}_{j=1}^k$. The definitions of $\hat{\varphi}$ and of the notions of "subspace generated by" in X_k and Y_k respectively imply easily that the subspace generated by $\{\hat{\varphi}(\mathbf{n}_i)\}_{i=1}^\infty$ in Y_k is monochromatic. $\square$

We are now ready to prove Theorem 13.18 for functions $g : S(c_0) \to \mathbb{R}$ which are unconditional; i.e., such that $g(\lambda_1, \lambda_2, \dots) = g(|\lambda_1|, |\lambda_2|, \dots)$. For such functions we can estimate the oscillation of the restriction $g_{|S(c_0) \cap F}$ (for a subspace F which is also a sublattice) by considering only vectors in $PS(c_0)$, the positive part of the unit sphere of c_0.

Proof of Theorem 13.18 for an unconditional function g. Fix $\varepsilon > 0$. By the uniform continuity of g there is a $\delta > 0$ such that $\|x - y\| \leq \delta$ implies $|g(x) - g(y)| < \varepsilon/2$. Choose k with $(1+\delta)^{-(k-1)} < \delta$, and consider the set Δ of all finitely supported the functions $f : \mathbb{N} \to \{1, (1+\delta)^{-1}, (1+\delta)^{-2}, \dots, (1+\delta)^{-(k-1)}, 0\}$

which take the value 1 at least once. The set Δ is δ-dense in $PS(c_0)$, and the map ψ defined by

$$\psi(f)(n-1) = \begin{cases} k + \log_{1+\delta} f(n) & \text{if } f(n) \neq 0 \\ 0 & \text{if } f(n) = 0 \end{cases}$$

for $n \geq 1$ is a one-one correspondence between Δ and Y_k.

Since g is bounded, its range is contained in a finite union $\bigcup_{j=1}^{r} I_j$, where the I_j's are intervals of length bounded by $\varepsilon/2$. Color Δ with r colors by assigning to $f \in \Delta$ an index j such that $g(f) \in I_j$. This coloring of Δ induces via ψ a coloring of Y_k. By Proposition 13.24 there is a sequence $\{h_i\}_{i=1}^{\infty}$ in Y_k which generates a monochromatic subspace. The sequence $\{\psi^{-1}h_i\}_{i=1}^{\infty}$ is a (positive) block basis of the unit vector basis of c_0, and we put $F = \overline{\text{span}}\{\psi^{-1}h_i\}_{i=1}^{\infty}$. Then F is a sublattice of c_0, and by their choice the $\psi^{-1}h_i$'s generate a monochromatic δ-dense set in $PS(F)$. By the definition of the coloring this gives that the oscillation of g on this δ-dense set is at most $\varepsilon/2$. By the choice of δ the oscillation of g on all of $PS(F)$ is at most ε. $\diamondsuit$

The natural approach to extend this proof to general uniformly continuous functions on $S(c_0)$ would be the following: Consider Z_k, the set of all finitely supported functions $f : \mathbb{N} \to \{-k, -(k-1), \dots, 0, \dots, k\}$ which take either the value $+k$ or the value $-k$ at least once. Define the operator T on Z_k by

$$Tf(n) = \big((|f(n)| - 1) \vee 0\big)\text{sign} f(n) \tag{13.10}$$

and if $\{f_i\}_{i=1}^{\infty}$ is a sequence of disjointly supported functions in Z_k, define the subspace generated by them to be the set of all the functions f of the form

$$f = \sum_{j=1}^{k} \sum_{i \in A_j} T^{k-j} f_i - \sum_{j=1}^{k} \sum_{i \in B_j} T^{k-j} f_i$$

where the finite sets $\{A_j\}_{j=1}^{k}$ and $\{B_j\}_{j=1}^{\infty}$ are all mutually disjoint, and either A_k or B_k (or both) are nonempty. Observe that any such f is in Z_k.

In analogy to Proposition 13.24 we would like to find, for any finite coloring of Z_k, a sequence $\{f_i\}_{i=1}^{\infty}$ in Z_k which generates a monochromatic subspace. This, however, is false: Color each $f \in Z_k$ by the sign of its first nonzero coordinate, then f and $-f$ will always have a different color. The assertion does not hold even for colorings which give f and $-f$ the same color. For example, color f red if its first and last nonzero coordinate have the same sign and blue otherwise, then for any $f_1, f_2 \in Z_k$ with $\max\text{supp} f_1 < \min\text{supp} f_2$ the two functions $f_1 + f_2$ and $f_1 - f_2$ will have different colors.

Thus the proof of Theorem 13.18 for unconditional functions does not generalize directly. There is, however, an approximate generalization of Proposition 13.24 which suffices for our purpose. For a subset A of Z_k put

$$\hat{A} = \Big\{ f \in Z_k : \text{ there is an } h \in A \text{ with } \|f - h\|_\infty \leq 1 \Big\}.$$

Proposition 13.25. *Let k be an integer, and color Z_k with finitely many colors. Then there is a monochromatic set A and a sequence $\{f_i\}_{i=1}^{\infty}$ of disjointly supported elements of Z_k such that the subspace W generated by $\{f_i\}_{i=1}^{\infty}$ satisfies $W \subset \hat{A}$.*

Observe that Proposition 13.25 suffices to complete the proof of Theorem 13.18. Indeed, the argument used in the proof for unconditional g (and using the same notation) will now yield a block basis $\{h_i\}_{i=1}^{\infty}$ of the unit vector basis of c_0, a δ-dense set in $S(F)$ (where $F = \overline{\text{span}}\{h_i\}_{i=1}^{\infty}$) and a color j such that for every h in this δ-dense set there is an $h' \in \Delta$ with $\|h' - h\|_{c_0} \leq \delta$ and with h' colored by j. Thus the oscillation of g on $S(F)$ is at most $3\varepsilon/2$.

The remainder of this section is devoted to the proof of Proposition 13.25.

Denote by $\mathcal{V}(Z_k)$ the set of filters α on Z_k which satisfy the following conditions:

(i) α is cofinite; i.e., for every integer n it contains the set of all the elements in Z_k supported in (n, ∞).

(ii) If $A \in \alpha$, then $-A \in \alpha$, where $-A = \{-f : f \in A\}$.

(iii) If Z_k is represented as a finite union $Z_k = \bigcup_{i=1}^{n} A_i$, then there is an i such that $\hat{A}_i \in \alpha$.

Observe that we do not require that the filters in $\mathcal{V}(Z_k)$ are ultrafilters, but condition (iii) says that they are "approximately maximal": The same condition (even for $n = 2$ only) with $\hat{A}_i$ replaced by A_i is equivalent to the requirement that α is an ultrafilter. Note also that $\mathcal{V}(Z_k)$ is a compact Hausdorff space under the product topology, because it is closed in $2^{2^{Z_k}}$. (That $\mathcal{V}(Z_k)$ is nonempty will be proved in Lemma 13.26 below).

We next extend the operation of addition (which was defined for ultrafilters before Lemma 13.23) to $\mathcal{V}(Z_k)$. For $\alpha \in \mathcal{V}(Z_j)$ and $\beta \in \mathcal{V}(Z_k)$ put

$$\alpha + \beta = \Big\{ A \in Z_{j \vee k} : \Lambda_\alpha f \big[\Lambda_\beta h [\operatorname{supp} f \cap \operatorname{supp} h = \emptyset \text{ and } f + h \in A] \big] \Big\}.$$

Since $\Lambda_\alpha f \big[\Lambda_\beta h [\operatorname{supp} f \cap \operatorname{supp} h = \emptyset] \big]$ holds for any two cofinite filters, we shall omit writing this in the definition of $\alpha + \beta$. Moreover, we shall always assume implicitly that the f and h appearing in the formula satisfy

$$\max \operatorname{supp} f < \min \operatorname{supp} h.$$

We now verify that $\alpha + \beta \in \mathcal{V}(Z_{j \vee k})$. Conditions (i) and (ii) are obvious. To check (iii) assume that $Z_{j \vee k} = \bigcup_{i=1}^{n} A_i$, and for every $f \in Z_j$ and $1 \leq i \leq n$ put $A_i^f = \{h \in Z_k : f + h \in A_i\}$. Then $\bigcup_{i=1}^{n} A_i^f = Z_k$ for every $f \in Z_j$, and there is at least one $1 \leq i \leq n$ such that $\hat{A}_i^f \in \beta$. Putting $B_i = \{f \in Z_j : \hat{A}_i^f \in \beta\}$, this means that $\bigcup_{i=1}^{n} B_i = Z_j$, and there is an i_0 such that $\hat{B}_{i_0} \in \alpha$. Note that if f, h, f', h' are such that $\max \operatorname{supp} f < \min \operatorname{supp} h$ and $\max \operatorname{supp} f' < \min \operatorname{supp} h'$, and if $\|f - f'\|_\infty \leq 1$ and $\|h - h'\|_\infty \leq 1$, then $\|(f + h) - (f' + h')\|_\infty \leq 1$. Consequently $\Lambda_\alpha f \big[\Lambda_\beta h [f + h \in \hat{A}_{i_0}] \big]$; i.e., $\hat{A}_{i_0} \in \alpha + \beta$.

It is easy to verify that for every fixed $\beta \in \mathcal{V}(Z_k)$ the map $\alpha \to \alpha + \beta$ from $\mathcal{V}(Z_j)$ to $\mathcal{V}(Z_{j \vee k})$ is continuous.

For every $\alpha \in \mathcal{V}(Z_k)$ put $\sigma(\alpha) = \{A \subset Z_{k-1} : \Lambda_\alpha f[Tf \in A]\}$ (where T is defined by (13.10)), and we check that $\sigma(\alpha) \in \mathcal{V}(Z_{k-1})$. As above, we only need to check condition (iii), so assume that $Z_{k-1} = \bigcup_{i=1}^{n} A_i$. Then $Z_k = \bigcup_{i=1}^{n} T^{-1} A_i$ and there is an i_0 with $(T^{-1} A_{i_0})\hat{} \in \alpha$. Since $\|Tf - Tf'\|_\infty \leq \|f - f'\|_\infty$, it follows that if $Tf' \in A_i$ and $\|f - f'\|_\infty \leq 1$, then $Tf \in \hat{A}_i$, and consequently $\hat{A}_{i_0} \in \sigma(\alpha)$.

It is easy to check that the map $\sigma : \mathcal{V}(Z_k) \to \mathcal{V}(Z_{k-1})$ is continuous and that $\sigma(\alpha + \beta) = \sigma(\alpha) + \sigma(\beta)$. Note, however, that unlike the analogous definition in $\mathcal{U}(X_k)$ we do not claim here that σ is surjective.

The proof of Proposition 13.25 requires the following lemma which resembles Lemma 13.23 but has a somewhat different proof.

Lemma 13.26. *For every integer $k \geq 1$ there is a filter $\alpha \in \mathcal{V}(Z_k)$ such that*

$$\sigma^j(\alpha) + \alpha = \alpha + \sigma^j(\alpha) = \alpha \text{ for all } 0 \leq j \leq k-1.$$

Proof. The first task is to prove that $\mathcal{V}(Z_k)$ is nonempty. Let γ be any cofinite ultrafilter on Z_k, and let β be the filter

$$(-1)^{k-1}\sigma^{k-1}(\gamma) + (-1)^{k-2}\sigma^{k-2}(\gamma) + \cdots - \sigma(\gamma) + \gamma - \sigma(\gamma) + \ldots (-1)^{k-1}\sigma^{k-1}(\gamma).$$

(If α is a filter on some Z_j, then $-\alpha = \{-A : A \in \alpha\}$.) Then β is also a cofinite ultrafilter on Z_k, and it is easy to check that if $A \in \beta$, then a suitable translate of A differs from an element of $-\beta$ by a vector whose $\|\cdot\|_\infty$-norm is at most one. Thus $A \in \beta$ implies that $-\hat{A} \in \beta$. Let now β_0 be the filter consisting of all sets A such that both A and $-A$ belong to β. Then the filter β_0 satisfies (i) and (ii), and it also satisfies (iii) because β is an ultrafilter.

We now proceed with the proof of the lemma. By Lemma 13.20 the nonempty compact semigroup $\sigma^{k-1}(\mathcal{V}(Z_k))$ contains an element $\alpha_1 \in \sigma^{k-1}(\mathcal{V}(Z_k))$ which satisfies $\alpha_1 + \alpha_1 = \alpha_1$. Inductively, fix $1 \leq j < k-1$ and assume that there is an $\alpha_j \in \sigma^{k-j}(\mathcal{V}(Z_k))$ satisfying

$$\sigma^i(\alpha_j) + \alpha_j = \alpha_j + \sigma^i(\alpha_j) = \alpha_j \text{ for all } 0 \leq i \leq j-1.$$

The compact set

$$\Big\{\beta : \beta \in \sigma^{k-j-1}(\mathcal{V}(Z_k)) \text{ such that } \sigma(\beta) = \alpha_j\Big\} + \alpha_j$$

is a subsemigroup. Indeed, if $\beta', \beta'' \in \sigma^{k-j-1}(\mathcal{V}(Z_k))$ satisfy $\sigma(\beta') = \sigma(\beta'') = \alpha_j$, then $\beta' + \alpha_j + \beta'' \in \sigma^{k-j-1}(\mathcal{V}(Z_k))$ and

$$\sigma(\beta' + \alpha_j + \beta'') = \sigma(\beta') + \sigma(\alpha_j) + \sigma(\beta'') = \alpha_j + \sigma(\alpha_j) + \alpha_j = \alpha_j.$$

(If $j = 1$, then $\sigma(\alpha_1) = 0$ because $Tf = 0$ for all $f \in Z_1$.) By Lemma 13.20 there is a $\gamma \in \sigma^{k-j-1}(\mathcal{V}(Z_k))$ satisfying $\sigma(\gamma) = \alpha_j$ and $\gamma + \alpha_j + \gamma + \alpha_j = \gamma + \alpha_j$. Put

$$\alpha_{j+1} = \alpha_j + \gamma + \alpha_j \in \sigma^{k-j-1}(\mathcal{V}(Z_k))$$

and one easily verifies that α_{j+1} satisfies the requirements of the lemma for all $i \leq j$. The filter $\alpha = \alpha_k$ has the desired property. □

Proof of Proposition 13.25. Let α be a filter in $\mathcal{V}(Z_k)$ as in Lemma 13.26. For every $A \in \alpha$ and every $0 \leq j \leq k$

$$\Lambda_\alpha f\big[\Lambda_\alpha h[f + T^j h \in A]\big] \text{ and } \Lambda_\alpha f\big[\Lambda_\alpha h[T^j f + h \in A]\big].$$

By condition (ii) in the definition of $\mathcal{V}(Z_k)$ we also have

$$\Lambda_\alpha f\big[\Lambda_\alpha h[\pm f \pm T^j h \in A]\big] \text{ and } \Lambda_\alpha f\big[\Lambda_\alpha h[\pm T^j f \pm h \in A]\big].$$

In other words, $A \in \alpha$ implies that $\Lambda_\alpha f\big[\Lambda_\alpha h[\langle f, h\rangle \subset A]\big]$, where $\langle f, h\rangle$ denotes the subspace generated by f and h.

Assume that Z_k is colored by finitely many colors. Then there is a monochromatic set $A \in Z_k$ such that $A_1 = \hat{A} \in \alpha$. The argument above shows that this implies that $\Lambda_\alpha f\big[\Lambda_\alpha h[\langle f, h\rangle \subset A_1]\big]$. Choose f_1 such that $\Lambda_\alpha h[\langle f_1, h\rangle \subset A_1]$, and put $A_2 = \{h \in A_1 : \langle f_1, h\rangle \subset A_1\}$. Then $A_2 \subset A_1$ and $A_2 \in \alpha$, and we continue inductively and choose sets $\{A_n\}_{n=1}^\infty$ in α and elements $f_n \in A_n$ such that

$\Lambda_\alpha f\big[\Lambda_\alpha h[\langle f,h\rangle \subset A_n]\big]$ and $\Lambda_\alpha h[\langle f_n,h\rangle \subset A_n]$. The set $A_{n+1} \subset A_n$ is defined by $A_{n+1} = \{h \in A_n : \langle f_n, h\rangle \subset A_n\}$.

Since $f_n \in A_n$, it follows from the definition of f_{n-1} that $\langle f_{n-1}, f_n\rangle \subset A_{n-1}$. Thus $\langle f_{n-2}, f\rangle \subset A_{n-2}$ for every $f \in \langle f_{n-1}, f_n\rangle$, and $\langle f_{n-2}, f_{n-1}, f_n\rangle \subset A_{n-2}$. Continuing inductively we deduce that $\langle f_1, f_2, \ldots, f_n\rangle \subset A_1 = \hat{A}$ for every n. $\diamondsuit$

As observed earlier, this also concludes the proof of Theorem 13.18.

4. Asymptotic Biorthogonal Systems

In Section 1 we observed that the existence of a distorted norm on E (or of a uniformly continuous function on $S(E)$ which is not oscillation stable) is equivalent to the existence of two asymptotic sets in $S(E)$ which are at a positive distance from each other. In this section we study the properties and existence of a richer system of asymptotic sets.

Definition 13.27. *A Banach space E is said to admit an asymptotic biorthogonal system with constant δ (where $0 < \delta < 1/2$) if there are two sequences $\{A_n\}_{n=1}^\infty$ and $\{A_n^*\}_{n=1}^\infty$ of subsets of the unit sphere $S(E)$ of E and of the unit ball $B(E^*)$ of E^* respectively such that*

(i) Each A_n is an asymptotic set (i.e., $\operatorname{dist}(A_n, F) = 0$ for every infinite-dimensional subspace F of E).

(ii) For every n and for every $x \in A_n$ there is an $x^* \in A_n^*$ with $\langle x^*, x\rangle > 1 - \delta$.

(iii) If $n \neq m$, then $|\langle x^*, x\rangle| < \delta$ for every $x \in A_n$ and $x^* \in A_m^*$.

We start with a simple observation.

Proposition 13.28. *If a Banach space E admits an asymptotic biorthogonal system for every $0 < \delta < 1/2$, then E is arbitrarily distortable.*

Proof. We only use the existence, for every $\delta > 0$, of sets A_1, A_2 and A_1^* as above. Indeed, fix $\varepsilon > 0$ and define an equivalent norm on E by

$$|||x||| = \varepsilon\|x\| + \sup\big\{|\langle x^*, x\rangle| : \ x^* \in A_1^*\big\}.$$

Then $|||y||| \geq \varepsilon + 1 - \delta$ for every $y \in A_1$ and $|||z||| \leq \varepsilon + \delta$ for $z \in A_2$. Since every infinite-dimensional subspace F of E contains points which are arbitrarily close to such y and z, the new norm is $(1+\varepsilon-\delta)/(\varepsilon+\delta)$-distorted, and ε and δ are arbitrary. $\square$

Theorem 13.29. *Let E be a separable Banach space which admits an asymptotic biorthogonal system with $\delta < 1/36$. Then there is an equivalent norm on E such that the unconditional constant (with respect to the new norm) of every basic sequence in E is at least $(36\delta)^{-1/2}$.*

Proof. Since E is separable, we may assume that all the sets A_n^* are countable. Then $A^* = \bigcup A_n^*$ is also countable, and there is a one-one map σ from the collection of all finite sequences of elements of A^* onto the set of positive integers.

Let $r = [\delta^{-1/2}]$. A sequence $\{z_i\}_{i=1}^r \subset E^*$ is called a special sequence of functionals if $z_1^* \in A_1^*$ and $z_{i+1}^* \in A^*_{\sigma(z_1^*, z_2^*, \ldots, z_i^*)}$ for every $1 \leq i \leq r-1$.

Let Γ_r be the set of all vectors of the form $z^* = \sum_{i=1}^r z_i^*$, where $\{z_i^*\}_{i=1}^r \subset E^*$ is a special sequence of functionals, and define an equivalent norm on E by

$$|||x||| = \max\Big(\|x\|, r\sup\big\{|\langle z^*, x\rangle| : \ z^* \in \Gamma_r\big\}\Big).$$

We shall show that if $\{x_j\}$ is any infinite basic sequence in E, then there is an arbitrarily small perturbation $\{z_i\}_{i=1}^r$ of a block basis of the x_j's such that $|\!|\!| \sum_{i=1}^r (-1)^i z_i |\!|\!| < 4r$ and $|\!|\!| \sum_{i=1}^r z_i |\!|\!| > r(r-1)$. By the choice of r and since $\delta < 1/36$ this will prove the theorem.

The z_i's are chosen as follows. Since A_1 is an asymptotic set, there is a $z_1 \in A_1$ which is close to some finite linear combination of the x_j's, say of $\{x_j\}_{j=1}^{n_1}$. Choose $z_1^* \in A_1^*$ such that $\langle z_1^*, z_1 \rangle > 1-\delta$, and let z_2 be any element of $A_{\sigma(z_1^*)}$ which is close to a finite linear combination of $\{x_j\}_{j=n_1+1}^{n_2}$ for some n_2. Next pick $z_2^* \in A_{\sigma(z_1^*)}$ such that $\langle z_2^*, z_2 \rangle > 1 - \delta$. Proceeding in an obvious manner, we obtain small perturbations $\{z_i\}_{i=1}^r$ of a block basis of the x_j's and vectors $z_i^* \in E^*$ such that

$$z_i \in A_{\sigma(z_1^*,\ldots,z_{i-1}^*)}; \quad z_i^* \in A^*_{\sigma(z_1^*,\ldots,z_{i-1}^*)}; \quad \langle z_i^*, z_i \rangle > 1-\delta$$

for all i. The injectivity of σ implies that $|\langle z_i^*, z_k \rangle| < \delta$ for $i \neq k$.

By the construction $\{z_i^*\}_{i=1}^r$ is a special sequence of functionals. Hence

$$|\!|\!| \sum_{i=1}^r z_i |\!|\!| \geq r \Big\langle \sum_{i=1}^r z_i^*, \sum_{i=1}^r z_i \Big\rangle > r\big(r(1-\delta) - r(r-1)\delta\big) \geq r(r-1)$$

by the choice of r.

Let now $\{w_i^*\}_{i=1}^r$ be any special sequence of functionals of length r. If there is an i such that $w_i^* = z_i^*$, denote the largest such i by k, otherwise put $k=0$. Write

$$\begin{aligned} &\Big|\sum_{i=1}^r (-1)^i \langle w_i^*, z_i \rangle\Big| \\ &\qquad = \Big|\sum_{i=1}^k (-1)^i \langle z_i^*, z_i \rangle\Big| + \big|\langle w_{k+1}^*, z_{k+1} \rangle\big| + \Big|\sum_{i=k+2}^r (-1)^i \langle w_i^*, z_i \rangle\Big| \end{aligned}$$

and we estimate each term separately. Since σ is injective, w_i^* and z_j^* belong to different A_n's whenever $i \neq j$ or $i = j > k+1$. Hence $\big|\sum_{i=k+2}^r (-1)^i \langle w_i^*, z_i \rangle\big| \leq r\delta$, and since $1 - \delta \leq \langle z_i^*, z_i \rangle \leq 1$ for $i \leq k$ we obtain

$$\Big|\sum_{i=1}^r (-1)^i \langle w_i^*, z_i \rangle\Big| \leq (1 + k\delta/2) + 1 + \delta r \leq 2(1+\delta r)$$

and consequently

$$\Big|\Big\langle \sum_{i=1}^r w_i^*, \sum_{i=1}^r (-1)^i z_i \Big\rangle\Big| \leq 2(1+\delta r) + r(r-1)\delta < 4$$

by the choice of r.

By the triangle inequality $\| \sum_{i=1}^r (-1)^i z_i \| \leq r$; hence $|\!|\!| \sum_{i=1}^r (-1)^i z_i |\!|\!| \leq 4r$ as required. $\square$

Remark. If E has a basis which is 1-unconditional under its original norm (and this will be the case below and in the next section), then Theorem 13.29 produces a norm on E which is $(36\delta)^{-1}$ distorted. It is simpler, however, to obtain a norm with a large distortion by using Proposition 13.28 above.

Our aim is to produce asymptotic biorthogonal systems in a Hilbert space. At present there is no reasonably explicit way to do so, and the strategy we follow is to first produce such a system in some other Banach space, which we construct for this purpose, and then transfer the sets to Hilbert space by a suitable map between their spheres.

The space that we construct for this purpose will be called the Schlumprecht space, and is denoted by $\mathcal{S}$. It is closely related to the Tsirelson space T which was defined and studied in Section 5 of Chapter 10, but the discussion here will be self-contained and will not depend on the results of Chapter 10.

Let $\mathcal{F}$ be the class of functions $f : [1, \infty) \to [1, \infty)$ which satisfy

(i) $f(1) = 1$ and $f(x) < x$ for $x > 1$.

(ii) f is strictly increasing, the right derivative $f'(1)$ of f at 1 is strictly positive, and $\lim_{x\to\infty} f(x) = \infty$.

(iii) $\lim_{x\to\infty} x^{-q} f(x) = 0$ for every $q > 0$.

(iv) The function $x/f(x)$ is concave and nondecreasing.

(v) $f(xy) \le f(x)f(y)$ for all x and y.

(vi) $f(x^2) \le 2f(x)$ for every x.

These requirements are obviously not independent of each other. We listed them all since each will be used at some point.

A simple example of a function in $\mathcal{F}$ is $f(x) = \log_2(x + 1)$. It will suffice for our purposes to work with this example only, but we prefer to work with general functions in $\mathcal{F}$ because this requires no additional effort, and it actually simplifies the notations at some points. (The class $\mathcal{F}$ is quite large, and contains, for example, $\sqrt{f}$ for every $f \in \mathcal{F}$, but we shall not use this fact.)

Let $\{e_n\}$ be a basis in some normed space, and let I be a finite subset of the integers. If $x = \sum a_n e_n$ we put $Ix = \sum_{n\in I} a_n e_n$.

For each $f \in \mathcal{F}$ we now define a Schlumprecht space $\mathcal{S}_f$. Let $\{e_n\}$ be the unit vector basis of c_0, and define inductively a sequence of norms $\|\cdot\|_k$ on the linear span of $\{e_n\}$ as follows. Let $x = \sum a_n e_n$. If $k = 0$ put $\|x\|_0 = \max |a_n|$, and then inductively

$$\|x\|_{k+1} = \max \left(\|x\|_0, \sup \left\{ f(m)^{-1} \sum_{j=1}^{m} \|I_j x\|_k \right\} \right)$$

where the supremum is taken over all integers m and all choices of intervals I_j of integers such that $I_1 \prec I_2 \prec \dots$ (i.e., $\max\{i : i \in I_j\} < \min\{i : i \in I_{j+1}\}$ for all j). Since $\|x\|_k \le \|x\|_{k+1} \le \sum |a_n|$, it follows that $\|x\| = \lim_{k\to\infty} \|x\|_k$ exists for every x in the linear span of $\{e_n\}$. The space $\mathcal{S}_f$ is the completion of c_0 with respect to this norm. (To simplify the notation we denote $\mathcal{S}_f$ simply by $\mathcal{S}$ when f is understood.) The vectors $\{e_n\}$ form a 1-subsymmetric basis for $\mathcal{S}$, and the norm in $\mathcal{S}$ satisfies

$$\|x\| = \max \left(\|x\|_0, \sup \left\{ f(m)^{-1} \sum_{j=1}^{m} \|I_j x\| \right\} \right) \tag{13.11}$$

where the supremum is again taken over all integers m and intervals I_j as above.

Remark. We shall use below the fact that when $\|x\| > \|x\|_0$, then the supremum in (13.11) can actually be taken over $m \ge 2$; i.e., over nontrivial decompositions. Indeed, fix any $\varepsilon < \|x\| - \|x\|_0$, and let k be minimal such that $\|x\|_k > \|x\| - \varepsilon$. It follows that $\|x\|_k > \|x\|_{k-1}$, and therefore the supremum in the definition of $\|x\|_k$

can actually be taken over nontrivial decompositions; hence there is an $l \geq 2$ and intervals $\{I'_j\}_{j=1}^{l}$ such that

$$\|x\|_k \leq f(l)^{-1} \sum_{j=1}^{l} \|I'_j x\|_{k-1} + \varepsilon \leq f(l)^{-1} \sum_{j=1}^{l} \|I'_j x\| + \varepsilon.$$

Thus $\|x\| \leq \|x\|_k + \varepsilon \leq \sup_{m \geq 2}\{f(m)^{-1} \sum_{j=1}^{m} \|I_j x\|\} + 2\varepsilon$, and the assertion follows because ε was arbitrary.

Note also that if I is an interval containing the support of x, then the the intervals $\{I_j\}$ can all be taken to be contained in I.

Let us just point out that (13.11) differs from the norm of the Tsirelson space T in that the factor $f(m)^{-1}$ here replaces the factor $1/2$ in T, and we do not require here that I_1 starts after the mth place. This latter requirement was needed in the definition of T to ensure that it is not isomorphic to l_1. This is achieved here by the factor $f(m)^{-1}$ which tends to 0 as $m \to \infty$ (see Corollary 13.31 below).

The rest of the section is devoted to the proof of the following theorem.

Theorem 13.30. *For every $f \in \mathcal{F}$ the space $\mathcal{S}_f$ admits asymptotic biorthogonal systems for every $\delta > 0$.*

The proof (or more precisely, the construction) will be accomplished by a sequence of lemmas, but we first point out the following corollary which shows that $\mathcal{S}_f$ shares the main property of T.

Corollary 13.31. *For every $f \in \mathcal{F}$ the space $\mathcal{S}_f$ is reflexive and does not have a subspace isomorphic to l_p for any $1 < p < \infty$.*

Proof. $\mathcal{S}$ does not contain a subspace isomorphic to c_0 or l_p for $1 < p < \infty$ because if $\{z_j\}$ is any infinite normalized block basic sequence of $\{e_n\}$, then (13.11) implies that $\|z_1 + \ldots + z_m\| \geq m/f(m)$ for every integer m. By property (iii) of functions in $\mathcal{F}$ this expression tends to infinity faster than $m^{1/p}$ for any $p > 1$.

That $\mathcal{S}$ does not contain a subspace isomorphic to l_1 follows from Proposition 13.6 because it is distortable by Theorem 13.30 and Proposition 13.28.

The reflexivity follows from Theorem E.3. □

A vector in $\mathcal{S}$ is called an l_{1+}^n-vector with constant λ (where n is an integer and $\lambda > 1$) if it can be written as $x = \sum_{j=1}^{n} x_j$ where the x_j's satisfy $x_1 \prec x_2 \ldots \prec x_n$ (i.e., the maximal index in the support of x_j precedes the minimal index in the support of x_{j+1}) and $0 < \|x_j\| \leq \lambda \|x\|/n$ for every j. If in addition $\|x\| = 1$, then x is called an l_{1+}^n-average with constant λ.

Lemma 13.32. *Fix $\lambda > 1$ and an integer n. Then every infinite-dimensional subspace F of $\mathcal{S}$ spanned by a block basis of the unit vector basis contains l_{1+}^n-averages with constant λ.*

Proof. Suppose that F contains no such averages; i.e., every normalized block basic sequence $\{x_j\}_{j=1}^{n} \subset F$ of the e_i's satisfies $\|\sum_{j=1}^{n} x_j\| \leq n/\lambda$. If $\{x_j\}_{j=1}^{n^2}$ is such a block basic sequence of length n^2, then dividing it to n blocks of n consecutive vectors and using the estimate twice gives $\|\sum_{j=1}^{n^2} x_j\| \leq n^2/\lambda^2$. By induction we see that $\|\sum_{j=1}^{n^k} x_j\| \leq n^k/\lambda^k$ for each k and for any normalized block basic sequence $\{x_j\}_{j=1}^{n^k} \subset F$. On the other hand $\|\sum_{j=1}^{n^k} x_j\| \geq n^k/f(n^k)$ by (13.11). But this is impossible for large enough k in view of property (iii) in the definition of $\mathcal{F}$. □

Lemma 13.33. *Let m and n be integers, and fix $\lambda > 1$. Then for every l_{1+}^n-vector $x \in \mathcal{S}$ with constant λ and every choice of m intervals $I_1 \prec \ldots \prec I_m$*

$$\sum_{k=1}^{m} \|I_k x\| \leq \lambda(1 + 2mn^{-1})\|x\|.$$

Proof. Assume that $x = \sum_{j=1}^{n} x_j$ with $\|x\| = n$ where $x_1 \prec \ldots \prec x_n$ and $\|x_j\| \leq \lambda$ for all j. For each k put $A_k = \{j : \operatorname{supp} x_j \subset I_k\}$ and $B_k = \{j : I_k x_j \neq 0\}$. Then $|B_k| \leq |A_k| + 2$ and thus

$$\|I_k x\| = \Big\| \sum_{j \in B_k} x_j \Big\| \leq \lambda(|A_k| + 2).$$

Since $\sum_k |A_k| \leq n$, we obtain

$$\sum_{k=1}^{m} \|I_k x\| \leq \sum_{k=1}^{m} \lambda(|A_k| + 2) \leq \lambda(n + 2m) = \lambda(1 + 2mn^{-1})\|x\|.$$

□

Let $\varphi(x) = \varphi_f(x) = f^{-1}(6x)$ (where f^{-1} is the inverse of the function f). A sequence of vectors $x_1 \prec x_2 \ldots \prec x_N$ in $\mathcal{S}$ is called a rapidly increasing sequence (RIS) of l_{1+}-averages of length N with constant $\lambda = 1 + \varepsilon$ if each x_k is an $l_{1+}^{n_k}$-averages with constant $1 + \varepsilon$ and

(i) $n_1 \geq (1 + \varepsilon)\varphi(N/\varepsilon)/\varepsilon f'(1)$.

(ii) $\varepsilon f(n_k) > 2|\operatorname{supp}(x_{k-1})|$ for all $1 < k \leq N$.

The point in the definition is that n_k is very large compared with the size of the support of x_{k-1}. Conditions (i) and (ii) are just the quantitative formulation of this property, suitable for the computations below. Note in particular that $\{n_k\}$ is an increasing sequence.

A functional $x^* \in \mathcal{S}^*$ is said to be an (M, f)-form if $x^* = \sum_{j=1}^{M} x_j^*$ where $x_1^* \prec \ldots \prec x_M^*$ and $\|x_j^*\| \leq 1/f(M)$ for all j. Note that $\|x^*\| \leq 1$ for every (M, f)-form by (13.11).

Lemma 13.34. *Let $\varepsilon > 0$, and let $\{x_i\}_{i=1}^N$ be an RIS in $\mathcal{S}$ with constant $1 + \varepsilon$ and $x = \sum x_i$. If $M \geq \varphi(N/\varepsilon)$ and x^* is an (M, f)-form, then $|\langle x^*, x\rangle| \leq 1 + 2\varepsilon$.*

Proof. Represent $x^* = \sum_{j=1}^{M} x_j^*$ as in the definition of an (M, f)-form, and let I_j be the minimal interval containing the support of x_j^*.

Since $\|x^*\| \leq 1$ and $\|x_i\| = 1$ (as an $l_{1+}^{n_i}$-average), we obtain for every i that

$$|\langle x^*, x_i\rangle| \leq 1. \tag{13.12}$$

We shall need two more estimates for $|\langle x^*, x_i\rangle|$. By the definition of an (M, f)-form and by Lemma 13.33 we obtain the estimate

$$\begin{aligned} |\langle x^*, x_i\rangle| &= \Big|\sum_{j=1}^{M} \langle x_j^*, I_j x_i\rangle\Big| \leq f(M)^{-1} \sum_{j=1}^{M} \|I_j x_i\| \\ &\leq (1 + \varepsilon)(1 + 2Mn_i^{-1})f(M)^{-1}. \end{aligned} \tag{13.13}$$

By (13.11) we obtain that $\sum_{j=1}^{M} \|I_j x_i\| \leq \sum_{k \in \operatorname{supp} x_i} |x_i(k)| \leq f(|\operatorname{supp} x_i|)$, and using this inequality in (13.13) instead of Lemma 13.33 gives the other estimate, namely

$$|\langle x^*, x_i\rangle| \le f\big(|\operatorname{supp} x_i|\big) f(M)^{-1}. \tag{13.14}$$

Let t be the largest integer such that $n_t \le M$. By (ii) in the definition of RIS and the monotonicity of f

$$f\big(|\operatorname{supp} x_{t-1}|\big) \le \big|\operatorname{supp} x_{t-1}\big| \le \varepsilon f(n_t)/2 \le \varepsilon f(M)/2.$$

It also follows that if $i < t$, then

$$\begin{aligned} f\big(|\operatorname{supp} x_i|\big) &\le \big|\operatorname{supp} x_i\big| \le \varepsilon f(n_{i+1})/2 \\ &\le \varepsilon f\big(|\operatorname{supp} x_{i+1}|\big)/2 \le \dots \\ &\le f\big(|\operatorname{supp} x_{t-1}|\big)(\varepsilon/2)^{t-1-i} \le f(M)(\varepsilon/2)^{t-i}. \end{aligned} \tag{13.15}$$

Writing

$$|\langle x^*, x\rangle| \le \sum_{i=1}^{N} |\langle x^*, x_i\rangle| = \sum_{i=1}^{t-1} |\langle x^*, x_i\rangle| + |\langle x^*, x_t\rangle| + \sum_{i=t+1}^{N} |\langle x^*, x_i\rangle|$$

we now estimate the first term by using (13.14) and (13.15), the second by (13.12) and the third by (13.13) and obtain

$$\begin{aligned} |\langle x^*, x\rangle| &\le f(M)^{-1} f(M) \sum_{k=1}^{t-1} (\varepsilon/2)^k + 1 + 3(1+\varepsilon) f(M)^{-1}(N-t-1) \\ &\le \varepsilon + 1 + \varepsilon. \end{aligned}$$

(In estimating the last term we used that $1 + 2Mn_i^{-1} \le 3$ for $i > t$ by the choice of t, and that $3(1+\varepsilon)f(M)^{-1}N \le \varepsilon$ because $f(M) \ge f\big(\varphi(N/\varepsilon)\big) \ge 6N/\varepsilon$.) □

For the next lemma we introduce the notion of the length $l(I)$ of an interval with respect to a given RIS $x_1 \prec \ldots \prec x_N$. By definition each x_i has a representation $x_i = \sum_{j=1}^{n_i} x_{i,j}$ for some $x_{i,1} \prec \ldots \prec x_{i,n_i}$. Denote the smallest (respectively largest) integer i such that $Ix_i \ne 0$ by i_I (respectively k_I). Let also r_I (respectively s_I) be the smallest (respectively largest) integer j such that $Ix_{i_I,j} \ne 0$ (respectively $Ix_{k_I,j} \ne 0$). Then $l(I)$ is defined as to be $k_I - i_I + s_I/n_{k_I} - r_I/n_{i_I}$ (if $\operatorname{supp} x_{i_I,r_I} \subset I$ we replace r_I in the formula by $r_I - 1$); it is the number of x_i's whose support is contained in I allowing fractional terms for the proportions of the "parts" of the two extremal x_i's whose supports intersect I. It follows from the definition that if $I_1 \prec \ldots \prec I_M$ and if $I \supset \bigcup_{m=1}^{M} I_m$, then $l(I) \ge \sum_{m=1}^{M} l(I_m)$.

Lemma 13.35. *Let $0 < \varepsilon \le 1$, and let $x_1 \prec \ldots \prec x_N$ be an RIS in $\mathcal{S}$ with constant $1+\varepsilon$. Then*

$$\Big\|\sum_{i=1}^{N} x_i\Big\| \le (1+2\varepsilon)N/f(N).$$

This is the crucial lemma in the proof of Theorem 13.30. If $x_1 \prec \ldots \prec x_N$ is any normalized block basis of the unit vectors in $\mathcal{S}$, then $\|\sum_{i=1}^{N} x_i\| \ge N/f(N)$ by (13.11). The lemma says that for RIS's this inequality is essentially an equality.

Proof. Let $g(t) = t$ for $0 < t < 1$ and $g(t) = t/f(t)$ for $t \ge 1$. Then the assumptions on $\mathcal{F}$ give that g is increasing and concave on $(0,\infty)$, $g(1) = 1$, and $g(ts) \ge g(t)g(s)$. Its right derivative at 1 satisfies $0 \le g'(1) = 1 - f'(1) \le 1$.

Put $x = \sum_{i=1}^{N} x_i$. We shall show that if I is an interval with $l(I) \geq (1+\varepsilon)/\varepsilon n_1$ (note that $(1+\varepsilon)/\varepsilon n_1 < 1$ by the choice of n_1), then

$$\|Ix\| \leq (1+2\varepsilon) g\big(l(I)\big) \tag{13.16}$$

Taking in (13.16) any interval I containing the support of x (hence $l(I) = N$) will give the desired result.

We shall use several times the following estimate which holds for any interval J and follows directly from the triangle inequality.

$$\|Jx\| \leq (1+\varepsilon)\big(l(J) + n_1^{-1}\big). \tag{13.17}$$

Note that if $l(I) \leq 1$, then $\|Ix\| \leq (1+2\varepsilon)l(I) = (1+2\varepsilon)g\big(l(I)\big)$ by (13.17) and since $l(I) \geq (1+\varepsilon)/\varepsilon n_1$. Thus, assuming that (13.16) does not hold, let I be a minimal interval for which (13.16) fails, and then $l(I) \geq 1$. Since the absolute value of each coordinate of x is bounded by 1, and since the failure of (13.16) implies that $\|Ix\| > 1$, it follows that $\|Ix\|_0 \leq 1 < \|Ix\|$. By the remark following (13.11) there are $M \geq 2$ proper subintervals $I_1 \prec \ldots \prec I_M$ of I such that

$$\|Ix\| = f(M)^{-1} \sum_{j=1}^{M} \|I_j x\|.$$

(The supremum in (13.11) is attained here because I is finite, and the supremum is therefore taken over a finite set.)

Put $A = \{j : l(I_j) \leq (1+\varepsilon)/\varepsilon n_1\}$, and let B be its complement. Put $l_j = l(I_j)$ and $k = |B|$. It follows from the minimality of I that $\|I_j x\| \leq (1+2\varepsilon)g(l_j)$ for every $j \in B$.

For each j choose $x_j^* \in S^*$ supported in I_j with $\|x_j^*\| = 1$ and $\langle x_j^*, I_j x\rangle = \|I_j x\|$. Then $x^* = f(M)^{-1} \sum_{j=1}^{M} x_j^*$ is an (M, f)-form and

$$\langle x^*, x\rangle = f(M)^{-1} \sum_{j=1}^{M} \|I_j x\| = \|Ix\|.$$

Since $\|Ix\| > (1+2\varepsilon)g\big(l(I)\big) \geq 1+2\varepsilon$, it follows from Lemma 13.34 that $M < \varphi(N/\varepsilon)$.

Assume first that $k = 0$; i.e., every j belongs to A. Then by (13.17)

$$\begin{aligned} \|Ix\| &= f(M)^{-1} \sum_{j=1}^{M} \|I_j x\| \leq (1+\varepsilon) f(M)^{-1} \sum_{j=1}^{M} (l_j + n_1^{-1}) \\ &\leq M f(M)^{-1} (1+\varepsilon)(1+2\varepsilon)/\varepsilon n_1. \end{aligned}$$

By (i) in the definition of an RIS

$$n_1 \geq (1+\varepsilon)\varphi(N/\varepsilon)/\varepsilon f'(1) \geq (1+\varepsilon)M/\varepsilon$$

and thus $\|Ix\| \leq f(M)^{-1}(1+2\varepsilon) < (1+2\varepsilon)$, contradicting the choice of I.

Assume now that $k > 0$. Since (13.16) holds for each I_j with $j \in B$, the concavity and monotonicity of g and $\sum_{j\in B} l_j \leq \sum_{j=1}^{M} l_j \leq l(I)$ give

$$\sum_{j\in B} \|I_j x\| \leq (1+2\varepsilon) \sum_{j\in B} g(l_j) \leq (1+2\varepsilon) k g\big(l(I)/k\big).$$

Consequently

$$\begin{aligned}\|Ix\| &= f(M)^{-1}\Big(\sum_{j\in A}\|I_jx\|+\sum_{j\in B}\|I_jx\|\Big)\\ &\le f(M)^{-1}\Big((M-k)(1+\varepsilon)(1+2\varepsilon)/\varepsilon n_1+k(1+2\varepsilon)g\big(l(I)/k\big)\Big).\end{aligned}$$

By the definition of g and by property (v) of functions in $\mathcal{F}$

$$f(M)^{-1}g\big(l(I)/k\big)=M^{-1}g(M)g\big(l(I)/k\big)\le M^{-1}g\big(Ml(I)/k\big)$$

and by using also the estimate above for n_1, we obtain

$$\|Ix\|\le(1+2\varepsilon)\Big((1-\frac{k}{M})f'(1)+\frac{k}{M}g\big(Ml(I)/k\big)\Big).$$

The concavity of g gives

$$\begin{aligned}g(\lambda) &\ge g(\lambda+1-t)-(1-t)g'(1)\ge tg(\lambda/t)+(1-t)\big(g(1)-g'(1)\big)\\ &= tg(\lambda/t)+(1-t)f'(1)\end{aligned}$$

for every $0<t\le 1$ and $\lambda\ge 1$. Taking $t=k/M$ we get $\|Ix\|\le(1+2\varepsilon)g\big(l(I)\big)$, which contradicts again the choice of I. □

Proof of Theorem 13.30. Fix $\delta>0$, and let $\{N_j\}_{j=1}^\infty$ be an increasing sequence of squares with $N_j\delta/2>f(N_j)>8/\delta$ for all j and $N_j>\varphi(2N_{j-1})$ for $j>1$.

For each k let A_k be the set of all norm-one vectors $x=\sum_{i=1}^{N_k}x_i$, where $\{x_i\}_{i=1}^{N_k}$ is a multiple of an RIS of length N_k and constant $1+\delta/2$. By (13.11) this multiple is at most $f(N_k)N_k^{-1}$. Let A_k^* be the set of all (N_k,f)-forms; i.e., norm-one functionals of the form $x^*=f(N_k)^{-1}\sum_{i=1}^{N_k}x_i^*$ where $x_1^*\prec\ldots\prec x_{N_k}^*$ and $\|x_i^*\|\le 1$ for every i.

By Lemma 13.32 the A_k's are asymptotic sets, and we shall show that together with the A_k^*'s they form an asymptotic biorthogonal system with constant δ.

Assume first that $j>k$, then $N_j>\varphi(2N_k)$. By Lemma 13.34 with $N=N_k$, $M=N_j$ and $\varepsilon=1/2$, and taking into account the normalization factor, we obtain that $|\langle x^*,x\rangle|<2f(N_k)/N_k<\delta$ for every $x^*\in A_j^*$ and $x\in A_k$.

Fix now $j<k$ and pick $x=\sum_{i=1}^{N_k}x_i\in A_k$ and $x^*\in A_j^*$. For any subset $B\subset\{1,\ldots,N_k\}$ Lemma 13.35 (with $\varepsilon=1/2$) and the normalization give

$$\|\sum_{i\in B}x_i\|\le 2\big(|B|/f(|B|)\big)\cdot\big(f(N_k)/N_k\big).$$

If $|B|=N_k^{1/2}$, it then follows from property (vi) in the definition of $\mathcal{F}$ that $\|\sum_{i\in B}x_i\|\le 4N_k^{-1/2}$. Decomposing $\{1,\ldots,N_k\}$ into $N_k^{1/2}$ consecutive blocks of length $N_k^{1/2}$ we can represent x as an $l_{1+}^{N_k^{1/2}}$-average with constant 4. Write $x^*=f(N_j)^{-1}\sum_{i=1}^{N_j}x_i^*$ and choose $I_1\prec\ldots\prec I_{N_j}$ such that x_i^* is supported in I_i. Using Lemma 13.33 with $m=N_j$, $n=N_k^{1/2}$ and $\lambda=4$ we deduce that

$$\begin{aligned}|\langle x^*,x\rangle| &= f(N_j)^{-1}\Big|\sum_{i=1}^{N_j}\langle x_i^*,I_ix\rangle\Big|\le f(N_j)^{-1}\sum_{i=1}^{N_j}\|I_ix\|\\ &\le 4f(N_j)^{-1}(1+2N_jN_k^{-1/2})\le 8f(N_j)^{-1}\le\delta.\end{aligned}$$

Finally, if $x=\sum_{i=1}^{N_k}x_i\in A_k$, pick for each i a norm-one functional $x_i^*\in\mathcal{S}^*$ with the same support as x_i and $\langle x_i^*,x_i\rangle=\|x_i\|$. Then $x^*=f(N_k)^{-1}\sum_{i=1}^{N_k}x_i^*\in A_k^*$. By the definition of A_k the norm $\|x_i\|$ does not depend on i, say it is equal to α. Lemma 13.35 then gives $1 = \|\sum_{i=1}^{N_k} x_i\| \le (1+\delta)\alpha N_k f(N_k)^{-1}$, or $\alpha \ge f(N_k)/N_k(1+\delta)$. Hence

$$\langle x^*, x\rangle = f(N_k)^{-1}\sum_{i=1}^{N_k}\|x_i\| = N_k f(N_k)^{-1}\alpha \ge 1/(1+\delta) \ge 1-\delta.$$

$\diamondsuit$

Remark. An obvious modification of the proof of the theorem yields the following stronger result: If $\delta_k \searrow 0$ then there are asymptotic sets A_k in $S(\mathcal{S})$ and sets A_k^* in $B(\mathcal{S}^*)$ so that for every $x \in A_k$ there is an $x^* \in A_k^*$ with $\langle x^*, x\rangle \ge 1-\delta_k$, while $|\langle x^*, x\rangle| \le \delta_{\min(k,j)}$ for $x \in A_k$ and $x^* \in A_j^*$ when $j \ne k$. Such a system is called an asymptotic biorthogonal system with vanishing constants.

5. Asymptotic Biorthogonal Systems in l_p

We now transfer the results obtained in the previous section for $\mathcal{S}$ to l_p for $1 < p < \infty$.

Let E be a Banach space with an unconditional basis, and we shall define a map ψ_E from the finitely supported vectors in l_1 to E. We start with an auxiliary function, the entropy function. It is defined for two finitely supported vectors u and x with coordinates x_i and u_i receptively by $e(u,x) = \sum |u_i|\log|x_i|$, where $0\cdot\log 0$ is understood as 0. This function has the property that for every finitely supported vector u there is a unique $x \in S(E)$ such that

(i) $e(u,x) = \max\{e(u,y) : y \in S(E)\} = \max\{e(u,y) : y \in B(E)\}$.

(ii) $\operatorname{supp} x = \operatorname{supp} u$.

(iii) $\operatorname{sign} x_i = \operatorname{sign} u_i$ for every $i \in \operatorname{supp} u$.

The compactness of $\{y \in S(E) : \operatorname{supp} y \subset \operatorname{supp} u\}$ implies existence of x. If $u_i = 0$ and $x_i \ne 0$, then $e(u,x) = -\infty$, i.e., (ii) holds. The unconditionality of the basis in E implies that x can be chosen to satisfy (iii). Uniqueness of x follows from the strict concavity of the logarithm which implies that if $|x| \ne |y|$, then $e\big(u, \frac{1}{2}(|x|+|y|)\big) > \frac{1}{2}e(u,|x|) + \frac{1}{2}e(u,|y|)$.

We denote this unique vector x by $\psi_E(u)$, and the number $e\big(u,\psi_E(u)\big)$ by $e_E(u)$. We emphasize again that $\psi_E(u)$ and $e_E(u)$ are only defined for finitely supported u. The first lemma gives a quantitative version of the uniqueness of $\psi_E(u)$. The estimate depends on the following parameter $\eta(\varepsilon)$ which measures the concavity of log. For $0 < \varepsilon < 1$ put

$$\begin{aligned}\eta(\varepsilon) &= \varepsilon\log\Big(\frac{1}{2}(1+\varepsilon)^{1/2} + \frac{1}{2}(1+\varepsilon)^{-1/2}\Big)\\ &= \varepsilon\min\Big\{\log\frac{1}{2}\big(t^{1/2}+t^{-1/2}\big) : |1-t| \ge \varepsilon\Big\}\end{aligned}$$

which is strictly positive for $\varepsilon > 0$. If u is a vector and H is a subset of $\mathbb{N}$, we denote the vector which agrees with u on H and vanishes outside H by Hu. Inequalities for vectors are understood to hold coordinatewise. The set of nonnegative vectors in $B(E)$ is denoted by $B(E)^+$.

Lemma 13.36. *Let u be a nonnegative vector in l_1 with $\|u\|_1 = 1$ and finite support. Let $0 < \varepsilon < 1$, and let $y \in B(E)^+$ satisfy $e(u,y) \ge e_E(u) - \eta(\varepsilon)$. Then* *there is a subset $H \subset \operatorname{supp} u$ such that $(1-\varepsilon)H\psi_E(u) \le Hy \le (1+\varepsilon)H\psi_E(u)$ and $\|Hu\|_1 \ge 1-\varepsilon$.*

Proof. Let $x = \psi_E(u)$ and $D = \operatorname{supp} u = \operatorname{supp} x$. Since u is nonnegative so is x, and $(x+y)/2 \in B(E)^+$, hence $e_E(u) \ge e_E\big(u, (x+y)/2\big)$. Passing to coordinates we obtain

$$\begin{aligned}
\eta(\varepsilon) &\ge e_E\big(u,(x+y)/2\big) - e(u,y) = \sum_{i\in D} u_i\big(\log\frac{1}{2}(x_i+y_i) - \log y_i\big)\\
&= \sum_{i\in D} u_i\Big(\frac{1}{2}\log x_i + \frac{1}{2}\log y_i + \log\frac{1}{2}(x_i+y_i) - \log\sqrt{x_iy_i} - \log y_i\Big)\\
&= \frac{1}{2}\sum_{i\in D} u_i(\log x_i - \log y_i) + \sum_{i\in D} u_i \log\frac{1}{2}\big(\sqrt{x_i/y_i} + \sqrt{y_i/x_i}\big)\\
&\ge \sum_{i\in D} u_i \log\frac{1}{2}\big(\sqrt{x_i/y_i} + \sqrt{y_i/x_i}\big).
\end{aligned}$$

Put $H = \{i \in D : (1-\varepsilon)x_i \le y_i \le (1+\varepsilon)x_i\} = \{i \in D : |y_i/x_i - 1| \le \varepsilon\}$. Then $\log\frac{1}{2}\big(\sqrt{x_i/y_i} + \sqrt{y_i/x_i}\big) \ge \eta(\varepsilon)/\varepsilon$ for $i \in D \setminus H$, and hence

$$\sum_{i\in D\setminus H} u_i \le \frac{\varepsilon}{\eta(\varepsilon)} \sum_{i\in D} u_i \log\frac{1}{2}\big(\sqrt{x_i/y_i} + \sqrt{y_i/x_i}\big) \le \varepsilon$$

i.e., $\|Hu\|_1 \ge 1-\varepsilon$. □

From now on we shall take E to be the space $\mathcal{S}^*$, where $\mathcal{S}$ is the Schlumprecht space considered in the previous section. (As in the previous section, we omit the function $f \in \mathcal{F}$ from the notation.) We also denote $\psi_{\mathcal{S}^*}(u)$ by $\psi(u)$ and $e_{\mathcal{S}^*}(u)$ by $e(u)$.

The next lemma resembles Lemma 13.32 in its statement, and in particular in its proof.

Lemma 13.37. *Let U be an infinite-dimensional subspace of l_1 spanned by a normalized nonnegative block basic sequence $\{u_k\}$. Let m be a positive integer, and let $0 < \tau < 1$. Then there is a normalized block basic sequence $\{v_j\}_{j=1}^m$ of $\{u_k\}$ such that*

$$\sum_{j=1}^m e(v_j) - e\Big(\sum_{j=1}^m v_j\Big) \le \tau m.$$

Proof. If not, then in particular $\sum_{j=1}^m e(u_{k_j}) - e\big(\sum_{j=1}^m u_{k_j}\big) \ge \tau m$ for every subsequence $\{u_{k_j}\}$ of $\{u_k\}$. Put $v_j = \sum_{i=1}^m u_{jm+i}$. Then $\|v_j\|_1 = m$ for every j.

Since $e(\alpha w) = \alpha e(w)$ for every vector w and $\alpha > 0$, the failure of the lemma gives

$$\sum_{j=0}^{m-1} e\Big(\sum_{i=1}^m u_{mj+i}\Big) - e\Big(\sum_{k=1}^{m^2} u_k\Big) \ge \tau m^2.$$

Hence

$$\begin{aligned}\sum_{k=1}^{m^2} e(u_k) - e\Big(\sum_{k=1}^{m^2} u_k\Big)\\ &= \sum_{j=0}^{m-1}\Big(\sum_{i=1}^{m} e(u_{mj+i}) - e\Big(\sum_{i=1}^{m} u_{mj+i}\Big)\Big) + \sum_{j=0}^{m-1} e\Big(\sum_{i=1}^{m} u_{mj+i}\Big) - e\Big(\sum_{k=1}^{m^2} u_k\Big)\\ &\geq m\tau m + \tau m^2 = 2\tau m^2\end{aligned}$$

and by induction $\sum_{k=1}^{m^p} e(u_k) - e\big(\sum_{k=1}^{m^p} u_k\big) \geq p\tau m^p$ for every p.

Let now $x_k^* = \psi(u_k) \in \mathcal{S}^*$. Then $f(m^p)^{-1}\sum_{k=1}^{m^p} x_k^* \in B(\mathcal{S}^*)$, and thus, since the cross terms below vanish and $\|u_k\|_1 = 1$

$$\begin{aligned} e\Big(\sum_{k=1}^{m^p} u_k\Big) &\geq e\Big(\sum_{k=1}^{m^p} u_k, f(m^p)^{-1}\sum_{k=1}^{m^p} x_k^*\Big)\\ &= \sum_{k=1}^{m^p}\Big(e(u_k, x_k^*) - \log f(m^p)\Big) = \sum_{k=1}^{m^p} e(u_k) - m^p \log f(m^p).\end{aligned}$$

Comparing the two estimates gives a contradiction if $p\tau > \log f(m^p)$. (Such p exists by property (iii) of functions in $\mathcal{F}$.) $\square$

We now come to the main lemma of this section. If $y \in \mathcal{S}$ and $y^* \in \mathcal{S}^*$ we denote the vector obtained by multiplying their respective coordinates by $y^* \cdot y$. Note that $y^* \cdot y \in l_1$ and $\|y^* \cdot y\|_1 = \sum |y_i^* y_i| = \langle |y^*|, |y|\rangle \leq \|y^*\|\,\|y\|$.

Lemma 13.38. *Let U be as in the statement of Lemma 13.37. Let $\varepsilon > 0$, and let m be an integer. Then there are an l_{1+}^m-average y in $\mathcal{S}$ with constant $1+\varepsilon$ and $y^* \in B(\mathcal{S}^*)$ such that $\operatorname{dist}\big(y \cdot y^*, S(U)\big) \leq \varepsilon$.*

Proof. Let $0 < \tau < \eta(\varepsilon)/m$, and let $\{v_j\}_{j=1}^m$ be as in Lemma 13.37. Write $x^* = \psi(\sum_{j=1}^m v_j) \in \mathcal{S}^*$ as $\sum_{j=1}^m x_j^*$ where $\operatorname{supp} x_j^* = \operatorname{supp} v_j$. Let $y_j^* = \psi(v_j)$, and choose $y_j \in S(\mathcal{S})$ with $\operatorname{supp} y_j = \operatorname{supp} v_j$ such that $y_j^* \cdot y_j = v_j$. (The existence of such y_j's follows from Lemma 9.5 and from the observations related to this lemma in the Notes and Remarks section of Chapter 9.) By the definition of the entropy function

$$\begin{aligned}\tau m &\geq \sum_{j=1}^m e(v_j) - e\Big(\sum_{j=1}^m v_j\Big) = \sum_{j=1}^m e(v_j, y_j^*) - e\Big(\sum_{j=1}^m v_j, \sum_{j=1}^m x_j^*\Big)\\ &= \sum_{j=1}^m \big(e(v_j, y_j^*) - e(v_j, x_j^*)\big).\end{aligned}$$

Since all the terms in the last sum are nonnegative, it follows from the choice of τ that $e(v_j, y_j^*) - e(v_j, x_j^*) \leq \eta(\varepsilon)$ for every j.

By Lemma 13.36 there are subsets H_j of $\operatorname{supp} v_j$ such that $\|v_j - H_j v_j\|_1 \leq \varepsilon$ and $(1-\varepsilon)H_j y_j^* \leq H_j x_j^* \leq (1+\varepsilon)H_j y_j^*$. Thus

$$\begin{aligned}\|v_j - H_j x_j^* \cdot y_j\|_1 &\leq \varepsilon + \|H_j v_j - H_j x_j^* \cdot y_j\|_1\\ &= \varepsilon + \|H_j(y_j^* - x_j^*) \cdot y_j\|_1 \leq 2\varepsilon\end{aligned}$$

because $\|H_j y_j^* - H_j x_j^*\| \le \varepsilon$ and $\|y_j\| = 1$. It follows that $\langle H_j x_j^*, y_j\rangle \ge 1 - 2\varepsilon$, and thus

$$\Big\|\sum_{j=1}^m y_j\Big\| \ge \langle x^*, \sum_{j=1}^m y_j\rangle \ge \sum_{j=1}^m \langle H_j x_j^*, y_j\rangle \ge m(1-2\varepsilon)$$

i.e., $\{y_j\}_{j=1}^m$ is an l_{1+}-sequence in $\mathcal{S}$ with constant $(1-2\varepsilon)^{-1}$, and the l_{1+}^m-average $y = (\sum_{j=1}^m y_j)/\|\sum_{j=1}^m y_j\|$ satisfies $\|y - \tilde{y}\| \le 2\varepsilon$, where $\tilde{y} = m^{-1}\sum_{j=1}^m y_j$. Hence

$$\begin{aligned}\Big\|m^{-1}\sum_{j=1}^m v_j - \Big(\bigcup_{j=1}^m H_j\Big)x^* \cdot y\Big\|_1 &\\ \le \Big\|m^{-1}\sum_{j=1}^m v_j - \Big(\bigcup_{j=1}^m H_j\Big)x^* \cdot \tilde{y}\Big\|_1 + \|y - \tilde{y}\| &\\ \le m^{-1}\sum_{j=1}^m \|v_j - H_j x_j^* \cdot y_j\|_1 + \|y - \tilde{y}\| \le 4\varepsilon.&\end{aligned}$$

Taking $y^* = \big(\bigcup_{j=1}^m H_j\big)x^*$ concludes the proof because $m^{-1}\sum_{j=1}^m v_j \in S(U)$. $\square$

We are now ready to transfer to l_p the specific asymptotic systems in $\mathcal{S}$ constructed in the proof of Theorem 13.30. In the following lemma these sets are mapped to asymptotic sets in l_1. Since l_1 is not distortable, these images are only asymptotic in l_1 and cannot form an asymptotic biorthogonal system there. We shall show, however, that their images in l_p and l_p^* via the respective Mazur maps are such a system for any $1 < p < \infty$.

A set $D \in l_1$ is called unconditional if $\sum \pm\alpha_i e_i \in D$ whenever $\sum \alpha_i e_i \in D$. It is called spreading if $\sum \alpha_i e_{n_i} \in D$ for every increasing sequence $\{n_i\}$ and $\sum \alpha_i e_i \in D$.

Fix $\delta > 0$, and let $A_j \subset \mathcal{S}$ and $A_j^* \subset \mathcal{S}^*$ be the biorthogonal asymptotic system with constant $\delta > 0$ constructed in the previous section. Put

$$D_j = \Big\{x^* \cdot x/\|x^* \cdot x\|_1 : x^* \in A_j^*;\ x \in A_j;\ \|x^* \cdot x\|_1 = \langle |x^*|, |x|\rangle \ge 1 - 3\delta\Big\}$$

then $D_j \subset S(l_1)$.

Lemma 13.39. *The sets D_j are unconditional, spreading and asymptotic.*

Proof. That each D_j is unconditional and spreading is evident from the definitions of the A_j's and A_j^*'s and from the fact that the standard basis of $\mathcal{S}$ is 1-subsymmetric.

To prove that D_j is asymptotic it suffices to show, since D_j is unconditional, that $\operatorname{dist}(D_j, S(U)) = 0$ for subspaces U as in Lemma 13.37. Fix $0 < \varepsilon < \delta$ such that $(1+2\varepsilon)^{-1}(1+2\delta)^{-1} > 1 - 3\delta$. By Lemma 13.38 there are an RIS $\{x_i\}_{i=1}^{N_j}$ in $S(\mathcal{S})$ with constant smaller than $1+\delta$, vectors $\{x_i^*\}_{i=1}^{N_j} \subset B(\mathcal{S}^*)$ and a finite normalized block basic sequence $\{w_i\}_{i=1}^{N_j}$ in U such that $\|x_i^* \cdot x_i - w_i\|_1 < \varepsilon$ and $\operatorname{supp} x_i^* = \operatorname{supp} x_i$ for all i. (In particular $x_i^* \cdot x_k = 0$ for $i \ne k$.)

Put $\lambda_i = \|x_i^* \cdot x_i\|_1^{-1}$. Then $1 \le \lambda_i \le 1 + 2\varepsilon$ (because $\|x_i^* \cdot x_i\|_1 \ge 1 - \varepsilon$). Fix $y = \sum_{i=1}^{N_j} x_i/\|\sum_{i=1}^{N_j} x_i\| \in A_j$ and $y^* = (1+2\varepsilon)^{-1} f(N_j)^{-1}\sum_{i=1}^{N_j}\lambda_i x_i^* \in A_j^*$, then

Lemma 13.35 gives $\|\sum_{i=1}^{N_j} x_i\| \le N_j(1+2\delta)f(N_j)^{-1}$, hence

$$\begin{aligned}\|y^*\cdot y\|_1 &= \sum_{i=1}^{N_j}\lambda_i\|x_i^*\cdot x_i\|_1\Big/\Big((1+2\varepsilon)f(N_j)\Big\|\sum_{i=1}^{N_j}x_i\Big\|\Big)\\ &\ge (1+2\varepsilon)^{-1}(1+2\delta)^{-1}\ge 1-3\delta\end{aligned}$$

and $y^*\cdot y/\|y^*\cdot y\|_1 \in D_j$. Since $y^*\cdot y/\|y^*\cdot y\|_1 = N_j^{-1}\sum_{i=1}^{N_j}\lambda_i x_i^*\cdot x_i$, we obtain

$$\Big\|y^*\cdot y/\|y^*\cdot y\|_1 - N_j^{-1}\sum_{i=1}^{N_j} w_i\Big\|_1 \le N_j^{-1}\sum_{i=1}^{N_j}\|\lambda_i x_i^*\cdot x_i - w_i\|_1 \le 3\varepsilon$$

and thus $\operatorname{dist}\big(D_j, S(U)\big)\le 3\varepsilon$. □

Theorem 13.40. *For every $\delta>0$ and $1<p<\infty$ the space l_p admits an asymptotic biorthogonal system with constant δ.*

Proof. The case $p=2$ is somewhat simpler because $l_2^* = l_2$, and we consider it first.

For every j put $C_j = \{x\in S(l_2) : |x|^2 \in D_j\}$. The set C_j is the image of D_j under the Mazur map from l_1 to l_2. Since the Mazur map maps block sequences to block sequences and is a uniform homeomorphism between the respective unit spheres, it follows from Lemma 13.39 that every C_j is an asymptotic set. To see that the C_j's are almost orthogonal fix $y_j\in C_j$ and $y_k \in C_k$ with $j\ne k$. Then $|y_j|^2 = x_j^*\cdot x_j/\|x_j^*\cdot x_j\|_1$ where $x_j\in A_j$ and $x_j^*\in A_j^*$, and $\|x_j^*\cdot x_j\|_1\ge 1-3\delta$. Similarly $|y_k|^2 = x_k^*\cdot x_k/\|x_k^*\cdot x_k\|_1$. Hence

$$\begin{aligned}\big\langle |y_j|,|y_k|\big\rangle &\le (1-3\delta)^{-1}\sum_{i=1}^{\infty}\big|x_j^*(i)x_j(i)x_k^*(i)x_k(i)\big|^{1/2}\\ &\le (1-3\delta)^{-1}\Big(\sum_{i=1}^{\infty}|x_j^*(i)x_k(i)|\Big)^{1/2}\Big(\sum_{i=1}^{\infty}|x_k^*(i)x_j(i)|^{1/2}\Big)^{1/2}\\ &\le (1-3\delta)^{-1}\big\langle|x_j^*|,|x_k|\big\rangle^{1/2}\big\langle|x_k^*|,|x_j|\big\rangle^{1/2}\le(1-3\delta)^{-1}\delta.\end{aligned}$$

For $p\ne 2$ we proceed similarly, but define for every j two sets $C_j\subset l_p$ and $C_j^*\subset l_q = l_p^*$. These are again the images of D_j under the respective Mazur maps to l_p and l_q. Assuming that $p>2$, say, and using Hölder's inequality with exponents $p/2$ and $p/(p-2)$, it follows that for every $k\ne j$ if $y_k\in C_k$ and $y_j^*\in C_j^*$, then $(1-3\delta)\big\langle|y_j|,|y_k|\big\rangle$ is bounded by

$$\begin{aligned}&\sum_{i=1}^{\infty}\big|x_j^*(i)x_j(i)\big|^{1/p}\big|x_k^*(i)x_k(i)\big|^{1/q}\\ &=\sum_{i=1}^{\infty}\big|x_j^*(i)x_j(i)x_k^*(i)x_k(i)\big|^{1/p}\big|x_k^*(i)x_k(i)\big|^{1/q-1/p}\\ &\le\Big(\sum_{i=1}^{\infty}|x_j^*(i)x_j(i)x_k^*(i)x_k(i)|^{1/2}\Big)^{2/p}\Big(\sum_{i=1}^{\infty}|x_k^*(i)x_k(i)|\Big)^{(p-2)/p}\\ &\le\delta^{2/p}\end{aligned}$$

□

Remark. If we start with an asymptotic biorthogonal system in $\mathcal{S}$ with vanishing constants (see the remark at the end of the previous section), then the procedure above yields a similar system in any l_p with $1 < p < \infty$.

6. Notes and Remarks

The first result on distortion in infinite-dimensional spaces was Proposition 13.6, due to James [**273**]. The natural question what happens in l_p for $1 < p < \infty$, and in particular in Hilbert space, became to be known as "the distortion problem".

The study of oscillation stability of general uniformly continuous or Lipschitz functions on the unit sphere of infinite-dimensional Banach spaces (and not just of equivalent norms) was initiated by Milman [**415**]. He was motivated by the finite-dimensional results Proposition 12.3 and Corollary 12.11. Milman introduced the notion of asymptotic sets (without naming them explicitly), and proved Propositions 13.3 and 13.5.

Proposition 13.8 is due to Olevskiĭ [**445**], who stated a somewhat weaker form. The present formulation appears in Lindenstrauss and Pełczyński [**365**]. In the proof of the proposition in [**365**] the splitting of the sets $A_{n,j}$ was done by Liapounoff's theorem (Theorem D.11). Propositions 13.9 and 13.10 are also due to Lindenstrauss and Pełczyński [**365**]. We refer to this paper for the proof of Proposition 13.9.

In contrast to Proposition 13.10 (for L_1), we mention that Dowling, Randianantoanina and Turret [**156**] proved that if a dual space E contains a subspace isomorphic to L_1, then for every $\varepsilon > 0$ it contains a subspace which is $(1+\varepsilon)$-isometric to L_1. Note that such an E cannot be separable because L_1 fails the RNP property.

Milman [**416**] studied the question whether there is any Banach space with a distortable norm. He conjectured that no such space exists, and hoped that this will constitute the proper infinite-dimensional generalization of Dvoretzky's theorem. In this connection he proved Theorem 13.11, and hoped that it will be used to show that every infinite-dimensional Banach space contains a subspace isomorphic to c_0 or to l_p for some $1 \le p < \infty$. All the examples of Banach spaces known at that time contained such subspaces. In some spaces this was obvious, but in some cases (like Orlicz spaces, see Lindenstrauss and Tzafriri [**369**]) this required a proof. Thus this latter question was very natural, and it was raised and studied also by other mathematicians. Note also that the spaces c_0 and l_p are not isomorphic to a subspace of each other, and therefore the list cannot be reduced.

There have been major advances on this problem (including Tsirelson's example which will be discussed later on in this section), and currently there are quite strong general results ensuring the existence of a subspace isomorphic to c_0 or to l_p in a large class of Banach spaces. Krivine and Maurey [**339**] proved that stable Banach spaces (see Section 5 in Chapter 9) contain, for every $\varepsilon > 0$, a subspace which is $(1+\varepsilon)$-isometric to l_p for some p. (In the important special case of L_1 this was proved earlier by Aldous [**7**].) It was this result that motivated Krivine and Maurey to introduce the class of stable Banach spaces.

Garling [**197**] and Argyros, Negrepontis and Zachariades [**21**] introduced the related class of weakly stable Banach spaces. A space E is weakly stable if whenever K is a weakly compact set in E and $\{x_n\}$ and $\{y_n\}$ are two sequences in K, then

$$\lim_{n\in\mathcal{U}}\lim_{m\in\mathcal{V}}\|x_n+y_m\| = \lim_{m\in\mathcal{V}}\lim_{n\in\mathcal{U}}\|x_n+y_m\|$$

for any two ultrafilters $\mathcal{U}$ and $\mathcal{V}$ on $\mathbb{N}$. A reflexive space is weakly stable if and only if it is stable, and the spaces c_0 or c are simple examples of weakly stable Banach spaces which are not stable. In these two papers the theorem of Krivine and Maurey was generalized to say that a weakly stable Banach space contains, for every $\varepsilon > 0$, a subspace which is $(1+\varepsilon)$-isometric to c_0 or l_p. For more information on stable spaces and on the Krivine-Maurey theorem see the book [**239**] by Guerre-Delabriére.

The proof we presented here of Theorem 13.11 is due to Haydon, Odell, Rosenthal and Schlumprecht [**250**]. (This unpublished proof appears in the survey article [**442**] by Odell and Schlumprecht.) The original proof of Milman is quite close in spirit to this proof. The undistorted norms he uses are close to the ones we used here. Instead of Krivine's theorem (which was not known at that time) he uses Zippin's characterization [**594**] of perfectly homogeneous bases.

Proposition 13.12 is due to Krivine and Maurey. They introduced in [**339**] the notion of "type" on a Banach space E, i.e., a function of the form $\psi(x) = \lim \|x + x_j\|$, where $\{x_j\}$ is a sequence for which the limit exists for every $x \in E$. Types of the form appearing in the statement of Proposition 13.12 were called by them l_p-types.

Proposition 13.13 was observed by several mathematicians; see, e.g., Lindenstrauss [**356**], page 95. Borsuk's antipodal theorem appeared in [**78**], and the proof can be found in most introductory texts on algebraic topology.

Theorem 13.19 is due to Hindman [**254**]. Lemma 13.20 is due to Namakara [**433**]. Lemma 13.21 and the proof we presented of Theorem 13.19 are due to Glazer; see Graham, Rothchild and Spencer [**227**]. This book also contains two other proofs of Hindman's theorem.

The remainder of Section 3, and in particular Theorem 13.18, is due to Gowers [**224**]. The proof for an unconditional function g follows Gowers' proof very closely. The proof of the general case (Proposition 13.25 and Lemma 13.26) differs somewhat from the presentation in [**224**]. The simpler proof here was communicated to us by Gowers.

The conjecture that there are no distortable norms (hence, in view of Theorem 13.11, that every infinite-dimensional Banach space contains a copy of c_0 or l_p) was disproved when Tsirelson [**563**] produced a counterexample (see also Proposition 10.31).

The norm on Tsirelson's space T is given by an implicit formula (10.9), and is generally hard to compute except for very specific simple vectors. (This implicit definition of the norm is in the heart of the novelty of Tsirelson's construction, and is the source of the many remarkable properties of this space.)

Tsirelson's space was studied extensively very soon after the appearance of [**563**], and many new variants with remarkable properties were constructed; see Section 5 of Chapter 10, and in particular the book [**112**] by Casazza and Shura. Surprisingly little was done on examining the space in the context of distortable norms. By Theorem 13.11 T admits a distortable norm, but only some 15 years after its discovery Odell constructed explicit distorted norms on T. He checked that T admits λ-distortable norms for any $\lambda < 2$, but was unable to decide what happens for $\lambda \geq 2$. (This question is still open, and will be discussed further at the end of this section.)

Motivated by Odell's work on T Schlumprecht [**518**] constructed the first example of a Banach space which is arbitrarily distortable, the space $\mathcal{S}$ of Section 4. The definition of $\mathcal{S}$ is closely related to that of T, but there are significant differences between them. The proof in [**518**] that $\mathcal{S}$ is arbitrarily distortable is closely related to some of the arguments presented in Section 4.

The results and proofs of the paper [**518**] by Schlumprecht strongly influenced two major developments — the article of Gowers and Maurey [**226**] and the solution of the distortion problem by Odell and Schlumprecht [**443**] (see also [**444**]). Our presentation of the results in Section 4 follows the arguments in Sections 1 and 2 of [**226**] very closely. They introduced the notion of an asymptotic biorthogonal system and proved Theorem 13.29. The proof of this theorem is a modification of the argument used by Maurey and Rosenthal [**403**] to construct a normalized basic sequence without an unconditional subsequence.

Odell and Schlumprecht [**443**] proved the following stronger version of Theorem 13.29 (with a similar proof).

Theorem 13.41. *Let E be a Banach space which has an asymptotic biorthogonal system with vanishing constants and a basis $\{e_i\}$. Then for every integer n and $\varepsilon > 0$ there is an equivalent norm on E with the following property. For any normalized monotone basic sequence $\{y_j\}$ in any Banach space and for every block basic sequence $\{x_k\}$ of the e_i's there is a normalized sequence $\{z_j\}$ of the x_k's which is $(1+\varepsilon)$-equivalent (in the new norm) to $\{y_j\}$.*

Theorem 13.29 is the special case where the y_j's are the "summing basis", i.e., when $\|\sum_{j=1}^n a_j y_j\| = \max_{k\leq n} |\sum_{j=1}^k a_j|$.

Note that Theorem 13.41 applies also to l_2, and we obtain the surprising result that for every n and $\varepsilon > 0$ and for every $1 \leq p \leq \infty$ there is an equivalent norm on l_2 such that any infinite-dimensional subspace of l_2 contains a subspace which is $(1+\varepsilon)$-isometric to l_p^n under the new norm.

The results of Section 4 imply that for every C there is a renorming of $\mathcal{S}$ such that no basic sequence in $\mathcal{S}$ is C-unconditional under the new norm. (It follows from the results in Section 5 that the same is true for l_p when $1 < p < \infty$.) However, since $\mathcal{S}$ has an unconditional basis, it also has one under any equivalent norm. The main part of the paper [**226**] of Gowers and Maurey is devoted to modifying the definition of $\mathcal{S}$ in order to obtain the first example of a Banach space which contains no unconditional basic sequence.

Actually (as pointed out by Johnson), the space they constructed is hereditarily indecomposable (HI) in the sense that no closed subspace F of the space can be decomposed as $F = F_1 \oplus F_2$ with $\dim F_1 = \dim F_2 = \infty$. Of course any Banach space with an unconditional basis $\{x_i\}$ is decomposable — it is the direct sum of the closed spans of the even and of the odd basis vectors. The existence of an HI space was completely unexpected in Banach space theory. There is now a considerable amount of information on HI spaces (and, of course, many open problems). We mention just one result, the important dichotomy discovered by Gowers [**225**]: Every Banach space contains a subspace which is either HI or has an unconditional basis.

The results of Section 5 are due to Odell and Schlumprecht [**443**]. They form the solution of (a strong version of) the distortion problem mentioned in the beginning of this section. We followed the proof in [**443**]. This proof already contains some modification of the original proof suggested by Maurey in view of the presentation of the results of Schlumprecht [**518**] in Gowers and Maurey [**226**].

The function $e(u, x)$ and the mapping ψ_E originate from Gillespie [**208**]. Odell and Schlumprecht used the mapping ψ_E in [**443**] to prove Theorem 9.4. The arguments in Section 5 do not use the fact that for certain Banach spaces ψ_E is a uniform homeomorphism of the unit balls. In fact we use this map here for $E = \mathcal{S}^*$ which contains l_∞^n's uniformly, and therefore its unit ball is not uniformly homeomorphic to that of l_1. We note also that for $1 \leq p < \infty$ and $E = l_p$ the map ψ_E is the Mazur map.

Comparing the solution of the distortion problem with the finite-dimensional situation as described in Chapter 12 reveals a basic difference and a new genuine infinite-dimensional phenomenon, which should be investigated further and understood better.

The construction in Sections 4 and 5 is not simple, and does not simplify significantly if we just want to find a distortable norm in Hilbert space, i.e., two asymptotic sets in the unit sphere of l_2 which are at a positive distance from each other. One naturally wonders if there is a direct proof of this simply stated geometric property of Hilbert space. Perhaps one of the many different realizations of Hilbert space could be useful in this direction. (Of course the notions of "simple" and "direct" are not well defined.)

The sets C_j constructed in the proof of Theorem 13.40 are spreading and unconditional. Maurey [**401**] and, independently, Schlumprecht (unpublished) proved that one can replace them by sets which are even symmetric. More precisely, Maurey proved

Theorem 13.42. *There is a sequence of sets U_j in l_2 with the following properties*

(i) *U_j is asymptotic.*

(ii) *U_j is symmetric in the sense that if $\sum a_i e_i \in U_j$, then $\sum \pm a_{\pi(i)} e_i \in U_j$ for every choice of signs and every permutation π of the integers.*

(iii) *There is a constant C such that $C^{-1} \leq \|x\| \leq C$ for every $x \in \bigcup U_j$.*

(iv) *There is a sequence $\varepsilon_j \searrow 0$ such that $|\langle x, y\rangle| \leq \varepsilon_j$ whenever $x \in U_j$ and $y \in U_k$, and $k > j$.*

It follows that there are arbitrarily distorted norms on l_2 for which the unit vector basis is a symmetric basis; just put for every j

$$|\!|\!|x|\!|\!|_j = \sup\left\{|\langle x, y\rangle| : y \in U_j\right\} + \varepsilon_j \|x\|/C.$$

Then $|\!|\!|x|\!|\!|_j \geq \|x\|$ for $x \in U_j$, and if $k > j$ and $x \in U_k$, then $|\!|\!|x|\!|\!|_j \leq \varepsilon_j(1 + \|x\|/C)$. Since both U_k and U_j are asymptotic, we deduce that $|\!|\!| \cdot |\!|\!|$ is at least $C^{-4}/2\varepsilon_j$-distorted.

A similar result is proved in [**401**] for l_p, $1 < p < \infty$.

In Sections 1 and 2 we discussed distorted norms, while in Sections 4 and 5 we showed that certain spaces are actually arbitrarily distortable. It is a natural question whether every Banach space with a distorted norm is arbitrarily distortable or has, at least, an arbitrarily distortable subspace. This question is still open, but there are several known results in this direction and we quote now two such results without proof and a third will be proved.

Before stating these results, we make the following observation concerning the concept of arbitrary distortion: Let E be a Banach space with no arbitrarily distortable infinite-dimensional subspace, then there are a constant $\mu < \infty$ and an infinite-dimensional subspace F of E such that the distortion of any equivalent norm on any infinite-dimensional subspace of F is at most μ. Indeed, if there is no such μ, then there is a decreasing sequence $F_1 \supset F_2 \supset \ldots$ of infinite-dimensional subspaces of E and equivalent norms $\|\!|\cdot|\!\|_n$ on F_n whose distortion is at least n. Normalize the norms so that $\|\!|x|\!\|_n \leq \|x\|$ for every $x \in F_n$. Let $\tilde{F}$ be a subspace of E such that $\tilde{F} \cap F_n$ has finite codimension in $\tilde{F}$ for every n. For each n define an equivalent norm on $\tilde{F}$ by taking as its unit ball the convex hull of $B(\tilde{F}) \cup \{x \in \tilde{F} \cap F_n : \|\!|x|\!\|_n \leq 1\}$. The distortion of this norm is at least n; hence $\tilde{F}$ is arbitrarily distortable.

A Banach space F is called asymptotically l_p (respectively c_0) if it has a basis $\{e_i\}$ and if there is a constant C such that for every n there is an integer $N(n)$ such that every block basis $\{x_j\}_{j=1}^n$ of $\{e_i\}_{i>N(n)}$ is C-equivalent to the unit vector basis of l_p^n (respectively l_∞^n).

Theorem 13.43. *Let E be a Banach space which contains no arbitrarily distortable infinite-dimensional subspace. Then E contains a subspace which is either asymptotically l_p for some $1 \leq p < \infty$ or is asymptotically c_0.*

This theorem is due to Milman and Tomczak-Jaegermann [**422**], and a somewhat simplified proof is given in Maurey [**400**]. It is closely related to Theorem 13.11, and the proof of these theorems have several points in common.

The following theorem is due to Maurey [**400**]. The proof is obtained by a careful adaptation of the arguments presented in Section 5 to the setting of asymptotic l_p spaces.

Theorem 13.44. *Let E be an asymptotic l_p space for some $1 < p < \infty$ with an unconditional basis and of type r for some $r > 1$. Then E admits an asymptotic biorthogonal system with constant δ for every $\delta > 0$.*

The next theorem is due to Tomczak-Jaegermann [**558**]

Theorem 13.45. *Let E be a Banach space, and assume that there is a constant μ such that no infinite-dimensional subspace of E is μ-distortable. Then E has a subspace with an unconditional basis.*

Combining Theorems 13.43, 13.44 and 13.45 gives immediately the following result.

Corollary 13.46. *An infinite-dimensional Banach space with type $r > 1$ has an arbitrarily distortable subspace.*

Theorems 13.43, 13.44 and 13.45 do not give any new information on the Tsirelson space. This space is an asymptotic l_1-space with an unconditional basis and does not have nontrivial type. It is thus a natural candidate for a distortable space which is not arbitrarily distortable. (Note, however, that there are asymptotic l_1-space with an unconditional basis which are arbitrarily distortable, see Argyros and Deliyanni [**20**] and Androulakis and Odell [**17**].) In view of Theorem 13.44 the convexified Tsirelson spaces $\mathcal{T}^{(p)}$ for $1 < p < \infty$ (see Section 5 of Chapter 10) are arbitrarily distortable.

We shall now prove Theorem 13.45, following the proof in [**558**]. It is based on a special case of the Kunen-Martin theorem in descriptive set theory. Proofs of this type were introduced to Banach space theory by Bourgain, who used them in several papers. We only mention one of these, [**84**], where it was used in the study of the existence of certain universal Banach spaces.

The Kunen-Martin theorem is discussed in detail in Kechris' book [**323**], pages 239–241. We only need a simple special case which we shall prove after introducing some notation and terminology.

Let E be a set, and let E^n denote its n-fold cartesian product. A tree T on E is a subset of $\bigcup_{n=1}^{\infty} E^n$ such that whenever $(x_1, \dots, x_n, x_{n+1}) \in T$ for some n, then the initial segment $(x_1, \dots, x_n)$ also belongs to T. A tree T is said to be well-founded if there is no infinite sequence $\{x_n\}_{n=1}^{\infty}$ in E such that $(x_1, \dots, x_n) \in T$ for every n.

Let T be a well-founded tree. For every countable ordinal α define by transfinite induction a tree T^α as follows. Start with $T^1 = T$. An n-tuple $(x_1, \dots, x_n)$ belongs to $T^{\alpha+1}$ if there is an $x \in E$ such that $(x_1, \dots, x_n, x) \in T^\alpha$. If β is a limit ordinal, then $T^\beta = \bigcap_{\alpha<\beta} T^\alpha$. Since T is well-founded, the transfinite sequence of trees T^α is strictly decreasing until the minimal ordinal α (which is called the order of T and is denoted by $o(T)$) for which $T^\alpha = \emptyset$. If $T^\alpha \neq \emptyset$ for every countable ordinal α, we put $o(T) = \omega_1$.

Assume that E is a topological space. We say that a tree T on E is closed if $T(n) = T \bigcap E^n$ is a closed subset of E^n (in the product topology) for every n. The next proposition is the special case of the Kunen-Martin theorem which we need.

Proposition 13.47. *Let E be a complete separable metric space, and let T be a closed and well-founded tree on E. Then $o(T) < \omega_1$.*

Proof. We show first that if $\tilde{T}$ is a nonempty closed tree on E such that $\tilde{T}(n) = \overline{\pi_n \tilde{T}(n+1)}$ for every n, then $\tilde{T}$ is not well-founded. (Here π_n is the natural projection from E^{n+1} onto the first n coordinates, which we identify with E^n.) Indeed, by the assumption there are n-tuples $(x_{1,n}, x_{2,n}, \dots, x_{n,n}) \in \tilde{T}(n)$ for each n such that $d(x_{i,n}, x_{i,n+1}) < 2^{-n}$ for every $i \leq n$. Since E is complete the sequence $\{x_{i,n}\}_{n \geq i}$ converges for every i to a limit y_i, say. Since $\tilde{T}$ is closed, $(y_1, \dots, y_n) \in \tilde{T}$ for every n; i.e., $\tilde{T}$ is not well-founded.

A strictly decreasing chain of closed sets in a separable metric space is necessarily countable. Thus for each n there is a countable ordinal $\eta(n)$ such that $\overline{T^\alpha(n)} = \overline{T^{\eta(n)}(n)}$ for every $\alpha > \eta(n)$. It follows that there is a countable ordinal η such that $\overline{T^{\eta+1}(n)} = \overline{T^\eta(n)}$ for all n. Since $T^{\alpha+1}(n) = \pi_n T^\alpha(n+1)$ for all α and n by the definition of T^α, it follows that $\overline{T^\eta(n)} = \overline{\pi_n T^\eta(n+1)}$ for all n.

The tree $\tilde{T} = \bigcup_{n=1}^{\infty} \overline{T^\eta(n)}$ is closed and is well-founded because T is. Also, for every n

$$\tilde{T}(n) = \overline{T^\eta(n)} = \overline{\pi_n T^\eta(n+1)} = \overline{\pi_n \tilde{T}(n+1)}.$$

It follows from the first part of the proof that $\tilde{T}$ is empty, i.e., $T^\eta(n) = \emptyset$ for all n. Thus $o(T) \leq \eta$. □

Let E be a Banach space with a basis $\{e_i\}$. Let Σ be a family of finite collections of finite intervals $\{I_j\}_{j=1}^n$ in $\mathbb{N}$ with $I_1 \prec \ldots \prec I_n$ satisfying

(i) Σ contains all single intervals.

(ii) If $\{I_j\}_{j=1}^n \in \Sigma$ and if $\{J_j\}_{j=1}^n$ are intervals (some of which could be empty) with $J_j \subset I_j$ for every j, then $\{J_j\}_{j=1}^n \in \Sigma$.

To each collection $\{I_j\}_{j=1}^n \in \Sigma$ as above we associate 2^n linear operators on E defined by $\varphi(x) = \sum_{j=1}^n \theta_j I_j x$, where $\theta_j = \pm 1$. The collection of all these operators φ for all $\{I_j\}_{j=1}^n \in \Sigma$ will be denoted by M_Σ. Let F be a subspace of E spanned by a block basis $\{y_m\}_{m=1}^\infty$ of the e_i's. We denote by $M_\Sigma(F)$ the operators $\varphi \in M_\Sigma$ for which $I_j y_m$ is either 0 or y_m for every j and m. Clearly $\varphi(F) \subset F$ for every $\varphi \in M_\Sigma(F)$.

The distortion assumption in Theorem 13.45 is used in the following lemma.

Lemma 13.48. *Let E be as in Theorem 13.45, and assume also that it has a basis $\{e_i\}$. Let Σ be a family as above, and let F be a subspace of E spanned by a block basis $\{y_m\}_{m=1}^\infty$ of the e_i's. Assume further that $\sup\{\|\varphi\| : \varphi \in M_\Sigma(F)\} < \infty$. Then there is a subspace G of F, spanned by a block basis of the y_m's, such that $\|\psi\| \le 4\mu + 1$ for every $\psi \in M_\Sigma(G)$.*

Proof. Introduce a new norm on F by

$$|||x||| = \sup\{\|\varphi(x)\| : \varphi \in M_\Sigma(F)\}.$$

Then $|||x||| \ge \|x\|$ for every $x \in F$ by (i), and hence $|||\cdot|||$ is equivalent to the original norm. Using (ii), we see that if $\varphi, \psi \in M_\Sigma(F)$, then $\varphi \circ \psi = \xi_1 + \xi_2 + \xi_3 + \xi_4$, where each ξ_i belongs to $M_\Sigma(F)$.

By the distortion assumption on E there is a subspace G of F, spanned by a block basis of the y_m's and a constant $0 < \lambda < \infty$ such that

$$\lambda |||x||| \le \|x\| \le (\mu + \frac{1}{4})\lambda |||x|||$$

for every $x \in G$. By the remark above we see that for every $\psi \in M_\Sigma(F)$

$$|||\psi(x)||| = \sup\{\|\varphi \circ \psi(x)\| : \varphi \in M_\Sigma(F)\} \le 4|||x|||.$$

If in addition $\psi \in M_\Sigma(G)$, then ψ leaves G invariant, and for every $x \in G$

$$\|\psi(x)\| \le (\mu + \frac{1}{4})\lambda |||\psi(x)||| \le (4\mu + 1)\lambda |||x||| \le (4\mu + 1)\|x\|.$$

□

Proof of theorem 13.45. Passing to a subspace and renorming we may assume that E has a bimonotone basis $\{e_i\}$ (see appendix E). Let T be the tree on E consisting of the tuples of the form $(y_1, \dots, y_n)$, where n is arbitrary and $\{y_j\}_{j=1}^n$ is a normalized block of the e_i's such that $\|\sum_{j=1}^n \pm a_j y_j\| \le (4\mu + 1)\|\sum_{j=1}^n a_j y_j\|$ for all choices of signs and scalars $\{a_j\}$. Then T is a closed tree, and we shall show that it is not well-founded. This will prove the theorem.

By Proposition 13.47 it suffices to show that $o(T) = \omega_1$. To this end we define by transfinite induction for each countable ordinal α a family Σ_α satisfying (i) and (ii) as above and a block subspace F_α of E so that the following conditions hold.

(iii) If $\gamma < \alpha$, then $F_\alpha \cap F_\gamma$ is of finite codimension in F_α.

(iv) $\|\varphi(x)\| \le (4\mu + 1)\|x\|$ for every $x \in F_\alpha$ and $\varphi \in M_{\Sigma_\alpha}(F_\alpha)$.

(v) For every block subspace G of E, let $T_{\Sigma_\alpha}(G)$ be the tree on G consisting of all tuples $(z_1, \dots, z_n)$ with $z_j \in G$ and $\|z_j\| = 1$, and such that $(I_1, \dots, I_n) \in \Sigma_\alpha$, where I_j is the minimal interval containing the support of z_j. Then $o(T_{\Sigma_\alpha}(G)) \ge \alpha$.

Once this construction is accomplished, then by (iv) the tree $T_{\Sigma_\alpha}(F_\alpha)$ is a subset of T and by (v) $o(T) \geq o\big(T_{\Sigma_\alpha}(F_\alpha)\big) \geq \alpha$. Since α is an arbitrary countable ordinal, the desired result follows.

We now pass to the construction of Σ_α and F_α. For $\alpha = 1$ we take Σ_1 to be the family consisting of all single intervals, and F_1 is the space given by Lemma 13.48 for Σ_1. (Actually, since the basis is bimonotone, the norm of every element in M_{Σ_1} is 1, and we could take $F_1 = E$.) Condition (iii) is vacuous in this case, (iv) holds by Lemma 13.48, and (v) is evident.

Passing from α to $\alpha+1$, we take $\Sigma_{\alpha+1}$ to consist of all collections $\{I_j\}_{j=1}^n$ where $\{I_j\}_{j=2}^n \in \Sigma_\alpha$. Then (i) and (ii) hold for $\Sigma_{\alpha+1}$ because they hold for Σ_α. Each operator $\varphi \in M_{\Sigma_{\alpha+1}}(F_\alpha)$ is of the form $\varphi = I_1 \pm \psi$ where $\psi \in M_{\Sigma_\alpha}(F_\alpha)$. Since the operators in $M_{\Sigma_\alpha}(F_\alpha)$ are uniformly bounded on F_α (by $4\mu + 1$), it follows that $M_{\Sigma_{\alpha+1}}(F_\alpha)$ is also uniformly bounded on F_α. By Lemma 13.48 there is a subspace $F_{\alpha+1}$ of F_α, spanned by blocks, such that $\|\varphi\| \leq 4\mu+1$ for every $\varphi \in M_{\Sigma_{\alpha+1}}(F_{\alpha+1})$. Condition (iii) holds because $F_{\alpha+1} \subset F_\alpha$ and because (iii) holds up to α. Condition (iv) is obvious, and (v) follows from the following observation: Let G be any block subspace of E, and let z_1 be any normalized block in G. Let G_0 be the subspace of G spanned by vectors starting after z_1. If $\{z_j\}_{j=2}^n$ belongs to $T_{\Sigma_\alpha}(G_0)$, then $\{z_j\}_{j=1}^n \in T_{\Sigma_{\alpha+1}}(G)$. Hence $o\big(T_{\Sigma_{\alpha+1}}(G)\big) \geq o\big(T_{\Sigma_\alpha}(G_0)\big) + 1 \geq \alpha + 1$.

Finally we consider a limit ordinal β, and we choose a sequence $\alpha_n \nearrow \beta$. We start with the definition of the subspace F_β. Choose a normalized block basis $\{y_j\}_{j=1}^\infty$ such that $y_j \in F_{\alpha_m}$ for $m \leq j$. This is possible because $\{F_\alpha\}_{\alpha<\beta}$ satisfy (iii). We then put $F_\beta = \overline{\operatorname{span}\{y_j\}}$. Then the codimension of F_β in F_{α_m} is at most m, hence $\{F_\alpha\}_{\alpha\leq\beta}$ also satisfies (iii).

Let now Σ_β be the family consisting of all single intervals as well as of collections $\{I_j\}_{j=1}^n$ such that there is an m for which $\{I_j\}_{j=1}^n \in \Sigma_{\alpha_m}$ and $\operatorname{supp} y_{m-1} \prec I_1$. It is clear that Σ_β satisfies (i) and (ii). If $\varphi(x) = \sum_{j=1}^n \pm I_j x$ where $\operatorname{supp} y_{m-1} \prec I_1$ and $x \in F_\beta$, then there is an $\tilde{x} \in F_{\alpha_m}$ with $\varphi(x) = \varphi(\tilde{x})$ and $\|\tilde{x}\| \leq \|x\|$ (by the bimonotonicity of the basis). Hence $\|\varphi(x)\| \leq (4\mu+1)\|x\|$ for every $\varphi \in M_{\Sigma_\beta}(F_\beta)$ and $x \in F_\beta$.

It remains to check (v) for β. Let G be a block subspace of E, and for every m put $G_m = \{z \in G : y_{m-1} \prec z\}$. Then

$$o\big(T_{\Sigma_\beta}(G)\big) \geq o\big(T_{\Sigma_\beta}(G_m)\big) \geq o\big(T_{\Sigma_{\alpha_m}}(G_m)\big) \geq \alpha_m.$$

Since this holds for every m, we see that $o\big(T_{\Sigma_\beta}(G)\big) \geq \beta$. ◊

A nice overview of much of the material in this chapter and of related developments in infinite-dimensional Banach space theory is given in Maurey [**399**].

CHAPTER 14

Perturbations of Local Isometries

1. Isometries

In this and the next chapter we study maps which nearly preserve distances and their special properties. In particular, how close are they to true isometries? In this chapter we study "local" questions; i.e., the maps will be defined only on some bounded subset, often just a ball. In the next chapter we shall consider "global" questions, i.e., for maps defined on the entire space. The questions, as well as the methods, are somewhat different in the two setups, and in the global case one usually investigates the behavior of $f(x)$ as $\|x\| \to \infty$.

In the first section we consider isometries, and we then introduce, in Section 2, the class of quasi-isometries. These maps are, by their definition, locally one-one, and we discuss under what conditions they are actually injective on large subsets. This is done in the framework of general Banach spaces.

In Section 3 we study the approximation of quasi-isometries by isometries mainly, but not exclusively, in the setting of finite-dimensional Euclidean spaces.

In Section 4 we study the approximation of the differentials of quasi-isometries on finite-dimensional Euclidean spaces (rather than of the quasi-isometries themselves). Historically, BMO functions were first introduced in this framework.

A map f from an open subset of a Banach space E into a Banach space F is called open if it maps open sets into open sets. It is called a local isometry if every point has a neighborhood U so that the restriction of f to U is an isometry.

Theorem 14.1. *Let f be an open local isometry from an open connected subset Ω of a Banach space E onto an open subset of a Banach space F. Then f is the restriction of an affine isometry of E onto F. In particular, any isometry from E onto F is affine.*

Proof. We show that if f is an isometry from $B(z, 2r)$ onto $B(f(z), 2r)$, then it is affine on $B(z, r)$. Since Ω can then be covered by small balls so that f is affine on each of them, and since two affine maps that agree on an open set are the restrictions of the same affine map, the result follows from the connectedness of Ω.

Fix $x, y \in B(z, r)$, and define inductively a decreasing family of closed sets by

$$K_0 = \{u : \|u - x\| = \|u - y\| = \|x - y\|/2\}$$
$$K_{n+1} = \{u \in K_n : K_n \subseteq B(u, d_n/2)\}$$

where $d_n = \operatorname{diam}(K_n)$. Note that f is an isometry on K_0 because $K_0 \subset B(z, 2r)$.

We shall show that $\bigcap K_n = \{(x + y)/2\}$. Since the definition of the K_n's is purely metric, this gives a metric characterization of $(x + y)/2$ (and similarly for $(f(x) + f(y))/2$). Since f is an isometry of $B(z, 2r)$ onto $B(f(z), 2r)$, it follows that $(f(x) + f(y))/2 = f((x + y)/2)$, and hence that f is affine in $B(z, r)$.

To simplify the notation, translate so that $y = -x$, and hence $(x+y)/2 = 0$. It is clear that each K_n is symmetric (i.e., $u \in K_n$ if and only if $-u \in K_n$), and we prove by induction that $0 \in K_n$ and that the diameter of K_{n+1} is at most $d_n/2$.

Fix $u \in K_n$. Then $2\|u\| = \|u-(-u)\| \le d_n$ because $-u \in K_n$ by the symmetry of K_n. Thus $\|u\| \le d_n/2$, and $K_n \subseteq B(0, d_n/2)$, i.e., $0 \in K_{n+1}$. Also, if $u, v \in K_n$, then $v \in B(u, d_n/2)$. Thus $\|u - v\| \le d_n/2$.

Finally $\bigcap K_n = \{0\}$ because $\operatorname{diam}(K_n) \to 0$ and they all contain 0. □

For strictly convex F Theorem 14.1 is trivial even for into maps, but this is false in general. Indeed, if $g : \mathbb{R} \to \mathbb{R}$ is any 1-Lipschitz function, then $f(t) = \big(t, g(t)\big)$ is an isometry from $\mathbb{R}$ into l_2^∞. Note, however, that in this case the linear map $T : l_2^2 \to \mathbb{R}$ defined by $T(t,s) = t$ is a norm-one operator so that $T \circ f$ is the identity on $\mathbb{R}$. It turns out that this is the structure of a general globally defined into isometry. (For onto isometries we thus obtain a different proof of Theorem 14.1.)

Theorem 14.2. *Let f be an isometry from a Banach space E into a Banach space F with $f(0) = 0$. Then there is a unique linear operator T from the closed linear span of $f(E)$ into E such that $\|T\| = 1$ and $Tf(x) = x$ for every $x \in E$.*

Proof. Uniqueness is obvious, and it implies that it suffices to prove existence for finite-dimensional E. Indeed, for each finite-dimensional subspace $D \subset E$ we then obtain an operator T_D on $\overline{\operatorname{span}} f(D) \subset F$, so that $T_D f(x) = x$ for $x \in D$. By the uniqueness $D \subset C$ implies that T_D is the restriction of T_C to $\overline{\operatorname{span}} f(D)$; hence the T_D's define an operator as required on all of $\overline{\operatorname{span}} f(E)$.

Put $n = \dim E < \infty$, and assume that F is the closed linear span of $f(E)$.

Let $n = 1$ (i.e., $f : \mathbb{R} \to F$). For each k find a norm-one $x_k^* \in F^*$ so that

$$\langle x_k^*, f(k) - f(-k)\rangle = \|f(k) - f(-k)\| = 2k.$$

The fact that f is an isometry and that $f(0) = 0$ implies that $\langle x_k^*, f(\lambda)\rangle = \lambda$ for all $|\lambda| \le k$. Let x^* be a w^*-limit point of the sequence $\{x_k^*\}_{k=1}^\infty$. Then $\langle x^*, f(\lambda)\rangle = \lambda$ for all λ, so $T = x^*$ works, and clearly $\|x^*\| = 1$.

Now let $n > 1$. By the above, for every $x \in E$ with $\|x\| = 1$ there is a $u_x^* \in F^*$ such that $\|u_x^*\| = 1$ and $\langle u_x^*, f(\lambda x)\rangle = \lambda$ for all real λ. We claim that if x is a smooth point in the unit sphere of E, then $u_x^* f$ is linear: it is the unique norm-one functional which supports the unit ball of E at x. Denote this functional by $v_x^* \in E^*$, and assume that for some $y \in E$, $\langle u_x^*, f(y)\rangle \ne \langle v_x^*, y\rangle$. The vector $y - \langle u_x^*, f(y)\rangle x$ is not in the kernel of v_x^*; hence, as x is a smooth point, there is a scalar t such that $\|x - t\big(y - \langle u_x^*, f(y)\rangle x\big)\| < 1$ or alternatively $\|y - \big(\langle u_x^*, f(y)\rangle + 1/t\big)x\| < 1/|t|$. A contradiction follows because the homogeneity of $u_x^* f$ in the direction of x gives

$$\begin{aligned}
1/|t| &= \Big|\langle u_x^*, f(y)\rangle - \big(\langle u_x^*, f(y)\rangle + 1/t\big)\Big| \\
&= \Big|\langle u_x^*, f(y)\rangle - \Big\langle u_x^*, f\Big(\big(\langle u_x^*, f(y)\rangle + 1/t\big)x\Big)\Big\rangle\Big| \\
&= \Big|\Big\langle u_x^*, f(y) - f\big((\langle u_x^*, f(y)\rangle + 1/t)x\big)\Big\rangle\Big| \\
&\le \Big\|f(y) - f\Big(\big(\langle u_x^*, f(y)\rangle + 1/t\big)x\Big)\Big\| \\
&= \big\|y - \big(\langle u_x^*, f(y)\rangle + 1/t\big)x\big\| < 1/|t|.
\end{aligned}$$

By Theorem 4.17 the smooth points are dense in the unit sphere of E. Hence there is a basis $\{v_i^*\}_{i=1}^n$ of E^* such that each v_i^* supports the unit sphere of E at a smooth point. Let $\{u_i^*\} \subset F^*$ satisfy $\|u_i^*\| = 1$ and $u_i^* f = v_i^*$ for all i. Choose $\{z_i\}_{i=1}^n$ in E such that $\langle v_i^*, z_j \rangle = \delta_{i,j}$, and define a linear operator $T : F \to E$ by

$$Ty = \sum_{i=1}^{n} \langle u_i^*, y \rangle z_i.$$

Then $T \circ f$ is the identity on E. We shall show that $\|y\| \geq 1$ whenever Ty is a smooth point of the unit sphere of E. Since the smooth points are dense in the unit sphere of E, and since T is linear and surjective, this proves that $\|T\| = 1$.

Let v^* be the (unique) supporting functional at Ty, and let $u^* \in F^*$ satisfy $\|u^*\| = 1$ and $v^* = u^* f$. Then the linear functional $u^* \circ f \circ T$ is equal to u^*. Indeed, since $f(E)$ spans a dense subspace of F, it is enough to note that $u^* \circ f \circ T \circ f = u^* \circ f$ (because $T \circ f$ is the identity on E). Hence

$$1 = \|Ty\| = \langle v^*, Ty \rangle = \langle u^* \circ f, Ty \rangle = \langle u^* \circ f \circ T, y \rangle = \langle u^*, y \rangle \leq \|y\|.$$

□

Remark. It is worthwhile to note that there is a nonexpansive retraction from the closed linear span of $f(E)$ onto $f(E)$. Indeed, $f \circ T$ is such a retraction.

2. Quasi-Isometries, Injectivity

Let f be a map from an open subset Ω of a Banach space E into a Banach space F. The local distortion of distances by f will be measured by the two functions

$$D^+ f(x) = \limsup_{y \to x} \|f(y) - f(x)\| / \|y - x\|,$$
$$D^- f(x) = \liminf_{y \to x} \|f(y) - f(x)\| / \|y - x\|.$$

Definition 14.3. *A map f from an open subset Ω of a Banach space E into a Banach space F is called an (m, M)-quasi-isometry if it satisfies the following two conditions:*

(i) *f is a local homeomorphism; i.e., every point $x \in \Omega$ has an open neighborhood V such that f is a homeomorphism of V onto an open subset of F.*

(ii) *f satisfies $m \leq D^- f(x) \leq D^+ f(x) \leq M$ for every $x \in \Omega$.*

The map f is called a quasi-isometry if it is an (m, M)-quasi-isometry for some m and M.

The distance of m and M to 1 will serve as a measure of how close f is to being an isometry. In some questions the important parameter is the ratio $\mu = M/m$, which measures how far f is from a multiple of an isometry.

Our aim is to study the global geometric properties that follow from the local constraints imposed in the definition. For example: What geometric conditions on Ω and on the size of m and M force the mapping f to actually be injective on all of Ω? When does f necessarily satisfy global upper and lower Lipschitz conditions on all of Ω or on some "large" subset of it? Can f be approximated by an isometry?

The physical motivation for these questions comes from the theory of elasticity. Imagine Ω as an elastic solid (in three-, or in n-dimensional Euclidean space) being deformed by the mapping f. Being an (m, M)-quasi-isometry means that we only allow limited local deformations, and this is a constraint on the strain at each point. The questions we ask are an attempt to understand the possible global geometry of the deformation. All of these geometric properties have a physical interpretation: the global injectivity of f means that no buckling occurs. Control on the global upper and lower Lipschitz constants of f mean that we can infer global information on the distortion of the deformation from the local strain. Finally, the possibility to approximate f by an isometry means that small strain at every point necessarily leads to deformations which are close to rigid motions.

Global upper estimates of the Lipschitz constant follow automatically from local ones when the set is convex.

Lemma 14.4. *Let f be a map from a convex subset Ω of a Banach space E into a Banach space F. If $D^+f(x) \leq M$ for all $x \in \Omega$, then $\|f(x) - f(y)\| \leq M\|x - y\|$ for all x and y in Ω.*

Proof. If $\|f(y) - f(x)\|/\|y - x\| = N > M$, put $z = (x + y)/2$. By the triangle inequality either $\|f(z) - f(x)\|/\|z - x\| \geq N$ or $\|f(z) - f(y)\|/\|z - y\| \geq N$. Repeating the argument, we find intervals $[x_n, y_n]$ with a unique common point u such that $\|x_n - y_n\| \to 0$, and such that $\|f(y_n) - f(x_n)\|/\|y_n - x_n\| \geq N$. By the triangle inequality again, and by passing to a subsequence, we can assume that $\|f(u) - f(x_n)\|/\|u - x_n\| \geq N$. But then $D^+f(u) \geq N > M$, a contradiction. □

From now on we assume that Ω is convex. Thus the upper Lipschitz condition holds globally, and we look for lower estimates. We start with a simple observation. Assume that f is injective on an arc γ in Ω and that $D^-f(x) \geq m$ for every x in γ. If the image $f(\gamma)$ is an interval $[f(u), f(v)]$, then $\|f(y) - f(z)\|/\|y - z\| \geq m$ for all $y, z \in \gamma$. Indeed, under these circumstances f^{-1} is defined as a map from the interval $[f(u), f(v)]$ onto γ, and it satisfies the conditions of Lemma 14.4 on this interval with $M = 1/m$. This observation motivates the following definition:

Definition 14.5. *Let Ω be an open connected set in a Banach space E, and let $f : \Omega \to F$ be a local homeomorphism. The star S_x of the point $x \in \Omega$ is the set of all points $y \in \Omega$ such that the interval $[f(x), f(y)]$ is contained in $f(\Omega)$ and such that there is an arc $\gamma \subset \Omega$ connecting x and y with the property that the restriction of f to γ is a homeomorphism from γ onto the interval $[f(x), f(y)]$.*

The next lemma summarizes the properties of the stars S_x.

Lemma 14.6. *Let f be an (m, M)-quasi-isometry from an open convex subset Ω in a Banach space E into a Banach space F. Then*

(i) *$\|f(x) - f(y)\| \geq m\|x - y\|$ for all $y \in S_x$.*

(ii) *$y \in S_x$ if and only if $x \in S_y$.*

(iii) *S_x and $f(S_x)$ are open, and f is a homeomorphism from S_x onto its image. We denote its inverse on $f(S_x)$ by f_x^{-1}.*

(iv) *Assume $[f(x), u)$ is a maximal interval in $f(S_x)$, i.e., $u \notin f(S_x)$ and $v \in f(S_x)$ for all $v \in [f(x), u)$. Then $z = \lim_{v \to u} f_x^{-1}(v)$ exists, and z lies on the boundary of Ω. (The limit is taken over points $v \in [f(x), u)$.)*

Proof. (i) follows from the remarks preceding the definition of the star, and (ii) is obvious because the same arc that connects x to y connects y to x.

(iii) Fix $y \in S_x$, and let γ be the arc connecting x and y as in the definition of the star. We first show that there is an open neighborhood U of γ such that f is a homeomorphism when restricted to U. If no such U exists, we could find pairs of points $u_n \neq v_n$ such that $d(u_n, \gamma), d(v_n, \gamma) \to 0$ and such that $f(u_n) = f(v_n)$. By the compactness of γ, we can assume that $u_n \to u$ and $v_n \to v$ for some $u, v \in \gamma$. Since f is one-one on γ, it follows that $u = v$. Restricting to a small neighborhood of this common limit, the equality $f(u_n) = f(v_n)$ contradicts the assumption that f is a local homeomorphism.

The set $f(U)$ is open because f is a local homeomorphism. It thus contains intervals connecting $f(x)$ to all points in some neighborhood of $f(y)$; hence S_x and $f(S_x)$ are open. To show that f is injective on S_x, assume that $y \neq y_1$ satisfy $f(y) = f(y_1) = u$, and find arcs γ and γ_1 connecting x with y and with y_1 respectively so that their images are the segment $[f(x), u]$. Thus f is not a homeomorphism in a neighborhood of the last point where the two arcs intersect, contradicting the assumption that f is a local homeomorphism.

(iv) Note that f_x^{-1} is a Lipschitz map on the half-open interval $[f(x), u)$. Thus the limit $z = \lim_{v \to u} f_x^{-1}(v)$ exists. That $z \notin \Omega$ follows from the fact that f is a local homeomorphism and from the maximality of $[f(x), u)$. □

Put $E(x, y, k) = \{z \in E : \|x - z\| + \|y - z\| \leq k\|y - x\|\}$. In Hilbert space, this is an ellipsoid of revolution with foci at x and y and eccentricity $1/k$.

Theorem 14.7. *Let f be an (m, M)-quasi-isometry from a convex domain Ω in a Banach space E into a Banach space F. Then*

(i) *Assume that Ω is the open ball $B_E(z, r)$. Then $f(S_z) \supset B_F(f(z), rm)$ and $S_z \supset B_F(z, rm/M)$. The function f satisfies $M \geq \|f(y) - f(x)\|/\|y - x\| \geq m$ for all $x, y \in B_E(z, rm/M)$.*

(ii) *If Ω contains $E(x, y, M/m)$, then y is in the star S_x of x. In particular, $\|f(x) - f(y)\| \geq m\|x - y\|$.*

Proof. (i) We may assume that $z = 0$ and that $f(0) = 0$. We denote, here and later on, the ball of radius t centered at the origin by $B(t)$.

Fix u in the unit sphere of F, and let $[0, \eta u)$ be a maximal interval in $f(S_0)$. (Of course, η depends on u.) By Lemma 14.6(iv) the limit $x = \lim_{t \to \eta^-} f_0^{-1}(tu)$ exists and lies on the boundary of $B_E(r)$. By (i) of the same lemma $\|f_0^{-1}(tu)\| \leq t/m$ for every $0 < t < \eta$. Taking $t \to \eta$, we see that $r = \|x\| \leq \eta/m$. Thus $\eta \geq rm$ for every u, and hence $B_F(rm) \subseteq f(S_0)$ as claimed.

Choose a point x in $B_E(rm/M)$, and fix $0 \leq t \leq 1$. Since f is Lipschitz with constant M, it follows that $\|f(tx)\| = \|f(tx) - f(0)\| \leq M\|tx\| < rm$. By the first part of the proof there is a unique $y_t \in S_0$ such that $f(tx) = f(y_t)$. We now show that $y_t = tx$. The set $T = \{t \in [0, 1] : y_t = tx\}$ is nonempty ($0 \in T$), closed (f_0^{-1} is continuous), and open (S_0 is open). Thus $y_t = tx$ for all t and $x \in S_0$.

The argument above actually shows that f_0^{-1} is a homeomorphism of the ball $B_F(mr)$ onto a set containing $B_E(rm/M)$. Since $B_F(mr)$ is convex, it follows that $\|f(y) - f(x)\|/\|y - x\| \geq m$ in $B_E(rm/M)$. The upper estimate is obvious.

(ii) Assume that the claim is false. Replacing, if necessary, y by the point in the interval nearest to x for which the claim is false, we shall assume that $z \in S_x$ for every $z \in [x, y)$. For $0 \leq t \leq 1$ and $0 \leq \lambda \leq 1$, put

$$x_t = (1 - t)x + ty \ ; \quad z_{\lambda,t} = (1 - \lambda)f(x) + \lambda f(x_t).$$

We first analyze the situation for $t < 1$. Since $x_t \in S_x$ for such t, it follows that f_x^{-1} is well-defined on the interval $[f(x), f(x_t)]$, and we put $x_{\lambda,t} = f_x^{-1}(z_{\lambda,t})$. By its definition $x_{\lambda,t} \in S_x \cap S_{x_t}$; hence $\|f(x_{\lambda,t}) - f(x)\| \geq m\|x_{\lambda,t} - x\|$ and also $\|f(x_{\lambda,t}) - f(x_t)\| \geq m\|x_{\lambda,t} - x_t\|$.

Since $f(x), f(x_t)$ and $f(x_{\lambda,t})$ are collinear, and since $\|f(x)-f(x_t)\| \le M\|x-x_t\|$ by Lemma 14.4, we obtain

$$(14.1)\qquad \begin{aligned} M\|x-x_t\| &\ge \|f(x)-f(x_t)\| \\ &= \|f(x)-f(x_{\lambda,t})\| + \|f(x_{\lambda,t})-f(x_t)\| \\ &\ge m\big(\|x_{\lambda,t}-x\| + \|x_{\lambda,t}-x_t\|\big). \end{aligned}$$

We now consider $t = 1$. Since $x_1 = y \notin S_x$, there is a minimal $\mu < 1$ such that f_x^{-1} is not defined at $z_{\mu,1} = (1-\mu)f(x) + \mu f(y)$.

By Lemma 14.6(iv), the limit $x_{\mu^-,1} = \lim_{\lambda\to\mu^-} x_{\lambda,1}$ exists and lies on the boundary of Ω. It follows that $x_{\mu^-,1}$ does not belong to $E(x,y,M/m)$, and hence

$$\|x_{\mu^-,1}-x\| + \|x_{\mu^-,1}-y\| > M\|x-y\|/m.$$

Since $x_{\mu^-,1} = \lim_{\lambda\to\mu^-} x_{\lambda,1}$, we can fix $\lambda < \mu$ so that

$$\|x_{\lambda,1}-x\| + \|x_{\lambda,1}-y\| > M\|x-y\|/m.$$

Letting $t \to 1$ in (14.1) for this λ leads to a contradiction. □

Corollary 14.8. *Let f be a local homeomorphism from a connected open set Ω in a Banach space E onto an open subset of a Banach space F satisfying*

$$\lim_{y\to x} \|f(y)-f(x)\|/\|y-x\| = 1$$

for all $x \in \Omega$. Then f is the restriction of an affine isometry from E onto F.

Proof. By Theorem 14.7(i), f is an isometry in every ball contained in Ω, and the corollary follows from Theorem 14.1. □

We now pass to the main result in this section, and we first need a lemma.

Lemma 14.9. *Let f be an (m,M)-quasi-isometry, defined on the (open) ball $B(R)$ in E, and assume that f is not injective. Then there is an $r < R$, an $\eta > 0$ and sequences $\{x_n\}$ and $\{y_n\}$ in $B(R)$ such that*

(i) *f is injective on $B(r)$.*

(ii) *$f(x_n) = f(y_n)$ for all n.*

(iii) *$\|x_n - y_n\| \ge \eta$ for all n and* $\lim\|x_n\| = \lim\|y_n\| = r$.

Proof. Let $r = \inf\{s : f \text{ is not injective on } B(s)\}$. Clearly $0 < r < R$. By the definition of r it is possible to find, for each n, points $x_n \ne y_n$ in $B(r+1/n)$ with $f(x_n) = f(y_n)$. If $1/n < (R-r)/2$, then $B(x_n,(R-r)/2) \subset B(R)$, and hence $\|x_n - y_n\| \ge (R-r)m/2M$ by Theorem 14.7(i). Put $\eta = (R-r)m/2M$. Passing to a subsequence, assume that $\lim\|x_n\|$ and $\lim\|y_n\|$ both exist. One of them, say $\lim\|y_n\|$, must be equal to r, and we show that $\alpha = \lim\|x_n\|$ is also equal to r.

Assume that $\alpha < r$. If $\delta_1 < (r-\alpha)/3$, then $B(x_n,\delta_1) \subset B\big(r-(r-\alpha)/2\big)$, provided n is large enough. Similarly, if $\delta_2 < (r-\alpha)/3$, and if n is large enough, then $B(y_n,\delta_2)$ intersects $B(r)$ and is disjoint from $B(x_n,\delta_1)$. Pick $z_n \in B(y_n,\delta_2)$ such that $\|z_n\| < r$. Then $\|f(z_n)-f(x_n)\| = \|f(z_n)-f(y_n)\| < M\delta_2$, and hence, by Theorem 14.7(i), $f(z_n) \in f\big(B(x_n,\delta_1)\big)$ for small enough δ_2. As $z_n \notin B(x_n,\delta_1)$, this contradicts the injectivity of f on $B(r)$. □

Theorem 14.10. *There is a number $\mu_0 > 1$ with the following property: Let f be an (m,M)-quasi-isometry from a ball in a Banach space E into a Banach space F. If $M/m < \mu_0$, then f is injective.*

The number μ_0 found in the proof below is the unique real root of the equation

$$\lambda = \frac{\lambda + (25\lambda^2 - 8\lambda)^{1/2}}{2(3\lambda^2 - \lambda)} \tag{14.2}$$

which is 1.114.... It is unknown what the best possible value is. The following simple example shows that it is not larger than 2. Let $\Omega \subset \mathbb{R}^2$ be the open disk with radius 1 centered at $(1,0)$. For each $\alpha > 2$ the map on Ω, given in polar coordinates by $(r,\theta) \to (r, \alpha\theta)$, has $m = 1$ and $M = \alpha$, and it is not injective.

Proof. Let f be a noninjective (m, M)-quasi-isometry defined on $B(R)$, and put $\mu = M/m$. Let r and η be as in the lemma, and assume by rescaling that $r = 1$. Fix ε such that

$$0 < \varepsilon < \eta/2, \quad \text{and} \quad \frac{2\mu(1+\varepsilon)}{2\mu + \eta - 2\varepsilon} < \frac{4\mu + \eta}{4\mu + 2\eta}. \tag{14.3}$$

Choose n large enough so that $\big|\|x_n\| - 1\big|, \big|\|y_n\| - 1\big| < \varepsilon$, and put $x = x_n/\|x_n\|$ and $y = y_n/\|y_n\|$. Then $\|x - x_n\|, \|y - y_n\| < \varepsilon$, and hence $\|x - y\| > \eta - 2\varepsilon$ and also $\|f(x) - f(y)\| \le \|f(x) - f(x_n)\| + \|f(y_n) - f(y)\| < 2M\varepsilon$.

Fix $t < 1$, and assume (for reasons to become clear later) that

$$\frac{4\mu + \eta}{4\mu + 2\eta} < t < 1. \tag{14.4}$$

The number $s = \sup\Big\{\tau : 0 < \tau < t \text{ and } [f(\tau x), f(\tau y)] \subset f\big(B(t)\big)\Big\}$ is positive because $f\big(B(t)\big)$ is open, and if $\tau < s$, then

$$\|f(\tau x) - f(\tau y)\| \ge m\|\tau x - \tau y\| = \tau m\|x - y\| > \tau m(\eta - 2\varepsilon). \tag{14.5}$$

On the other hand,

$$\begin{aligned} \|f(\tau x) - f(\tau y)\| &\le \|f(x) - f(y)\| + \|f(\tau x) - f(x)\| + \|f(\tau y) - f(y)\| \\ &\le 2M\varepsilon + 2M(1-\tau) = 2M(1 - \tau + \varepsilon). \end{aligned} \tag{14.6}$$

Comparing with (14.5), we see that $\tau < 2\mu(1+\varepsilon)/(2\mu + \eta - 2\varepsilon)$, and hence the choice of ε and t in (14.3) and in (14.4) gives

$$0 < s \le \frac{4\mu + \eta}{4\mu + 2\eta} < t. \tag{14.7}$$

By the maximality of s there is a point u with $\|u\| = t$ such that $f(u)$ belongs to the interval $[f(sx), f(sy)]$. Since $\|u - sx\|, \|u - sy\| \ge t - s$ there is a $\lambda > 1$ with

$$\|u - sx\| + \|u - sy\| = 2\lambda(t - s). \tag{14.8}$$

Since $[f(sx), f(u)]$ and $[f(sy), f(u)]$ are contained in $f\big(B(1)\big)$, Lemma 14.6(i) gives

$$\begin{aligned} \|f(sx) - f(sy)\| &= \|f(sx) - f(u)\| + \|f(u) - f(sy)\| \\ &\ge m(\|sx - u\| + \|sy - u\|) = 2m\lambda(t - s). \end{aligned}$$

Comparing with (14.6) (for $\tau = s$), we obtain

$$\mu \ge \lambda(t - s)/(1 - s + \varepsilon). \tag{14.9}$$

We now derive another inequality involving μ and λ which, when combined with (14.9), will give the desired estimate $\mu \ge \mu_0$. Put $z = (sx + sy)/2$. Then $\|z\| \le s$, and by (14.8) and the triangle inequality

$$t - s \le \|z - u\| \le \lambda(t - s). \tag{14.10}$$

Put

$$v = z + (u - z)/\lambda\mu \qquad w = z - (u - z)/\lambda\mu.$$

Then $v, w \in B(z, (t-s)/\mu)$ by (14.10). Since also $B(z, t-s) \subset B(t)$, Theorem 14.7(i) gives that $\|f(v) - f(w)\| \geq m\|v - w\| \geq 2m(t-s)/\lambda\mu$. Subtracting the inequality

$$\|f(v) - f(u)\| \leq M\|v - u\| = M(1 - 1/\lambda\mu)\|z - u\| \leq M(1 - 1/\lambda\mu)\lambda(t - s)$$

we obtain

$$\|f(w) - f(u)\| \geq (t - s)\big(2m/\lambda\mu - \lambda M(1 - 1/\lambda\mu)\big). \tag{14.11}$$

On the other hand, $f(u) \in [f(sx), f(sy)]$ implies that

$$\begin{aligned} \|f(w) - f(u)\| &\leq \max\big(\|f(w) - f(sx)\|, \|f(w) - f(sy)\|\big) \\ &\leq \max\big(M\|w - sx\|, M\|w - sy\|\big). \end{aligned} \tag{14.12}$$

To estimate $\|w - sx\|$, we use $w \in [z, 2z - u]$ and estimate (using (14.8))

$$\begin{aligned} \|z - sx\| &= \|(sx - sy)/2\| \leq \|(sx - u)/2\| + \|(sy - u)/2\| \\ &= \lambda(t - s) \leq (2\lambda - 1)(t - s) \end{aligned}$$

(because $\lambda \geq 1$), and

$$\|2z - u - sx\| = \|sy - u\| = 2\lambda(t - s) - \|sx - u\| \leq (2\lambda - 1)(t - s).$$

Thus $\|w - sx\| \leq (2\lambda - 1)(t - s)$, with a similar estimate for $\|w - sy\|$. Using these estimates together with (14.11) and (14.12) yields another inequality involving λ and μ, namely

$$M(2\lambda - 1) \geq 2m/\lambda\mu - \lambda M(1 - 1/\lambda\mu)$$

or (since $\mu = M/m$) that $(3\lambda^2 - \lambda)\mu^2 - \lambda\mu - 2 \geq 0$, which implies, in turn, that

$$\mu \geq \frac{\lambda + (25\lambda^2 - 8\lambda)^{1/2}}{2(3\lambda^2 - \lambda)}. \tag{14.13}$$

This inequality contains only μ and λ, while the inequality (14.9) contains the additional parameters s, t and ε (and λ depends on them of course). The quantities μ and η (in (14.4)) are fixed. In the choice of t and ε we were restricted only by $\varepsilon < \eta/2$ and by (14.4), so ε and $1 - t$ can be chosen as small as we please. The only control we have on s is (14.7), which shows that $t - s$ is bounded from below as $t \to 1$. Letting $\varepsilon \to 0$ and $t \to 1$, we see that there is a λ with $\mu \geq \lambda$ for which (14.13) holds. Since the right-hand side of (14.13) decreases with λ, we deduce that $\mu \geq \mu_0$, where μ_0 is the unique solution of (14.2). □

3. Approximation of Quasi-Isometries by Isometries

In the main results of this and the next section $E = F$ is the n-dimensional Euclidean space $\mathbb{R}^n$. The constants appearing in the statements and proofs depend on the dimension. We give explicit estimates that follow from the chosen proofs, but are certainly not best possible. It may even be possible that some of them can be replaced by absolute constants independent of n.

We shall denote by K absolute constants, which may have different values even in the same formula. The unit vectors in $\mathbb{R}^n$ will be denoted by $\{e_i\}$, and the identity matrix by I. An (m, M)-quasi-isometry with $(1 + \varepsilon)^{-1} \leq m \leq M \leq 1 + \varepsilon$ will be called an ε-quasi-isometry.

Observe that if $\{v_i\}$ are "nearly orthogonal" vectors in $\mathbb{R}^n$ in the sense that $|\delta_{ij} - \langle v_i, v_j\rangle| \leq \eta$ for some $\eta > 0$, then there is an orthogonal matrix U such that $\|v_i - Ue_i\| \leq n\eta$. Indeed, write the matrix V, whose columns are the v_j's, as $V = UP$ where P is positive definite and U orthogonal. Replacing the v_i's by $U^{-1}v_i$, we can assume that $U = I$, i.e., $V = P$; hence $\|(I+V)^{-1}\| \leq 1$. We then obtain $\|v_i - e_i\| = \|(V-I)e_i\| \leq n\eta$ because

$$\begin{aligned}\|V - I\| &\leq \|V^2 - I\|\,\|(V+I)^{-1}\| \leq \|V^2 - I\| \leq \|V^2 - I\|_{HS} \\ &= \Big(\sum_{i\neq j} |\langle v_i, v_j\rangle|^2 + \sum |\langle v_i, v_i\rangle - 1|^2\Big)^{1/2} \leq n\eta\end{aligned}$$

where $\|A\|_{HS}$ denotes the Hilbert-Schmidt norm of the matrix A.

Theorem 14.11. *Let Ω be a convex domain in $\mathbb{R}^n$ which contains the ball $B(z,r)$ of radius r and is contained in the concentric ball $B(z,R)$ of radius R. Let $0 < \varepsilon < 1$, and let f be an ε-quasi-isometry in Ω. Then there is an orthogonal matrix U such that $\|f(x) - f(z) - U(x-z)\| \leq Kn^{3/2}\varepsilon R^2/r$ for every $x \in \Omega$, where K is an absolute constant.*

Proof. Note first that the right-hand side is properly scaled: R/r is a geometric invariant measuring the "regularity" of Ω, and the extra R is the scaling factor.

We first assume that $\Omega = B(z,r)$ is a ball and find an orthogonal matrix U so that $\|f(x) - f(z) - U(x-z)\| \leq Kn^{3/2}r\varepsilon$ for every x in the smaller concentric ball $B(z,(1+\varepsilon)^{-2}r)$. By translation and scaling we can assume that $z = 0$, $r = (1+\varepsilon)^2$, and that $f(0) = 0$. We then need to find an orthogonal matrix which approximates f in the unit ball (which we denote by B). We shall use repeatedly Theorem 14.7(i) in the following form:

If $B(u,\rho) \subseteq \Omega$, then $(1+\varepsilon)^{-1} \leq \|f(x)-f(y)\|/\|x-y\| \leq 1+\varepsilon$ on $B(u,(1+\varepsilon)^{-2}\rho)$.

In particular, we see that $(1+\varepsilon)^{-1} \leq \|f(x) - f(y)\|/\|x-y\| \leq 1+\varepsilon$ in B. It follows that $|\langle f(x), f(y)\rangle - \langle x,y\rangle| \leq 10\varepsilon$ for all $x, y \in B$. Indeed,

$$\begin{aligned}&|\langle f(x), f(y)\rangle - \langle x, y\rangle| \\ &= \Big|\|f(x)\|^2 - \|x\|^2 + \|f(y)\|^2 - \|y\|^2 - \big(\|f(x) - f(y)\|^2 - \|x-y\|^2\big)\Big|/2 \\ &\leq \big((1+\varepsilon)^2 - 1\big)\big(\|x\|^2 + \|y\|^2 + \|x-y\|^2\big)/2 \leq 10\varepsilon.\end{aligned}$$

Taking $v_i = f(e_i)$, the observation preceding the theorem implies that there is an orthogonal matrix U so that $\|f(e_i) - Ue_i\| \leq 10n\varepsilon$. Replacing f by $U^{-1}\circ f$, we can assume that $U = I$. We shall show that

$$\|f(x) - x\| = \big(\sum \langle f(x) - x, e_i\rangle^2\big)^{1/2} \leq Kn^{3/2}\varepsilon$$

for every $x \in B$. Indeed, $\|e_i - f(e_i)\| \leq 10n\varepsilon$ implies that

$$\begin{aligned}&|\langle f(x) - x, e_i\rangle| \\ &\leq |\langle f(x), e_i - f(e_i)\rangle| + |\langle f(x), f(e_i)\rangle - \langle x, e_i\rangle| \\ &\leq (1+\varepsilon)10n\varepsilon + 10\varepsilon \leq Kn\varepsilon.\end{aligned}$$

Passing to the general case, translate so that $z = 0$ and $f(0) = 0$. Let U_0 be the matrix found in the first part of the proof, i.e., $\|f(x) - U_0x\| \leq Kn^{3/2}r\varepsilon$ on the ball $B(r(1+\varepsilon)^{-2})$. We shall show that U_0 actually approximates f well on all of Ω. This will be done by first considering f on sub-balls, in each of which we shall use the estimate from the first part to find an approximating isometry, and then showing that all these isometries can be replaced by U_0.

To estimate the difference between two such isometries, we shall use the following simple fact: if A_1 and A_2 are two matrices such that the affine transformations $L_i x = y_i + A_i x$ satisfy $\|L_1 x - L_2 x\| \le \alpha$ on some (not necessarily central) ball of radius β in $\mathbb{R}^n$, then $\|A_1 - A_2\| \le 2\alpha/\beta$.

Put $q = 1/(1 + rR^{-1}(1+\varepsilon)^{-2}) < 1$. Fix any point $x \in \Omega$, and consider the points $x_j = (1-q^j)x$. Since Ω is convex and contains the ball $B(r)$, the distance of x_j to the boundary of Ω is larger than $q^j r$. Thus if we put $r_j = q^j(1+\varepsilon)^{-2} r$, then Theorem 14.7(i), when applied to each of the balls $B_j = B(x_j, r_j)$, implies that $(1+\varepsilon)^{-1} \le \|f(z) - f(y)\|/\|z - y\| \le 1 + \varepsilon$ there. By the first part of the proof we can find, for each j, an orthogonal matrix U_j such that $\|f(y) - L_j y\| \le Kn^{3/2}\varepsilon r_j$ on B_j, where L_j is the isometry given by $L_j y = f(x_j) + U_j(y - x_j)$.

It follows from the the choice of q that $\big((1-q^j) - (1-q^{j-1})\big)R = r_j$. Thus the radius of B_j is larger than the distance, $\big((1-q^j) - (1-q^{j-1})\big)\|x\|$, between x_j and x_{j-1}. It follows that the points x_j and x_{j-1} belong to the intersection $B_j \cap B_{j-1}$, and hence this intersection contains a ball of radius $r_j/2$ around $(x_j + x_{j-1})/2$.

Since $\|(L_j - L_{j-1})y\| \le Kn^{3/2}\varepsilon(r_j + r_{j-1})$ in $B_j \cap B_{j-1}$, hence in a ball of radius $r_j/2$, it follows that $\|U_j - U_{j-1}\| \le 4Kn^{3/2}\varepsilon(r_j + r_{j-1})/r_j \le Kn^{3/2}\varepsilon$. Since $x_j \in B_j \cap B_{j-1}$, we also obtain

$$\|(L_j - L_{j-1})x_j\| \le Kn^{3/2}\varepsilon(r_j + r_{j-1}) \le Kn^{3/2}\varepsilon q^j r.$$

It follows that for each $k \ge j$,

$$\|(L_j - L_{j-1})x_k\| \le \|U_j - U_{j-1}\|\,\|x_j - x_k\| + \|(L_j - L_{j-1})x_j\| \le KRn^{3/2}\varepsilon q^j$$

because $\|x_j - x_k\| = (q^j - q^k)\|x\| \le q^j R$. Since $f(x_k) = L_k(x_k)$ and $L_0 = U_0$, summing from 1 to k and using the choice of q gives

$$\begin{aligned}\|f(x_k) - U_0 x_k\| &= \|(L_k - L_0)x_k\| \le \sum \|(L_j - L_{j-1})x_k\| \\ &\le KRn^{3/2}\varepsilon(1-q)^{-1} \le Kn^{3/2}\varepsilon R^2/r.\end{aligned}$$

Letting $k \to \infty$ yields $\|f(x) - U_0 x\| \le Kn^{3/2}\varepsilon R^2/r$ as claimed. □

Many natural questions related to Theorem 14.11 seem to be open. One such question is whether a theorem of this nature holds with estimates independent of the dimension. Another question is whether the Euclidean norm can be replaced by other norms. For general norms this question may not make sense. Given $\varepsilon > 0$, it is evident that there is a norm $|||\cdot|||$ on l_2^n with $\|x\| \le |||x||| \le (1+\varepsilon)\|x\|$ for every x so that $(l_2^n, |||\cdot|||)$ has only two isometries, namely, $\pm I$. Every nontrivial unitary operator on $(l_2^n, \|\cdot\|)$ is an ε-quasi-isometry on $(l_2^n, |||\cdot|||)$ which cannot be approximated by isometries on this space. However, there are reasonable questions which one can ask about classical spaces which are rich with isometries. There are known positive results of this kind even for isometries into. For example, we mention the following result without proof. Recall that $C(K)$ denotes the space of continuous real-valued functions on K with the supremum norm.

Theorem 14.12. *Let K_i (for $i = 1, 2$) be compact metric spaces, and denote the ball with radius r centered at the origin in $C(K_i)$ by $B_i(r)$. Let $0 < \varepsilon < 1/15$, and let f be a map from $B_1(1)$ into $B_2(1)$ with $f(0) = 0$ so that*

$$(1-\varepsilon)\|u - v\| \le \|f(u) - f(v)\| \le (1+\varepsilon)\|u - v\|, \qquad u, v \in B_1(1). \tag{14.14}$$

Then there is an isometry F from $B_1(1-8\varepsilon)$ into $B_2(1)$ with $\|f(u)-F(u)\| \le 8\varepsilon$ for $u \in B_1(1-8\varepsilon)$.

There is no analogous result for l_1. We denote the ball of radius r in l_1 centered at the origin by $B(r)$.

Example 14.13. *Fix $M \ge 3$. For every $\varepsilon > 0$ there is a map f from $B(1)$ into itself satisfying (14.14), such that whenever F is an isometry from a subset of l_1 containing $B(6/M)$ into l_1, then $\|F(x)-f(x)\| \ge 1/M^2$ for some $x \in B(3/M)$.*

Denote the unit vector basis of l_1 by $\{e_i\}$, and assume, as we may, that $\varepsilon^{-1} = n$ is an integer. The map f is defined as a composition, $f = g \circ h$, where g and h are defined on all of l_1. For $x = \sum_{i=1}^{\infty} a_i e_i$ put

$$g(x) = \begin{cases} x & \text{if } a_{n+1} \le 0 \\ x + a_{n+1}\left(\varepsilon \sum_{i=1}^{n} e_i - e_{n+1}\right)/M & \text{if } a_{n+1} \ge 0 \end{cases}$$

and $h(x) = \sum_{i=1}^{\infty} h(a_i e_i)$ where

$$h(a_i e_i) = \begin{cases} a_i e_{n+1+i} & i \le n \text{ and } a_i < 2/M \\ (a_i - 2/M)e_i + 2/M\, e_{n+1+i} & i \le n \text{ and } a_i \ge 2/M \\ a_{n+1} e_{n+1} & i = n+1 \\ a_i e_{n+1+i} & i > n+1 \end{cases}$$

We check that this map has the desired properties.

The map h is clearly an isometry from l_1 into itself. For every $u \in l_1$ denote the set of those indices $i \le n$ which are in the support of u by S_u. If $u = h(x)$ for some $x = \sum a_1 e_i \in B(1)$, then the coordinates of u are nonnegative for $i \le n$, and the ith coordinate (for $i \le n$) is strictly positive only if $a_i > 2/M$. Since $\|x\| \le 1$, it follows that $\mathrm{Card}(S_u) \le M/2$. Thus, to prove that f satisfies (14.14), it suffices to prove the following:

Assume that $u = \sum_{i=1}^{\infty} b_i e_i$ and $v = \sum_{i=1}^{\infty} c_i e_i$ satisfy $\mathrm{Card}(S_u) \le M/2$ and $\mathrm{Card}(S_v) \le M/2$, and that b_i and c_i are nonnegative for $1 \le i \le n$. Then

$$\|u-v\| \ge \|g(u)-g(v)\| \ge (1-\varepsilon)\|u-v\|.$$

Since $g(u) = u$ if $b_{n+1} \le 0$, this is obvious if $b_{n+1} \le 0$ and $c_{n+1} \le 0$. It remains to check the cases $b_{n+1} \ge c_{n+1} \ge 0$ and $b_{n+1} \ge 0 \ge c_{n+1}$. This is straightforward, and we check, e.g., the case $b_{n+1} \ge c_{n+1} \ge 0$.

$$\begin{aligned} \|g(u)-g(v)\| = & \sum_{\substack{i \le n \\ i \notin S_u \cup S_v}} \varepsilon(b_{n+1}-c_{n+1})/M + \sum_{i \in S_u \setminus S_v} (b_i + \varepsilon(b_{n+1}-c_{n+1})/M) \\ & + \sum_{i \in S_v \setminus S_u} |c_i - \varepsilon(b_{n+1}-c_{n+1})/M| + \sum_{i \in S_u \cap S_v} |b_i - c_i + \varepsilon(b_{n+1}-c_{n+1})/M| \\ & + (1-1/M)(b_{n+1}-c_{n+1}) + \sum_{i=n+2}^{\infty} |b_i - c_i|. \end{aligned}$$

An inspection of this expression (use $\varepsilon = n^{-1}$) shows that $\|g(u)-g(v)\| \le \|u-v\|$. The sum of the first, second and fifth summands is

$$b_{n+1} - c_{n+1} - \sum_{i \in S_v} \varepsilon(b_{n+1}-c_{n+1})/M + \sum_{i \in S_u \setminus S_v} b_i$$

and using also that $\mathrm{Card}(S_v) \le M/2$, we get

$$\|g(u)-g(v)\| \ge \|u-v\| - 2\varepsilon(b_{n+1}-c_{n+1})\,\mathrm{Card}(S_v)/M \ge (1-\varepsilon)\|u-v\|.$$

To finish the proof, we note first that if F is an isometry from $B(2r)$ (for some $r > 0$) into l_1 with $F(0) = 0$, then F takes disjointly supported vectors to disjointly supported vectors. Indeed, if $x, y \in B(r)$ have disjoint supports, then $\|x \pm y\| = \|x\| + \|y\|$; hence

$$\|F(x) - F(y)\| = \|x - y\| = \|x\| + \|y\| = \|F(x)\| + \|F(y)\| \tag{14.15}$$

and it follows from

$$\begin{aligned} \|F(x+y)\| &\leq \|F(x+y) - F(x)\| + \|F(x)\| \\ &= \|y\| + \|x\| = \|x+y\| = \|F(x+y)\| \end{aligned}$$

that the coordinates of $F(x)$ and $F(x+y)$ have the same signs. Similarly, the coordinates of $F(y)$ and $F(x+y)$ have the same signs. Thus the coordinates of $F(x)$ and $F(y)$ have the same signs, and by (14.15) they must have disjoint supports.

Now let F be an isometry defined on $B(6/M)$, and assume first that $F(0) = 0$. If there is an $i \leq n$ so that the ith coordinate of $F(3e_i/M)$ is 0, then for this i

$$\|F(3e_i/M) - f(3e_i/M)\| \geq 1/M \geq 3/M^2.$$

If the ith coordinate of $F(3e_i/M)$ is different from 0 for all $1 \leq i \leq n$, then the disjointness assertion implies that the support of $F(3e_{n+1}/M)$ is disjoint from $\{1, 2, \dots, n\}$ and again $\|F(3e_{n+1}/M) - f(3e_{n+1}/M)\| \geq 3/M^2$.

To remove the assumption that $F(0) = 0$ note that if $\|F(0)\| \geq 1/M^2$, we are done, since $f(0) = 0$. If $\|F(0)\| \leq 1/M^2$, we apply the argument above to $F(x) - F(0)$ and get a point x in $B(3/M)$ so that $\|F(x) - f(x)\| \geq 2/M^2$.

4. Approximation of the Derivative

By Rademacher's theorem (Proposition 6.41) a quasi-isometric map on an open subset Ω of $\mathbb{R}^n$, being Lipschitz, is differentiable almost everywhere. The derivative $D_f(x)$ of f at the point x gives the infinitesimal deformation by f at x.

Write the polar decomposition of D_f in the form $D_f(x) = C(x)\big(I + E(x)\big)$, where $I + E(x)$ is a positive definite matrix and $C(x)$ is orthogonal. The matrix $E(x)$ is called the "strain matrix". Its eigenvectors describe the main directions of the strain, and its eigenvalues give the stretching coefficients along these directions. The matrix $C(x)$ is called the rotation matrix. It describes the rotational part of the deformation f at x. The norms $\|E(x)\|$ at the points of Ω give a measure of how close f is to an isometry: $\sup \|E(x)\| \leq \varepsilon$ if and only if f is an ε-quasi-isometry.

Consideration of the degree of the map f gives that the determinant $\det\big(D_f(x)\big)$ has a constant sign on each connected component of Ω. (See Lemma 6.7 in Reshetnyak [**489**], p.190.) Since $I + E(x)$ is positive definite, it follows that $\det\big(C(x)\big)$ is also constant on each connected component of Ω.

Theorem 14.11 gives a good answer to the problem of the approximation of an ε-quasi-isometry f by an isometry. Is it also possible to obtain good global control on the rotation? Is it possible, for example, to find one orthogonal matrix U such that $\|C(x) - U\|$ is small for all $x \in \Omega$? Note that $D_f = C(I + E) = C + CE$ implies that $\|D_f(x) - C(x)\| = \|C(x)E(x)\| = \|E(x)\| \leq \varepsilon$ for all $x \in \Omega$; hence approximation of $C(x)$ by a fixed rotation is equivalent to approximating D_f.

The naive answer to these questions is no: Fix $\varepsilon > 0$, and consider the map f from the unit disk in the plane to itself, given in polar coordinates by

$$f(r, \theta) = (r, \theta + \varepsilon \log r).$$

Direct computations show that f is an ε-quasi-isometry and that the rotation matrix $C(x)$ at the point $x = (r, \theta)$ is essentially rotation by $\varepsilon \log r$, which changes very fast near the origin. Hence $C(x)$ cannot be approximated by one fixed rotation.

Note, however, that the oscillations of $C(x)$ are restricted to a "small" subset and that the identity matrix approximates $C(x)$ well on "most" of the disk: Given $M > 0$, the set where $\|I - C(x)\| > M\varepsilon$ is essentially a disk of radius e^{-M} (because essentially $\|I - C(x)\| \le \varepsilon|\log r|$). Its measure, πe^{-2M}, is very small if M is large. The discussion here will show that this example is typical and that the rotations can always be approximated by one orthogonal matrix on "large" sets.

Recall that a Banach space-valued function φ defined in a domain Ω in $\mathbb{R}^n$ is said to be of Bounded Mean Oscillation (BMO) if there is a constant K such that for every subcube Q of Ω there is a vector u_Q such that $|Q|^{-1} \int_Q \|\varphi(x) - u_Q\| \le K$. ($|Q|$ denotes the volume of Q.) The smallest possible constant K is called the BMO norm of φ and is denoted by $\|\varphi\|_{BMO}$.

We refer the reader to [**542**] for an exposition of the theory of BMO functions. Here we shall use only the fundamental inequality of John and Nirenberg: Let Q be a cube in $\mathbb{R}^n$, and denote the average of φ on Q by $\varphi_Q = |Q|^{-1} \int_Q \varphi$. Then

$$\mu\{x \in Q : \|\varphi(x) - \varphi_Q\| > t\} \le 2|Q| \exp\big(-2^{-n-2}\|\varphi\|_{BMO}^{-1}\, t\big)$$

for all $t > 0$, where μ is the Lebesgue measure $\mathbb{R}^n$.

Before stating the main result of this section, we recall some standard notation and facts about matrices. A^* will denote the transpose of the matrix A, and $\operatorname{tr}[A]$ its trace. Clearly $|\operatorname{tr}[A]| \le n\|A\|$. When A is positive definite, $\|A\|$ is its largest eigenvalue, hence smaller than the sum of the eigenvalues, i.e., $\|A\| \le \operatorname{tr}[A]$. In general the matrix A^*A is positive definite; hence $\|A\|^2 = \|A^*A\| \le \operatorname{tr}[A^*A]$ for every A. Finally $\operatorname{tr}[A] = \operatorname{tr}[A^*]$, and $\operatorname{tr}[AB] = \operatorname{tr}[BA]$.

Theorem 14.14. *Let f be an ε-quasi-isometry in a cube Q in $\mathbb{R}^n$. Then*

(i) *There is an orthogonal matrix U such that $|Q|^{-1} \int_Q \|D_f(x) - U\| \le Kn^{7/2}\varepsilon$.*

(ii) *$D_f(x)$ and $C(x)$ are in BMO. Their BMO norms are bounded by $Kn^{7/2}\varepsilon$.*

(iii) *There is an orthogonal matrix W so that*

$$\mu\{x : \|C(x) - W\| > M\varepsilon\} \le 4|Q| \exp(-K2^{-n}n^{-7/2}M)$$

for every constant $M > 0$.

Proof. Parts (ii) and (iii) follow from (i). Indeed, the estimates in (i) also apply to every subcube of Q. This is the definition of the BMO functions and norm, and (ii) follows. To see (iii), apply the John-Nirenberg inequality to obtain

$$\begin{aligned} \mu \otimes \mu\{(x, y) : \|C(x) - C(y)\| > t\} &\le 2\mu\{x : \|C(x) - C_Q\| > t/2\} \\ &\le 4 \exp\big(-2^{-n-2} t \|C\|_{BMO}^{-1}/2\big). \end{aligned}$$

Taking $t = M\varepsilon$, and using the estimate of (ii) for $\|C\|_{BMO}$, we obtain

$$\mu \otimes \mu\{(x, y) : \|C(x) - C(y)\| > M\varepsilon\} \le 4 \exp\big(-K2^{-n}n^{-7/2}M\big).$$

By Fubini's theorem there is a y_0 such that

$$\mu\{x : \|C(x) - C(y_0)\| > M\varepsilon\} \le 4 \exp\big(-K2^{-n}n^{-7/2}M\big).$$

Thus $W = C(y_0)$ satisfies (iii).

We now prove (i). By scaling and translation we may assume that Q is the cube of volume 1 centered at 0 and that $f(0) = 0$. Since $B(1/2) \subseteq Q \subseteq B(n^{1/2}/2)$, Theorem 14.11 gives an orthogonal matrix, which we shall assume to be the identity, so that $\|f(x) - x\| \leq Kn^{5/2}\varepsilon$. We remark that it follows from these normalizations that $\det\big(C(x)\big) > 0$ at any point where f is differentiable. To see this, assume that f is differentiable at 0 (otherwise, take a point arbitrarily close to 0). Since the sign of $\det\big(C(x)\big)$ is constant, it suffices to check $\det\big(C(0)\big)$. The homotopy $f_t(x) = (1-t)x + tf(x)$ connects I to f. Also $f_t(x) \neq 0$ for any $x \in \partial Q$, because otherwise $1/2 \leq \|x\| \leq t\|x - f(x)\| \leq Kn^{5/2}\varepsilon$, which is impossible for small ε. By the invariance of the degree under such homotopies (see Property I in Reshetnyak [**489**], p. 88), the two maps have the same degree at 0. The claim now follows because $\det\big(C(0)\big)$ and $\det\big(D_f(0)\big)$ have the same sign, and since the latter is positive by the definition of the degree.

We shall show that under these normalizations the theorem holds with $U = I$. Since the L^1-norm is smaller than the L^2-norm, this will follow once we show that $(\int_Q \|D_f(x) - I\|^2)^{1/2} \leq Kn^{7/2}\varepsilon$.

The proof will be done in two steps. We shall first show that

$$\left(\int_Q \|D_f(x) - I\|^2\right)^{1/2} \leq Kn^{7/4}\varepsilon^{1/2} \tag{14.16}$$

and then improve the dependence on ε from $\varepsilon^{1/2}$ to ε.

Put $H(x) = I - D_f(x)$, and note that

$$H^*H = (I - D_f)^*(I - D_f) = {D_f}^*D_f - I + (H + H^*)$$

and that for every x

$$\|{D_f}^*D_f - I\| = \|(I+E)C^*C(I+E) - I\| = \|2E + E^2\| \leq 3\varepsilon.$$

We thus obtain the two estimates

$$\|H^*H - (H + H^*)\| = \|{D_f}^*D_f - I\| \leq 3\varepsilon \tag{14.17}$$

$$\big|\mathrm{tr}\,[H^*H] - 2\mathrm{tr}\,[H]\big| \leq n\|{D_f}^*D_f - I\| \leq 3n\varepsilon. \tag{14.18}$$

By (14.18)

$$\int_Q \|D_f(x) - I\|^2 = \int_Q \|H\|^2 \leq \int_Q \mathrm{tr}\,[H^*H] \leq 3n\varepsilon + 2\int_Q |\mathrm{tr}\,[H]| \tag{14.19}$$

and thus what we need to show is that

$$\int_Q |\mathrm{tr}\,[H]| \leq Kn^{7/2}\varepsilon. \tag{14.20}$$

Fix i, and let $J = [a, b]$ be the intersection of a line parallel to the x_i-axis and Q. (Thus a and b differ only in their ith coordinates which are $\pm 1/2$ respectively.) Denote the ith component of f by f^i, and the ith partial derivative of the ith component of $x - f(x)$ by H_i^i; i.e., H_i^i is the (i,i)th term in the matrix H.

Since f^i is Lipschitz, hence absolutely continuous, it is differentiable a.e. on J and is the integral of its derivative there. Also, the fact that f is an ε-quasi-isometry implies that $|\partial f^i/\partial x_i(x)| \leq 1 + \varepsilon$ whenever the derivative exists. It follows that

$H_i^i = 1 - \partial f^i/\partial x_i \geq -\varepsilon$, and hence that $|H_i^i| \leq H_i^i + 2\varepsilon$ a.e. on J. Since also $\|f(x) - x\| \leq Kn^{5/2}\varepsilon$ we obtain

$$\begin{aligned}(14.21)\quad \int_J |H_i^i| &\leq \int (H_i^i + 2\varepsilon) = \big(1/2 - f^i(b)\big) - \big(-1/2 - f^i(a)\big) + 2\varepsilon \\ &\leq \|f(a) - a\| + \|f(b) - b\| + 2\varepsilon \leq Kn^{5/2}\varepsilon.\end{aligned}$$

Integrating this estimate with respect to the other $(n-1)$ coordinates, we obtain $\int_Q |H_i^i| \leq Kn^{5/2}\varepsilon$. Summing over i proves equation (14.20), hence also (14.16).

To improve the dependence on ε, we now strengthen (14.20) to

$$\int_Q \operatorname{tr}[H^*H] \leq Kn^7\varepsilon^2.$$

Using (14.17) we obtain

$$\begin{aligned}&\big|\operatorname{tr}[(H^*H)^2] - 2\operatorname{tr}[H^2] - 2\operatorname{tr}[H^*H]\big| \\ &= \big|\operatorname{tr}[(H^*H)^2] - \operatorname{tr}[(H + H^*)^2]\big| \\ &= \operatorname{tr}\big[(H^*H + H + H^*)(H^*H - H - H^*)\big] \\ &\leq n\big\|(H^*H + H + H^*)(H^*H - H - H^*)\big\| \leq Kn\varepsilon\big(\|H^*H\| + \varepsilon\big)\end{aligned}$$

and (14.18) implies that

$$\begin{aligned}\big|\operatorname{tr}^2[H^*H] - 4\operatorname{tr}^2[H]\big| &= \big|\operatorname{tr}[H^*H] - 2\operatorname{tr}[H]\big|\,\big|\operatorname{tr}[H^*H] + 2\operatorname{tr}[H]\big| \\ &\leq 3n\varepsilon\big(3n\varepsilon + 4|\operatorname{tr}[H]|\big).\end{aligned}$$

Integrating and using equations (14.19) and (14.20), we obtain

$$\int_Q \big|\operatorname{tr}[(H^*H)^2] - 2\operatorname{tr}[H^2] - 2\operatorname{tr}[H^*H]\big| \leq Kn^{9/2}\varepsilon^2$$

and

$$\int_Q \big|\operatorname{tr}^2[H^*H] - 4\operatorname{tr}^2[H]\big| \leq Kn^{9/2}\varepsilon^2.$$

Using these estimates, we can represent $\int_Q \operatorname{tr}[H^*H]$ as the sum of three terms:

$$\int_Q \Big(\operatorname{tr}[(H^*H)^2] - \operatorname{tr}^2[H^*H]/2 - \operatorname{tr}[H^*H]\Big) + 2\int_Q \Big(\operatorname{tr}^2[H] - \operatorname{tr}[H^2]\Big) + \rho$$

where the last term, ρ, is an error term, smaller than $Kn^{9/2}\varepsilon^2$. We shall show that

(a) The integrand in the first term is nonpositive.

(b) The second term is bounded by $Kn^7\varepsilon^2$.

The theorem then follows because $\operatorname{tr}[H^*H]$ is nonnegative.

Proof of (a). We distinguish two cases. If $\operatorname{tr}[H^*H] < 2$, then

$$\begin{aligned}&\operatorname{tr}[(H^*H)^2] - \operatorname{tr}^2[H^*H]/2 - \operatorname{tr}[H^*H] \\ &= \Big(\operatorname{tr}[(H^*H)^2] - \operatorname{tr}^2[H^*H]\Big) - \Big(1 - \operatorname{tr}[H^*H]/2\Big)\operatorname{tr}[H^*H] \leq 0\end{aligned}$$

because H^*H is positive definite, and hence the first term is negative by the Cauchy-Schwartz inequality, and the second is positive since $\operatorname{tr}[H^*H] < 2$.

If $\operatorname{tr}[H^*H] \geq 2$, write $H = I - D_f = I - C - CE$ and $A = (I - C^*)(I - C)$. It follows that $H^*H = A + B$, where $B = E^2 - (I - C)^*CE - EC^*(I - C)$, so that $\|B\| \leq K\varepsilon$. Similarly $(H^*H)^2 = A^2 + C$, where $\|C\| \leq K\varepsilon$. It follows that $|\mathrm{tr}\,[B]| \leq Kn\varepsilon$ and that $|\mathrm{tr}\,[C]| \leq Kn\varepsilon$. Thus

$$\mathrm{tr}\,[(H^*H)^2] - \mathrm{tr}^2[H^*H]/2 - \mathrm{tr}\,[H^*H] = \mathrm{tr}\,[A^2] - \mathrm{tr}^2[A]/2 - \mathrm{tr}\,[H^*H] + \tau$$

where the error term, τ, is smaller than $Kn\varepsilon$. Using $\mathrm{tr}\,[H^*H] \geq 2$, we see that if $\varepsilon < (Kn)^{-1}$, we need to check only that $\mathrm{tr}\,[A^2] - \mathrm{tr}^2[A]/2 \leq 0$. (If ε is larger, there is nothing to prove anyhow.) This is done by an explicit computation: Under our normalization the determinant of the orthogonal matrix $C(x)$ is identically one. Thus -1 appears as an eigenvalue of $C(x)$ an even number of times. Also, the nonreal eigenvalues of $C(x)$ appear in conjugate pairs. It follows that the nonzero eigenvalues of A are pairs of positive numbers, say $\mu_1, \ldots, \mu_N$ (where each μ appears twice), and hence $\mathrm{tr}^2[A]/2 = (2\sum \mu_i)^2/2 \geq 2\sum \mu_i^2 = \mathrm{tr}\,[A^2]$.

Proof of (b). Write explicitly $\mathrm{tr}\,[H^2] - \mathrm{tr}^2[H] = \sum(H^i_k H^k_i - H^i_i H^k_k)$, where H^i_k is the derivative of the ith component of $x - f(x)$ with respect to x_k (and similarly for other choices of indices). We shall estimate each of the terms separately.

Fix i and k, and let S be the square obtained by intersecting the cube Q with a plane parallel to the (x_i, x_k)-plane. By Gauss' theorem (which reduces in the simple case of a square to integration by parts on segments parallel to its sides), and by using the absolute continuity of f^i and f^k on such segments, we obtain

$$\iint_S (H^i_k H^k_i - H^i_i H^k_k) = \int_{\partial S} (x_i - f^i(x))H^k_i dx_i + (x_i - f^i(x))H^k_k dx_k.$$

The right-hand side is a sum of one-dimensional integrals. Two of these are of the form $\int_J (x_i - f^i(x))H^k_i dx_i$, where J is an interval parallel to x_i. For the other combination of indices the integration is also taken over an interval in the same direction as the partial derivative. Since $\|x - f(x)\| \leq Kn^{5/2}$, this estimate holds also for its components, the $(x_i - f^i(x))$'s, and as in (14.21) we obtain

$$\begin{aligned}\int_J \left|(x_i - f^i(x))H^k_i(x)\right| dx_i &\leq \max_{x\in J} \left|(x_i - f^i(x))\right| \int_J |H^k_i(x)| dx_i \\ &\leq \left(Kn^{5/2}\varepsilon\right)^2 = Kn^5\varepsilon^2.\end{aligned}$$

Summing over the four terms on the right-hand side, we see that

$$\left|\iint_S (H^i_k H^k_i - H^i_i H^k_k)\right| \leq Kn^5\varepsilon^2.$$

Integrating with respect to the other $(n-2)$ coordinates and summing over the $n(n-1)$ pairs of i and k yield (b). □

5. Notes and Remarks

Theorem 14.1 is a local version (with the same proof) of a classical theorem of Mazur and Ulam [**407**].

Theorem 14.2 is due to Figiel [**187**]. Another proof appears in Holsztyński [**256**].

The theory of quasi-isometric maps was initiated by F. John in [**281**]. Motivated by problems in elasticity, he embarked in [**281**] and later in [**282**]–[**284**] on a purely geometric study. Most of the results of this chapter are due to him.

The main injectivity result (Theorem 14.10) was proved by John only in Hilbert space, and he used special properties of Hilbert space geometry. The generalization to Banach spaces is due to J. Gevirtz [**202**], and we presented here his proof.

There are injectivity results for functions defined on more general sets. A domain Ω is called a uniform domain if there is a constant $b > 0$ so that for each $u, v \in \Omega$ there is a path $\phi(t)$ connecting u and v and parameterized by arc length so that $B(\phi(t), b\min(t, L-t)) \subset \Omega$ for all t. Gevirtz [**201**] (improving results and estimates of John in the Hilbert space setting) proved that there is a constant $\mu(b) > 0$, independent of the Banach space, such that if f is a quasi-isometry on a uniform domain with constant b, and if $M/m < \mu(b)$, then f is injective.

Questions of this type were examined in much more detail in finite-dimensional Euclidean spaces, and in particular in the plane. See the survey [**198**] by Gehring.

Another study of injectivity was done in the setting of quasi-conformal mappings. A local homeomorphism f from a domain Ω in a Banach space E into a Banach space F is called κ-quasi-conformal if

$$\limsup_{\delta \to 0} \frac{\sup\{\|f(y) - f(x)\| : \|x - y\| = \delta\}}{\inf\{\|f(y) - f(x)\| : \|x - y\| = \delta\}} \leq \kappa$$

for every $x \in \Omega$. Every (m, M)-quasi-isometry is m/M-quasi-conformal.

For a presentation of the theory of quasi-conformal mappings in $\mathbb{R}^n$, see the book [**568**] by Väisälä. Martio, Rickman and Väisälä proved in [**385**] the following injectivity theorem: For every $\kappa \geq 1$ and every integer $n \geq 3$, there is an $r(\kappa, n) > 0$ so that every κ-quasi-conformal mapping from the unit ball of $\mathbb{R}^n$ into $\mathbb{R}^n$ is injective on the ball of radius $r(\kappa, n)$ centered at the origin. It seems to be unknown if there are infinite-dimensional generalizations of the theorem of Martio, Rickman and Väisälä even in Hilbert space. See Zorich [**596**] for a recent survey and discussion of this topic.

Let f be an ε-quasi-isometry in a bounded convex domain Ω of $\mathbb{R}^n$. John [**283**] proved that if ε is small enough, then f is injective in Ω. In fact, there is a $\delta > 0$ (which depends on the ratio between the radii of a ball contained in Ω and a ball containing it, and not on n) so that $\|f(x) - f(y)\| \geq (1-\delta)\|x - y\|$. Denote the smallest possible value of δ by $\delta_\Omega(\varepsilon)$. The function $R_\Omega(\varepsilon) = \delta_\Omega(\varepsilon)/\varepsilon$ is a measure of the "rigidity" of Ω. As 0-quasi-isometries are isometries (Corollary 14.8), one expects that $\lim_{\varepsilon\to 0} R_\Omega(\varepsilon) = 1$. The estimates in [**283**] give only that $\limsup_{\varepsilon \to 0} R_\Omega(\varepsilon) \leq 3$. Whether $\lim_{\varepsilon\to 0} R_\Omega(\varepsilon) = 1$ seems to be still open.

Theorem 14.11 is due to John [**281**]. As remarked already in the text, it is unknown if this perturbation result generalizes to infinite dimensions.

The asymptotics as $\varepsilon \to 0$ in Theorem 14.11 is studied by Semenov in [**528, 529**]. Put $\lambda(\varepsilon) = \sup_f \inf_U \|f - U\|$, where the f's are ε-quasi-isometries and the U's are isometries. Semenov gives improved estimates on $\lambda(\varepsilon)$ by a completely different method than that of John [**281**] which we presented here. In particular he shows that $1 \leq \limsup_{\varepsilon\to 0} \lambda(\varepsilon)/\varepsilon \leq 3(r + 2R)^2/r$ whenever Ω is a star-shaped domain which contains a ball of radius r and is contained in a ball of radius R.

Theorem 14.12 is due to Lövblom [**374**]. Jarosz [**279**] obtained a result on the stability of quasi-isometries on $C(K)$ spaces about the same time as Lövblom. His result, however, treats only surjective maps. On the other hand, his proof works for maps between certain uniform algebras. Example 14.13 is due to Lövblom [**375**].

Theorem 14.14 was proved in John [**281**], and functions of Bounded Mean Oscillation made their first appearance in this article. The celebrated John-Nirenberg inequality was proved in the next article in the same issue of *Comm. Pure Appl.*

Math. [**286**]. The first application to PDE appeared in an article of Moser "On Harnack's theorem for elliptic differential equations" in the same volume.

Reshetnyak [**491**], Section 5.2, proves the analog of Theorem 14.14 for general domains (with appropriately regular boundaries). This requires quite sophisticated analytic tools, such as singular integrals and elliptic differential operators.

The direct applications of the theory to elasticity are purely geometric. They consider only the geometry, and not the forces causing the deformation. In John [**285**] this aspect is taken into consideration, and he studies bodies in equilibrium, i.e., when f deforms the body in such a way that after the deformation there are no more elastic forces causing further deformation. Using BMO functions and Theorem 14.14, he obtains a uniqueness result for ε-quasi-isometric deformations in equilibrium and with prescribed boundary values.

In [**284**] John studies mappings whose strain matrix is "small" in the L_p sense rather than the L_∞ sense. He does require some a priori L_∞ estimate, and assumes that there is a constant $\delta = \delta(p, n)$ so that $\|E\|_\infty \leq \delta$. Under this additional assumption, he shows all the results from [**281**] are true under the condition that $\|E\|_p \leq \varepsilon$ and studies the behavior of the mappings as $\varepsilon \to 0$.

Delicate L_q estimates of the degree of approximation of f by an isometry in terms of $\|E\|_p$, without any apriori assumptions, appear in Kohn [**331**].

Reshetnyak [**490**] studied another notion of "nearness" to isometry. Instead of measuring the distortion of f by the biggest eigenvalue of ${D_f}^* D_f$, he uses other spectral invariants. Given a smooth function f on a domain Ω in $\mathbb{R}^n$, define

$$R_p(f) = \left\| n - \operatorname{tr}\left[{D_f}^* D_f\right] \right\|_{L_p(\Omega)}^{1/2} + \left\| 1 - \det(D_f) \right\|_{L_p(\Omega)}.$$

Fix $p > n$. Reshetnyak shows that there is a positive function $\theta(t)$, converging to zero as $t \to 0$, so that if $f \in W_p^1(\Omega)$ and if $R_p(f) \leq t$, then there is an isometry U with $\|f - U\|_{W_p^1(\Omega)} \leq \theta(t)$. (The introduction of Sobolev spaces is not just a technicality. This is the appropriate setup for continuity of functionals and compactness arguments. This is studied and applied by Reshetnyak and others in many related problems.) In some of the proofs, Reshetnyak applies Theorem 14.14.

There are many interesting open questions on quasi-isometries. For example, it seems to be unknown if, for small ε, an ε-quasi-isometry is conjugate to an isometry by a bi-Lipschitz map. Another problem of a similar nature is whether an ε-quasi-isometry can always be represented as the composition of δ-quasi-isometries for some $\delta < \varepsilon$; see Freedman [**196**].

The structure of quasi-isometries with a small ε and the possibility to extend such maps from their domain of definition to larger domains have been studied by several authors. See, e.g., Partanen and Väisälä [**447**] and its references.

Stability problems very close in spirit to those of this section have also been studied in the theory of quasi-conformal mappings. See the books [**489, 491**] by Reshetnyak for information on this subject.

CHAPTER 15

Perturbations of Global Isometries

1. The Hyers-Ulam Problem

In this chapter we study surjective maps between Banach spaces which nearly preserve distances. It is important here that the maps are not restricted to a bounded subset, and a crucial role is played by the limit (if it exists) of $f(\lambda x)/\lambda$ as $\lambda \to \infty$.

We first consider surjective maps f for which $\big|\|f(x)-f(y)\|-\|x-y\|\big|$ is bounded by a constant independent of x and y. Hyers and Ulam asked whether such maps are close to (onto) isometries, and we present the solution to this question in this section. In the next section we study the situation when the expression above is allowed to tend to infinity as $\|x\|, \|y\| \to \infty$. We give a precise estimate on the rate in which this expression may tend to infinity so that the answer to the natural generalization of the Hyers-Ulam problem is still positive.

Before treating almost isometries we present a result on perturbation of almost additive mappings. This is related to the Hyers-Ulam problem but is much simpler.

Proposition 15.1. *Let f be a mapping from a Banach space E into a Banach space F, and assume that there is an $\varepsilon > 0$ so that*

$$\|f(x+y) - f(x) - f(y)\| \leq \varepsilon \tag{15.1}$$

for all $x, y \in E$. Then there is a unique additive map $T : E \to F$ which satisfies $\|f(x) - Tx\| \leq \varepsilon$ for all $x \in E$. If, in addition to (15.1), $f(0) = 0$ and f is also assumed to be bounded on the unit ball, then the map T is a bounded linear operator.

Proof. It follows from (15.1) by a straightforward induction that

$$\|f(2^n x) - 2^n f(x)\| \leq \varepsilon(2^n - 1) \tag{15.2}$$

for all $n \geq 1$ and for all $x, y \in E$. Thus

$$\|f(2^m x) - 2^{m-n} f(2^n x)\| \leq \varepsilon \cdot 2^{m-n}$$

for $m > n$, or

$$\|f(2^m x)/2^m - f(2^n x)/2^n\| \leq \varepsilon/2^n.$$

Consequently,

$$Tx = \lim_{n\to\infty} f(2^n x)/2^n \tag{15.3}$$

exists for all $x \in X$. By using (15.1) for $2^n x$ and $2^n y$ instead of x and y and by letting $n \to \infty$, we deduce that $T(x+y) = Tx + Ty$, i.e., that T is additive. From (15.2) we deduce that

$$\|Tx - f(x)\| \leq \varepsilon$$

for all $x \in E$. Conversely, an additive T which satisfies this inequality must be given by (15.3); i.e., T is unique. The last assertion of the proposition is obvious. $\square$

We now state the main result of this section.

Theorem 15.2. *Let E and F be Banach spaces, and let f be a surjective map from E onto F so that $f(0) = 0$ and so that there is an $\varepsilon > 0$ with*

$$\Big| \|f(x) - f(y)\| - \|x - y\| \Big| \le \varepsilon \tag{15.4}$$

for all $x, y \in E$. Then there is a surjective isometry U from E onto F so that $U0 = 0$ and so that

$$\|f(x) - Ux\| \le 2\varepsilon \tag{15.5}$$

for all $x \in E$.

Before passing to the proof of the theorem we make some comments. Note that f is not assumed to be injective or continuous. The surjectivity assumption is, however, important even in finite-dimensional Euclidean spaces. For example, the map f from $\mathbb{R}$ into l_2^2 given by $f(x) = \big(x, g(x)\big)$, where $g(x) = \log x$ for $x \ge 1$ and $g(x) = 0$ for $x \le 1$, satisfies (15.4) for a suitable ε, but it is not a bounded perturbation of an isometry from $\mathbb{R}$ into l_2^2.

The constant 2 appearing in (15.5) is the best possible. Let f be the (discontinuous) map from $\mathbb{R}$ to $\mathbb{R}$ defined by $f(t) = -3t$ for $t \in [0, 1/2]$ and $f(t) = t - 1$ elsewhere. Then f is surjective and satisfies (15.4) with $\varepsilon = 1$. The only isometry (with $U0 = 0$) which can appear in (15.5) is $Ux = x$, but then $|f(1/2) - U(1/2)| = 2$.

There are evidently no homeomorphisms of $\mathbb{R}$ onto itself which would give such an example, but already in the Euclidean plane there is a homeomorphism f of $\mathbb{R}^2$ onto itself satisfying (15.4) with $\varepsilon = 1$ and so that $\sup \|f(x) - U(x)\| = 2$ for every isometry U of $\mathbb{R}^2$ satisfying $U0 = 0$ (see [**446**] pp. 620–621).

The first step in the proof of Theorem 15.2 will be to show that maps which satisfy (15.4) are almost affine. More precisely

Lemma 15.3. *There is an absolute constant K so that if f is as in the statement of Theorem 15.2, then*

$$\left\| f\left(\frac{x+y}{2}\right) - \frac{f(x) + f(y)}{2} \right\| \le K\Big(\|x - y\| \cdot 2^{-n} + n\varepsilon\Big) \tag{15.6}$$

for every positive integer n and for all $x, y \in E$.

Proof. We may assume that f is injective. Indeed, by (15.4) f is one-one on any maximal 2ε-separated set in E. Thus E and F have the same density character. It follows that the cardinality of E is the same as that of F or of any nonempty ball in F. Given any $\delta > 0$ we may partition F into disjoint sets $\{F_\alpha\}_{\alpha \in A}$ with nonempty interior and with diameters smaller than δ. (For example, we may pick a maximal $\delta/2$-separated set $\{u_\alpha\}_{\alpha \in A}$ in F and choose F_α so that $F_\alpha \supset B_F(u_\alpha, \delta/4)$ and so that a point v not in the union of these balls is assigned to some F_α with $\|u_\alpha - v\| \le \delta/2$.) Let $f_0 : E \to F$ be any map whose restriction to $f^{-1}(F_\alpha)$ is a one-one map onto F_α. Then f_0 is both injective and surjective, and $\|f(x) - f_0(x)\| \le \delta$ for all x. Once (15.6) is established for f_0 (with $\varepsilon + 2\delta$ instead of ε) the desired result follows for f because δ was arbitrary.

We may thus assume that f is both injective and surjective. Fix $x, y \in E$, and put $d = \|f(x) - f(y)\|/2$, $p = (x + y)/2$ and $q = \big(f(x) + f(y)\big)/2$. We need to estimate the distance between q and $f(p)$. We shall do this by using reflections around q and $f(p)$ (somewhat in the spirit of the proof of the Mazur-Ulam theorem). Since we want to consider both reflections on the same space, say F, we use f to carry the reflection around p to F and define for $u \in F$

$$g_0(u) = f\big(2p - f^{-1}(u)\big), \qquad g_1(u) = 2q - u.$$

For $n \geq 2$ we define inductively a sequence of maps g_n from F onto itself by putting $g_{n+1} = g_{n-1} \circ g_n \circ g_{n-1}^{-1}$. A straightforward induction gives

$$\Big| \|g_n(u) - g_n(v)\| - \|u - v\| \Big| \leq 2^{n+1}\varepsilon \tag{15.7}$$

for all $u, v \in F$. Notice also that for all n

$$g_n\big(f(x)\big) = f(y), \qquad g_n\big(f(y)\big) = f(x). \tag{15.8}$$

Define a sequence of points in F by $q_1 = q$ and $q_{n+1} = g_{n-1}(q_n)$ for $n \geq 1$. It follows from (15.7) and (15.8), by induction on n, that

$$\|q_n - f(x)\|, \|q_n - f(y)\| \leq d + 2^n\varepsilon$$

and hence that

$$\|q_n - q_{n-1}\| \leq 2d + 2^{n+1}\varepsilon. \tag{15.9}$$

Let us prove next, again by induction on n, that

$$\|g_n(u) - u\| \geq 2\|q_n - u\| - 3 \cdot 2^n\varepsilon \tag{15.10}$$

for all $u \in F$. This is obvious for $n = 1$. Also by (15.7) and the induction hypothesis

$$\begin{aligned}
\|g_{n+1}(u) - u\| &= \|g_{n-1} \circ g_n \circ g_{n-1}^{-1}(u) - u\| \\
&\geq \|g_n\big(g_{n-1}^{-1}(u)\big) - g_{n-1}^{-1}(u)\| - 2^n\varepsilon \\
&\geq 2\|q_n - g_{n-1}^{-1}(u)\| - 4 \cdot 2^n\varepsilon \\
&= 2\|g_{n-1}^{-1}\big(g_{n-1}(q_n)\big) - g_{n-1}^{-1}(u)\| - 4 \cdot 2^n\varepsilon \\
&\geq 2\|g_{n-1}(q_n) - u\| - 6 \cdot 2^n\varepsilon \\
&= 2\|q_{n+1} - u\| - 3 \cdot 2^{n+1}\varepsilon.
\end{aligned}$$

By substituting $u = q_{n+1}$ in (15.10) we get

$$\|q_{n+2} - q_{n+1}\| \geq 2\|q_{n+1} - q_n\| - 3 \cdot 2^n\varepsilon$$

and hence

$$\|q_n - q_{n-1}\| \geq 2^{n-2}\|q_2 - q_1\| - 3n2^n\varepsilon.$$

Combining the last inequality with (15.9) gives

$$\|q_1 - q_2\| \leq 8d \cdot 2^{-n} + 4(3n + 2)\varepsilon$$

and this inequality implies (15.6) because

$$\begin{aligned}
\|q_1 - q_2\| &= \|q - f\big(x + y - f^{-1}(q)\big)\| \geq \|2f^{-1}(q) - (x + y)\| - \varepsilon \\
&\geq 2\left\|q - f\left(\frac{x+y}{2}\right)\right\| - 3\varepsilon = 2\left\|\frac{f(x) + f(y)}{2} - f\left(\frac{x+y}{2}\right)\right\| - 3\varepsilon
\end{aligned}$$

and $2d = \|f(x) - f(y)\| \leq \|x - y\| + \varepsilon$. □

Proof of Theorem 15.2. Let $z \in E$, and let n be a positive integer. Using (15.6) with $x = 2^n z$ and $y = 0$ gives

$$\|2^{-(n-1)} f(2^{n-1} z) - 2^{-n} f(2^n z)\| \leq 2^{-(n-1)} K(\|z\| + n\varepsilon).$$

It follows that $Uz = \lim 2^{-n} f(2^n z)$ exists. Applying (15.6) with $x = 2^m z$ and $y = 0$ and taking $n = m + l$ for fixed l, we get

$$\|2^{-(m-1)} f(2^{m-1} z) - 2^{-m} f(2^m z)\| \leq K\big(2^{-l}\|z\| + (m+l)\varepsilon\big)2^{-(m+1)}.$$

Summing over m from 1 to ∞, we get

$$\|Uz - f(z)\| \leq K_1\big(2^{-l}\|z\| + l\varepsilon\big).$$

By taking $l \sim \log \|z\|$ we get for $\|z\| \geq 1$

$$\|Uz - f(z)\| \leq K_2\big(1 + \varepsilon \log \|z\|\big). \tag{15.11}$$

(Actually all that we need below is that $\|Uz - f(z)\| = o(\|z\|)$ as $z \to \infty$.) It follows from (15.6) that U is linear and from (15.4) that it is an isometry. To see that U is surjective choose an arbitrary $w \in F$. For each positive integer n choose $x_n \in E$ such that $f(x_n) = nw$. Then $\|x_n\| \leq n\|w\| + \varepsilon$, and by (15.11) with $z = x_n$

$$\|w - n^{-1} U x_n\| \leq n^{-1} K_2\big(1 + \varepsilon \log(n\|w\| + \varepsilon)\big) \to 0.$$

Since U is an isometry, its range is closed and hence contains w.

Since U is an onto isometry, we can identify E and F via U; that is, we may assume that $E = F$ and that U is the identity. To complete the proof we need to show that $\|x - f(x)\| \leq 2\varepsilon$ for all $x \in E$. Let $x \in E$, and put $a = \|x - f(x)\|$ and $y = a^{-1}(f(x) - x)$. (If $a = 0$, there is nothing to prove.) Define $z_n \in E$ by the equation $f(x + z_n) = x + (n + a)y$. By (15.11) (with $U = I$)

$$\|z_n - (n+a)y\| = \|x + z_n - f(x + z_n)\| = o(n)$$

and hence $y = \lim z_n/\|z_n\|$. Put $\lambda = \limsup(\|x + z_n\| - \|z_n\|)$. For every real α and for large n (so that $1 - \alpha\|z_n\|^{-1} \geq 0$)

$$\begin{aligned}
&\|x + z_n\| - \|z_n\| + \alpha \\
&= \Big\|x + \alpha\|z_n\|^{-1} z_n + \big(1 - \alpha\|z_n\|^{-1}\big) z_n\Big\| - \Big\|\big(1 - \alpha\|z_n\|^{-1}\big) z_n\Big\| \\
&\leq \Big\|x + \alpha\|z_n\|^{-1} z_n\Big\|.
\end{aligned}$$

Consequently $\|x + \alpha y\| \geq \lambda + \alpha$. Using this inequality with $\alpha = n + a$, the fact that $f(0) = 0$, and the inequality

$$\|z_n\| \leq \|f(x + z_n)\| - \|f(x)\| + \varepsilon = \|x + (n+a)y\| - \|ay + x\| + \varepsilon \leq n + \varepsilon$$

we obtain

$$\begin{aligned}
\varepsilon &\geq \|f(x + z_n)\| - \|x + z_n\| = \|x + (n+a)y\| - \|x + z_n\| \\
&\geq \lambda + a + n - \|x + z_n\| \geq \lambda - (\|x + z_n\| - \|z_n\|) + a - \varepsilon.
\end{aligned}$$

Passing to the limit gives $\varepsilon \geq a - \varepsilon$; i.e., $a \leq 2\varepsilon$ as required. ◊

2. Larger Perturbations

Let f be a map from a Banach space E into a Banach space F. Put

$$\varphi_f(t) = \sup\Big\{\Big|\|f(x) - f(y)\| - \|x - y\|\Big|\Big\}$$

for $t \geq 0$, where the supremum is taken over all pairs of points x and y so that either $\|x - y\| \leq t$ or $\|f(x) - f(y)\| \leq t$.

In the first section we studied surjective maps f (normalized by $f(0) = 0$) for which $\varphi_f(t) \leq \varepsilon$ for all t. In this section we allow $\varphi_f(t)$ to be unbounded, and we give a rather precise result on the possible growth of $\varphi_f(t)$ as $t \to \infty$, which still forces f to be close to an actual isometry. Just as in the proof of Lemma 15.3 we may also assume here that f is injective.

Assume that $\varphi_f(t) = o(t)$ as $t \to \infty$, and let $\rho_n \to \infty$. Let $\mathcal{U}$ be a free ultrafilter on $\mathbb{N}$. The map

$$\tilde{f}(x_1, \dots, x_n, \dots) = \big(\rho_1^{-1} f(\rho_1 x_1), \dots, \rho_n^{-1} f(\rho_n x_n), \dots\big)$$

induces an isometry from the ultrapower $E_{\mathcal{U}}$ onto the ultrapower $F_{\mathcal{U}}$. By the Mazur-Ulam theorem this isometry must be affine. It follows that

$$\lim_{\mathcal{U}} \rho_n^{-1}\Big(f\big(\rho_n(x + y)\big) - f(\rho_n x) - f(\rho_n y)\Big) = 0 \tag{15.12}$$

for every $x, y \in E$, and the convergence is even uniform on bounded sets. It turns out (see Example 15.6) that this does not ensure, e.g., that $\lim 2^{-n} f(2^n x)$ exists for all x, or even that E and F are actually isometric. We shall give here a precise estimate on the possible growth of $\varphi_f(t)$ that ensures that this happens.

Note first that $\varphi_f(t)$ is monotone increasing in t and that

$$\varphi_f(t + s) \leq \varphi_f(t) + \varphi_f(s) \qquad t, s \geq 0. \tag{15.13}$$

Indeed, let $x, y \in E$, and assume first that $\|f(x) - f(y)\| \geq \|x - y\|$ and that $\|x - y\| \leq t + s$. Pick $z \in [x, y]$ so that $\|x - z\| \leq t$ and $\|z - y\| \leq s$. Then

$$\begin{aligned}\Big|\|f(x) - f(y)\| - \|x - y\|\Big| &= \|f(x) - f(y)\| - \|x - y\| \\ &\leq \|f(x) - f(z)\| - \|x - z\| + \|f(z) - f(y)\| - \|z - y\| \\ &\leq \varphi_f(t) + \varphi_f(s).\end{aligned}$$

If $\|f(x) - f(y)\| \leq t + s$, we proceed similarly but pick z so that $f(z) \in [f(x), f(y)]$ and $\|f(x) - f(z)\| \leq t$, $\|f(z) - f(y)\| \leq s$.

It follows from (15.13) that $\varphi_f(t)/t$ is essentially decreasing in t. More precisely if $t \geq s > 0$, then

$$\frac{\varphi_f(t)}{t} \leq \left(\left[\frac{t}{s}\right] + 1\right) \frac{s}{t} \frac{\varphi_f(s)}{s} \leq 2\frac{\varphi_f(s)}{s}. \tag{15.14}$$

Theorem 15.4. *Let f be a surjective map from the Banach space E onto the Banach space F. Assume that $f(0) = 0$, and let $\varphi_f(t)$ be as above. If*

$$\int_1^\infty \frac{\varphi_f(t)}{t^2} dt < \infty \tag{15.15}$$

then there is a (linear) isometry U from E onto F so that

$$\|f(x) - Ux\| = o(\|x\|) \text{ as } \|x\| \to \infty. \tag{15.16}$$

More precisely

$$\|f(x) - Ux\| \le K\Big(\|x\| \int_{\|x\|}^{\infty} \frac{\varphi(t)}{t^2}\,dt + \int_1^{\|x\|} \frac{\varphi(t)}{t}\,dt\Big) \tag{15.17}$$

where K is an absolute constant and $\varphi(t) = \max\big(1, \varphi_f(t)\big)$.

We shall show later that Theorem 15.4 is sharp and that assumption (15.15) cannot be relaxed. Note that by the monotonicity of φ_f condition (15.15) implies that $\varphi(t) = o(t)$; hence (15.17) implies (15.16).

As in the proof of Theorem 15.2, the proof of Theorem 15.4 is based on an estimate of how close f is to an affine map.

Lemma 15.5. *Let f be as in Theorem 15.4, and let φ be any monotone non-decreasing function so that $\varphi(t) \ge \varphi_f(t)$ and $\varphi(2t) \le 2\varphi(t)$ for all $t > 0$. Then*

$$\Big\|f\Big(\frac{x+y}{2}\Big) - \frac{f(x)+f(y)}{2}\Big\| \le 19\varphi(d) + \varphi(d/2) + \varphi(d/4) + \cdots + \varphi(d/2^k)$$

for all $x, y \in E$, where $d = \|x-y\|$ and k is an integer so that $2^k \ge d$.

The proof of Lemma 15.5 is along the same lines as that of Lemma 15.3, but it is more involved since we have to keep track of all the error estimates. We start with the proof that Lemma 15.5 implies the theorem, which is also similar to the argument in the previous section and is even simpler.

Proof of Theorem 15.4. Let f be as in the statement of the theorem, and put $\varphi(t) = \max(\varphi_f(t), 1)$. Then $\varphi(2t) \le 2\varphi(t)$ by (15.13). Let $z \in E$ with $\|z\| \ge 1$, and let $m \ge 0$ be such that $2^{m-1} \le \|z\| \le 2^m$. Applying Lemma 15.5 with $x = 2^n z$ and $y = 0$, we obtain

$$\|f_n(z)\| \le 2^{-n+1}\Big(19\varphi(2^{n+m}) + \varphi(2^{n+m-1}) + \varphi(2^{n+m-2}) + \cdots + \varphi(1)\Big)$$

for every n, where $f_n(z) = 2^{-n}f(2^n z) - 2^{-(n-1)}f(2^{n-1}z)$. Thus there is an absolute constant K so that

$$\begin{aligned}
\sum_{n=1}^{\infty} \|f_n(z)\| &\le K \sum_{n=1}^{\infty}\sum_{j=0}^{n+m} 2^{-n}\varphi(2^j) \\
&= K \sum_{j=0}^{\infty} \varphi(2^j) \sum_{n=1\vee(j-m)}^{\infty} 2^{-n} \\
&= K\Big(2^{m+1} \sum_{j=m+1}^{\infty} 2^{-j}\varphi(2^j) + \sum_{j=0}^{m+1} \varphi(2^j)\Big).
\end{aligned}$$

From (15.15), the monotonicity of φ, and the usual relation between integrals and sums, it follows that the sum on the right in the expression above is finite. Hence

$$Uz = \lim 2^{-n} f(2^n z) = f(z) + \sum_{n=1}^{\infty} f_n(z)$$

exists and moreover (15.17) holds. It follows from Lemma 15.5 that U is additive, and from the definition of φ_f that it is an isometry. The proof that U is surjective is the same as in Theorem 15.2. ♢

Proof of Lemma 15.5. As in the proof of Lemma 15.3, the argument is based on the interplay of the two reflections; the one in E around $(x+y)/2$ and the one in F around $\big(f(x)+f(y)\big)/2$. By transferring from one space to the other they can be iterated, and the quantitative behavior of these iterates will yield the result. The bookkeeping is a little tedious, and unlike in the proof of Lemma 15.3 we now choose to work in E (but we also define a sequence of points in F). As in the proof of Lemma 15.3 we may assume that f is both injective and surjective.

We pass to the calculations. Let $x, y \in E$, and define $\{p_j\}_{j=-\infty}^{\infty}$ in E and $\{q_j\}_{j=-\infty}^{\infty}$ in F inductively by $p_0 = (x+y)/2$, $q_0 = \big(f(x)+f(y)\big)/2$, and

$$(15.18)\qquad \begin{aligned} p_{j+1} &= f^{-1}(q_{-j}), \quad & q_{j+1} &= f(p_{-j}),\\ p_{-j}+p_j &= x+y, \quad & q_{-j}+q_j &= f(x)+f(y) \end{aligned}$$

for $j \geq 0$. Thus, the order in which the points are chosen is $p_0, q_0, p_1, q_1, p_{-1}, q_{-1}, p_2, q_2, \ldots$ and so on. Note that for every $j = 0, \pm1, \pm2, \ldots$

$$f(p_j) = q_{-j+1}, \quad f^{-1}(q_j) = p_{-j+1}.$$

We need to estimate $\|f(p_0) - q_0\|$, i.e., $\|q_1 - q_0\| = \frac{1}{2}\|q_1 - q_{-1}\|$ in our new notation. To this end we estimate the norms of the differences $q_i - q_j$ (as well as those of $p_i - p_j$) with $i - j$ even. For each $j \geq 0$ put

$$(15.19)\qquad \begin{aligned} \delta_{2j} &= \max\big\{\|p_{2k} - p_{2l}\| : \ |k-l| = j, \ -j \leq k, l \leq j\big\}\\ \Delta_j &= \max\big\{\|p_{2i} - x\|, \ \|p_{2i} - y\|; \ 0 \leq i \leq j\big\}. \end{aligned}$$

Sublemma 1. *With the notation above and with φ as in the statement of the lemma, $\delta_{2^{m+1}} \geq 2\delta_{2^m} - 2^m\varphi(\delta_{2^m})$ for every $m \geq 1$.*

Proof of Sublemma 1. It follows from (15.18) that $\|p_r - p_s\| = \|p_{-r} - p_{-s}\|$ for all r and s, and similarly for the q's. If k, l and j are such that $|k-l| = j$ and $-j+1 \leq k, l \leq j$, then by (15.18), the remark above, and the assumption on φ

$$\begin{aligned} &\Big|\|p_{2k-2} - p_{2l-2}\| - \|p_{2k} - p_{2l}\|\Big|\\ &= \Big|\|p_{-2k+2} - p_{-2l+2}\| - \|p_{2k} - p_{2l}\| - \|q_{2k-1} - q_{2l-1}\| + \|q_{-2k+1} - q_{-2l+1}\|\Big|\\ &\leq \Big|\|p_{-2k+2} - p_{-2l+2}\| - \|f(p_{-2k+2}) - f(p_{-2l+2})\|\Big|\\ &\quad + \Big|\|f(p_{2k}) - f(p_{2l})\| - \|p_{2k} - p_{2l}\|\Big| \leq 2\varphi(\delta_{2j}). \end{aligned}$$

Given m, choose k and l so that $|k-l| = 2^{m-1}$ for $-2^{m-1} \leq k, l \leq 2^{m-1}$ and $\delta_{2^m} = \|p_{2k} - p_{2l}\|$. Since we can replace k, l by $-k$, $-l$ we may also assume that $k \geq 2^{m-2}$. Applying the inequality above $2^{m-1} - k$ times, we obtain

$$\Big|\|p_{2k} - p_{2l}\| - \|p_{2^m} - p_0\|\Big| \leq 2(2^{m-1} - k)\varphi(\delta_{2^m}) \leq 2^{m-1}\varphi(\delta_{2^m})$$

and this proves the sublemma since $2\|p_{2^m} - p_0\| = \|p_{2^m} - p_{-2^m}\| \leq \delta_{2^{m+1}}$. ◇

Sublemma 2. *Let $d = \|x - y\|$, and let Δ_j be given by (15.19). Let φ be as above. Then $\Delta_j \leq d$ for every $j \leq d/8\varphi(d)$.*

Proof of Sublemma 2. Since

$$\|f(p_{-2j}) - f(y)\| = \|q_{2j+1} - f(y)\| = \|q_{-(2j+1)} - f(x)\| = \|f(p_{2j+2}) - f(x)\|$$

we obtain the estimate

$$\begin{aligned}\Big|\|p_{2j+2}-x\|-\|p_{2j}-x\|\Big| &= \Big|\|p_{2j+2}-x\|-\|p_{-2j}-y\|\Big| \\ &\le \Big|\|f(p_{-2j})-f(y)\|-\|p_{-2j}-y\|\Big| + \Big|\|p_{2j+2}-x\|-\|f(p_{2j+2})-f(x)\|\Big|.\end{aligned}$$

Since $\|p_{-2j}-y\| \le \Delta_j$, it follows that

$$\|f(p_{2j+2})-f(x)\| = \|f(p_{-2j})-f(y)\| \le \Delta_j + \varphi(\Delta_j)$$

and we deduce that

$$\begin{aligned}\|p_{2j+2}-x\| &\le \|p_{2j}-x\| + \varphi(\Delta_j) + \varphi\big(\Delta_j + \varphi(\Delta_j)\big) \\ &\le \|p_{2j}-x\| + 2\varphi\big(\Delta_j + \varphi(\Delta_j)\big).\end{aligned}$$

A similar estimate holds for $\|p_{2j+2}-y\|$, and hence

$$\Delta_{j+1} \le \Delta_j + 2\varphi\big(\Delta_j + \varphi(\Delta_j)\big).$$

Repeated applications of this estimate and $\Delta_0 = d/2$ give for every $k \ge 0$

$$\Delta_k \le d/2 + 2k\varphi\big(\Delta_{k-1} + \varphi(\Delta_{k-1})\big). \tag{15.20}$$

Assume now that $j \le d/8\varphi(d)$. We prove by induction on k that $\Delta_k \le 3d/4$ for $0 \le k \le j$. This is clear for $k = 0$, hence also whenever $\varphi(d) \ge d/4$, and we may assume that $\varphi(d) < d/4$. The induction hypothesis and (15.20) give

$$\Delta_k \le d/2 + 2k\varphi(3d/4 + d/4) \le d/2 + 2j\varphi(d) \le d/2 + d/4 = 3d/4.$$

◇

We return to the proof of Lemma 15.5. Since both $\|f\big((x+y)/2\big) - f(x)\|$ and $\|f\big((x+y)/2\big) - f(y)\|$ are smaller than $d/2 + \varphi(d/2)$, the lemma is obvious if $d \le 16\varphi(d)$. So assume that $d \ge 16\varphi(d)$, and find an integer $n \ge 1$ so that

$$2^n \le d/8\varphi(d) < 2^{n+1}. \tag{15.21}$$

By their definitions, $\delta_{2j} \le 2\Delta_j$ for every j. It thus follows from Sublemma 2 that $\delta_{2^{n+1}} \le 2\Delta_{2^n} \le 2d$. We claim next that

$$\delta_{2^{n-k}} \le 2^{(-k+1)/2} d \tag{15.22}$$

for $-1 \le k \le n$. For $k = -1$ we have just checked this. For other values of k Sublemma 1 gives $\delta_{2^{n-k}} \le \frac{1}{2}\delta_{2^{n-(k-1)}} + 2^{n-k-1}\varphi(\delta_{2^{n-k}})$.

By induction, the first term is bounded by $2^{-k/2}d$. By the choice of n in (15.21), and since $\delta_{2^{n-k}} \le \delta_{2^{n+1}} \le 2d$ when $k \ge -1$, the second term is bounded by $2^{-k-3}\big(d/2\varphi(d)\big)\varphi(2d)$. Since also $\varphi(2d) \le 2\varphi(d)$, we obtain

$$\delta_{2^{n-k}} \le (2^{-k/2} + 2^{-k-3})d \le 2^{(-k+1)/2}d$$

and this proves (15.22). Repeated use of Sublemma 1 gives

$$\delta_{2^{n+1}} \ge 2^n\Big(\delta_2 - \big(\varphi(\delta_{2^n}) + \varphi(\delta_{2^{n-1}}) + \cdots + \varphi(\delta_2)\big)\Big). \tag{15.23}$$

By (15.22), the typical element on the right satisfies

$$\varphi(\delta_{2^{n-k}}) \le \varphi(2^{(-k+1)/2}d).$$

By taking $k = 2j+1$, we obtain $\varphi(\delta_{2^{n-k}}) + \varphi(\delta_{2^{n-(k+1)}}) \le 2\varphi(2^{-j}d)$. Since also $2^{-n}\delta_{2^{n+1}} \le \frac{16\varphi(d)}{d} \cdot 2d$ (because $2^{-n} \le \frac{16\varphi(d)}{d}$ by (15.21)), substitution in (15.23) gives

$$\delta_2 \le 34\varphi(d) + 2\sum_{j=1}^{[n/2]+1} \varphi(2^{-j}d).$$

Since $\varphi(\delta_2) \le \varphi(2d) \le 2\varphi(d)$ (because $\delta_2 \le 2d$ by (15.22)), we deduce that

$$\delta_2 + \varphi(\delta_2) \le 38\varphi(d) + 2\sum_{j=1}^{n} \varphi(2^{-j}d).$$

This concludes the proof since $2^n \le d/8\varphi(d) \le d/8 \le d$ and

$$2\|q_0 - f(p_0)\| = 2\|q_1 - q_0\| = \|q_1 - q_{-1}\| = \|f(p_0) - f(p_2)\| \le \delta_2 + \varphi(\delta_2).$$

◇

We turn to showing that Theorem 15.4 is sharp, and we first give an example that shows that its conclusion may fail when $\int_1^\infty \varphi_f(t)t^{-2}dt = \infty$ even for maps of the Euclidean plane onto itself.

Let φ be a monotone increasing function which satisfies (15.14) (for φ) and such that $\int_1^\infty \varphi(t)t^{-2}dt = \infty$. Put $\Phi(t) = \frac{1}{2}\int_1^t \varphi(s)s^{-2}ds$ for $t > 1$ and $\Phi(t) = 0$ for $0 \le t \le 1$. For $t \ge 0$ let V_t denote the clockwise rotation by an angle t. Then $\|V_t x - V_s x\| \le (t-s)\|x\|$ whenever $t \ge s \ge 0$. Put

$$f(x) = V_{\Phi(\|x\|)}x.$$

Then f is a surjective homeomorphism of $\mathbb{R}^2$, and there is no isometry U of the plane with $\|f(x) - Ux\| = o(\|x\|)$ because $\lim_{t\to\infty}\Phi(t) = \infty$. We now check that $\varphi_f(t) \le \varphi(t)$ for all t. Indeed, if $1 \le \|y\| \le \|x\|$, then $\|x-y\| = \|V_{\Phi(\|x\|)}x - V_{\Phi(\|x\|)}y\|$ and (15.14) give

$$\begin{aligned}
\Big|\|f(x) - f(y)\| - \|x-y\|\Big| &\le \Big\|V_{\Phi(\|x\|)}y - V_{\Phi(\|y\|)}y\Big\| \le \frac{1}{2}\|y\|\int_{\|y\|}^{\|x\|} \varphi(s)s^{-2}ds \\
&\le \frac{1}{2}\|y\|\varphi(\|x\|)\int_{\|y\|}^{\|x\|} s^{-2}ds \\
&= \big(\|x\| - \|y\|\big)\varphi\big(\|x\|\big)/2\|x\| \\
&\le \varphi\big(\|x\| - \|y\|\big) \le \varphi\big(\|x-y\|\big).
\end{aligned}$$

For finite-dimensional spaces, the existence of a map f from E onto F with $\varphi_f(t) = o(t)$ ensures that E is isometric to F. Indeed, if $\rho_n \to \infty$, then there is a subsequence $\{\rho_{n(m)}\}$ so that $Ux = \lim \rho_{n(m)}^{-1} f(\rho_{n(m)}x)$ exists for all $x \in E$. Clearly $\|Ux\| = \|x\|$, and U is linear by (15.12). Thus U is an isometry.

The example above shows, however, that when $\int_1^\infty \varphi_f(t)t^{-2}dt = \infty$, then U may depend on the particular subsequence $\{\rho_{n(m)}\}$, and that $\lim_{t\to\infty} t^{-1}f(tx)$ may fail to exist even for a single $x \ne 0$.

In the example below we use infinite-dimensional spaces to construct a map with $\varphi_f(t) = o(t)$ when E and F are not even isometric. The general idea is similar to the construction above, but the details are naturally more complicated.

Example 15.6. *Let $\varphi(t)$ be a positive increasing function on $(0,\infty)$ so that*

$$\frac{\varphi(t)}{t} \le 2\frac{\varphi(s)}{s}, \qquad t \ge s > 0 \tag{15.24}$$

and

$$\int_1^\infty \frac{\varphi(t)}{t^2}\,dt = \infty. \tag{15.25}$$

Then there exist two nonisometric Banach spaces E and F, and a homeomorphism f from E onto F with $f(0)=0$ so that $\varphi_f(t) \le \varphi(t)$ for all $t>0$.

We shall essentially work in the Hilbert space l_2. Define the function ψ : $[0,\infty) \to [0,\infty)$ by $\psi(t) = 2$ for $0 \le t \le 1$ and by

$$\psi(t) = 2 + \frac{1}{10}\int_1^t \frac{\tilde{\varphi}(s)}{s^2}\,ds$$

for $t \ge 1$, where $\tilde{\varphi}(s) = \min\{s/2, \varphi(s/2)\}$. Then (15.25) implies that $\psi(n) \to \infty$.

Denote the unit vector basis in l_2 by $\{e_i\}_{i=0}^\infty$. The usual norm on l_2 and the norm of operators on l_2 with respect to this norm are denoted by $\|\!|\cdot|\!\|$. (This distinguishes them from the new norms to be defined below.) Consider the one-parameter family U_t of unitary operators on l_2 defined by

$$U_{n+\theta}(e_i) = \begin{cases} e_i & \text{for } i \ge n+1 \\ -e_{i-1} & \text{for } 1 \le i \le n-1 \\ \cos\frac{\theta\pi}{2}e_{n-1} + \sin\frac{\theta\pi}{2}e_n & \text{for } i = 0 \\ -\sin\frac{\theta\pi}{2}e_{n-1} + \cos\frac{\theta\pi}{2}e_n & \text{for } i = n \end{cases}$$

for $n = 1,2,\dots$ and $0 \le \theta \le 1$. It is easy to verify that if $t \ge 1$ and $s > 0$, then

$$\|\!|U_{t+s} - U_t|\!\| \le \pi s/2 \le 2s. \tag{15.26}$$

The spaces E and F will be renormings of l_2. Define t_n by

$$\psi(t_n/3) = n+2 \tag{15.27}$$

and choose a decreasing sequence of positive numbers λ_n so that

$$0 < \lambda_n < \tilde{\varphi}(t_n)/12t_n, \quad \Big(\sum_{n=1}^\infty \lambda_n^2\Big)^{1/2} \le 1/4 \tag{15.28}$$

$$\Big(\lambda_n^2 + 2\sum_{m=n+1}^\infty \lambda_m^2\Big)^{1/2} \le 2\lambda_n. \tag{15.29}$$

We let E, respectively F, be the space l_2 with the norms

$$\|x\|_E = \|\!|x|\!\| + \sum_{i=1}^\infty \lambda_i|x_{i+1}|\,; \quad \|x\|_F = \|\!|x|\!\| + \sum_{i=1}^\infty \lambda_i|x_i|$$

where $x = (x_0, x_1, x_2, \dots) \in l_2$. Then E and F are not isometric. Indeed, span$\{e_0, e_1\}$ in E is isometric to l_2^2, while F does not have any smooth two-dimensional subspace because when a sum of convex functions is differentiable at a point, then so are all the summands.

Consider the map f from E onto F given by

$$f(x) = U_{\psi(\|\!|x|\!\|)}x.$$

Clearly f is a surjective homeomorphism, and we turn to proving that $\varphi_f(t) \leq \varphi(t)$. We start with some remarks on ψ and on the new norms. If $s, t > 0$, then

$$t\big(\psi(t+s) - \psi(t)\big) \leq \tilde{\varphi}(s)/5. \tag{15.30}$$

Indeed, (15.30) holds trivially for $0 < t < 1$ because $\psi(t) = \psi(1)$, and for $t \geq 1$

$$\begin{aligned}\psi(t+s) - \psi(t) &= 10^{-1}\int_t^{t+s} u^{-2}\tilde{\varphi}(u)du \leq 10^{-1}\tilde{\varphi}(t+s)\int_t^{t+s} u^{-2}du \\ &= \tilde{\varphi}(t+s)/(t+s)\cdot s/10t \leq 2\tilde{\varphi}(s)/s \cdot s/10t.\end{aligned}$$

From (15.28) we obtain for every $x \in l_2$

$$|||x||| \leq \|x\|_E \leq 5|||x|||/4, \quad |||x||| \leq \|x\|_F \leq 5|||x|||/4. \tag{15.31}$$

Now let $x, y \in l_2$ with $|||x||| \geq |||y|||$, and put $t = |||y|||$ and $s = |||x||| - |||y|||$. Then $\|x-y\|_E \geq |||x-y||| \geq s$, and we get by (15.26) and (15.30)

$$\begin{aligned}\|U_{\psi(t+s)}y - U_{\psi(t)}y\|_E &\leq 5|||U_{\psi(t+s)}y - U_{\psi(t)}y|||/4 \\ &\leq 10\big(\psi(t+s) - \psi(t)\big)|||y|||/4 \\ &= 10t\big(\psi(t+s) - \psi(t)\big)/4 \\ &\leq \tilde{\varphi}(s)/2 \leq s/4 \leq \|x-y\|_E/4\end{aligned} \tag{15.32}$$

(recall that by definition $\tilde{\varphi}(s) \leq s/2$). From (15.31), the computation in (15.32) and the fact that the U's are unitary, we deduce that

$$\begin{aligned}\|f(x) - f(y)\|_F &\geq |||f(x) - f(y)||| = |||U_{\psi(t+s)}x - U_{\psi(t)}y||| \\ &\geq |||x-y||| - |||U_{\psi(t+s)}y - U_{\psi(t)}y||| \\ &\geq 4\|x-y\|_E/5 - \|x-y\|_E/5 = 3\|x-y\|_E/5.\end{aligned} \tag{15.33}$$

We now proceed to estimate the difference

$$\begin{aligned}&\Big|\|f(x) - f(y)\|_F - \|x-y\|_E\Big| \\ &\leq \|U_{\psi(t+s)}y - U_{\psi(t)}y\|_F + \Big|\|U_{\psi(t+s)}x - U_{\psi(t+s)}y\|_F - \|x-y\|_E\Big|.\end{aligned} \tag{15.34}$$

The first summand was estimated in (15.32). To estimate the second, fix n so that $n+1 \leq \psi(t+s) < n+2$. Inspection of the definition of U_r and of the norms in E and F shows that if $z = (z_0, z_1, z_2, \ldots)$, then $\|U_{\psi(t+s)}z\|_F - \|z\|_E$ is equal to

$$\lambda_n|\alpha z_0 - \beta z_{n+1}| + \lambda_{n+1}|\beta z_0 + \alpha z_{n+1}| - \lambda_{n+1}|z_{n+1}| + \sum_{i=n+2}^{\infty}|z_i|(\lambda_{i-1} - \lambda_i)$$

for suitable α, β with $\alpha^2 + \beta^2 = 1$. It follows from this and from (15.29) that

$$\Big|\|U_{\psi(t+s)}z\|_F - \|z\|_E\Big| \leq 3\lambda_n|||z||| \leq 3\lambda_n\|z\|_E. \tag{15.35}$$

By (15.27) we have

$$t_n/3 \geq t+s = |||x||| \geq |||x-y|||/2 \geq \|x-y\|_E/3$$

so $\|x-y\|_E \leq t_n$. Hence, by (15.28), (15.35) and (15.24) (which holds also for $\tilde{\varphi}$),

$$\begin{aligned}\Big|\|U_{\psi(t+s)}x - U_{\psi(t+s)}y\|_F - \|x-y\|_E\Big| &\leq 3\lambda_n\|x-y\|_E \\ &\leq \tilde{\varphi}(t_n)\|x-y\|_E/4t_n \\ &\leq \tilde{\varphi}(\|x-y\|_E)/2.\end{aligned}$$

This, together with the computation in (15.32) and (15.34), yields

$$\Big| \|f(x) - f(y)\|_F - \|x - y\|_E \Big| \le \tilde{\varphi}(\|x - y\|_E).$$

By (15.33) $\|x - y\|_E \le 2\|f(x) - f(y)\|_F$; hence, going back to the definition of φ_f, we deduce that $\varphi_f(u) \le \tilde{\varphi}(2u) \le \varphi(u)$. This concludes the verification that f has all the required properties.

We now give a sharp generalization of Proposition 15.1 in the spirit of Theorem 15.4 and Example 15.6. The proofs are, however, much easier. For a map $f : E \to F$ let

$$\alpha_f(t) = \sup \big\{ \|f(x + y) - f(x) - f(y)\| : \ \|x\|, \|y\| \le t \big\}.$$

Note that $\alpha_f(t)$ is a monotone nondecreasing function of t.

Proposition 15.7. (i) *Let E and F be Banach spaces. If a map f from E to F satisfies $\int_1^\infty t^{-2}\alpha_f(t) < \infty$, then there is an additive map $T : E \to F$ so that $\|f(x) - Tx\| = o(\|x\|)$ as $\|x\| \to \infty$.*

(ii) *Let $\alpha(t)$ be a positive increasing function on $(1, \infty)$. If $\int_1^\infty t^{-2}\alpha(t)dt = \infty$, then there is a map $f : \mathbb{R} \to \mathbb{R}$ so that $\alpha_f(t) \le \alpha(t)$ for all $t > 1$ and so that there is no linear $T : \mathbb{R} \to \mathbb{R}$ with $|f(x) - Tx| = o(|x|)$ as $|x| \to \infty$.*

Proof. Part (i) is a trivial extension of Proposition 15.1. Indeed, by definition $\|f(2y) - 2f(y)\| \le \alpha_f(\|y\|)$. Fixing $x \in E$ and applying this inequality with $y = 2^n x$, we find that $Tx = \lim 2^{-n} f(2^n x)$ exists and satisfies $\|Tx - f(x)\| = o(\|x\|)$. That T is additive follows from the definition of α_f (note that $\alpha_f(t) = o(t)$).

(ii) Define $\psi : \mathbb{R} \to \mathbb{R}$ by $\psi(x) = 0$ if $|x| \le 2$ and $\psi(x) = \frac{1}{3}\int_2^{|x|} \alpha(u/2)u^{-2}du$ if $|x| \ge 2$, and let $f(x) = x\psi(x)$. Then $\psi(x) \to \infty$ implies that there is no linear $T : \mathbb{R} \to \mathbb{R}$ with $|f(x) - Tx| = O(|x|)$ (let alone $o(|x|)$).

To estimate α_f assume that $2 \le s = |y| \le |x| = t$, and put $r = |x + y|$. Assume also that $r \ge 2$. Then

$$\begin{aligned} \big|f(x+y) - f(x) - f(y)\big| &= \big|(x+y)\psi(x+y) - x\psi(x) - y\psi(y)\big| \\ &\le |x+y|\big|\psi(x+y) - x\psi(x)\big| + |y|\big|\psi(x) - \psi(y)\big| \\ &= \frac{r}{3}\Big|\int_r^t \alpha(u/2)u^{-2}du\Big| + \frac{s}{3}\int_s^t \alpha(u/2)u^{-2}du. \end{aligned}$$

For $t \ge r$ the first integral is positive, and the right-hand side is bounded by $2\alpha(t/2)/3$. For $t \le r$ we use $r \le 2t$ and obtain the bound $\alpha(t)$. Minor changes show that the same estimates also hold when either r or s (or both) is smaller than 2. Hence $\alpha_f(t) \le \alpha(t)$ for all $t \ge 1$. $\square$

3. Notes and Remarks

Proposition 15.1 is due to Hyers [**261**]. The literature contains several related results; see e.g. the expository paper [**262**].

Hyers and Ulam [**264**] proved a result on convex functions which is related to Proposition 15.1 but which depends on the dimension of the space. They proved that there is a constant C_n, depending only on the dimension n, such that if D is a convex subset of $\mathbb{R}^n$ and if $f : D \to \mathbb{R}$ satisfies

$$f\big(\lambda x + (1 - \lambda)y\big) \le \lambda f(x) + (1 - \lambda)f(y) + \varepsilon \tag{15.36}$$

for all $x, y \in D$ and $0 < \lambda < 1$, then there is a convex function $g : D \to \mathbb{R}$ so that $|g(x) - f(x)| \le C_n\varepsilon$ for all $x \in D$. This does not extend to infinite-dimensional spaces because the constants C_n really depend on n. Cholewa [**119**] proved that $C_n \le C\log(n+1)$ for some absolute constant C. Laczkovich [**345**] proved that the result is sharp in the sense that there is a constant $\gamma > 0$ such that whenever $D \subset \mathbb{R}^n$ is a convex set with nonempty interior there is a function $f : D \to \mathbb{R}$ which satisfies (15.36) for some $\varepsilon > 0$ (say $\varepsilon = 1$) so that

$$\sup\bigl\{|g(x) - f(x)| : x \in D\bigr\} \ge \gamma\log(n+1)\varepsilon.$$

A somewhat weaker result was proved earlier by Casini and Papini [**113**]. Dilworth, Howard and Roberts [**150**] prove similar results for functions satisfying (15.36) for $\lambda = 1/2$ only. They also use these results to study approximately convex sets in $\mathbb{R}^n$, i.e., sets A for which $\operatorname{dist}\bigl((x+y)/2, A\bigr) \le 1$ for every $x, y \in A$.

Instead of characterizing functions that are close to linear functions as in Proposition 15.1, one can also look for characterizations of functions that are close to polynomials. A continuous function f on a linear space is a polynomial of degree at most $m-1$ if and only if $\Delta^m(f) \equiv 0$, where Δ^m is the mth order difference operator

$$\Delta^m(h; x, y) = \sum_{j=0}^{m}(-1)^{m-j}\binom{m}{j}h(x+jy).$$

(For example, $\Delta^2(h; x, y) = h(x+2y) - 2h(x+y) + h(x)$.)

Brudnyi and Kalton [**103**] investigate the stability of this characterization. Let D be a convex subset in a Banach space E. For each $m \ge 2$ let $w_m(D)$ be the smallest constant K with the following property: If f is a continuous real-valued function on D, then there is a polynomial P of degree at most $m-1$ so that

$$\sup\bigl\{|f(x) - P(x)| : x \in D\bigr\} \le K\sup\bigl\{|\Delta^m(f; x, y)| : x, x+my \in D\bigr\}$$

($w_m(D)$ could be infinite). For $m = 2$ this gives a measure of approximation by linear functions. Since f is nearly linear when both f and $-f$ are nearly convex, this case is close to the material discussed in the preceding paragraph. They also consider the case $m \ge 3$. They study in particular the situation where D is the unit ball of E. In this case they relate $w_m(D)$ to Banach space properties of E (such as type). They show that if D is an ellipsoid in $\mathbb{R}^n$, then there is a constant C, independent of the dimension, so that $w_2(D) \le C$ and $w_3(D) \le C\log(n+1)$ and that this estimate is sharp (except for the value of C). They also give estimates of $w_2(D)$ and $w_3(D)$ when D is the unit ball of l_p^n for $1 \le p \le \infty$ $(p \ne 2)$. They also point out that the investigation of the constants $w_m(D)$ is closely related to the subject matter of the next chapter.

The question whether Theorem 15.2 is valid (with a constant K instead of the exact constant 2 in (15.5)) was posed by Hyers and Ulam [**263**]. They were motivated by concrete problems of construction of models for configurations where the experimental data (e.g. aerial photos) for the distances between two points contain an error which is uniformly bounded for all pairs of points.

They proved that the answer is positive (with $K = 10$) if $E = F = l_2$. Additional partial contributions were made by Hyers and Ulam, D.G. Bourgin and R.D. Bourgin. The decisive steps in the solution of the general case were done by Gruber [**235**], who verified that if there is an isometry U with $\|f(x) - Ux\| = o(\|x\|)$ as

$x \to \infty$, then (15.5) is valid with 5 instead of 2, and in particular by Gevirtz [**203**], who proved that such a U really exists. The proof of Lemma 15.3 given here is a (somewhat simplified) presentation of the argument of Gevirtz. The proof that Lemma 15.3 implies Theorem 15.2 is due to Omladič and Šemrl [**446**], who streamlined Gruber's argument and obtained the best constant 2 in (15.5).

Theorem 15.2 and the examples presented after its statement required the isometries to satisfy $U0 = 0$. In both examples, if this requirement is dropped, then it is possible to find affine isometries with $\sup \|f(x) - Ux\| \leq 1$. Šemrl [**530**] gives an example which shows that even without assuming $U0 = 0$ it is generally impossible to reduce the constant 2 in (15.5). He notes, however, that when the assumption $U0 = 0$ is removed, then (15.5) holds with 2 replaced by the Jung constant γ_E of E. (Recall that the Jung constant of E is the infimum of the r's such that every subset of E of diameter 1 is contained in a closed ball of radius $r/2$. Evidently $1 \leq \gamma_E \leq 2$, and it is easily verified that $\gamma_E = 1$ if and only if E is a $\mathcal{P}_1$ space.)

Although the surjectivity assumption in Theorem 15.2 cannot be removed, it can be relaxed. This is discussed, e.g., in Dilworth [**149**] (and in its references). In particular, no surjectivity assumption is needed when E and F are finite-dimensional spaces of the same dimension.

Theorem 15.4 and Example 15.6 are due to Lindenstrauss and Szankowski [**368**]. As they point out, the proof of Lemma 15.5 (and for that matter also Lemma 15.3) is valid in the setting of general metric spaces provided we postulate the existence of involutions which are isometries in E and F. (These involutions replace the reflections such as $x \to 2p - x$ which were used in the proofs here.)

Let E and F be Banach spaces so that there is a surjective map $f : E \to F$ with $\varphi_f(t) = o(t)$ as $t \to \infty$. As pointed out in Section 2, E and F have isometric ultrapowers, but they need not be isometric. It seems to be unknown whether for each ε there must be an isomorphism T_ε from E onto F so that $\|T_\varepsilon\| \, \|T_\varepsilon^{-1}\| \leq 1+\varepsilon$. This is easily seen to be the case for the spaces constructed in Example 15.6.

We originally constructed in Proposition 15.7(ii) a map $f : l_2 \to l_2$ (using the operators U_t from Example 15.6). J. Castillo pointed out to us that the same argument can be used to give an example on the line, and this is what we presented here.

CHAPTER 16

Twisted Sums

1. Quasi-Linear Functions

In this chapter we introduce a class of nonlinear functions, the quasi-linear functions, and use them for a construction with important consequences in linear functional analysis. The essential property of quasi-linear functions (for a precise definition see Definition 16.1 below) is that the quantity $\alpha_f(t)$, defined before Proposition 15.7, satisfies $\alpha_f(t) = O(t)$ as $t \to \infty$ instead of $\alpha_f(t) = o(t)$ as in Chapter 15. In this section we shall discuss the general construction, and in Section 2 we shall consider in detail one special case which is of much interest from the point of view of Banach space theory.

The natural context for the study of quasi-linear functions is not that of Banach spaces but the more general class of quasi-Banach spaces. We refer the reader to Appendix H for information on quasi-Banach spaces.

We consider the following problem: Let E and F be quasi-Banach spaces. What are the quasi-Banach spaces G which contain a subspace E_0 isomorphic to E such that G/E_0 is isomorphic to F, that is, the spaces G such that there is a short exact sequence (with continuous linear operators)

$$0 \longrightarrow E \longrightarrow G \longrightarrow F \longrightarrow 0.$$

The direct sum $G = E \oplus F$ satisfies these conditions trivially. It is usual to call G a twisted sum of E and F (note that the order of E and F is important) if it satisfies these conditions and if it is not such a direct sum (i.e., when the image of E in G is not complemented). We find it convenient, however, to also call the direct sum $E \oplus F$ a (trivial) twisted sum.

Two twisted sums G_1 and G_2 of E and F are called equivalent if there is an isomorphism T from G_1 onto G_2 such that the diagram

$$\begin{array}{ccccccccc} 0 & \longrightarrow & E & \longrightarrow & G_1 & \longrightarrow & F & \longrightarrow & 0 \\ & & \big\downarrow{\scriptstyle I_E} & & \big\downarrow{\scriptstyle T} & & \big\downarrow{\scriptstyle I_F} & & \\ 0 & \longrightarrow & E & \longrightarrow & G_2 & \longrightarrow & F & \longrightarrow & 0 \end{array}$$

commutes, where I_E and I_F denote the identity maps on E and F respectively.

Definition 16.1. *A function f from a quasi-Banach space F to a quasi-Banach space E is called quasi-linear if there is a constant M such that*

(i) *$f(tx) = tf(x)$ for all $t \in \mathbb{R}$ and $x \in F$.*

(ii) *$\|f(x+y) - f(x) - f(y)\| \le M(\|x\| + \|y\|)$ for all $x, y \in F$.*

Note that there is no continuity assumption on f. In fact, the functions f which are usually employed are actually discontinuous.

Two quasi-linear functions f_1 and f_2 from F to E are called equivalent if there is a linear map $T : F \to E$ (not necessarily continuous!) and a constant K such that

$$\|f_1(x) - f_2(x) - Tx\| \le K\|x\|$$

for all $x \in F$.

To every quasi-linear map f from F to E we assign a twisted sum of E and F, which we denote by $E \oplus_f F$. This is the space of pairs (x, y), where $x \in E$ and $y \in F$, equipped with the quasi-norm

$$\|(x, y)\| = \|x - f(y)\| + \|y\|.$$

To see that this formula defines a quasi-norm, let M be the quasi-linearity constant of f, and let k_1 and k_2 be the concavity constants of E and F respectively. Then

$$\begin{aligned}
\big\|(x_1 + x_2, y_1 + y_2)\big\| &= \|x_1 + x_2 - f(y_1 + y_2)\| + \|y_1 + y_2\| \\
&\le k_1\|x_1 - f(y_1) + x_2 - f(y_2)\| \\
&\qquad + k_1\|f(y_1) + f(y_2) - f(y_1 + y_2)\| + k_2\big(\|y_1\| + \|y_2\|\big) \\
&\le k_1^2\big(\|x_1 - f(y_1)\| + \|x_2 - f(y_2)\|\big) \\
&\qquad + k_1 M\big(\|y_1\| + \|y_2\|\big) + k_2\big(\|y_1\| + \|y_2\|\big) \\
&\le C\big(\|(x_1, y_1)\| + \|(x_2, y_2)\|\big)
\end{aligned}$$

where $C = \max(k_1^2, k_2 + k_1 M)$. The subspace $E_0 = \{(x, 0) : x \in E\}$ of $E \oplus_f F$ is isometric to E, and the quotient $(E \oplus_f F)/E_0$ is isometric to F; hence $E \oplus_f F$ is a twisted sum of E and F.

We now check that $E \oplus_f F$ is complete. (This is not completely obvious because f is not assumed to be continuous.) Let $\{z_n\}$ be a Cauchy sequence in $E \oplus_f F$, and let $q : E \oplus_f F \to F$ be the quotient map. Then there is a $z \in E \oplus_f F$ such that $qz_n \to qz$, and hence there are $x_n \in E$ such that $\|z - z_n - (x_n, 0)\| \to 0$. Consequently, $\{x_n\}$ is a Cauchy sequence in E; hence convergent to some $x \in E$. Thus $\lim z_n = z - (x, 0)$.

Theorem 16.2. *The correspondence $f \leftrightarrow E \oplus_f F$ is a one-one correspondence between the equivalence classes of quasi-linear function from F to E and the equivalence classes of twisted sums of E and F. In particular, $E \oplus_f F$ is a trivial twisted sum if and only if f is equivalent to a linear function.*

Proof. We first show that if G is a twisted sum of E and F, then there is a quasi-linear $f : F \to E$ such that G is equivalent to $E \oplus_f F$. Let j be a linear into-isomorphism, and let q be a surjective linear map yielding the exact sequence

$$0 \longrightarrow E \xrightarrow{\;j\;} G \xrightarrow{\;q\;} F \longrightarrow 0.$$

Let $\varphi : F \to G$ be a linear (not necessarily continuous) right inverse of q (i.e., $q\big(\varphi(y)\big) = y$ for all $y \in F$). Such a φ exists by trivial linear algebra (use a Hamel basis in F). Let ψ be a homogeneous (not necessarily continuous) right inverse of q for which there is a constant K with $\|\psi(y)\| \le K\|y\|$ for all $y \in F$. To construct such a ψ, define it first on the unit sphere (by the open mapping theorem), and then extend by homogeneity. Then the map $f : F \to E$, defined by $f(y) = j^{-1}\big(\psi(y) - \varphi(y)\big)$, is well defined (because $\psi(y) - \varphi(y)$ is in the kernel of q)

and is quasi-linear. Indeed,

$$\begin{aligned}\|f(x+y)-f(x)-f(y)\| &= \|j^{-1}\big(\psi(x+y)-\psi(x)-\psi(y)\big)\| \\ &\le \|j^{-1}\|\,k_1^2\Big(\|\psi(x+y)\|+\|\psi(x)\|+\|\psi(y)\|\Big) \\ &\le \|j^{-1}\|\,Kk_1^2(k_2+1)\big(\|x\|+\|y\|\big)\end{aligned}$$

where k_1 and k_2 are the moduli of concavity of G and F respectively.

Consider the linear map $Tz = \big(j^{-1}\big(z-\varphi(qz)\big), qz\big)$ from G to $E \oplus_f F$. The formula $\|Tz\| = \|j^{-1}\big(z-\psi(qz)\big)\| + \|qz\|$ shows that T is a bounded linear operator and that it is also bounded from below. Indeed, this is clear if $\|qz\| \ge \|z\|/2K$; otherwise, since $\|\psi(qz)\| \le K\|qz\|$, it follows that

$$\|Tz\| \ge \big\|j^{-1}\big(z-\psi(qz)\big)\big\| \ge \big(\|z\| - \|\psi(qz)\|\big)/\|j\| \ge \|z\|/2\|j\|.$$

It is also easy to check that T is onto and that the relevant diagram commutes. Thus T is an equivalence between G and $E \oplus_f F$.

Assume next that $E \oplus_{f_1} F$ and $E \oplus_{f_2} F$ are equivalent twisted sums, and let $T : E \oplus_{f_1} F \to E \oplus_{f_2} F$ be the isomorphism which gives this equivalence. Then T must be of the form $T(x, y) = (x + Ly, y)$ where L is a linear map from F to E. Applying this to $(f_1(y), y)$, we get that $\|f_1(y) - f_2(y) + Ly\| \le \|T\|\,\|y\|$ and thus f_1 is equivalent to f_2.

It is trivial to verify that the equivalence of f_1 and f_2 implies that of $E \oplus_{f_1} F$ and $E \oplus_{f_2} F$, and that $E \oplus_f F$ is a trivial twisted sum when f is linear. □

Since quasi-linear maps may fail to be continuous, it is difficult (if at all possible) to represent them explicitly on the entire space F. Often the function is defined explicitly only on a dense linear subspace. The surprising fact is that in spite of their discontinuity, these functions can be extended from any dense subspace to the whole space and in an essentially unique way. This is a consequence of Theorem 16.2 and the usual procedure of completion of a quasi-normed space.

Proposition 16.3. *Let E and F be quasi-Banach spaces, and let F_0 be a dense linear subspace of F. Let $f_0 : F_0 \to E$ be a quasi-linear function. Then f_0 can be extended to a quasi-linear function $f : F \to E$, and the function f is uniquely determined up to equivalence.*

Proof. Consider the quasi-normed space $E \oplus_{f_0} F_0$ (whose norm is given by $\|(x, y)\| = \|x - f_0(y)\| + \|y\|$ for $x \in E$ and $y \in F_0$), and let G be its completion. The space G is a twisted sum of E and F. Indeed, $jx = (x, 0)$ is an embedding of E into G, and the map $(x, y) \to y$ extends to a quotient map q from G onto F. To check that the kernel of q coincides with $j(E)$, notice that $j(E) \subset \ker(q)$, and fix $z \in G$ with $qz = 0$. Then $z = \lim(x_n, y_n)$ with $y_n \in F_0$ and $y_n \to 0$. Since $\|(f_0(y_n), y_n)\| = \|y_n\| \to 0$, it follows that $z = \lim(x_n - f_0(y_n), 0)$; i.e., z belongs to $j(E)$. It follows from Theorem 16.2 that there is a quasi-linear map $g : F \to E$ such that G is equivalent to $E \oplus_g F$. This means that there is an isomorphism T from G onto $E \oplus_g F$ such that the restriction of T to the subspace $E \oplus_{f_0} F$ of G is given by $T(x, y) = (x + Ly, y)$ for some linear map $L : F_0 \to E$. Hence

$$\|f_0(y) - g(y) + Ly\| \le \|T\|\,\|y\|, \qquad y \in F_0.$$

Define the function $f : F \to E$ to be f_0 on F_0 and by $f(y) = g(y) - Ly$ for $y \notin F_0$. Then f is a quasi-linear extension of f_0 to all of F. The uniqueness of f up to equivalence follows from the uniqueness of the completion. □

We now give an example of a nontrivial quasi-linear map f from l_1 to the real line. In view of Proposition 16.3 it suffices to define it on the subspace F_0 of finitely supported sequences in l_1. Put (with the logarithm taken to the base e)

$$f(x) = \sum x_n \log |x_n| - \Big(\sum x_n\Big) \log \Big|\sum x_n\Big|, \qquad x = (x_1, x_2, \dots) \in F_0$$

where $0 \log 0$ is taken as 0. That f is quasi-linear follows from the inequality $\big|s \log |s| + t \log |t| - (s+t) \log |s+t|\big| \leq |s| + |t|$ which holds for all real s, t.

The function f is not trivial, i.e., not equivalent to 0. Indeed, $f(e_n) = 0$ for all n (where e_n denotes the n'th unit vector), and thus any linear function φ on l_1 which satisfies $|f(x) - \varphi(x)| \leq M\|x\|$ has to be bounded on the unit sphere of F_0, but $f(n^{-1} \sum_{i=1}^{n} e_i) = -\log n$.

It follows that $\mathbb{R} \oplus_f l_1$ is not locally convex (i.e., not a Banach space). Indeed, in a locally convex space G, the Hahn-Banach theorem implies that every one-dimensional space E is complemented; i.e., $G = E \oplus (G/E)$ (a usual direct sum).

This example justifies the remark made in the beginning of the section that the natural context for the discussion of quasi-linear functions is that of quasi-normed spaces. Even when E and F are Banach spaces, the quasi-norm in $E \oplus_f F$ may fail to even be equivalent to a norm. This cannot happen for spaces with nontrivial type. (See Appendix G for information on type and on related notions.)

Theorem 16.4. *Let E and F be Banach spaces, and let G be a twisted sum of E and F.*

(i) *If both E and F have nontrivial type, then the space G is isomorphic to a Banach space.*

(ii) *If both E and F have type p for some $1 < p \leq 2$, then the Banach space G has type r for all $r < p$.*

(iii) *If both E and F have equivalent norms whose moduli of smoothness satisfy $\rho(\varepsilon) = O(\varepsilon^p)$ for some $1 < p \leq 2$, then for every $r < p$ the Banach space G can be renormed to have a modulus of smoothness satisfying $\rho(\varepsilon) = O(\varepsilon^r)$. In particular, G is superreflexive whenever both E and F are.*

Before passing to the proof of this theorem, we record one consequence regarding quasi-linear functions. It is an immediate corollary of part (i) and of the fact that a one-dimensional subspace of a Banach space is complemented.

Corollary 16.5. *Let f be a real-valued quasi-linear function on a Banach space F with nontrivial type. Then there is a (not necessarily continuous) linear function $\varphi : F \to \mathbb{R}$ and a constant M such that $|f(x) - \varphi(x)| \leq M\|x\|$ for all $x \in E$.*

Recall that the type-p constants of a quasi-Banach space E are denoted by $a_{n,p}(E)$. The next lemma generalizes Proposition G.3(i).

Lemma 16.6. *Let G be a twisted sum of E and F, and let k be the modulus of concavity of the quasi-norm in G. Then*

$$a_{nm,p}(G) \leq k^2 \Big(a_{n,p}(E) a_{m,p}(G) + a_{n,p}(E) a_{m,p}(F) + a_{n,p}(G) a_{m,p}(F) \Big)$$

for all integers n and m and for all $1 \leq p \leq 2$.

Proof. Let $\{x_j\}_{j=1}^{mn} \subset G$, and let $\{r_j\}$ be the Rademacher functions. Put

$$X_i(t) = \sum_{j=(i-1)m+1}^{im} r_j(t)x_j$$

for $1 \le i \le n$, and

$$Z(t) = \Big(\int_0^1 \Big\| \sum_{i=1}^{n} r_i(s)X_i(t) \Big\|^p ds \Big)^{1/p}.$$

By the definition of the Rademacher functions, $\{r_i(s)r_j(t)\}$ have the same joint distribution as $\{r_j(t)\}$; hence

$$\int_0^1 Z(t)^p dt = \int_0^1 \Big\| \sum_{j=1}^{mn} r_j(t)x_j \Big\|^p dt. \tag{16.1}$$

Let q be the quotient map from G onto F. Given $\varepsilon > 0$, we use the definition of the quotient quasi-norm to find for each $i \le n$ an E-valued function $Y_i(t)$ such that

$$\|X_i(t) + Y_i(t)\| \le \|qX_i(t)\| + \varepsilon$$

for all $t \in [0,1]$. Since

$$\Big\| \sum r_i(s)X_i(t) \Big\| \le k\Big(\Big\| \sum r_i(s)Y_i(t) \Big\| + \Big\| \sum r_i(s)\big(Y_i(t) + X_i(t)\big) \Big\| \Big)$$

we get

$$\begin{aligned} Z(t) &\le k\Big(\int \Big\| \sum r_i(s)Y_i(t) \Big\|^p ds \Big)^{1/p} + k\Big(\int \Big\| \sum r_i(s)\big(X_i(t) + Y_i(t)\big) \Big\|^p ds \Big)^{1/p} \\ &\le k\, a_{n,p}(E)\big(\sum \|Y_i(t)\|^p \big)^{1/p} + k\, a_{n,p}(G)\Big(\sum \|X_i(t) + Y_i(t)\|^p \Big)^{1/p}. \end{aligned}$$

We now use the fact that $\|Y_i\| \le k\big(\|X_i\| + \|X_i + Y_i\|\big)$ and the definition of Y_i to obtain

$$\begin{aligned} Z(t) &\le k^2\, a_{n,p}(E)\big(\sum \|X_i(t)\|^p \big)^{1/p} \\ &+ k^2\, \big(a_{n,p}(E) + a_{n,p}(G)\big)\Big(\big(\sum \|qX_i(t)\|^p \big)^{1/p} + n^{1/p}\varepsilon \Big). \end{aligned}$$

Since $\int \|X_i(t)\|^p \le a_{m,p}(G)^p \sum_{j=(i-1)m+1}^{im} \|x_j\|^p$ and an analogous estimate also holds for $\int \|qX_i(t)\|^p$, integration of the last inequality with respect to t gives that $\big(\int Z(t)^p \big)^{1/p}$ is bounded by

$$k^2 \Big(a_{n,p}(E)a_{m,p}(G) + \big(a_{n,p}(E) + a_{n,p}(G)\big)a_{m,p}(F) \Big) \Big(\sum_{j=1}^{nm} \|x_j\|^p \Big)^{1/p} + O(\varepsilon)$$

and the lemma follows from (16.1) because ε was arbitrary. □

Proof of Theorem 16.4.(i) It follows from Lemma 16.6 with $m = n$ and $p = 2$, and from $a_{n,2}(E) \le a_{n,2}(G)$ that

$$a_{n^2,2}(G)/n \le k^2 \gamma_n a_{n,2}(G)/n^{1/2}$$

where $\gamma_n = \big(a_{n,2}(E) + 2a_{n,2}(F)\big)/n^{1/2}$, and $\gamma_n \to 0$ by Proposition G.3(ii). Thus $n^{-1}a_{n^2,2}(G) \to 0$, and since $a_{n,2}(G)$ is a monotone function of n, it follows that $n^{-1/2}a_{n,2}(G) \to 0$. By Propositions G.3(ii) and H.3, G is a Banach space.

(ii) Since $\sup a_{n,p}(E) < \infty$ and $\sup a_{n,p}(F) < \infty$, it follows from Lemma 16.6 with $n = m$ that $a_{n^2,p}(G) \leq C a_{n,p}(G)$ for some constant C and all n. Put $2^\gamma = C$ and $\alpha_n = a_{n,p}(G)/(\log n)^\gamma$. Then $\alpha_{n^2} \leq \alpha_n$, and the monotonicity of $a_{n,p}(G)$ as a function of n implies that $\sup \alpha_n < \infty$.

By (i) we may assume that $\|\cdot\|$ is a norm, and we shall show that for any Banach space G, the boundedness of $\{\alpha_n\}$ implies that G is of type r for every $r < p$. Fix such r and x_i's with $\sum_{i=1}^n \|x_i\|^r = 1$. Let n_j be the cardinality of the set $A_j = \{i : 2^{-j-1} < \|x_i\| \leq 2^{-j}\}$. Then $n_j \leq 2^{r(j+1)}$ for all j, and

$$\begin{aligned}\Big(\int \Big\|\sum_{i=1}^n r_i(t)x_i\Big\|^p\Big)^{1/p} &\leq \sum_j \Big(\int \Big\|\sum_{i\in A_j} r_i(t)x_i\Big\|^p\Big)^{1/p} \\ &\leq K_1 \sum_j (\log n_j)^\gamma n_j^{1/p} 2^{-j} \\ &\leq K_2 \sum_j j^\gamma 2^{-j(1-r/p)} = K_3.\end{aligned}$$

In other words,

$$\Big(\int \|\sum r_i(t)x_i\|^p dt\Big)^{1/p} \leq K_3\big(\sum \|x_i\|^r\big)^{1/r}$$

for all $\{x_i\}$ in G, and G has type r by Kahane's inequality.

(iii) The proof of this part is similar to that of (ii), and we only sketch it. It is based on the martingale analogues of the type constants. Fix $1 < p \leq 2$ and an integer n, and let $\alpha_{n,p}(E)$ be the smallest number c such that for every E-valued martingale $\{X_i\}_{i=0}^n$ with $X_0 = 0$ we have

$$\Big(\int \|X_n\|^p\Big)^{1/p} \leq c \sum_{i=1}^n \Big(\int \|X_i - X_{i-1}\|^p\Big)^{1/p}.$$

Pisier [**462**] proved that if $\sup \alpha_{n,p}(E) < \infty$, then E has an equivalent norm with modulus of smoothness $\rho(t)$ satisfying $\rho(t) \leq Kt^p$. Conversely, every Banach space E with such a modulus of smoothness satisfies $\sup \alpha_{n,p}(E) < \infty$. A Banach space E is superreflexive if and only if $n^{-1/2}\alpha_{n,2}(E) \to 0$ as $n \to \infty$.

What is needed for the argument in (ii) to be carried out for the constants $\alpha_{n,p}(E)$ is the following variant of Lemma 16.6.

Lemma 16.7. *Let the Banach space G be the twisted sum of E and F. Then for all $1 < p \leq 2$ and integers n and m*

$$\alpha_{nm,p}(G) \leq \alpha_{n,p}(E)\alpha_{m,p}(G) + 2\alpha_{n,p}(E)\alpha_{m,p}(F) + 2\alpha_{n,p}(G)\alpha_{m,p}(F).$$

The proof of this lemma is similar to (but slightly more involved than) the proof of Lemma 16.6. ◇

To strengthen Theorem 16.4(i) we first introduce the following definition.

Definition 16.8. *A Banach space F is called a $\mathcal{K}$-space if every twisted sum of $\mathbb{R}$ and F is trivial.*

In this language the statement of Corollary 16.5 is that every Banach space with a nontrivial type is a $\mathcal{K}$-space.

Proposition 16.9. *Let F be a $\mathcal{K}$-space. Then the twisted sum of any Banach space E with F is isomorphic to a Banach space.*

Proof. Let G be the twisted sum of E and F, and assume, as we may, that the embedding of E into G is an isometry. We first show that every continuous linear functional x^* on E extends to a continuous linear functional on G. Indeed, let H be the kernel of x^*. Then E/H is one-dimensional. It is easy to check that G/H is a twisted sum of the one-dimensional space E/H and F and hence, by our assumption, a Banach space. Identifying x^* with a functional on E/H, the Hahn-Banach theorem gives a continuous linear functional y^* on G/H which extends it. The required extension of x^* to G is the composition $y^* \circ q$, where $q : G \to G/H$ is the natural quotient map.

By the open mapping theorem there is a constant γ such that the extension can be chosen to have norm at most γ whenever $\|x^*\| \leq 1$. Consequently, if $C(r)$ denotes the convex hull of the ball $B(r)$ of radius r and center 0 in G, then

$$C(r) \cap E \subset B(\gamma r). \tag{16.2}$$

(Use the separation theorem and that $B(\gamma r) \cap E$ is convex.)

To finish the proof, we show that there is a constant η such that $C = C(1)$ is contained in $B(\eta)$. Let $q : G \to F$ be the quotient map. Since F is a Banach space, there is a constant λ such that $qC \subset \lambda\, qB(1) = qB(\lambda)$. Thus for every $x \in C$ there is a $y \in B(\lambda)$ with $q(x) = q(y)$. In particular, $x - y \in E$ and (16.2) shows that

$$x - y \in \big(C + B(\lambda)\big) \cap E \subset B(\gamma) + B(\lambda) \subset B\big(k(\gamma + \lambda)\big)$$

where k is the modulus of concavity of G. Finally,

$$x = (x - y) + y \in B\big(k(\gamma + \lambda)\big) + B(\lambda) \subset B(\eta)$$

where $\eta = k\big(k(\gamma + \lambda) + \lambda\big)$. □

For some further comments on $\mathcal{K}$-spaces, see the Notes and Remarks section.

We conclude this section by presenting a simple general construction of quasi-linear functions from a Banach space with an unconditional basis into itself. These special functions give rise (as we shall see in the next section) to very interesting examples of twisted sums.

Proposition 16.10. *Let E be a Banach space with an unconditional basis $\{e_i\}$, and denote the subspace of finite linear combinations $y = \sum_i y_i e_i$ by E_0. Let $\varphi : \mathbb{R}^+ \to \mathbb{R}$ be a Lipschitz function, and define $f : E_0 \to E_0$ by $f(y) = \sum_i v_i e_i$, where $v_i = 0$ if $y_i = 0$ and*

$$v_i = y_i \varphi\big(\log(\|y\|/|y_i|)\big), \qquad y_i \neq 0.$$

Then f is quasi-linear.

Proof. Note first that for all scalars a and b

$$\Big|(a+b)\varphi\big(\log|a+b|\big) - a\varphi\big(\log|a|\big) - b\varphi\big(\log|b|\big)\Big| \leq 2L\big(|a| + |b|\big) \tag{16.3}$$

where L is the Lipschitz constant of φ, the log is taken to the base e, and $0 \cdot \varphi(\log 0)$ is understood as 0. Indeed, we may assume that $\varphi(0) = 0$. If $|a| + |b| = 1$, then (16.3) follows from the fact that $\big|u \log |u|\big| \leq 1/e$ for $0 < u < 1$. For $|a|+|b| \neq 1$ just put $t = |a|+|b|$, and apply (16.3) to a/t, b/t, replacing $\varphi(s)$ by $\varphi(s+\log t) - \varphi(\log t)$.

Define $g : E_0 \to E_0$ by $g(\sum y_i e_i) = \sum w_i e_i$, where $w_i = 0$ if $y_i = 0$ and $w_i = y_i\varphi(\log(1/|y_i|))$ if $y_i \neq 0$. It follows from (16.3) that

$$\|g(x+y) - g(x) - g(y)\| \leq 2CL(\|x\| + \|y\|) \tag{16.4}$$

where C is the unconditionality constant of the basis. We shall show that

$$\sup\left\{\big\|\|u\|g(u/\|u\|) - g(u)\big\| : \|u\| \leq 1\right\} < \infty.$$

Since $f(x) = \|x\|g(x/\|x\|)$, this, together with (16.4), will imply that there is a constant K such that $\|f(x+y) - f(x) - f(y)\| \leq K$ when $\|x\| + \|y\| = 1$. Since f is homogeneous, it then follows that it is quasi-linear.

Fix $u \in E_0$ with $\|u\| \leq 1$. The ith coordinate of $\|u\|g(u/\|u\|) - g(u)$ is bounded by $L|u_i \log\|u\||$; hence

$$\big\|\|u\|g(u/\|u\|) - g(u)\big\| \leq CL\|u\|\big|\log\|u\|\big| \leq CL/e.$$

□

Remarks. (i) By Proposition 16.3, f has a unique extension (up to equivalence) to a quasi-linear function from E to E.

(ii) We worked with real scalars, but all the results, including Proposition 16.10, hold verbatim in the complex case. (The function φ will then be a Lipschitz function from $\mathbb{R}^+$ to the complex plane.) This will be used in the next section.

2. Twisted Sums of Hilbert Spaces

In this section we examine twisted sums of Hilbert space with itself. These spaces are, perhaps, the most interesting examples of nontrivial twisted sums, and we use them to illustrate several important points in Banach space theory. In particular, we use them to distinguish between some natural notions of unconditional structure of Banach spaces and to exhibit two nonisomorphic complex Banach spaces which are real-isomorphic as real spaces.

We shall not make here a systematic study of general twisted sums of Hilbert space with itself, and will only consider some examples which arise from the quasi-linear functions of Proposition 16.10 above.

Let φ be a Lipschitz function on $(0, \infty)$. We denote by $Z_2(\varphi)$ the completion of the space of pairs of finitely supported sequences $x = \sum x_n e_n$ and $y = \sum y_n h_n$ endowed with the quasi-norm

$$\|(x,y)\| = \left(\sum_{n=1}^{\infty}(x_n - y_n\varphi(\log(\|y\|_2/|y_n|))^2\right)^{1/2} + \|y\|_2 \tag{16.5}$$

where $\|y\|_2 = (\sum |y_n|^2)^{1/2}$, and where $\{e_n\}_{n=1}^{\infty}$ and $\{h_n\}_{n=1}^{\infty}$ denote the standard unit vector bases in the respective Hilbert spaces. (We take $0 \cdot \varphi(\log\infty)$ as 0.) Clearly $\|(x,0)\| = \|x\|_2$, and thus $E = \overline{\operatorname{span}}\{(e_n, 0)\}_{n=1}^{\infty}$ is isometric to l_2. Also, $Z_2(\varphi)/E$ is isometric to l_2. It follows from the results of the previous section that the quasi-norm (16.5) is equivalent to a norm, that the Banach space $Z_2(\varphi)$ is superreflexive, and that for every $\varepsilon > 0$ $Z_2(\varphi)$ has type $2-\varepsilon$ and cotype $2+\varepsilon$ and can be renormed to have modulus of smoothness $\rho(t) \leq c(\varepsilon)t^{2-\varepsilon}$ and modulus of convexity $\delta(t) \geq k(\varepsilon)t^{2+\varepsilon}$.

Proposition 16.11. *The spaces $Z_2(\varphi)$ and $Z_2(\psi)$ are equivalent twisted sums of l_2 with itself if and only if* $\sup_{t>0}|\varphi(t)-\psi(t)|<\infty$. *In particular $Z_2(\varphi)$ is isomorphic to l_2 (i.e., the twisted sum is trivial) if and only if* $\sup_{t>0}|\varphi(t)|<\infty$.

Proof. It is trivial that if $\varphi-\psi$ is bounded, then $Z_2(\varphi)$ is equivalent to $Z_2(\psi)$. We prove the converse. Let f_φ and f_ψ be the quasi-linear functions associated to φ and ψ via Proposition 16.10, and assume that there is a (perhaps unbounded) linear map T from l_2 into l_2 such that

$$\|f_\varphi(x)-f_\psi(x)-Tx\|\leq K\|x\|. \tag{16.6}$$

By definition, $f_\varphi(h_n)=\varphi(0)e_n$ and $f_\psi(h_n)=\psi(0)e_n$ for all n; thus the sequence $\{Th_n\}_{n=1}^\infty$ is bounded and has a weakly convergent subsequence. By a standard argument we can pass to a further subsequence $\{n(k)\}_{k=1}^\infty$ for which there are disjointly supported vectors u_k in l_2 such that if $x_k=h_{n(2k)}-h_{n(2k-1)}$, then their images satisfy $\|Tx_k-u_k\|\leq 2^{-k}$ for all k. Put $M=\sup\|Tx_k\|$.

For each positive integer m put $z_m=\sum_{k=1}^m x_k$. Then $\|z_m\|=(2m)^{1/2}$ and $\|Tz_m\|\leq m^{1/2}M+2$. By the definition of f_φ and f_ψ we get

$$\|f_\varphi(z_m)-f_\psi(z_m)\|=(2m)^{1/2}\Big|\varphi\big(\log\left((2m)^{1/2}\right)\big)-\psi\big(\log\left((2m)^{1/2}\right)\big)\Big|$$

and it follows from (16.6) that the sequence $\big\{\varphi(\frac{1}{2}\log 2m)-\psi(\frac{1}{2}\log 2m)\big\}$ is bounded. Since φ and ψ are Lipschitz functions, it follows that $\sup_{t>0}|\varphi(t)-\psi(t)|<\infty$. $\square$

Proposition 16.12. *The space $Z_2(\varphi)$ is isomorphic to its dual.*

Proof. Note first that the spaces $Z_2(\varphi)$ and $Z_2(-\varphi)$ are isometric via the map $(x,y)\to(-x,y)$ (although Proposition 16.11 implies that $Z_2(\varphi)$ and $Z_2(-\varphi)$ are, in general, not equivalent as twisted sums). We shall show that the bilinear form

$$B\big((x,y),(u,v)\big): Z_2(\varphi)\times Z_2(-\varphi)\to\mathbb{R}$$

defined on finitely supported vectors by $B\big((x,y),(u,v)\big)=\langle x,v\rangle+\langle y,u\rangle$ (where $\langle\cdot,\cdot\rangle$ is the usual inner product in l_2) is bounded and is bounded from below in the sense that there is an $\alpha>0$ for which

$$\sup\Big\{\big|B\big((x,y),(u,v)\big)\big|:\ \|(u,v)\|=1\Big\}\geq\alpha\|(x,y)\|$$

for each (x,y), and a similar inequality holds when the roles of (x,y) and (u,v) are interchanged. Since $Z_2(\varphi)$ is reflexive (even superreflexive), this implies that $Z_2(\varphi)^*$ is isomorphic to $Z_2(-\varphi)$.

For any choice of finitely many scalars a_i and b_i

$$\Big|\sum a_ib_i\log\Big|\frac{a_i}{b_i}\Big|\Big|\leq\sum(a_i^2+b_i^2)/e.$$

Indeed, by summing over $I=\{i:|a_i|<|b_i|\}$, we get (using $|t\log t|<1/e$ for $t\in(0,1)$)

$$\Big|\sum_{i\in I}a_ib_i\log\Big|\frac{a_i}{b_i}\Big|\Big|\leq\Big(\sum_{i\in I}b_i^2\Big)/e$$

and an analogous estimate holds on the complement of I.

It follows that for all finitely supported vectors y and v

$$\begin{aligned}\left|\langle f_\varphi(y), v\rangle + \langle y, f_{-\varphi}(v)\rangle\right| &= \left|\sum y_i v_i \Big(\varphi\big(\log \|y\|_2/|y_i|\big) - \varphi\big(\log \|v\|_2/|v_i|\big)\Big)\right| \\ &\le C\|y\|_2\|v\|_2 \left|\sum \frac{y_i}{\|y\|_2}\frac{v_i}{\|v\|_2}\log\frac{\|y\|_2|v_i|}{|y_i|\,\|v\|_2}\right| \\ &\le 2Ce^{-1}\|y\|_2\|v\|_2\end{aligned}$$

where C is the Lipschitz constant of φ. Consequently

$$\begin{aligned}&\big|B\big((x,y),(u,v)\big)\big| \\ &\quad = \big|\langle x - f_\varphi(y), v\rangle + \langle y, u - f_{-\varphi}(v)\rangle + \langle f_\varphi(y), v\rangle + \langle y, f_{-\varphi}(v)\rangle\big| \\ &\quad \le \big\|x - f_\varphi(y)\big\|_2\|v\|_2 + \|y\|_2\big\|u - f_{-\varphi}(v)\big\|_2 + 2Ce^{-1}\|y\|_2\|v\|_2 \\ &\quad \le K\|(x,y)\|\,\|(u,v)\|\end{aligned}$$

i.e., B is bounded. To see that it is bounded from below, fix (x,y). Choose v with $\|v\|_2 = 1$ and $\langle x - f_\varphi(y), v\rangle = \|x - f_\varphi(y)\|_2$, and let $u = f_{-\varphi}(v) + (2Ce^{-1}+1)y/\|y\|_2$. Then $\|(u,v)\| = 2Ce^{-1} + 2$ and

$$\begin{aligned}\big|B\big((x,y),(u,v)\big)\big| &= \|x - f_\varphi(y)\|_2 + (2Ce^{-1}+1)\|y\|_2 + \langle f_\varphi(y), v\rangle + \langle y, f_{-\varphi}(v)\rangle \\ &\ge \|x - f_\varphi(y)\|_2 + (2Ce^{-1}+1)\|y\|_2 - 2Ce^{-1}\|y\|_2 = \|(x,y)\|.\end{aligned}$$

The other lower estimate is proved similarly. □

We examine next the unconditional structure of $Z_2(\varphi)$. The decomposition of $Z_2(\varphi)$ as a sum of the two-dimensional subspaces $E_n = \operatorname{span}\{(e_n,0),(0,h_n)\}$ is 1-unconditional; i.e., $\|\sum u_n\| = \|\sum \theta_n u_n\|$ for every choice of finitely many vectors $u_n \in E_n$ and signs $\theta_n = \pm 1$. It follows in particular that the sequence $(e_1,0),(0,h_1),(e_2,0),(0,h_2),\ldots$ is a Schauder basis for $Z_2(\varphi)$. This basis is generally not unconditional. In fact, $Z_2(\varphi)$ may fail to have an unconditional basis at all, and we shall prove an even stronger assertion. For the purpose of this discussion we specialize to $\varphi(t) = t$ and denote the space $Z_2(\varphi)$ with this particular φ by Z_2.

Denote the unconditional constant of a space F with an unconditional basis by $\chi(F)$. A Banach space E is said to have local unconditional structure (l.u.st) if there is a constant K such that for every finite-dimensional subspace $B \subset E$, the inclusion map I_B of B into E factors through a space with an unconditional basis and the factorizations are bounded by K. More precisely, for each such B there is a space F with an unconditional basis and operators $U : B \to F$ and $V : F \to E$ such that $VU = I_B$ and $\|U\| \cdot \|V\| \cdot \chi(F) \le K$. This is a very weak notion of unconditionality. A space with an unconditional basis certainly has l.u.st, and so does any Banach lattice. A complemented subspace of a space with l.u.st also has l.u.st. It is quite surprising that the space Z_2, which has an unconditional decomposition into two-dimensional spaces, fails to have even this rather weak property of unconditionality.

Theorem 16.13. *The space Z_2 does not have local unconditional structure.*

For the proof we need two lemmas. The first is a technical result concerning the notion of l.u.st, and we omit its proof. (The proof is given in Johnson, Lindenstrauss and Schechtman [**294**]. In our context, the reader may just consider the conclusion of the lemma as the definition of l.u.st.)

Lemma 16.14. *Let E be a superreflexive Banach space with an unconditional decomposition into finite-dimensional subspaces $\{E_k\}_{k=1}^{\infty}$. Then E has local unconditional structure if and only if there is a Banach space $F \supset E$ with an unconditional basis $\{z_{n,k}\}_{k<\infty, n\le m_k}$, such that $E_k \subset \operatorname{span}\{z_{n,k}\}_{n=1}^{m_k}$ for all k, and such that there is a bounded linear projection P from F onto E with $Pz_{n,k} \in E_k$ for all k and $n \le m_k$.*

Lemma 16.15. *Let A be the 2×2 matrix $\left(\begin{smallmatrix}\alpha & \beta\\ \gamma & \delta\end{smallmatrix}\right)$. Then the map T defined by*

$$T(x_n e_n, y_n h_n) = \big((\alpha x_n + \beta y_n)e_n, (\gamma x_n + \delta y_n)h_n\big)$$

for all n extends to a bounded linear operator on Z_2 if and only if $\alpha = \delta$ and $\gamma = 0$.

Proof. The "if" part is obvious. To prove the other direction, consider first $u_n = n^{-1/2}\sum_{k=1}^n (e_k, 0)$. Then $\|u_n\| = 1$, and there is a constant $C > 0$ such that

$$\|Tu_n\| = n^{-1/2}\big\|(\alpha \sum e_k, \gamma \sum h_k)\big\| \ge \gamma C \log n$$

i.e., $\gamma = 0$. If $\alpha \ne \delta$, put $u_n = n^{-1/2}\big((\log\sqrt{n})\sum_{k=1}^n e_k, \sum_{k=1}^n h_k\big)$. Then $\|u_n\| = 1$, and a direct computation shows that $\|Tu_n\| \ge C\log n$. □

Proof of the Theorem. Assume that Z_2 has l.u.st. We shall construct a bounded linear operator $\tilde{T}$ from Z_2 into itself such that $\tilde{T}E_k \subset E_k$ for all k, and such that the matrices A_k of the restrictions of $\tilde{T}$ to E_k with respect to the natural bases satisfy

$$\inf_k \operatorname{dist}\left(A_k, \left\{\left(\begin{smallmatrix}\alpha & \beta\\ 0 & \alpha\end{smallmatrix}\right) : \alpha, \beta \in \mathbb{R}\right\}\right) > 0$$

where the distance is taken with respect to any fixed norm on the space of 2×2 matrices. Once such a $\tilde{T}$ is defined, find a subsequence $\{k_i\}$ of the integers such that all the matrices A_{k_i} are small perturbations of a fixed matrix which is not of the form $\left(\begin{smallmatrix}\alpha & \beta\\ 0 & \alpha\end{smallmatrix}\right)$. This contradicts Lemma 16.15, because $\overline{\operatorname{span}}\{E_{k_i}\}_{i=1}^{\infty}$ is naturally isometric to Z_2, and the lemma clearly continues to hold when all the matrices A_{k_i} are small enough perturbations of one fixed matrix.

To produce $\tilde{T}$, assume that F, the $z_{n,k}$'s and P are as in Lemma 16.14, and let $z_{n,k}^* \in F^*$ be the biorthogonal vectors to $\{z_{n,k}\}$. Since $\{z_{n,k}\}$ is an unconditional basis, the operators

$$T_J u = P\Big(\sum_{k,n\in J} \langle z_{n,k}^*, u\rangle z_{n,k}\Big) \qquad u \in Z_2$$

are bounded for any set of indices J. We shall find subsets $J_k \subset \{1,2,\dots,m_k\}$ such that the 2×2 matrices A_k corresponding to the restrictions of $P\big(\sum_{n\in J_k}\langle z_{n,k}^*, \cdot\rangle z_{n,k}\big)$ to E_k respectively satisfy

$$\operatorname{dist}\left(A_k, \left\{\left(\begin{smallmatrix}\alpha & \beta\\ 0 & \alpha\end{smallmatrix}\right) : \alpha, \beta \in \mathbb{R}\right\}\right) \ge \mu \tag{16.7}$$

for all k and for some fixed $\mu > 0$. We then take $\tilde{T} = T_J$ for $J = \bigcup J_k$.

Fix k. For each $1 \le n \le m_k$ consider the rank-one operator $P\big(\langle z_{n,k}^*, \cdot\rangle z_{n,k}\big)$ restricted to E_k. Since it is rank-one, its matrix is either of the form $\left(\begin{smallmatrix}a_n & b_n\\ \alpha_n a_n & \alpha_n b_n\end{smallmatrix}\right)$ or of the form $\left(\begin{smallmatrix}0 & 0\\ a_n & b_n\end{smallmatrix}\right) = \left(\begin{smallmatrix}0 & 0\\ \alpha_n a_n & \alpha_n b_n\end{smallmatrix}\right)$ where, to unify the notation, we just take $\alpha_n = 1$ in the second case. The sum of all these m_k matrices is the identity matrix; hence $\sum_{n=1}^{m_k} \alpha_n b_n = 1$. Also by the unconditionality of the $z_{n,k}$'s, there is an absolute constant K such that $\sum_{n=1}^{m_k}|b_n| \le K$. (For each k sum the absolute values of the upper right entries of the matrices of the first form, and the lower right entries of the matrices of the second form.) Consequently, if $I_k = \{n : |\alpha_n| \geq 1/2K\}$, then $\sum_{n\in I_k} \alpha_n b_n \geq 1/2$. Fix $n \in I_k$, and assume that for this n the matrix is of the first form. Then one of the following four possibilities must occur:

(i) $\alpha_n a_n \geq (1+2K)^{-1}|\alpha_n b_n|$,

(ii) $-\alpha_n a_n \geq (1+2K)^{-1}|\alpha_n b_n|$,

(iii) $a_n - \alpha_n b_n \geq (1+2K)^{-1}|\alpha_n b_n|$,

(iv) $\alpha_n b_n - a_n \geq (1+2K)^{-1}|\alpha_n b_n|$.

Indeed, if none of the four holds, then summing the two inequalities

$$|a_n| < (1+2K)^{-1}|b_n| \leq 2K(1+2K)^{-1}|\alpha_n b_n|$$
$$|a_n - \alpha_n b_n| < (1+2K)^{-1}|\alpha_n b_n|$$

we obtain $|\alpha_n b_n| < |\alpha_n b_n|$, a contradiction.

If n is such that the matrix is of the second form, the analogue of (iii) and (iv), namely, $|\alpha_n b_n| \geq (1+2K)^{-1}|\alpha_n b_n|$, holds trivially.

Hence there is a set $J_k \subset I_k$ such that $\sum_{n\in J_k} \alpha_n b_n \geq 1/8$ and such that one of the four possibilities above holds for all $n \in J_k$ simultaneously. With this choice of J_k condition (16.7) holds. ♢

The next result on the structure of the complemented subspaces of Z_2, when combined with Theorem 16.13, shows that Z_2 does not have a complemented subspace with local unconditional structure.

Theorem 16.16. *Every infinite-dimensional complemented subspace W of Z_2 contains a complemented subspace isomorphic to Z_2.*

Proof. Note first that if $\{v_n\}_{n=1}^\infty$ is a normalized block basis of the unit vector basis of l_2, then $V = \overline{\text{span}}\{(v_n, 0), (f(v_n), v_n)\}_{n\geq 1}$ is isometric to Z_2 and is complemented in Z_2. (Here f is the quasi-linear function corresponding to $\varphi(t) = t$.) Indeed, direct computation shows that the linear operator $T_1 : Z_2 \to V$ defined by $T_1(e_n, 0) = (v_n, 0)$ and by $T_1(0, h_n) = (f(v_n), v_n)$ is an isometry. To see that it is complemented, recall that Z_2^* is identified with $Z_2(-t)$ via the bilinear form B of Proposition 16.12. The map T_2 on $Z_2(-t)$, defined by $T_2(e_n, 0) = (v_n, 0)$ and $T_2(0, h_n) = (-f(v_n), v_n)$, is again an into-isometry. When we consider T_2 as an operator on Z_2^*, then T_2^* is an operator on Z_2 which satisfies $T_2^*(v_n, 0) = (e_n, 0)$ and $T_2^*(f(v_n), v_n) = (0, h_n)$ for all n. Consequently $T_1 T_2^*$ is a bounded linear projection from Z_2 onto V.

Now let W be an infinite-dimensional complemented subspace of Z_2. Note first that like any infinite-dimensional subspace of a twisted sum of l_2 with itself, W has a subspace isomorphic to l_2. Indeed, let q be the quotient map of Z_2 onto l_2. If $W \cap \ker(q)$ is infinite-dimensional, this is clear. Otherwise there is a subspace W_1 of W of finite codimension on which q is an isomorphism, and we are done again.

Let $\{w_n\}_{n=1}^\infty$ be a sequence in W, equivalent to the unit vector basis in l_2. Since $w_n \to 0$ weakly, we may assume (by passing to a subsequence) that there is an increasing sequence of integers p_n and vectors $x_n \in \text{span}\{e_k : p_n \leq k < p_{n+1}\}$ and $u_n \in \text{span}\{h_k : p_n \leq k < p_{n+1}\}$ such that $\|w_n - (x_n, u_n)\| \leq 2^{-n}$. We claim that $\|u_n\|_2 \to 0$. Indeed, otherwise we may assume that $\|u_n\|_2 > \eta > 0$ for all n, and hence $\big\|f\big(\sum_{i=1}^n u_i\big) - \sum_{i=1}^n f(u_i)\big\|_2 \geq \rho n^{1/2} \log n$ for some $\rho > 0$ by the definition of f. Thus

$$\begin{aligned}\Big\|\Big(\sum_{i=1}^{n} x_i, \sum_{i=1}^{n} u_i\Big)\Big\| &\geq \Big\|\sum_{i=1}^{n} x_i - f\Big(\sum_{i=1}^{n} u_i\Big)\Big\|_2 \\ &\geq \Big\|f\Big(\sum_{i=1}^{n} u_i\Big) - \sum_{i=1}^{n} f(u_i)\Big\|_2 - \Big\|\sum_{i=1}^{n} \big(x_i - f(u_i)\big)\Big\|_2 \\ &\geq \rho n^{1/2} \log n - \Big\|\sum_{i=1}^{n} \big(x_i - f(u_i)\big)\Big\|_2\end{aligned}$$

which is impossible, because $\big\| \sum_{i=1}^{n} \big(x_i - f(u_i)\big) \big\|_2^2 = \sum_{i=1}^{n} \big\|x_i - f(u_i)\big\|_2^2 \leq n$, while $\big\|\big(\sum_{i=1}^{n} x_i, \sum_{i=1}^{n} u_i\big)\big\| \sim n^{1/2}$ because the vectors (x_n, u_n) are a small perturbation of $\{w_n\}$. We may thus assume that $\|u_n\|_2 \leq 2^{-n}$, and hence

$$\|w_n - (v_n, 0)\| \leq 2^{-n+1}$$

where $v_n = x_n - f(u_n)$. Replacing w_n by $\lambda_n w_n$, where the sequence $\{\lambda_n\}$ satisfies $\inf \lambda_n > 0$ and $\sup \lambda_n < \infty$, we may also assume that $\|v_n\|_2 = 1$ for all n.

With this choice of $\{v_n\}_{n=1}^{\infty}$, let T_1 and T_2 be the operators defined in the first part of the proof. Let Q be a bounded linear projection from Z_2 onto W. Then

$$\big\|Q(v_n, 0) - (v_n, 0)\big\| \leq \big\|Q(v_n, 0) - w_n\big\| + \big\|w_n - (v_n, 0)\big\| \leq 2^{-n+1}\big(\|Q\| + 1\big).$$

Hence the operator $K : Z_2 \to Z_2$ defined by

$$K(e_n, 0) = Q(v_n, 0) - (v_n, 0) \quad \text{and} \quad K(0, h_n) = 0$$

is compact. The operator $S = QT_1 - T_1 - K$ vanishes on vectors of the form $(x, 0)$ and is strictly singular; i.e., S is not an isomorphism on any infinite-dimensional subspace. Indeed, as already shown above, any infinite-dimensional subspace of Z_2 contains a sequence which is equivalent to the unit vector basis in l_2, and this sequence has a subsequence which is arbitrarily close to vectors of the form $(x, 0)$.

It follows that the operator $QT_1 - T_1$, being a compact perturbation of a strictly singular operator, is also strictly singular, and the same is true, of course, for $T_2^*\big(QT_1 - T_1\big)$. Since $T_2^* T_1 = I$, it follows (cf. Lindenstrauss and Tzafriri [**371**], p. 79) that $T_2^* Q T_1$ is a Fredholm operator of index 0. In particular, the restriction of QT_1 to $\overline{\text{span}}\{(e_n, 0), (0, h_n)\}_{n \geq n_0}$ is an isomorphism provided n_0 is large enough. Replacing $\{v_n\}_{n=1}^{\infty}$ by $\{v_n\}_{n \geq n_0}$, we may thus assume without loss of generality that QT_1 itself is an isomorphism. Consequently there is a finite-rank operator A on Z_2 such that $(T_2^* + A)QT_1$ is invertible. Denote its inverse by U. Then $QT_1 Z_2$ is a subspace of W which is isomorphic to Z_2, and $QT_1 U(T_2^* + A)$ is a projection from Z_2 on this subspace. □

In the final result of this section we consider complex scalars. For every complex Banach space E we denote by $\tilde{E}$ the "complex-conjugate space" of E. This is the same vector space as E, with the same norm, but with a new rule for scalar multiplication: $\lambda * x = \bar{\lambda} x$. Of course $\tilde{E}$ is again a complex Banach space, and the identity map from E to $\tilde{E}$ is an isomorphism (actually isometry) when E and $\tilde{E}$ are regarded as real spaces. The next example shows that E is not necessarily complex isomorphic to $\tilde{E}$.

For real α we denote by $Z_2(\alpha)$ the space $Z_2(\varphi)$ with $\varphi(t) = t^{1+i\alpha}$, $t > 0$. Thus $Z_2(0)$ is the space denoted above by Z_2. It follows directly from the definitions that $\widetilde{Z_2(\alpha)}$ is isometric (as a complex space) to $Z_2(-\alpha)$. Thus the desired example follows from the next theorem.

Theorem 16.17. *If $\alpha \neq \beta$, then $Z_2(\alpha)$ is not isomorphic to $Z_2(\beta)$.*

Proof. We assume, as we may, that $\alpha \neq 0$. The key to the proof is the observation that if T is an operator defined, as in Lemma 16.15, by a 2×2 complex matrix $\left(\begin{smallmatrix}\lambda & \mu \\ \sigma & \eta\end{smallmatrix}\right)$ such that either $\lambda \neq 0$ or $\eta \neq 0$, then T is not bounded as an operator from $Z_2(\alpha)$ to $Z_2(\beta)$. (We shall actually only use that if T is bounded, then necessarily $\eta = 0$.) Indeed, exactly as in the proof of Lemma 16.15 we get $\sigma = 0$. Proceeding further as in Lemma 16.15, take

$$u_n = n^{-1/2}\left((\log\sqrt{n})^{1+i\alpha}\sum_{k=1}^{n} e_k, \sum_{k=1}^{n} h_k\right) \in Z_2(\alpha).$$

Then $\|u_n\|_{Z_2(\alpha)} = 1$ for all n, while

$$\begin{aligned}\|Tu_n\|_{Z_2(\beta)} &= \Big\|n^{-1/2}\Big(\big(\lambda(\log\sqrt{n})^{1+i\alpha}+\mu\big)\sum_{k=1}^{n} e_k, \eta\sum_{k=1}^{n} h_k\Big)\Big\|_{Z_2(\beta)}\\ &= \big|\lambda(\log\sqrt{n})^{1+i\alpha}+\mu-\eta(\log\sqrt{n})^{1+i\beta}\big| + |\eta|\end{aligned}$$

which is unbounded when either $\lambda \neq 0$ or $\eta \neq 0$ because $\alpha \neq \beta$.

We shall show that if there is an isomorphism from $Z_2(\alpha)$ into $Z_2(\beta)$, then there is also a bounded operator between these spaces corresponding to a 2×2 matrix with $\eta \neq 0$, which is impossible.

Assume that there is an operator $S : Z_2(\alpha) \to Z_2(\beta)$ with $\|S\| \leq 1$ and such that $\|S(e_n, 0)\|_{Z_2(\beta)} \geq c$ for some $c > 0$ and for all n. Since both $\{(e_n, 0)\}$ and $\{(0, h_n)\}$ converge weakly to zero, we may assume without loss of generality (by passing to a subsequence and by a small perturbation) that there is an increasing sequence of integers $p_1 < p_2 < \dots$ such that if we put $I_n = \{k : p_n \leq k < p_{n+1}\}$, then

$$S(e_n, 0) = (u_n, y_n)\ ; \quad S(0, h_n) = (v_n, w_n)$$

where $u_n, v_n \in \operatorname{span}\{e_k : k \in I_n\}$ and $y_n, w_n \in \operatorname{span}\{h_k : k \in I_n\}$. We verify next that $\|y_n\|_2 \to 0$. Indeed, otherwise we may assume by passing to a subsequence that $\|y_n\|_2 > \delta$ for some $\delta > 0$ and for all n. Note that $\|(u_n, y_n)\| = \|S(e_n, 0)\| \leq 1$, and thus

$$\sum_{k \in I_n}\left|u_n(k) - y_n(k)\Big(\log\frac{\|y_n\|_2}{|y_n(k)|}\Big)^{1+i\beta}\right|^2 \leq 1 \tag{16.8}$$

(where $y_n(k)$ and $u_n(k)$ denote the k's component of y_n and u_n respectively), and also $\|y_n\| \leq 1$.

Fix an integer N. Then

$$\Big\|\Big(\sum_{n=1}^{N} u_n, \sum_{n=1}^{N} y_n\Big)\Big\| = \Big\|S\Big(\sum_{n=1}^{N} e_n, 0\Big)\Big\| \leq \Big\|\Big(\sum_{n=1}^{N} e_n, 0\Big)\Big\| = N^{1/2}$$

and thus

$$\sum_{n=1}^{N}\sum_{k \in I_n}\left|u_n(k) - y_n(k)\Big(\log\frac{\|\sum_{n=1}^{N} y_n\|_2}{|y_n(k)|}\Big)^{1+i\beta}\right|^2 \leq N. \tag{16.9}$$

Using (16.8) for each n and the triangle inequality in Hilbert space, and then (16.9), we find that

$$\sum_{n=1}^{N}\sum_{k\in I_n}\left|y_n(k)\left(\left(\log\frac{\|\sum_{n=1}^{N}y_n\|_2}{|y_n(k)|}\right)^{1+i\beta}-\left(\log\frac{\|y_n\|_2}{|y_n(k)|}\right)^{1+i\beta}\right)\right|^2\leq 4N.$$

Since $|t^{1+i\beta}-s^{1+i\beta}|\geq|t^{1+i\beta}|-|s^{1+i\beta}|=t-s$ for $t>s>0$, it follows that

$$\sum_{n=1}^{N}\sum_{k\in I_n}|y_n(k)|^2\Big(\log\frac{\|\sum_{n=1}^{N}y_n\|_2}{\|y_n\|_2}\Big)^2\leq 4N$$

and we obtain a contradiction for large enough N because the left-hand side is larger than or equal to $\sum_{n=1}^{N}\|y_n\|_2^2\big(\log\delta N^{1/2}\big)^2\geq N\delta^2\big(\log\delta N^{1/2}\big)^2$.

Since $\|y_n\|_2\to 0$, we may assume by passing to a subsequence and a small perturbation that $y_n=0$ for all n, and consequently that $\|u_n\|_2\geq c$ for all n.

We shall show that $\|w_n\|_\infty\not\to 0$. Note first that this will complete the proof. Indeed, assume that there are a $\delta>0$ and integers n_k and r_k such that the absolute value of the r_kth coordinate of w_{n_k} is larger than δ. By using diagonalization (cf. Lindenstrauss and Tzafriri [**371**], p. 20) with respect to the canonical unconditional decomposition of $Z_2(\beta)$ into two-dimensional spaces and by relabeling the indices, we get from S a bounded operator $\tilde{S}$ from $Z_2(\alpha)$ to $Z_2(\beta)$ which corresponds to a sequence of 2×2 matrices $\left(\begin{smallmatrix}\lambda_k & \mu_k\\ 0 & \eta_k\end{smallmatrix}\right)$ (a different matrix for every coordinate) such that $|\eta_k|\geq\delta$ for all k. Passing to a subsequence, we may assume that the matrix is fixed. As noted in the beginning of the proof, this leads to a contradiction.

Thus, assume that $\|w_n\|_\infty\to 0$. Let $K>0$ be an arbitrary constant, and find n_0 such that $\|w_n\|_\infty\leq e^{-K}$ for $n>n_0$. Let m_1 and m_2 be arbitrary integers, $M=4\max\{m_1^2,m_2^2\}$, and put $\sigma_i=\frac{1}{2}\log m_i$. Put $J_M=\{k:n_0+1\leq k\leq n_0+M\}$.

Let $A\subset J_M$ be an arbitrary set of cardinality m_1. Since $\|S\|=1$ and since $\big\|\big(\sum_{n\in A}\sigma_1^{1+i\alpha}e_n,\sum_{n\in A}h_n\big)\big\|_{Z_2(\alpha)}=m_1^{1/2}$, the norm of its image under S satisfies $\big\|\big(\sum_{n\in A}(\sigma_1^{1+i\alpha}u_n+v_n),\sum_{n\in A}w_n\big)\big\|_{Z_2(\beta)}\leq m_1^{1/2}$. Hence, in particular,

$$\sum_{n\in A}\sum_{k\in I_n}\left|\sigma_1^{1+i\alpha}u_n(k)+v_n(k)+w_n(k)\Big(\log\frac{\|\sum_{n\in A}w_n\|}{|w_n(k)|}\Big)^{1+i\beta}\right|^2\leq m_1. \tag{16.10}$$

Since $|t^{1+i\beta}-s^{1+i\beta}|\leq|\beta|(t-s)$ for $t>s>0$, and since $|t\log t|<1/e$ for $0<t<1$, it follows that

$$\begin{aligned}
&\sum_{n\in A}\sum_{k\in I_n}\left|w_n(k)\Big(\log\frac{\|\sum_{n\in A}w_n\|}{|w_n(k)|}\Big)^{1+i\beta}-w_n(k)\Big(\log\frac{m_1^{1/2}}{|w_n(k)|}\Big)^{1+i\beta}\right|^2\\
&\leq\sum_{n\in A}\sum_{k\in I_n}|w_n(k)|^2|\beta|^2\Big(\log\frac{\|\sum_{n\in A}w_n\|}{m_1^{1/2}}\Big)^2\\
&=\sum_{n\in A}\|w_n\|^2|\beta|^2\Big(\log\frac{\|\sum_{n\in A}w_n\|}{m_1^{1/2}}\Big)^2\\
&=|\beta|^2\,\Big\|\sum_{n\in A}w_n\Big\|^2\Big(\log\frac{\|\sum_{n\in A}w_n\|}{m_1^{1/2}}\Big)^2\leq|\beta|^2m_1/e^2.
\end{aligned}$$

From this, (16.10) and the triangle inequality in Hilbert space, we infer that

$$\sum_{n\in A}\sum_{k\in I_n}\left|\sigma_1^{1+i\alpha}u_n(k)+v_n(k)+\big(\sigma_1-\log|w_n(k)|\big)^{1+i\beta}w_n(k)\right|^2\le Cm_1$$

where $C=(1+|\beta|/e)^2$. Averaging the inequality over all subsets A of J_M with cardinality m_1, we deduce that

$$\sum_{n=n_0+1}^{n_0+M}\sum_{k\in I_n}\left|\sigma_1^{1+i\alpha}u_n(k)+v_n(k)+\big(\sigma_1-\log|w_n(k)|\big)^{1+i\beta}w_n(k)\right|^2\le CM. \tag{16.11}$$

Similarly, averaging over subsets of J_M of cardinality m_1^2, we get that

$$\sum_{n=n_0+1}^{n_0+M}\sum_{k\in I_n}\left|(2\sigma_1)^{1+i\alpha}u_n(k)+v_n(k)+\big(2\sigma_1-\log|w_n(k)|\big)^{1+i\beta}w_n(k)\right|^2 \tag{16.12}$$

is also bounded by MC. Using next (twice) the mean value estimate

$$\left|t^{1+i\beta}-s^{1+i\beta}-(1+i\beta)s^{i\beta}(t-s)\right|\le|1+i\beta|^2(t-s)^2/2s$$

which holds for $t>s>0$ and that $-\log|w_n(k)|\ge K$ for $n\ge n_0$ and all k, we see that whenever $w_n(k)\ne 0$ the expression

$$\left|\big(2\sigma_1-\log|w_n(k)|\big)^{1+i\beta}-\big(\sigma_1-\log|w_n(k)|\big)^{1+i\beta}-(1+i\beta)\big(-\log|w_n(k)|\big)^{i\beta}\sigma_1\right|$$

is bounded by $|1+i\beta|^2 4\sigma_1^2/K$.

By the estimates on the expressions in (16.11) and (16.12), the triangle inequality in Hilbert space, and the fact that $\|w_n\|_2\le\|S\|\le 1$ for all n, we obtain

$$\begin{aligned}\sum_{n=n_0+1}^{n_0+M}\sum_{k\in I_n}&\left|\frac{(2\sigma_1)^{1+i\alpha}-\sigma_1^{1+i\alpha}}{\sigma_1}u_n(k)-(1+i\beta)\big(-\log|w_n(k)|\big)^{i\beta}w_n(k)\right|^2\\ &\le M\Big(\frac{2C}{\sigma_1^2}+\frac{16\sigma_1^2\,|1+i\beta|^4}{K^2}\Big).\end{aligned}$$

Using sets of cardinality m_2 and m_2^2, we obtain the same inequality with σ_1 replaced by σ_2. By combining the inequalities for σ_1 and σ_2 and applying the triangle inequality in Hilbert space, we get

$$\begin{aligned}\sum_{n=n_0+1}^{n_0+M}\sum_{k\in I_n}&\left|\sigma_1^{i\alpha}(2^{1+i\alpha}-1)-\sigma_2^{i\alpha}(2^{1+i\alpha}-1)\right|^2|u_n(k)|^2\\ &\le M\left(\frac{2C}{\sigma_1^2}+\frac{2C}{\sigma_2^2}+\frac{16\,|1+i\beta|^4}{K^2}(\sigma_1^2+\sigma_2^2)\right).\end{aligned}$$

On the other hand, since $\|u_n\|\ge c$ for all n, the sum on the left is bounded from below by $Mc^2|\sigma_1^{i\alpha}-\sigma_2^{i\alpha}|^2|2^{1+i\alpha}-1|^2$. Since K was arbitrary, it follows that

$$|\sigma_1^{i\alpha}-\sigma_2^{i\alpha}|^2\le C_1\left(\frac{1}{\sigma_1^2}+\frac{1}{\sigma_2^2}\right)$$

for some absolute constant C_1, and this is impossible for appropriate large σ_1 and σ_2 because $\alpha\ne 0$. □

Remark. The proof actually shows that if $\alpha\ne 0$, then $\lim S(e_n,0)=0$ for every bounded operator $S:Z_2(\alpha)\to Z_2(\beta)$. In particular, $Z_2(\alpha)$ is not even isomorphic to a subspace of $Z_2(\beta)$ when $0\ne\alpha\ne\beta$.

3. Notes and Remarks

Although questions related to twisted sums were studied earlier, the serious study of this notion was motivated by the problem, posed by Palais, whether there are nontrivial twisted sums of l_2 with itself. Enflo, Lindenstrauss and Pisier [**178**] gave the first example, which was based on the existence of a homogeneous function $f : l_2 \to l_2$ such that $\| \sum f(x_i)\| \leq \sum \|x_i\|$ whenever $\{x_i\} \subset l_2$ satisfy $\sum x_i = 0$, and such that $\sup\{\|f(x) - Tx\| : \|x\| \leq 1\} = \infty$ for every linear map $T : l_2 \to l_2$.

The notion of a quasi-linear function and its relation to twisted sums is due to Ribe [**494**]. He used it to construct a nontrivial twisted sum of $\mathbb{R}$ with l_1 (the example presented here after Proposition 16.3). Examples of this type were constructed independently and at about the same time by Kalton [**310**] (with a similar approach to that of Ribe) and Roberts [**499**]. A detailed study of quasi-linear functions and of their connection to twisted sums was done by Kalton and Peck [**318**]. This paper initiated considerable further research (including most of the results reported here), mainly by Kalton and several of his co-authors.

Theorem 16.2 and Proposition 16.3 are from Kalton and Peck [**318**]. Lemmas 16.6 and 16.7, as well as parts (ii) and (iii) of Theorem 16.4, are from Enflo, Lindenstrauss and Pisier [**178**]. Part (i) of this theorem, due to Kalton, appears in [**312**], where a more thorough study of questions involving type, convexity and their relation to twisted sums in the setting of quasi-Banach spaces is done.

Proposition 16.9 is due to Dierolf [**145**]; see also Kalton [**312**]. We proved in the text that l_1 fails to be a $\mathcal{K}$-space and that every space with a nontrivial type is a $\mathcal{K}$-space. Another interesting result is that c_0 (and actually every quotient of an $\mathcal{L}_\infty$ space) is a $\mathcal{K}$-space. This is proved in Kalton and Roberts [**321**] and is a consequence of the following result: Let ν be a real-valued function on an algebra $\mathcal{A}$ of sets, such that $|\nu(A \cup B) - (\nu(A) + \nu(B))| \leq 1$ for all disjoint $A, B \in \mathcal{A}$. Then there is a finitely additive measure μ on $\mathcal{A}$ such that $|\mu(A) - \nu(A)| \leq M$ for all $A \in \mathcal{A}$, where M is some absolute constant. It is of interest that the proof of this result uses the existence of concentrator graphs.

Other examples of $\mathcal{K}$-spaces were constructed in Cabelló Sanchez and Castillo [**107**], where it is also proved that if a Banach space E contains a subspace F such that both F and E/F are $\mathcal{K}$-spaces, then so is E.

Kalton, Peck and Roberts [**319**] conjecture that a Banach space is a $\mathcal{K}$-space if and only if its dual has nontrivial cotype.

Not much is known on pairs of infinite-dimensional spaces E and F whose only twisted sums are trivial. By Proposition 16.9 and Sobczyk's theorem (see Lindenstrauss and Tzafriri [**371**], page 106), this is the case if F is a separable $\mathcal{K}$-space and E is c_0. Kalton and Pełczyński [**320**] study this question when E is a Hilbert space. Their results are "formal" consequences of the fact that there is a nontrivial twisted sum of a Hilbert space with itself. In Kalton and Peck [**318**], where Proposition 16.10 is proved, there are also some additional results on the case $E = F$ when E is a sequence space. Function spaces $E = F$ are treated in Kalton [**315**]. Brudnyi and Kalton [**103**] is closely related to the case $E^* = F$ of this question. Additional information can be found in Cabelló Sanchez and Castillo [**107**].

Section 2 is devoted to properties of twisted sums of l_2 with itself. As mentioned before, the existence of such spaces was first proved in Enflo, Lindenstrauss and

Pisier [**178**]; however, here we follow the approach of Kalton and Peck [**318**]. The results extend easily to more general settings (e.g. if l_2 is replaced by l_p, $1 < p < \infty$).

The results in Section 2, and in particular their proofs, belong naturally to linear Banach space theory and are not really in the spirit of the other material in this book. We chose to include them in order to exhibit the striking applications of the nonlinear construction of Section 1 to the linear theory.

Propositions 16.11 and 16.12 are from Kalton and Peck [**318**].

There are several notions of local unconditional structure. The one used here is usually called the Gordon-Lewis (or GL) l.u.st. Since we do not use the other notions, we have omitted the prefix GL.

Theorem 16.13 is from Johnson, Lindenstrauss and Schechtman [**294**]. The method of proof of this theorem was later refined and extended to general contexts by several authors. In particular, it was an important ingredient in the work of Komorowski and Tomczak-Jaegermann [**334**], which, when combined with Gowers' dichotomy principle [**225**] shows that a separable Hilbert space is the only Banach space isomorphic to all of its infinite-dimensional subspaces (see [**334**] for further references).

Casazza and Kalton [**110**] generalized Theorem 16.13 considerably, and showed that if a Banach space E with l.u.st. has an unconditional decomposition into finite-dimensional spaces of bounded dimension, then it actually has an unconditional basis. They also proved that a large family of twisted sums of Hilbert spaces fails to have l.u.st. They conjecture that if a twisted sum of a Hilbert space with itself has l.u.st, then it must be trivial, i.e., isomorphic to a Hilbert space.

Theorem 16.16 is due to Kalton and Peck [**318**] and to Johnson, Lindenstrauss and Schechtman [**294**]. It is unknown if Z_2 is actually "prime", i.e., whether every infinite-dimensional complemented subspace of Z_2 is isomorphic to Z_2. There are several other open problems concerning the structure of the space Z_2. For example, it is conjectured in Johnson, Lindenstrauss and Schechtman [**294**] that Z_2 is not isomorphic to its hyperplanes (in which case it is certainly not prime!).

The first example of a complex Banach space which is not isomorphic to its complex conjugate space was constructed by Bourgain [**86**]. He used random Banach spaces (following a general method of Gluskin). A similar construction was done independently by Szarek [**550**]. The examples $Z_2(\alpha)$ presented here are explicit and much simpler. Theorem 16.17 is taken from Kalton [**317**].

There is much further study related to the material in this chapter. Kalton did an extensive study of the relation of twisted sums to nonlinear commutators and to interpolation theory. He used this study to show, e.g., that there is a twisted sum of l_2 with itself which fails to be a U.M.D. space. (E is U.M.D if E-valued martingale differences form an unconditional sequence in $L_2(\mu, E)$.) See Kalton [**314**] and its references. In another study [**316**] Kalton produced a quasi-Banach space which is a twisted sum of $\mathbb{R}$ and l_1 and which contains no infinite basic sequence. His construction relies heavily on the construction of Gowers and Maurey [**226**] of Banach spaces which do not contain an unconditional basic sequence.

The recent book of Castillo and Gonzalez [**114**] surveys all the aspects related to twisted sums of Banach spaces.

CHAPTER 17

Group Structure on Banach Spaces

1. Banach Groups

In this chapter we consider a more general structure than that of a Banach space, namely that of a metric topological group uniformly homeomorphic to a Banach space. Using the homeomorphism, we can consider the Banach space as equipped with an additional group operation. If E is a Banach space and ϕ is a uniform homeomorphism of E onto a metric topological group F (or a linear metric space, or even another Banach space) which carries 0 to 0, then E is such a group under the product defined by

$$x \cdot y = \phi^{-1}\big(\phi(x)\phi(y)\big).$$

The group operation is unrelated (at least a priori) to the vector space operations, and it is only assumed to be uniformly continuous with respect to the original norm topology.

In this section we consider groups defined on the whole Banach space, and we look for conditions that ensure that the group is actually the additive group of a Banach space. In the next section we consider local group structure on Banach spaces, i.e., a neighborhood of zero in the Banach space which is uniformly homeomorphic to a neighborhood of the identity in a group. The questions examined in the next section are infinite-dimensional analogues of Hilbert's fifth problem.

Definition 17.1. *A uniform Banach group G is a Banach space equipped with a group operation $(x, y) \to x \cdot y$ which is uniformly continuous as a function of two variables and so that the identity element of the group is zero.*

It should, perhaps, be mentioned at the outset that there are no known examples of noncommutative Banach groups.

We shall use multiplicative notation, *including the dot* $\cdot$, for the group operation (except for denoting the identity of the group by 0). Additive notation will be preserved for the vector space addition, and tx will denote the multiplication of a vector x by a scalar t. The map $x \to x^{-1}$ is also uniformly continuous. Indeed, for every $\varepsilon > 0$ the uniform continuity of the group multiplication gives a $\delta > 0$ such that $\|x - y\| < \delta$ implies $\|w \cdot x \cdot z - w \cdot y \cdot z\| < \varepsilon$ for every $w, z \in G$. Now take $w = x^{-1}$ and $z = y^{-1}$.

Let us also remark that for every R there is a constant $C(R)$ so that $\|x\| \le R$ implies that $\|w \cdot x \cdot z - w \cdot z\| \le C(R)$ for every w and z. Indeed, take δ_0 which corresponds to $\varepsilon = 1$ above, and let $n = [R/\delta_0] + 1$. Put $x_k = kx/n$. Then

$$\|w \cdot x \cdot z - w \cdot z\| \le \sum_{k=0}^{n-1} \|w \cdot x_{k+1} \cdot z - w \cdot x_k \cdot z\| \le n.$$

A uniform Banach group admits a G-invariant metric which is uniformly equivalent to the norm (i.e., the identity map is uniformly continuous as a map between G equipped with the new metric and with the norm distance). One such metric is given by $d(x, y) = \sup \|w \cdot x \cdot z - w \cdot y \cdot z\|$. It satisfies $d(x, y) \geq \|x - y\|$, and it is uniformly equivalent to the norm because the group multiplication is uniformly continuous. Note that every G-invariant metric ρ on G satisfies a triangle inequality of the form $\rho(x \cdot y, 0) \leq \rho(x, 0) + \rho(y, 0)$. Indeed

$$\rho(x \cdot y, 0) = \rho(x, y^{-1}) \leq \rho(x, 0) + \rho(0, y^{-1}) = \rho(x, 0) + \rho(y, 0).$$

The key to detecting and reconstructing a linear structure on G from its group operation is the existence of "well-behaved" square roots for its elements. Taking square roots in G corresponds to the map $z \to z/2$ in a linear space, and this map clearly suffices to reconstruct the whole linear structure of a Banach space.

If multiplication in G comes from the additive group of a Banach space, then $\|x^n\| \to \infty$ for every $x \neq 0$, and uniformly so when x is outside a neighborhood of zero. It turns out that this property also controls the operation of taking square roots in more general groups.

Definition 17.2. *A uniform Banach group G is called dissipative if $\|x^n\| \to \infty$ for every $x \neq 0$. It is uniformly dissipative if for every $\varepsilon > 0$ and for every $K > 0$ there is an $N = N(\varepsilon, K)$ such that $\|x^n\| > K$ whenever $\|x\| \geq \varepsilon$ and $n \geq N$.*

Lemma 17.3. (i) *Let G be a dissipative uniform Banach group. Then $x = y$ whenever $x^2 = y^2$; i.e., the map $z \to z^{1/2}$ is single-valued on its domain.*

(ii) *If G is uniformly dissipative, then for every R and $\varepsilon > 0$ there is a $\delta > 0$ such that if $\|x^2 - y^2\| < \delta$ and if $\|x\|, \|y\| \leq R$, then $\|x - y\| < \varepsilon$. Thus the map $z \to z^{1/2}$ is uniformly continuous on sets of the form $\{x^2 : \|x\| \leq R\}$.*

(iii) *If G is abelian and uniformly dissipative, then the map $z \to z^{1/2}$ is uniformly continuous on its entire domain of definition.*

Proof. (i) Assume that $x \neq y$ and $x^2 = y^2$. We shall show that the even powers of $x^{-1} \cdot y$ are bounded, contradicting the dissipativity of G.

Since multiplication is uniformly continuous there is a constant $C = C(R)$, depending only on R, such that

$$\|x^{-1} \cdot g \cdot x \cdot g - g^2\| \leq C$$

for all $g \in G$ and for all x satisfying $\|x\| \leq R$. Fix $n > 0$, and set $g = (x^{-1} \cdot y)^n$. Then $x^2 = y^2$ implies that $x^{-1} \cdot g \cdot x \cdot g = 0$. Hence if $\|x\| \leq R$, then the choice of C gives $\|(x^{-1} \cdot y)^{2n}\| = \|g^2\| \leq \|x^{-1} \cdot g \cdot x \cdot g\| + C = C$.

(ii) To prove the second assertion, choose $C = C(R)$ as above. The uniform continuity of the multiplication gives an $\eta = \eta(\varepsilon) > 0$ so that $\|x^{-1} \cdot y\| \geq \eta$, and hence uniform dissipativity gives an $N = N(\varepsilon, R)$ so that $\|(x^{-1} \cdot y)^{2N}\| > C + 2$ for all x and y with $\|x - y\| \geq \varepsilon$.

We now repeat the argument in (i) with $g = (x^{-1} \cdot y)^N$. By the choice of C and by the uniform continuity of the multiplication there is a $\delta > 0$ so that if $\|x\|, \|y\| \leq R$ and if $\|x^2 - y^2\| < \delta$, then $\|(x^{-1} \cdot y)^{2N}\| < C + 1$. But then $\|x - y\| \geq \varepsilon$ is impossible by the choice of N, so $\|x - y\| < \varepsilon$ as required.

(iii) Fix $\varepsilon > 0$, and choose N such that $\|(x^{-1} \cdot y)^{2N}\| > 1$ whenever $\|x - y\| \geq \varepsilon$. Choose $\delta = \delta(N) > 0$ so that $\|(x^2)^{-N} \cdot (y^2)^N\| < 1$ whenever $\|x^2 - y^2\| < \delta$. Then $\|x^2 - y^2\| < \delta$ implies that $\|x - y\| < \varepsilon$. □

A one-parameter subgroup of a topological group is a continuous homomorphism from the real line onto a nontrivial subgroup H. We shall denote the homomorphism by $X(t)$.

Corollary 17.4. *Let G be a uniformly dissipative uniform Banach group. Assume that for every $L > 0$ there is an $R > 0$ so that the set*

$$\{x^2 : \|x\| \le R \text{ and } \|x^2\| \le L\}$$

is dense in the ball $\{y : \|y\| \le L\}$. Then

(i) *Every element in G is a square, and the map $z \to z^{1/2}$ is uniformly continuous on bounded sets.*

(ii) *For every $x \ne 0$ in G there is a unique one-parameter subgroup $X(t)$ with $X(1) = x$. The homomorphism X is actually an isomorphism, and $\|X(t)\| \to \infty$ uniformly as $t \to \infty$ for x outside any fixed neighborhood of zero.*

(iii) *If G is abelian, then the denseness condition can be relaxed, and it is enough to assume that the set of all squares in G is dense in G with no estimates on the norms. In this case the map $z \to z^{1/2}$ is uniformly continuous on all of G, and G is the additive group of a quasi-normed space.*

Proof. (i) By Lemma 17.3(ii) the map $z \to z^{1/2}$ is uniformly continuous on $\{x^2 : \|x\| \le R\}$ for every R. Since this set is dense in $\{y : \|y\| \le L\}$ for an appropriate R, the map extends by continuity to all of $\{y : \|y\| \le L\}$.

(ii) Each $x \ne 0$ has a unique 2^nth root for every n by (i), and these roots commute. For a dyadic rational $t = \sum 2^{-m_i}$ set x^t to be the product of the 2^{m_i}th roots of x. The map $t \to x^t$ is a homomorphism of the dyadic rationals into G. The heart of the proof is that this homomorphism is uniformly continuous. Once this is proved, the required one-parameter subgroup is obtained when $X(t) = x^t$ is extended to all of $\mathbb{R}$ by continuity. Uniform dissipativity gives $\|X(t)\| \to \infty$ uniformly, and hence $X(t)$ is actually an isomorphism of $\mathbb{R}$ into G.

Note also that $\|x^t - y^t\| \to 0$ when $\|x - y\| \to 0$ for each fixed t, and uniformly so for t's in bounded sets.

Since $t \to x^t$ is a homomorphism, to show that it is uniformly continuous we need to show only that $\lim_{t\to 0} x^t = 0$. We shall show that this holds, in fact, uniformly for x in bounded sets.

Let $W_n = \{y^{2^{-j}} : \|y\| < 1,\ j \ge n\}$. Note that each W_n is open and that the uniform dissipativity implies that $\operatorname{diam}(W_n) \to 0$.

By the uniform continuity of multiplication there is a k such that if $u, v \in W_k$, then $u \cdot v \in W_0$. It follows that for all n, if u and v are two commuting elements in W_{n+k}, then their product belongs to W_n. (Write $u = (y^{2^{-m}})^{2^{-n}}$ and $u = (z^{2^{-l}})^{2^{-n}}$, where $l, m \ge k$ and y and z commute. Then $y^{2^{-m}} \cdot z^{2^{-l}} \in W_0$.)

Writing $t = \sum_i 2^{-m_i} = \sum_{j=0}^{k-1} \sum_i 2^{-kl_i + j} = \sum_{j=0}^{k-1} t_j$, we find that it suffices to show that $\lim_{t \to 0} y^t = 0$ uniformly on $\{y : \|y\| < 1\}$ when the t's are restricted to be finite sums of the form $\sum_i 2^{-kl_i+j}$ for some fixed $j \le k-1$. Without loss of generality we may assume that $j = 0$, so t is of the form $t = \sum_{i=1}^M 2^{-kl_i}$ (and we need estimates that are independent of M).

For $p \ge 1$ put $s_p = \sum_{i=p}^M 2^{-kl_i}$. By the choice of k every y with $\|y\| < 1$ satisfies

$$y^{s_{M-1}} = y^{2^{-kl_M}} \cdot y^{2^{-kl_{M-1}}} \in W_{kl_{M-1}-k} \subset W_{kl_{M-2}}.$$

Inductively we see that $y^{s_p} \in W_{kl_{p-1}}$ for $1 \leq p < M$; hence $y^t = y^{s_M} \in W_{kl_0}$ where $l_0 = l_1 - 1$. Since $l_0 \to \infty$ as $t \to 0$ and since the diameters of the W_n's converge to zero as $n \to \infty$, the assertion follows.

(iii) The map $z \to z^{1/2}$ extends by continuity to a uniformly continuous map defined on all of G by Lemma 17.3(iii). Define vector space operations as follows: Addition is the group operation $(x, y) \to x \cdot y$, and multiplication of x by the scalar t is given by x^t. That these are vector space operations follows for dyadic t's by definition, hence by continuity for all of $\mathbb{R}$. The continuity of multiplication by scalars as a function of two variables follows from $\|x^t - y^s\| \leq \|x^t - y^t\| + \|y^t - y^s\|$.

To see that G with this linear structure is a quasi-normed space, we show that the open neighborhood $B = \{x : \|x\| < 1\}$ of 0 is bounded; that is, for every neighborhood V of 0 there is a $t > 0$ so that $B^t \subseteq V$ (see Appendix H). Indeed, t is chosen by uniform dissipativity so that $\|x^{1/t}\| \geq 1$ whenever $x \notin V$. □

In addition to conditions on the group, we shall also impose geometric conditions on the Banach space on which the group is defined. We first introduce the necessary definitions and notation.

Let X be a metric space. An n-cube in X is a set of 2^n points $\{x_\varepsilon\}$ indexed by the vertices of the unit cube $\varepsilon = (\varepsilon_1, \ldots \varepsilon_n)$ where each ε_i is either 0 or 1. (Alternatively, they can be indexed as $\{x_A\}$ over the subsets A of $\{1, \ldots, n\}$.)

A diagonal is a pair $(x_\varepsilon, x_\delta)$ where ε and δ differ in all n coordinates. There are 2^{n-1} diagonals, and we denote their lengths by d's.

An edge is a pair $(x_\varepsilon, x_\delta)$ where ε and δ differ in exactly one coordinate. There are $n2^{n-1}$ edges, and their lengths are denoted by e's.

We say that X has roundness p if for every 2-cube $\sum d_\alpha^p \leq \sum e_\beta^p$, where the first sum is taken over the lengths of all diagonals of the 2-cube and the second over all the lengths of its edges.

If X has roundness p, then a simple induction shows that the inequality

$$\sum d_\alpha^p \leq \sum e_\beta^p \tag{17.1}$$

holds for all n-cubes for every n.

Every metric space has roundness 1 by the triangle inequality. Hilbert space has roundness 2 by the parallelogram law. Since the roundness of the real line is exactly 2, the roundness of every normed space is bounded by 2. Direct computation shows that for $1 \leq p \leq 2$ the roundness of L_p is exactly p.

We denote the minimal length of a diagonal in an n-cube by $d_{\min}$ and the maximal length of an edge by $e_{\max}$. Roundness p will be used in the following way:

If X has roundness p, then every n-cube satisfies $d_{\min} \leq n^{1/p} e_{\max}$.

This follows from (17.1) by counting the edges and the diagonals in an n-cube.

Proposition 17.5. *Let G be an abelian uniform Banach group modeled on a Banach space with roundness $p > 1$. Assume that*

$$\|x_1 \cdot y - x_2 \cdot y\| = o(\|x_1 - x_2\|^{1/p})$$

uniformly in x_1, x_2, y as $\|x_1 - x_2\| \to 0$. Then the set $S = \{u^2 : u \in G\}$ of all squares in G is dense in G.

In particular, if G is uniformly dissipative, then it is the additive group of a quasi-normed space.

Proof. Let $z \neq 0$. To approximate it by squares, fix an integer n and set $y_j = (jz/n) \cdot \big((j-1)z/n\big)^{-1}$ for $1 \leq j \leq n$. Clearly $z = y_1 \cdot \ldots \cdot y_n$.

For each subset A of $\{1, \ldots, n\}$ let y_A be the product of the y_j's with $j \in A$, and consider the n-cube whose vertices are the 2^n partial products y_A. Since G is abelian, each edge is of the form $(y_j \cdot y_A, y_A)$ for some j and for some set A not containing j. Set $y = \big((j-1)z/n\big)^{-1} \cdot y_A$, and write the edge in the form $\big((jz/n) \cdot y, \big((j-1)z/n\big) \cdot y\big)$. By the condition of the proposition its length is $o(\|z/n\|^{1/p})$, and roundness p gives $d_{\min} \leq n^{1/p} e_{\max} = o(1)$ as $n \to \infty$.

Let (u, v) be a diagonal of minimal length. Being a diagonal, u and v are complementary products of y_j's; hence $v \cdot u = z$. Since $d_{\min} = \|u - v\|$ can be made arbitrarily small by choosing n large, we obtain

$$\|u^2 - z\| = \|u \cdot u - v \cdot u\| = o(\|u - v\|^{1/p}) \to 0$$

as $n \to \infty$, and we have found squares arbitrarily close to z.

The last claim follows from part (iii) of Corollary 17.4. □

We shall also need information on the location of the square roots. This is best accomplished in terms of an "intrinsic" invariant metric, which we now define.

Let d be any G-invariant metric, uniformly equivalent to the norm, and such that $d(x, y) \geq \|x - y\|$ (see the beginning of this section). An admissible chain between two points $x, y \in G$ is a sequence $x = x_0, \ldots, x_n = y$ where $d(x_i, x_{i+1}) \leq 1$ for all i. The intrinsic (or chain) metric d_I on G is defined by

$$d_I(x, y) = \inf \left\{ \sum d(x_i, x_{i+1}) \right\}$$

where the infimum is taken over all admissible chains $\{x_i\}$ between x and y.

The metric d_I is G-invariant because d is. It is uniformly equivalent to the norm because d is and because $d_I(x, y) = d(x, y)$ whenever $d(x, y) \leq 1$. Clearly $d_I(x, y) \geq d(x, y) \geq \|x - y\|$ for all x and y.

As for any metric which is uniformly equivalent to the norm, there is a constant C such that $d_I(x, y) \leq C\|x - y\|$ whenever $\|x - y\| \geq 1$. Indeed, choose C such that $d_I(u, v) \leq C/2$ whenever $\|u - v\| \leq 1$. If $\|x - y\| \geq 1$, then choose $n \leq 2\|x - y\|$ and $x = x_0, \ldots, x_n = y$ in the interval $[x, y]$ so that $\|x_i - x_{i+1}\| \leq 1$. Then the triangle inequality, the estimate on n and the definition of C give

$$d_I(x, y) \leq \sum d_I(x_i, x_{i+1}) \leq 2\|x - y\| \cdot C/2 = C\|x - y\|.$$

We also note that there is a constant C_1 so that $d_I(x, y) \leq C_1\|x - y\|$ whenever $d_I(x, y) \geq 1$. Indeed, we need to check only pairs of points x, y with $\|x - y\| < 1$, so $d_I(x, y) \leq C/2$ by the choice of C. If there were no C_1 as claimed, we would find sequences x_n, y_n so that $1 \leq d_I(x_n, y_n) \leq C/2$ while $\|x_n - y_n\| \to 0$, contradicting the uniform equivalence of d_I and the norm.

The property that makes d_I so convenient for our purposes is that there is essential equality in the triangle inequality along nearly "optimal" chains. More precisely, if $\{x_i\}$ is a chain that essentially realizes the d_I distance between z and w, then each segment of the chain essentially realizes the distance between its end points. Indeed, if there were a more efficient chain for realizing the distance between the end points of the segment, it could be substituted instead of the original segment and improve the estimate on the distance between z and w. In particular we obtain essential equality in the triangle inequality along the chain.

The next proposition finds for every $z \in G$ a square u^2 which is close to z.

Proposition 17.6. *Let G be an abelian uniform Banach group modeled on a Banach space with roundness $p > 1$, and let d_I be as above. Then there is a constant $K > 0$ so that for every $z \in G$ with $d_I(z, 0) \geq 1$ there is a $u \in G$ satisfying*

$$d_I(u^2, z) \leq K \quad \textit{and} \quad \left|2d_I(u, 0) - d_I(u^2, 0)\right| \leq K d_I^{1/p}(z, 0).$$

If every element of G has a unique square root and if the map $z \to z^{1/2}$ is uniformly continuous, then there is a constant C so that every v with $d_I(v, 0) \geq 1$ satisfies

$$\left|2d_I(v, 0) - d_I(v^2, 0)\right| \leq C d_I^{1/p}(v, 0).$$

Proof. The idea of the proof is similar to that of Proposition 17.5. Choose K_1 such that $d_I(x, y) \leq K_1 \|x - y\|$ whenever $d_I(x, y) \geq 1$, and fix $z \in G$ with $d_I(z, 0) \geq 1$. We shall use an iterative procedure, so we denote z by z_1. Let $\{x_i\}_{i \leq n}$ be an admissible chain between 0 and z_1 that essentially realizes $d_I(z_1, 0)$, and set $y_j = x_j \cdot x_{j-1}^{-1}$.

We shall make repeated use of the fact that there are other nearly optimal chains between 0 and z_1 with the same set of distances and where the y_j's are just permutations of those of the original chain. Indeed, fix k and replace x_k by $x_{k+1} \cdot x_k^{-1} \cdot x_{k-1}$, keeping the other x_i's unchanged. The new chain is also admissible (by the invariance of d_I), it has the same set of distances, and y_k and y_{k+1} are interchanged.

As a first application we show that the chain can be replaced by another, for which there is at most one j with $d_I(y_j, 0) \leq 1/2$. If this is not already the case, permute the y_j's so that there are two consecutive j's with $d_I(y_j, 0) \leq 1/2$, say y_k and y_{k+1}. But then dropping x_k gives another admissible chain (by the d-triangle inequality), which also nearly gives the distance between 0 and z_1 and with at least one less j with $d_I(y_j, 0) \leq 1/2$. The desired chain is obtained by repeating this argument. The final chain will then satisfy $n \geq d_I(z_1, 0) \geq (n-1)/2$.

We now start the iterative procedure. Consider the n-cube $\{y_A\}$ of all partial products of the y_j's. The d-length of each edge is at most 1, hence also its norm length. Since E has roundness p there is a diagonal (u_1, v_1) with $\|u_1 - v_1\| \leq n^{1/p}$. By the choice of K_1 this implies that $d_I(u_1, v_1) \leq K_1 n^{1/p} \leq K_2 d_I^{1/p}(z_1, 0)$.

Changing the chain so that the y_j's are permuted, we may assume that u_1 is the product of an initial segment of the chain and that v_1 is the product of its last terms. The two distances $d_I(u_1, 0)$ and $d_I(u_1, z_1)$ are essentially realized by the appropriate parts of the chain; hence $d_I(u_1, 0) + d_I(v_1, 0) = d_I(0, u_1) + d_I(u_1, z_1) \simeq d_I(z_1, 0)$. Together with $|d_I(u_1, 0) - d_I(v_1, 0)| \leq d_I(u_1, v_1) \leq K_2 d_I^{1/p}(z_1, 0)$ this implies that

$$\begin{aligned} d_I(u_1^2, z_1) &= d_I(u_1^2, u_1 \cdot v_1) = d_I(u_1, v_1) \leq K_2 d_I^{1/p}(z_1, 0). \\ d_I(u_1^2, 0) &\leq 2d_I(u_1, 0) \leq d_I(z_1, 0) + K_2 d_I^{1/p}(z_1, 0) \leq K_3 d_I(z_1, 0). \end{aligned}$$

If $d_I(u_1^2, z_1) < 1$, we are finished. Indeed,

$$\begin{aligned} d_I(z_1, 0) - 1 &\leq d_I(z_1, 0) - d_I(u_1^2, z_1) \leq d_I(u_1^2, 0) \\ &\leq 2d_I(u_1, 0) \leq d_I(z_1, 0) + K_2 d_I^{1/p}(z_1, 0) \end{aligned}$$

i.e., $0 \leq 2d_I(u_1, 0) - d_I(u_1^2, 0) \leq K d_I^{1/p}(z_1, 0)$.

If $d_I(u_1^2, z_1) \geq 1$, put $z_2 = u_1^{-2} z_1$, and then define u_2 by the procedure above. Proceeding inductively we obtain points $z_m = u_{m-1}^{-2} z_{m-1}$ so that

$$\begin{aligned} d_I(u_m^2, z_m) &\leq K_2 d_I^{1/p}(z_m, 0), \\ d_I(u_m^2, 0) &\leq 2d_I(u_m, 0) \leq K_3 d_I(z_m, 0). \end{aligned}$$

Let N be the smallest integer so that either $d_I(u_N^2, z_N) < 1$ or $d_I^{1/p^N}(z_1, 0) < 2$. Since this inductive choice is possible as long as $d_I(u_{m-1}^2, z_{m-1}) \geq 1$, we make it for $m \leq N$. We now define the vector u by

$$u = u_1 \cdot \ldots \cdot u_N.$$

Noting that

$$d_I(z_m, 0) = d_I(u_{m-1}^2, z_{m-1}) \leq K_2 d_I^{1/p}(z_{m-1}, 0) \leq \ldots \leq K_4 d_I^{1/p^{m-1}}(z_1, 0)$$

(where $K_4 = K_2^{(1+1/p+1/p^2 \ldots)}$) and recalling that $z = z_1$, we see that

$$d_I(u^2, z) = d_I(u_N^2, z_N) < K.$$

Also, $d_I^{1/p^{N-1}}(z_1, 0) \geq 2$ implies that $\sum_{m=2}^N d_I^{1/p^{m-1}}(z_1, 0) \leq K_p d_I^{1/p}(z_1, 0)$. Thus finally

$$\begin{aligned} d_I(z, 0) - K &\leq d_I(u^2, 0) \leq 2d_I(u, 0) \leq 2d_I(u_1, 0) + 2\sum_{m=2}^N d_I(u_m, 0) \\ &\leq d_I(z_1, 0) + K_2 d_I^{1/p}(z_1, 0) + K_3 \sum_{m=2}^N d_I(z_m, 0) \\ &\leq d_I(z_1, 0) + K_2 d_I^{1/p}(z_1, 0) + K_5 \sum_{m=2}^N d_I^{1/p^{m-1}}(z_1, 0) \\ &\leq d_I(z, 0) + K d_I^{1/p}(z, 0) \end{aligned}$$

hence $0 \leq 2d_I(u, 0) - d_I(u^2, 0) \leq K d_I^{1/p}(z, 0)$.

For the second part of the proposition assume that $z \to z^{1/2}$ is well-defined and uniformly continuous. With K as above, find a constant L so that $d_I(u^2, v^2) \leq K$ implies that $d_I(u, v) \leq L$.

Fix $v \in G$, and use the first part of the proof to find $u \in G$ which corresponds to the choice $z = v^2$. Then (note that if $d_I(v^2, 0) < 1$, the assertion is obvious)

$$\begin{aligned} \left|2d_I(v, 0) - d_I(v^2, 0)\right| &\leq 2d_I(v, u) + \left|2d_I(u, 0) - d_I(u^2, 0)\right| + d_I(v^2, u^2) \\ &\leq 2L + K d_I^{1/p}(z, 0) + K \leq C d_I^{1/p}(v, 0) \end{aligned}$$

because $d_I(z, 0) \leq 2d_I(v, 0)$, and $d_I(v, 0) \geq 1$. □

Example 17.7. The group G of all integer-valued functions in $L^2(\mathbb{R})$ is closed and connected, and it is homeomorphic to a Hilbert space (see Dobrowolski [**152**]). We now deduce from Proposition 17.6 that G is not uniformly homeomorphic to a Banach space with roundness bigger than 1. Indeed, the squares in G are functions whose values are even integers; hence the characteristic function of an interval of length T cannot be approximated better than $T^{1/2}$ by squares in G. Since T is arbitrary, there is no K such that every element in G can be K-approximated by a square.

Proposition 17.8. *Let G be a uniform Banach group modeled on a Banach space with roundness $p > 1$. Assume that G is the additive group of a quasi-normed space; then G is normable. That is, there is a G-invariant norm on G uniformly equivalent to the given quasi-norm.*

Proof. Since G is the additive group of a linear space, z^t is well-defined for every real t. Let $|||\cdot|||$ be the quasi-norm on G. By Proposition H.2 there is an $0 < r < 1$ such that $|||\cdot|||^r$ satisfies the triangle inequality. Thus $d(x, y) = |||x-y|||^r$ is a metric on G which induces the same uniform structure as $|||\cdot|||$. With this metric let d_I be the invariant intrinsic metric introduced before Proposition 17.6, and set

$$N(z) = \limsup_{t\to\infty} d_I(z^t, 0)/t.$$

We shall show that $N(\cdot)$ is the required norm.

$N(\cdot)$ is positive homogeneous. The invariance of d_I implies that $N(\cdot)$ satisfies the triangle inequality and that $d_I(z^t, 0) \le t d_I(z, 0)$ for positive integers t. Hence for each z there is a constant C_z such that $d_I(z^t, 0) \le t d_I(z, 0) + C_z$ for all t. Thus $N(\cdot)$ is a semi-norm that satisfies $N(z) \le d_I(z, 0)$ for every z.

By Proposition 17.6 there is a constant C so that

$$\left|d_I(v^2, 0) - 2d_I(v, 0)\right| \le C\big(1 \vee d_I(v, 0)\big)^{1/p}$$

for all $v \in G$. It follows that if $d_I(v, 0) \ge 1$, then for every n

$$\begin{aligned}
\left|d_I(v^{2^n}, 0) - 2^n d_I(v, 0)\right| &\le \sum 2^{n-j}\left|d_I(v^{2^j}, 0) - 2d_I(v^{2^{j-1}}, 0)\right| \\
&\le \sum 2^{n-j} C\big(1 \vee d_I(v^{2^{j-1}}, 0)\big)^{1/p} \\
&\le 2^n C \sum 2^{-j}\big(2^{j-1} d_I(v, 0)\big)^{1/p} \\
&\le 2^n K d_I^{1/p}(v, 0)
\end{aligned}$$

for some constant $K > 0$. Thus

$$N(v) \ge \limsup_{n\to\infty} 2^{-n} d_I(v^{2^n}, 0) \ge d_I(v, 0) - K d_I^{1/p}(v, 0) \ge d_I(v, 0)/2$$

provided $d_I(v, 0)$ is large enough, say $d_I(v, 0) \ge L$.

If $|||v||| = L^{1/r}$, then $d_I(v, 0) \ge L$ and thus $N(v) \ge d_I(v, 0)/2 \ge L/2$. As both $|||\cdot|||$ and N are homogeneous, it follows that $N(v) \ge \frac{1}{2} L^{1-\frac{1}{r}} |||v|||$ for all $v \in G$. Since also $N(v) \le d_I(v, 0)$ for all v, it follows that N is uniformly equivalent to the given quasi-norm. □

It follows from Proposition 17.8 that $L_r(\mu)$ is not uniformly homeomorphic to $L_p(\nu)$ for $r < 1 < p < \infty$. Indeed, such a homeomorphism would make $L_r(\mu)$ a uniform Banach group modeled on the Banach space $L_p(\nu)$ whose roundness is bigger than 1. By the proposition $L_r(\mu)$ would be normable, which it is not.

Finally we put all the results together and obtain

Theorem 17.9. *Let G be an abelian uniformly dissipative Banach group modeled on a Banach space with roundness $p > 1$. Assume that*

$$\|x_1 \cdot y - x_2 \cdot y\| = o(\|x_1 - x_2\|^{1/p})$$

uniformly in x_1, x_2, y as $\|x_1 - x_2\| \to 0$. Then G is the additive group of a Banach space.

Proof. By Proposition 17.5 G is the additive group of a quasi-normed space, and it is normable by Proposition 17.8. Since it is complete, it is a Banach space. □

2. Hilbert's Fifth Problem in Infinite Dimensions

Hilbert's fifth problem was whether a topological group which is locally Euclidean is a Lie group, i.e., whether one can introduce local coordinates so that the group operation becomes real analytic (in the commutative case even just vector addition). Many authors contributed to the solution of this problem, and a very nice exposition of the positive solution (which actually goes much further than the original question) is given in the book of Montgomery and Zippin [**426**].

We are concerned here with the infinite-dimensional version of Hilbert's question. Let G be a metric topological group which is locally homeomorphic to an open set in a Banach space E (the model space) and in which the multiplication is uniformly continuous (as a function of two variables). Is G (in an obvious sense) a Lie group? The proof given in [**426**] is very finite-dimensional in character and cannot be carried over to the infinite-dimensional setting. Moreover, not only do the methods of proof need to be modified, there is also an inherent new problem which does not exist in the finite-dimensional case. As we saw in Chapter 9 there are many pairs of Banach spaces which are not isomorphic but which have uniformly homeomorphic balls. Thus, for example, L_2 with its usual addition is a nice topological group modeled on any L_p space ($1 \le p < \infty$). If we want to introduce "good" coordinates in the group, we have to change the model space and replace it by L_2. Moreover, in Chapter 7 we saw that there are nonisomorphic Banach spaces E and F which are Lipschitz equivalent via a map $\varphi : E \to F$. The product $x \cdot y = \varphi^{-1}\big(\varphi(x) + \varphi(y)\big)$ on E is, of course, globally defined and commutative, and it satisfies a Lipschitz condition (with constant $C = \|\varphi\|_{Lip}\|\varphi^{-1}\|_{Lip}$):

$$\|x \cdot y - u \cdot v\| \le C(\|x - u\| + \|y - v\|) \qquad x, y, u, v \in E.$$

This product cannot, however, be represented as a vector sum without changing the model space from E to F.

The extent to which Hilbert's fifth problem has a positive answer in the infinite-dimensional case when we allow a change of the model space seems to be wide open. One partial positive result (which is closely related to the material of Section 1) will be quoted at the end of this section. Most of this section is devoted, however, to the question of what can be done without changing the model space. The first main result shows that Lie theory can be done without changing the model space if we make assumptions (17.2) and (17.3) below, which are stronger than the assumption that the group action is Lipschitz. Assumptions (17.2) and (17.3) are satisfied if the group action, as a function of two variables, is C^1 near the origin. (See also the discussion before Proposition 17.11.) The second main result of this section is that one-sided differentiability suffices: If $x \cdot y$ is C^1 in x for every fixed y, then it is also C^1 in y for every fixed x (see below for a precise statement).

The problems discussed here are local ones, and we shall not work with groups, but rather with local groups G with a model space E. We assume that an associative operation is defined from a neighborhood of $(0, 0)$ in $E \times E$ to E so that $x \cdot 0 = 0 \cdot x = x$. We shall impose continuity conditions which imply, besides the uniform continuity of the operation, also the existence and uniform continuity of the inverse operation x^{-1}. All results and constructions are supposed to hold in a neighborhood of the origin only, and the size of the neighborhood will change during the proofs (we shall not specify it). In the discussion below we shall use the expression $\psi(t)$ to describe quantities which are $o(t)$. Thus ψ is any continuous, increasing, nonnegative function on $[0,\infty)$ with $\lim_{t\to 0^+}\psi(t)/t = 0$. Unless specifically said otherwise, the same symbol ψ will be used for different functions at different places.

Theorem 17.10. *Let G be a local group modeled on a Banach space E, and assume that for $x,u,v\in G$*

$$\|x\cdot u - x\cdot v - (u-v)\| \le \psi\big(\|x\|+\|u\|+\|v\|\big)\|u-v\| \tag{17.2}$$

$$\|u\cdot x - v\cdot x - (u-v)\| \le \psi\big(\|x\|+\|u\|+\|v\|\big)\|u-v\|. \tag{17.3}$$

Then there is a homeomorphism φ between neighborhoods of 0 in E so that the new local group $\tilde G$ (with E as a model space) defined by

$$x*y = \varphi^{-1}\big(\varphi(x)\cdot\varphi(y)\big)$$

satisfies (17.2) and (17.3) (with $*$ instead of $\cdot$) and so that $x*x = 2x$ for all x in $\tilde G$. Moreover, if G is abelian, then $x*y = x+y$ for all x and y in $\tilde G$.

Proof. By (17.2) and (17.3) the maps $y\to y - x\cdot y$ and $y\to y - y\cdot x$ are contractions when $\|x\|$ is small and when we work on a sufficiently small ball. Hence Banach's fixed point theorem ensures the existence of x^{-1} (as well as the uniform continuity of the map $x\to x^{-1}$).

Taking $u=0$ in (17.2), we obtain

$$\|x\cdot y - (x+y)\| \le \psi\big(\|x\|+\|y\|\big)\|y\| \tag{17.4}$$

for all x and y. Hence

$$x+y = \lim_{\lambda\to 0}\lambda^{-1}(\lambda x)\cdot(\lambda y). \tag{17.5}$$

We prove now by induction on r that

$$\Big\|\prod_{k=1}^{r} x_k - \sum_{k=1}^{r} x_k\Big\| \le \psi\Big(\sum_{k=1}^{r}\|x_k\|\Big)\sum_{k=1}^{r}\|x_k\|. \tag{17.6}$$

In the inductive argument below it is important to work with a fixed function ψ. We shall show that the function appearing in (17.4) works provided we assume, as we may, that it is concave. For $r=2$ inequality (17.6) follows from (17.4). For the induction step put $s=\sum_{k=1}^{r-1}\|x_k\|$ and $t=\|x_r\|$. By (17.4) and by the induction hypothesis

$$\begin{aligned}
\Big\|\prod_{k=1}^{r} x_k - \sum_{k=1}^{r} x_k\Big\| &\le \Big\|\prod_{k=1}^{r-1} x_k\cdot x_r - \Big(\prod_{k=1}^{r-1} x_k + x_r\Big)\Big\| + \Big\|\prod_{k=1}^{r-1} x_k - \sum_{k=1}^{r-1} x_k\Big\| \\
&\le t\psi\Big(\Big\|\prod_{k=1}^{r-1} x_k\Big\| + t\Big) + s\psi(s) \\
&\le t\psi(s+s\psi(s)+t) + s\psi(s) \\
&= t\big(\psi(s+s\psi(s)+t) - \psi(s+t)\big) + t\psi(s+t) + s\psi(s) \\
&\le s\psi(s)\big(\psi(s+t)-\psi(s)\big) + t\psi(s+t) + s\psi(s)
\end{aligned}$$

because the concavity of ψ gives

$$\Big(\psi(s+s\psi(s)+t) - \psi(s+t)\Big)/s\psi(s) \le \Big(\psi(s+t)-\psi(s)\Big)/t.$$

Assuming, as we may, that $\psi(s) \le 1$, inequality (17.6) follows for r.

Allowing ψ to vary again, we shall show that if both $\psi(m\|u\|) \le 1$ and $\psi(m\|v\|) \le 1$, then

$$\|u^m - v^m - (mu - mv)\| \le m\|u - v\|\psi\big(m\|u\| + m\|v\|\big). \tag{17.7}$$

Indeed, note that (17.2) and (17.3) combine to give

$$\|x \cdot u \cdot y - x \cdot v \cdot y - (u - v)\| \le \psi\big(\|x\| + \|y\| + \|u\| + \|v\|\big)\|u - v\|.$$

With $1 \le l \le m$, $x = u^{l-1}$ and $y = v^{m-l}$ we obtain

$$\begin{aligned}
&\|u^l \cdot v^{m-l} - u^{l-1} \cdot v^{m-l+1} - (u - v)\| \\
&= \|u^{l-1} \cdot u \cdot v^{m-l} - u^{l-1} \cdot v \cdot v^{m-l} - (u - v)\| \\
&\le \psi\big(\|u^{l-1}\| + \|v^{m-l}\| + \|u\| + \|v\|\big)\|u - v\| \le \psi\big(m\|u\| + m\|v\|\big)\|u - v\|
\end{aligned}$$

where the last inequality follows from $\psi(m\|u\|), \psi(m\|v\|) \le 1$ and from (17.6). Summing over l, the triangle inequality gives (17.7).

As a special case of (17.7) we get for small enough x and y and for all n that

$$\left\|\left(\frac{x}{n}\right)^n - \left(\frac{y}{n}\right)^n - (x - y)\right\| \le \psi\big(\|x\| + \|y\|\big)\|x - y\|. \tag{17.8}$$

In particular, taking $y = 0$ gives

$$\left\|\left(\frac{x}{n}\right)^n - x\right\| \le \psi(\|x\|)\|x\|.$$

Replacing n by m and x by x/n gives $\left\|\left(\frac{x}{nm}\right)^m - \frac{x}{n}\right\| \le \psi\big(\frac{\|x\|}{n}\big)\frac{\|x\|}{n}$; hence

$$\left\|n\left(\frac{x}{nm}\right)^m - nm\frac{x}{nm}\right\| \le \psi\big(\|x\|/n\big)\|x\| \le \|x\|$$

for all positive integers m and n. In particular, $\|n\left(\frac{x}{nm}\right)^m\| \le 2\|x\|$. Using (17.8) with $y = n\left(\frac{x}{nm}\right)^m$ gives

$$\begin{aligned}
&\left\|\left(\frac{x}{n}\right)^n - \left(\frac{x}{nm}\right)^{nm} - \left(x - n\left(\frac{x}{nm}\right)^m\right)\right\| \\
&\le \psi\big(\|x\| + \|y\|\big)\left\|x - n\left(\frac{x}{nm}\right)^m\right\| \le \psi(3\|x\|)\left\|x - n\left(\frac{x}{nm}\right)^m\right\|
\end{aligned}$$

and if we assume that $\psi(3\|x\|) \le 1$, then

$$\left\|\left(\frac{x}{nm}\right)^{nm} - \left(\frac{x}{n}\right)^n\right\| \le 2n\left\|\left(\frac{x}{nm}\right)^m - \frac{x}{n}\right\| \le 2\psi\left(\frac{\|x\|}{n}\right)\|x\|.$$

Since $\psi(0) = 0$ this proves that the sequence $\left(\frac{x}{n}\right)^n$ is a Cauchy sequence; hence $\varphi(x) = \lim_{n\to\infty}\left(\frac{x}{n}\right)^n$ exists. Clearly $\varphi(0) = 0$, and by (17.8)

$$\|\varphi(x) - \varphi(y) - (x - y)\| \le \psi\big(\|x\| + \|y\|\big)\|x - y\|. \tag{17.9}$$

It follows from (17.9) and from Banach's fixed point theorem that φ is a uniform (even Lipschitz) homeomorphism from some neighborhood of 0 in E onto another such neighborhood. It also follows from (17.9) that

$$x * y = \varphi^{-1}\big(\varphi(x) \cdot \varphi(y)\big)$$

satisfies (17.2) and (17.3) (when $\cdot$ is replaced by $*$). It is clear from the definition of φ that $\varphi(2x) = \varphi(x) \cdot \varphi(x)$, i.e., that $x * x = 2x$. If G is commutative, then

$$\left(\frac{x}{n}\right)^n \cdot \left(\frac{y}{n}\right)^n = \left(\frac{x}{n} \cdot \frac{y}{n}\right)^n$$

and (17.4) implies that $n\left\|\frac{x}{n}\cdot\frac{y}{n}-\frac{x+y}{n}\right\|\to 0$ as $n\to\infty$. By an argument similar to the one used to show that $\left(\frac{x}{n}\right)^n$ is a Cauchy sequence, it follows from (17.8) that

$$\left(\frac{x}{n}\right)^n\cdot\left(\frac{y}{n}\right)^n-\left(\frac{x+y}{n}\right)^n\to 0 \qquad \text{as } n\to\infty$$

i.e., that $\varphi(x)\cdot\varphi(y)=\varphi(x+y)$, or that $x*y=x+y$. □

Remarks. (i) The map φ is called the exponential map. As an example let G be an open ball of radius smaller than one in the space of bounded linear operators on a Banach space E, centered at the identity operator I. Then G is a local group under the usual multiplication of operators and $\varphi(I+T)=\lim\left(I+\frac{T}{n}\right)^n=e^T$ is the usual exponential map.

(ii) Any continuous homomorphism L from $\tilde{G}$ into itself is induced by a bounded linear operator on E. Indeed, since with respect to the $*$ multiplication $x^n=nx$, we conclude that $Lnx=Lx^n=(Lx)^n=nLx$ whenever n is such that $x^n\in\tilde{G}$. By a routine argument $L\lambda x=\lambda Lx$ for every small enough real λ. By using (17.5) (with $*$ instead of $\cdot$) we deduce that $L(x+y)=Lx+Ly$.

(iii) We can go further and prove the real analyticity of the group operation $*$ in the noncommutative case. We shall not do this here since this is not really in the main line of this book. We just indicate that the main ingredient of Lie theory, the bracket operation $[x,y]$, is defined by the formula

$$[x,y]=\lim_{\lambda,\mu\to 0}(\lambda\mu)^{-1}\Big((\lambda x)^{-1}*(\mu y)^{-1}*(\lambda x)*(\mu y)\Big).$$

The fact that this limit exists follows quite easily from the inequalities above. These inequalities also yield that $[x,y]$ is a bilinear function of x and y, that $[x,y]=-[y,x]$ and that $[[x,y],z]+[[y,z],x]+[[z,x],y]=0$.

We return now to conditions (17.2) and (17.3) and examine the differentiability assumptions which imply them. (Recall that we always assume that the operation $x\cdot y$ is uniformly continuous in both variables.) It will be convenient to introduce the notation $f(x;y)=x\cdot y-y$. By $f'(x;y)$ we denote the Fréchet derivative of f with respect to x at the point (x,y). It is a bounded linear operator from E to E. Since $x\cdot 0=x$ it follows that $f'(x;0)=I$. If $x\cdot y$ is differentiable in x for fixed y (i.e., $f'(x;y)$ exists and, as usual, just for small x and y), and if $f'(x;y)$ is continuous as a function of x and y at (0,0) (i.e. $\lim_{x,y\to 0}f'(x;y)=I$), then (17.3) holds. This follows directly from the mean value theorem

$$\begin{aligned}\|x\cdot y-z\cdot y-(x-z)\| &= \|f(x;y)-f(z;y)-(x-z)\|\\ &\le \sup_{0\le\theta\le 1}\|f'(\theta x+(1-\theta)z;y)-I\|\,\|x-z\|.\end{aligned}$$

Thus the assumptions of Theorem 17.10 hold when $x\cdot y$ is differentiable with respect to both x and y and when the derivatives are continuous at the point $(0,0)$. We shall weaken the differentiability assumptions in two steps. First we show that if the left multiplication, say, is C^1, then the derivative is continuous in a stronger sense. This is used in the second, and deeper, step to show that differentiability of left multiplication also implies that of right multiplication.

Proposition 17.11. *Let G be a local group modeled on a Banach space E. Assume that $f'(x;y)$ exists and is continuous as a function of x for each fixed y. Then there is a smaller neighborhood G_0 of 0 in E in which the map $(x,y)\to f'(x;y)$*

is continuous as a function of two variables, i.e., as a map from $G_0 \times G_0$ into the space $L(E,E)$ of bounded linear operators on E with the operator norm.

Proof. We show first that for every $\varepsilon > 0$ and for every ball B in G there is a smaller ball $B_0 \subset B$ such that

$$\limsup_{z\to y} \|f'(0;z) - f'(0;y)\| \leq 4\varepsilon \tag{17.10}$$

for every $y \in B_0$.

Put $r_y(x) = x \cdot y - y - f'(0;y)(x)$. It follows from Fréchet differentiability that for every $y \in B$ there is an integer $n(y)$ such that $\|r_y(x)\| \leq \varepsilon\|x\|$ whenever $\|x\| \leq 1/n(y)$. By Baire's category theorem there is an n_0 and a set D whose closure contains a ball $B_0 \subset B$ such that $n(y) \leq n_0$ for all $y \in D$. Let y be an interior point of B_0, and let $u \in D$. If $\|x\| \leq \min\big(1/n_0, 1/n(y)\big)$, then

$$\|\big(f'(0;y) - f'(0;u)\big)(x)\| \leq \|x\cdot y - x\cdot u\| + \|y-u\| + 2\varepsilon\|x\|.$$

Fix $\tau > 0$. By the uniform continuity of the multiplication there is a $\delta > 0$ so that if we also assume that $\|u-y\| < \delta$, then

$$\|\big(f'(0;y) - f'(0;u)\big)(x)\| \leq 2(\varepsilon+\tau)\|x\|$$

and hence

$$\|f'(0;y) - f'(0;u)\| \leq 2(\varepsilon+\tau).$$

Now fix another interior point $z \in B_0$ so that $\|z-y\| < \delta$. By the same argument as above, there is a $\delta_1 > 0$ such that

$$\|f'(0;z) - f'(0;u)\| \leq 2(\varepsilon+\tau)$$

whenever $u \in D$ satisfies $\|u-z\| < \delta_1$. Since D is dense in B_0 there is a point $u \in D$ satisfying both $\|u-y\| < \delta$ and $\|u-z\| < \delta_1$; hence

$$\|f'(0;z) - f'(0;y)\| \leq 4(\varepsilon+\tau)$$

and (17.10) follows.

Using (17.10) let B_n be a decreasing sequence of balls whose radii tend to 0 and such that the oscillation of $f'(0;y)$ at any point in B_n is less than $1/n$. Then the map $y \to f'(0;y)$ from G into $L(E,E)$ is continuous at $y_0 = \bigcap B_n$.

The definition of $f(x;y)$ and the associative law for the group operation yield the identity

$$f\big(y + f(x;y); z\big) = f(y;z) + f(x; y\cdot z)$$

for every $x, y, z \in G$. Differentiating with respect to x and putting $x = 0$ give

$$f'(y;z) \circ f'(0;y) = f'(0; y\cdot z).$$

Since $f'(0;y)$ is continuous at y_0, the equation above and the continuity of $f'(y;z)$ as a function of y give that $f'(0;y)$ is continuous at $y = y_0 \cdot z$ for every $z \in G$, hence in some neighborhood $\tilde{G}$ of 0. Since $f'(0;0) = I$ we may also assume that $\tilde{G}$ is chosen so that $f'(0;y)^{-1}$ exists in $\tilde{G}$ and that it depends continuously on y. The result follows from $f'(y;z) = f'(0;y\cdot z) \circ f'(0;y)^{-1}$. □

Theorem 17.12. *Let G be a local group modeled on a Banach space (with a uniformly continuous group operation). Assume that $x \cdot y$ is continuously differentiable as a function of x for each fixed y. Then there is a neighborhood G_0 of 0 so that $x \cdot y$ is continuously differentiable in G_0 as a function of y for each fixed x.*

We start with two lemmas. The first is purely technical.

Lemma 17.13. *Let $v \in G$, and let $\eta > 0$ be such that the open rectangle*

$$U = \{(u_1, u_2) : \|v - u_1\| < \eta \text{ and } \|u_2\| < \eta\}$$

contains $(v, 0)$ and is contained in $G \times G$. Assume that $\|f'(u_1; u_2) - I\| \leq \varepsilon$ for $(u_1, u_2) \in U$ and that $(v \cdot u, u^k) \in U$ for some u, some integer n, and for all $0 \leq k \leq n-1$. Then

$$\|v \cdot u^n - v - n(v \cdot u - v)\| \leq n\varepsilon \|v \cdot u - v\|.$$

Proof. By the mean value theorem

$$\begin{aligned} \|v \cdot u^{k+1} - v \cdot u^k - (v \cdot u - v)\| &= \|f(v \cdot u; u^k) - f(v; u^k) - (v \cdot u - v)\| \\ &\leq \varepsilon \|v \cdot u - v\| \end{aligned}$$

and the lemma follows from the triangle inequality by summing over k. □

The next lemma shows that under the assumptions of Theorem 17.12 the multiplication $x \cdot y$ is a Lipschitz function in y.

Lemma 17.14. *Under the assumptions of Theorem 17.12 there is an open neighborhood $\tilde{G}$ of 0 and a constant C such that $\|x \cdot y - x \cdot z\| \leq C\|y - z\|$ for every $x, y, z \in \tilde{G}$.*

Proof. In view of Proposition 17.11 we can choose $0 < \varepsilon < 1/2$ and $\delta > 0$ so that $\|f'(x; z) - I\| \leq \varepsilon$ whenever $\|z\| \leq 2\delta$ and $\|x\| \leq 3\delta$. We first claim that $\|x \cdot w\| < 3\delta$ whenever $\|w\|, \|x\| \leq \delta$. Indeed

$$\|x \cdot w - w\| = \|f(x; w) - f(0; w)\| \leq \sup_{0 \leq \theta \leq 1} \|f'(\theta x; w)\| \, \|x\| \leq (1 + \varepsilon)\|x\|.$$

Let $\|w\|, \|x\| \leq \delta$, and let $n \geq 0$ be the largest integer so that $(1 + \varepsilon)n\|w\| \leq \delta$. By Lemma 17.13 (with $u = w$, $v = 0$) we get by induction that $\|w^k\| \leq \delta$ for every $0 \leq k \leq n$. Using Lemma 17.13 again (with $u = w$, $v = x$) we get

$$\|x \cdot w^n - x\| \geq n(1 - \varepsilon)\|x \cdot w - x\|$$

and it follows from $\|x \cdot w^n - x\| \leq 3\delta + \delta$ that $n(1 - \varepsilon)\|x \cdot w - x\| \leq 4\delta$. By the maximality of n

$$\|x \cdot w - x\| \leq \frac{4(1 + \varepsilon)(n + 1)\|w\|}{n(1 - \varepsilon)} \leq 24\|w\|. \tag{17.11}$$

Assume next that $\|x\|$, $\|y\|$ and $\|z\|$ are all smaller than $\delta/4$, and put $w = z \cdot y^{-1}$. By the mean value theorem

$$\|w\| = \|z \cdot y^{-1} - y \cdot y^{-1}\| \leq 2\|y - z\| \leq \delta.$$

Using the mean value theorem again and then (17.11), we get

$$\|x \cdot y - x \cdot z\| = \|x \cdot y - x \cdot w \cdot y\| \leq 2\|x \cdot w - x\| \leq 48\|w\| \leq 96\|y - z\|.$$

□

Proof of Theorem 17.12. We note first that it suffices to prove the differentiability of $x \cdot y$ as a function of y (with x fixed) at $y = 0$. Indeed, assume that T is a bounded linear operator so that $x \cdot z = x + Tz + o(\|z\|)$. For small x, y and z the relation

$$y + z = (y + z) \cdot y^{-1} \cdot y = \big(f'(y; y^{-1})(z) + o(\|z\|)\big) \cdot y$$

gives

$$\begin{aligned} x \cdot (y+z) &= x \cdot \big(f'(y;y^{-1})(z) + o(\|z\|)\big) \cdot y \\ &= \big(x + T \circ f'(y;y^{-1})(z) + o(\|z\|)\big) \cdot y \\ &= x \cdot y + f'(x;y) \circ T \circ f'(y;y^{-1})(z) + o(\|z\|). \end{aligned}$$

In other words $x \cdot y$ is differentiable as a function of y at every small y, and its derivative $f'(x;y) \circ T \circ f'(y;y^{-1})$ is continuous in y by Proposition 17.11.

To prove the existence of T, we prove first the existence of the directional derivatives. Fix $x, y \in G$ and $\varepsilon > 0$. If $t > 0$ is small enough and if $s > 0$ is small enough with respect to t, then Lemma 17.13 (with $u = sy$, $v = 0$) gives

$$\|(sy)^{[t/s]} - ty\| \leq \varepsilon \|ty\|$$

and thus by Lemma 17.14

$$\|x \cdot (sy)^{[t/s]} - x \cdot ty\| \leq Ct\varepsilon \|y\|.$$

Another application of Lemma 17.13 (with $u = sy$, $v = x$) and of Lemma 17.14 gives

$$\|x \cdot (sy)^{[t/s]} - x - t/s(x \cdot sy - x)\| \leq 2Ct\varepsilon \|y\|.$$

The last two inequalities combine to give

$$\left\| \frac{x \cdot ty - x}{t} - \frac{x \cdot sy - x}{s} \right\| \leq 3C\varepsilon \|y\|.$$

Since $\varepsilon > 0$ was arbitrary, it follows that

$$Ty = \lim_{t \to 0} \frac{x \cdot ty - x}{t}$$

exists and the convergence is uniform on spheres of G, i.e.,

$$x \cdot ty = x + tTy + o(t\|y\|).$$

Since T is a Lipschitz map by Lemma 17.14 it remains to check its linearity. Since $f'(v;u) \to I$ as $u, v \to 0$ it follows that $ty + tz = (ty) \cdot (tz) + o(t)$. Using this fact, Lemma 17.14 and Proposition 17.11, we obtain

$$\begin{aligned} x \cdot (ty + tz) &= x \cdot (ty) \cdot (tz) + o(t) = (x + tTy) \cdot (tz) + o(t) \\ &= x \cdot (tz) + tTy + o(t) = x + t(Tz + Ty) + o(t) \end{aligned}$$

i.e., $T(y+z) = Tz + Ty$. ◊

We conclude this chapter by stating without proof one result in the spirit of Hilbert's fifth problem where a change of the model space is allowed. The proof of the theorem is close to some arguments presented in Section 1.

Theorem 17.15. *Let G be a commutative local group modeled on a Banach space E which has roundness p with $p > 1$. Assume that*

$$\|x \cdot y - x \cdot z\| \leq C\|y - z\|$$

and that square roots in G are determined uniquely (i.e., that $x^2 = y^2$ implies that $x = y$). Then there is a Banach space F so that G is isomorphic to a neighborhood $\tilde{G}$ of 0 in F and so that the operation in $\tilde{G}$ is just vector addition.

3. Notes and Remarks

The entire chapter is motivated by Hilbert's fifth problem and its solution. All the results in Section 1 are due to Enflo. Several ideas in the proofs have their origin in the solution of Hilbert's problem as presented in the book of Montgomery and Zippin [**426**]. Enflo examined systematically to what extent the results and the arguments in this book can be extended to the infinite-dimensional case.

The book [**426**] also contains much material on topological transformation groups. Enflo [**170**] and Tian [**556**] contain results of this nature in the infinite-dimensional setting.

Local versions and variations on Lemma 17.3 and Corollary 17.4 appear in Section 2.2 of Enflo [**171**]. Some results also appear in Enflo [**170**]. Note that the proof of part (i) of Lemma 17.3 actually shows that $x^{-1} \cdot y$ generates a bounded subgroup. For odd powers we obtain after the cancellation $\|x^{-1} \cdot y\| + C$, which does not depend on n, and for negative powers conjugate with y instead of x.

The relations between local boundedness of a metric linear space, dissipativity, and the fact that there are neighborhoods of 0 which do not contain a nontrivial subgroup (as well as uniform versions of these conditions) are studied in Section 3.1 of Enflo [**171**].

Section 5.1 of Enflo [**172**] contains Propositions 17.5 and 17.6. As with the other results here it is unclear whether the assumptions imposed in their statements are really necessary. For example, in Proposition 17.5, does the assumption that G is uniformly dissipative follow from the other assumptions?

Theorem 5.1.5 in [**172**] gives analogous results when the multiplication is locally Lipschitz: If, in addition, the Banach space on which the abelian group is locally modeled has roundness $p > 1$, then there is an invariant metric d_1 which is locally Lipschitz equivalent to the norm and such that for every $\varepsilon > 0$ and for every z in some neighborhood of 0 there is a v with $v^2 = z$ and $|d_1(z,0) - 2d_1(v,0)| < \varepsilon$. Theorem 17.15 is an easy consequence of this fact.

Example 17.7 and Proposition 17.16 below appear in Section 5.2 of Enflo [**172**]. We do not know if G is uniformly homeomorphic to a Banach space (which would necessarily have no roundness $p > 1$). Note that G is generated in $L^2(\mathbb{R})$ under addition and closure by the curve $X(t) = \chi_{[0,t]}$, the characteristic function of $[0,t]$ (or $\chi_{[t,0]}$ if t is negative). This raises the question of finding conditions on a curve $X(t)$ in a Banach space E so that the closed additive subgroup it generates is actually a linear subspace. One such condition is given in

Proposition 17.16. *Let E be a Banach space with roundness $p > 1$, and let $X(t)$ be a curve in E with $X(0) = 0$ and $\|X(t) - X(s)\| = o(|s-t|^{1/p})$ as $s - t \to 0$. Then the closed group generated by the curve is a linear subspace.*

Proof. It suffices to show that for every t the point $X(t)/2$ can be approximated arbitrarily well by elements in the group generated by the curve.

Put $y_j = X\big(jt/n\big) - X\big((j-1)t/n\big)$, and form the n-cube of all possible sums of the y's. Then $e_{\max} = o(n^{-1/p})$, and since the space has roundness p there is a diagonal (u, v) of length $d_{\min} \le n^{1/p} e_{\max} = o(1)$. Thus $u + v = X(t)$, and $\|u - v\|$ can be made arbitrarily small if n is large; hence $\|u - X(t)/2\|$ is arbitrarily small. Of course u belongs to the group generated by the curve. □

Proposition 17.16 is sharp. For $1 < p \le 2$ the space $L^p(\mathbb{R})$ has roundness p exactly, and the curve $X(t) = \chi_{[0,t]}$ satisfies $\|X(t) - X(s)\| = |t-s|^{1/p}$ in $L^p(\mathbb{R})$ (instead of $o(|t-s|^{1/p})$). The group it generates is the group of integer-valued function, which is not a subspace. Alspach and Benyamini (unpublished) constructed analogous sharp examples in l^p and in some other spaces.

Proposition 17.8 is clearly related to the fact that a quasi-normed space with type $p > 1$ is isomorphic to a Banach space (see Proposition H.3).

As a consequence of Proposition 17.8 the uniform classification of L^p spaces is now nearly complete: We show in Chapter 10 that L^p and l^p are not uniformly homeomorphic for $p \ge 1$. For $p \le 1$ this is proved by Weston in [**583**]. Combining the remark after Proposition 17.8 with the results in Chapter 10 also shows that L^p and L^q (or l^p and l^q) are not uniformly homeomorphic if $\max(p,q) > 1$ and $p \neq q$. The only remaining open cases are whether L^p and L^q (and l^p and l^q) are uniformly homeomorphic when $p, q \le 1$.

The question whether L_p for $p < 1$ and L_1 are uniformly homeomorphic is a special case of a more general question. We do not know if a nonnormable quasi-Banach space can be uniformly homeomorphic to a Banach space. Partial results on this question are contained in Naor [**435**]. In this connection we mention the following observation of Bessaga [**58**].

Proposition 17.17. *If a locally convex space F is uniformly homeomorphic to a normed space E, then F is a normed space (i.e., its topology is defined by a norm).*

Proof. Let $\phi : E \to F$ be an onto uniform homeomorphism with $\phi(0) = 0$. Let B be the open unit ball in E, and put $U = \phi(B)$. Then U is an open neighborhood of 0 in F, and we shall show that it is bounded (i.e., every neighborhood of 0 in F has a multiple which contains U). Since F is locally convex the existence of such a bounded neighborhood implies that F is normable. (A norm is given by the Minkowski functional of any symmetric convex neighborhood of 0 contained in U.) Let V be a convex neighborhood of 0 in F. Since ϕ is uniformly continuous, there is a neighborhood W of 0 in E such that $x - y \in W$ implies that $\phi(x) - \phi(y) \in V$. There is an n such that every $x \in B$ can be written as $x = \sum_{i=1}^{n} w_i$ with $w_i \in W$ for all i. Thus $\phi\big(\sum_{i=1}^{m+1} w_i\big) - \phi\big(\sum_{i=1}^{m} w_i\big) \in V$ for all $1 \le m < n$, and the choice of W gives $\phi(x) \in V + \ldots + V = nV$, i.e., $U \subset nV$. □

Theorem 17.10 and its proof are due to G. Birkhoff [**65**]. Birkhoff carries the theory further in his article, including the proof that multiplication is real analytic.

Proposition 17.11 and Theorem 17.12 are due to Enflo [**170**]. In the setting of finite-dimensional groups Theorem 17.12 was proved earlier by Segal [**527**]. At the time it was an important step in the work towards the solution of Hilbert's fifth problem. Segal's proof is very much finite-dimensional, and Enflo's argument (presented here) is completely different.

To illustrate the importance of the assumption made here that multiplication is uniformly continuous, consider the following example from Enflo [**170**]. Let G be the group of continuously differentiable maps f from $[0,1]$ into itself so that $f(0) = 0$, $f(1) = 1$ and $f'(t) > 0$ for all t. Multiplication on G is composition, and the distance is given by $d(f,g) = \max_t |f'(t) - g'(t)|$. Then G is a topological group, and the map $f(t) \to f(t) - t$ maps G isometrically onto a neighborhood of 0 in the Banach space of continuous differentiable functions on $[0, 1]$ satisfying $h(0) = h(1) = 0$ with the norm $\|h\| = \max_t |h'(t)|$. For a fixed g the map $f \to f \circ g$ is linear in f, so certainly Fréchet differentiable and uniformly continuous. On the other hand, the map $g \to f \circ g$ for fixed f is not uniformly continuous and not differentiable.

Another example in [**170**] shows that without uniform continuity of the multiplication, square roots may fail to be unique in any neighborhood of 0 even in the commutative case. These examples are studied further in Tian [**555**].

As we already mentioned, Theorem 17.15 is proved in Enflo [**172**]. The proof uses the "cube technique", which gives information only on the existence of square roots and on their location, not on uniqueness. It is not known whether the uniqueness assumption in the statement of Theorem 17.15 is really necessary.

Appendices

A. Convexity

A Banach space E is said to be strictly convex if $\|x+y\|/2 < 1$ for every $x \neq y$ in E satisfying $\|x\|, \|y\| \leq 1$. It is said to be smooth if for every $x \neq 0$ there is a unique supporting functional at x, i.e., a unique functional $x^* \in E^*$ such that $\|x^*\| = 1$ and $\langle x^*, x\rangle = \|x\|$. This supporting functional is denoted by $j(x)$, and the map $x \to j(x)$ is called the duality map.

It follows directly from the definitions that if E^* is strictly convex (respectively, smooth), then E is smooth (respectively, strictly convex), and that if E is smooth, then the norm is Gâteaux differentiable at every point $x \neq 0$. (The Gâteaux derivative of the norm at the point x is given by $j(x)$.)

A Banach space E is said to be uniformly convex if for every $\varepsilon > 0$ there is a $\delta = \delta(\varepsilon) > 0$ such that $\|x+y\|/2 \leq 1-\delta(\varepsilon)$ whenever $\|x\|, \|y\| \leq 1$ and $\|x-y\| \geq \varepsilon$. The function

$$\delta_E(\varepsilon) = \inf\left\{1 - \|x+y\|/2 : \ \|x\|, \|y\| \leq 1 \ \text{ and } \ \|x-y\| \geq \varepsilon\right\}$$

is called the modulus of convexity of E, and E is uniformly convex if and only if $\delta_E(\varepsilon) > 0$ for every $\varepsilon > 0$.

E is called uniformly smooth if $\|x+y\| + \|x-y\| - 2 = o(\|y\|)$ as $\|y\| \to 0$. The function

$$\rho_E(\tau) = \sup\left\{\|x+y\|/2 + \|x-y\|/2 - 1 : \ \|x\| \leq 1 \ \text{ and } \ \|y\| \leq \tau\right\}$$

is called the modulus of smoothness of E, and E is uniformly smooth if and only if $\rho_E(\tau)/\tau \to 0$ as $\tau \to 0$.

We remark that δ_E is unchanged if we only take $\|x\| = \|y\| = 1$ in its definition, and similarly we can take only $\|x\| = 1$ and $\|y\| = \tau$ in the definition of ρ_E. (See, for example, Lindenstrauss and Tzafriri [**372**], page 60.)

The modulus of convexity of a Hilbert space is $\delta_H(\varepsilon) = 1 - \left(1 - \varepsilon^2/4\right)^{1/2}$. Indeed, if $\|x\| = \|y\| = 1$, then $\|x+y\|^2 = 4 - \|x-y\|^2$ by the parallelogram law. A similar computation shows that $\rho_H(\tau) = \left(1+\tau^2\right)^{1/2} - 1$.

The next proposition shows that Hilbert space is the "most" convex Banach space in the sense that $\delta_E \leq \delta_H$ for every Banach space E. The elegant geometric argument is due to Nördlander [**439**]. We shall use the following notation. Let S and B be the unit sphere and the unit ball respectively of a norm on $\mathbb{R}^2$. For $0 \leq t \leq 2\pi$ let $x(t) = \big(x_1(t), x_2(t)\big)$ be the point in S which makes an angle t with the positive x-axis. Fix $0 < \varepsilon < 2$. For each t let $y(t) = \big(y_1(t), y_2(t)\big)$ be the unique point in S in the clockwise direction from $x(t)$ such that $\|x(t) - y(t)\| = \varepsilon$. The convexity of S implies that the functions $x_i(t)$ and $y_i(t)$ are of bounded variation and that the curve $\big(x(t) + y(t)\big)/2$ is simple and bounds a convex subset B_ε of B. Denote the areas of B and B_ε by $|B|$ and $|B_\varepsilon|$ respectively.

Proposition A.1. *Let E be a two-dimensional Banach space, and let $\varepsilon > 0$. Then, with the notation above, $|B_\varepsilon| = (1-\varepsilon^2/4)|B|$; hence $\delta_E(\varepsilon) \le \delta_H(\varepsilon)$.*

Proof. Application of Green's theorem gives $|B| = \oint x_1 dx_2 = \oint y_1 dy_2$ and $|B_\varepsilon| = \oint \big((x_1+y_1)/2\big) d\big((x_2+y_2)/2\big)$. Also the curve $x(t) - y(t)$ traces the sphere εS of radius ε exactly once; hence $|\varepsilon B| = \oint (x_1 - y_1) d(x_2 - y_2)$. Thus

$$\begin{aligned}(1-\varepsilon^2/4)|B| &= \frac{1}{2}\Big(\oint x_1 dx_2 + \oint y_1 dy_2\Big) - \frac{1}{4}\oint (x_1-y_1)d(x_2-y_2) \\ &= \oint \Big((x_1+y_1)/2\big) d\big((x_2+y_2)/2\Big) = |B_\varepsilon|.\end{aligned}$$

To see that $\delta_E(\varepsilon) \le \delta_H(\varepsilon)$, note that the set B_ε cannot be strictly contained in $(1-\varepsilon^2/4)^{1/2}B$ because the two sets have the same area. Thus there are points $\|x\| = \|y\| = 1$ with $\|x - y\| = \varepsilon$ such that $\|x + y\| \ge (1-\varepsilon^2/4)^{1/2}$. □

The equality $\delta_E \equiv \delta_H$ actually characterizes inner product spaces. This follows by combining the previous proposition with the following result, due to Day [**136**] and to Schoenberg [**523**], which characterizes inner product spaces by "parallelogram inequalities". For a systematic presentation of the various characterizations of inner product spaces, see the book [**15**] by Amir.

Proposition A.2. *Let E be a normed space such that $\|x+y\|^2 + \|x-y\|^2 \ge 4$ for every $\|x\| = \|y\| = 1$. Then E is an inner product space. The same result holds when $\|x+y\|^2 + \|x-y\|^2 \le 4$ for every $\|x\| = \|y\| = 1$.*

Proof. We may assume that E is two-dimensional. Assume that it satisfies the first inequality, and let $\mathcal{E}$ be the ellipsoid of maximal area contained in $B(E)$. Thus the norm $\|\!|\cdot|\!\|$ induced by $\mathcal{E}$ satisfies $\|\!|u|\!\| \ge \|u\|$ for every $u \in E$, and we shall show that $B(E) = \mathcal{E}$, i.e., that the two norms coincide. Indeed, assume that this is not so. Then the boundaries of $B(E)$ and $\mathcal{E}$ do not coincide and touch in at least four points; hence there are points x, y with $x \ne \pm y$ such that $\|x\| = \|\!|x|\!\| = \|y\| = \|\!|y|\!\| = 1$ and $\|\!|u|\!\| > \|u\|$ for every u on the short arc connecting x and y. In particular $\|\!|x+y|\!\| > \|x+y\|$, and the parallelogram law for $\|\!|\cdot|\!\|$ gives $\|x+y\|^2 + \|x-y\|^2 < \|\!|x+y|\!\|^2 + \|\!|x-y|\!\|^2 = 4$, a contradiction.

The proof of the second part is similar, using the ellipsoid of minimal area containing $B(E)$. □

It follows by duality from Proposition A.3 below that also $\rho_E \ge \rho_H$ for every Banach space and that if $\rho_E \equiv \rho_H$, then E is an inner product space.

The spaces L_p are uniformly convex and uniformly smooth for $1 < p < \infty$. The asymptotic behavior of their moduli were computed by Hanner [**249**] and are given by

$$\delta_p(\varepsilon) = \begin{cases} (p-1)\varepsilon^2/8 + o(\varepsilon^2) & \text{if } 1 < p \le 2 \\ \varepsilon^p/p2^p + o(\varepsilon^p) & \text{if } 2 \le p < \infty \end{cases}$$

and

$$\rho_p(\tau) = \begin{cases} \tau^p/p + o(\tau^p) & \text{if } 1 < p \le 2 \\ (p-1)\tau^2/2 + o(\tau^2) & \text{if } 2 \le p < \infty. \end{cases}$$

It follows from the definition that if E is uniformly smooth, then the norm on E is Fréchet differentiable at every $x \ne 0$.

Proposition A.3. (i) *Uniformly convex (and uniformly smooth) spaces are reflexive.*

(ii) *E is uniformly convex if and only if E^* is uniformly smooth. More precisely*

$$\rho_E^*(\tau) = \sup\left\{\varepsilon\tau/2 - \delta_E(\varepsilon) : 0 < \varepsilon < 2\right\}$$

for every $\tau > 0$.

Part (i) is due to D. P. Milman [**413**] and Pettis [**458**]. The first part of (ii) is due to Day [**135**], and the second to Lindenstrauss [**355**]; see also Lindenstrauss and Tzafriri [**372**], page 61.

The following proposition is due to Figiel [**188**]; see also page 66 in Lindenstrauss and Tzafriri [**372**].

Proposition A.4. *The function $\delta_E(\varepsilon)/\varepsilon$ is nondecreasing.*

The restriction of the duality map to the unit sphere of a uniformly smooth space is uniformly continuous. More precisely

Proposition A.5. *Let E be a uniformly smooth Banach space. Then*

$$\|j(x) - j(y)\| \le \rho_E\left(2\left\|\frac{x}{\|x\|} - \frac{y}{\|y\|}\right\|\right) \Big/ \left\|\frac{x}{\|x\|} - \frac{y}{\|y\|}\right\|$$

for every $x, y \in E$.

Proof. The norm is a convex function, and its derivative at x is $j(x)$; hence

$$\langle j(u), v\rangle \le \|u + v\| - \|u\|$$

for every $u \ne 0$ and for every v.

Fix $x, y \in E$, and we may assume that $\|x\| = \|y\| = 1$ (because $j(\frac{x}{\|x\|}) = j(x)$ and similarly for y). Let z be any vector with $\|z\| = \|x - y\|$. Then

$$\begin{aligned}
\langle j(y), z\rangle - \langle j(x), z\rangle &\le \|y + z\| - 1 - \langle j(x), z\rangle \\
&\le \|y + z\| - 1 + \langle j(x), x - y - z\rangle \\
&\le \|y + z\| - 1 + \|2x - y - z\| - 1 \\
&= \|x + (y - x + z)\| + \|x - (y - x + z)\| - 2 \\
&\le \rho_E(\|y - x + z\|) \le \rho_E(2\|y - x\|)
\end{aligned}$$

because $\|z\| = \|y - x\|$. Since z was arbitrary, the lemma follows.

□

Note that since E is uniformly smooth, the right-hand side of the inequality of the proposition tends to zero as $\|x - y\| \to 0$. Thus j is, indeed, uniformly continuous on the unit sphere

We now discuss the class of Banach spaces that can be given an equivalent uniformly convex or uniformly smooth norm. We refer to Appendix F for the definitions of superreflexivity and finite representation.

An (n, δ)-tree in E (where $\delta > 0$, and n is a positive integer) is a set of the form

$$T_n = \left\{x_{\epsilon_1,\dots,\epsilon_k} : 1 \le k \le n, \quad \epsilon_i = \pm 1\right\}$$

where the points $\{x_{\epsilon_1,\dots,\epsilon_k}\}$ satisfy $x_{\epsilon_1,\dots,\epsilon_k} = (x_{\epsilon_1,\dots,\epsilon_k,-1} + x_{\epsilon_1,\dots,\epsilon_k,1})/2$, and $\|x_{\epsilon_1,\dots,\epsilon_k,-1} - x_{\epsilon_1,\dots,\epsilon_k,1}\| \ge \delta$ for all $\varepsilon_1, \dots, \varepsilon_k$. (A variation of this definition is used for infinite trees in Chapter 5.)

We say that E has the finite tree property for some $\delta > 0$ if its unit ball contains (n, δ)-trees for every n.

Theorem A.6. *Let E be a Banach space. Then the following are equivalent:*

(i) *E is superreflexive.*

(ii) *E does not have the finite tree property for some $0 < \delta < 2$.*

(ii′) *E does not have the finite tree property for any $0 < \delta < 2$.*

(iii) *E admits an equivalent uniformly convex norm.*

(iii′) *E admits an equivalent uniformly convex norm whose modulus of convexity satisfies $\delta(\varepsilon) \geq C\varepsilon^q$ for some $q \geq 2$.*

(iv) *E admits an equivalent uniformly smooth norm.*

(iv′) *E admits an equivalent uniformly smooth norm whose modulus of smoothness satisfies $\rho(\tau) \leq C\tau^p$ for some $1 < p \leq 2$.*

The equivalence of (i), (ii) and (ii′) is due to James [**274**], who used it to show that E is superreflexive if and only if E^* is. Since (ii) is preserved by isomorphisms, it follows that so is superreflexivity. That (iii) implies (i) follows from this observation, Theorem A.3, and the fact that a Banach space which is finitely representable in a uniformly convex space E is itself uniformly convex.

Enflo [**173**] proved that (ii) implies (iii). Since E is superreflexive if and only if E^* is, and since uniform convexity and uniform smoothness are dual properties, this also proves that (i) implies (iv).

The strengthening of (iii) and (iv) to (iii′) and (iv′), giving power type estimates of the moduli, is due to Pisier [**462**].

Remark. If E is not superreflexive, it contains trees with stronger separation between its nodes. To see this we use a result of James [**274**] that a Banach space E is not superreflexive if and only if for every $\delta > 0$ and for every integer m there are norm-one vectors $\{z_i\}_{i=1}^m$ such that for every $1 \leq k \leq m$

$$\Big\|\sum_{i=1}^{k} z_i - \sum_{i=k+1}^{m} z_i\Big\| > m - \delta.$$

Note that if $z_k^* \in E^*$ satisfy $\|z_k^*\| = 1$ and $\langle z_k^*, \sum_{i=1}^k z_i - \sum_{i=k+1}^m z_i\rangle > m - \delta$, then $\langle z_k^*, z_i\rangle > 1 - \delta$ for $i \leq k$ and $\langle z_k^*, -z_i\rangle > 1 - \delta$ for $i > k$.

We now construct arbitrarily large trees in E whose nodes are well separated from each other. Take for example $m = 8$, and consider the 3-tree given by

$$\begin{aligned} x_\emptyset &= (z_1 + \ldots + z_8)/8 \\ x_{-1} &= (z_1 + \ldots + z_4)/4\,, \quad x_1 = (z_5 + \ldots + z_8)/4 \\ x_{-1,-1} &= (z_1 + z_2)/2\,, \quad x_{-1,1} = (z_3 + z_4)/2\,, \quad x_{1,-1} = (z_5 + z_6)/2, \\ &\quad\; x_{1,1} = (z_7 + z_8)/2 \\ x_{-1,-1,-1} &= z_1\,, \quad x_{-1,-1,1} = z_2, \ldots\,. \end{aligned}$$

Translating by $x_\emptyset$ gives a tree of height 3 in the 2-ball in E. Similarly, if we take $m = 2^n$, we obtain in the same manner an n-tree. These trees have the additional property (which is easily verified by applying the z_k^*'s) that the distance between any two vectors in the tree is at least $1 - \delta$.

For an exposition of the various characterizations of superreflexivity, see also Beauzamy [**44**] and Deville, Godefroy and Zizler [**144**].

The following theorem is due to Figiel and Pisier [**191**]. See also Lindenstrauss and Tzafriri [**372**], page 78. The proof here is taken from Deville, Godefroy and Zizler [**144**], page 164. Combined with parts (iii′) and (iv′) of the previous theorem, it implies that superreflexive spaces have nontrivial type and cotype (see Appendix G for the definitions).

Theorem A.7. (i) *Let E be a uniformly smooth Banach space whose modulus of smoothness satisfies $\rho_E(\tau) \leq K\tau^p$ for some $1 < p \leq 2$. Then E has type p.*

(ii) *Let E be a uniformly convex Banach space whose modulus of convexity satisfies $\delta_E(\varepsilon) \geq K\varepsilon^q$ for some $q \geq 2$. Then E has cotype q.*

Proof. (i) We first show that there is a constant C such that

$$\|u+v\|^p + \|u-v\|^p - 2\|u\|^p \leq C\|v\|^p. \tag{A.1}$$

It will then follow by induction that

$$\int_0^1 \Big\|\sum_{i=1}^n r_i(t)x_i\Big\|^p dt \leq C\sum_{i=1}^n \|x_i\|^p$$

for every n; i.e., E has type p.

Proposition A.5 and the standard inequality

$$\begin{aligned}\left\|\frac{x}{\|x\|} - \frac{y}{\|y\|}\right\| &\leq \left\|\frac{x}{\|x\|} - \frac{y}{\|x\|}\right\| + \left\|\frac{y}{\|x\|} - \frac{y}{\|y\|}\right\| \\ &\leq \|x-y\|/\|x\| + \big|\|x\| - \|y\|\big|/\|x\| \\ &\leq 2\|x-y\|/\|x\|\end{aligned}$$

give

$$\begin{aligned}\|j(x) - j(y)\| &\leq \rho_E\Big(2\Big\|\frac{x}{\|x\|} - \frac{y}{\|y\|}\Big\|\Big)\Big/\Big\|\frac{x}{\|x\|} - \frac{y}{\|y\|}\Big\| \\ &\leq K_1\Big\|\frac{x}{\|x\|} - \frac{y}{\|y\|}\Big\|^{p-1} \leq K_2\Big(\|x-y\|/\|x\|\Big)^{p-1}.\end{aligned}$$

It follows that the Gâteaux derivative of the function $\psi(x) = \|x\|^p$, namely $D_\psi(x) = p\|x\|^{p-1}j(x)$, satisfies (recall that $0 < p-1 \leq 1$)

$$\begin{aligned}\|D_\psi(x) - D_\psi(y)\| &\leq p\|x\|^{p-1}\|j(x) - j(y)\| + p\|j(y)\|\,\big|\|x\|^{p-1} - \|y\|^{p-1}\big| \\ &\leq pK_2\|x-y\|^{p-1} + p\|x-y\|^{p-1} = K_3\|x-y\|^{p-1}.\end{aligned}$$

By the mean value theorem there are $0 < \theta_1, \theta_2 < 1$ such that

$$\|u+v\|^p - \|u\|^p = D_\psi(u+\theta_1 v)v \quad\text{and}\quad \|u-v\|^p - \|u\|^p = -D_\psi(u-\theta_2 v)v.$$

Adding the left-hand sides gives the left-hand side of (A.1), and the previous estimate shows that norm of the sum of the right-hand sides is bounded by

$$K_3\|\theta_1 v + \theta_2 v\|^{p-1}\|v\| \leq C\|v\|^p$$

where $C = 2^{p-1}K_3$.

Part (ii) follows from part (i) by duality (use Proposition A.3). □

We denote the set of all equivalent norms on the Banach space E, equipped with the metric $d(\rho_1, \rho_2) = \sup\{|\rho_1(x) - \rho_2(x)| : \|x\| = 1\}$, by $N(E)$. Note that $N(E)$ is open in the space of all continuous semi-norms on E and the latter is a complete metric space under the same metric. It follows that the Baire category theorem holds in $N(E)$.

Proposition A.8. *A superreflexive space admits an equivalent norm which is both uniformly convex and uniformly smooth.*

Proof. By Theorem A.6 we may assume that the original norm is uniformly convex. Denote the modulus of convexity of a norm ρ by δ_ρ.

For every $\rho \in N(E)$ the norms $(\rho + \frac{1}{m}\|\cdot\|)$ converge to ρ in $N(E)$ as $m \to \infty$, and they are uniformly convex. Also the set $G_n = \{\rho : \delta_\rho(1/n) > 0\}$ is open in $N(E)$ for every n. Thus the set of all equivalent uniformly convex norms on E, which is just $\bigcap G_n$, is a dense G_δ-set in $N(E)$.

Since the map $\rho \to \rho^*$ is a homeomorphism of $N(E)$ onto $N(E^*)$ (as is the case for every reflexive space), it follows by duality that the same is true for the set of all equivalent uniformly smooth norms. The theorem now follows from the Baire category theorem. □

Asplund [**26**] gave a general averaging method for constructing an equivalent norm that shares two properties, initially known to hold separately by two distinct norms. In particular, he showed that a space which admits two equivalent norms, one of them uniformly convex and the other uniformly smooth, also admits a third equivalent norm with both properties. (This last result also follows from the computation, due to Beauzamy [**43**], of the modulus of convexity of interpolation spaces; see also Lindenstrauss and Tzafriri [**372**], page 229.)

The idea of applying the Baire category theorem is due to Fabian, Zajîček and Zizler [**181**]; see also section II.4 in Deville, Godefroy and Zizler [**144**]. The same method is applied there to mix other possible properties of norms. For example, if the sets of strictly convex or of locally uniformly convex norms on E are nonempty, then they are residual.

It seems to be an open problem whether a space that admits a uniformly convex norm whose modulus of convexity satisfies $\delta(\varepsilon) \geq K\varepsilon^q$ and a uniformly smooth norm whose modulus of smoothness satisfies $\rho(\tau) \leq K\tau^p$ has a third equivalent norm whose moduli are of power type with the same exponents p and q. All the methods above give weaker estimates. (For $p = q = 2$ such a space has type 2 and cotype 2 by Theorem A.7. By a theorem of Kwapien (see Appendix G), it is isomorphic to a Hilbert space. This gives an indirect positive answer in this special case.)

A bounded convex set K is said to have a normal structure if every convex subset S of K which does not reduce to a single point contains a nondiametral point, i.e., a point x such that $\sup_{y \in S} \|x - y\| < \operatorname{diam}(S)$. If there is a constant $\gamma < 1$ such that $\inf_{x \in E} \sup_{y \in S} \|x - y\| \leq \gamma \operatorname{diam}(S)$ for every such S, we say that K has a uniform normal structure. We say that a Banach space has a uniform normal structure if its unit ball does.

Proposition A.9. *Let E be a uniformly convex or a uniformly smooth Banach space. Then E has a uniform normal structure.*

Proof. Let K be a bounded convex subset of a uniformly convex space, and assume that $\operatorname{diam}(K) = 1$ and $u, v \in K$. Put $x = (u+v)/2$. Then every $z \in K$ satisfies $\|z-u\|, \|z-v\| \leq 1$ and $\|(z-u)-(z-v)\| = \|u-v\|$. By the definition of uniform convexity $\|z-x\| = \|(z-u)+(z-v)\|/2 \leq 1 - \delta_E(\|u-v\|)$. Since u, v were arbitrary, the result follows with $\gamma = 1 - \delta_E(1)$.

We pass now to uniformly smooth spaces. The main step is to show that every bounded and convex set K has a nondiametral point.

Let K be a bounded convex set with $\operatorname{diam}(K) = 1$, and assume that every point in K is diametral. We may assume that K is separable, and let $\{y_j\}$ be a dense sequence in K. For each n choose $x_n \in K$ with $\|x_n - \frac{1}{n}\sum_{j=1}^n y_j\| > 1 - 1/n^2$. It follows from $\|x_n - y_j\| \leq 1$ and the triangle inequality that $\|x_n - y_j\| > 1 - 1/n$ for every $j \leq n$. Since uniformly smooth spaces are reflexive, we may assume, by passing to a subsequence, that $\{x_n\}$ is weakly convergent and by translation that it converges weakly to zero. In particular, $0 \in K$, and since 0 can be approximated arbitrarily well by the y_j's, it follows that $\|x_n\| \to 1$.

For each n let x_n^* be the norm-one functional with $\langle x_n^*, x_n\rangle = \|x_n\|$. By passing to a further subsequence, assume that $\{x_n^*\}$ is w^*-convergent to $x^* \in E^*$.

Fix $\varepsilon > 0$. We claim that there are i, j such that $\|x_i\|, \|x_j\| > 1 - \varepsilon$ and such that $|\langle x_i^*, x_j\rangle|, |\langle x_j^*, x_i\rangle| < \varepsilon$. Indeed, choose i such that $\|x_n\| > 1 - \varepsilon$ for all $n \geq i$ and $|\langle x^*, x_i\rangle| < \varepsilon/2$. Then choose $j \geq i$ large enough such that $|\langle x_i^*, x_j\rangle| < \varepsilon$ and also $|\langle x_j^* - x^*, x_i\rangle| < \varepsilon/2$. Hence $|\langle x_j^*, x_i\rangle| \leq |\langle x_j^* - x^*, x_i\rangle| + |\langle x^*, x_i\rangle| < \varepsilon$.

Since also $\|x_i - x_j\| \leq 1$, we obtain

$$\begin{aligned}
\rho_E(\tau) &\geq \frac{1}{2}\Big(\|x_i - x_j + \tau x_i\| + \|x_i - x_j - \tau x_i\|\Big) - 1 \\
&\geq \frac{1}{2}\Big(\langle x_i^*, (1+\tau)x_i - x_j\rangle + \langle x_j^*, x_j - (1-\tau)x_i\rangle\Big) - 1 \\
&\geq \frac{1}{2}\Big((1+\tau)(1-\varepsilon) - \varepsilon + (1-\varepsilon) - (1-\tau)\varepsilon\Big) - 1 = \tau/2 - 2\varepsilon
\end{aligned}$$

for every $\tau > 0$. Since ε was arbitrary, $\rho_E(\tau)/\tau \geq 1/2$, contradicting the uniform smoothness of E.

That convex bounded subsets of uniformly smooth spaces have a uniform normal structure, with γ depending only on the modulus ρ of smoothness of the space, now follows by a routine application of the ultraproducts technique (see Appendix F). Indeed, otherwise we could find a sequence of uniformly smooth Banach spaces E_n with a common modulus of smoothness ρ, and for each n a convex and bounded set $K_n \subset E_n$ with $\operatorname{diam}(K_n) = 1$ such that $\inf_{x\in E}\sup_{y\in K_n}\|x-y\| \to 1$. Let E be an ultraproduct of the E_n's, and let $K \subset E$ be the product of the K_n's. Then E, being an ultraproduct of uniformly smooth spaces with a common modulus of smoothness ρ, is also such a space, K is a convex bounded subset of E, and every point of K is diametral, contradicting the first part of the proof. □

The notions of a diametral point and of normal structure and the fact that uniformly convex spaces have normal structure are all due to Brodskiĭ and Milman [**93**]. That uniformly smooth spaces have normal structure was first proved by Turett [**565**]. We followed the proof in Prus [**478**].

A space with a uniformly normal structure is reflexive; see Maluta [**381**]. It is an open problem whether such a space has to be superreflexive. For further partial results in this direction, see Naor [**435**].

B. Partitions of Unity

Let X be a Hausdorff space. An open cover $\mathcal{V}$ of X is a family of open subsets of X whose union is all of X. It is called locally finite if every $x \in X$ has a neighborhood U that intersects only finitely many sets in $\mathcal{V}$. An open cover $\mathcal{V}'$ is a refinement of $\mathcal{V}$ if every $V' \in \mathcal{V}'$ is contained in some $V \in \mathcal{V}$.

The space X is said to be paracompact if every open cover of X admits a locally finite open refinement. The class of paracompact spaces was introduced by Dieudonné [**148**]. It is the "correct" setting for many problems in topology, analysis and geometry. On one hand, Proposition B.1 below shows that it is rich enough to contain both compact and metric spaces. On the other hand, the existence of partitions of unity (Proposition B.2 below) makes it the natural class in which it is possible to "glue" together local constructions in order to obtain a global one.

Proposition B.1. (i) *Paracompact spaces are normal.*

(ii) *Compact Hausdorff spaces are paracompact.*

(iii) *Metrizable spaces are paracompact.*

Parts (i) and (ii) follow directly from the definitions. Part (iii) is due to A. H. Stone [**545**]. A simpler proof was given by M. E. Rudin [**505**]; see also Engelking [**179**], page 349. (That separable metric spaces are paracompact is trivial; see Proposition B.3 below.)

A locally finite partition of unity on X is a family $\{f_\alpha : \alpha \in A\}$ of continuous functions from X to $[0,1]$ such that

(i) the sets $\{x \in X : f_\alpha(x) > 0\}$ form a locally finite open cover of X,

(ii) $\sum_\alpha f_\alpha(x) = 1$ for all $x \in X$.

Note that each point $x \in X$ has a neighborhood U where the sum in (ii) is a finite sum. We shall say that the partition of unity $\{f_\alpha\}$ is subordinated to the cover $\mathcal{V}$ if each of the sets $\{x \in X : f_\alpha(x) > 0\}$ is contained in some $V \in \mathcal{V}$; i.e., when the cover consisting of these sets is a refinement of $\mathcal{V}$.

Proposition B.2. *Let X be a paracompact space. Then for every open cover $\mathcal{V}$ of X there is a locally finite partition of unity, subordinated to $\mathcal{V}$.*

Proof. We first refine $\mathcal{V}$ to a locally finite open cover $\mathcal{W}$ with the property that for each $W \in \mathcal{W}$ there is a closed set $F_W \subset W$ such that

$$\bigcup\{F_W : W \in \mathcal{W}\} = X.$$

To this end we may assume (since X is paracompact) that the original cover $\mathcal{V}$ is already locally finite. For each $x \in X$ choose an open set U_x containing it so that $\overline{U_x} \subset V$ for some $V \in \mathcal{V}$, and let $\mathcal{U}$ be a locally finite cover of X which refines the cover $\{U_x : x \in X\}$.

By the construction, we can choose for every $U \in \mathcal{U}$ a set $V_U \in \mathcal{V}$ such that the closure of U satisfies $\overline{U} \subset V_U$, and we put $\mathcal{W} = \{W = V_U : U \in \mathcal{U}\}$.

As a subfamily of the locally finite cover $\mathcal{V}$, the cover $\mathcal{W}$ is itself locally finite. For each $W \in \mathcal{W}$, put $F_W = \bigcup\{\overline{U} : U \in \mathcal{U} \text{ and } V_U = W\}$. Since $\mathcal{V}$ is locally finite, every $x \in X$ has an open neighborhood that intersects only finitely many of the closed sets $\{\overline{U} : U \in \mathcal{U} \text{ and } V_U = W\}$. Thus F_W is, indeed, closed.

By Urysohn's Lemma we can find, for each $W \in \mathcal{W}$, a continuous function $g_W : X \to [0,1]$ such that $g_W(x) = 1$ for $x \in F_W$ and $g_W(x) = 0$ for $x \notin W$. Since $\mathcal{W}$ is a locally finite cover, the function $g = \sum_{W\in\mathcal{W}} g_W$ is a well-defined strictly positive and continuous function on X. The required partition of unity is then $\{f_W = g_W/g : W \in \mathcal{W}\}$. □

For more information on paracompact spaces and on partitions of unity, see Chapter 5 in Engelking [**179**].

Proposition B.2 cannot give partitions of unity with some additional prescribed properties because there is no control on the geometry of the sets W and F_W. We now present another approach which yields such partitions of unity on separable metric spaces. (The example to keep in mind is of partitions of unity with C^1 or C^k functions in case differentiation on X makes sense.)

Let X be a separable metric space. If $B = B(x,r)$ is an open ball in X, we denote the ball $B(x, 2r)$ by $2B$. Let $\mathcal{F}$ be an algebra of continuous functions on X such that

(i) For each ball B in X there is a function $f : X \to [0,1]$ in $\mathcal{F}$ with $f(x) = 1$ for $x \in B$ and $f(x) = 0$ for $x \notin 2B$.

(ii) If $f, g \in \mathcal{F}$, and if $g(x) > 0$ for every $x \in X$, then $f/g \in \mathcal{F}$.

(iii) If $f_i \in \mathcal{F}$, and if the family $\{x : f_i(x) \neq 0\}$ is locally finite, then the function $f = \sum_{i=1}^{\infty} f_i$ belongs to $\mathcal{F}$.

Proposition B.3. *Let X be a separable metric space, and let $\mathcal{F}$ be a family of functions on X as above. Then for every open cover $\mathcal{V}$ of X there is a locally finite partition of unity, consisting of functions from $\mathcal{F}$, subordinated to $\mathcal{V}$.*

Proof. For each $x \in X$ choose a ball B containing x such that there is a $V \in \mathcal{V}$ with $2B \subset V$. Since X is separable, there is a countable family $\{B_i\}_{i=1}^{\infty}$ of these balls such that $\bigcup B_i = X$.

For each i choose $h_i : X \to [0,1]$ in $\mathcal{F}$ with $h_i(x) = 1$ on B_i and $h_i(x) = 0$ for $x \notin 2B_i$, and put

$$g_n = h_n \prod_{i<n} (1 - h_i).$$

Since $g_n(x) = 0$ for $x \in B_i$ whenever $i < n$, it follows that $\{x : g_n(x) \neq 0\}$ is locally finite; hence $g = \sum g_n \in \mathcal{F}$, and it is clearly strictly positive. The functions $f_n = g_n/g$ are the required partition of unity. □

C. Invariant Means

Let G be a semigroup. A left-invariant mean on G is a linear functional M on $l_\infty(G)$ such that

(i) $M(1) = 1$.

(ii) $M(f) \geq 0$ for every $f \geq 0$.

(iii) $M(f_g) = M(f)$ for every $g \in G$, where f_g is the left-translation of f by g; i.e., $f_g(h) = f(gh)$ for every $h \in G$.

Note that (i) and (ii) are equivalent to (i) and $\|M\| = 1$.

We shall usually denote the invariant mean by M_x or $M_x(\cdot)$, emphasizing that the mean is taken with respect to the variable x.

The first examples of invariant means are due to Banach [**36**], who constructed what we now call Banach limits on $\mathbb{R}$ and $\mathbb{Z}$. We shall need invariant means only on abelian semigroups. For more information on invariant means and on their applications, see Greenleaf [**228**], Wagon [**579**] and Paterson [**449**].

Theorem C.1. *Let G be an abelian semigroup. Then G admits an invariant mean.*

Proof. We start with the observation that if W is a convex and compact set in a linear topological space, and if $T : W \to W$ is affine and continuous, then the set F_T of fixed points of T is nonempty. To see this, fix any point $x \in W$, and let $x_n = (\sum_{i=0}^{n-1} T^i x)/n$. Since $Tx_n - x_n = (T^n x - x)/n \to 0$, it follows that any accumulation point of the sequence $\{x_n\}$ is fixed by T; hence F_T is nonempty, and it is clearly convex and compact.

Observe that F_T is invariant under any affine continuous map S of W which commutes with T. Indeed, if $x \in F_T$, then $T(Sx) = S(Tx) = Sx$. Thus Sx is fixed by T and belongs to F_T. Since by the first argument S, as a map on the closed and convex set F_T, has a fixed point, we see that $F_S \cap F_T \neq \emptyset$. By induction $\bigcap F_T \neq \emptyset$ for any finite collection of commuting continuous affine maps on W.

Now let $W = \{0 \leq \varphi \in l_\infty^*(G) : \langle \varphi, 1 \rangle = 1\}$. The set W is convex and w^*-compact. The elements of G act on W as affine continuous maps by translation: If $\varphi \in W$, we define $g(\varphi)$ by the formula $\langle g(\varphi), f \rangle = \langle \varphi, f_g \rangle$ for every $f \in l_\infty(G)$.

By the observations above, the fixed points sets $\{F_g : g \in G\}$ in W are nonempty, and since G is abelian, they have the finite intersection property. Since W is compact, they have a common intersection, and any point M in this intersection is fixed under all of G; hence it is an invariant mean. □

Remark. In locally convex spaces F_T is nonempty also for nonlinear maps; see Theorem 3.1. The affine case is much simpler, and therefore we gave here the direct argument (which, by the way, does not require the space to be locally convex).

We shall also need to apply invariant means to bounded Banach space-valued functions. It follows from the theorem that this is possible when the space is a dual space.

Corollary C.2. *Let G be a semigroup that admits a left-invariant mean M, and let E be a dual Banach space. Then M induces an operator (which we also denote by M) from $l_\infty(G, E)$, the space of bounded E-valued functions on G, into E, satisfying*

(i) *$M(\hat{x}) = x$ for every $x \in E$, where $\hat{x}$ is the function that takes the constant value x.*

(ii) *$\|M\| = 1$.*

(iii) *$M(f_g) = M(f)$ for every $g \in G$.*

Proof. Assume that $E = F^*$, and define $M(f)$ by the formula

$$\langle M(f), y \rangle = M_x\big(\langle f(x), y \rangle\big)$$

for every $y \in F$. □

D. Measure and Probability

Let S be a complete separable metric space. By a measure on S we shall always mean a scalar-valued σ-additive measure defined on the Borel sets of S, i.e., on the σ-algebra generated by the open subsets of S. The set of probability measures on S is denoted by $\mathcal{P}(S)$. The algebra of Borel subsets of the product $S_1 \times S_2$, where S_1 and S_2 are separable metric spaces, is the product of the Borel algebras of the factors. This is often used implicitly, e.g., when Fubini's theorem is applied.

Proposition D.1. *Every measure $\mu \in \mathcal{P}(S)$ is regular and tight; i.e., for every Borel set A and for every $\varepsilon > 0$ there are an open set $G \supset A$ and a compact set $K \subset A$ such that $\mu(G \setminus K) < \varepsilon$.*

Proof. We first show that there are $F \subset A \subset G$, with F closed and G open, such that $\mu(G \setminus F) < \varepsilon$. Since the family of sets satisfying the claim is a σ-algebra, and since the σ-algebra generated by the closed sets is the algebra of all Borel sets, it suffices to check this for a closed set A. But we can then take $F = A$, and since $A = \bigcap_{\delta>0}\{x : d(x, A) < \delta\}$, we can take $G = \{x : d(x, A) < \delta\}$ for small enough δ.

We can thus assume that A in the proposition is closed and we need to find K. By the separability of S we can cover A, for each n, by a countable collection of closed balls $A_{i,n}$ of radius $1/n$. Put $B_{i,n} = A \cap A_{i,n}$. By the σ-additivity of μ there is an i_n such that $\mu\big(\bigcup_{i\leq i_n} B_{i,n}\big) > \mu(A) - \varepsilon/2^n$. Thus $\mu\big(\bigcap_{n\geq 1}\bigcup_{i\leq i_n} B_{i,n}\big) > \mu(A) - \varepsilon$. This latter set is closed and is totally bounded by the construction, since S is complete, it is actually compact. □

Let μ be a nonnegative finitely additive measure, defined on an algebra $\mathcal{F}$ of subsets of S, with $\mu(S) = 1$. Then μ extends to a σ-additive measure on the σ-algebra generated by $\mathcal{F}$ if and only if $\mu(A_n) \to 0$ whenever $\{A_n\}$ is a decreasing sequence of sets in $\mathcal{F}$ with empty intersection.

We now discuss measures on product spaces. Let I be an infinite set, and consider a collection $\{\mu_A : A \text{ is a finite subset of } I\}$ of probability measures, where each μ_A is a measure on $\prod_{i\in A} S_i$. Assume that the collection is consistent in the sense that whenever $A \subset B$, the measure μ_A is the projection of μ_B. Then there is a probability measure on the infinite product $\prod_{i\in I} S_i$ whose projection on each $\prod_{i\in A} S_i$ is μ_A. This is Kolmogorov's consistency theorem (see, e.g., Billingsley [**64**], page 433). The finitely additive measures μ, defined on the algebra generated by the finite products, is well defined by the consistency condition, and the key to the proof of Kolmogorov's theorem is that it satisfies the criterion above. Note that the existence of infinite product measures is a special case of this theorem.

The signed Borel measures on S induce bounded linear functionals on the space $C_B(S)$ of real-valued bounded continuous functions on S, and the norm of the functional induced by μ is its variation. Following the common terminology in the probability literature, we shall say that a sequence of probability measures $\mu_n \in \mathcal{P}(S)$ converges weakly to a probability measure μ if $\int f d\mu_n \to \int f d\mu$ for every bounded continuous function f on S, i.e., in the ω^*-topology on the dual of $C_B(S)$.

In the topology of weak convergence $\mathcal{P}(S)$ is a complete metric space. One possible complete metric on $\mathcal{P}(S)$ is the so-called Lévy metric

$$\rho(\mu, \nu) = \inf\big\{\delta : \mu(A) \leq \nu(A_\delta) + \delta \text{ and } \nu(A) \leq \mu(A_\delta) + \delta\big\}$$

where $A_\delta = \{x \in S : d(x, A) \leq \delta\}$.

The next theorem is due to Prohorov.

Theorem D.2. *Let S be a complete separable metric space. A set $M \subset \mathcal{P}(S)$ is relatively compact in the topology of weak convergence of measures if and only if it is tight; i.e., for every $\varepsilon > 0$ there is a compact set $K \subset S$ such that $\mu(K) > 1 - \varepsilon$ for every $\mu \in M$.*

Proof. Assume that $\{\mu_n\}$ is a tight sequence of measures. Let $\{K_m\}$ be an increasing sequence of compact sets such that $\mu_n(K_m) \geq 1 - 1/m$ for all n and m.

For each fixed m the ω^*-compactness of the unit ball of $C(K_m)^*$ gives a subsequence of the restrictions of the measures to K_m such that $\mu_{n_j|K_m}$ is ω^*-convergent in $C(K_m)^*$ to a measure ν_m on K_m. By a standard diagonal argument, we can pass to a subsequence (and assume that it is the original sequence) so that $\mu_{n|K_m}$ converges ω^* in $C(K_m)^*$ to a measure ν_m on K_m for every m. If $m < l$, then $\|\mu_{n|K_m} - \mu_{n|K_l}\| \leq 1/m$ for every n, hence also $\|\nu_m - \nu_l\| \leq 1/m$. It follows that $\{\nu_m\}$ is a norm-Cauchy sequence, and hence norm-convergent to some probability measure ν on S. One checks easily that the μ_n's converge weakly to ν.

The converse is a modification of the proof of Proposition D.1. Assume that M is relatively compact. Since S is separable, it can be covered, for each n, by a countable collection of open balls $B_{i,n}$ of radius $1/n$. We claim that for every $\varepsilon > 0$ there is an i_n (depending only on ε) such that $\mu\big(\bigcup_{i\leq i_n} B_{i,n}\big) > 1 - \varepsilon/2^n$ for every $\mu \in M$.

Indeed, if this were not so, we could find for each j a measure $\mu_j \in M$ such that $\mu_j\big(\bigcup_{i\leq j} B_{i,n}\big) \leq 1 - \varepsilon/2^n$. Since M is relatively compact, we can assume that the μ_j's converge weakly to some probability measure μ. The measure μ is tight, and thus there is a compact set $K \subset S$ such that $\mu(K) > 1 - \varepsilon/2^n$. Choose j_0 such that K is contained in $\bigcup_{i\leq j_0} B_{i,n}$ and a continuous function $0 \leq f \leq 1$, supported in $\bigcup_{i\leq j_0} B_{i,n}$, such that $f \equiv 1$ on K. Then $\int f d\mu > 1 - \varepsilon/2^n$, hence also $\int f d\mu_j > 1 - \varepsilon/2^n$ for large j, a contradiction.

Thus $\mu\big(\bigcap_{n\geq 1}\bigcup_{i\leq i_n} B_{i,n}\big) > 1 - \varepsilon$ for every $\mu \in M$. By the construction this latter set is totally bounded, and since S is complete, its closure is compact. □

For more information on probability measures on metric spaces, see Billingsley [**63**]; Shiryayev [**532**], Chapter 3; and Stroock [**547**], Section 3.1.

We shall need information on the structure of Borel sets in metric spaces. In most cases we shall need only the fact that a general Borel set can be obtained from the open (or from the closed) sets by countable transfinite iteration of the operations of complementation and countable unions. (Or, alternatively, by countable transfinite iteration of the operations of countable intersection of decreasing sequences of sets and countable unions of increasing sequences of sets.) At one point we shall need, however, deeper facts on Borel sets in complete metric spaces.

Consider the space $\mathbb{N}^{\mathbb{N}}$ of infinite sequences of positive integers. Taking the discrete topology on $\mathbb{N}$, the product is a complete metric space, which is actually homeomorphic to the set of irrational real numbers (via the expansion as continued fractions). We denote the elements of $\mathbb{N}^{\mathbb{N}}$ by $\overline{m} = (m_1, m_2, \ldots)$. The following theorem can be found, e.g., in Kechris [**323**], page 89.

Theorem D.3. *A subset A of a complete separable metric space is Borel if and only if there is a continuous surjective map φ from some closed subset B of $\mathbb{N}^{\mathbb{N}}$ onto A.*

We next use this theorem to give a very convenient representation (via the so-called Souslin operation) of Borel sets in a complete metric space.

Let A be a Borel set, and let B and φ be as in the theorem. Given a finite set of positive integers $\{n_i\}_{i=1}^k$, put

$$A_{n_1,\ldots,n_k} = \varphi\{\overline{m} : \ m_i \leq n_i, \ \ i \leq k\}.$$

By the theorem each of these sets is again a Borel set, and clearly

$$A_{n_1,\dots,n_k} = \bigcup_{n=1}^{\infty} A_{n_1,\dots,n_k,n}$$

and

$$A = \bigcup_{\overline{n}\in\mathbb{N}^{\mathbb{N}}} \bigcap_{k=1}^{\infty} A_{n_1,\dots,n_k}.$$

Note that this is a representation as an uncountable union. The crucial property of this representation for our purposes is that if $F_{n_1,\dots,n_k}$ denotes the closure of $A_{n_1,\dots,n_k}$, then $\bigcap_{k=1}^{\infty} F_{n_1,\dots,n_k} \subset A$ for each $\overline{n} = (n_1, n_2, \dots) \in \mathbb{N}^{\mathbb{N}}$. Indeed, if x belongs to the intersection, then for each k there is a point $\overline{m}^k \in B$ such that $m_i^k \leq n_i$ for $i \leq k$ and such that $\rho\big(x, \varphi(\overline{m}^k)\big) \leq 1/k$. The sequence $\{\overline{m}^k\}$ has a subsequence which converges to some point $\overline{m}$, and since B is closed, $\overline{m} \in B$. Clearly $\varphi(\overline{m}) = x$.

Note that our description of the Souslin operation differs slightly from the standard description. Usually the sets $A_{n_1,\dots,n_k}$ in the representation are required to be closed and are not required to be contained in A.

The book by Kechris mentioned above contains a very detailed treatment of Borel sets and related material.

Let X be a random variable on a probability space $(\Omega, \mathcal{B}, \mu)$. Its distribution (or law) $\nu = \nu_X$ is the measure $\nu(B) = \mu\big(X^{-1}(B)\big)$ on $\mathbb{R}$ (or on the space E when X is E-valued). Sometimes it is convenient to specify the probability space and the random variable, but usually we shall only be interested in its distribution.

In this appendix we shall usually discuss only scalar-valued random variables. It will be clear from the context (and we shall occasionally indicate this explicitly) which notions and results can be easily generalized to $\mathbb{R}^n$-valued or, more generally, to Banach space-valued variables.

We shall say that a sequence of random variables X_n converges in distribution to a random variable X if their distributions, ν_{X_n}, converge weakly to the distribution ν_X of X.

Note that convergence in distribution does not require the random variables to even be defined on the same probability space. Even when they are, convergence in distribution does not give information on the random variables themselves. One exceptional case (when the random variables are defined on the same space) is that X_n converge to a constant in distribution if and only if they converge to it in probability. (In the text we use the terms convergence in probability and convergence in measure interchangeably.)

The expectation of X is denoted by $\mathbb{E}X = \int_{\mathbb{R}} x d\nu_X(x)$. We shall also use the notation $\int_\Omega X d\mu$ when Ω and μ are specified.

A family of random variables $\{X_i\}_{i\in I}$ is called independent if

$$\mu\Big(\bigcap_{i\in J}\{X_i \in B_i\}\Big) = \prod_{i\in J}\mu\Big(\{X_i \in B_i\}\Big)$$

for every finite subset $J \subset I$ and for every choice of Borel sets $B_i \subset \mathbb{R}$. A sequence of independent and identically distributed random variables $\{X_i\}$ is said to be an i.i.d. sequence.

If $\{X_i\}_{i \in I}$ are random variables on the probability spaces $(\Omega_i, \mathcal{B}_i, \mu_i)$, then the functions Y_i, defined on the product space $\prod \Omega_i$ by $Y_i(\omega) = X_i(\omega_i)$ are independent random variables with respect to the product measure. They are independent copies of the X_i's in the sense that each Y_i has the same distribution as the corresponding X_i. Every independent family can be realized in this way.

Let $(\Omega, \mathcal{B}, \mu)$ be a probability space, and let $\{\mathcal{B}_i\}$ be σ-subalgebras of $\mathcal{B}$. The tail algebra of the sequence is $\mathcal{B}_0 = \bigcap_n \mathcal{A}_n$, where $\mathcal{A}_n$ is the σ-algebra generated by $\{\mathcal{B}_i\}_{i>n}$. The zero-one law says that if the algebras $\mathcal{B}_i$ are independent (i.e., $\mu(\bigcap_{i=1}^n A_i) = \prod_{i=1}^n \mu(A_i)$ for all $A_i \in \mathcal{B}_i$ and $i \leq n$), and if $A \in \mathcal{B}_0$, then $\mu(A) = 0$ or $\mu(A) = 1$. Indeed, if B belongs to the algebra generated by $\{\mathcal{B}_i\}_{i \leq n}$, then, since $A \in \mathcal{A}_n$, it follows from independence that $\mu(A \cap B) = \mu(A)\mu(B)$. But A itself is a limit of such B's; hence $\mu(A) = \mu(A)^2$.

Assume that $\{X_i\}$ are random variables on some probability space. Typical examples of tail events are the set where the X_i's converge and when they do converge to a limit X, say, sets of the form $\{X > \alpha\}$. It follows from the zero-one law that if the X_i are independent, then they either converge almost everywhere or almost nowhere and in the first case the limit is a constant almost everywhere.

Let μ and ν be two probability measures on the same space. Let λ be a nonnegative measure such that both μ and ν are absolutely continuous with respect to λ (e.g., $\lambda = \mu + \nu$). The Hellinger integral of μ and ν is

$$H(\mu, \nu) = \int \sqrt{d\mu/d\lambda}\sqrt{d\nu/d\lambda}\, d\lambda.$$

It is easy to check that H does not depend on the choice of λ. By Cauchy-Schwartz $0 \leq H(\mu, \nu) \leq 1$ and $H(\mu, \nu) = 1$ if and only if $\mu = \nu$. Also $H(\mu, \nu) = 0$ if and only if μ and ν are mutually singular. The following theorem is due to Kakutani [**309**].

Theorem D.4. *For each i let μ_i and ν_i be two equivalent probability measures. Then the two product measures $\mu = \prod \mu_i$ and $\nu = \prod \nu_i$ are mutually equivalent if and only if $\prod H(\mu_i, \nu_i) > 0$. If $\prod H(\mu_i, \nu_i) = 0$, then μ and ν are singular with respect to each other.*

The characteristic function of an $\mathbb{R}^n$-valued random variable X is the function $\varphi_X(x) = \mathbb{E}e^{i\langle x, X\rangle} = \int_{\mathbb{R}^n} e^{i\langle x, y\rangle} d\nu_X(y)$ on $\mathbb{R}^n$. Up to conventional normalizations, it is the Fourier transform of the measure ν_X. The characteristic function of a random variable depends only on its distribution, and if the latter is ν, we shall sometime denote the characteristic function by φ_ν. The characteristic function φ_X is positive definite and uniformly continuous. It satisfies $|\varphi_X(x)| \leq 1$ for every x, $\varphi_X(0) = 1$ and $\varphi_{aX}(x) = \varphi_X(ax)$ for $a \in \mathbb{R}$ and for every $x \in \mathbb{R}^n$. (Recall that a function φ is positive definite if $\sum \varphi(x_i - x_j) c_i \bar{c}_j \geq 0$ for all choices of finitely many complex numbers $\{c_j\}$ and finitely many vectors $\{x_j\}$ in $\mathbb{R}^n$. See Section 1 of Chapter 8 for more information on positive definite functions.) Also, if ν is a measure on $\mathbb{R}^n$, and if μ is its projection on $\mathbb{R}^k$ for some $k < n$, then φ_ν is an extension of φ_μ.

The next proposition summarizes some of the properties of characteristic functions.

Theorem D.5. (i) *The distribution ν_X is uniquely determined by its characteristic function.*

(ii) *Every continuous positive definite function φ on $\mathbb{R}^n$ with $\varphi(0) = 1$ is the characteristic function of some $\mathbb{R}^n$-valued random variable. In fact, it is enough to assume that the real part of φ is continuous at 0.*

(iii) X *and* Y *are independent if and only if* $\varphi_{X+Y} = \varphi_X \varphi_Y$.

(iv) $\varphi_{\nu_n} \to \varphi_\nu$ *pointwise if and only if* $\{\nu_n\}$ *converges weakly to* ν. *In particular, since weak convergence and convergence in measure agree when the limit is constant,* X_n *converge in probability to a constant* c *if and only if* $\varphi_{X_n}(t) \to e^{ict}$ *for every* $t \in \mathbb{R}$.

Part (ii) is a theorem of Bochner; see Rudin [**506**], page 19; Feller [**185**], page 585; or Chung [**122**], page 179. For proofs and more details on the other parts, see Chapter 15 in Feller's book, Chapter 6 in Chung's book, and Section 26 in Billingsley [**64**].

The Rademacher functions on $[0,1]$ are defined by $r_j(t) = \operatorname{sign} \sin 2^j \pi t$. Considered as random variables on $[0,1]$ with the Lebesgue measure, they are a model of an i.i.d. sequence, with each variable taking the values ± 1 with probability $\frac{1}{2}$.

We now estimate the tail of the distribution of $|\sum a_j r_j|$ and also give a lower estimate on the probability that the sum will take relatively small values. Denote $\left(\sum a_j^2\right)^{1/2}$ by $\|a\|_2$, and notice that $\|\sum a_j r_j\|_2 = \|a\|_2$.

Proposition D.6. (i) $m\{t : |\sum a_j r_j| \geq \lambda \|a\|_2\} \leq 2e^{-\lambda^2/2}$ *for every* $\lambda > 0$.

(ii) $m\{t : |\sum a_j r_j| \geq \lambda \|a\|_2\} \geq (1-\lambda^2)^2/3$ *for every* $0 < \lambda < 1$. *In particular, taking* $\lambda = 1/4$, *we find that if* $m\{t : |\sum a_j r_j| \leq K/4\} \geq 3/4$ *for some constant* K, *then* $\|a\|_2 \leq K$.

Proof. (i) We can assume that $\|a\|_2 = 1$. For every $x \in \mathbb{R}$ we have the estimate $\mathbb{E}e^{xr_j} = \frac{1}{2}(e^x + e^{-x}) \leq e^{x^2/2}$. By the symmetry and independence of the r_j's and by Chebyshev's inequality in the form $m\{e^X > t\} \leq t^{-1}\mathbb{E}e^X$, we obtain

$$\begin{aligned}
m\left\{t : \left|\sum a_j r_j\right| \geq \lambda\right\} &= 2m\left\{t : \sum a_j r_j \geq \lambda\right\} \\
&= 2m\left\{t : \exp\left(\lambda \sum a_j r_j\right) \geq e^{\lambda^2}\right\} \\
&\leq 2e^{-\lambda^2} \mathbb{E} \exp\left(\lambda \sum a_j r_j\right) \\
&= 2e^{-\lambda^2} \prod \mathbb{E} e^{\lambda a_j r_j} \\
&\leq 2\exp\left(-\lambda^2 + \sum \lambda^2 a_j^2/2\right) = 2e^{-\lambda^2/2}.
\end{aligned}$$

(ii) Assume that $\|a\|_2 = 1$, and put $A = \{t : |\sum a_j r_j| \geq \lambda\}$ and $B = [0,1] \setminus A$. Then

$$\begin{aligned}
1 - \lambda^2 &\leq \int_0^1 \left|\sum a_j r_j\right|^2 dt - \int_B \left|\sum a_j r_j\right|^2 dt \\
&= \int_A \left|\sum a_j r_j\right|^2 dt \leq m(A)^{1/2} \Big(\int_0^1 \left|\sum a_j r_j\right|^4 dt\Big)^{1/2}.
\end{aligned}$$

Since the Rademacher functions are orthonormal, we obtain

$$\int_0^1 \left|\sum a_j r_j\right|^4 dt = \sum a_j^4 + 6 \sum_{i<j} a_i^2 a_j^2 \leq 3\left(\sum a_j^2\right)^2 = 3.$$

Substituting in the previous inequality gives the result. □

Part (i) is a standard and commonly used estimate. Part (ii) is due to Paley and Zygmund.

The next result is Khintchine's inequality.

Proposition D.7. *For every $0 < p < \infty$ there are constants A_p and B_p such that*

$$A_p\Big(\int_0^1 |\sum a_j r_j|^p\Big)^{1/p} \le \Big(\int_0^1 |\sum a_j r_j|^2\Big)^{1/2} \le B_p\Big(\int_0^1 |\sum a_j r_j|^p\Big)^{1/p}.$$

Proof. This follows immediately from the tail estimate in part (i) of Proposition D.6 and from the formula $\mathbb{E}|X|^p = \int_{\mathbb{R}} |x|^p d\nu_X(x)$. □

Remark. Since $\int_0^1 |\sum a_j r_j|^2 = \sum |a_j|^2$, another way to state the proposition is that for every $0 < p < \infty$ the Rademacher functions are equivalent in L_p to the unit basis vectors in l_2. It is easy to check that it also follows that the orthogonal projection from L_2 onto the span of the Rademacher functions is bounded on L_p for every $1 < p < \infty$; hence the space L_p contains for these values of p a complemented subspace isomorphic to a Hilbert space. It also follows that l_p contains uniformly complemented copies of l_2^n. It follows that for $1 < p < \infty$ every $L_p(\mu)$ space either contains an infinite-dimensional Hilbert space or, at least, uniformly complemented copies of l_2^n.

There are many other proofs of Khintchine's inequality. One such proof actually follows from the method of estimating $\|\sum a_j r_j\|_4$ in the proof of part (ii) of Proposition D.6. The estimate we gave proves the inequality for $p = 4$, and the same method works for every p which is an even integer. The general case follows from the monotonicity of the L_p norm and from Hölder's inequality.

The best value of the constant A_1 is $\sqrt{2}$ (of course $B_1 = 1$). This was proved by Szarek [**549**]. For other values of p the exact constants were computed by Haagerup [**242**].

The standard one-dimensional Gaussian (or normal) distribution is the measure on $\mathbb{R}$ whose density is $(2\pi)^{-1/2}e^{-x^2/2}$. The measure γ_n on $\mathbb{R}^n$ whose density is $(2\pi)^{-n/2}e^{-\|x\|^2/2}$ is called the standard Gaussian measure on $\mathbb{R}^n$. More generally, if S is a symmetric positive invertible operator on $\mathbb{R}^n$, and if m is a vector in $\mathbb{R}^n$, then the measure with density

$$(2\pi)^{-n/2}|S^{-1}|e^{-\frac{1}{2}\big\langle (x-m), S^{-1}(x-m)\big\rangle}$$

(where $|S|$ denotes the determinant of the matrix of S) is called a nondegenerate Gaussian measure on $\mathbb{R}^n$ with mean m and covariance operator S. The characteristic function of this measure is $\varphi(y) = e^{i\langle y,m\rangle - \langle Sy,y\rangle/2}$. In the one-dimensional case the density is usually written in the form $(2\pi)^{-1/2}\sigma^{-1}e^{-(x-m)^2/2\sigma^2}$, and σ is called the standard deviation of the distribution. It is equal, when $m = 0$, to the L^2 norm of the random variable. A distribution which is concentrated in some proper affine subspace of $\mathbb{R}^n$ and which is a nondegenerate Gaussian distribution on this subspace is called a degenerate Gaussian distribution on $\mathbb{R}^n$. The degenerate Gaussian distributions on the line are thus just the dirac measures (and we conventionally treat them as Gaussian measures with $\sigma = 0$).

An $\mathbb{R}^n$-valued random variable is called Gaussian if its distribution is a Gaussian distribution. When $\mathbb{E}X = 0$, we shall say that X is a centered, or symmetric, Gaussian variable.

It follows from the definition that the standard Gaussian measure is invariant under the orthogonal group on $\mathbb{R}^n$. It also follows easily, by considering the characteristic functions, that a linear combination of independent Gaussian variables is also Gaussian.

Let X be a centered Gaussian variable with covariance operator S. Since S is symmetric, there are orthogonal coordinates in $\mathbb{R}^n$ with respect to which its matrix is diagonal. In this case the coordinate functionals are independent one-dimensional centered Gaussian variables. By translation, we find that any $\mathbb{R}^n$-valued Gaussian variable can be represented as the sum of n independent one-dimensional Gaussian variables.

If $T : \mathbb{R}^n \to \mathbb{R}^m$ is a linear transformation, and if γ is a Gaussian measure on $\mathbb{R}^n$ with covariance operator S, then $T\gamma$ is a Gaussian measure on $\mathbb{R}^m$, and its covariance operator is $TS^{-1}T^*$, where T^* is the adjoint of T. (If $m > n$, then $T\gamma$ is, of course, degenerate and is concentrated in the image of T.)

A measure on $\mathbb{R}^n$ is Gaussian if and only if all its one-dimensional projections are Gaussian. (See Proposition 6.18 for a proof that holds even in infinite-dimensional spaces.)

Let $\{g_j\}$ be i.i.d. standard one-dimensional Gaussian variables on some probability space (Ω, μ). Then $\sum a_j g_j$ is a symmetric Gaussian variable with standard deviation $(\sum a_j^2)^{1/2}$. Hence $\left(\mathbb{E}|\sum a_j g_j|^p\right)^{1/p} = (\sum a_j^2)^{1/2}\left(\mathbb{E}|g_1|^p\right)^{1/p}$; i.e., the g_j's span a subspace of $L_p(\mu)$ which is isometric to a Hilbert space (and $\{g_j\}$ is an orthogonal basis for this Hilbert space).

For more information on Gaussian variables, see Feller [**185**] and Billingsley [**64**].

A nondegenerate random variable X is called a symmetric stable variable if for every two independent random variables X_1 and X_2 with the same distribution as X and for every $a_1, a_2 \in \mathbb{R}$, there is a $b \in \mathbb{R}$ such that $a_1X_1 + a_2X_2$ has the same distribution as bX. (Taking $a_1 = 1$ and $a_2 = -1$, we see that X is indeed symmetric.) The next theorem summarizes the facts we need on symmetric stable distributions.

Theorem D.8. (i) *For each $0 < p \leq 2$ there is a symmetric stable random variable whose characteristic function is $e^{-|t|^p}$.*

(ii) *Every symmetric stable variable is a multiple of one of the variables in* (i). *We call a distribution with characteristic function $e^{-c|t|^p}$, for some constant $c > 0$, a p-stable law.*

(iii) *If $p < 2$ and X is p-stable, then $\mathbb{E}|X|^r < \infty$ if and only if $r < p$.*

(iv) *For $p > 1$ denote by X_p a p-stable variable normalized such that $\mathbb{E}|X_p| = 1$, and write its characteristic function as $e^{-c_p|t|^p}$. Then $\lim_{p\to 1} c_p = 0$.*

Proof. The Gaussian distribution satisfies the requirements for $p = 2$. We shall thus be interested only in the case $p \neq 2$.

(i) We need to show that $e^{-|t|^p}$ is a characteristic function. It is clear that it is then the characteristic function of a symmetric stable distribution.

Let $\{Z_i\}_{i=1}^n$ be i.i.d. variables which are uniformly distributed in the interval $[-n, n]$, and let

$$Y_n = \sum_{i=1}^{n} |Z_i|^{-1/p}\text{sign}Z_i.$$

Write the characteristic function of $|Z_i|^{-1/p}\text{sign}Z_i$ in the form

$$\begin{aligned}\frac{1}{2n}\int_{-n}^{n}\exp\{it|x|^{-1/p}\text{sign}x\}dx &= 1-\frac{1}{n}\int_0^n(1-\cos tx^{-1/p})dx\\ &= 1-\frac{1}{n}\int_0^\infty(1-\cos tx^{-1/p})dx+o(1/n).\end{aligned}$$

(The integral converges because $p<2$.) Then the characteristic function of Y_n is

$$\Big(1-\frac{1}{n}\int_0^\infty(1-\cos tx^{-1/p})dx+o(1/n)\Big)^n\to\exp\Big\{-\int_0^\infty(1-\cos tx^{-1/p})dx\Big\}$$

which is equal to $e^{-c_p|t|^p}$ by the identity

$$|t|^r=\alpha_r\int_0^\infty(1-\cos ty)y^{-r-1}dy \tag{D.1}$$

which holds for every $0<r<2$ and which we use with $r=p$, substituting $y=x^{-1/p}$.

Being a limit of positive definite function, $e^{-|t|^p}$ is also positive definite, hence a characteristic function by Proposition D.5(ii). For another proof that $e^{-|t|^p}$ is positive definite, see Chapter 8, Section 1.

(ii) By the definition of the symmetric stable distribution, its characteristic function satisfies the functional equation $\varphi(a_1t)\varphi(a_2t)=\varphi\big(f(a_1,a_2)t\big)$ for some function f. Since we do not need this part explicitly, we refer to Lamperti [**346**], page 76, for the details of the proof that the only solutions of this functional equation are functions of the form $\varphi(t)=e^{-c|t|^p}$. Since φ is continuous and nonconstant, it follows that $p>0$. Finally, p cannot be bigger than 2, for otherwise φ would be twice differentiable with $\varphi''(0)=0$. If it were the characteristic function of a random variable X, this variable would satisfy $\mathbb{E}X^2=0$; i.e., X would be constant, which is impossible.

(iii) Since X is symmetric, its characteristic function is $\mathbb{E}\cos(tX)$. By (D.1) we obtain

$$\begin{aligned}\mathbb{E}|X|^r &= \alpha_r\mathbb{E}\int_0^\infty\big(1-\cos(tX)\big)t^{-r-1}dt=\alpha_r\int_0^\infty\big(1-\mathbb{E}\cos(tX)\big)t^{-r-1}dt\\ &= \alpha_r\int_0^\infty\big(1-e^{-c|t|^p}\big)t^{-r-1}dt\end{aligned}$$

and this integral is finite if and only if $r<p$.

(iv) Rather than estimate the value of c_p from the proofs above, we just note that if we could find $p_j\to1$ such that $c_{p_j}\to c\neq0$, then $e^{-c|t|}$ would be the characteristic function of a random variable X satisfying $E|X|\le1$. But this contradicts (iii) above. □

Stable distributions were introduced and studied by Paul Lévy in the early twenties. For another proof of (iii), based on estimation of the tail of the p-stable distribution, see exercise 4, page 186, in Chung [**122**].

Our main interest in p-stable laws is that they give a concrete embedding of l_p into L_r for $0<r<p\le2$. Let X_i be i.i.d. with p-stable distribution, and assume that $(\sum|a_i|^p)^{1/p}=1$. It follows that the characteristic function of $\sum a_iX_i$ is $\prod e^{-c|a_it|^p}=\exp\{-c\sum|a_it|^p\}=e^{-c|t|^p}$; i.e., the sum has the same distribution as the original sequence. It follows that if $r<p$ (so that X_i are in L_r), and if

they are normalized so that their norm in L_r is one, then they are isometrically equivalent in L_r to the unit vectors of l_p.

Let $\mathcal{U}$ be an ultrafilter such that $(l_p)_{\mathcal{U}}$ contains a subspace isometric to $L_p(0,1)$. Since $(L_r(0,1))_{\mathcal{U}}$ is isometric to $L_r(\nu)$ for some nonatomic measure ν (see Appendix F), and since every separable subspace of $L_r(\nu)$ is contained in some subspace isometric to $L_r(0,1)$, it follows that $L_p(0,1)$ is isometric to a subspace of $L_r(0,1)$ for every $0 < r < p \leq 2$.

We remark that there is a difference between the Gaussian case (i.e., $p = 2$) and $p < 2$. Gaussian i.i.d. variables span a Hilbertian subspace H in L_r for any $1 \leq r < \infty$, and the orthogonal projection on H is bounded for all $1 < r < \infty$. On the other hand, if $p < 2$, then l_p does not embed isomorphically into L_r for $r > p$. Moreover, if $r < p$, then any subspace of L_r which is isomorphic to l_p is not complemented. See Kadec and Pełczyński [**304**] for details.

Let $\mathcal{B}_0 \subset \mathcal{B}$ be two σ-algebras of subsets of Ω, and let μ be a probability measure on $(\Omega, \mathcal{B})$. The conditional expectation of $f \in L_1(\Omega, \mathcal{B}, \mu)$ with respect to $\mathcal{B}_0$, denoted by $\mathbb{E}_{\mathcal{B}_0} f$, is the function $g \in L_1(\Omega, \mathcal{B}_0, \mu)$ which satisfies

$$\int_A f d\mu = \int_A g d\mu$$

for every $A \in \mathcal{B}_0$. This function g, which, by definition, has to be $\mathcal{B}_0$ measurable, is unique (up to a set of measure 0). The conditional expectation exists by the Radon-Nikodým theorem, and it satisfies $\|\mathbb{E}_{\mathcal{B}_0} f\|_1 \leq \|f\|_1$. Thus $\mathbb{E}_{\mathcal{B}_0}$ is a contractive projection from $L_1(\Omega, \mathcal{B}, \mu)$ onto $L_1(\Omega, \mathcal{B}_0, \mu)$. Similarly, $\mathbb{E}_{\mathcal{B}_0}$ is a contractive projection from $L_p(\Omega, \mathcal{B}, \mu)$ onto $L_p(\Omega, \mathcal{B}_0, \mu)$ for every $1 \leq p \leq \infty$.

Assume that $\{\mathcal{B}_n\}_{n=1}^\infty$ is an increasing sequence of σ-algebras of subsets of Ω, and that μ is a probability measure on a σ-algebra $\mathcal{B}$ containing all the $\{\mathcal{B}_n\}_{n=1}^\infty$. (Usually $\mathcal{B}$ will be the algebra generated by the $\mathcal{B}_n$'s.) A sequence of functions $f_n \in L_1(\Omega, \mathcal{B}_n, \mu)$ is called a ($\mathcal{B}_n$-adapted) martingale if $f_m = \mathbb{E}_{\mathcal{B}_m} f_n$ for all $m < n$. If $f \in L_1(\Omega, \mathcal{B}, \mu)$, then the sequence $f_n = \mathbb{E}_{\mathcal{B}_n} f$ forms a $\mathcal{B}_n$-adapted martingale, and if $\mathcal{B}$ is the σ-algebra generated by $\{\mathcal{B}_n\}_{n=1}^\infty$, then $\|f_n - f\|_1 \to 0$. For simple functions f this follows directly from the definition, and the density of simple functions gives the general case. The convergence is also pointwise:

Proposition D.9. *A martingale which converges in the L_1 norm converges almost everywhere (to the same limit).*

The proof is based on the following "maximal" lemma.

Lemma D.10. *Let f_n be a $\mathcal{B}_n$-adapted martingale, and let $r > 0$. Then*

$$\limsup_n \int_{A_r} \big(|f_n(w)| - r\big) d\mu \geq 0$$

where $A_r = \{\omega : \sup_n |f_n(\omega)| > r\}$. In particular $r\mu(A_r) \leq \sup_n \|f_n\|_1$.

Proof. Put

$$A_r^m = \big\{w : \ |f_k(w)| \leq r \ \text{ for all } k < m \ \text{ and } |f_m(w)| > r\big\}.$$

Then $A_r^m \in \mathcal{B}_m$ and A_r is the disjoint union of the sets $\{A_r^m\}_{m=1}^\infty$. If $n > m$, then since $\mathbb{E}_{\mathcal{B}_m} |f_n| \geq |f_m|$ by the martingale property, we obtain

$$\int_{A_r^m} \big(|f_n| - r\big) d\mu \geq \int_{A_r^m} \big(|f_m| - r\big) d\mu$$

and thus

$$\int_{\bigcup_{m=1}^{n} A_r^m} (|f_n| - r)d\mu \geq \sum_{m=1}^{n} \int_{A_r^m} (|f_m| - r)d\mu \geq 0.$$

Since, on the other hand, $\big||f_n| - r\big| \leq r$ on A_r^m for $m > n$, it follows that

$$\Big|\int_{\bigcup_{m>n} A_r^m} (|f_n| - r)d\mu\Big| \leq r\mu\Big(\bigcup_{m>n} A_r^m\Big) \to 0,$$

as $n \to \infty$, and the lemma follows. □

Proof of Proposition D.9. Assume that the martingale $\{f_n\}$ converges to f in L_1, and fix $\varepsilon > 0$. It follows that there is an $m = m(\varepsilon)$ such that $\|f_n - f_m\|_1 \leq \varepsilon^2$ for every $n \geq m$. The sequence $\{f_n - f_m\}_{n=m}^{\infty}$ is a $\mathcal{B}_n$-adapted martingale, and applying Lemma D.10 to this martingale with $r = \varepsilon$ gives

$$\varepsilon\mu\Big\{w : \sup_{n\geq m} |f_n(w) - f_m(w)| > \varepsilon\Big\} < \sup_{n\geq m} \|f_n - f_m\|_1 < \varepsilon^2.$$

Since ε was arbitrary, it follows that $\{f_n\}$ converges almost everywhere. ♢

For more details on martingales, see, e.g., Neveu [**436**]; Shiryayev [**532**], Chapter 7; and Stroock [**547**], Chapter 5.

We conclude the appendix by quoting two well-known results from measure theory. The first is Liapounoff's theorem on the range of vector-valued measures. A positive measure μ is said to be nonatomic if every measurable set A with $\mu(A) > 0$ contains a measurable subset $B \subset A$ such that $\mu(A) > \mu(B) > 0$. A signed measure is nonatomic if its positive and negative parts are.

Theorem D.11. *Let $\{\mu_i\}_{i=1}^n$ be n finite nonatomic signed measures on the same measure space S. Then its range*

$$\{(\mu_1(A), \dots, \mu_1(A)) : A \subset S\}$$

is a compact and convex subset of $\mathbb{R}^n$.

See Lindenstrauss [**360**] for a simple proof.

The second result is the Brunn-Minkowski theorem, one of the basic tools in finite-dimensional geometry. We denote the Lebesgue measure on $\mathbb{R}^n$ by λ_n.

Theorem D.12. *Let A and B be two Lebesgue measurable subsets of $\mathbb{R}^n$ with finite measure. Then*

$$\lambda_n(A + B)^{1/n} \geq \lambda_n(A)^{1/n} + \lambda_n(B)^{1/n}$$

and

$$\lambda_n\big((A + B)/2\big) \geq \big(\lambda_n(A)\lambda_n(B)\big)^{1/2}.$$

For a proof see Milman and Schechtman [**420**], page 134, or Schneider [**521**], page 309. The second inequality follows immediately from the first because

$$\begin{aligned} \lambda_n\big((A + B)/2\big)^{1/n} &\geq \lambda_n(A/2)^{1/n} + \lambda_n(B/2)^{1/n} \\ &= \big(\lambda_n(A)^{1/n} + \lambda_n(B)^{1/n}\big)/2 \\ &\geq \big(\lambda_n(A)\lambda_n(B)\big)^{1/2n}. \end{aligned}$$

E. Bases and Lattices

All the Banach spaces in this appendix are assumed to be infinite-dimensional unless stated otherwise. Some parameters (e.g., the basis constant) have an obvious meaning also in the finite-dimensional case.

A sequence $\{x_n\}$ in a Banach space E is called a basis if every $x \in E$ has a unique expansion as a convergent series $x = \sum a_n x_n$, where $\{a_n\}$ are scalar coefficients. The basis is called normalized if $\|x_n\| = 1$ for every n. If the space E contains a sequence of finite-dimensional subspaces $\{E_n\}$ such that every $x \in E$ has a unique representation as $x = \sum x_n$, where $x_n \in E_n$ for every n, we say that E has a finite-dimensional decomposition (FDD).

A basis $\{x_n\}$ is called unconditional if the convergence of each of the expansions $x = \sum a_n x_n$ is unconditional; i.e., $x = \sum \lambda_n a_n x_n$ converges for every bounded sequence $\{\lambda_n\}$. Equivalently, $\sum a_{\pi(n)} x_{\pi(n)}$ converges for every permutation π of the natural numbers, or $\sum_{n \in M} a_n x_n$ converges for every subset M of $\mathbb{N}$. Unconditional FDD are defined in a similar way.

A basis $\{x_n\}$ is called symmetric if for every vector $x = \sum a_n x_n$ all the series $x = \sum \lambda_n a_n x_{\pi(n)}$, where $\{\lambda_n\}$ is a bounded sequence and π is a permutation of the natural numbers, converge.

A basis $\{x_n\}$ is called K-spreading if

$$K^{-1}\Big\|\sum a_n x_n\Big\| \leq \Big\|\sum a_n x_{m_n}\Big\| \leq K\Big\|\sum a_n x_n\Big\|$$

for all scalars $\{a_n\}$ and for all $m_1 < m_2 < \dots$. If the basis is spreading and unconditional, it is called subsymmetric.

A sequence $\{x_n\}$ is called a basic sequence if it is a basis for the closed linear subspace of E that it spans. Unconditional and symmetric basic sequences are defined similarly.

Let $\{x_n\}$ be a basis of the space E. It follows from the closed graph theorem that the basis projections $P_N x = \sum_{n \leq N} a_n x_n$, where $x = \sum a_n x_n$, are uniformly bounded. The basis constant of $\{x_n\}$ is $K = \sup_N \|P_N\|$. The functionals x_n^*, assigning to each $x = \sum a_n x_n$ the coefficient a_n, are called the biorthogonal functionals. They are clearly bounded by $2K/\|x_n\|$.

The basis is called monotone if $\|P_N\| = 1$ for all N, it is called bimonotone if $\|P_N\| = \|I - P_N\| = 1$ for all N. A space with a basis can be renormed by $|||x||| = \sup_N \|P_N x\|$ so that the basis is monotone under the new norm. Similarly $|||x||| = \sup_{N>M} \|(P_N - P_M)x\|$ defines an equivalent norm under which the basis is bimonotone.

If $\{x_n\}$ is an unconditional basis of the space E, then all the operators of the form $P_\theta x = \sum \theta_n a_n x_n$, where $\theta = \{\theta_n\}$ is a sequence of ± 1's, and all the projections $P_A x = \sum_{n \in A} a_n x_n$, where A is a subset of the natural numbers, are uniformly bounded. The unconditional basis constant of $\{x_n\}$ is $\sup_\theta \|P_\theta\|$. In this case the formula $|||x||| = \sup_\theta \|P_\theta x\|$ gives an equivalent norm on E. The basis $\{x_n\}$ is 1-unconditional with respect to this new norm, and $|||P_A||| \leq 1$ for every subset A of $\mathbb{N}$.

Similar projections and constants are defined for spaces with FDD. For more details and a systematic presentation of all these notions, see Lindenstrauss and Tzafriri [**371**].

Let $\{x_n\}$ be a basis of the Banach space E. A sequence of vectors $y_k \neq 0$ of the form $y_k = \sum_{n=m_k}^{m_{k+1}-1} a_n x_n$, where $1 \leq m_1 < m_2 < \dots$, is called a block basis of the basis $\{x_n\}$ or, simply, a block basic sequence. Such a sequence is always a basic sequence, and its basis constant (respectively, its unconditional basis constant) is not larger than that of $\{x_n\}$.

Two basic sequences $\{x_n\}$ and $\{y_n\}$ are said to be equivalent if the map $T\big(\sum a_n x_n\big) = \sum a_n y_n$ is a well-defined isomorphism from span$\{x_n\}$ onto span$\{y_n\}$. If T is an isometry, the bases are said to be isometrically equivalent. For example, any normalized basic sequence of the unit vector basis of c_0 or of l_p for $1 \le p < \infty$ is isometrically equivalent to the basis itself. (Actually, the unit vector bases of c_0 and l_p are the only bases which are equivalent to all their block bases; see the Notes and Remarks section in Chapter 12.) A basic sequence which is equivalent to an unconditional basis is, of course, unconditional itself.

Proposition E.1. (i) *Let $\{x_n\}$ be a basis of a Banach space E, and let F be an infinite-dimensional subspace of E. Then for every sequence $\{\varepsilon_n\}$ of positive numbers there are a block basis $\{y_k\}$ of $\{x_n\}$ and vectors $z_k \in F$ with $\|z_k\| = 1$ such that $\|z_k - y_k\| < \varepsilon_k$ for all k.*

(ii) *Let E be a Banach space, and let $\{y_n\}$ be a normalized basic sequence in E with basis constant K. Assume that $\{\varepsilon_n\}$ are positive numbers with $\sum \varepsilon_n < 1/4K$. If $z_n \in E$ satisfy $\|z_n - y_n\| < \varepsilon_n$ for every n, then $\{z_n\}$ is a basic sequence which is equivalent to $\{y_n\}$.*

Proof. (i) Denote the biorthogonal functionals to $\{x_n\}$ by $\{x_n^*\}$, and define a sequence of integers $1 < m_1 < m_2 < \dots$ and vectors z_k and y_k inductively as follows.

Fix $z_1 = \sum a_i^1 x_i \in F$ with $\|z_1\| = 1$, and choose m_1 with $\| \sum_{i=m_1+1}^{\infty} a_i^1 x_i \| < \varepsilon_1$. Then take $y_1 = \sum_{i=1}^{m_1} a_i^1 x_i$.

Assume that m_{k-1}, z_{k-1} and y_{k-1} have already been chosen. Then the subspace $F_k = \{z \in F : \langle x_i^*, z\rangle = 0 \text{ for all } i \le m_{k-1}\}$ is of finite codimension in F, hence nontrivial. Choose $z_k = \sum_{i>m_{k-1}} a_i^k x_i \in F_k$ with $\|z_k\| = 1$, and choose m_k such that $\| \sum_{i=m_k+1}^{\infty} a_i^k x_i \| < \varepsilon_k$. Then take $y_k = \sum_{i=m_{k-1}+1}^{m_k} a_i^k x_i$.

(ii) Let T be the linear operator mapping y_n to z_n, and fix $y = \sum a_n y_n$. Then

$$\|y - Ty\| = \Big\| \sum a_n (y_n - z_n) \Big\| \le \sum |a_n| \varepsilon_n \le \|y\|/2$$

because $|a_n| \le 2K\|y\|$. Thus $\|y\|/2 \le \|Ty\| \le 3\|y\|/2$. □

As an immediate consequence, we obtain

Corollary E.2. *Every subspace of a space with an unconditional basis contains an unconditional basic sequence.*

All the classical separable Banach spaces have bases. Enflo [**174**] constructed the first example of a separable Banach space without a basis; see also Section 2.d in Lindenstrauss and Tzafriri [**371**].

We now describe the Haar system, which is a basis in many function spaces. Denote the dyadic intervals of order n in $[0,1]$ by $\Delta_{n,j} = [j/2^n, (j+1)/2^n]$ for $0 \le j \le 2^n - 1$. The Haar functions (normalized in the L_∞ norm) are the functions $h_0 \equiv 1$ and $h_{n,j} = 1_{\Delta_{n,2j}} - 1_{\Delta_{n,2j+1}}$ for $n = 1, 2\dots$ and $0 \le j \le 2^{n-1} - 1$, where 1_A denotes the characteristic function of the set A.

The Haar system is a monotone basis in $L_p[0,1]$ for any $1 \le p < \infty$. The basis projections are just the conditional expectations with respect to suitable algebras of dyadic intervals. For $1 < p < \infty$ the Haar basis is unconditional in $L_p(0,1)$; see Section 2.c in Lindenstrauss and Tzafriri [**372**].

Schauder introduced a monotone basis for $C(0,1)$. It consists of the functions 1 and $F_{n,j}(t) = \int_0^t h_{n,j}(s)ds$ for $0 \le j \le 2^n - 1$; see Lindenstrauss and Tzafriri [**371**], page 3. As we remark after the proof of Proposition 13.8, $L_1(0,1)$ and $C(0,1)$ do not even embed in a space with an unconditional basis.

Theorem E.3. (i) *Every Banach space contains a subspace with a basis. In fact, every normalized sequence that converges weakly to zero contains a basic subsequence.*

(ii) *Let E be a superreflexive space. Then there are $1 < p < q < \infty$ and a constant $C > 0$ such that every normalized basic sequence $\{x_n\}$ in E satisfies*

$$C^{-1}\big(\sum |a_n|^q\big)^{1/q} \le \big\|\sum a_n x_n\big\| \le C\big(\sum |a_n|^p\big)^{1/p}$$

for every choice of scalars $\{a_n\}$ for which $\sum a_n x_n$ converges.

(iii) *Let E be a superreflexive space with an unconditional basis. Then every subspace of a quotient of E contains an unconditional basic sequence.*

(iv) *Let E be a Banach space with an unconditional basis. Then E is reflexive if and only if it does not contain a subspace isomorphic to c_0 or to l_1.*

(v) *Every separable Banach space E has a quotient space with a basis. More precisely, if $\{x_n^*\}$ is a normalized sequence in E^* which is weak*-convergent to zero, then there is a Banach space F with a basis $\{y_j\}$ and a quotient map T from E onto F such that the isometry T^* from F^* into E^* maps the biorthogonal functionals $\{y_j^*\}$ to a subsequence $\{x_{n_j}^*\}$ of $\{x_n^*\}$.*

The first part of (i) was known to Banach, and the second part is due to Bessaga and Pełczyński [**59**]. For proofs see Lindenstrauss and Tzafriri [**371**], pages 4 and 7. Part (ii) is due to Gurarii and Gurarii [**241**] and to James [**275**]. For a proof, see also Theorem 352 in Habala, Hájek and Zizler [**243**]. Part (iii) is a special case of Theorem II.7 in Johnson [**290**]. Part (iv) is due to James [**271**]; see also Theorem 1.c.12(a) in Lindenstrauss and Tzafriri [**371**]. Part (v) is due to Johnson and Rosenthal [**297**].

Corollary E.2 and part (iii) above, as well as Proposition E.8 below, are early positive partial results on the question whether every Banach space contains an unconditional basic sequence, i.e., whether the claim in part (i) could be strengthened to the existence of a subspace with an unconditional basis. For general Banach spaces the answer is negative, as was shown by Gowers and Maurey [**226**].

The spaces l_1 and c_0 play a very important role in the theory of Banach spaces. The two parts of the following theorem characterize, respectively, Banach spaces which do not contain a subspace isomorphic to l_1 and the subspaces of c_0. Recall that a sequence $\{x_n\}$ in a Banach space E is called weak Cauchy if $\lim\langle x^*, x_n\rangle$ exists for every $x^* \in E^*$.

Theorem E.4. (i) *A Banach space E does not contain a subspace isomorphic to l_1 if and only if every bounded sequence in E contains a weak Cauchy subsequence.*

(ii) *Let E be a separable Banach space, and assume that there is a constant $\lambda > 0$ such that*

$$\limsup \|x_n^* + x^*\| \ge \|x^*\| + \lambda \limsup \|x_n^*\|$$

for every x^* *and* x_n^* *in* E^* *such that* $w^*\text{-}\lim x_n^* = 0$. *Then* E *is isomorphic to a subspace of* c_0.

Part (i) is a somewhat weaker form of the celebrated theorem of Rosenthal [**504**]; see also Theorem 2.e.5 in Lindenstrauss and Tzafriri [**371**]. Part (ii) is due to Godefroy, Kalton and Lancien [**210**]; a simpler proof appears in Johnson, Lindenstrauss, Preiss and Schechtman [**292**].

A basis as in the following lemma is called an Auerbach basis. Its existence in arbitrary finite-dimensional Banach spaces is a very convenient tool in their study.

Lemma E.5. *Let* F *be an* n*-dimensional Banach space. Then there are* n *linearly independent vectors* $\{y_i\}_{i=1}^n$ *with biorthogonal functionals* $\{y_i^*\}_{i=1}^n$ *such that* $\|y_i\| = \|y_i^*\| = 1$ *for every* i. *In particular,* $\|\sum a_i y_i\| \geq \max |a_i|$ *for all* $\{a_i\}_{i=1}^n$.

Proof. Fix a basis $\{e_i\}_{i=1}^n$ in F. Given n vectors $\{x_i\}_{i=1}^n$ in the unit ball of F, let $V(x_1, \dots, x_n)$ be the determinant of the matrix $(a_{i,j})$, where $x_j = \sum a_{i,j} e_i$. The required y_i's are chosen so that V attains its maximum at $\{y_i\}_{i=1}^n$. Then $y_i^*(x) = V(y_1, \dots, y_{i-1}, x, y_{i+1}, \dots)/V(y_1, \dots, y_n)$. □

Note that if F is a subspace of E, and if $z_i^* \in E^*$ is a Hahn-Banach extension of y_i^* for every i, then $Pz = \sum \langle z_i^*, z\rangle y_i$ is a projection from E onto F with $\|P\| \leq n$. Thus every n-dimensional space is n-complemented in any space containing it. (Actually, there is a projection P from E onto F with $\|P\| \leq \sqrt{n}$. See Proposition G.1.)

The next lemma is useful when we need to control the structure of a general finite-dimensional subspace of a space with an unconditional basis. It was proved by Pełczyński and Rosenthal [**453**] for L_p-spaces, and Johnson [**289**] noticed that the proof works for general spaces.

Lemma E.6. *For every* n *there is an* $N = N(n)$ *with the following property: Let* E *be a Banach space with a* 1*-unconditional basis* $\{e_j\}$, *and let* F *be an* n*-dimensional subspace of* E. *Then* F *is contained in an* N*-dimensional subspace* G *of* E *which is* 2*-isomorphic to the span of* N *disjointly supported vectors.*

Proof. We shall show that there are a subspace H spanned by N disjoint vectors and an isomorphism T of F into H such that $\|y - Ty\| \leq \|y\|/4n$ for every $y \in F$. Since TF is n-dimensional, there is a projection $P : H \to TF$ with $\|P\| \leq n$, and we denote its kernel by H_0. The required subspace is $G = \operatorname{span}\{F, H_0\}$. Indeed, write the elements of H as $x = x_0 + x_1$, where $x_1 = Px \in TF$ and $x_0 \in H_0$, and define $S : H \to G$ by $Sx = x_0 + T^{-1}x_1$. Since clearly $\|T^{-1}\| \leq 2$, it follows that

$$\begin{aligned} \|x - Sx\| &= \|x_1 - T^{-1}x_1\| \leq \|T^{-1}x_1\|/4n \\ &\leq \|T^{-1}\|\,\|P\|\,\|x\|/4n \leq \|x\|/2 \end{aligned}$$

and hence S is a 2-isomorphism.

Let $\{y_i\}_{i=1}^n$ be a normalized Auerbach basis for F, and write $y_i = \sum_{j=1}^\infty a_{i,j} e_j$. Then $|y_i| = \sum_{j=1}^\infty |a_{i,j}| e_j$ are also normalized, and hence

$$w = \sum_{i=1}^n |y_i| = \sum_{j=1}^\infty \Big(\sum_{i=1}^n |a_{i,j}|\Big) e_j = \sum_{j=1}^\infty \omega_j e_j$$

satisfies $\|w\| \leq n$. The algebra generated by the sets

$$A_{i,k} = \big\{ j : \ (k-1)\omega_j/4n^3 < a_{i,j} \leq k\omega_j/4n^3 \big\}$$

for $1 \le i \le n$ and $-4n^3 \le k \le 4n^3$ is generated by at most $N = (8n^3+1)^n$ disjoint sets, which we denote by $\{A_l\}_{l=1}^N$.

Let H be the linear span of the disjointly supported vectors $z_l = \sum_{j\in A_l} w_j e_j$. By the construction we can find for each $i \le n$ a vector $x_i = \sum_{j=1}^\infty b_{i,j} e_j \in H$ such that $|b_{i,j} - a_{i,j}| \le \omega_j/4n^3$ for every j. Thus $\|x_i - y_i\| \le \|w\|/4n^3 \le 1/4n^2$.

We now define $T : F \to H$ by $T\big(\sum_{i=1}^n a_i y_i\big) = \sum_{i=1}^n a_i x_i$. If $y = \sum_{i=1}^n a_i y_i$, then $|a_i| \le \|y\|$ for every i, hence $\|y - Ty\| \le \sum_{i=1}^n |a_i|\|y_i - x_i\| \le \|y\|/4n$ as required. □

The following useful characterization of reflexivity is not really connected to bases or lattices, but this seems to be the most natural place to mention it. It is due to James [**272**] and, independently, to D. P. and V. D. Milman [**414**]. See also Theorem I.6.1(iii) in Guerre-Delabriére [**239**]. The characterization is closely related to the characterizations of superreflexivity in Theorem A.6 and the remark following it. It should be pointed out, however, that Theorem A.6 is deeper and that Proposition E.7 is quite easy to prove.

Proposition E.7. *Let E be a Banach space. Then the following properties are equivalent:*

(i) *E is nonreflexive.*

(ii) *For every $0 < \delta < 1$ there are normalized vectors $\{x_n\} \subset E$ and $\{x_n^*\} \subset E^*$ such that $\langle x_n^*, x_m\rangle = 0$ for $n > m$ and $\langle x_n^*, x_m\rangle \ge \delta$ for $n \le m$.*

(iii) *E contains a normalized sequence $\{x_i\}$ such that*

$$\Big\|\sum_{i=1}^{l} a_i x_i - \sum_{i=l+1}^{n} a_i x_i\Big\| \ge \frac{1}{2}\sum_{i=1}^{n} a_i$$

for every $1 \le l \le n$ and for all nonnegative coefficients $\{a_i\}$.

A partially ordered Banach space E is called a Banach lattice if it satisfies the following properties:

(i) $x \le y$ implies that $x + z \le y + z$ for every $x, y, z \in E$.

(ii) $\lambda x \ge 0$ whenever $x \ge 0$ in E and λ is a nonnegative scalar.

(iii) Each $x, y \in E$ admit a least upper bound $x \vee y$ and a greatest lower bound $x \wedge y$.

(iv) $\|x\| \le \|y\|$ whenever $|x| \le |y|$. (The absolute value of $x \in E$ is defined by $|x| = x \vee (-x)$.)

Two elements x, y in a Banach lattice E are called disjoint if $|x| \wedge |y| = 0$. An ideal in E is a (not necessarily closed) linear subspace F of E such that if $y \in F$ and $|x| \le |y|$, then $x \in F$. An element $e \ge 0$ in E is called a weak unit if the only $x \in E$ for which $x \wedge e = 0$ is $x = 0$. It is easy to check that every separable lattice has a weak unit.

A space with a 1-unconditional basis is a lattice under the natural order obtained when we consider it as a space of sequences (when we identify $x = \sum a_n x_n$ with the sequence (a_n)). Thus the results below also hold for spaces with a 1-unconditional basis and, by passing to an equivalent norm, they also apply to spaces with a general unconditional basis.

The same Banach space may very well admit different order structures. For example, as mentioned above the Haar functions form an unconditional basis in $L_p(0,1)$ for $1 < p < \infty$. The order induced by this basis is, of course, completely different from its order structure as a function space.

A set $\{x_\alpha : \alpha \in A\}$ in a Banach lattice E is said to be order bounded if there is an $x \in E$ such that $|x_\alpha| \leq |x|$ for every $\alpha \in A$. The lattice E is called order (respectively, σ-order) complete if every order bounded set (respectively, sequence) has a least upper bound $\vee_{\alpha\in A} x_\alpha$ and a greatest lower bound $\wedge_{\alpha \in A} x_\alpha$. The lattice E is called order (respectively, σ-order) continuous if every downward directed net $\{x_\alpha : \alpha \in A\}$ (respectively, sequence) in E with $\wedge_{\alpha\in A} x_\alpha = 0$ satisfies $\|x_\alpha\| \to 0$.

If the lattice E is not σ-complete or σ-order continuous, then it necessarily contains a subspace isomorphic to c_0; see Lindenstrauss and Tzafriri [**372**], pages 6–7. In particular, if E is reflexive or if it does not contain l_∞^n's uniformly, then it is both σ-complete and σ-order continuous, hence (by [**372**], Proposition 1.a.8) also order continuous.

Proposition E.8. *Let E be an order continuous Banach lattice. Then every subspace of E contains an unconditional basic sequence.*

For the proof, see Proposition 1.c.9 in Lindenstrauss and Tzafriri [**372**].

The following representation of "nice" lattices as function spaces is very useful.

Proposition E.9. *Let E be an order continuous lattice with a weak unit. Then there are a probability space (Ω, Σ, μ), an ideal $\tilde{E}$ in $L_1(\Omega, \Sigma, \mu)$, and a lattice norm $\|\cdot\|_{\tilde{E}}$ on $\tilde{E}$, such that*

(i) *E is isometric and order isomorphic to $\tilde{E}$*

(ii) *$\tilde{E}$ is dense in $L_1(\Omega, \Sigma, \mu)$ and $L_\infty(\Omega, \Sigma, \mu)$ is dense in $\tilde{E}$*

(iii) *$\|f\|_1 \leq \|f\|_{\tilde{E}} \leq 2\|f\|_\infty$ whenever $f \in L_\infty(\Omega, \Sigma, \mu)$*

(iv) *the dual of the isometry given in* (i) *maps E^* onto the Banach lattice $\tilde{E}^*$ of all measurable functions g for which*

$$\|g\|_{\tilde{E}^*} = \sup\Big\{ \int fg\, d\mu : \ \|f\|_{\tilde{E}} \leq 1 \Big\} < \infty.$$

The value of the functional corresponding to g at $f \in \tilde{E}$ is given by $\int fg\, d\mu$.

For the proof see Lindenstrauss and Tzafriri [**372**], page 25.

A Banach lattice E of functions on a measure space Ω is called a rearrangement invariant function space if the two functions $|f|$ and $g(t) = f\big(\sigma(t)\big)$ belong to E for every $f \in E$ and for every measure preserving map σ of Ω onto itself, and if $\|g\|_E = \|f\|_E = \||f|\|_E$.

In the next discussion we shall write expressions like $\big(\sum |x_i|^p\big)^{1/p}$, where the x_i's are elements of the Banach lattice E. Such expressions are defined pointwise when E is represented as a function space as above, but it is also possible to define them via a functional calculus for general Banach lattices; see Lindenstrauss and Tzafriri [**372**], pages 40–42.

Let $1 < p < \infty$. A Banach lattice E is said to be p-convex if there is a constant M such that

$$\Big\|\big(\sum |x_i|^p\big)^{1/p}\Big\| \leq M\big(\sum \|x_i\|^p\big)^{1/p}$$

for every choice of vectors $\{x_i\}$ in E. It is said to be q-concave ($1 < q < \infty$) if there is a constant M such that

$$\big(\sum \|x_i\|^q\big)^{1/q} \leq M\Big\|\big(\sum |x_i|^q\big)^{1/q}\Big\|$$

for every choice of vectors $\{x_i\}$ in E. The optimal constants in these inequalities are called the p-convexity and q-concavity constants of E, and they are denoted, respectively, by $M^{(p)}(E)$ and by $M_{(q)}(E)$.

In analogy to the relation between the different $L_p(\mu)$ spaces, where the map $f \to |f|^{1/s}\text{sign}f$ maps $L_r(\mu)$ onto $L_{rs}(\mu)$, there is a construction of a Banach lattice $E^{(r)}$, the r-convexification of E, for any given Banach lattice E. The new lattice $E^{(r)}$ is r-convex with $M^{(r)}(E^{(r)}) = 1$. If E is p-convex and q-concave, then $E^{(r)}$ is pr-convex and qr-concave.

If E is assumed apriori to be p-convex for some $p > 1$, then there is also an analogous construction for the r-concavification $E_{(r)}$ of E. It is r-concave with $M_{(r)}(E_{(r)}) = 1$ and if E is p-convex and q-concave, then $E_{(r)}$ is p/r-convex and q/r-concave.

In particular, every p-convex and q-concave lattice can be renormed so that with the new norm (and with the same order) its p-convexity and q-concavity constants are equal to one.

For details on the construction and properties of the convexification and concavification of Banach lattices, see Section 1.d in Lindenstrauss and Tzafriri [**372**], and especially pages 53–55.

Proposition E.10. (i) *Let $1 < p \leq 2 \leq q < \infty$, and assume that E is a p-convex and q-concave Banach lattice. Then E can be renormed by a lattice norm which is uniformly convex and uniformly smooth and such that the moduli of uniform convexity and of uniform smoothness are of power types q and p respectively. Moreover, if $M^{(p)}(E) = M_{(q)}(E) = 1$, then the original norm is already uniformly convex and uniformly smooth, with moduli of power type as above.*

(ii) *If E is either p-convex for some $p > 1$ or has cotype s for some $s < \infty$, then it is q-concave for some $q < \infty$.*

Part (i) is proved in Lindenstrauss and Tzafriri [**372**], page 80, and part (ii) on page 88.

Let $1 < p < \infty$. A Banach lattice E is said to satisfy an upper p-estimate if there is a constant M such that

$$\big\|\sum x_i\big\| \leq M\big(\sum \|x_i\|^p\big)^{1/p}$$

for every choice of *disjoint* vectors $\{x_i\}$ in E. If the reverse inequality holds, we say that E satisfies a lower p-estimate. The optimal constants in these inequalities are called the upper (lower) p-estimate constants of E.

Proposition E.11. (i) *If the Banach lattice E is p-convex (p-concave), then it satisfies an upper p-estimate (lower p-estimate). Conversely, if E satisfies an upper p-estimate (lower p-estimate) for some $1 < p < \infty$, then it is r-convex (q-concave) for every $1 < r < p < q < \infty$. Moreover, there is an absolute constant K such that if $p < 2$ and if E satisfies a lower p-estimate with constant M, then it is q-convex with constant KM for $q = 1/(2-p)$.*

(ii) *E satisfies an upper (lower) p-estimate for some $p > 1$ if and only if E^* satisfies a lower (upper) p'-estimate, where $1/p + 1/p' = 1$.*

The first part of (i) is trivial; for proofs of the second part and of (ii), see Lindenstrauss and Tzafriri [**372**], pages 85 and 83 respectively. The claim on the constants follows from tracing the constants in the proof.

The first chapter of Lindenstrauss and Tzafriri [**372**] gives a systematic presentation of Banach lattices and their structure. See, in particular, the diagrams on pages 100–101 which summarize the connections between the various parameters associated with them.

F. Local Structure of Infinite-Dimensional Spaces

In this appendix we discuss several aspects of the way in which the structure of an infinite-dimensional Banach space is determined by its finite-dimensional subspaces and by the way in which these subspaces are embedded into each other.

The Banach-Mazur distance between two Banach spaces E and F is defined by $d(E,F) = \inf\{\|T\|\,\|T^{-1}\|\}$, where the infimum is taken over all isomorphisms T from E onto F.

A Banach space F is said to be finitely representable in a Banach space E if for every $\varepsilon > 0$ and for every finite-dimensional subspace $F_1 \subset F$ there is a subspace $E_1 \subset E$ such that $d(E_1, F_1) < 1 + \varepsilon$.

The following proposition is called the principle of local reflexivity. It gives a stronger version of the well-known fact that E^{**} is finitely representable in E for every Banach space E. We shall consider E as a subspace of E^{**}.

Proposition F.1. *Let E be a Banach space, and let G be a finite-dimensional subspace of E^{**}. Then for every $\varepsilon > 0$ and for every finite-dimensional subspace F of E^* there is a linear operator $T : G \to E$ such that*

(i) $Tx^{**} = x^{**}$ *for every* $x^{**} \in G \cap E$,

(ii) $(1-\varepsilon)\|x^{**}\| \le \|Tx^{**}\| \le (1+\varepsilon)\|x^{**}\|$ *for every* $x^{**} \in G$,

(iii) $\langle x^*, Tx^{**}\rangle = \langle x^{**}, x^*\rangle$ *for every* $x^{**} \in G$ *and* $x^* \in F$,

(iv) *if G is complemented in E^{**} by a projection P, then T can be chosen so that there is a projection $Q : E \to TG$ satisfying $\|Q\| \le (1+\varepsilon)\|P\|$.*

The principle of local reflexivity is due to Lindenstrauss and Rosenthal [**367**] and to Johnson, Rosenthal and Zippin [**298**].

A Banach space E is called superreflexive if every Banach space which is finitely representable in E is reflexive. Superreflexive spaces can be renormed to be uniformly convex and uniformly smooth, and a Banach space which is isomorphic to a superreflexive space is also superreflexive; see Appendix A.

Let $1 \le p \le \infty$ and $\lambda \ge 1$. A Banach space E is called an $\mathcal{L}_{p,\lambda}$-space if for every finite-dimensional subspace F of E there is a finite-dimensional subspace G containing F such that $d(G, l_p^m) \le \lambda$ (where m is the dimension of G). We say that E is an $\mathcal{L}_p$-space if it is an $\mathcal{L}_{p,\lambda}$-space for some λ.

Theorem F.2. (i) *$L_p(\mu)$ (respectively $C(K)$) is an $\mathcal{L}_{p,1+\varepsilon}$-space (respectively $\mathcal{L}_{\infty,1+\varepsilon}$-space) for every $\varepsilon > 0$.*

(ii) *E is an $\mathcal{L}_{2,\lambda}$-space if and only if it is λ-isomorphic to a Hilbert space.*

(iii) *E is an $\mathcal{L}_p$-space if and only if E^* is an $\mathcal{L}_q$-space, where $p^{-1} + q^{-1} = 1$.*

(iv) *Let E be a complemented subspace of $L_p(\mu)$. Then it is either an $\mathcal{L}_p$-space or (only when $1 < p < \infty$) isomorphic to a Hilbert space. Conversely, if $1 < p < \infty$*

and if E is an $\mathcal{L}_p$-space, then E is isomorphic to a complemented subspace of $L_p(\mu)$ for some measure μ.

(v) *E is an $\mathcal{L}_\infty$-space if and only if E^{**} is an $\mathcal{L}_\infty$-space, and if and only if E^{**} is injective (i.e., complemented in every Banach space containing it). E is an $\mathcal{L}_1$-space if and only if E^* is injective.*

(vi) *If E is a separable $\mathcal{L}_p$-space $(1 < p < \infty)$ which does not contain an isomorphic copy of l_2, then E is isomorphic to l_p.*

(vii) *Let E be a Banach space that does not contain an isomorphic copy of l_2, and let $2 < p < \infty$. Then every operator $T : E \to L_p$ factors through l_p; i.e., there are operators $A : E \to l_p$ and $B : l_p \to L_p$ such that $T = BA$.*

(viii) *c_0 is a minimal $\mathcal{L}_\infty$-space in the sense that if E is a $\mathcal{L}_\infty$-subspace of c_0, then E is isomorphic to c_0.*

$\mathcal{L}_{p,\lambda}$-spaces were introduced by Lindenstrauss and Pełczyński [**364**], where many of their properties, as well as their special role in the general theory of Banach spaces, are discussed. Parts (iii)–(v) were proved by Lindenstrauss and Rosenthal [**367**]. Part (vi) is due to Johnson and Odell [**296**]. Part (vii) is due to Johnson [**288**]. It generalizes (vi) (for $p > 2$) because complemented subspaces of l_p are isomorphic to l_p. Part (viii) is due to Johnson and Zippin [**299**].

Bourgain [**85**] contains a presentation of the theory of $\mathcal{L}_p$-spaces and, in particular, a discussion of many interesting examples.

A family $\mathcal{U}$ of subsets of a set I is called a filter if it is closed under finite intersections, $\emptyset \notin \mathcal{U}$, and $A \in \mathcal{U}$ whenever $A \supset B$ and $B \in \mathcal{U}$. A maximal filter is called an ultrafilter. By Zorn's lemma every filter is contained in an ultrafilter, and $\mathcal{U}$ is an ultrafilter if and only if for every $A \subset I$ either $A \in \mathcal{U}$ or $I \setminus A \in \mathcal{U}$. We shall always assume without further mention that the ultrafilters are free, i.e., that there is no point which is common to all the sets in $\mathcal{U}$.

We shall say that a set of points $\{x_i : i \in I\}$ in a topological space X converges to a point x with respect to the ultrafilter $\mathcal{U}$ on I (and write $\lim_{\mathcal{U}} x_i = x$) if $\{i \in I : x_i \in V\}$ belongs to $\mathcal{U}$ for every neighborhood V of x. If X is a compact Hausdorff space, then every set $\{x_i : i \in I\}$ in X converges to a unique point with respect to any ultrafilter on I.

An ultrafilter $\mathcal{U}$ on I is said to be countably incomplete if there is a decreasing sequence of sets $I_1 \supset I_2 \supset \dots$ in $\mathcal{U}$ such that $\bigcap I_n = \emptyset$. For most of what we do we can use the set of natural numbers $\mathbb{N}$ as the set I. In this case every ultrafilter is countably incomplete. We do not, however, make any assumptions on the cardinality of I, and we shall also need ultrafilters on uncountable sets.

Let $\{E_i : i \in I\}$ be a family of Banach spaces, and let $\mathcal{U}$ be an ultrafilter on I. The space $l_\infty(E_i)$ of all bounded families of points $x = (x_i)$, where $x_i \in E_i$ for all i, equipped with the norm $\|x\| = \sup \|x_i\|$, is a Banach space, and its subspace $N = \{x : \lim_{\mathcal{U}} \|x_i\| = 0\}$ is closed. The ultraproduct of the spaces $\{E_i\}$ is the quotient space $l_\infty(E_i)/N$ with the usual quotient norm. The ultraproduct of the spaces E_i with respect to the ultrafilter $\mathcal{U}$ will be denoted $(E_i)_{\mathcal{U}}$. (Another common notation is $\prod E_i /\mathcal{U}$.) The elements of $(E_i)_{\mathcal{U}}$ are denoted by $(x_i)_{\mathcal{U}}$, and their norm can be computed directly by the formula $\|(x_i)_{\mathcal{U}}\| = \lim_{\mathcal{U}} \|x_i\|$.

Let $T_i : E_i \to F_i$ be a uniformly bounded set of linear operators. The ultraproduct $(T_i)_{\mathcal{U}} : (E_i)_{\mathcal{U}} \to (F_i)_{\mathcal{U}}$ of the operators is defined by $(T_i)_{\mathcal{U}}\big((x_i)_{\mathcal{U}}\big) = (T_i x_i)_{\mathcal{U}}$. It is a well-defined bounded linear operator because $\lim_{\mathcal{U}} \|x_i\| = 0$ implies that

$\lim_{\mathcal{U}} \|T_i x_i\| = 0$ by the uniform boundedness of the T_i's. If each of the T_i's is an into isomorphism, and if the T_i^{-1}'s are also uniformly bounded, then $(T_i)_{\mathcal{U}}$ is an into isomorphism.

The ultraproduct construction preserves many classes of spaces. For example

Proposition F.3. (i) *If all the E_i's are Banach lattices, then so is $(E_i)_{\mathcal{U}}$.*

(ii) *Let $1 \leq p \leq \infty$. If $(\Omega_i, \Sigma_i, \mu_i)$ are measure spaces, then there is a measure space (Ω, Σ, μ) such that $\big(L_p(\mu_i)\big)_{\mathcal{U}}$ is isometric and order isomorphic to $L_p(\mu)$.*

(iii) *If $\{K_i\}$ are compact Hausdorff spaces, then there is a compact Hausdorff space K such that $\big(C(K_i)\big)_{\mathcal{U}}$ is isometric and order isomorphic to $C(K)$.*

(iv) *If $1 \leq p \leq \infty$, and if all the E_i's are $\mathcal{L}_{p,\lambda}$-spaces for some $\lambda \geq 1$, then so is $(E_i)_{\mathcal{U}}$.*

See Heinrich [**251**] for proofs.

Part (ii) has the following important corollary.

Corollary F.4. *Let E be a Banach space, and let $1 \leq p < \infty$. If every finite-dimensional subspace of E is isometric to a subspace of L_p, then E is isometric to a subspace of $L_p(\mu)$ for some measure μ.*

Proof. For simplicity we shall assume that E is separable, so fix an increasing sequence F_n of finite-dimensional subspaces of E whose union is dense in E, and isometries T_n from F_n to L_p. Then $(T_n)_{\mathcal{U}}$ is an isometry of $(F_n)_{\mathcal{U}}$ into the $L_p(\mu)$ space $(L_p)_{\mathcal{U}}$, and E is naturally isometric to a subspace of $(F_n)_{\mathcal{U}}$. Indeed, assign to each $x \in \bigcup F_n$ the class of the constant sequence $(x, x, \dots)$. (The sequence is in $l_\infty(F_n)$ except for the first finitely many coordinates; hence its class in $(F_n)_{\mathcal{U}}$ is well defined.) □

Ultraproducts give a convenient way to treat finite representability and to study properties of Banach spaces that are determined by their "local structure", i.e., by the structure of their finite-dimensional subspaces. The next proposition (see Proposition 6.2 in Heinrich [**251**]) formalizes one such connection.

Proposition F.5. *Let E be a Banach space, and let E_i be a family of Banach spaces such that for each finite-dimensional subspace G of E there is an i such that G is K-isomorphic to a subspace of E_i. Then E is K-isomorphic to a subspace of some ultraproduct of the E_i's.*

An ultraproduct in which all the spaces E_i are equal to one space E is called an ultrapower of E and denoted by $(E)_{\mathcal{U}}$. There is a natural linear isometric embedding of E into $(E)_{\mathcal{U}}$, which we denote by J_E. It maps $x \in E$ to the equivalence class of the family with the constant value x, i.e., $J_E(x) = (x)_{\mathcal{U}} \in (E)_{\mathcal{U}}$.

Let $\mathcal{U}$ and $\mathcal{V}$ be two ultrafilters on sets I and J respectively. The product ultrafilter is the family of all sets $A \subset I \times J$ such that $\big\{j : \{i : (i,j) \in A\} \in \mathcal{U}\big\} \in \mathcal{V}$. It follows directly from the definitions that $\big((E)_{\mathcal{U}}\big)_{\mathcal{V}} = (E)_{\mathcal{U}\times\mathcal{V}}$

The next proposition summarizes the relations between finite representability in E, the second dual E^{**}, and the ultrapowers $(E)_{\mathcal{U}}$. See Sections 6 and 7 in Heinrich [**251**] for details and references.

Proposition F.6. (i) *$(E)_{\mathcal{U}}$ is finitely representable in E.*

(ii) *There is an ultrafilter* $\mathcal{U}$ *such that the natural map* $Q_E : (E)_{\mathcal{U}} \to E^{**}$, *given by*

$$Q_E\big((x_i)_{\mathcal{U}}\big) = w^* - \lim_{\mathcal{U}} x_i$$

is a norm-one quotient map and $Q_E J_E$ *is the identity on* E.

(iii) *A Banach space* F *is finitely representable in* E *if and only if there is an ultrafilter* $\mathcal{U}$ *such that* F *is isometric to a subspace of* $(E)_{\mathcal{U}}$. *In particular this is true for* $F = E^{**}$, *and in this case the isometry, which we denote by* T, *can be chosen so that its restriction to* E *is* J_E *and* $Q_E T$ *is a norm-one projection from* $(E)_{\mathcal{U}}$ *onto* E^{**}.

(iv) *The natural embedding* $I : (E^*)_{\mathcal{U}} \to \big((E)_{\mathcal{U}}\big)^*$ *given by*

$$\langle I\big((x_i^*)_{\mathcal{U}}\big), (x_i)_{\mathcal{U}}\rangle = \lim_{\mathcal{U}} \langle x_i^*, x_i\rangle$$

is an into isometry. If $\mathcal{U}$ *is countably incomplete, then this map is onto if and only if* E *is superreflexive.*

(v) *If* $F \subset (E)_{\mathcal{U}}$ *and* $G \subset \big((E)_{\mathcal{U}}\big)^*$ *are finite-dimensional subspaces, then for every* $\varepsilon > 0$ *there is a subspace* H *of* $(E^*)_{\mathcal{U}}$ *and a* $(1+\varepsilon)$*-isomorphism of* H *onto* G *such that* $\langle Iu, x\rangle = \langle Tu, x\rangle$ *for every* $u \in H$ *and* $x \in F$. *In particular, every finite-dimensional* λ*-complemented subspace of* $(E)_{\mathcal{U}}$ *is* $(1+\varepsilon)$*-isomorphic to a* $(\lambda+\varepsilon)$*-complemented subspace of* E *for every* $\varepsilon > 0$.

It follows from (i) and (iii) that E is superreflexive if and only if each ultrapower of E is reflexive and that if E is reflexive, then E is 1-complemented in $(E)_{\mathcal{U}}$. Part (v), which is a strengthening of part (i), is called the ultrapower version of local duality.

All $\mathcal{L}_p$-spaces for a fixed p have the same local structure. We shall need the stronger result (Stern [**543**]) that if E is an $\mathcal{L}_p$-space $(1 < p < \infty)$, then there is a countably incomplete ultrafilter such that $(E)_{\mathcal{U}}$ is isomorphic to $(l_p)_{\mathcal{U}}$.

The following proposition summarizes the structure of linear operators to and from finite-dimensional subspaces of an ultrapower.

Proposition F.7. (i) *Let* G *be a finite-dimensional Banach space, and let* S *be a linear operator from* G *to the ultraproduct* $(F_i)_{\mathcal{U}}$. *Then there are linear operators* $S_i : G \to F_i$ *such that* $S = (S_i)_{\mathcal{U}}$.

(ii) *Let* H_i *be* n*-dimensional Banach spaces. Then* $(H_i)_{\mathcal{U}}$ *is also* n*-dimensional. If* $\{x_{i,j} : j \le n\}$ *is a basis for* H_i *for each* i, *and if there is a constant* $d > 0$ *such that* $\|x_{i,j}\| = 1$ *and* $\operatorname{dist}\big(x_{i,j}, \operatorname{span}\{x_{i,k} : k \ne j\}\big) \ge d$ *for all* i *and* j, *then* $x_j = (x_{i,j})_{\mathcal{U}}$ *is a basis for* $(H_i)_{\mathcal{U}}$.

(iii) *Let* H_i *and the bases* $\{x_{i,j}\}$ *and* $\{x_j\}$ *be as in part (ii), and let* T *be a linear operator from* $(H_i)_{\mathcal{U}}$ *to a Banach space* E. *Then there are linear operators* $T_i : H_i \to E$ *such that* $T_i x_{i,j} = T x_j$ *for all* i *and* j.

For a proof see Proposition 6.1 in Heinrich [**251**] and the arguments in the proof of Theorem 5.2 in Heinrich and Mankiewicz [**252**].

The following result is Proposition 3.4 in Heinrich and Mankiewicz [**252**] (which we only formulate for separable subspaces). The reflexive case was proved earlier by Lindenstrauss [**359**]. Sims and Yost [**535**] showed how to derive the general result from the argument in [**359**] without using model theory (which was employed in [**252**]).

Proposition F.8. *Let F be a separable subspace of the Banach space E. Then there is a separable subspace $F \subset G \subset E$ such that G^* is isometric to a norm-one complemented subspace of E^*. In particular, if E is reflexive, then G is complemented in E.*

A "decomposition scheme" is any method that enables one to deduce that E and F are isomorphic to each other from the assumption that they are isomorphic to complemented subspaces of each other. The next proposition lists two such situations.

Proposition F.9. *Let E and F be two Banach spaces which are isomorphic to complemented subspaces of each other. Assume that they satisfy one of the following two conditions:*

(i) F is isomorphic to $F \oplus F$ and E is isomorphic to $E \oplus E$

or

(ii) E contains a complemented copy of $c_0(E)$ or of $l_p(E)$ for some $1 \le p < \infty$. Then they are isomorphic to each other.

Proof. We denote isomorphism by equality, and represent $E = F \oplus F_1$ and $F = E \oplus E_1$. Assuming (i), we then obtain

$$E \oplus F = E \oplus (E \oplus E_1) = (E \oplus E) \oplus E_1 = E \oplus E_1 = F$$

and $E \oplus F = E$ by a similar computation; hence $E = F$.

We now check (ii). Note that the assumption here is only on one of the spaces and not on both. We consider only the case of $l_p(E)$. Write $E = W \oplus l_p(E)$. Then

$$\begin{aligned} E &= W \oplus l_p(E) = W \oplus \big(l_p(F_1) \oplus l_p(F)\big) \\ &= \big(W \oplus l_p(F_1) \oplus l_p(F)\big) \oplus F = E \oplus F \end{aligned}$$

and

$$F = E_1 \oplus E = E_1 \oplus \big(W \oplus l_p(E)\big) = \big(E_1 \oplus W \oplus l_p(E)\big) \oplus E = F \oplus E.$$

□

The decomposition method first appeared in the context of Banach space theory in Pełczyński [**450**].

Proposition F.10. *Assume that E is isomorphic to a complemented subspace of $(F)_{\mathcal{U}_1}$ and that F is isomorphic to a complemented subspace of $(E)_{\mathcal{U}_2}$. Then there is an ultrafilter $\mathcal{U}$ such that $(F)_{\mathcal{U}}$ and $(E)_{\mathcal{U}}$ are isomorphic to complemented subspaces of each other. If F and E satisfy one of the "decomposition scheme" assumptions above, then in fact $(F)_{\mathcal{U}}$ and $(E)_{\mathcal{U}}$ are isomorphic to each other.*

The proof of the proposition is contained in the proofs of Proposition 4.4 and of Theorem 4.5 in Heinrich and Mankiewicz [**252**].

For more details and references on ultraproducts of Banach spaces, see Heinrich [**251**] and the Preliminaries section of Heinrich and Mankiewicz [**252**].

G. Quantitative Theory of Banach Spaces

To any Banach space one can associate several natural numerical parameters. The study of the behavior of these parameters for families of finite-dimensional spaces, as the dimension of the spaces grows to infinity, is an important and well-developed theory. We present here very few quantitative results of this nature.

The Banach-Mazur distance between two isomorphic Banach spaces E and F is defined to be

$$\inf\left\{\|T\|\,\|T^{-1}\|\right\}$$

where the infimum is taken over all isomorphisms T from E onto F.

The relative projection constant of a subspace E of a Banach space F is defined by

$$\lambda(E,F)=\inf\left\{\|P\|\right\}$$

where P is a projection from F onto E. The projection constant of E is

$$\lambda(E)=\sup\left\{\lambda(E,F)\right\}$$

where the supremum is taken over all Banach spaces F containing E.

By the Hahn-Banach theorem $\lambda(l^n_\infty)=1$; see also Lemma 1.1(ii). The spaces l^n_∞ are the only finite-dimensional spaces whose projection constant is equal to 1.

Proposition G.1. *Let E be an n-dimensional Banach space. Then*
(i) $d(E,l^n_2)\le\sqrt{n}$,
(ii) $\lambda(E)\le\sqrt{n}$.

Part (i) is due to John [**280**]; see also Milman and Schechtman [**420**], page 10. Part (ii) is due to Kadec and Snobar [**305**]. See also Chapter 3 of Pisier [**467**] for another approach to both parts. For the weaker estimate $\lambda(E)\le n$, see Lemma E.5.

A Banach space E is said to have type p if there is a constant C such that

$$\text{(G.1)}\qquad \Big(\int_0^1\big\|\sum r_i(t)x_i\big\|^p dt\Big)^{1/p}\le C\big(\sum\|x_i\|^p\big)^{1/p}$$

for every choice of finitely many vectors $\{x_i\}$ in E, where $\{r_i(t)\}$ are the Rademacher functions on $[0,1]$. The best constant C in the inequality is called the type-p constant of E.

If a Banach space satisfies (G.1), then every finite family of independent E-valued random variables $\{X_j\}$ with mean zero satisfies

$$\mathbb{E}\big\|\sum X_j\big\|^p\le(2C)^p\sum\mathbb{E}\,\|X_j\|^p.$$

Indeed, let $\{Y_j\}$ be independent variables which are also independent from the X_j's and such that each Y_j has the same distribution as X_j. Then X_j-Y_j are symmetric, and since $\mathbb{E}(X_j-Y_j)=0$ we obtain

$$\begin{aligned}\mathbb{E}\big\|\sum X_j\big\|^p &\le \mathbb{E}\big\|\sum(X_j-Y_j)\big\|^p=\mathbb{E}\int_0^1\big\|\sum r_i(t)(X_j-Y_j)\big\|^p dt\\ &\le C^p\mathbb{E}\sum\|X_j-Y_j\|^p\le(2C)^p\sum\mathbb{E}\,\|X_j\|^p\end{aligned}$$

as claimed.

By Khintchine's inequality

$$\Big(\int_0^1\big|\sum_{i=1}^n r_i(t)\big|^p dt\Big)^{1/p}\sim\Big(\int_0^1\big|\sum_{i=1}^n r_i(t)\big|^2 dt\Big)^{1/2}=n^{1/2}.$$

Thus by taking all the x_i's in the definition of type to be equal to one fixed vector, it follows that no Banach space can have type $p>2$. By the triangle inequality, every Banach space has type 1; i.e., $1\le p\le 2$ for every Banach space. The notion of type makes sense also for quasi-normed spaces (see Appendix H), and in this case any $0 < p \leq 2$ is possible.

A Banach space E is said to have cotype q if there is a constant C such that

$$\Big(\int_0^1 \|\sum r_i(t)x_i\|^q dt\Big)^{1/q} \geq C\big(\sum \|x_i\|^q\big)^{1/q}$$

for every choice of finitely many vectors $\{x_i\}$ in E. The best constant C is called the cotype q constant of E. Again, by Khintchine's inequality no Banach space can have cotype $q < 2$.

It is an important fact that the left-hand sides in the inequalities defining type and cotype are actually equivalent for all p and on q. This follows from Kahane's inequality, which generalizes Khintchine's inequality to Banach spaces:

For every $0 < p < q < \infty$ there is a constant C, depending only on p and q, such that

$$\Big(\int_0^1 \|\sum r_i(t)x_i\|^q dt\Big)^{1/q} \leq C\Big(\int_0^1 \|\sum r_i(t)x_i\|^p dt\Big)^{1/p}$$

for every choice of finitely many vectors $\{x_i\}$ in any Banach space.
(The reverse inequality is evident with constant 1.)

Kahane's inequality appears in Kahane [**306**]; see also page 74 of Lindenstrauss and Tzafriri [**372**] and page 134 of Milman and Schechtman [**420**]. See also Kalton [**312**] for the minor changes needed in the proof for quasi-normed spaces.

The type and cotype are known for many concrete Banach spaces. For example, $L_p(\mu)$ has type p and cotype 2 for $1 \leq p \leq 2$, and type 2 and cotype p for $2 \leq p < \infty$, and these results are sharp when the space is infinite-dimensional; see Lindenstrauss and Tzafriri [**372**], page 73. Tomczak-Jaegermann [**557**] showed that the same is true for the Schatten classes C_p.

We now state several results on spaces with type and/or cotype 2.

Theorem G.2. (i) *A Banach space E is isomorphic to a Hilbert space if and only if it has type 2 and cotype 2.*

(ii) *Let E be a Banach space with type 2, and let F be a subspace of E. Let H be a Hilbert space. Then every bounded linear operator $T : F \to H$ extends to an operator $S : E \to H$ with $\|S\| \leq C\|T\|$, where C is the type-2 constant of E. In particular, every Hilbertian subspace of E is complemented.*

(iii) *A Banach lattice E has type 2 if and only if it is 2-convex and q-concave for some $q < \infty$. In particular, E has type 2 when it is a superreflexive lattice which satisfies an upper p-estimate for some $p > 2$.*

That a Hilbert space has type 2 and cotype 2 follows from the parallelogram law. The contents of (i) is the converse, which is due to Kwapien [**343**]; see also Chapter 3 of Pisier [**466**]. Part (ii) is due to Maurey [**397**]. For a proof of (iii) see Proposition 1.f.17 in Lindenstrauss and Tzafriri [**372**]. The second part of (iii) follows from the first by Proposition E.11(i).

The sequence of type-p constants of E, denoted by $a_{n,p} = a_{n,p}(E)$, is the sequence of the smallest constants for which the inequality in the definition of type holds for a fixed n and for all choices of $\{x_i\}_{i=1}^n$ in E. It follows from the inequality $\sum_{i=1}^n \|x_i\| \leq n^{1/2}\big(\sum_{i=1}^n \|x_i\|^2\big)^{1/2}$ that $a_{n,p}(E) \leq n^{1/2}$ for every Banach space E. The sequence $b_{n,q} = b_{n,q}(E)$ of cotype-q constants of E is defined in an analogous way. The following proposition and its corollary are due to Maurey and Pisier; see, e.g., [**402**].

Proposition G.3. (i) *The type-p and cotype-q constants of E are submultiplicative; i.e.,*

$$a_{nm,p} \le a_{n,p}a_{m,p} \quad \textit{and} \quad b_{nm,q} \le b_{n,q}b_{m,q}$$

for all integers n and m and for all $1 \le p \le 2 \le q < \infty$.

(ii) *E has type p for some* $p > 1$ *if and only if* $\lim n^{-1/2}a_{n,2} = 0$ *(if and only if there is a natural number k such that* $a_{k,2} < \sqrt{k}$*).*

(iii) *E has cotype q for some* $q < \infty$ *if and only if* $\lim n^{-1/2}b_{n,2} = 0$ *(if and only if there is a natural number k such that* $b_{k,2} < \sqrt{k}$*).*

Proof. We shall consider only the type-p constants. The cotype constants are treated similarly.

(i) Let $\{x_j\}_{j=1}^{mn} \subset G$, and put

$$X_i(t) = \sum_{j=(i-1)m+1}^{im} r_j(t)x_j \,, \quad 1 \le i \le n.$$

Then

$$\int_0^1 \|X_i(t)\|^p dt \le a_{m,p}^p \sum_{j=(i-1)m+1}^{im} \|x_j\|^p.$$

The products $\{r_i(s)r_j(t)\}$ have the same joint distribution as $\{r_j(t)\}$; hence

$$\begin{aligned}\int_0^1 \Big\|\sum_{j=1}^{mn} r_j(t)x_j\Big\|^p dt &= \int_0^1\int_0^1 \Big\|\sum_{i=1}^{n} r_i(s)X_i(t)\Big\|^p ds dt \\ &\le a_{n,p}^p \int_0^1 \sum_{i=1}^{n} \|X_i(t)\|^p dt \le a_{n,p}^p a_{m,p}^p \sum_{j=1}^{nm} \|x_j\|^p\end{aligned}$$

i.e., $a_{nm,p} \le a_{n,p}a_{m,p}$.

(ii) Kahane's inequality and $\big(\sum_{i=1}^n \|x_i\|^p\big)^{1/p} \le n^{-1/2+1/p}\big(\sum_{i=1}^n \|x_i\|^2\big)^{1/2}$ imply that $a_{n,2} = O(n^{-1/2+1/p}a_{n,p})$. Hence $n^{-1/2}a_{n,2} = O(n^{-1+1/p}) \to 0$ whenever $p > 1$ and E has type p.

Conversely, assume that there is a k with $a_{k,2} < \sqrt{k}$, and fix $s > 2$ such that $a_{k,2} \le k^{1/2-1/s}$. By the monotonicity and submultiplicativity of the type-2 constants, there is a constant C such that $a_{n,2} \le Cn^{1/2-1/s}$ for every n. We shall show that E has type p for any $p < s'$, where $1/s + 1/s' = 1$, i.e., $1 - 1/s - 1/p < 0$.

Fix $\{x_i\}_{i=1}^n$ in E, and assume that $\|x_1\| \ge \ldots \ge \|x_n\|$. It then follows that $\|x_i\| \le i^{-1/p}\big(\sum_{i=1}^n \|x_i\|^p\big)^{1/p}$ for every i, and hence

$$\begin{aligned}\Big(\int \Big\|\sum_{i=1}^n r_i x_i\Big\|^2\Big)^{1/2} &\le a_{n,2}\Big(\sum_{i=1}^n \|x_i\|^2\Big)^{1/2} \\ &\le Cn^{1/2-1/s}\Big(\sum_{i=1}^n i^{-2/p}\Big)^{1/2}\Big(\sum_{i=1}^n \|x_i\|^p\Big)^{1/p}.\end{aligned}$$

The choice of p shows that $n^{1/2-1/s}\big(\sum_{i=1}^n i^{-2/p}\big)^{1/2}$ is bounded, and the result follows from Kahane's inequality. □

A Banach space E is said to contain l_p^n's uniformly if for every n there is an n-dimensional subspace E_n of E such that $\sup d(E_n, l_p^n) < \infty$. We shall use the notation

$$p_E = \sup\{p:\ E \text{ has type } p\},$$
$$q_E = \sup\{q:\ E \text{ has cotype } q\}.$$

Corollary G.4. *A Banach space E contains l_1^n's (respectively l_∞^n's) uniformly if and only if $p_E = 1$ (respectively $q_E = \infty$).*

Proof. We prove only the l_1^n's case. One direction is evident, so assume that E does not have type p for any $p > 1$. Fix n. By part (ii) of the previous lemma $a_{n,2} \geq \sqrt{n}$. It follows that we can choose for every $\varepsilon > 0$ vectors $\{x_i\}_{i=1}^n$ such that $\big(\int \|\sum r_i x_i\|^2\big)^{1/2} \geq (n-\varepsilon)^{1/2}\big(\sum \|x_i\|^2\big)^{1/2}$. (All the sums in this proof are taken over $1 \leq i \leq n$.)

Since also $\big(\int \|\sum r_i x_i\|^2\big)^{1/2} \leq \sum \|x_i\| \leq n^{1/2}\big(\sum \|x_i\|^2\big)^{1/2}$, it follows from the identity $\sum(\|x_i\| - \|x_j\|)^2 = n\sum\|x_i\|^2 - \big(\sum\|x_i\|\big)^2$ that all the $\|x_i\|$'s are essentially equal. By renormalizing we can assume that $\|x_i\| \leq 1+\varepsilon$ for all i and $\big(\int \|\sum r_i x_i\|^2\big)^{1/2} > n$.

Let f be a linear functional on $L_2([0,1], E)$ with $\|f\| = 1$ and $\langle f, \sum r_i x_i\rangle \geq n$. Since $\langle f, r_i x_i\rangle \leq 1+\varepsilon$ for every i, it follows that $\langle f, r_i x_i\rangle \geq 1 - n\varepsilon$ for every i. We shall show that if ε is small enough, then the x_i's are $(1+\varepsilon)^2$-equivalent to the unit vectors of l_1^n. Indeed, fix $\{a_i\}_{i=1}^n$. Then

$$\begin{aligned}\Big(\int \|\sum a_i r_i x_i\|^2\Big)^{1/2} &= \Big(\int \|\sum |a_i| r_i x_i\|^2\Big)^{1/2}\\ &\geq \sum |a_i|\langle f, r_i x_i\rangle \geq (1-n\varepsilon)\sum|a_i|.\end{aligned}$$

Hence

$$(1-n\varepsilon)\sum|a_i| \leq m(A)(1+\varepsilon)\sum|a_i| + \big(1-m(A)\big)(1+\varepsilon)^{-1}\sum|a_i|$$

where $A = \big\{t:\ \|\sum a_i r_i(t) x_i\| \geq (1+\varepsilon)^{-1}\sum|a_i|\big\}$. It follows that if ε is small enough, then $m(A) \geq \big(1 - n\varepsilon - (1+\varepsilon)^{-1}\big)/\big(1+\varepsilon-(1+\varepsilon)^{-1}\big) > 1 - 2^{-n}$.

Since $\|\sum a_i r_i(t) x_i\|$ is constant on the dyadic intervals of length 2^{-n}, it follows that A is all of $[0,1]$; i.e., $(1+\varepsilon)^{-1}\sum|a_i| \leq \|\sum a_i x_i\| \leq (1+\varepsilon)\sum|a_i|$. □

The following, much deeper theorem, also due to Maurey and Pisier [**402**] (see also Milman and Schechtman [**420**], Section 13), generalizes the simple cases $p_E = 1$ and $q_E = \infty$ to all values of these parameters.

Theorem G.5. *Let E be an infinite-dimensional Banach space. Then E contains $l_{p_E}^n$'s and $l_{q_E}^n$'s uniformly. In fact, for every n and for every $\varepsilon > 0$ there are n-dimensional subspaces F_n and G_n of E such that $d(F_n, l_{p_E}^n), d(G_n, l_{q_E}^n) \leq 1+\varepsilon$.*

A basic tool used to prove this theorem is Krivine's theorem, which is discussed and proved in the second section of Chapter 12.

H. Quasi-Normed Spaces

Definition H.1. *A quasi-norm on a real vector space E is a function $x \longrightarrow \|x\|$ from E to $[0, \infty)$ which satisfies*

(i) $\|x\| > 0$ for every $x \neq 0$ in E,

(ii) $\|tx\| = |t|\,\|x\|$ for every $t \in \mathbb{R}$ and $x \in E$,

(iii) there is a $k \geq 1$ such that $\|x+y\| \leq k(\|x\| + \|y\|)$ for every $x, y \in E$.

The smallest possible k is called the modulus of concavity of $\|\cdot\|$.

The balls with respect to $\|\cdot\|$ define a linear topology on E, and it is possible to define an invariant metric d on E which determines the same topology. (Proposition H.2 below shows that, in fact, there is an $0 < r \leq 1$ such that $d(x,y) = \|x-y\|^r$ is such a metric.) A quasi-normed space E is called a quasi-Banach space if every $\|\cdot\|$-Cauchy sequence in E converges; i.e., E is complete with respect to the invariant metric d. The most important class of quasi-Banach spaces which are not already Banach spaces is the class of $L_p(\mu)$ spaces for $0 < p < 1$ with the usual quasi-norm $\|g\|_p = (\int |g|^p)^{1/p}$. In this case

$$\|g+h\|_p \leq 2^{\frac{1}{p}-1}\big(\|g\|_p + \|h\|_p\big)$$

i.e., the modulus of concavity of $L_p(\mu)$ is $2^{\frac{1}{p}-1}$.

A linear operator T between quasi-Banach spaces E and F is continuous if and only if it is bounded, and we put, as usual, $\|T\| = \sup\{\|Tx\| : \|x\| \leq 1\}$. The standard results depending on Baire category, like the open mapping theorem, are valid also in the context of quasi-Banach spaces (with the same proof). On the other hand, quasi-normed spaces are not necessarily locally convex, and the Hahn-Banach theorem and results depending on it are in general false in this context. If E is a quasi-Banach space and if F a closed linear subspace of E, then the quotient space E/F is a quasi-Banach space with the usual definition of the quotient quasi-norm. The notion of equivalence of quasi-norms is defined just as for norms.

A subset W of a linear topological space E is called bounded if every neighborhood of 0 in E contains a multiple of W. The topology of a linear topological space is induced by a quasi-norm if and only if it is locally bounded, i.e., if and only if it contains a bounded open set. Indeed, assume that W is a bounded neighborhood of 0 in E. The continuity of multiplication by scalars implies that there is an $\varepsilon > 0$ such that $V = \bigcup\{tW : |t| < \varepsilon\}$ is contained in W. Replacing W by V, we can thus assume that W is a symmetric star-shaped bounded neighborhood of 0. Similarly, using the continuity of addition and the boundedness of W, we can also assume that there is a constant k such that $W + W \subset kW$. Under these conditions a quasi-norm on E that induces the original topology is given by the Minkowski functional of W, namely, $\|x\| = \inf\{t > 0 : t^{-1}x \in W\}$.

A quasi-Banach space is isomorphic to a Banach space if and only if it is locally convex, i.e., when 0 has a convex bounded neighborhood.

The following proposition is due to Aoki [**18**] and Rolewicz [**500**]. We follow the proof in the book [**319**] by Kalton, Peck and Roberts.

Proposition H.2. *For every quasi-normed space E there is an $0 < r \leq 1$ and an equivalent quasi-norm $\|\cdot\|$ satisfying*

$$\|x+y\|^r \leq \|x\|^r + \|y\|^r$$

for all $x, y \in E$.

Proof. Denote the original quasi-norm on E by $\|\!|\cdot|\!\|$, and let k be its concavity constant. Let r be such that $2^{1/r} = 2k$, and define a new quasi-norm by

$$\|x\| = \inf\Big\{\Big(\sum \|\!|x_i|\!\|^r\Big)^{1/r} : x = \sum x_i\Big\}.$$

It is evident that $\|tx\| = |t|\,\|x\|$, that $\|x\| \le \|\!|x|\!\|$, and that $\|\cdot\|$ satisfies the required inequality. We shall show that

$$\text{(H.1)} \qquad \Big\|\!\Big|\sum_{i=1}^n x_i\Big|\!\Big\| \le 4^{1/r}\Big(\sum_{i=1}^n \|\!|x_i|\!\|^r\Big)^{1/r}$$

for all $\{x_i\}_{i=1}^n$. This will show that $\|\!|x|\!\| \le 4^{1/r}\|x\|$ and will conclude the proof of the proposition.

For every $x \in E$ put $N(x) = 2^{m/r}$, where the integer m is chosen so that $2^{(m-1)/r} < \|\!|x|\!\| \le 2^{m/r}$. With this notation (H.1) will follow once we show that

$$\text{(H.2)} \qquad \Big\|\!\Big|\sum_{i=1}^n x_i\Big|\!\Big\| \le 2^{1/r}\Big(\sum_{i=1}^n N(x_i)^r\Big)^{1/r}.$$

If there are two x_i's, say x_1 and x_2, such that $N(x_1) = N(x_2) = 2^{m/r}$, then we can replace the pair x_1, x_2 in the proof of (H.2) by their sum. Indeed,

$$\|\!|x_1 + x_2|\!\| \le k\big(\|\!|x_1|\!\| + \|\!|x_2|\!\|\big) \le 2k2^{m/r} = 2^{(m+1)/r}$$

by the choice of r. Thus

$$N(x_1 + x_2)^r \le 2^{m+1} = N(x_1)^r + N(x_2)^r.$$

Using this remark and rearranging the x_i's, we may assume that the sequence $\{N(x_i)\}_{i=1}^n$ is strictly decreasing; hence

$$\|\!|x_i|\!\| \le N(x_i) \le 2^{-(i-1)/r}N(x_1)$$

for $1 \le i \le n$.

Repeated use of the definition of the concavity modulus and the definition of r then give

$$\begin{aligned}\Big\|\!\Big|\sum_{i=1}^n x_i\Big|\!\Big\| &\le k\|\!|x_1|\!\| + k^2\|\!|x_2|\!\| + \ldots + k^n\|\!|x_n|\!\| \\ &\le \sum_{i=1}^n N(x_1)k^i 2^{-(i-1)/r} = 2^{1/r}N(x_1)\sum_{i=1}^n 2^{-i} \\ &\le 2^{1/r}N(x_1) \le 2^{1/r}\Big(\sum_{i=1}^n N(x_i)^r\Big)^{1/r}.\end{aligned}$$

□

The next proposition is due to N. Kalton; see Theorem 2.5(ii) in [**310**] and Theorem 2.1. in [**312**]. The reader is referred to Appendix G for information on type and on type-p constants.

Proposition H.3. *If a quasi-Banach space E has type p for some $p > 1$, then E is a Banach space (i.e. there is a norm equivalent to the given quasi-norm).*

Proof. For every integer n put

$$b_n = \sup\Big\{\big\|\sum_{i=1}^{n} x_i\big\| : \ \|x_i\| \le 1\Big\}.$$

Then E is a Banach space if and only if there is a constant C such that $b_n \le Cn$. One direction follows from the triangle inequality. Conversely, let $\lambda_i = m_i/l$ be positive rationals whose sum is 1, and let $\|x_i\| \le 1$ for $1 \le i \le n$. Then $l = \sum m_i$, and thus

$$\big\|\sum_{i=1}^{n} \lambda_i x_i\big\| = l^{-1}\big\|\sum_{i=1}^{n} m_i x_i\big\| \le l^{-1} b_l \le C.$$

By continuity the same estimate holds for all $\lambda_i \ge 0$ with $\sum \lambda_i = 1$. This shows that the convex hull of the unit ball of E is bounded; i.e., E is a Banach space.

We now show that if E has type $p > 1$, then b_n/n is bounded. Fix vectors $\{x_i\}_{i=1}^{2n}$ with $\|x_i\| \le 1$ for all i. By the definition of the type-2 constants $a_{2n,2}$ of E there are signs $\theta_i = \pm 1$ such that

$$\big\|\sum_{i=1}^{2n} \theta_i x_i\big\| \le (2n)^{1/2} a_{2n,2}.$$

Assume, as we may, that the cardinality of the set A of the indices i with $\theta_i = +1$, is at most n and choose $0 < r < 1$ (by Proposition H.2) such that $\|u+v\|^r \le \|u\|^r + \|v\|^r$ for all $u, v \in E$. Since

$$\sum_{i=1}^{2n} x_i = 2\sum_{i\in A} x_i - \sum_{i=1}^{2n} \theta_i x_i$$

it follows that

$$\big\|\sum_{i=1}^{2n} x_i\big\|^r \le 2^r\big\|\sum_{i\in A} x_i\big\|^r + \big\|\sum_{i=1}^{2n} \theta_i x_i\big\|^r$$

and hence

$$\big(b_{2n}\big)^r \le 2^r b_n^r + \big((2n)^{1/2} a_{2n,2}\big)^r.$$

Putting

$$\alpha_j = (a_{2^j,2}/2^{j/2})^r, \quad \beta_j = (b_{2^j}/2^j)^r$$

the last inequality translates into

$$\beta_{j+1} - \beta_j \le \alpha_{j+1}.$$

Since $\{\alpha_j\}$ is submultiplicative and $\lim_{j\to\infty} \alpha_j = 0$ (by Lemma G.3), it follows that there are a $\lambda < 1$ and an index k such that $\alpha_{kj} \le \lambda^j$; consequently $\sum \alpha_j < \infty$. Summing the last inequality over j, we obtain $\sup \beta_j < \infty$; hence $\sup b_n/n < \infty$. $\square$

For systematic presentations of quasi-normed spaces we refer the reader to the books by Kalton, Peck and Roberts [**319**] and by Rolewicz [**501**].

I. The Complex Interpolation Method

In this appendix we describe the complex interpolation method and present the results from this theory which are needed in the main text. Occasionally we impose some simplifying conditions which provide, however, a sufficiently general framework for our purposes. All the Banach spaces in this appendix will be over the complex field.

Denote the strip $\{z : 0 < \Re(z) < 1\}$ in the complex plane by $\mathcal{S}$, and denote its closure by $\bar{\mathcal{S}}$ and its boundary by $\partial\mathcal{S}$.

For $\xi \in \mathcal{S}$ let μ_ξ be the harmonic measure for the point ξ; i.e., μ_ξ is the unique probability measure on $\partial\mathcal{S}$ such that

$$f(\xi) = \int_{\partial\mathcal{S}} f(z) d\mu_\xi(z) \tag{I.1}$$

for every scalar-valued bounded continuous function on $\bar{\mathcal{S}}$ which is harmonic in $\mathcal{S}$. The measure μ_ξ is absolutely continuous with respect to the Lebesgue measure on $\partial\mathcal{S}$, and its Radon-Nikodým derivative is given by the Poisson kernel. For each fixed $\xi, \zeta \in \mathcal{S}$, the Radon-Nikodým derivative $d\mu_\xi/d\mu_\zeta$ (which is just the quotient of the appropriate Poisson kernels) is a bounded function on $\partial\mathcal{S}$.

The representation (I.1) holds under less restrictive assumptions on the function f. If $f \in H_p(\mathcal{S})$ for some $1 \le p \le \infty$, then its nontangential boundary values exist almost everywhere on $\partial\mathcal{S}$, and (I.1) still holds when these boundary values are used as the definition of f on $\partial\mathcal{S}$. The representation also holds for Banach space-valued functions.

An interpolation pair (E_0, E_1) is a pair of Banach spaces which are linearly embedded in a linear space W so that $E_0 + E_1 = W$. The space W is a Banach space when normed by

$$\|x\|_W = \inf\big\{\|u\|_0 + \|v\|_1 : x = u + v,\ u \in E_0, v \in E_1\big\}.$$

It is just the quotient of $(E_0 \oplus E_1)_1$ modulo the subspace $\{(u, -u) : u \in E_0 \cap E_1\}$. In particular, W is reflexive whenever E_0 and E_1 are.

When we also need to consider dual spaces, it is convenient to assume that $V = E_0 \cap E_1$ is dense in both E_0 and E_1. The space V is a Banach space under the norm

$$\|x\|_V = \max\big\{\|x\|_0, \|x\|_1\big\}$$

and it is just the diagonal subspace of $(E_0 \oplus E_1)_\infty$.

Since V is dense in E_0, the identity map from E_0^* to V^* is one-one. We identify E_0^* with its image, which is a (not necessarily closed) subspace of V^*. Similarly, E_1^* is identified with a subspace of V^*, and (E_0^*, E_1^*) is an interpolation pair with V^* as the containing linear space. The subspace $V^\perp = \{(u^*, -u^*) : u^* \in E_0^* \cap E_1^*\}$ is the annihilator of V in $(E_0 \oplus E_1)_\infty^* = (E_0^* \oplus E_1^*)_1$, and standard duality shows that $V^* = (E_0^* \oplus E_1^*)_1/V^\perp$; i.e., the natural dual norm on V^* is exactly the standard norm as defined above on the containing space of the interpolation pair (E_0^*, E_1^*).

When we use duality, we shall always make the tacit assumption that $E_0 \cap E_1$ is dense in E_0 and in E_1 (and that $E_0^* \cap E_1^*$ is dense in E_0^* and E_1^* when we also need to talk of their duals).

Let (E_0, E_1) be an interpolation pair, and denote by $\mathcal{F} = \mathcal{F}(E_0, E_1)$ the space of all bounded continuous functions $f : \bar{\mathcal{S}} \to W$ which are analytic in $\mathcal{S}$, satisfying $f(it) \in E_0$ and $f(1 + it) \in E_1$ for every $t \in \mathbb{R}$, and such that the restrictions of f

to the lines $i\mathbb{R}$ and $1 + i\mathbb{R}$ are bounded and continuous as maps into E_0 and E_1 respectively. The space $\mathcal{F}$ is a Banach space under the norm

$$\|f\|_{\mathcal{F}} = \max\left\{\sup\|f(it)\|_{E_0}, \sup\|f(1+it)\|_{E_1}\right\} < \infty$$

and it follows from (I.1) that $\|f(\xi)\|_W \leq \|f\|_{\mathcal{F}}$ for every $\xi \in \mathcal{S}$.

For $0 < \theta < 1$ we define the interpolation space $E_\theta = [E_0, E_1]_\theta$ as the space of all $x \in W$ such that $x = f(\theta)$ for some $f \in \mathcal{F}$, equipped with the norm

$$\|x\|_\theta = \inf\left\{\|f\|_{\mathcal{F}} : \ f \in \mathcal{F} \text{ satisfies } f(\theta) = x\right\}.$$

The following proposition summarizes several facts on interpolation spaces.

Proposition I.1. (i) *If $f \in \mathcal{F}(E_0, E_1)$ and $g \in \mathcal{F}(E_0^*, E_1^*)$, then the scalar-valued function $\psi(z) = \langle f(z), g(z)\rangle$ is analytic in $\mathcal{S}$ and bounded by $\|f\|\,\|g\|$.*

(ii) *Let (E_0, E_1) be an interpolation pair such that at least one of the spaces E_0 and E_1 is reflexive, and let $0 < \theta < 1$. Then E_θ is reflexive, and $E_\theta^* = [E_0^*, E_1^*]_\theta$.*

(iii) *The reiteration theorem: If $0 < \theta_1, \theta_2, \tau < 1$, then $[E_{\theta_0}, E_{\theta_1}]_\tau = E_\eta$, where $\eta = (1-\tau)\theta_0 + \tau\theta_1$.*

(iv) *If one of the spaces E_0 or E_1 is uniformly convex, then E_θ is uniformly convex for every $0 < \theta < 1$.*

(v) *Fix $1 \leq p_0 < p_1 \leq \infty$. Then $[L_{p_0}(\mu), L_{p_1}(\mu)]_\theta = L_p(\mu)$ where $\frac{1}{p} = \frac{1-\theta}{p_0} + \frac{\theta}{p_1}$.*

(vi) *With p, p_0, p_1 and θ as above, $[C_{p_0}, C_{p_1}]_\theta = C_p$ (where the C's are the Schatten classes of compact operators on a Hilbert space; see* Appendix J*).*

The equalities in (ii)-(vi) *mean that the spaces in question can be identified so that their natural norms coincide.*

Part (i) is immediate if f and g are finite linear combinations of functions of the form $\varphi(z)x$ and $\eta(z)x^*$, where $x \in E_0 \cap E_1$, $x^* \in E_0^* \cap E_1^*$, and the functions φ and η are scalar analytic functions. The general case follows because such functions are dense in $\mathcal{F}(E_0, E_1)$ and in $\mathcal{F}(E_0^*, E_1^*)$ respectively (see Lemma 4.2.3 in Bergh and Löfström [**56**]).

Part (ii) is due to Calderón [**108**] (and see also Corollary 4.5.2 in Bergh and Löfström [**56**]).

Part (iii) was proved under additional assumptions by Calderón [**108**]. The general case was proved by Cwikel [**128**].

Part (iv) is due to Cwikel and Reisner [**129**].

For a proof of part (v), see Theorem 5.1.1 in Bergh and Löfström [**56**].

For part (vi) see Gohberg and Krein [**219**], page 137, or Simon [**534**], page 32. (The proofs in these references are formulated as estimates of the norm of an operator from C_p to itself in terms of its norms as an operator on C_{p_0} and on C_{p_1}. They are based on Hadamard's three lines theorem and can be interpreted as identifying $[C_{p_0}, C_{p_1}]_\theta$ with C_p.)

The development of the complex interpolation method (mainly by Calderón in [**108**]) is motivated by Thorin's classical proof of the Riesz interpolation theorem. In the language of interpolation spaces, the "interpolation theorem" is that if (E_0, E_1) and (F_0, F_1) are two interpolation pairs, and if T is a linear operator from $E_0 \cap E_1$ to $F_0 \cap F_1$ which extends to a bounded operator from E_i to F_i (with norms $\|T\|_i$ respectively), then it is also bounded as an operator from E_θ to F_θ, and its norm satisfies $\|T\|_\theta \leq \|T\|_0^{1-\theta}\|T\|_1^\theta$.

It will be very useful to introduce another representation of $\|x\|_\theta$, due to Calderón [**108**]. Let $\mathcal{F}_2(\theta)$ be the completion of the space $\mathcal{F}$ under the norm

$$\|f\|_{\mathcal{F}_2(\theta)} = \Big(\int \|f(z)\|^2 d\mu_\theta(z)\Big)^{1/2}$$

where the norm under the integral sign is that of E_0 on the line $i\mathbb{R}$ and that of E_1 on the line $1+i\mathbb{R}$.

If we denote the normalized restrictions of μ_θ to the lines $i\mathbb{R}$ and $1+i\mathbb{R}$ by μ_θ^0 and μ_θ^1 respectively, then $\mathcal{F}_2(\theta)$ is a subspace of $\big(L^2(\mu_\theta^0, E_0)\oplus L^2(\mu_\theta^1, E_1)\big)_2$. Hence $\mathcal{F}_2(\theta)$ is reflexive, or uniformly convex, whenever E_0 and E_1 are.

Proposition I.2. *The norm on E_θ can be computed by*

$$\|x\|_\theta = \inf\big\{\|f\|_{\mathcal{F}_2(\theta)} :\ f(\theta) = x\big\}. \tag{I.2}$$

Proof. It was shown by Calderón [**108**] (see also Lemma 4.3.2 in Bergh and Löfström [**56**]) that for every $f\in\mathcal{F}$

$$\log\|f(\theta)\|_\theta \le \int \log\|f(z)\|d\mu_\theta(z).$$

Hence $\|f(\theta)\|_\theta \le \big(\int\|f(z)\|^2 d\mu_\theta(z)\big)^{1/2} = \|f\|_{\mathcal{F}_2(\theta)}$ by the convexity of e^{2t}.

By the definition of the norm in E_θ, it follows that for every $x\in E_\theta$

$$\|x\|_\theta \le \inf\big\{\|f\|_{\mathcal{F}_2(\theta)} :\ f\in\mathcal{F} \text{ and } f(\theta) = x\big\}. \tag{I.3}$$

But this infimum coincides with that of (I.2) because $\mathcal{F}$ is dense in $\mathcal{F}_2(\theta)$.

The reverse inequality is obvious. □

Proposition I.3. *Let (E_0, E_1) be an interpolation pair, and let $0<\theta<1$. If E_0 and E_1 are reflexive, then for every $x\in E_\theta$ with $\|x\|_\theta = 1$ there is a function $f_x\in\mathcal{F}_2(\theta)$ such that $\|f_x\|_{\mathcal{F}_2(\theta)} = 1$ and $f_x(\theta) = x$. If E_0 and E_1 are uniformly convex, then this function is unique and satisfies*

(i) $\|f_x(z)\| = 1$ *almost everywhere on $\partial\mathcal{S}$ with respect to μ_θ,*

(ii) $\|f_x(\tau)\|_\tau = 1$ *for every $0<\tau<1$.*

Proof. Since E_0 and E_1 are reflexive, so is $\mathcal{F}_2(\theta)$. Thus the infimum in (I.2) is actually a minimum. If the two spaces are uniformly convex, so is $\mathcal{F}_2(\theta)$, and this minimum is attained at a unique element f_x.

To prove (i) choose $f_n\in\mathcal{F}$ with $f_n(\theta) = x$ and $\|f_n\|_{\mathcal{F}}\to 1$. (This is possible because the infima in (I.2) and (I.3) coincide.) Since $1\le\|f_n\|_{\mathcal{F}_2(\theta)}\le\|f_n\|_{\mathcal{F}}\to 1$, it follows that $\mu_\theta\{\|f_n(z)\|<1-\varepsilon\}\to 0$ as $n\to\infty$ for every $\varepsilon>0$. Thus (i) follows, since $\|f_n - f_x\|_{\mathcal{F}_2(\theta)}\to 0$ by the uniqueness of f_x.

We now prove (ii). Note first that since μ_τ is equivalent to μ_θ, it follows that $\|f_x(z)\| = 1$ almost everywhere with respect to τ; hence $\|f_x(\tau)\|_\tau\le 1$.

To see that actually $\|f_x(\tau)\|_\tau = 1$, we use duality. Fix $x^*\in[E_0,E_1]_\theta^* = [E_0^*, E_1^*]_\theta$ with $\|x^*\|_\theta = 1 = \langle x, x^*\rangle$. Denote by $\mathcal{F}_2(\theta)^\#$ the space analogous to $\mathcal{F}_2(\theta)$, with E_0 and E_1 replaced by E_0^* and E_1^*. Since E_0^* and E_1^* are reflexive, there is a function $g\in\mathcal{F}_2(\theta)^\#$ such that $\|g\|_{\mathcal{F}_2(\theta)^\#} = 1$ and $g(\theta) = x^*$. The scalar analytic function $\psi(z) = \langle f_x(z), g(z)\rangle$ is then bounded by 1 on $\mathcal{S}$ and attains this value at the interior point θ. By the maximum principle it is identically equal to 1. In particular, this holds at the point $z=\tau$; hence $\|f_x(\tau)\|_\tau\ge\langle f_x(\tau), g(\tau)\rangle = 1$. □

Remark. If we define the spaces $\mathcal{F}_p(\theta)$ for $1 \le p < \infty$ in the obvious way, the same proofs show that a formula analogous to I.2 holds for all such p and that f_x is simultaneously optimal in $\mathcal{F}_p(\theta)$ for all p.

In the rest of the appendix we shall discuss Banach lattices. For information on lattices and on p-convexity and q-concavity, see Appendix E. All our lattices will be lattices of measurable functions on some measure space $(\Omega, \mathcal{B}, \mu)$. By Proposition E.9 many abstract Banach lattices have such a representation. If E is such a lattice, we shall also use the same letter E to denote its complexification, namely, the space of measurable functions f on Ω such that $|f| \in E$ with the norm $\|f\| = \||f|\|$.

Let E_0 and E_1 be two Banach lattices on the same measure space, and let $0 < \theta < 1$. We denote by $E_0^{1-\theta}E_1^\theta$ the Banach lattice of all measurable functions f such that $|f| = |f_0|^{1-\theta}|f_1|^\theta$ with $f_i \in E_i$ under the norm $\|f\| = \inf\{\|f_0\|_{E_0}^{1-\theta}\|f_1\|_{E_1}^\theta\}$, where the infimum is taken over all such representations. We shall use repeatedly the fact, proved in §13.6 of Calderón [**108**], that if at least one of the spaces E_0 or E_1 is reflexive, then $[E_0, E_1]_\theta = E_0^{1-\theta}E_1^\theta$.

As a first application of this representation we check that if E_0 is p_0-convex and E_1 is p_1-convex, then E_θ is p-convex where $1/p = (1-\theta)/p_0 + \theta/p_1$. Indeed, fix $f_i \in E_\theta$, and factor $f_i = x_i^{1-\theta}y_i^\theta$ with $x_i \in E_0$ and $y_i \in E_1$. To simplify the notation, assume that the convexity constants of E_0 and E_1 are equal to 1 and that the representation of f_i is such that we have the exact equality $\|f_i\|_\theta = \|x_i\|_0^{(1-\theta)}\|y_i\|_1^\theta$.

Put $\lambda_i = \left(\|y_i\|_1^{p_1}/\|x_i\|_0^{p_0}\right)^{\frac{\theta(1-\theta)}{\theta p_0 + (1-\theta)p_1}}$. By Hölder's inequality, the definition of the norm in $E_0^{1-\theta}E_1^\theta$, and the definition of p-convexity we obtain

$$\begin{aligned}
\left\|\left(\sum |f_i|^p\right)^{1/p}\right\|_\theta &= \left\|\left(\sum |\lambda_i^p x_i^{(1-\theta)p}\lambda_i^{-p}y_i^{\theta p}|\right)^{1/p}\right\|_\theta \\
&\le \left\|\left(\sum |\lambda_i^{1/(1-\theta)}x_i|^{p_0}\right)^{(1-\theta)/p_0}\left(\sum |\lambda_i^{-1/\theta}y_i|^{p_1}\right)^{\theta/p_1}\right\|_\theta \\
&\le \left\|\left(\sum |\lambda_i^{1/(1-\theta)}x_i|^{p_0}\right)^{1/p_0}\right\|_0^{1-\theta}\left\|\left(\sum |\lambda_i^{-1/\theta}y_i|^{p_1}\right)^{1/p_1}\right\|_1^\theta \\
&\le \left(\sum \|\lambda_i^{1/(1-\theta)}x_i\|_0^{p_0}\right)^{(1-\theta)/p_0}\left(\sum \|\lambda_i^{-1/\theta}y_i\|_1^{p_1}\right)^{\theta/p_1} \\
&= \left(\sum \left(\|x_i\|_0^{1-\theta}\|y_i\|_1^\theta\right)^p\right)^{1/p} = \left(\sum \|f_i\|_\theta^p\right)^{1/p}.
\end{aligned}$$

A similar result holds by duality for the interpolation of q-concave lattices.

The following result of Pisier (Theorem 2.2 and Remark 2.5 in [**464**]) is a converse to these direct computations, in the sense that it represents an r-convex and s-concave lattice as an interpolation space between L_2 and a lattice F which has the "correct" convexity and concavity. We denote the conjugate exponent of p by p'; i.e., $1/p + 1/p' = 1$.

Proposition I.4. *Let E be a Banach lattice of measurable functions on some measure space $(\Omega, \mathcal{B}, \mu)$, and assume that E is r-convex and s-concave for some $1 < r < s < \infty$ with $M^r(E) = M_s(E) = 1$. Assume that $1 < p < 2$ satisfies $p \le r$ and $p' \ge s$. Then there is a Banach lattice F of measurable functions on Ω such that $E = [L_2(\mu), F]_\theta$, where $1/p = (1-\theta)/2 + \theta/1$; i.e., $\theta = 2/p - 1$.*

The lattice F is p_0-convex and q_0-concave where $1/r = (1-\theta)/2 + \theta/p_0$ and $1/s = (1-\theta)/2 + \theta/q_0$. Explicitly $p_0 = 2(p'-p)r/\big((2-r)pp' + (p'-p)r\big)$, and $q_0 = 2(p'-p)s/\big((2-s)pp' + (p'-p)s\big)$.

We shall use Proposition I.4 for representing lattices which are r-convex and "almost" r-concave as being "close" to L_r in an interpolation scale between two appropriately chosen lattices. This representation is used in Chapters 9 and 10. In these applications it is important to give quantitative estimates for the various parameters. The technical formulation of Corollary I.5 below stems from the need to give these estimates.

Let F be a reflexive Banach lattice of measurable functions on $(\Omega, \mathcal{B}, \mu)$. It follows from the Lozanovskii factorization (see the Notes and Remarks section in Chapter 9) that $[F, F^*]_{1/2} = L_2$. Indeed, since F is reflexive, every $f_1 \geq 0$ in L_1 can be factored as $f_1 = gh$, where $g, h \geq 0$ are in F and F^* respectively, and $\|f_1\|_1 = \|g\|_F \, \|h\|_{F^*}$. If $f \in L_2$, factor $|f|^2 = gh$ as above, and then write

$$f = \big(g^{1/2}\mathrm{sign}(f)\big) \cdot h^{1/2} \in F^{1/2}(F^*)^{1/2} = [F, F^*]_{1/2}.$$

That $[F, F^*]_{1/2} \subset L_2$ is obvious.

Corollary I.5. *Let E be a Banach lattice of measurable functions on some measure space $(\Omega, \mathcal{B}, \mu)$, and let $r_0 > 1$ and $s_0 < \infty$. Assume that E is r-convex and s-concave for some $r_0 < r < s < s_0$, with $M^r(E) = M_s(E) = 1$. Then there are uniformly convex lattices G and H of measurable functions on Ω whose moduli of convexity depend only on r_0 and on s_0, and $0 < \eta < 1$ (which depends on r and on s) such that $\eta \to 1/2$ as $s \to r$, and such that $L_r = [G, H]_{1/2}$ and $E = [G, H]_\eta$.*

Proof. We shall construct G and H which are p_0-convex and q_0-concave for some p_0 and q_0 which are bounded away from 1 and ∞ by functions of r_0 and s_0. By Proposition E.10, G and H are uniformly convex as required.

Consider first the case $r = 2$. Let $p < s'$, and write $E = [F, L_2]_\theta$ as in Proposition I.4, where $\theta/2 = 1/p - 1/2$. We then take $G = F$ and $H = F^*$. The formulas for θ, p_0, and q_0 in the proposition show that $p_0 = 2$; and since we can choose $p \to 2$ when $s \to 2$, we also obtain $\theta \to 0$. (In the proof for general r below, we shall need the fact that by appropriate choice of p we can also ensure that $q_0 \to 2$ as $s \to 2$.)

As remarked above $[F, F^*]_{1/2} = L_2$. Also, by the reiteration theorem

$$E = [F, L_2]_\theta = [F, [F, F^*]_{1/2}]_\theta = [F, F^*]_{1/2-\theta/2}$$

and $\eta = 1/2 - \theta/2$ satisfies the requirements.

Assume now that $r < 2$. (The case $r > 2$ follows by duality.)

The $2/r$-convexification $E^{(2/r)}$ of E is a 2-convex and $2s/r$-concave lattice. Fix $q > 2$ such that $q' > 2/r$. It follows from the case $r = 2$ above that $E^{(2/r)} = [F, F^*]_\eta$ and $L_2 = [F, F^*]_{1/2}$, where $\eta \to 1/2$ as $s \to r$, and also that F is 2-convex and q concave. Hence F^* is q'-convex. Since $2 > q' > 2/r$, the $2/r$-concavification of all the spaces involved is well defined. Taking this concavification, we find that the result follows with $G = F_{(2/r)}$ and $H = \big(F^*\big)_{(2/r)}$. (That concavification commutes with the interpolation follows directly from the formula $[E_0, E_1]_\theta = E_0^{1-\theta} E_1^\theta$.) □

For a systematic presentation of the theory of interpolation and for historical notes, see Bergh and Löfström [**56**].

J. Operators on Hilbert Space

A bounded linear operator T on a complex Hilbert space H is said to be nonnegative if $T = T^*$ and $\langle Tx, x\rangle \geq 0$ for every $x \in H$. For every bounded linear operator T on H there is a unique nonnegative operator $|T|$ (the absolute value of T) such that $|T|^2 = T^*T$. It satisfies $\big\||T|\big\| = \big\|T\big\|$, and there is an isometry U from the closure of the range of $|T|$ onto the closure of the range of T such that $T = U|T|$. The operator U is extended to a partial isometry on H by defining it to be zero on $(|T|H)^{\perp}$, and the representation $T = U|T|$ is called the polar decomposition of T. The operator T is compact if and only if $|T|$ is.

Every nonnegative compact operator S on a Hilbert space H can be represented as $S = \sum \lambda_j e_j \otimes e_j$, where $\lambda_j \searrow 0$ is the sequence (which may be finite) of nonzero eigenvalues of S (with their multiplicities), $\{e_j\}$ is the orthonormal system of eigenvectors of S, and $e \otimes f$ denotes the rank-one operator $(e \otimes f)(x) = \langle e, x\rangle f$. It follows from the polar decomposition that every compact operator T on H has a representation of the form

$$T = \sum \mu_j e_j \otimes f_j$$

where $\mu_j \to 0$ is the sequence (which may be finite) of nonzero eigenvalues of $|T|$ (with multiplicity), and $\{e_j\}$ and $\{f_j\}$ are orthonormal systems in H. The μ_j's are called the singular numbers of T. They are positive, and we arrange them in a nonincreasing sequence.

For $1 \leq p < \infty$, we say that the compact operator T belongs to the Schatten class C_p if $\sum \mu_j^p < \infty$. The class C_p is a Banach space under the natural norm $\|T\|_p = (\sum \mu_j^p)^{1/p}$. The class of all compact operators with the operator norm $\|T\| = \|T\|_\infty = \mu_1 = \max \mu_i$ is denoted by C_∞. The C_p norms satisfy the following identities and inequalities:

(i) $\|T\|_p = \|T^*\|_p$.

(ii) $\|T\|_p = \|UTV\|_p$ for any two unitary operators U and V.

(iii) $\|S_1 T S_2\|_p \leq \|S_1\|\,\|S_2\|\,\|T\|_p$ for any two bounded operators S_1 and S_2.

(iv) $\|T_1 T_2\|_r \leq \|T_1\|_p \|T_2\|_q$ for $1/r = 1/p + 1/q$.

The space C_2 is a Hilbert space, and its elements are called Hilbert-Schmidt operators. The norm in C_2 is called the Hilbert-Schmidt norm. If $T \in C_2$, and if $\{e_j\}$ and $\{f_i\}$ are any two complete orthonormal systems, then

$$\|T\|_2^2 = \sum_{n=1}^{\infty} \|Te_n\|^2 = \sum_{n=1}^{\infty} \|T^* f_n\|^2 = \sum_{n,m=1}^{\infty} \big|\langle Te_n, f_m\rangle\big|^2.$$

The operators in C_1 are called trace-class operators. The trace of an operator $T \in C_1$ is defined as $\operatorname{tr} T = \sum \langle Te_n, e_n\rangle$, where $\{e_n\}$ is a complete orthonormal system in H (and the definition does not depend on the choice of $\{e_n\}$). The series converges absolutely, and the trace is a norm-one linear functional on C_1. One has that $\operatorname{tr} T^* = \overline{\operatorname{tr} T}$ and that $\operatorname{tr} ST = \operatorname{tr} TS$ for every bounded linear operator S. A deep theorem of Lidskii (which we do not use in this book) says that the trace of $T \in C_1$ is given by $\operatorname{tr} T = \sum \lambda_j$, where $\{\lambda_j\}$ are the eigenvalues of T counted with multiplicity, and the series converges absolutely.

From the inequalities mentioned above it follows that if $T \in C_p$ and $S \in C_q$, with $1/p + 1/q = 1$, then

$$|\mathrm{tr}\, TS| \leq \|T\|_p \|S\|_q.$$

As in the case of l_p spaces, the dual of C_p (for $1 < p < \infty$) can be identified with C_q via trace duality: $\langle T, S\rangle = \mathrm{tr}\,(TS)$. (If $p = q = 2$, then, as is usual, we use the antisymmetric inner product $\langle T, S\rangle = \mathrm{tr}\,(TS^*)$.) The dual of C_∞ is identified by trace duality with C_1 and the dual of C_1 with the space $L(H)$ of all bounded linear operators on H with the usual operator norm. The natural embedding of C_∞ in $L(H)$ is the canonical embedding of a Banach space in its second dual.

The books by Schatten [**515**], Gohberg and Krein [**219**] and Simon [**534**] contain basic information and further references on the spaces C_p. The following proposition summarizes the Banach space properties of these spaces that are needed in the main text.

Proposition J.1. (i) *For $1 < p < \infty$ the spaces C_p are uniformly convex and uniformly smooth, and their moduli of convexity and of smoothness behave as those of the corresponding L_p spaces. Thus $\delta_{C_p}(\varepsilon)$ behaves as ε^2 for $1 < p \leq 2$ and as ε^p for $2 \leq p < \infty$, and $\rho_{C_p}(\tau)$ behaves as τ^p for $1 < p \leq 2$ and as τ^2 for $2 \leq p < \infty$.*

(ii) *The spaces C_p have the same type and cotype as the corresponding L_p spaces. Thus the type of C_p is p for $1 < p \leq 2$ and 2 for $2 \leq p < \infty$, and the cotype is 2 for $1 \leq p \leq 2$ and p for $2 \leq p < \infty$.*

(iii) *The spaces C_p form an interpolation scale for the complex interpolation method: If $r < p < s$, then $C_p = [C_r, C_s]_\theta$ where $1/p = \theta/r + (1-\theta)/s$.*

(iv) *The spaces C_p have a Schauder basis, but (except for $p = 2$) they do not have an unconditional basis. Moreover, for $p \neq 2$ they do not even embed isomorphically into any Banach lattice with a nontrivial cotype.*

Parts (i) and (ii) are due to Tomczak-Jaegermann [**557**]. For a proof of (iii) see Gohberg and Krein [**219**], page 137, or Simon [**534**], page 32. For information on bases in C_p and the fact that they do not have an unconditional basis, see Kwapien and Pełczyński [**344**]. That C_p does not embed in a Banach lattice with a finite cotype is due to Pisier [**463**].

Bibliography

1. I. Aharoni, *Every separable metric space is Lipschitz equivalent to a subset of* c_0, Israel J. Math. **19** (1974), 284–291. (cited on pp 184, 196)
2. ———, *Lipschitz maps and uniformly continuous functions between Banach spaces* (Hebrew), Ph.D. thesis, Hebrew University, 1978. (cited on p 184)
3. I. Aharoni and J. Lindenstrauss, *Uniform equivalence between Banach spaces*, Bull. Amer. Math. Soc. **84** (1978), 281–283. (cited on pp 33, 184)
4. ———, *An extension of a result of Ribe*, Israel J. Math. **52** (1985), 59–64. (cited on p 257)
5. I. Aharoni, B. Maurey, and B. S. Mityagin, *Uniform embeddings of metric spaces and of Banach spaces into Hilbert spaces*, Israel J. Math. **52** (1985), 251–265. (cited on p 195)
6. Ya. I. Alber, *A bound for the modulus of continuity for metric projections in a uniformly convex and uniformly smooth Banach spaces*, J. Approx. Th. **85** (1996), 237–249. (cited on p 58)
7. D. Aldous, *Subspaces of* L^1 *via random measures*, Trans. Amer. Math. Soc. **267** (1981), 445–463. (cited on pp 258, 333)
8. D. E. Alspach, *The dual of the Bourgain-Delbaen space*, Israel J. Math., to appear. (cited on p 255)
9. ———, *Quotients of* c_0 *are almost isometric to subspaces of* c_0, Proc. Amer. Math. Soc. **76** (1979), 285–288. (cited on p 255)
10. ———, *A fixed point free nonexpansive map*, Proc. Amer. Math. Soc. **82** (1981), 423–424. (cited on p 80)
11. D. E. Alspach and Y. Benyamini, $C(K)$ *quotients of separable* $\mathcal{L}_\infty$ *spaces*, Israel J. Math. **32** (1979), 145–160. (cited on p 255)
12. A. Ambrosetti and G. Prodi, *A Primer of non Linear Analysis*, Cambridge studies in advanced mathematics, vol. 34, Cambridge University Press, 1995. (cited on p 7)
13. D. Amir, *Continuous functions' spaces with the bounded extension property*, Bull. Res. Council of Israel **10F** (1962), 133–138. (cited on p 33)
14. ———, *Chebyshev centers and uniform convexity*, Pacific J. Math. **77** (1978), 1–6. (cited on p 34)
15. ———, *Characterizations of Inner Product Spaces*, Operator Theory: Advances and Applications, vol. 20, Birkhäuser, 1986. (cited on p 410)
16. R. D. Anderson, *Hilbert space is homeomorphic to the countable product of lines*, Bull. Amer. Math. Soc. **72** (1966), 515–519. (cited on p 1)
17. G. Androulakis and E. Odell, *Distorting mixed Tsirelson spaces*, Israel J. Math. **100** (1999), 125–149. (cited on p 337)
18. T. Aoki, *Locally bounded linear topological spaces*, Proc. Imp. Acad. Tokyo **18** (1942), 588–594. (cited on p 445)
19. R. Arens, *Extension of functions on fully normal spaces*, Pacific J. Math. **2** (1952), 11–22. (cited on p 33)
20. S. Argyros and I. Deliyanni, *Examples of asymptotic* l_1 *Banach spaces*, Trans. Amer. Math. Soc. **349** (1997), 973–995. (cited on p 337)

21. S. Argyros, S. Negrepontis, and Th. Zachariades, *Weakly stable Banach spaces*, Israel J. Math. **57** (1987), 68–88. (cited on p 333)
22. J. Arias-De-Reyna, K. Ball, and R. Villa, *Concentration of the distance in finite-dimensional normed spaces*, Mathematika **45** (1998), 245–252. (cited on p 299)
23. N. Aronszajn, *Differentiability of Lipschitzian mappings between Banach spaces*, Studia Math. **57** (1976), 147–190. (cited on pp 163, 164, 183)
24. N. Aronszajn and P. Panitchpakdi, *Extensions of uniformly continuous transformations and hyperconvex metric spaces*, Pacific J. Math. **6** (1956), 405–439. (cited on p 59)
25. Z. Artstein, *Extension of Lipschitz selections and an application to differential inclusions*, Nonlinear Anal. **16** (1991), 701–704. (cited on p 33)
26. E. Asplund, *Averaged norms*, Israel J. Math. **5** (1967), 227–233. (cited on p 414)
27. ———, *Fréchet differentiability of convex functions*, Acta Math. **121** (1968), 31–47. (cited on p 97)
28. P. Assouad, *Remarques sur un article de Israel Aharoni sur les prolongements Lipschitziens dans c_0*, Israel J. Math. **31** (1978), 97–100. (cited on p 184)
29. J. M. Ayerbe Toledano, T. Dominguez Benavides, and G. López Acedo, *Measures of Noncompactness in Metric Fixed Point Theory*, Birkhäuser, 1997. (cited on p 81)
30. J. B. Baillon, *Un théorème de type ergodique pour les contractions non linéaires dans un espace de Hilbert*, C.R. Acad. Sci. Paris **280** (1975), A1511–A1514, Ser. A-B. (cited on p 81)
31. ———, *Comportement asymptotique des itérés de contractions non linéaires dans les espaces L^p*, C.R. Acad. Sci. Paris **286** (1978), A157–A159, Ser. A-B. (cited on p 81)
32. ———, *Quelques aspects de la théorie des points fixes dans les espaces de Banach I*, Séminaire d'analyse fonctionnelle, École Polytechnique, 1978–79. (cited on p 80)
33. J. B. Baillon, R. E. Bruck, and S. Reich, *On the asymptotic behavior of nonexpansive mappings and semigroups in Banach spaces*, Houston J. Math. **4** (1978), 1–9. (cited on p 81)
34. J. W. Baker, *Dispersed images of topological spaces and uncomplemented subspaces of $C(X)$*, Proc. Amer. Math. Soc. **41** (1973), 309–314. (cited on p 256)
35. S. Banach, *Sur les opérations dans les ensembles abstraits et leur application aux équations intégrales*, Fund. Math. **3** (1922), 133–181. (cited on p 79)
36. ———, *Sur le problème de la mesure*, Fund. Math. **4** (1923), 7–23. (cited on p 417)
37. R. G. Bartle, N. Dunford, and J. T. Schwartz, *Weak compactness and vector measures*, Canad. J. Math. **7** (1955), 289–305. (cited on p 121)
38. R. G. Bartle and L. M. Graves, *Mappings between function spaces*, Trans. Amer. Math. Soc. **72** (1952), 400–413. (cited on p 33)
39. S. M. Bates, *On the image size of singular maps II*, Duke Math. J. **68** (1992), 463–476. (cited on p 277)
40. ———, *Towards a precise smoothness hypothesis in Sard's theorem*, Proc. Amer. Math. Soc. **117** (1993), 279–283. (cited on p 277)
41. ———, *On smooth, nonlinear surjections of Banach spaces*, Israel J. Math. **100** (1997), 209–220. (cited on p 277)
42. S. M. Bates, W. B. Johnson, J. Lindenstrauss, D. Preiss, and G. Schechtman, *Affine approximation of Lipschitz functions and nonlinear quotients*, Geom. Funct. Anal. (1999). (cited on pp 123, 279)
43. B. Beauzamy, *Espaces d'Interpolation Réels: Topologie et Géométrie*, Lect. Notes in Math., vol. 666, Springer, 1978. (cited on p 414)
44. ———, *Introduction to Banach Spaces and their Geometry*, Math. Studies, vol. 86, North-Holland, 1982. (cited on p 413)
45. B. Beauzamy and J. T. Lapresté, *Modèles étalés des Espaces de Banach*, Hermann, Paris, 1984. (cited on p 299)

46. B. Beauzamy and B. Maurey, *Points minimaux et ensembles optimaux dans les espaces de Banach*, J. Funct. Anal. **24** (1977), 107–139. (cited on p 82)
47. Y. Benyamini, *Constants of simultaneous extension of continuous functions*, Israel J. Math. **16** (1973), 258–262. (cited on p 34)
48. ———, *An extension theorem for separable Banach spaces*, Israel J. Math. **29** (1978), 24–30. (cited on p 255)
49. ———, *Small into isomorphisms between spaces of continuous functions*, Proc. Amer. Math. Soc. **83** (1981), 479–485. (cited on p 60)
50. ———, *Simultaneously continuous retractions on the unit ball of a Banach space*, J. Approx. Th. **38** (1983), 28–42. (cited on pp 55, 60)
51. ———, *Two point symmetrizations, the isoperimetric inequality on the sphere and some applications*, Longhorn notes, The Univ. of Texas at Austin Functional Analysis Seminar (E. Odell and H. P. Rosenthal, eds.), 1983-4, pp. 53–76. (cited on p 299)
52. ———, *The uniform classification of Banach spaces*, Longhorn notes, The Univ. of Texas at Austin Functional Analysis Seminar (E. Odell and H. P. Rosenthal, eds.), 1984-5, pp. 15–38. (cited on p 257)
53. Y. Benyamini and Y. Sternfeld, *Spheres in infinite-dimensional normed spaces are Lipschitz contractible*, Proc. Amer. Math. Soc. **88** (1983), 439–445. (cited on p 79)
54. C. Berg, J. P. R. Christensen, and P. Ressel, *Harmonic Analysis on Semigroups; Theory of positive definite and related functions*, Graduate texts in Math., vol. 100, Springer, 1984. (cited on p 195)
55. M. S. Berger, *Nonlinearity and Functional Analysis. Lectures on Nonlinear Problems in Mathematical Analysis*, Pure and Applied Math., vol. 74, Academic Press, 1977. (cited on p 7)
56. J. Bergh and J. Löfström, *Interpolation Spaces. An Introduction*, Springer, 1976. (cited on pp 33, 449, 450, 452)
57. M. Besbes, *Extension des théorèmes de points fixes de Maurey aux fonctions à valeurs vectorielles*, Séminaire d'Initiation à l'Analyse, University of Paris VI, 1993/4. (cited on p 80)
58. C. Bessaga, *On topological classification of complete linear metric spaces*, Fund. Math. **56** (1965), 251–288. (cited on p 407)
59. C. Bessaga and A. Pełczyński, *On bases and unconditional convergence of series in Banach spaces*, Studia Math. **17** (1958), 151–164. (cited on p 431)
60. ———, *Spaces of continuous functions IV: On isomorphical classification of spaces of continuous functions*, Studia Math. **19** (1960), 53–62. (cited on p 255)
61. ———, *On extreme points in separable conjugate spaces*, Israel J. Math. **4** (1966), 262–264. (cited on p 122)
62. ———, *Selected Topics in Infinite Dimensional Topology*, PWN, Warsaw, 1975. (cited on pp 7, 80, 215)
63. P. Billingsley, *Convergence of Probability Measures*, Wiley, 1968. (cited on pp 162, 420)
64. ———, *Probability and Measure*, Wiley, 1979. (cited on pp 419, 423, 425)
65. G. Birkhoff, *Analytical groups*, Trans. Amer. Math. Soc. **43** (1938), 61–101. (cited on p 407)
66. E. Bishop and R. R. Phelps, *A proof that every Banach space is subreflexive*, Bull. Amer. Math. Soc. **67** (1961), 97–98. (cited on p 96)
67. ———, *The support functionals of a convex set*, Convexity (V. Klee, ed.), Proc. Symp. Pure Math., vol. 7, Amer. Math. Soc., 1963, pp. 27–35. (cited on p 96)
68. B. O. Björnestål, *Local Lipschitz continuity of the metric projection operator*, Approximation Theory (Warsaw) (Z. Ciesielski, ed.), Banach Center Publications, vol. 4, PWN, 1979, pp. 43–53. (cited on p 58)
69. S. Bochner, *Integration von Funktionen deren Werte die Elemente eines Vectorraumes sind*, Fund. Math. **20** (1933), 262–276. (cited on p 121)

70. S. Bochner and A. E. Taylor, *Linear functionals on certain spaces of abstractly valued functions*, Ann. of Math. **39** (1938), 913–944. (cited on p 123)
71. V. I. Bogachev, *Negligible sets in locally convex spaces* (Russian), Mat. Zametki **36** (1984), 51–64, English translation: Math. Notes**3** 6 (1984), 519–526. (cited on p 163)
72. ———, *Some results on differentiable measures*, Math. USSR-Sb. **55** (1986), 335–349. (cited on p 163)
73. ———, *Gaussian Measures*, Math. Surveys, vol. 62, Amer. Math. Soc, 1998. (cited on pp 162, 163)
74. V. I. Bogachev and S. A. Shkarin, *Differentiable and Lipschitizian mappings of Banach spaces* (Russian), Mat. Zametki **44** (1988), 567–583, English translation: Math. Notes**44** (1988), 790–798. (cited on p 164)
75. H. F. Bohnenblust, *An axiomatic characterization of L_p spaces*, Duke Math. J. **6** (1940), 627–640. (cited on p 300)
76. B. Bollobás, I. Leader, and A. J. Radcliffe, *Reverse Kleitman inequalities*, Proc. London Math. Soc. **58** (1989), 153–168. (cited on p 216)
77. R. Bonic and J. Frampton, *Smooth functions on Banach manifolds*, J. Math. Mech. **15** (1966), 877–898. (cited on pp 97, 259)
78. K. Borsuk, *Drei Sätze über die n-dimensionale euklidische Sphäre*, Fund. Math. **20** (1933), 177–190. (cited on p 334)
79. ———, *Über Isomorphie der Funktionalräume*, Bull. Acad. Polon. Sci. Sér. Sci. Math. Astrono. Phys. **1/3** (1933), 1–10. (cited on p 33)
80. J. M. Borwein and D. Noll, *Second order differentiability of convex functions in Banach spaces*, Trans. Amer. Math. Soc. **342** (1994), 43–81. (cited on p 161)
81. J. M. Borwein and D. Preiss, *A smooth variational principle with applications to subdifferentiability and to differentiability of convex functions*, Trans. Amer. Math. Soc. **303** (1987), 517–527. (cited on p 97)
82. J. Bourgain, *On dentability and the Bishop-Phelps property*, Israel J. Math. **28** (1977), 265–271. (cited on pp 97, 122)
83. ———, *La propriété de Radon Nikodým*, Pub. Math. de l'Univ. Pierre et Marie Curie n. 36, Paris, 1980. (cited on pp 122, 123, 124)
84. ———, *On separable Banach spaces universal for all separable reflexive spaces*, Proc. Amer. Math. Soc. **79** (1980), 241–246. (cited on p 338)
85. ———, *New Classes of $\mathcal{L}_p$ Spaces*, Lect. Notes in Math., vol. 889, Springer, 1981. (cited on p 437)
86. ———, *Real isomorphic complex Banach spaces need not be complex isomorphic*, Proc. Amer. Math. Soc. **96** (1986), 221–226. (cited on p 390)
87. ———, *Remarks on the extension of Lipschitz maps defined on discrete sets and uniform homeomorphisms*, Israel GAFA Seminar 1985-86 (J. Lindenstrauss and V. D. Milman, eds.), Lect. Notes in Math., vol. 1267, Springer, 1987, pp. 157–167. (cited on p 254)
88. J. Bourgain and F. Delbaen, *A class of special $\mathcal{L}_\infty$ spaces*, Acta Math. **145** (1980), 155–176. (cited on p 124)
89. J. Bourgain and H. P. Rosenthal, *Martingales valued in certain subspaces of L_1*, Israel J. Math. **37** (1980), 54–75. (cited on p 124)
90. J. Bourgain and M. Talagrand, *Dans un espace de Banach réticulé solide, la propriété de Radon Nikodým et celle de Krein-Milman sont équivalentes*, Proc. Amer. Math. Soc. **81** (1981), 93–96. (cited on p 122)
91. R. D. Bourgin, *Geometric Aspects of Convex Sets with the Radon Nikodým Property*, Lect. Notes in Math., vol. 993, Springer, 1983. (cited on pp 122, 123, 124)
92. J. Bretagnolle, D. Dacunha-Castelle, and J. L. Krivine, *Lois stables et espaces L^p*, Ann. Inst. H. Poincaré **2** (1966), 231–259. (cited on p 195)
93. M. S. Brodskiĭ and D. P. Milman, *On the center of a convex set*, Dokl. Akad. Nauk SSSR **59** (1948), 837–840. (cited on pp 80, 415)

94. F. E. Browder, *Non expansive nonlinear operators in a Banach space*, Proc. Nat. Acad. Sci. U.S.A **54** (1965), 1041–1044. (cited on p 80)
95. ———, *Semicontractive and semiaccretive nonlinear mappings in Banach spaces*, Bull. Amer. Math. Soc. **74** (1968), 660–665. (cited on p 81)
96. ———, *Nonlinear Operators and Nonlinear Equations of Evolution in Banach Spaces*, Proc. Symp. Pure Math., vol. 18/2, American Math. Soc., 1976. (cited on pp 7, 80)
97. A. L. Brown, *A rotund reflexive space having a subspace of codimension two with a discontinuous metric projection*, Michigan Math. J. **21** (1974), 145–151. (cited on p 58)
98. R. E. Bruck, *Properties of fixed-point sets of nonexpansive mappings in Banach spaces*, Trans. Amer. Math. Soc. **179** (1973), 251–262. (cited on p 82)
99. ———, *A simple proof of the mean ergodic theorem for nonlinear contractions in Banach spaces*, Israel J. Math. **32** (1979), 107–116. (cited on pp 81, 82)
100. ———, *On the convex approximation property and the asymptotic behavior of nonlinear contractions in Banach spaces*, Israel J. Math. **38** (1981), 304–314. (cited on pp 81, 82)
101. ———, *Asymptotic behavior of nonexpansive mappings*, Contemp. Math. **18** (1983), 1–47. (cited on p 81)
102. ———, *A simple proof that rate of asymptotic regularity of $(I+T)/2$ is $O(1/n^{1/2})$*, Recent advances in Metric Fixed Point Theory (T. Dominguez Benavides, ed.), Univ. de Sevilla, 1996, pp. 11–18. (cited on p 81)
103. Y. A. Brudnyi and N. J. Kalton, *Polynomial approximation on convex subsets of $\mathbb{R}^n$*, Constr. Approx., (to appear). (cited on pp 371, 389)
104. A. Brunel and L. Sucheston, *On B convex Banach spaces*, Math. Systems Th. **7** (1974), 294–299. (cited on p 299)
105. ———, *On J convexity and some ergodic super properties of Banach spaces*, Trans. Amer. Math. Soc. **204** (1975), 79–90. (cited on p 299)
106. D. Burago and B. Kleiner, *Separated nets in Euclidean space and Jacobians of biLipschitz maps*, Geom. Funct. Anal. **8** (1998), 273–282. (cited on p 257)
107. F. Cabelló Sanchez and J. M. F. Castillo, *The long homology sequence for quasi-Banach spaces, with applications*, preprint. (cited on p 389)
108. A. Calderón, *Intermediate spaces and interpolation, the complex method*, Studia Math. **24** (1964), 113–190. (cited on pp 449, 450, 451)
109. P. G. Casazza, W. B. Johnson, and L. Tzafriri, *On Tsirelson's space*, Israel J. Math. **47** (1984), 81–98. (cited on p 258)
110. P. G. Casazza and N. J. Kalton, *Unconditional bases and unconditional finite-dimensional decompositions in Banach spaces*, Israel J. Math. **95** (1996), 349–373. (cited on p 390)
111. P. G. Casazza and E. Odell, *Tsirelson's space and minimal subspaces*, Longhorn Notes, Texas Functional Analysis Seminar, Univ. of Texas at Austin, 1982-83, pp. 61–72. (cited on p 258)
112. P. G. Casazza and T. J. Shura, *Tsirelson's Space*, Lect. Notes in Math., vol. 1363, Springer, 1989. (cited on pp 248, 249, 258, 334)
113. E. Casini and P. L. Papini, *A counterexample to the infinity version of the Hyers and Ulam stability theorem*, Proc. Amer. Math. Soc. **118** (1993), 885–890. (cited on p 371)
114. J. M. F. Castillo and M. Gonzalez, *Three-Space Problems in Banach Space Theory*, Lect. Notes in Math., vol. 1667, Springer, 1997. (cited on pp 9, 390)
115. M. Cepedello-Boiso, *Approximation of Lipschitz functions by Δ-convex functions in Banach spaces*, Israel J. Math. **106** (1998), 269–284. (cited on p 98)
116. F. Chaatit, *On the uniform homeomorphisms of the unit spheres of certain Banach lattices*, Pacific J. Math. **168** (1995), 11–31. (cited on p 216)

117. K. C. Chang and L. Shujie, *A remark on expanding maps*, Proc. Amer. Math. Soc. **85** (1982), 583–586. (cited on p 217)
118. S. D. Chatterji, *Martingale convergence and the Radon Nikodým theorem in Banach spaces*, Math. Scand. **22** (1968), 21–41. (cited on p 122)
119. P. W. Cholewa, *Remarks on the stability of functional equations*, Aequationes Math. **27** (1984), 76–86. (cited on p 371)
120. J. P. R. Christensen, *On sets of Haar measure zero in abelian Polish groups*, Israel J. Math. **13** (1972), 255–260. (cited on p 161)
121. ———, *Measure theoretic zero sets in infinite dimensional spaces and applications to differentiability of Lipschitz mappings II*, Publ. Dep. Math. Lyon **10** (1973), 29–39. (cited on pp 164, 183)
122. K. L. Chung, *A Course in Probability Theory*, second ed., Academic Press, 1974. (cited on pp 423, 426)
123. H. B. Cohen, *Injective envelopes of Banach spaces*, Bull. Amer. Math. Soc. **70** (1964), 723–726. (cited on p 32)
124. H. H. Corson, *The weak topology of a Banach space*, Trans. Amer. Math. Soc. **101** (1961), 1–15. (cited on p 60)
125. H. H. Corson and V. L. Klee, *Topological classification of convex sets*, convexity (V. Klee, ed.), Proc. Symp. Pure Math., vol. 7, Amer. Math. Soc., 1963, pp. 37–51. (cited on p 33)
126. H. H. Corson and J. Lindenstrauss, *On simultaneous extensions of continuous functions*, Bull. Amer. Math. Soc. **71** (1965), 542–545. (cited on pp 34, 53)
127. M. Csörnyei, *Aronszajn null and Gaussian null sets coincide*, Israel J. Math. **111** (1999), 191–202. (cited on p 163)
128. M. Cwikel, *Complex interpolation spaces, a discrete definition and reiteration*, Indiana Univ. Math. J. **27** (1978), 1005–1009. (cited on p 449)
129. M. Cwikel and S. Reisner, *Interpolation of uniformly convex Banach spaces*, Proc. Amer. Math. Soc. **84** (1982), 555–559. (cited on p 449)
130. M. Daher, *Homémorphismes uniformes entre les sphères unité des espaces d'interpolation*, C.R. Acad. Sci. Paris **316** (1993), 1051–1054, série I. (cited on p 216)
131. G. David, *Morceaux de graphes lipschitziens et integrales singulières sur un surface*, Revista Matemática Iberoamericana **4** (1988), 73–114. (cited on p 280)
132. G. David and S. Semmes, *Regular mappings between dimensions*, preprint. (cited on p 280)
133. W. J. Davis and P. Enflo, *Contractive projections on subsets of ℓ_p spaces*, Analysis in Urbana (E. Berkson, T. Peck, and J. J. Uhl, eds.), London Math. Soc. Lecture Notes, vol. 137, 1989, pp. 151–161. (cited on p 82)
134. W. J. Davis and R. R. Phelps, *The Radon Nikodým property and dentable sets in Banach spaces*, Proc. Amer. Math. Soc. **45** (1974), 119–122. (cited on p 122)
135. M. M. Day, *Uniform convexity in factor and conjugate spaces*, Ann. of Math. **45** (1944), 375–385. (cited on p 411)
136. ———, *Some characterizations of inner product spaces*, Trans. Amer. Math. Soc. **62** (1947), 320–337. (cited on p 410)
137. ———, *Strict convexity and smoothness of normed spaces*, Trans. Amer. Math. Soc. **78** (1955), 516–528. (cited on p 97)
138. ———, *Normed Linear Spaces*, third ed., Ergebnisse, vol. 21, Springer Verlag, 1973. (cited on p 31)
139. K. Deimling, *Nonlinear functional analysis*, Springer, 1985. (cited on p 7)
140. R. Deville, *Geometrical implications of existence of very smooth bump functions in Banach spaces*, Israel J. Math. **67** (1989), 1–22. (cited on p 96)

141. R. Deville, G. Godefroy, D. E. G. Hare, and V. Zizler, *Differentiability of convex functions and the convex point of continuity property in Banach spaces*, Israel J. Math. **59** (1987), 245–255. (cited on p 98)
142. R. Deville, G. Godefroy, and V. Zizler, *The three-space problem for smooth partitions of unity and $C(K)$ spaces*, Math. Ann. **288** (1990), 613–625. (cited on p 184)
143. ______, *A smooth variational principle with applications to Hamilton-Jacobi equations in infinite dimensions*, J. Funct. Anal. **111** (1993), 197–212. (cited on p 97)
144. ______, *Smoothness and Renormings in Banach Spaces*, Pitman Monographs, vol. 64, Pitmamn, 1993. (cited on pp 9, 59, 96, 98, 260, 413, 414)
145. S. Dierolf, *Über Vererbbarkeitseigenschaften in topologischen Vectorräumen*, Ph.D. thesis, Univ. of Munich, 1974. (cited on p 389)
146. J. Diestel and J. J. Uhl, *Vector Measures*, Math. Surveys, vol. 15, Amer. Math. Soc., 1977. (cited on pp 9, 122, 123, 124)
147. ______, *Progress in vector measures 1977–83*, Measure theory and its applications (J. M. Belley et al., ed.), Lect. Notes in Math., vol. 1033, Springer, 1983, pp. 144–192. (cited on p 122)
148. J. Dieudonné, *Une généralisation des espaces compacts*, J. Math Pures Appl. **23** (1944), 65–76. (cited on p 416)
149. S. J. Dilworth, *Approximate isometries on finite dimensional normed spaces*, Bull. London Math. Soc., to appear. (cited on p 372)
150. S. J. Dilworth, R. Howard, and J. W. Roberts, *Extremal approximately convex functions and estimating the size of convex hulls*, Advances in Math., to appear. (cited on p 371)
151. J. Dixmier, *Étude sur les variétés et les opérateurs de Julia, avec quelques applications*, Bull. Soc. Math. France **77** (1949), 11–101. (cited on p 163)
152. T. Dobrowolski, *Examples of topological groups homeomorphic to l_2^f*, Proc. Amer. Math. Soc. **98** (1986), 303–311. (cited on p 397)
153. A. Le Donne and M. V. Marchi, *Representation of Lipshitzian compact-convex valued mappings*, Atti. Accad. Naz. Lincei. **68** (1980), 278–280. (cited on p 33)
154. R. Dougherty, *Examples of non shy sets*, Fund. Math. **144** (1994), 73–88. (cited on p 161)
155. P. N. Dowling and C. J. Lennard, *Every nonreflexive subspace of $L_1[0,1]$ fails the fixed point property*, Proc. Amer. Math. Soc. **125** (1997), 443–446. (cited on p 80)
156. P. N. Dowling, N. Randianantoanina, and B. Turret, *Remarks on James' distortion theorems*, Bull. Austral. Math. Soc. **57** (1998), 49–54. (cited on p 333)
157. J. Dugundji, *An extension of Tietze's theorem*, Pacific J. Math. **1** (1951), 353–367. (cited on p 33)
158. N. Dunford and B. J. Pettis, *Linear operations on summable functions*, Trans. Amer. Math. Soc. **47** (1940), 323–392. (cited on pp 32, 122)
159. N. Dunford and J. T. Schwartz, *Linear Operators. Part I: General Theory*, Interscience, 1958. (cited on pp 32, 67, 122)
160. A. Dvoretzky, *Some results on convex bodies and Banach spaces*, Proc. Int. Symp. on Linear space, Jerusalem Academic Press, 1961, pp. 123–160. (cited on p 298)
161. A. Dvoretzky and C. A. Rogers, *Absolute and unconditional convergence in normed linear spaces*, Proc. Nat. Acad. Sci. U.S.A **36** (1950), 192–197. (cited on p 299)
162. I. S. Edelstein and P. Wojtaszczyk, *On projections and unconditional bases in direct sums of Banach spaces*, Studia Math. **56** (1976), 263–276. (cited on p 255)
163. M. Edelstein and R. C. O'Brien, *Nonexpansive mappings, asymptotic regularity and successive approximations*, J. London Math. Soc. **17** (1978), 547–554. (cited on p 81)
164. G. A. Edgar, *Extremal integral representations*, J. Funct. Anal. **23** (1976), 145–161. (cited on p 123)
165. I. Ekeland, *On the variational principle*, J. Math. Anal. Appl. **47** (1974), 324–353. (cited on pp 96, 97)

166. ———, *Non convex minimization problems*, Bull. Amer. Math. Soc. **1** (1979), 443–474. (cited on pp 96, 97)
167. J. Elton, P. K. Lin, E. Odell, and S. Szarek, *Remarks on the fixed point problem for nonexpansive maps*, Contemp. Math. **18** (1983), 87–120. (cited on p 80)
168. P. Enflo, *On a problem of Smirnov*, Ark. Mat. **8** (1969), 107–109. (cited on pp 196, 201, 216)
169. ———, *On the nonexistence of uniform homeomorphism between L_p spaces*, Ark. Mat. **8** (1969), 103–105. (cited on pp 183, 253)
170. ———, *Topological groups in which multiplication on one side is differentiable or linear*, Math. Scand. **24** (1969), 195–207. (cited on pp 406, 407, 408)
171. ———, *Uniform structures and square roots in topological groups I*, Israel J. Math. **8** (1970), 230–252. (cited on pp 183, 406)
172. ———, *Uniform structures and square roots in topological groups II*, Israel J. Math. **8** (1970), 253–272. (cited on pp 406, 408)
173. ———, *Banach spaces which can be given an equivalent uniformly convex norm*, Israel J. Math. **13** (1973), 281–288. (cited on p 412)
174. ———, *A counterexample to the approximation property in Banach spaces*, Acta Math. **130** (1973), 309–317. (cited on p 430)
175. ———, *Uniform homeomorphism between Banach spaces*, Sém. Maurey-Schwartz, École Polytechnique, Paris, 1975-76, Exposé no. 18. (cited on p 253)
176. ———, *Contractive projections onto subsets of $L_1(0,1)$*, Analysis in Urbana (E. Berkson, T. Peck, and J. J. Uhl, eds.), London Math. Soc. Lecture Notes, vol. 137, 1989, pp. 162–184. (cited on p 82)
177. ———, *Contractive projections onto subsets of L_p-spaces*, Function Spaces (K. Jarosz, ed.), Lecture notes in pure and applied Mathematics, vol. 136, Marcel Dekker, 1992, pp. 79–95. (cited on p 82)
178. P. Enflo, J. Lindenstrauss, and G. Pisier, *On the three space problem*, Math. Scand. **36** (1975), 199–210. (cited on pp 389, 390)
179. R. Engelking, *General Topology*, PWN, Warszawa, 1977. (cited on pp 416, 417)
180. M. J. Fabian, *Gâteaux Differentiability of Convex Functions and Topology. Weak Asplund Spaces*, Canadian Math. Soc. Monographs, Wiley-Interscience, 1997. (cited on p 98)
181. M. J. Fabian, L. Zajíček, and V. Zizler, *On the residuality of the set of rotund norms on a Banach space*, Math. Ann. **258** (1981/82), 349–351. (cited on p 414)
182. M. J. Fabian and D. Preiss, *On intermediate differentiability of Lipschitz functions on certain Banach space*, Proc. Amer. Math. Soc. **113** (1991), 733–740. (cited on p 279)
183. M. J. Fabian and V. Zizler, *An elementary approach to some questions in higher order smoothness in Banach space*, Extracta Math., to appear. (cited on p 96)
184. J. Feldman, *Equivalence and perpendicularity of gaussian processes*, Pacific J. Math. **8** (1958), 699–708. (cited on p 162)
185. W. Feller, *An Introduction to Probability Theory and its Applications*, vol. 2, Wiley, 1971. (cited on pp 423, 425)
186. X. Fernique, *Intégrabilité des vecteurs gaussiens*, C.R. Acad. Sci. Paris **270** (1970), A1698–A1699. (cited on p 163)
187. T. Figiel, *On non linear isometric embedding of normed linear spaces*, Bull. Acad. Polon. Sci. Sér. Sci. Math. Astrono. Phys. **16** (1968), 185–188. (cited on p 356)
188. ———, *On the moduli of convexity and smoothness*, Studia Math. **56** (1976), 121–155. (cited on p 411)
189. T. Figiel and W. B. Johnson, *A uniformly convex Banach space which contains no l_p*, Compositio Math. **29** (1974), 179–190. (cited on p 257)
190. T. Figiel, J. Lindenstrauss, and V. D. Milman, *The dimension of almost spherical sections of convex bodies*, Acta Math. **139** (1977), 53–94. (cited on pp 298, 299)

191. T. Figiel and G. Pisier, *Séries aléatoires dans les espaces uniformément convexes ou uniformément lisses*, C.R. Acad. Sci. Paris **279** (1974), 611–614, Serie A. (cited on pp 58, 413)
192. D. G. de Figueiredo and L. A. Karlovitz, *The extension of contractions and the intersection of balls in Banach spaces*, J. Funct. Anal. **11** (1972), 168–178. (cited on p 59)
193. ______, *On the extension of contractions on normed spaces*, Proc. Symp. Pure Math., vol. 18/1, Amer. Math. Soc., 1976, pp. 95–104. (cited on p 82)
194. P. A. Fillmore and J. P. Williams, *On operator ranges*, Advances in Math **7** (1971), 254–281. (cited on p 163)
195. V. P. Fonf, *Semi-embeddings and G_δ-embeddings* (Russian), Mat. Zametki **39** (1986), 550–561, English translation: Math. Notes**39** (1986), 302–307. (cited on p 124)
196. M. H. Freedman, *Working and playing with the 2-disk*, Mathematics into the twenty-first century (F. Browder, ed.), vol. 2, Amer. Math. Soc., 1992, pp. 37–47. (cited on p 358)
197. D. J. H. Garling, *Stable Banach spaces, random measures and Orlicz function spaces*, Lect. Notes in Math., vol. 928, Springer, 1982, pp. 121–175. (cited on p 333)
198. F. W. Gehring, *Uniform domains and the ubiquitous quasidisk*, Jahresber. Deutch. Math. Verein. **89** (1987), 88–103. (cited on p 357)
199. I. M. Gelfand, *Abstracte Funktionen und lineare Operatoren*, Mat. Sbornik **(4)46** (1938), 235–284. (cited on p 122)
200. A. Genel and J. Lindenstrauss, *An example concerning fixed points*, Israel J. Math. **22** (1975), 81–86. (cited on p 81)
201. J. Gevirtz, *Injectivity in Banach spaces and the Mazur-Ulam theorem on isometries*, Trans. Amer. Math. Soc. **274** (1982), 307–318. (cited on p 357)
202. ______, *Injectivity of quasi-isometric mappings of balls*, Proc. Amer. Math. Soc. **85** (1982), 345–349. (cited on p 356)
203. ______, *Stability of isometries on Banach spaces*, Proc. Amer. Math. Soc. **89** (1983), 633–636. (cited on p 372)
204. N. Ghoussoub, G. Godefroy, B. Maurey, and W. Schachermayer, *Some topological and geometrical structures in Banach spaces*, Mem. Amer. Math. Soc., vol. 378, AMS, 1987. (cited on p 122)
205. N. Ghoussoub and B. Maurey, *G_δ embeddings in Hilbert space*, J. Funct. Anal. **61** (1985), 72–97. (cited on p 124)
206. ______, *H_δ embeddings in Hilbert space and optimization on G_δ sets*, Mem. Amer. Math. Soc., vol. 349, AMS, 1986. (cited on pp 122, 124)
207. I. I. Gihman and A. V. Skorohod, *The Theory of Stochastic Processes III*, Grundlehren # 232, Springer Verlag, 1979. (cited on p 163)
208. T. A. Gillespie, *Factorization in Banach function spaces*, Indag. Math. **43** (1981), 287–300. (cited on pp 216, 336)
209. I. V. Girsanov and B. S. Mityagin, *Quasi invariant measures and linear topological spaces*, Naučn. Dokl. Vys. Škol **2** (1959), 5–10. (cited on p 161)
210. G. Godefroy, N. J. Kalton, and G. Lancien, *Subspaces of $c_0(\mathbb{N})$ and Lipschitz isomorphisms*, to appear. (cited on pp 256, 432)
211. ______, *Szlenk indices and uniform homeomorphisms*, to appear. (cited on p 255)
212. ______, *L'espace de Banach c_0 est déterminé par sa métrique*, C.R. Acad. Sci. Paris **327** (1998), 817–822. (cited on pp 254, 256)
213. K. Goebel, *On the structure of the minimal invariant sets for nonexpansive mappings*, Ann. Univ. Mariae Curie-Sklodowska, Sect. A **29** (1975), 73–77. (cited on p 80)
214. ______, *On a problem concerning involutions of Banach spaces*, Optimization and nonlinear analysis (A. Ioffe, M. Marcus, and S. Reich, eds.), Pitman Research Notes, vol. 244, Pitman, 1992, pp. 142–144. (cited on p 217)

215. K. Goebel and W. Kaczor, *Remarks on failure of Schauder's theorem in noncompact settings*, Ann. Univ. Mariae Curie-Sklodowska, Sect. A **51** (1997), 99–108. (cited on p 79)
216. K. Goebel and W. A. Kirk, *A fixed point theorem for transformations whose iterates have uniform Lipschitz constant*, Studia Math. **47** (1973), 135–140. (cited on p 81)
217. ———, *Topics in Metric Fixed Point Theory*, Cambridge studies in advanced mathematics, vol. 28, Cambridge University Press, 1990. (cited on pp 9, 79, 80, 81)
218. K. Goebel and S. Reich, *Uniform Convexity, Hyperbolic Geometry, and Nonexpansive Mappings*, Marcel Dekker, 1984. (cited on pp 80, 82)
219. I. C. Gohberg and M. G. Krein, *Introduction to the Theory of Linear Nonselfadjoint Operators*, Translation of Mathematical Monographs, vol. 18, Amer. Math. Soc., 1969. (cited on pp 449, 454)
220. D. Göhde, *Zum Prinzip der kontraktiven Abbildung*, Math. Nachr. **30** (1965), 251–258. (cited on p 80)
221. Y. Gordon, *Some inequalities for Gaussian processes and applications*, Israel J. Math. **50** (1985), 265–289. (cited on p 298)
222. E. Gorelik, *The uniform nonequivalence of L_p and l_p*, Israel J. Math. **87** (1994), 1–8. (cited on p 254)
223. W. T. Gowers, *Symmetric block bases of sequences with large average growth*, Israel J. Math. **69** (1990), 129–151. (cited on p 97)
224. ———, *Lipschitz functions on classical spaces*, European J. Combin. **13** (1992), 141–151. (cited on p 334)
225. ———, *A new dichotomy for Banach spaces*, Geom. Funct. Anal. **6** (1996), 1083–1093. (cited on pp 335, 390)
226. W. T. Gowers and B. Maurey, *The unconditional basic sequence problem*, J. Amer. Math. Soc. **6** (1993), 851–874. (cited on pp 183, 207, 335, 336, 390, 431)
227. R. L. Graham, B. L. Rothschild, and J. H. Spencer, *Ramsey Theory*, John Wiley, 1980. (cited on p 334)
228. F. Greenleaf, *Invariant Means on Topological Groups and Their Applications*, Van Nostrand, 1969. (cited on p 417)
229. M. Gromov, *Asymptotic invariants of infinite groups*, Geometric Group Theory (G. A. Niblo and M. A. Roller, eds.), vol. 2, Cambridge University Press, 1993. (cited on pp 196, 257)
230. ———, *Metric Structures for Riemannian and non-Riemannian Spaces*, Progress in Math., vol. 152, Birkhäuser, 1999. (cited on pp 33, 279, 298)
231. M. Gromov and V. D. Milman, *Generalization of the spherical isoperimetric inequality to uniformly convex Banach spaces*, Compositio Math. **62** (1987), 263–282. (cited on p 299)
232. A. Grothendieck, *Sur les applications linéaires faiblement compactes d'espace du type $C(K)$*, Canad. J. Math. **5** (1953), 129–173. (cited on p 32)
233. ———, *Une caractérisation vectorielle métrique des espaces L^1*, Canad. J. Math. **7** (1955), 552–561. (cited on p 59)
234. ———, *Erratum au mémoire: Produits tensoriels topologiques et espaces nucléaires*, Ann. Inst. Fourier (Grenoble) **6** (1956), 117–120. (cited on p 253)
235. P. M. Gruber, *Stability of isometries*, Trans. Amer. Math. Soc. **245** (1978), 263–277. (cited on p 371)
236. B. Grünbaum, *On a theorem of Kirszbraun*, Bull. Res. Council Israel **F7** (1958), 129–132. (cited on p 59)
237. ———, *Projection constants*, Trans. Amer. Math. Soc. **95** (1960), 451–465. (cited on pp 32, 59)
238. F. Grünbaum and E. H. Zarantonello, *On the extension of uniformly continuous mappings*, Michigan Math. J. **15** (1968), 65–74. (cited on p 33)

239. S. Guerre-Delabriére, *Classical Sequences in Banach Spaces*, Marcel Dekker, 1992. (cited on pp 218, 334, 433)
240. S. P. Gul'ko, *Uniformly homeomorphic spaces of functions*, Proc. Steklov Inst. Math. **3** (1993), 87–93. (cited on p 256)
241. V. I. Gurarii and N. I. Gurarii, *Bases in uniformly convex and uniformly smooth Banach spaces*, Izv. Akad. Nauk **35** (1971), 210–215. (cited on p 431)
242. U. Haagerup, *The best constants in the Khintchine inequality*, Studia Math. **70** (1982), 231–283. (cited on p 424)
243. P. Habala, P. Hájek, and V. Zizler, *Introduction to Banach Spaces, Volumes I and II*, MatFizPress, Univ. Karlovy, Prague, 1996. (cited on p 431)
244. J. Hájek, *On a property of normal distributions of any stochastic process* (Russian), Czechoslovak Math. J. **8** (1958), 610–618, English translation: Select. Transl. Math. Stat. Prob., Amer. Math. Soc., **1** (1961), 245–252. (cited on p 162)
245. P. Hájek, *Smooth functions on* c_0, Israel J. Math. **104** (1998), 17–27. (cited on p 278)
246. ———, *Smooth functions on* $C(K)$, Israel J. Math. **107** (1998), 237–252. (cited on p 278)
247. P. Halmos, *Naive Set Theory*, Springer, 1974. (cited on p 241)
248. O. Hanner, *Intersections of translates of convex bodies*, Math. Scand. **4** (1956), 65–87. (cited on p 59)
249. ———, *On the uniform convexity of* L^p *and* l^p, Ark. Mat. **3** (1956), 239–244. (cited on p 410)
250. R. Haydon, E. Odell, H. P. Rosenthal, and Th. Schlumprecht, *On distorted norms in Banach spaces and existence of* l_p*-types*, unpublished manuscript. (cited on p 334)
251. S. Heinrich, *Ultraproducts in Banach space theory*, J. Reine Angew. Math. **313** (1980), 72–104. (cited on pp 438, 439, 440)
252. S. Heinrich and P. Mankiewicz, *Applications of ultrapowers to the uniform and Lipschitz classification of Banach spaces*, Studia Math. **73** (1982), 225–251. (cited on pp 183, 253, 439, 440)
253. G. M. Henkin, *Impossibility of uniform homeomorphism between spaces of smooth functions of one and of n variables* ($n \geq 2$), Math. USSR-Sb. **3** (1967), 551–561. (cited on p 253)
254. N. Hindman, *Finite sums from sequences within cells of a partition of N*, J. Combin. Theory, Ser. A **17** (1974), 1–11. (cited on p 334)
255. A. Hohti, *On Lipschitz homogeneity of the Hilbert cube*, Trans. Amer. Math. Soc. **291** (1985), 75–86. (cited on p 218)
256. W. Holsztyński, *Linearization of isometric embeddings of Banach spaces. Metric envelopes*, Bull. Acad. Polon. Sci. Sér. Sci. Math. Astrono. Phys. **16** (1968), 189–193. (cited on p 356)
257. R. E. Huff, *Dentability and the Radon-Nikodým property*, Duke Math. J. **41** (1974), 111–114. (cited on p 122)
258. R. E. Huff and P. D. Morris, *Dual spaces with the Krein Milman property have the Radon Nikodým property*, Proc. Amer. Math. Soc. **49** (1975), 104–108. (cited on p 122)
259. B. R. Hunt, T. Sauer, and J. A. Yorke, *Prevalence: A translation-invariant "almost every" on infinite dimensional spaces*, Bull. Amer. Math. Soc. **27** (1992), 217–238. (cited on p 161)
260. ———, *Prevalence: An addendum*, Bull. Amer. Math. Soc. **28** (1993), 306–307. (cited on p 161)
261. D. H. Hyers, *On the stability of the linear functional equation*, Proc. Nat. Acad. Sci. U.S.A **27** (1941), 222–224. (cited on p 370)
262. ———, *The stability of homomorphisms and related topics*, Global analysis-Analysis on manifolds (T. M. Rassias, ed.), Teubner-Texte zur mathematik, vol. 57, 1983, pp. 140–153. (cited on p 370)

263. D. H. Hyers and S. Ulam, *On approximate isometries*, Bull. Amer. Math. Soc. **51** (1945), 288–292. (cited on p 371)
264. ———, *Approximately convex functions*, Proc. Amer. Math. Soc. **3** (1952), 821–828. (cited on p 370)
265. J. R. Isbell, *Uniform neighborhood retracts*, Pacific J. Math. **11** (1961), 609–648. (cited on p 33)
266. ———, *Injective envelopes of Banach spaces are rigidly attached*, Bull. Amer. Math. Soc. **70** (1964), 727–729. (cited on p 32)
267. ———, *Six theorems about injective metric spaces*, Comment. Math. Helv. **39** (1964), 65–76. (cited on p 32)
268. S. Ishikawa, *Fixed points and iteration of nonexpansive mappings in a Banach space*, Proc. Amer. Math. Soc. **59** (1976), 65–71. (cited on p 81)
269. D. J. Ives and D. Preiss, *Not too well differentiable Lipschitz isomorphisms*, Israel J. Math., to appear. (cited on p 184)
270. I. M. James, *Introduction to Uniform Spaces*, Cambridge University Press, 1990. (cited on p 279)
271. R. C. James, *Bases and reflexivity of Banach spaces*, Ann. of Math. **52** (1950), 518–527. (cited on p 431)
272. ———, *Characterizations of reflexivity*, Studia Math. **23** (1964), 205–216. (cited on p 433)
273. ———, *Uniformly non-square Banach spaces*, Ann. of Math. **80** (1964), 542–550. (cited on p 333)
274. ———, *Some self dual properties of normed linear spaces*, Symp. on inf. dim. topology (R. D. Anderson, ed.), Ann. of Math. Studies., vol. 69, 1972, pp. 159–175. (cited on p 412)
275. ———, *Super reflexive spaces with bases*, Pacific J. Math. **41** (1972), 409–419. (cited on p 431)
276. ———, *A non-reflexive Banach space that is uniformly nonoctahedral*, Israel J. Math. **18** (1974), 145–155. (cited on pp 215, 218)
277. R. C. James and J. Lindenstrauss, *The octahedral problem for Banach spaces*, Proceedings of a seminar on Random series, Convex sets and Geometry of Banach spaces, Aarhus Univ. (Denmark), 1974, pp. 100–120. (cited on p 218)
278. R. E. Jamison and W. H. Ruckle, *Factoring absolutely convergent series*, Math. Ann. **224** (1976), 143–148. (cited on p 216)
279. K. Jarosz, *Non linear generalization of the Banach Stone theorem*, Studia Math. **93** (1989), 97–107. (cited on p 357)
280. F. John, *Extremum problems with inequalities as subsidiary conditions*, Courant Anniversary Volume, Interscience, 1948, [Collected papers Volume 2, J. Moser, ed., Birkhäuser, (1985), 543–560], pp. 187–204. (cited on pp 299, 441)
281. ———, *Rotation and strain*, Comm. Pure Appl. Math. **14** (1961), 391–413, [Collected papers Volume 2, J. Moser, ed., Birkhäuser, (1985), 643–665]. (cited on pp 356, 357, 358)
282. ———, *Quasi isometric mappings*, Seminari dell Instituto Nazionale di Alta Mathematica (1964), 462–473, [Collected papers Volume 2, J. Moser, ed., Birkhäuser, (1985), 678–689]. (cited on p 356)
283. ———, *Distance changes in deformations with small strain*, Studies and Essays presented to Yu-Why-Chen, New York Univ., 1970, [Collected papers Volume 2, J. Moser, ed., Birkhäuser, (1985), 204–218], pp. 1–15. (cited on p 357)
284. ———, *Bounds for deformations in terms of average strains*, Inequalities Vol. III (O. Shisha, ed.), Academic Press, 1972, [Collected papers Volume 2, J. Moser, ed., Birkhäuser, (1985), 617–632], pp. 129–144. (cited on pp 356, 358)
285. ———, *Uniqueness of nonlinear elastic equilibrium for prescribed boundary displacements and sufficiently small strains*, Comm. Pure Appl. Math. **25** (1972), 617–634,

[Collected papers Volume 2, J. Moser, ed., Birkhäuser, (1985), 261–278]. (cited on p 358)

286. F. John and L. Nirenberg, *On functions of bounded mean oscillation*, Comm. Pure Appl. Math. **14** (1961), 415–426, [Collected papers Volume 2, J. Moser Ed., Birkhäuser, (1985), 666–677]. (cited on p 358)
287. J. Johnson and J. Wolfe, *Norm attaining operators and simultaneously continuous retractions*, Proc. Amer. Math. Soc. **86** (1982), 609–612. (cited on pp 60, 97)
288. W. B. Johnson, *Operators into L_p which factor through ℓ_p*, J. London Math. Soc. **14** (1976), 333–339. (cited on p 437)
289. ———, *A reflexive Banach space which is not sufficiently Euclidean*, Studia Math. **55** (1976), 201–205. (cited on pp 258, 432)
290. ———, *On quotients of L_p which are quotients of ℓ_p*, Compositio Math. **34** (1977), 69–89. (cited on p 431)
291. W. B. Johnson and J. Lindenstrauss, *Examples of $\mathcal{L}_1$ spaces*, Ark. Mat. **18** (1980), 101–106. (cited on p 124)
292. W. B. Johnson, J. Lindenstrauss, D. Preiss, and G. Schechtman, *Affine approximation of Lipschitz maps between infinite-dimensional Banach spaces*, to appear. (cited on p 432)
293. ———, *Uniform quotient mappings of the plane*, Michigan Math. J. **47** (2000), to appear. (cited on p 280)
294. W. B. Johnson, J. Lindenstrauss, and G. Schechtman, *On the relation between several notions of unconditional structure*, Israel J. Math. **37** (1980), 120–129. (cited on pp 382, 390)
295. ———, *Banach spaces determined by their uniform structure*, Geom. Funct. Anal. **6** (1996), 430–470. (cited on pp 254, 255, 256, 257, 258)
296. W. B. Johnson and E. Odell, *Subspaces of L_p which embed into ℓ_p*, Compositio Math. **28** (1974), 37–49. (cited on p 437)
297. W. B. Johnson and H. P. Rosenthal, *On w^*-basic sequences and their applications to the study of Banach spaces*, Studia Math. **43** (1972), 77–92. (cited on p 431)
298. W. B. Johnson, H. P. Rosenthal, and M. Zippin, *On bases, finite-dimensional decompositions and weaker structures in Banach spaces*, Israel J. Math. **9** (1971), 488–506. (cited on p 436)
299. W. B. Johnson and M. Zippin, *On subspaces of quotients of $(\sum G_n)_{l_p}$ and $(\sum G_n)_{c_0}$*, Israel J. Math. **13** (1972), 311–316. (cited on p 437)
300. ———, *Subspaces and quotient spaces of $(\sum G_n)_{l_p}$ and $(\sum G_n)_{c_0}$*, Israel J. Math. **17** (1974), 50–55. (cited on p 255)
301. P. Jones, *Lipschitz and biLipschitz functions*, Revista Matemática Iberoamericana **4** (1988), 115–122. (cited on p 280)
302. M. I. Kadec, *Conditions for differentiability of a norm in a Banach space* (Russian), Uspehi Math. Nauk. **20** (1965), 183–187. (cited on p 97)
303. ———, *A proof of topological equivalence of all separable infinite-dimensional Banach spaces* (Russian), Funk. Anal. i. Prilozen **1** (1967), 61–70. (cited on p 1)
304. M. I. Kadec and A. Pełczyński, *Bases, lacunary sequences and complemented subspaces in the spaces L_p*, Studia Math. **21** (1962), 161–176. (cited on pp 183, 258, 427)
305. M. I. Kadec and M. G. Snobar, *Certain functionals on the Minkowski compactum* (Russian), Mat. Zametki **10** (1971), 453–457, English translation: Math. Notes, **10** (1971) 694–696. (cited on p 441)
306. J. P. Kahane, *Some Random Series of Functions*, second ed., Cambridge University Press, 1985. (cited on p 442)
307. S. Kakutani, *Simultaneous extension of continuous functions considered as a positive linear operator*, Japanese J. Math. **17** (1940), 1–4. (cited on p 33)
308. ———, *Topological properties of the unit sphere of a Hilbert space*, Proc. Imp. Acad. Tokyo **19** (1943), 269–271. (cited on p 79)

309. ———, *On equivalence of infinite product measures*, Ann. of Math. **49** (1948), 214–224. (cited on p 422)
310. N. J. Kalton, *The three space problem for locally bounded F spaces*, Compositio Math. **37** (1978), 243–276. (cited on pp 389, 446)
311. ———, *Linear operators on L_p for $0 < p < 1$*, Trans. Amer. Math. Soc. **259** (1980), 319–355. (cited on p 195)
312. ———, *Convexity, type and the three space problem*, Studia Math. **69** (1981), 247–287. (cited on pp 389, 442, 446)
313. ———, *Banach spaces embedding into L_0*, Israel J. Math. **52** (1985), 305–319. (cited on p 195)
314. ———, *Nonlinear commutators in interpolation theory*, Mem. Amer. Math. Soc., vol. 385, AMS, 1988. (cited on p 390)
315. ———, *Differentials of complex interpolation processes for Köthe function spaces*, Trans. Amer. Math. Soc. **333** (1992), 479–529. (cited on p 389)
316. ———, *The basic sequence problem*, Studia Math. **116** (1995), 167–187. (cited on p 390)
317. ———, *An elementary example of a Banach space not isomorphic to its complex conjugate*, Canad. Math. Bull. **38** (1995), 218–222. (cited on p 390)
318. N. J. Kalton and N. T. Peck, *Twisted sums of sequence spaces and the three space problem*, Trans. Amer. Math. Soc. **255** (1979), 1–30. (cited on pp 389, 390)
319. N. J. Kalton, N. T. Peck, and J. W. Roberts, *An F-Space Sampler*, London Math. Soc. Lecture Notes, vol. 89, Cambridge Univ. Press, 1984. (cited on pp 389, 445, 447)
320. N. J. Kalton and A. Pełczyński, *Kernels of surjections from $\mathcal{L}_1$-spaces with an application to sidon sets*, Math. Ann. **309** (1997), 135–158. (cited on p 389)
321. N. J. Kalton and J. W. Roberts, *Uniformly exhaustive submeasures and nearly additive set functions*, Trans. Amer. Math. Soc. **278** (1983), 803–816. (cited on p 389)
322. L. Karlovitz, *Existence of fixed points of nonexpansive mappings in a space without normal structure*, Pacific J. Math. **66** (1976), 153–159. (cited on p 80)
323. A. S. Kechris, *Classical Descriptive Set Theory*, Graduate Texts in Math., vol. 156, Springer, 1995. (cited on pp 338, 420)
324. O. H. Keller, *Die Homeomorphie der kompakten konvexen Mengen in Hilbertschen Raum*, Math. Ann. **105** (1931), 748–758. (cited on p 215)
325. J. L. Kelley, *General Topology*, Van Nostrand, 1955. (cited on p 55)
326. W. A. Kirk, *A fixed point theorem for mappings which do not increase distances*, Amer. Math. Monthly **72** (1965), 1004–1006. (cited on p 80)
327. M. D. Kirszbraun, *Über die zusammenziehenden und Lipschitzchen Transformationen*, Fund. Math. **22** (1934), 77–108. (cited on p 33)
328. S. V. Kislyakov, *Sobolev imbedding operators and the nonisomorphism of certain Banach spaces*, Funct. Anal. Appl. **9** (1975), 290–294. (cited on p 254)
329. V. L. Klee, *Some topological properties of convex sets*, Trans. Amer. Math. Soc. **78** (1955), 30–45. (cited on p 79)
330. ———, *Mappings into normed linear spaces*, Fund. Math. **49** (1960-61), 25–34. (cited on p 97)
331. R. V. Kohn, *New integral estimates for deformations in terms of their nonlinear strains*, Arch. Rational Mech. Anal. **78** (1982), 131–172. (cited on p 358)
332. A. Koldobsky, *A Banach subspace of $L_{1/2}$ which does not embed in L_1 (isometric version)*, Proc. Amer. Math. Soc. **124** (1996), 155–160. (cited on p 195)
333. ———, *Positive definite functions, stable measures, and isometries on Banach spaces*, Interaction between functional analysis, harmonic analysis, and probability (Columbia Missouri 1994) (E. Saab N. J. Kalton and S. Montgomery-Smith, eds.), Lecture Notes in Pure and Applied Math., vol. 175, Dekker, 1996, pp. 275–290. (cited on p 195)

334. R. A. Komorowski and N. Tomczak-Jaegermann, *Banach spaces without local unconditional structure*, Israel J. Math. **89** (1995), 205–226, (See also Eratum, Israel J. Math., **105** (1998), 85–92.). (cited on p 390)
335. M. A. Krasnoselskii, *Two remarks about the method of successive approximations*, Uspehi. Mat. Nauk **10** (1955), 123–127. (cited on p 81)
336. ______, *Topological Methods in the Theory of non Linear Integral Equations*, Oxford Pergamon Press, 1964. (cited on p 7)
337. U. Krengel and M. Lin, *Order preserving nonexpansive operators in L_1*, Israel J. Math. **58** (1987), 170–192. (cited on p 81)
338. J. L. Krivine, *Sous espaces de dimension finie des espaces de Banach réticulés*, Ann. of Math. **104** (1976), 1–29. (cited on p 299)
339. J. L. Krivine and B. Maurey, *Espaces de Banach stables*, Israel J. Math. **39** (1981), 273–295. (cited on pp 212, 218, 258, 333, 334)
340. J. Kuelbs, *Gaussian measures on a Banach space*, J. Funct. Anal. **5** (1970), 354–367. (cited on p 163)
341. K. Kunen and H. P. Rosenthal, *Martingale proofs of some geometrical results in Banach space theory*, Pacific J. Math. **100** (1982), 153–175. (cited on p 123)
342. H. H. Kuo, *Gaussian Measures in Banach Spaces*, Lect. Notes in Math., vol. 463, Springer, 1975. (cited on p 162)
343. S. Kwapien, *Isomorphic characterizations of inner product spaces by orthogonal series with vector valued coefficients*, Studia Math. **44** (1972), 583–595. (cited on pp 58, 442)
344. S. Kwapien and A. Pełczyński, *The main triangle projection in matrix spaces and its applications*, Studia Math. **34** (1970), 43–68. (cited on p 454)
345. M. Laczkovich, *The local stability of convexity, affinity and of the Jensen equation*, preprint. (cited on p 371)
346. J. Lamperti, *Probability*, Benjamin, 1966. (cited on p 426)
347. J. M. Lasry and P. L. Lions, *A remark on regularization in Hilbert spaces*, Israel J. Math. **55** (1986), 257–266. (cited on p 98)
348. E. B. Leach and J. H. M. Whitfield, *Differentiable norms and rough norms on Banach spaces*, Proc. Amer. Math. Soc. **33** (1972), 120–126. (cited on p 97)
349. M. Ledoux and M. Talagrand, *Probability in Banach Spaces. Isoperimetry and Processes*, Ergebnisse Folge 3 # 23, Springer Verlag, 1991. (cited on p 163)
350. H. Lemberg, *Nouvelle démonstration d'un théorème de J. L. Krivine sur la finie représentation de l_p dans un espace de Banach*, Israel J. Math. **39** (1981), 341–348. (cited on p 299)
351. P. Lévy, *Problèmes Concrets d'Analyse Fonctionnelle*, Gauthier Villars, Paris, 1951. (cited on p 299)
352. A. Lima, *Intersection properties of balls and subspaces in Banach spaces*, Trans. Amer. Math. Soc. **227** (1977), 1–62. (cited on p 59)
353. P. K. Lin and Y. Sternfeld, *Convex sets with the Lipschitz fixed point property are compact*, Proc. Amer. Math. Soc. **93** (1985), 633–639. (cited on p 79)
354. J. Lindenstrauss, *On operators which attain their norm*, Israel J. Math. **1** (1963), 139–148. (cited on pp 97, 122)
355. ______, *On the modulus of smoothness and divergent series in Banach spaces*, Michigan Math. J. **10** (1963), 241–252. (cited on p 411)
356. ______, *Extensions of compact operators*, Mem. Amer. Math. Soc., vol. 48, AMS, 1964. (cited on pp 59, 334)
357. ______, *On nonlinear projections in Banach spaces*, Michigan Math. J. **11** (1964), 263–287. (cited on pp 33, 34, 183, 253)
358. ______, *On extreme points in ℓ_1*, Israel J. Math. **4** (1966), 59–61. (cited on p 122)
359. ______, *On nonseparable reflexive Banach spaces*, Bull. Amer. Math. Soc. **72** (1966), 967–970. (cited on p 439)

360. ______, *A short proof of Liapounoff's convexity theorem*, J. Math. Mech. **15** (1966), 971–972. (cited on p 428)
361. ______, *Almost spherical sections; their existence and their applications*, Jber. d. Dt. Math. Verein Jubiläumstagung (1990), 39–61. (cited on p 298)
362. ______, *Uniform embeddings, homeomorphisms and quotient maps between Banach spaces (a short survey)*, Topology Appl. **85** (1998), 265–279. (cited on p 6)
363. J. Lindenstrauss, E. Matoušková, and D. Preiss, *Lipschitz image of a measure-null set can have a null complement*, Israel J. Math., to appear. (cited on pp 162, 168, 184, 257)
364. J. Lindenstrauss and A. Pełczyński, *Absolutely summing operators in $\mathcal{L}_p$ spaces and their applications*, Studia Math. **29** (1968), 275–326. (cited on p 437)
365. ______, *Contributions to the theory of classical Banach spaces*, J. Funct. Anal. **8** (1971), 225–249. (cited on p 333)
366. J. Lindenstrauss and D. Preiss, *Almost Fréchet differentiability of finitely many Lipschitz functions*, Mathematika **43** (1996), 393–412. (cited on pp 258, 279)
367. J. Lindenstrauss and H. P. Rosenthal, *The $\mathcal{L}_p$ spaces*, Israel J. Math. **7** (1969), 325–349. (cited on pp 436, 437)
368. J. Lindenstrauss and A. Szankowski, *Non linear perturbations of isometries*, Astérisque **131** (1985), 357–371. (cited on p 372)
369. J. Lindenstrauss and L. Tzafriri, *On Orlicz sequence spaces*, Israel J. Math. **10** (1971), 379–390. (cited on p 333)
370. ______, *On the complemented subspaces problem*, Israel J. Math. **9** (1971), 263–269. (cited on p 44)
371. ______, *Classical Banach Spaces vol. I: Sequence spaces*, Ergebnisse, vol. 92, Springer, 1977. (cited on pp 183, 255, 385, 387, 389, 429, 430, 431, 432)
372. ______, *Classical Banach Spaces vol. II: Function spaces*, Ergebnisse, vol. 97, Springer, 1979. (cited on pp 58, 183, 409, 411, 413, 414, 431, 434, 435, 436, 442)
373. V. Lomonosov, *A counterexample to the Bishop Phelps theorem in complex spaces*, Israel J. Math., to appear. (cited on p 96)
374. G. M. Lövblom, *Isometries and almost isometries between spaces of continuous functions*, Israel J. Math. **56** (1986), 143–159. (cited on p 357)
375. ______, *Almost isometries on the unit ball of ℓ_1*, Israel J. Math. **63** (1988), 129–138. (cited on p 357)
376. ______, *Uniform homeomorphisms between unit balls in L_p-spaces*, Math. Scand. **62** (1988), 294–302. (cited on pp 215, 254)
377. ______, *Uniform homeomorphisms between the unit balls in L_p and ℓ_p*, Proc. Amer. Math. Soc. **123** (1995), 405–409. (cited on p 254)
378. G. Ya. Lozanovskii, *On some Banach lattices*, Siberian Math. J. **10** (1969), 584–599. (cited on p 216)
379. ______, *On some Banach lattices III*, Siberian Math. J. **13** (1972), 1304–1313. (cited on p 216)
380. O. Maleva, *A pathological example of a uniform quotient mapping between Euclidean spaces*, preprint. (cited on p 279)
381. E. Maluta, *Uniformly normal structure and related coefficients*, Pacific J. Math. **111** (1984), 357–369. (cited on p 415)
382. P. Mankiewicz, *On Lipschitz mappings between Fréchet spaces*, Studia Math. **41** (1972), 225–241. (cited on p 256)
383. ______, *On the differentiability of Lipschitz mappings in Fréchet spaces*, Studia Math. **45** (1973), 15–29. (cited on pp 163, 164, 183)
384. ______, *On topological, Lipschitz, and uniform classification of LF-spaces*, Studia Math. **52** (1974), 109–142. (cited on p 256)
385. O. Martio, S. Rickman, and J. Väisälä, *Topological and metric properties of quasiregular mappings*, Ann. Acad. Sci. Fenn. Ser. A I Math. (1971), 1–31. (cited on p 357)

386. J. Matoušek and E. Matoušková, *A highly nonsmooth norm on Hilbert space*, Israel J. Math. **112** (1999), 1–28. (cited on p 164)
387. E. Matoušková, *Extending of continuous and Lipschitz functions*, Canad. Math. Bull., to appear. (cited on p 60)
388. ———, *Lipschitz images of Haar null sets*, Bull. London Math. Soc., to appear. (cited on p 163)
389. ———, *Convexity and Haar null sets*, Proc. Amer. Math. Soc. **125** (1997), 1793–1799. (cited on p 162)
390. ———, *The Banach Saks property and Haar null sets*, Comment. Math. Univ. Carolinae **39** (1998), 71–80. (cited on p 162)
391. ———, *An almost nowhere Fréchet smooth norm on superreflexive spaces*, Studia Math. **133** (1999), 93–99. (cited on p 164)
392. E. Matoušková and C. Stegall, *A characterization of reflexive Banach spaces*, Proc. Amer. Math. Soc. **124** (1996), 1083–1090. (cited on p 162)
393. M. Matzeu and A. Vignoli (eds.), *Topological Nonlinear Analysis; Degree, singularity and variations I*, Progress in nonlinear differential equations and their applications, vol. 15, Birkhäuser, Boston, 1995. (cited on p 7)
394. ———, *Topological Nonlinear Analysis; Degree, singularity and variations II*, Progress in nonlinear differential equations and their applications, vol. 27, Birkhäuser, Boston, 1997. (cited on p 7)
395. B. Maurey, *Nouveaux théorèmes de Nikišhin*, Séminaire Maurey-Schwartz, École Polytechnique, Paris, 1973–1974, Exposé 4 and 5. (cited on p 195)
396. ———, *Théorèmes de factorisation pour les opérateurs linéaires à valeurs dans les espaces L^p*, Astérisque **11** (1974), 1–163. (cited on p 195)
397. ———, *Un théorème de prolongement*, C.R. Acad. Sci. Paris **279** (1974), 329–332. (cited on p 442)
398. ———, *Point fixes des contractions de certains faiblement compacts de L^1*, Séminaire d'Analyse Fonctionelle (Paris), École Polytechnique, Paris, 1980–1981. (cited on p 80)
399. ———, *Quelques progrés dans la compréhension de la dimension infinie*, Journée Annuelle, Soc. Math. de France (1994), 1–29. (cited on p 340)
400. ———, *A remark about distortion*, Gafa Seminar 92–94 (J. Lindenstrauss and V. D. Milman, eds.), Operator Theory, vol. 77, Birkhäuser, 1995, pp. 131–142. (cited on p 337)
401. ———, *Symmetric distortion in ℓ_2*, Gafa Seminar 92–94 (J. Lindenstrauss and V. D. Milman, eds.), Operator Theory, vol. 77, Birkhäuser, 1995, pp. 143–147. (cited on p 336)
402. B. Maurey and G. Pisier, *Séries de variables aléatoires vectorielles indépendantes et propriétés géométriques des espaces de Banach*, Studia Math. **58** (1976), 45–90. (cited on pp 443, 444)
403. B. Maurey and H. P. Rosenthal, *Normalized weakly null sequences with no unconditional subsequence*, Studia Math. **61** (1977), 77–98. (cited on p 335)
404. H. B. Maynard, *A geometrical characterization of Banach spaces with the Radon Nikodým property*, Trans. Amer. Math. Soc. **185** (1973), 493–500. (cited on p 122)
405. S. Mazur, *Une remarque sur l'homéomorphie des champs fonctionnels*, Studia Math. **1** (1929), 83–85. (cited on p 215)
406. ———, *Über konvexe Mengen in linearen normierten Räumen*, Studia Math. **4** (1933), 70–84. (cited on p 97)
407. S. Mazur and S. Ulam, *Sur les transformations isométriques d'espaces vectoriels normés*, C.R. Acad. Sci. Paris **194** (1932), 946–948. (cited on pp 1, 356)
408. S. Mazurkiewicz and W. Sierpinski, *Contribution à la topologie des ensembles dénombrables*, Fund. Math. **1** (1920), 17–27. (cited on p 255)

409. P. W. McCartney and R. C. O'Brien, *A separable space with the Radon Nikodým property that is not isomorphic to a subspace of a separable dual*, Proc. Amer. Math. Soc. **78** (1980), 40–42. (cited on p 124)
410. C. T. McMullen, *Lipschitz maps and nets in Euclidean space*, Geom. Funct. Anal. **8** (1998), 304–314. (cited on p 257)
411. W. J. Meyer, *Characterization of the Steiner point*, Pacific J. Math. **35** (1970), 717–725. (cited on p 59)
412. E. Michael, *Continuous selections I*, Ann. of Math. **63** (1956), 361–382. (cited on p 33)
413. D. P. Milman, *On some criteria for the regularity of spaces of the type* (B) (Russian), Dokl. Akad. Nauk SSSR **20** (1938), 243–246. (cited on p 411)
414. D. P. Milman and V. D. Milman, *Some geometric properties of non-reflexive spaces*, Soviet Math. Dokl. **4** (1963), 1250–1252. (cited on p 433)
415. V. D. Milman, *Spectrum of bounded continuous functions on the unit sphere of a Banach space*, Funct. Anal. Appl. **3** (1969), 67–79. (cited on pp 298, 333)
416. ______, *The geometric theory of Banach spaces part II: Geometry of the unit sphere* (Russian), Uspehi Mat. Nauk **26** (1971), 73–149, English translation: Russian Math. Surveys, **26** (1971), 79–163. (cited on p 333)
417. ______, *New proof of the theorem of A. Dvoretzky on sections of convex bodies*, Funct. Anal. Appl. **5** (1971), 28–37. (cited on p 298)
418. ______, *Almost Euclidean quotient spaces of subspaces of a finite-dimensional normed space*, Proc. Amer. Math. Soc. **94** (1985), 445–449. (cited on p 299)
419. ______, *Dvoretzky's Theorem – thirty years later (survey)*, Geom. Funct. Anal. **4** (1992), 455–479. (cited on p 298)
420. V. D. Milman and G. Schechtman, *Asymptotic Theory of Finite Dimensional Normed Spaces*, Lect. Notes in Math., vol. 1200, Springer, 1986. (cited on pp 9, 299, 428, 441, 442, 444)
421. ______, *An "isomorphic" version of Dvoretzky's theorem*, C.R. Acad. Sci. Paris **321** (1995), 541–544. (cited on p 299)
422. V. D. Milman and N. Tomczak-Jaegermann, *Asymptotic* ℓ_p *spaces and bounded distortion*, Contemp. Math. **144** (1993), 173–196. (cited on p 337)
423. R. A. Milnos, *Generalized random processes* (Russian), Trudy Mosk. Mat. Ob. **8** (1959), 497–518. (cited on p 162)
424. G. J. Minty, *On the extension of Lipschitz, Lipschitz-Hölder continuous, and monotone functions*, Bull. Amer. Math. Soc. **76** (1970), 334–339. (cited on p 33)
425. L. Mirsky, *Transversal Theory*, Academic Press, 1971. (cited on p 241)
426. D. Montgomery and L. Zippin, *Topological Transformation Groups*, Interscience, 1955. (cited on pp 399, 406)
427. E. H. Moore, *On properly positive hermitian matrices*, Bull. Amer. Math. Soc. **23** (1916), 66–67. (cited on p 195)
428. J. M. Morel and H. Steinlein, *On a problem of Nirenberg concerning expanding maps*, J. Funct. Anal. **59** (1984), 145–150. (cited on p 217)
429. J. Moser, *On the volume elements on a manifold*, Trans. Amer. Math. Soc. **120** (1965), 286–294. (cited on p 257)
430. R. Nahum, *On the Lipschitz classification of subsets of nonreflexive Banach spaces*, preprint. (cited on p 217)
431. ______, *On the Lipschitz equivalence of unit balls and of spheres in normed spaces*, preprint. (cited on p 216)
432. ______, *Lipschitz topology in Banach spaces*, Ph.D. thesis, University of Haifa, 1997. (cited on pp 33, 216, 217)
433. K. Namakara, *On bicompact semigroups*, Math. J. Okoyama Univ. **1** (1952), 99–108. (cited on p 334)

434. I. Namioka and R. R. Phelps, *Banach spaces which are Asplund spaces*, Duke Math. J. **42** (1975), 735–750. (cited on p 124)
435. A. Naor, *Geometric problems in non linear functional analysis*, Master's thesis, Hebrew University, 1998. (cited on pp 407, 415)
436. J. Neveu, *Discrete-Parameter Martingales*, North-Holland, 1975. (cited on p 428)
437. E. M. Nikišin, *Resonance theorems and expansions in eigenfunctions of the Laplace operator* (Russian), Izv. Akad. Nauk **36** (1972), 795–813, English translation: Math. USSR-Izv., **6** (1972), 788–806. (cited on p 195)
438. L. Nirenberg, *Topics in Nonlinear Functional Analysis*, Lecture Notes, Courant Inst., New York, 1974. (cited on pp 7, 217)
439. G. Nördlander, *The modulus of convexity in normed linear spaces*, Ark. Mat. **4** (1960), 15–17. (cited on p 409)
440. B. Nowak, *On the Lipschitzian retraction of the unit ball in infinite dimensional Banach spaces onto its boundary*, Bull. Acad. Polon. Sci. Sér. Sci. Math. Astrono. Phys. **27** (1979), 861–864. (cited on p 79)
441. E. Odell, H. P. Rosenthal, and Th. Schlumprecht, *On weakly null FDD's in Banach spaces*, Israel J. Math. **84** (1993), 333–351. (cited on p 300)
442. E. Odell and Th. Schlumprecht, *Distortion and asymptotic structure*, Hand-Book in Banach Spaces (W. B. Johnson and J. Lindenstrauss, eds.), North Holland, to appear. (cited on p 334)
443. ———, *The distortion problem*, Acta Math. **173** (1994), 259–281. (cited on pp 216, 335, 336)
444. ———, *Distortion and stabilized structure in Banach spaces. New geometric phenomena for Banach and Hilbert spaces*, Proc. 1994 ICM Zurich, Vol. II (S. D. Chatterji, ed.), Birkhäuser, 1995, pp. 955–965. (cited on p 335)
445. A. M. Olevskiĭ, *Fourier series and Lebesgue functions* (Russian), Uspehi Math. Nauk **22** (1967), 237–239. (cited on p 333)
446. M. Omladič and P. Šemrl, *On non-linear perturbations of isometries*, Math. Ann. **303** (1995), 617–628. (cited on pp 360, 372)
447. L. Partanen and J. Väisälä, *Extension of bi-Lipschitz maps of compact polyhedra*, Math. Scand. **72** (1993), 235–264. (cited on p 358)
448. J. R. Partington, *Norm attaining operators*, Israel J. Math. **43** (1982), 273–276. (cited on p 97)
449. A. L. T. Paterson, *Amenability*, Math. surveys and monographs, vol. 29, Amer. Math. Soc., 1988. (cited on p 417)
450. A. Pełczyński, *Projections in certain Banach spaces*, Studia Math. **19** (1960), 209–228. (cited on p 440)
451. ———, *Linear extensions, linear averagings, and their applications to linear topological classification of spaces of continuous functions*, Dissertationes Mathematicae, vol. 58, PWN, 1968. (cited on pp 34, 183, 257)
452. ———, *On almost diffeomorphic Banach spaces*, Indag. Math. **30** (1968), 202–208. (cited on p 259)
453. A. Pełczyński and H. P. Rosenthal, *Localization techniques in L_p spaces*, Studia Math. **52** (1975), 263–289. (cited on p 432)
454. A. Pełczyński and Z. Semadeni, *Spaces of continuous functions III. Spaces $C(\Omega)$ for Ω without perfect subsets*, Studia Math. **18** (1959), 211–222. (cited on p 256)
455. A. Pełczyński and K. Senator, *On isomorphisms of anisotropic Sobolev spaces with "classical Banach spaces" and a Sobolev type embedding theorem*, Studia Math. **84** (1986), 169–215. (cited on p 254)
456. B. J. Pettis, *Linear functionals and completely additive set functions*, Duke Math. J. **4** (1938), 552–565. (cited on p 121)
457. ———, *On integration in vector spaces*, Trans. Amer. Math. Soc. **44** (1938), 277–304. (cited on p 121)

458. ———, *A proof that every uniformly convex space is reflexive*, Duke Math. J. **5** (1939), 249–253. (cited on p 411)
459. R. R. Phelps, *Dentability and extreme points in Banach spaces*, J. Funct. Anal. **17** (1974), 78–90. (cited on pp 122, 123)
460. ———, *Gaussian null sets and differentiability of Lipschitz maps on Banach spaces*, Pacific J. Math. **77** (1978), 523–531. (cited on p 163)
461. ———, *Convex Functions, Monotone Operators and Differentiability*, second ed., Lect. Notes in Math., vol. 1364, Springer, 1993. (cited on pp 9, 98, 124)
462. G. Pisier, *Martingales with values in uniformly convex spaces*, Israel J. Math. **20** (1975), 326–350. (cited on pp 378, 412)
463. ———, *Some results on Banach spaces without local unconditional structure*, Compositio Math. **37** (1978), 3–19. (cited on p 454)
464. ———, *Some applications of the complex interpolation method to Banach lattices*, J. Analyse Math. **35** (1979), 264–281. (cited on p 451)
465. ———, *Remarques sur un résultat non publié de B. Maurey*, Seminaire d'Analyse Fonctionnelle, École Polytechnique, 1980-1981, expose 5. (cited on p 82)
466. ———, *Factorization of Linear Operators and Geometry of Banach Spaces*, CBMS Regional Conf. Series in Math., vol. 60, AMS, 1986. (cited on p 442)
467. ———, *The Volume of Convex Bodies and Banach Space Geometry*, Cambridge University Press, 1989. (cited on pp 299, 441)
468. G. Pisier and Q. Xu, *Random series in the real interpolation spaces between the spaces* v_p, Israel GAFA seminar 1985-86 (J. Lindenstrauss and V. D. Milman, eds.), Lect. Notes in Math., vol. 1267, Springer, 1987, pp. 185–209. (cited on p 218)
469. E. D. Positsel'skiĭ, *Lipschitzian mappings in the space of convex bodies* (Russian), Optimizacija **4(21)** (1971), 83–89. (cited on pp 59, 60)
470. ———, *Characterization of Steiner points* (Russian), Mat. Zametki **14** (1973), 243–247, English translation: Math. Notes, **14** (1973), 698–700. (cited on p 59)
471. D. Preiss, *Differentiability of Lipschitz functions on Banach spaces*, J. Funct. Anal. **91** (1990), 312–345. (cited on pp 6, 165, 274, 279)
472. D. Preiss and J. Tišer, *Two unexpected examples concerning differentiability of Lipschitz functions on Banach spaces*, GAFA Israel Seminar 92-94 (V. D. Milman and J. Lindenstrauss, eds.), Birkhäuser, 1995, pp. 219–238. (cited on p 164)
473. D. Preiss and J. Vilímovský, *In-between theorems in uniform spaces*, Trans. Amer. Math. Soc. **261** (1980), 483–501. (cited on p 33)
474. D. Preiss and L. Zajíček, *Directional derivatives of Lipschitz functions*, to appear. (cited on p 164)
475. ———, *Sigma-porous sets in products of metric spaces and sigma-directionally porous sets in Banach spaces*, Real Anal. Exchange, to appear. (cited on p 164)
476. ———, *Fréchet differentiation of convex functions in a Banach space with a separable dual*, Proc. Amer. Math. Soc. **91** (1984), 202–204. (cited on p 98)
477. ———, *Stronger estimates of smallness of sets of Fréchet non-differentiability of convex functions*, Suppl. Rend. Cir. Mat. Palermo II **3** (1984), 219–223, Proc. 11th Winter School. (cited on pp 98, 164)
478. S. Prus, *A remark on a theorem of Turett*, Bull. Acad. Polon. Sci. Sér. Sci. Math. Astrono. Phys. **36** (1988), 225–227. (cited on p 415)
479. K. Przesławski and D. Yost, *Continuity properties of selectors and Michael's theorem*, Michigan Math. J. **36** (1989), 113–134. (cited on p 60)
480. ———, *Lipschtiz retracts, selectors, and extensions*, Michigan Math. J. **42** (1995), 555–571. (cited on pp 33, 34)
481. S. Qian, *Nowhere differentiable Lipschitz maps and the Radon Nikodým property*, J. Math. Anal. Appl. **185** (1994), 613–616. (cited on p 123)
482. F. P. Ramsey, *On a problem of formal logic*, Proc. London Math. Soc. **30** (1930), 264–286. (cited on p 299)

483. Y. Raynaud, *Espaces de Banach superstables, distances stables et homéomorphismes uniformes*, Israel J. Math. **44** (1983), 33–52. (cited on p 218)

484. ———, *Ultrapowers of Köthe function spaces*, Collect. Math. **48** (1997), 733–742. (cited on p 216)

485. S. Reich, *Almost convergence and nonlinear ergodic theorems*, J. Approx. Th. **24** (1978), 269–272. (cited on p 82)

486. ———, *Strong convergence theorems for resolvents of accretive operators in Banach spaces*, J. Math. Anal. Appl. **75** (1980), 287–292. (cited on p 81)

487. ———, *Nonlinear semigroups, holomorphic mappings, and integral equations*, Nonlinear Functional Analysis and its Applications (F. E. Browder, ed.), Proc. Symp. Pure. Math., vol. 45/2, Amer. Math. Soc., 1986, pp. 307–324. (cited on p 80)

488. S. Reich and I. Shafrir, *Nonexpansive iterations in hyperbolic spaces*, Nonlinear Anal. **15** (1990), 537–558. (cited on p 80)

489. Yu. G. Reshetnyak, *Space Mappings with Bounded Distortion*, Amer. Math. Soc., Trans. Math. Monographs, vol. 73, Amer. Math. Soc. Math., 1989. (cited on pp 352, 354, 358)

490. ———, *On stability of isometric mappings*, Siberian Adv. Math. **1** (1991), 154–159. (cited on p 358)

491. ———, *Stability Theorems in Geometry and Analysis*, Kluwer, 1994. (cited on p 358)

492. M. Ribe, *On uniformly homeomorphic normed spaces*, Ark. Mat. **14** (1976), 237–244. (cited on p 253)

493. ———, *On uniformly homeomorphic normed spaces II*, Ark. Mat. **16** (1978), 1–9. (cited on pp 253, 256)

494. ———, *Examples for the nonlocally convex three space problem*, Proc. Amer. Math. Soc. **73** (1979), 351–355. (cited on p 389)

495. ———, *Existence of separable uniformly homeomorphic nonisomorphic Banach spaces*, Israel J. Math. **48** (1984), 139–147. (cited on p 257)

496. S. Rickman, *Quasiregular Mappings*, Ergebnisse, vol. 26, Springer, 1993. (cited on p 280)

497. M. A. Rieffel, *The Radon Nikodým theorem for the Bochner integral*, Trans. Amer. Math. Soc. **131** (1968), 466–487. (cited on p 122)

498. T. Rivière and D. Ye, *Resolutions of the prescribed volume from equation*, NoDEA Nonlinear Differential Equations Appl. **3** (1996), 323–369. (cited on p 257)

499. J. W. Roberts, *A nonlocally convex F space with the Hahn Banach approximation property*, Banach Spaces of Analytic Functions (J. Baker et al., ed.), Lect. Notes in Math., vol. 604, Springer, 1977, pp. 76–81. (cited on p 389)

500. S. Rolewicz, *On a certain class of linear metric spaces*, Bull. Acad. Polon. Sci. Sér. Sci. Math. Astrono. Phys. **5** (1957), 471–473. (cited on p 445)

501. ———, *Metric Linear Spaces*, second ed., D. Riedel and PWN, 1985. (cited on p 447)

502. H. P. Rosenthal, *On factors of $C[0,1]$ with non-separable dual*, Israel J. Math. **13** (1972), 361–378. (cited on p 255)

503. ———, *On subspaces of L^p*, Ann. of Math. **97** (1973), 344–373. (cited on pp 195, 221)

504. ———, *A characterization of Banach spaces containing l_1*, Proc. Nat. Acad. Sci. U.S.A **71** (1974), 2411–2413. (cited on p 432)

505. M. E. Rudin, *A new proof that metric spaces are paracompact*, Proc. Amer. Math. Soc. **20** (1969), 603. (cited on p 416)

506. W. Rudin, *Fourier Analysis on Groups*, Interscience, 1962. (cited on p 423)

507. G. T. Sallee, *A non continuous "Steiner point"*, Israel J. Math. **10** (1971), 1–5. (cited on p 59)

508. Z. Sasvari, *Positive Definite and Definitizable Functions*, Akademie-Verlag, 1994. (cited on p 195)
509. V. V. Sazonov, *On characteristic functionals* (Russian), Teor. Veryatnost. i Primenen. **3** (1958), 201–205. (cited on p 162)
510. W. Schachermayer, *Norm attaining operators on some classical Banach spaces*, Pacific J. Math. **105** (1983), 427–438. (cited on pp 60, 97)
511. ———, *For a Banach space isomorphic to its square the Radon-Nikodým property and the Krein-Milman property are equivalent*, Studia Math. **81** (1985), 329–339. (cited on p 122)
512. ———, *The Radon-Nikodým and the Krein-Milman properties are equivalent for strongly regular sets*, Trans. Amer. Math. Soc. **303** (1987), 673–687. (cited on p 122)
513. W. Schachermayer, A. Sersouri, and E. Werner, *Moduli of non-dentability and the Radon-Nikodým Property in Banach spaces*, Israel J. Math. **65** (1989), 225–257. (cited on p 123)
514. J. J. Schäffer, *Geometry of Spheres in Normed Spaces*, Marcel Dekker, 1976. (cited on p 33)
515. R. Schatten, *Norm Ideals of Completely Continuous Operators*, Springer, 1960. (cited on p 454)
516. J. Schauder, *Der Fixpunktsatz in Funktionalräumen*, Studia Math. **2** (1930), 171–180. (cited on p 79)
517. G. Schechtman, *On commuting families of nonexpansive operators*, Proc. Amer. Math. Soc. **84** (1982), 373–376. (cited on p 80)
518. T. Schlumprecht, *An arbitrarily distortable Banach space*, Israel J. Math. **76** (1991), 81–95. (cited on pp 335, 336)
519. R. Schneider, *On Steiner points of convex bodies*, Israel J. Math. **9** (1971), 241–249. (cited on p 59)
520. ———, *Equivariant endomorphisms of the space of convex bodies*, Trans. Amer. Math. Soc. **194** (1974), 53–78. (cited on p 59)
521. ———, *Convex Bodies: The Brunn-Minkowski Theory*, Cambridge University Press, 1993. (cited on pp 59, 428)
522. I. J. Schoenberg, *Metric spaces and positive definite functions*, Trans. Amer. Math. Soc. **44** (1938), 522–536. (cited on p 195)
523. ———, *A remark on M. M. Day's characterization of inner product spaces and a conjecture of L. M. Blumenthal*, Proc. Amer. Math. Soc. **3** (1952), 961–964. (cited on p 410)
524. S. O. Schönbeck, *On the extension of Lipschitz maps*, Ark. Mat. **7** (1967), 201–209. (cited on p 59)
525. I. Schur, *Bemerkungen zur Theorie der beschränkten Bilinearformen mit unendlisch vielen Veränderlichen*, J. Reine Angew. Math. **140** (1911), 1–29. (cited on p 195)
526. J. T. Schwartz, *Non Linear Functional Analysis*, Gordon and Breach, 1969. (cited on p 7)
527. I. E. Segal, *Topological groups in which multiplication on one side is differentiable*, Bull. Amer. Math. Soc. **52** (1946), 481–487. (cited on p 407)
528. V. I. Semenov, *Uniform stability estimates for isometries*, Siberian Math. J. **27** (1986), 466–472. (cited on p 357)
529. ———, *Estimates of stability for spatial quasiconformal mappings of a starlike region*, Siberian Math. J. **28** (1987), 946–959. (cited on p 357)
530. P. Šemrl, *Hyers-ulam stability of isometries*, to appear. (cited on p 372)
531. I. Shafrir, *The approximate fixed point property in Banach and hyperbolic spaces*, Israel J. Math. **71** (1990), 211–223. (cited on p 80)
532. A. N. Shiryayev, *Probability*, Graduate Texts in Math., vol. 95, Springer, 1984. (cited on pp 420, 428)

533. S. A. Shkarin, *On continuity points of Gâteaux differentiable mappings* (Russian), Siberian Math. J. **33** (1992), 176–185, English translation: Sib. J. Math, **33** (1992), 905–913. (cited on p 164)
534. B. Simon, *Trace Ideals and their Applications*, London Math. Soc. Lecture Notes Series, vol. 35, Cambridge Univ. Press, 1979. (cited on pp 449, 454)
535. B. Sims and D. Yost, *Linear Hahn-Banach extension operators*, Proc. Edinburgh Math. Soc. **32** (1989), 53–57. (cited on p 439)
536. A. G. Skaletskii, *Uniformly continuous selections in Fréchet spaces* (Russian), Vestnik Moskov Univ. Ser. I Mat. Mekh. **40** (1985), 24–28. (cited on p 34)
537. A. V. Skorohod, *Integration in Hilbert Space*, Ergebnisse, vol. 79, Springer, 1974. (cited on pp 161, 162)
538. V. L. Šmulian, *Sur la structure de la sphère unitaire dans l'espace de Banach*, Mat. Sbornik **9** (1941), 545–560. (cited on p 96)
539. S. Solecki, *On Haar null sets*, Fund. Math. **149** (1996), 205–210. (cited on p 161)
540. C. Stegall, *The Radon Nikodým property in conjugate Banach spaces*, Trans. Amer. Math. Soc. **206** (1975), 213–223. (cited on p 123)
541. ———, *Optimization of functions on certain subsets of Banach spaces*, Math. Ann. **236** (1978), 171–176. (cited on p 122)
542. E. M. Stein, *Harmonic Analysis: Real Variable Methods, Orthogonality, and Oscillatory Integrals*, Princeton Univ. Press, Princeton, New Jersey, 1993. (cited on p 353)
543. J. Stern, *Ultrapowers and local properties of Banach spaces*, Trans. Amer. Math. Soc. **240** (1978), 231–252. (cited on p 439)
544. S. Stoilow, *Lecons sur les Principes Topologiques de la Théorie des Fonctions Analytiques*, second ed., Gauthier-Villars, 1956. (cited on p 280)
545. A. H. Stone, *Paracompactness and product spaces*, Bull. Amer. Math. Soc. **54** (1948), 977–982. (cited on p 416)
546. T. Strömberg, *On regularization in Banach spaces*, Ark. Mat. **34** (1996), 383–406. (cited on p 98)
547. D. W. Stroock, *Probability Theory, an Analytic View*, Cambridge University Press, 1993. (cited on pp 420, 428)
548. V. N. Sudakov, *On quasi-invariant measures in linear spaces* (Russian), Vestnik Leningrad Univ **15** (1960), 5–8. (cited on p 161)
549. S. J. Szarek, *On the best constants in the Khintchine inequality*, Studia Math. **58** (1976), 197–208. (cited on p 424)
550. ———, *On the existence and uniqueness of complex structure and spaces with "few" operators*, Trans. Amer. Math. Soc. **293** (1986), 339–353. (cited on p 390)
551. W. Szlenk, *The non-existence of a separable reflexive Banach space universal for all separable reflexive Banach spaces*, Studia Math. **30** (1968), 53–61. (cited on p 255)
552. M. Talagrand, *Sur la propriété de Radon Nikodým dans les espaces de Banach réticulés*, C.R. Acad. Sci. Paris **288** (1979), no. ser A, 907–910. (cited on p 124)
553. ———, *La structure des espaces de Banach réticulés ayant la propriété de Radon Nikodým*, Israel J. Math. **44** (1983), 213–220. (cited on p 124)
554. ———, *Un espace de Banach réticulé qui a presque la propriété de Radon Nikodým*, Studia Math. **76** (1983), 255–260. (cited on p 124)
555. Y. Tian, *On one-parameter subgroups in a class of topological groups*, J. Math. Anal. Appl. **167** (1992), 482–504. (cited on p 408)
556. ———, *Uniform transformation groups on lower dimensional spaces*, Math. Scand. **79** (1996), 101–129. (cited on p 406)
557. N. Tomczak-Jaegermann, *The moduli of smoothness and convexity and the Rademacher averages of trace classes* S_p $(1 \leq p < \infty)$, Studia Math. **50** (1974), 163–182. (cited on pp 442, 454)

558. ———, *Banach spaces of type p have arbitrarily distortable subspaces*, Geom. Funct. Anal. **6** (1996), 1074–1082. (cited on pp 337, 338)
559. S. L. Troyanski, *On locally uniformly convex and differentiable norms in certain non separable Banach spaces*, Studia Math. **37** (1971), 173–180. (cited on p 122)
560. I. G. Tsařkov, *Theorems on global existence of an implicit function, and their application*, Russ. Acad. Sci. Dokl. Math. **45** (1992), 638–640. (cited on p 58)
561. ———, *On global existence of an implicit function* (Russian), Russian Acad. Sci. Sb. Math. **184** (1993), 79–116, English translation: **79** (1994), 287–313. (cited on p 58)
562. ———, *Smoothing of uniformly continuous mappings in L_p spaces* (Russian), Mat. Zametki **54** (1993), 123–140, English translation: Math. Notes, **54** (1993), 957–967. (cited on p 58)
563. B. S. Tsirelson, *Not every Banach space contains an imbedding of ℓ_p or c_0*, Funct. Anal. Appl. **8** (1974), 138–141. (cited on pp 257, 334)
564. P. Tukia and J. Väisälä, *Quasisymmetric embeddings of metric spaces*, Ann. Acad. Sci. Fenn. Ser. A I Math. **5** (1980), 97–114. (cited on p 258)
565. B. Turett, *A dual view of a theorem of Baillon*, Nonlinear Analysis and Applications (S. P. Singh and J. H. Burry, eds.), Lecture Notes in Pure and applied Mathematics, vol. 80, Marcel Dekker, 1982, pp. 279–286. (cited on p 415)
566. A. Tychonoff, *Ein Fixpunktsatz*, Math. Ann. **111** (1935), 767–776. (cited on p 79)
567. M. M. Vainberg, *Variational Methods for the Study of Nonlinear Operators*, Holden day, 1964. (cited on p 7)
568. J. Väisälä, *Lectures on n-Dimensional Quasiconformal Mappings*, Lect. Notes in Math., vol. 229, Springer, 1971. (cited on p 357)
569. ———, *Lipschitz homeomorphisms of the Hilbert cube*, Topology Appl. **11** (1980), 103–110. (cited on p 218)
570. ———, *Free quasiconformality in Banach spaces I*, Ann. Acad. Sci. Fenn. Ser. A I Math. **15** (1990), 355–379. (cited on p 258)
571. ———, *Free quasiconformality in Banach spaces II*, Ann. Acad. Sci. Fenn. Ser. A I Math. **16** (1991), 255–310. (cited on p 258)
572. ———, *Banach spaces and bilipschitz maps*, Studia Math. **103** (1992), 291–294. (cited on p 218)
573. ———, *Free quasiconformality in Banach spaces III*, Ann. Acad. Sci. Fenn. Ser. A I Math. **17** (1992), 393–408. (cited on p 258)
574. ———, *Free quasiconformality in Banach spaces IV*, Analysis and Topology (C. Andreian Cazacu et al, ed.), World Scientific, 1998, pp. 697–717. (cited on p 258)
575. ———, *The free quasiworld. Freely quasiconformal and related maps in Banach spaces*, Quasiconformal Geometry and Dynamics, Banach center Publications, vol. 48, Polish Academy of Sciences, 1999, pp. 55–118. (cited on p 258)
576. J. van Mill, *Infinite Dimensional Topology*, North-Holland, 1989. (cited on pp 7, 80, 215)
577. L. Veselý, *Metric projections after renorming*, J. Approx. Th. **66** (1991), 72–82. (cited on p 58)
578. R. A. Vitale, *The Steiner point in infinite dimensions*, Israel J. Math. **52** (1985), 245–250. (cited on p 60)
579. S. Wagon, *The Banach-Tarski Paradox*, Enc. Math and its applications, vol. 24, Cambridge Univ. Press, 1985. (cited on p 417)
580. A. Weil, *L'intégration dans les Groupes Topologiques et ses Application*, Hermann, Paris, 1940. (cited on p 161)
581. J. H. Wells and L. R. Williams, *Embeddings and Extensions in Analysis*, Ergebnisse, vol. 84, Springer, 1975. (cited on p 59)
582. A. Weston, *Some non linear problems in functional analysis*, Ph.D. thesis, Kent State University, 1993. (cited on pp 215, 254)

583. ———, *Some non-uniformly homeomorphic spaces*, Israel J. Math. **83** (1993), 375–380. (cited on p 407)
584. G. T. Whyburn, *Topological Analysis*, Princeton University Press, 1958. (cited on p 279)
585. R. Wittmann, *Mean ergodic theorems for nonlinear operators*, Proc. Amer. Math. Soc. **108** (1990), 781–788. (cited on p 81)
586. Y. Yamasaki, *Measures on Infinite-Dimensional Spaces*, Series in Pure Mathematics, vol. 5, World Scientific Singapore, 1985. (cited on p 161)
587. Z. Zahorski, *Sur l'ensemble des points de non-dérivabilité d'une fonction continue*, Bull. Soc. Math. France **74** (1946), 147–178. (cited on p 165)
588. L. Zajíček, *On the differentiability of convex functions in finite and infinite dimensional spaces*, Czechoslovak Math. J. **29** (1979), 340–348. (cited on p 98)
589. ———, *Differentiability of the distance function and points of multi-valuedness of the metric projection in Banach space*, Czechoslovak Math. J. **33** (1983), 292–308. (cited on pp 98, 164)
590. ———, *Porosity and σ-porosity*, Real Anal. Exchange **13** (1987-88), 314–350. (cited on pp 98, 167)
591. ———, *Fréchet differentiability, strict differentiability and subdifferentiability*, Czechoslovak Math. J. **41** (1991), 471–489. (cited on p 98)
592. E. H. Zarantonello, *Projections on convex sets in Hilbert space and spectral theory*, Contributions to Nonlinear Functional Analysis (E. H. Zarantonello, ed.), Academic Press, 1971, pp. 237–244. (cited on p 81)
593. E. Zeidler, *Nonlinear Functional Analysis and its Applications I, II, III, IV*, Springer, 1984–87. (cited on pp 7, 80)
594. M. Zippin, *On perfectly homogeneous bases in Banach spaces*, Israel J. Math. **4** (1966), 265–272. (cited on pp 300, 334)
595. ———, *On some subspaces of Banach spaces whose duals are L_1 spaces*, Proc. Amer. Math. Soc. **23** (1969), 378–385. (cited on p 59)
596. V. A. Zorich, *The global homeomorphism theorem for space quasi conformal mappings, its development and related open problems*, Quasi conformal space mappings (M. Vourinen, ed.), Lect. Notes in Math., vol. 1508, Springer, 1992, pp. 132–148. (cited on p 357)

Index

Selected Titles in This Series

(Continued from the front of this publication)